Background Modeling and Foreground Detection for Video Surveillance

Background Modeling and Foreground Detection for Video Surveillance

Edited by

Thierry Bouwmans

Université de La Rochelle
France

Fatih Porikli

Mitsubishi Electric Research Labs
Cambridge, Massachusetts, USA

Benjamin Höferlin

University of Osnabrück
Germany

Antoine Vacavant

Université d'Auvergne
Le Puy-en-Velay, France

CRC Press
Taylor & Francis Group
Boca Raton London New York

CRC Press is an imprint of the
Taylor & Francis Group, an **informa** business

CRC Press
Taylor & Francis Group
6000 Broken Sound Parkway NW, Suite 300
Boca Raton, FL 33487-2742

First issued in paperback 2020

ISBN-13: 978-1-4822-0537-4 (hbk)
ISBN-13: 978-0-367-65911-0 (pbk)

Visit the Taylor & Francis Web site at
http://www.taylorandfrancis.com

and the CRC Press Web site at
http://www.crcpress.com

Dedication

Background modeling has a long and well published history since it was commonly adopted as a means for object detection and tracking in video at the end of the last century. In the decade and a half since, there have been thousands of researchers and nearly 10,000 publications that have either contributed novel approaches or employed existing approaches in new computer vision and machine learning applications.

On behalf of the authors, I would like to dedicate this book to all the researchers who have contributed their experiences and understanding to this field. In particular, I would like to recognize researchers who have gone beyond explaining their algorithms to provide intuition and a greater understanding of the myriad approaches to background modeling. These researchers have impacted the field with their intuitive papers, comprehensive evaluations, landmark data sets, workshops, working groups, and books in background modeling. They have reduced redundancy and helped mature this field by conveying not just what they did, but also what they learned from their hundreds to thousands of hours toiling with videos. This book is intended as one such contribution, by bringing together a background modeling overview, history, applications, practical implementations, and evaluation practices in one resource. We hope you find it useful.

Christopher Stauffer, Ph.D.

on behalf of T. Bouwmans, F. Porikli, B. Höferlin, and A. Vacavant

Contents

Part I: Introduction and Background

Part II: Traditional and Recent Models

Part III: Applications in Video Surveillance

Part IV: Sensors, Hardware and Implementations

Part V: Benchmarking and Evaluation

Preface

Background modeling and foreground detection are important steps in the video processing field such as video-surveillance, optical motion capture, multimedia applications, teleconferencing and human-computer interface. Conventional background modeling methods exploit the temporal variation of each pixel to model the background while foreground detection applies change detection. The last decade witnessed very significant publications in this field and recently new applications in which background is not static, such as recordings taken from mobile devices or Internet videos, generate new developments to detect robustly moving objects in challenging environments. Thus, effective methods for robustness to deal both with dynamic backgrounds, illumination changes which occur in real-life scene with fixed cameras or mobile devices have been recently developed and so different strategies are used such as automatic feature selection, model selection or hierarchical models. Another feature of background modeling methods is that the use of advanced models has to be computed in real-time and low memory requirements. Algorithms have to be designed to meet these requirements.

In this context, this handbook solicited contributions to address this wide range of challenges met in background modeling and foreground detection for video-surveillance. Thus, it groups the works of the leading teams in this field over the recent years. By incorporating both existing and new ideas, this handbook gives a complete overview of the concepts, theories, algorithms, and applications related to background modeling and foreground detection. First, an introduction to background modeling and foreground detection for beginners is provided by surveying statistical models, clustering models, neural networks and fuzzy model. Furthermore, leading methods and algorithms for detecting moving objects in video surveillance are presented. A description of recent complete datasets and codes are given. Moreover, an accompanying website[1] is provided. This website contains the list of chapters, their abstracts and links to some software demonstrations. It allows the reader to have quick access to the main resources, datasets and codes in the field. Finally, with this handbook, we aim to bring a one-stop solution, *i.e.*, access to a number of different models, algorithms, implementations and benchmarking techniques in a single volume. The handbook consists of five parts.

Part I presents a comprehensive introduction to background modeling and foreground detection for video surveillance systems. Chapter 1 and Chapter 2 provide a first complete survey of the traditional and recent background models with a classification of them following the mathematical models used. Furthermore, these chapters present the available resources such as datasets and implementations, and thus facilitate comparison and evaluation of background subtraction algorithms. We provide an accompanying web site: the Background Subtraction Web Site[2]. This website contains a full list of the references in the field, links to available datasets and codes. In each case, the list is regularly updated and classified according to the sections of these chapters. Chapter 3 provides a first valuable overview of different background initialization strategies. An initial background model

---

[1]http://sites.google.com/site/backgroundmodeling/

[2]http://sites.google.com/site/backgroundsubtraction/Home

that describes the scene without foreground objects is the prerequisite in video-surveillance. Chapter 4 investigates background subtraction in the case of moving cameras. Indeed, most of today's videos are captured from moving platforms. Traditional background subtraction algorithms are not applicable for the moving-camera case. There have been some extensions of background subtraction algorithms to deal with mobile cameras where cameras motion is known or scene geometry is restricted, such as stationary mounted pan-tilt-zoom cameras. So, the authors review these extensions and present an effective solution for the case of videos captured from an online moving camera.

Part II concerns representative traditional and recent models. Chapter 5 gives a detailed overview of the statistical models. Chapter 6 describes a non-parametric method for background modeling/foreground detection in videos, specifically the well-known Pixel-Based Adaptive Segmenter (PBAS). The authors provide a detailed explanation of PBAS, as well as further optimization for the various parameters involved. Furthermore, PBAS is evaluated within a thorough experimental evaluation on the standard Change Detection dataset, including the assessment of how the choices of values for various parameters affect performance. Chapter 7 presents another famous background model named ViBe. ViBe achieved a fast initialization technique and a novel updating mechanism by introducing a memoryless update policy and spatial diffusion. The authors also discuss some improvements such as the distinction between the segmentation map and the updating mask, or a controlled diffusion mechanism. Concerning the computational cost, the notion of background subtraction complexity factor is introduced to express the speed of the algorithm, and show that ViBe has a low complexity factor to achieve real-time performance. Chapter 8 develops online learning models based on a stochastic approximation, which have an inherent robustness and low computational complexity. One of them models the background with a multivariate Gaussian distribution, while the other uses a probabilistic self-organizing map. Both of them associate a uniform distribution to the foreground to cope with any incoming objects adequately. Experimental results show that stochastic approximation methods perform well both in qualitative and quantitative terms when compared with a selection of state of the art algorithms. Chapter 9 presents group sparsity based methods for background and foreground separation and develops an algorithm called LRGS based on the group sparsity notion which is capable of handling video sequences recorded by both stationary and moving cameras. LRGS operates on matrices of tracked point trajectories and decomposes them into foreground and background trajectories by enforcing a low-rank constraint on the background matrix and applying a l_0-norm penalty to the l_2-norm of the rows (trajectories) of the foreground matrix thus encouraging row wise sparsity. The method is evaluated on both the synthetic data and several real videos and compared to other algorithms for trajectory labeling. Chapter 10 presents a robust detection through a rough set theory framework. Furthermore, an integrated 3D histon is proposed, where the histon distribution is calculated by considering the color value on three channels jointly. By determining the extent of similarity using Gaussian membership function, 3D fuzzy histon is subsequently used to compute 3D Fuzzy histon roughness index (3D FHRI). Foreground detection in a video sequence is performed by evaluating the Bhattacharyya distance between the model roughness index distribution and the roughness index distribution computed in the current frame in three types of histons (basic histon, 3D histon, and 3D fuzzy histon).

Part III focuses on several video-surveillance dynamic backgrounds and illumination changes. Chapter 11 introduces two efficient approaches for foreground detection and tracking for automated visual surveillance. One of them uses a non-parametric density estimation, while the other uses a single class background pixel model with Support Vector Regression (SVR). The density estimation is achieved either through an Adaptive Kernel

Density Estimation (AKDE) algorithm with a training phase to learn the kernel covariance and the foreground/background decision threshold per pixel or a Recursive Modeling (RM) algorithm with fixed per pixel memory requirements. For SVR, the Support Vector Data Descriptions (SVDD) are determined in the training phase and are used to calculate the radius of the hyper sphere that serves as the decision boundary during the classification stage. The SVR based background pixel modeling is augmented with a target tracking framework that handles multiple targets using size, color, motion and shape histogram information. For detecting moving objects in video sequences, Chapter 12 develops a novel approach based on a Beta-Liouville distribution, extended to the infinite case, and learned by means of incremental variational Bayes. This model is learned via variational Bayesian inference which is an efficient deterministic alternative to purely Bayesian inference. This method is robust and adaptive to dynamic background, and it has the ability to handle multi-modal background distributions. Moreover, thanks to the nature of non-parametric Bayesian models, the determination of the correct number of components is sidestepped by assuming that there are an infinite number of components. The results demonstrate the robustness of this approach in the presence of dynamic backgrounds on several real video surveillance sequences. Chapter 13 investigates three spatio-temporal background models for moving object detection. The first one is a combinational background model in which a statistical, a local feature-based and an extrapolation-based background models are combined adaptively. The second one is an integrated background model based on the feature SLDP where a statistical framework is applied to an illumination-invariant feature. The third one is an integrated background model StSIC realized by considering a similarity of intensity changes among pixels. Each background model uses a spatio-temporal feature to tackle various background changes robustly. The combinational model is suitable for close-up and low contrast scenes, and SLDP and StSIC are suitable for the scenes where illumination and dynamic changes affect the same region at the same time. Maritime environment represents a challenging application due to the complexity of the observed scene (waves on the water surface, boat wakes, weather issues). In this context, Chapter 14 presents a method for creating a discretization of an unknown distribution that can model highly dynamic background such as water background with varying light and weather conditions. A quantitative evaluation carried out on the recent MAR datasets demonstrates the effectiveness of this approach. Chapter 15 describes a spatial-color mixture of Gaussians based on a hierarchical scene model. It combines a global spatial modeling with a temporal pixel modeling of the scene which takes into account the spatial consistency between pixels. Global models permit to describe the whole scene, both background and foreground areas. The target model is dynamically created during the detection process, allowing for a better background/foreground separation and provides data information of a higher level than pixel. This may help a tracking step by introducing a natural combination between the detection and the tracking module. The detection results show that this approach outperforms several traditional methods under difficult conditions. Chapter 16 presents the Grassmannian Robust Adaptive Subspace Tracking Algorithm (GRASTA), an online algorithm for robust subspace tracking, and its variant t-GRASTA that incorporates geometric transforms on the data. For GRASTA, the authors use a low-rank model for data that may be corrupted by outliers and have missing data values. For t-GRASTA, a low-rank model for misaligned images that may be corrupted by outliers is proposed. Both GRASTA and t-GRASTA use the natural l_1-norm cost function for data corrupted by sparse outliers, and both perform incremental gradient descent on the Grassmannian, the manifold of all d-dimensional subspaces for fixed d. The two algorithms operate only one data vector at a time, making them faster than other state-of-the-art algorithms and suitable for streaming and real-time applications.

Part IV addresses sensors, hardware and implementations issues. Chapter 17 presents an overview of ubiquitous imaging sensors for presence detection. The main technologies considered are regular visible light cameras, thermal cameras, depth cameras and radar. The technologies are described, their operating principles explained and the performance analyzed. Chapter 18 presents a framework that is able to accurately segment foreground objects with RGB-D cameras. In particular, more reliable compact segmentations are obtained in the case of camouflage in color or depth. Chapter 19 details a GPU implementation of jointly modeling a scene background and separating the foreground. The authors propose a way of modeling that does not assume that background pixels are Gaussian random variables. They also incorporate an interesting periphery detector or a detector for short-term motion. Finally, a GPU implementation of this method provides a significant speed increase over the CPU implementation. Chapter 20 presents a GPU implementation for background-foreground-separation via Robust PCA and Robust Subspace Tracking. Starting with a brief introduction to some basic concepts of manifold optimization, the Robust PCA problem in the manifold context is discussed as well as the cost functions and how their optimization can practically be performed in batch and online mode. In order to keep the discussion close to real-world applications, pROST which performs online background subtraction on video sequences is presented. Its implementation on a graphics processing unit (GPU) achieves real-time performance at a resolution of 160×120 pixels. Experimental results show that the method succeeds in a variety of challenges such as camera jitter and dynamic backgrounds. Chapter 21 concerns two background subtraction methods for embedded hardware. Thus, the authors detail two FPGA architectures which is based on the Horprasert method, with a shadow detection extension, and the other one is based on the Codebook method. In each case, the architecture developed on reconfigurable hardware is explained and the performance is experimentally assessed. Chapter 22 presents a resource-efficient background modeling and foreground detection algorithm that is highly robust to illumination changes and dynamic backgrounds (swaying trees, water fountains, strong wind and rain). Compared to many traditional methods, the memory requirement for the data saved for each pixel is very small in the proposed algorithm. Moreover, the number of memory accesses and instructions are adaptive, and are decreased even more depending on the amount of activity in the scene and on a pixel's history. Furthermore, the authors present a feedback method to increase the energy efficiency of the foreground object detection even further. This way, foreground detection is performed in smaller regions as opposed to the entire frame. The feedback method significantly reduces the processing time of a frame. Finally, a detailed comparison of the feedback method and the sequential approach in terms of processing times and energy consumption is provided.

Part V covers the resources and datasets required for evaluation and comparison of background subtraction algorithms. Chapter 23 proposes the BGSLibrary which provides an easy-to-use C++ framework together with a couple of tools to perform background subtraction. First released in March 2012, the library provides 32 background subtraction algorithms. The source code is platform independent and available under the open source GNU GPL v3 license, thus the library is free for non-commercial use. The BGSLibrary also provides a Java based GUI (Graphical User Interface) allowing the users to configure the input video-source, regions of interest, and the parameters of each BS algorithm. Chapter 24 gives an overview of the most cited motion detection methods and datasets used for their evaluation. Then, the authors provide benchmarking results on different categories of videos, for different methods, different features and different post-processing methods. Thus, the chapter provides the reader with a broad overview on the most effective methods available today on different types of videos. All benchmark results are obtained on the changedetection.net dataset. Chapter 25 presents the Background Models Challenge

(BMC), which is a benchmark based on a set of both synthetic and real videos, together with several performance evaluation criteria. Finally this chapter presents the most recent results obtained by the BMC, such as the final ranking obtained in very first challenge. Finally, possible evolutions for this benchmark, according to recent advances in background subtraction and its potential applications are discussed.

The handbook is intended to be a reference for researchers and developers in industries, as well as graduate students, interested in background modeling and foreground detection applied to video surveillance and other related areas, such as optical motion capture, multimedia applications, teleconferencing, video editing and human-computer interfaces. It can be also suggested as reading text for teaching graduate courses in subjects such as computer vision, image processing, real-time architecture, machine learning and data mining.

The editors of this handbook would like to acknowledge with their sincere gratitude the contributors for their valuable chapters and the reviewers for the helpful comments concerning the chapters in this handbook. We also acknowledge the reviewers of the original handbook proposal. Furthermore, we are very grateful for the help that we have received from Sarah Chow, Marsha Pronin and others at CRC Press during the preparation of this handbook. Finally, we would like to acknowledge Shashi Kumar from Cenveo for his valuable support about the LaTeX issues.

About the Editors

Thierry Bouwmans (http://sites.google.com/site/thierrybouwmans/) is an Associate Professor at the University of La Rochelle, France. His research interests consist mainly in the detection of moving objects in challenging environments. He has recently authored 30 papers in the field of background modeling and foreground detection. These papers investigated particularly the use of fuzzy concepts, discriminative subspace learning models and robust PCA. They also develop surveys on mathematical tools used in the field. He has supervised Ph.D. students in this field. He is the creator and the administrator of the Background Subtraction Web Site. He has served as a reviewer for numerous international conferences and journals.

Benjamin Höferlin (http://www.vis.uni-stuttgart.de/nc/institut/mitarbeiter/benjam in-hoeferlin.html) received his M.Sc. degree and his Ph.D. degree in Computer Science from the University of Stuttgart, Germany (2013). He is currently associated with the Biologically Oriented Computer Vision Group, Institute of Cognitive Science, Osnabrück University, Germany. His research interests include computer vision, automated video analysis and visual analytics of video data. He is the author of the "Stuttgart Artificial Background Subtraction Dataset" (SABS) that allows precise challenge-based evaluation of background modeling techniques.

Fatih Porikli (http://www.porikli.com/) is currently a Professor of Computer Vision and Robotics at Australian National University and the Computer Vision Group Leader at NICTA since September 2013. Previously, he was a Distinguished Research Scientist at Mitsubishi Electric Research Labs (MERL) for 13 years. He received his Ph.D. from NYU Poly, NY. His work covers areas including computer vision, machine learning, video surveillance, multimedia processing, structured and manifold based pattern recognition, biomedical vision, radar signal processing, and online learning with over 100 publications and 60 patents. He has mentored more than 40 Ph.D. students and interns. He received Research and Developement 2006 Award in the Scientist of the Year category (select group of winners) in addition to 3 IEEE Best Paper Awards and 5 Professional Prizes. He serves as an Associate Editor of IEEE Signal Processing Magazine, SIAM Journal on Imaging Sciences, Springer Machine Vision Applications, Springer Real-time Image and Video Processing, and EURASIP Journal on Image and Video Processing. He served as the General Chair of IEEE Advanced Video and Signal based Surveillance Conference (AVSS) in 2010 and participated in the organizing committee of many IEEE events.

Antoine Vacavant (http://isit.u-clermont1.fr/ anvacava) obtained his Master's degree from the University Lyon 1, France, in 2005, and the Ph.D. degree in computer science from the University Lyon 2 in 2008. He is now associate professor at the University of Auvergne Clermont Ferrand 1. Head of the professional bachelor in 3D imaging, he gives lectures of image rendering, software engineering and object/event based programming in the IUT of Le Puy en Velay. Member of the ISIT lab, UMR 6284 UdA/CNRS / research team CaVITI, his main research topics are discrete and computational geometry, image analysis and computer vision. He has organized the first BMC (Background Models Challenge) at ACCV 2012, which addresses the evaluation of background subtraction algorithms thanks to a complete benchmark composed of real and synthetic videos.

List of Contributors

In alphabetical order

Catherine Achard, Univ. Pierre et Marie Curie, Paris, France

Laura Balzano, University of Michigan, Ann Arbor, USA

Olivier Barnich, EVS Broadcast, Belgium

George Bebis, Computer Vision Laboratory, University of Nevada, Reno, USA

Domenico Bloisi, Sapienza University of Rome, Italy

Nizar Bouguila, Concordia Institute for Information Systems Engineering (CIISE), Concordia University, Canada

Thierry Bouwmans, Laboratoire MIA, Univ. La Rochelle, La Rochelle, France

Massimo Camplani, ETSIT, Universidad de Madrid, Spain

Mauricio Casares, Syracuse University, USA

Thierry Chateau, Pascal Institute, Blaise Pascal University, Clermont-Ferrand, France

Chen Chen, University of Texas, Arlington, USA

Pojala Chiranjeevi, Indian Institute of Technology, Kharagpur, India

Xinyi Cui, Facebook, USA

Javier Diaz, CITIC, University of Granada, Spain

Ahmed Elgammal, Department of Computer Science, Rutgers University, USA

Ali Elqursh, Department of Computer Science, Rutgers University, USA

Wentao Fan, Concordia Institute for Information Systems Engineering (CIISE), Concordia University, Canada

Enrique J. Fernandez-Sanchez, CITIC, University of Granada, Spain

Christophe Gabard, CEA, LIST, France

Sadiye Guler, intuVision Inc., USA

Clemens Hage, Department of Electrical Engineering and Information Technology, Technische Universitat Munchen, Munchen, Germany

Jun He, Nanjing University of Information Science and Technology, China

Martin Hofmann, Institute for Human-Machine Communication, Technische Universitat Munchen, Munchen, Germany

Junzhou Huang, Department of Computer Science and Engineering, University of Texas, Arlington, USA

Pierre-Marc Jodoin, Université de Sherbrooke, Canada

Martin Kleinsteuber, Department of Electrical Engineering and Information, Technology Technische Universitat Munchen, Munchen, Germany

Ezequiel López-Rubio, University of Málaga, Spain

Laurent Lucat, CEA, LIST, France

Rafael M. Luque-Baena, University of Málaga, Spain

Lucia Maddalena, National Research Council, Institute for High-Performance Computing and Networking, Naples, Italy

Ashutosh Morde, intuVision Inc., USA

Hajime Nagahara, Laboratory for Image and Media Understanding, Kyushu University, Japan

Mircea Nicolescu, Computer Vision Laboratory, University of Nevada, Reno, USA

Yosuke Nonaka, Laboratory for Image and Media Understanding, Kyushu University, Japan

Alfredo Petrosino, Department of Applied Science, University of Naples Parthenope, Naples, Italy

Sébastien Piérard, Université de Liège, Belgium

Gerhard Rigoll, Institute for Human-Machine Communication, Technische Universitat Munchen, Munchen, Germany

Lionel Robinault, LIRIS/Foxstream, University Lyon 2, Lyon, France

Rafael Rodriguez-Gomez, CITIC, University of Granada, Spain

Eduardo Ros, CITIC, University of Granada, Spain

Luis Salgado, Video Processing and Understanding Lab, Universidad Autonoma de Madrid, Spain

Florian Seidel, Department of Informatics, Technische Universitat Munchen, Munchen, Germany

Somnath Sengupta, Indian Institute of Technology, Kharagpur, India

Atsushi Shimada, Laboratory for Image and Media Understanding, Kyushu University, Japan

Andrews Sobral, Laboratoire L3I, Univ. La Rochelle, La Rochelle, France

Arthur Szlam, City University of New York, USA

Rin-ichiro Taniguchi, Laboratory for Image and Media Understanding, Kyushu University, Japan

Alireza Tavakkoli, University of Houston-Victoria, USA

Philipp Tiefenbacher, Institute for Human-Machine Communication, Technische Universitat Munchen, Munchen, Germany

Laure Tougne, LIRIS, University Lyon 2, Lyon, France

Antoine Vacavant, ISIT, University of Auvergne, Clermont-Ferrand, France

Marc Van Droogenbroeck, University of Liège, Belgium

Senem Velipasalar, Syracuse University, USA

Satoshi Yoshinaga, Laboratory for Image and Media Understanding, Kyushu University, Japan

Junxian Wang, Microsoft Research, USA

Yi Wang, Université de Sherbrooke, Canada

I

Introduction and Background

1

Traditional Approaches in Background Modeling for Static Cameras

Thierry Bouwmans
Lab. MIA, Univ. La Rochelle, France

1.1 Introduction

Analysis and understanding of video sequences is an active research field. Many applications in this research area (video surveillance [60], optical motion capture [47], multimedia application [16]) need in the first step to detect the moving objects in the scene. So, the basic operation needed is the separation of the moving objects called foreground from the static information called the background. The process mainly used is the background subtraction and recent surveys can be found in [102] [77] [39]. The simplest way to model the background is to acquire a background image which doesn't include any moving object. In some environments, a static background is not available since the background dynamically changes due to varying illumination or moving objects. So, the background representation model must be robust and adaptive to address these challenges. Facing these challenges, many background subtraction methods have been designed over the last decade. Several surveys can be found in the literature but none of them address an overall review of this field. In 2000, Mc Ivor [218] surveyed nine algorithms allowing a first comparison of the models. However, this survey is mainly limited on a description of the algorithms. In 2004, Piccardi [238] provided a review on seven methods and an original categorization based on speed, memory requirements and accuracy. This review allows the readers to compare the complexity of the different methods and effectively helps them to select the most adapted method for their specific application. In 2005, Cheung and Kamath [60] classified several methods into non-recursive and recursive techniques. Following this classification, Elhabian et al. [102] provided a large survey in background modeling. However, this classification in terms of non-recursive and recursive techniques is more suitable for the background maintenance scheme than for the background modeling one. In their review in 2010, Cristiani et al. [77] distinguished the most popular background subtraction algorithms by means of their sensor utilization: single monocular sensor or multiple sensors. In 2010, Bouwmans et al. [39] provided a comprehensive survey on statistical background modeling methods for foreground detection classifying each approach following the statistical models used. Other recent surveys focus only on some models such as statistical background modeling [36], Mixture of Gaussians models [38], subspace learning models [35] and fuzzy models [37].

Considering all of this, we provide a first complete overview of the traditional and recent background models with a classification of them following the mathematical models used. This survey stretches across the first two chapters, i.e Chapter 1 and Chapter 2, respectively. Furthermore, these chapters present the available resources such as the datasets and libraries, and thus facilitates comparison and evaluation of background subtraction algorithms. Moreover, we provided an accompanying web site: the Background Subtraction Web Site[3]. This website contains a full list of the references in the field, links to available datasets and codes. In each case, the list is regularly updated and classified according to the sections of these chapters.

Two closely related problems to background subtraction are change detection [249] and salient motion detection [80]. Change detection addresses the detection of the changes between two images. So, background subtraction is a particular case when 1) one image is the background image and the other one is the current image, and 2) the changes are due to moving objects. On the other hand, salient motion detection aims at finding semantic regions and filtering out the unimportant areas. The idea of saliency detection is derived from the human visual system, where the first stage of human vision is a fast and simple pre-attentive process. So, salient motion detection can be viewed as a particular case of background subtraction.

The rest of Chapter 1 is organized as follows: In Section 1.2, we present the problem statement. The different steps and components in background subtraction are reviewed in Section 1.3. In Section 1.5, we present the challenges and issues related to background modeling and foreground detection. Then, in Section 1.6, we give an overview of the traditional background models classified according to the mathematical models. Then, measures for performance evaluation are presented in Section 1.7. In Section 1.8, we present the traditional datasets that are publicly available to test and evaluate background subtraction algorithms. Finally, we conclude the survey by providing possible directions.

---

[3]http://sites.google.com/site/backgroundsubtraction/Home

1.2   Problem Statement

Background subtraction is based on the assumption that the difference between the background image and a current image is caused by the presence of the moving objects. Pixels that have not changed are considered as "background" and pixels that have changed are considered as "moving objects". The scientific literature thus often denotes "moving objects" as "foreground" since moving objects are generally found in the foreground. Practically, pixels that have not changed could be part of moving objects, for example, if they have the same color as the background. Similarly, pixels that have changed could be part of the background when illumination changes occur for example. So that the idea can be perform in a general way, five main assumptions have to be respected:

- **Assumption 1:** The camera as well as its parameters are fixed.
- **Assumption 2:** The scene should not present illumination changes.
- **Assumption 3:** The background is visible and static, that is, the temporal luminance distribution of each pixel can be represented by a random process with uni-modal distribution. Furthermore, there are no moved or inserted background objects over time.
- **Assumption 4:** The initial background does not contain sleeping foreground objects.
- **Assumption 5:** The foreground and background can be easily separated by thresholding the difference between the background image and the current image. Here the choice of the feature is a key issue.

Practically, none of these asumptions is fully met due to many challenging situations which appear as well in indoor scenes as outdoor scenes. For example, Fig. 1.1 shows the original frame 309 of the sequence from [100], the generated background, the ground-truth and the foreground mask. We can see that several false positive detections are generated by the camera jitter.

1.3   Background Subtraction Steps and Issues

This section addresses the different steps and issues related to background subtraction. Table 1.1 gives an overview of these steps. Furthermore, Table 1.2 shows an overview of the foreground detection and the features and Table 1.3 groups the different key issues.

FIGURE 1.1 The first row presents the frame 309, the generated background, the ground truth (GT) and the foreground mask. The second row shows the same images for the frame 462. These frames come from the sequence [100].

1.3.1 Background Modeling

Background modeling (or representation) describes the kind of model used to represent the background. It essentially determines the ability of the model to deal with uni-modal or multi-modal backgrounds. In this chapter, the different background models used in the literature are classified in traditional and recent background models.

1.3.2 Background Initialization

Background initialization (generation, extraction or construction) regards the initialization of the model. In contrast to background model representation and model maintenance, the initialization of the background model was only marginally investigated. The main reason is that often the assumption made is that initialization can be achieved by exploiting some clean frames at the beginning of the sequence. Naturally, this assumption is rarely met in real scenarios, because of continuous clutter presence. Generally, the model is initialized using the first frame or a background model over a set of training frames, which contain or do not contain foreground objects. The main challenge is to obtain a first background model when more than half of the training contains foreground objects. This learning (or training) process can be done off-line and so the algorithm can be a batch one. However practically, some algorithms are : (1) batch ones using N training frames (consecutive or not) [222], (2) incremental with known N or (3) progressive ones with unknown N as the process generates partial backgrounds and continues until a complete background image is obtained [68] [73]. Furthermore, initialization algorithms depend on the number of modes and the complexity of their background models [247]. The main investigations use median [222], histogram [68] [73], stable intervals [198] [322] [125] [55], and SVM [191].

1.3.3 Background Maintenance

Background maintenance relies on the mechanism used for adapting the model to the changes of the scene over time. The background maintenance process has to be an incremental on-line algorithm, since new data is streamed and so dynamically provided. The key issues of this step are the following ones:

- **Maintenance schemes:** In the literature, three maintenance schemes are present: the blind, the selective, and the fuzzy adaptive schemes [18]. The blind background maintenance updates all the pixels with the same rules which is usually an IIR filter:

$$B_{t+1}(x, y) = (1 - \alpha) B_t(x, y) + \alpha I_t(x, y) \tag{1.1}$$

where α is the learning rate which is a constant in $[0, 1]$. B_t and I_t are the background and the current image at time t, respectively. The main disadvantage of this scheme is that the value of pixels classified as foreground are used in the computation of the new background and so polluted the background image. To solve this problem, some authors used a selective maintenance scheme that consists of updating the new background image with different learning rate depending on the previous classification of a pixel into foreground or background:

$$B_{t+1}(x, y) = (1 - \alpha) B_t(x, y) + \alpha I_t(x, y) \tag{1.2}$$
$$\qquad\qquad if \ (x, y) \ is \ background$$
$$B_{t+1}(x, y) = (1 - \beta) B_t(x, y) + \beta I_t(x, y) \tag{1.3}$$
$$\qquad\qquad if \ (x, y) \ is \ foreground$$

Here, the idea is to adapt very quickly a pixel classified as background and very slowly a pixel classified as foreground. For this reason, $\beta << \alpha$ and usually $\beta = 0$. So the Equation (1.3) becomes:

$$B_{t+1}(x, y) = B_t(x, y) \tag{1.4}$$

But the problem is that erroneous classification may result in a permanent incorrect background model. This problem can be addressed by a fuzzy adaptive scheme which takes into account the uncertainty of the classification. This can be achieved by graduating the update rule using the result of the foreground detection such as in [18].

- **Learning rate:** The learning rate determines the speed of the adaptation to the scene changes. It can be (1) fixed, or dynamically adjusted by (2) a statistical or (3) a fuzzy method. In the first case, the learning rate is fixed as the same value for all the sequence. Then, it is determined carefully such as in [370] or can be automatically selected by an optimization algorithm [337]. However, it can take one value for the learning step and one for the maintenance step [154]. Additionally, the rate may change over time following a tracking feedback strategy [166]. For the statistical case (2), Lee [181] used different learning rates for each Gaussian in the MOG model. The convergence speed and approximation results are significantly improved. Harville et al. [133] adapted the learning rate according to the pixel activity. Lindstrom [193] proposed a progressive learning rate where the relative update speed of each Gaussian component depends on how often the component has been observed, and thus integrates temporal information in the maintenance scheme. For the fuzzy case (3), Sigari et al. [278] [276] computed an adaptive learning rate at each pixel with respect to the fuzzy membership value obtained for each pixel during the fuzzy foreground detection. Recently, Maddalena and Petrosino [209] [208] improved the adaptivity by introducing spatial coherence information.

- **Maintenance mechanisms:** The learning rate determines the speed of adaptation to illumination changes but also the time a background change requires until it is incorporated into the model as well as the time a static foreground object can survive before being included in the model. So, the learning rate deals with different challenges which have different temporal characteristics. To decouple the adaptation mechanism and the incorporation mechanism, some authors [378] [320] used a set of counters which represents the number of times a pixel is classified as a foreground pixel. When this number is larger than a threshold, the pixel is considered as background. This gives a time limit on how long a pixel can be considered as a static foreground pixel. Another approach developed by Lindstrom et al. [193] uses a CUSUM detector to determine when components should be transfered from a foreground component to a background component. The CUSUM detector utilizes local counters for the foreground components. If the counter exceeds a threshold, the corresponding foreground component is added to the background model. If the maximum number of background components has been reached, it replaces the least likely background component. The foreground component remains in the model because other pixels might still be explained by it.

- **Frequency of the update:** The aim is to update only when it is needed. The maintenance may be done every frame but in absence of any significant changes, pixels are not required to be updated at every frame. For example, Porikli et al. [241] proposed adapting the time period of the maintenance mechanism with respect to an illumination score change. The idea is that no maintenance is needed if no illumination change is detected and a quick maintenance is necessary otherwise. In the same idea, Magee [212] used a variable adaptation frame rate following the activity of the pixel, which improves temporal history storage for slow changing pixels while running at high adaption rates for less stable pixels.

1.3.4   Foreground Detection

Foreground detection consists of comparing the background image with the current image to label pixels as background or foreground pixels. This task is a classification one, that can be achieved by crisp, statistical or fuzzy classification tools. For this, the different steps have to be achieved:

- **Preprocessing:** The preprocessing step avoids the detection of unimportant changes due to the motion of the camera or the illumination changes. This step may involve geometric and intensity adjustments [249]. As the scenes are usually rigid in nature and the camera jitter is small, geometric adjustments can often be performed using low-dimensional spatial transformations such as similarity, affine, or projective transformations [249]. On the other hand, there are several ways to achieve intensity adjustments. This can be done with intensity normalization [249]. The pixel intensity values in the current image are then normalized to have the same mean and variance as those in the background image. Another way consists in using an homomorphic filter based which is based on the shading model. This approach permits to separate the illumination and the reflectance. As only the reflectance component contains information about the objects in the scene, illumination-invariant foreground detection [311] [310] [231] can hence be performed by first filtering out the illumination component from the image.

- **Test:** The test which allows to classify pixels of the current image as background or foreground is usually the difference between the background image and the current image. This difference is then thresholded. Another ways to compare two images are the significance and hypothesis tests. The decision rule is then cast as a statistical hypothesis test. The decision as to whether or not a change has occurred at a given pixel corresponds to choosing one of two competing hypotheses: the null hypothesis H_0 or the alternative hypothesis H_1, corresponding to no-change and change decisions, respectively. Several significance tests can be found in [345] [281] [5] [4] [3] [220] [2] [6].

- **Threshold:** In literature, there are several types of threshold schemes. First, the threshold can be fixed and the same for all the pixels and the sequence. This scheme is simple but not optimal. Indeed, pixels present different activities and it needs an adaptive threshold. This can be done by computing the threshold via the local temporal standard deviation of intensity between the background and the current images, and by updating it using an infinite impulse response (IIR) filter such as in [71]. An adaptive threshold can be statistically obtained also from the variance of the pixel such as in [340]. Another way to adaptively threshold is to use fuzzy thresholds such as in [51].

- **Postprocessing:** The idea here is to enhance the consistency of the foreground mask. This can be done firstly by deleting isolated pixels with classical or statistical morphological operators [289]. Another way is to use fuzzy concepts such as fuzzy inference between the previous and the current foreground masks [282].

Moreover, foreground detection is a particular case of change detection when 1) one image is the background and the other one is the current image, and 2) the changes concern moving objects. So, all the techniques developed for change detection can be used in foreground detection. A survey concerning change detection can be found in [249] [254].

1.3.5   Size of the Picture's Element

This element may be a pixel [287], a block [107] or a cluster [30] as follows:

- **Pixel:** The background model is applied as an independent process on each pixel. This case is the most used but it does not take into account the spatial and temporal constraints.

- **Block:** Blocks can be overlapped or not [123]. A block is usually obtained as a vector of 3×3 neighbors of the current pixel. The advantage is to take into account the spatial dimension to improve the robustness and to reduce the computation time. Furthermore, blocks can be spatiotemporal ones [240]. Then, a dimensionality reduction technique is applied to obtain a compact vector representation of each block. These blocks provide a joint representation of texture and motion patterns. Their advantage is their robustness to noise and to movement in the background. The disadvantage is that the detection is less precise because only blocks are detected, making them unsuitable for applications that require detailed shape information.

- **Cluster:** First, the image in clusters which are generated using a color clustering mechanism of the nearest neighbor [30]. So, each cluster contains pixels that have similar features in the HSV space color. Then, the background model is applied on these clusters to obtain cluster of pixels classified as background or foreground. This cluster-wise approach gives less false alarms. Instead of the block-wise approach, the foreground detection is obtained with a pixel-wise precision.

Finally, the size of the element determines the robustness to the noise and the precision of the detection. A pixel-based method gives a pixel-based precision but it is less robust to noise than block-based or cluster-based methods.

1.3.6   Choice of the Features

The features characterize the picture's element. In the literature, there are five features commonly used: color features, edge features, stereo features, motion features and texture features. These features are generally crisp ones but some of them can be determined in a statistical or fuzzy way:

- **Crisp features:** Crisp features are the ones which do not use statistical or fuzzy concepts in their obtaining or their computation. The color features are the most used with RGB color space because it is the one directly available from the sensor or the camera. But the RGB color space has an important drawback: their three components are dependent which increase its sensitivity to illumination changes. For example, if a background point is covered by the shadow, the three component values at this point could be affected because the brightness and the chromaticity information are not separated. Alternative color spaces can be used such as YUV or YCrCb. Comparisons are presented in the literature [174] [252] [156] and usually YCrCb is selected as the most appropriate color space. Although color features are often very discriminative features of objects, they have several limitations in the presence of some challenges such as illumination changes, camouflage and shadows. To solve these problems, some authors proposed to use other features like edge, texture and stereo features in addition to the color features. For the edge features, they are computed using a gradient approach such as Canny edge detector. The edge handles the local illumination changes and the ghost leaved when waking foreground objects begin to move. For the stereo features, two cameras are needed to obtain the disparity or the depth. The stereo features deal with the camouflage. The motion features are usually obtained via optical flow but the main drawback is its computation time. Finally, the texture features are appropriate to illumination changes and to shadows. The texture features are generally obtained by using the Local Binary Pattern (LBP) [136] or the Local Ternary Pattern (LTP) [190]. Several variants of LBP and LTP can be found as can be seen in Table 1.2 at the part "Features Types".

- **Statistical features:** Texture features can be obtained by using statistical properties. The first work developed by Satoh et al. [259] proposed a Radial Reach Correlation (RRC) feature which has several variants: Bi-polar Radial Reach Correlation (BP-RRC) [258], Fast Radial Reach Correlation (F-RRC) [148], and Probabilistic Bi-polar Radial Reach Correlation (PrBP-RRC) [365]. In a similar way, Yokoi [364] used Peripheral Ternary Sign Correlation (PTESC). Recently, Yoshinaga et al. [368] proposed the Statistical Local Difference Pattern (SLDP). The aim of these statistical features is to be more robust to illumination changes.

- **Fuzzy features:** Texture features can be obtained by using fuzzy properties. For example, Chiranjeevi and Sengupta introduced fuzzy correlograms [62], fuzzy statistical texture features [63] and fuzzy 3D Histons [64]. The aim is to deal with illumination changes and dynamic backgrounds.

The features have different properties which allow to handle differently the critical situations (illumination changes, motion changes, structure background changes). If more than one feature is used, the operator which combines the results of each feature has to be choose. It may be a crisp operator (logical AND, logical OR), or a statistical one or a fuzzy one (Sugeno integral [373], Choquet integral [17]). The reader can see details about multi-features strategy in Section 2.3 of Chapter 2. Another way to take advantage of the properties of each feature is to make feature selection. The aim is to use the best feature or combination of features in a pixel [101].

1.3.7   Other Issues

Other issues can be summarized as follows:

- **Data Domains:** Background subtraction is usually achieved in the pixel domain but it can be also done in the measurement domain [336] [50] [185] [331] or in the compressed domain [329] [90].

- **Strategies:** Several strategies can be used to improve the robustness of a background model to the critical situations. It is important to note that these strategies can be generally applied with any background models (See Section 2.3).

- **Real time implementation:** Background subtraction is generally made in the context of real-time application on common PC or on smart embedded cameras. So, the computation time and memory requirement need to be reduced much as possible. This can be done by GPU implementations (See Section 2.4).

- **Performance evaluation:** Performance evaluation allows to compare the different models. Performance evaluation can be required in terms of time and memory

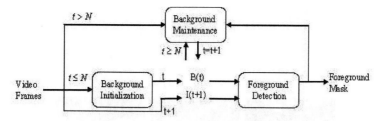

FIGURE 1.2 Background Subtraction Process. N is the number of frames that is used for the background initialization. B_t and I_t are the background and the current image at time t, respectively.

consuming or in terms of how well the algorithm detects the targets with less false alarms. Details are given in Section 1.7.

- **Datasets:** They are very important to allow a fair comparison of the algorithms on the same dataset. The traditional and recent datasets are reviewed in Section 1.8 and Section 2.7, respectively.

Figure 1.2 shows an overview of a background subtraction process which includes the following stages: (1) the background initialization module provides the first background image from N training frames, (2) Foreground detection that consists in classifying pixels as foreground or background, is achieved by comparing the background image and the current image. (3) Background maintenance module updates the background image by using the previous background, the current image and the foreground detection mask. The steps (2) and (3) are executed repeatedly as time progresses. Developing a background subtraction method, researchers must design each step and choose the features in relation to the critical situations (Section 1.5) they want to handle. These critical situations have different spatial and temporal properties.

1.4 Applications

Segmentation of moving foreground objects from a video stream is the fundamental step in many computer vision applications, such as:

- **Intelligent visual surveillance:** This is the main application of background modeling and foreground detection. The goal is to detect moving objects or abandoned objects to assure the security of the concern area, or to compute statistics on the traffic such as in road [316], airport [31] or maritime surveillance [32]. The objects of interest are very different such as vehicles, airplanes, boats, persons and baggages. Surveillance can be more specific such as to study consumer behavior in stores [184] [14] [182].
- **Intelligent visual observation of animals and insects:** Surveillance can also concern activities of animals in protected areas (river, ocean, etc.) or zoo for ethology. The objects of interest are then animals such as birds [171] [172] [215], fishes [285], honeybees [43] [170] [168] or hinds [214] [161].
- **Optical motion capture:** The aim is to obtain a full and precise capture of human with cameras [121]. The silhouette is generally extracted in each view by background subtraction. Then, the visual hull is obtained in three dimensions.
- **Human-machine interaction:** Several applications need interactions between human and machine through a video acquired in real-time by fixed cameras such as games (Microsoft's Kinect) and ludo-applications such as Aqu@theque [16].

FIGURE 1.3   The first row presents original frames in the following applications: Road Surveillance [120], Airport Surveillance [31] and Maritime Surveillance [32], the second row shows the ground truth (GT).

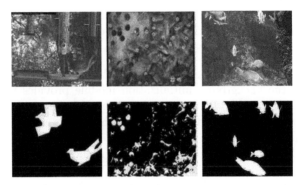

FIGURE 1.4   The first row presents original frames in animal surveillance: Birds [171], Honeybees [170] and Fish [16], the second row shows the corresponding ground truth (GT) or segmentation results.

- **Content based video coding:** To generate the video content, video has to be segmented into video objects and tracked as they transverse across the video frames. The registered background and the video object are then encoded separately. So, video coding needs an effective method for object detection from static and dynamic environments [61] [235].

Figure 1.3 and Figure 1.4 show samples of these applications, respectively. We can see that the objects to detect are very various in shape and in color such as cars, airplanes, boats, persons and animals (birds, honeybees, fishes). Therefore, there are generally no a priori on the shape and the color of the objects when background subtraction is applied.

Background subtraction can be used in applications in which cameras are slowly moving [266] [103] [291]. For example, Taneja et al. [298] proposed to model dynamic scenes recorded with freely moving cameras. In another way, Diaz et al. [91] presented a multi-view background subtraction for detecting dynamic objects in outdoor scenes.

All these applications show the importance of the moving object detection in video as it is the first step that is followed by tracking, recognition or behavior analysis. A study of the influence of background subtraction on these further steps can be found in [319]. Furthermore, these different applications present several specificities and need to deal with critical situations that are reviewed in Section 1.5 and some solutions are provided for each challenge.

TABLE 1.1 Background Subtraction Steps: An Overview

Categories	Sub-categories	Methods
Background Modeling	Traditional Models	*Basic:* Mean [180], Median [217], Histogram [384]
		Statistical: Gaussian (SG [340], SGG [162] MOG [287], MOGG [9], KDE [100])
		Support Vector (SVM [191], SVR [323], SVDD [302])
		Subspace Learning (PCA [230], ICA [357], INMF [41], IRT [187], LoPP [173])
		Clusters: K-means [42], Codebook [163], Basic Sequential Clustering [342]
		Neural Networks: General Regression [81] [82], Multivalued [203], Competitive [200], Dipolar Competitive [202]
		Neural Networks: Self Organizing (SOBS [206], SOBS-SC [210], 3dSOBS+ [211])
		Neural Networks: Growing Hierarchical Self Organizing [232], Adaptive Resonance Theory Neural Network [201]
		Estimation: Wiener Filter [312], Kalman Filter [157], Correntropy Filter [104], Chebychev Filter [53]
	Recent Models	*Advanced Statistical:* Mixture (Student's t [223], DMM [129], VarDMM [105], APMM [109])
		Advanced Statistical: Hybrid Models (KGMM [93] KGHM [196]), Non Parametric (ViBe [23], PBAS [138])
		Fuzzy: FCM [167], T2F-MOG [19]
		Discriminative and Mixed Subspace Models: IMMC [108], PCA-ILDA [213]
		RPCA: Principal Component Pursuit (PCP [46], SPCP [394], QPCP [25], BPCP [299], LPCP [338])
		RPCA: Outlier Pursuit [347], SpaCtrl [216], SpaCorr [141], LHR [89], IRLS [126], Bayesian RPCA [94]
		RPCA: Approximated RPCA (GoDec [391], Semi-Soft GoDec [391])
		Subspace Tracking: GRASTA [134], t-GRASTA [135], pROST [128] [261], GOSUS [348]
		Low Rank Minimization: DECOLOR [393], DRMF [346], DRMF-R [346], PRMF [359], BRMF [360]
		Sparse: Compressive Sensing [50], Structured Sparsity [145], Dynamic Group Sparsity [144]
		Sparse: Dictionary Learning [86], Sparse Error Estimation [92]
		Transform Domain: FFT [341], DCT [244], Walsh [309], Wavelet [115], Hadamard [22]
Background Initialization	Batch Models	*Basic Models:* Mean, Median [222]
	Progressive Models	*Basic Models:* Histogram [68] [73]
		Stable Intervals: Smoothness Detector (SD) [198], Adaptive Smoothness Detector (ASD) [198]
		Stable Intervals: Local Image Flow (LIF) [125], Optical Flow [55]
		Stable Intervals: Adaptive-Scale Sample Consensus (ASSC) [322]
		Stable Intervals: Dynamic Analysis [65], Spatio-Temporal Similarity [233]
		Statistical Models: SVM [191]
		Statistical Models: ECD based Mean Shift [195], Wronskian based SG [290]
		Statistical Models: Image Content Sensitivity based MOG [247]
		Transform Domain Models: Block-based DCT [251]
		Transform Domain Models: Block-based Hadamard Transform [22]
		Consecutive Frames: Block-based Sum of Absolute Differences (SAD) [307]
		Consecutive Frames: Patch-based Sum of Squared Distances (SSD) [72]
		Consecutive Frames: Moving Object Region Detection by Difference Product (DP) [127]
		Consecutive Frames: Evolutionary Algorithm (EA) [292]
		Consecutive Frames: Labeling Cost [70]
Background Maintenance	Strategical Models	Markov Random Fields (MRF) [251], Multi-Levels [76]
	Maintenance schemes	Blind update
		Selective update
		Fuzzy adaptive update [18]
	Learning rate	*Fixed learning rate:* One value (Good settings [370], Optimized settings [337])
		Variable learning rate: Two values (Initialization-Maintenance [154], Tracking feedback [166])
		Statistical learning rate: [181] [133] [193]
		Fuzzy learning rate: Exponential function [278] [276] [265], Adaptive exponential function [363]
		Fuzzy learning rate: Saturating linear function [207], Saturating linear function - Spatial coherence [209] [208]
	Maintenance mechanisms	Set of counters [378] [320]
		CUSUM detector [193]
	Frequence rate	Illumination changes score [241]
		Pixel activity [212]

TABLE 1.2 Foreground Detection and Features: An Overview

Categories	Sub-categories	Methods
Foreground Detection	Preprocessing	Geometric Adjustments [249]
		Intensity Adjustments: Intensity Normalization [249]
		Intensity Adjustments: Homomorphic Filtering [311] [310] [231]
		Tonal Adjustments: [288]
	Test	Difference Test
		Significance Test: Bhattacharyya Distance Test [345], Fast Significance Test [281]
		Significance Test: Probability Test [5] [4] [3]
		Significance Test: Total Least Squares Test [220] [2] [6]
		Linear Dependence Test: Linear dependence detector (LDD) [98], Wronskian Detector (WD) [99]
		Linear Dependence Test: Linear Approximation Model (LAM) [112]
	Threshold	Fixed Threshold
		Statistical Threshold: Adaptive Threshold [281] [4] [3] [220] [2] [6], Spatio-Temporal Adaptive Threshold [310]
		Statistical Threshold: Significance Invariance [1]
		Statistical Threshold: Standard Deviation [71], Variance [340]
		Fuzzy Threshold [51]
	Postprocessing	Morphological Operators
		Statistical Morphological Operators [289]
		Fuzzy Inference [282]
Features Size	Pixel	-
	Blocks	Blocks [123]
		Spatiotemporal Blocks [240]
	Clusters	Color Clusters [30]
Features Types	Crisp Features	*Color*: RGB, YUV [106], YCrCb [174], HSV [383], HSI [330]
		Color: Normalized RGB [350], Luv [361], Improved HLS Color Space [262]
		Color: Cylindrical RGB [163], Conical RGB [146]
		Histogram: LH [387], LDH [376]
		Edges: Gradient [150], Histogram of Oriented Gradient [151]
		Texture: LBP [136], STLBP [375], ELBP [326], SCS-LBP [355], HCS-LBP [354]
		Texture: SC-LBP [392], OC-LBP [183], SA-LBP [228]
		Texture: LTP [190], SILTP [380], MC-SILTP [380], SCS-LTP [204], CS-STLTP [351]
		Texture: Texton [286]
		Stereo: Disparity [119], Depth [133] [280] [178] [177]
		Motion: Optical flow [390]
	Statistical Features	*Texture*: PISC [257]
		Texture: RRC [259], BP-RCC [258], F-RRC [148], PrBP-RCC [365], PTESC [364]
		Texture: LDP [367], SLDP [368]
	Fuzzy Features	*Histogram*: Fuzzy 3D Histons [64]
		Texture: Fuzzy Correlograms [62], Fuzzy Statistical Texture Features [63]
Performance Evaluation	ROC Curves	True Positive (TP), True Negative (TN), False Positive (FP), False Negative (FN) [87]
		Detection Rate, F-Measure [87]
		F-Measure [87]
	Similarity	Similarity Measure [186]
	Perturbation Detection Rate	Perturbation Detection Rate [165]

TABLE 1.3 Background Subtraction Issues: An Overview

Categories	Sub-categories	Methods
Strategies	Multi-Features	*Two features:* Color and Edge [150], Color and Texture [67] [373] [17] *Two features:* Color and Disparity [119], Color and Depth [133] *Three features:* Color, Edge and Texture [15], Color, Gradient and Haar-like features [169] [130]
	Multi-Scales	Pixel and Block Scales [381] [85] [84] [83] [59] [122]
	Multi-Levels	*Two levels:* Pixel and Region [75] [78] [79] *Two levels:* Pixel and Frame [362] *Three levels:* Pixel, Region and Frame [150] [371] [312] [389]
	Multi-Resolutions	Multi-Resolutions [199] [385] [327] Hierarchical [56] [386] Coarse-to-Fine [29] [28] [114]
	Multi-Backgrounds	Short Term Background - Long Term Background *Same Background Models:* MOG-MOG [143] [242] [245] *Different Background Models:* Histogram (STLBP) - MOG [272] *Different Background Models:* SG (RRC) - KDE [297] [296] [295] *Different Background Models:* Predictive model - MOG [271] [273]
	Multi-Layers	Two Layers [358] Multiple Layers [243]
	Multi-Classifiers	Three Classifiers [250]
	Multi-Cues	*Two cues:* Bottom-up and Top-down Strategies [315] *Two cues:* Local and Global Cues [132] *Three cues:* Color, Edge and Intensity [146] *Three cues:* Color, Texture and Region [229]
	Multi-Criteria	Three Criteria [155]
	Map Selection	Two Areas [153]
	Markov Random Field	Markov Random Field [175] [260] Dynamic Hidden Markov Random Field [334] [335] [332] Hidden Conditional Random Fields [66]
Real-time Implementations	GPU Implementation	*Statistical:* SG [110], MOG [237] [176], SVM [118] [57] [58] *Clusters:* Codebook [110] *Neural Network:* SOBS [110]
	Embedded Implementation	Multimodal Mean [11] Light-Weight Foreground Detection [48] [49] CS-MOG [270]
	Dedicated Architecture	Digital Signal Processor (DSP) [13] [34] Very-large-Scale Integration (VLSI) [314] [236] Field-Programmable Gate Array (FPGA) [137] [339] [333] [12] [152] [188] [117] [253]
Sensors	Cameras	Monochromatic Cameras [279], Color CCD Cameras Web Cameras [284] [171] [172] High Definition Cameras [116]
	Infrared Cameras	Infrared Cameras [224] [74]
	RGB-D Cameras	Stereo Systems [119] [133] Time-of-Flight (ToF) Cameras [280] [178] [177] Microsoft's Kinect [45] [44]
Data Domain	Pixel Domain	-
	Measurement Domain	Compressive Sensing (CS) [336] [50] [185] [331]
	Compressed Domain	DCT [329], H.264 Streaming Video [90]
Datasets	Traditional Datasets	*Papers:* Wallflower [312], I2R [186], Carnegie Mellon [267] *Laboratories:* LIMU [367], USCD [52], SZTAKI [27] *Conference:* VSSN 2006, OTCBVS 2006, PETS *Project:* ATON [225]
	Recent Datasets	*Papers:* SABS [40] *Workshops:* ChangeDetection.net [120], Background Models Challenge (BMC) [318] *Projects:* MAR [32], RGB-D [44], CITIC RGB-D Dataset [111]

a) Low illum. b) Moderate illum. c) High illum. d) Foreground mask

FIGURE 1.5   From the left to the right: a) The first image presents an indoor scene with low illumination. b) The second image presents the same scene with a moderate illumination while the third image c) shows the scene with a high illumination. d) The fourth image shows the foreground mask obtained with MOG [287]. This sequence called "Time of Day" and comes from the Wallflower dataset [312].

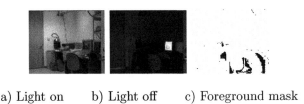

a) Light on   b) Light off   c) Foreground mask

FIGURE 1.6   From the left to the right: a) The first image presents an indoor scene with light on. b) The second image shows the same scene with light off. c) The third image shows the foreground mask obtained with MOG [287]. This sequence called "Light Switch" and comes from the Wallflower dataset [312].

1.5   Challenges

There are three main conditions which assure a good functioning of the background subtraction methods: the camera is fixed, the illumination is constant and the background is static, that is pixels have a unimodal distribution and no background objects are moved or inserted in the scene. In these ideal conditions, background subtraction gives good results. In practice, some critical situations may appear and perturb this process. In 1999, Toyama et al. [312] identified 10 challenging situations in the field of the video surveillance. In this Chapter, we extend this list to 13 which are the following ones :

- **Noise image:** It is due to a poor quality image source such as images acquired by a web cam or images after compression.
- **Camera jitter:** In some conditions, the wind may causes the camera to sway back and so it cause nominal motion in the sequence. Figure 1.8.a shows an illustration of camera jitter. The corresponding foreground mask shows false detections due to the motion.
- **Camera automatic adjustments:** Many modern cameras have auto focus, automatic gain control, automatic white balance and auto brightness control. These adjustments modify the dynamic in the color levels between different frames in the sequence.
- **Illumination changes:** They can be gradual such as ones in a day in an outdoor scene or sudden such as a light switch in an indoor scene. Figure 1.5 shows an indoor scene which presents a gradual illumination change. It causes false detections in several parts of the foreground mask as can be seen in Figure 1.5.d. Figure 1.6 shows the case of a sudden illumination change due to a light on/off. As all the pixels are affected by this change, a big amount of false detections is generated (see Figure 1.6.c).

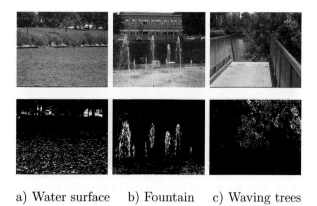

a) Water surface   b) Fountain   c) Waving trees

FIGURE 1.7 The first row presents original scenes containing dynamic backgrounds: a) Boats Sequence, b) Fountain Sequence and c) Overpass Sequence. These three sequences come from the ChangeDetection.net dataset [120]. The second row shows the corresponding foreground masks obtained by the MOG [287].

- **Bootstrapping:** During the training period, the background is not available in some environments. Then, it is impossible to compute a representative background image.

- **Camouflage:** A foreground object's pixel characteristics may be subsumed by the modeled background. Then, the foreground and the background cannot be distinguished. Figure 1.9.b shows an illustration of camouflage in color. A person enters the scene with a suitcase and leaves it on the floor. The difficulty is the similar color between the suitcase and the floor. Figure 1.11 and Figure 1.12 show a camouflage in color and in depth, respectively.

- **Foreground aperture:** When a moved object has uniform colored regions, changes inside these regions may not be detected. Thus, the entire object may not appear as foreground. Figure 1.9.c shows an illustration of foreground aperture. The foreground mask contains false negative detections.

- **Moved background objects:** Background objects can be moved. These objects should not be considered part of the foreground. Figure 1.8.b shows that both the initial and the new position of the box are detected without a robust maintenance mechanism.

- **Inserted background objects:** A new background object can be inserted. These objects should not be considered part of the foreground. Figure 1.8.c shows that the inserted box is detected without a robust maintenance mechanism.

- **Dynamic backgrounds:** Backgrounds can vacillate and this requires models which can represent disjointed sets of pixel values. Figure 1.7 shows three main types of dynamic backgrounds: Water surface, water rippling and waving trees. In each case, there is a big amount of false detections.

- **Beginning moving object:** When an object initially in the background moves, both it and the newly revealed parts of the background called "ghost" are detected. Figure 1.9.a shows a car which leaves a place. The corresponding foreground mask shows that both the initial and the new position of the car are detected without a robust maintenance mechanism.

- **Sleeping foreground object:** Foreground object that becomes motionless cannot be distinguished from a background object and then it will be incorporated in the background. How to manage this situation depends on the context. Indeed,

FIGURE 1.8   The first row presents original scenes containing camera jitter, moved background object, and inserted background object, respectively. a) Traffic, b) Abandoned Box and c) Sofa. These sequences come from the ChangeDetection.net dataset [120]. The second row shows the corresponding foreground masks obtained by the MOG [287].

FIGURE 1.9   The first row presents original scenes containing beginning moving object (a car), camouflage and foreground aperture, respectively. a) Parking (ChangeDetection.net dataset [120]), b) Camouflage (CITIC RGB-D Dataset [111]) and c) Foreground Aperture (Wallflower Dataset [312]). The second row shows the corresponding foreground masks obtained by the MOG [287].

in some applications, motionless foreground objects must be incorporated [147] and in others it is not the case [24].

- **Shadows:** Shadows can be detected as foreground and can come from background objects or moving objects (See Figure 1.10). Shadows detection is a research field itself. Complete studies and surveys can be found in [246] [377] [8] [256] [10].

The main difficulties come from the illumination changes and dynamic backgrounds. All the critical situations have different spatial and temporal properties. Table 1.4 gives an overview of which steps and issues are concerned to deal with them. The first column indicates the challenges and the second column the concerned step or issue with corresponding solutions. The reader is invited to read the following sections for the signification of each acronym. Furthermore, a similar table is available for the MOG improvements in [38] and for the KDE improvements in [36].

FIGURE 1.10   The first row presents original scenes containing shadows: a) Backdoor, b) Bungalows and c) PeopleInShade. These sequences come from the ChangeDetection.net dataset [120]. The second row shows the corresponding foreground masks obtained by the MOG [287].

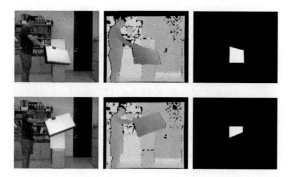

FIGURE 1.11   From the left to the right: The first image presents an indoor scene with a situation of camouflage in color. The second image presents the depth map which allows to deal with this camouflage and the third image shows the ground-truth. This sequence called "ColCamSeq" comes from the RGB-D Dataset [44].

FIGURE 1.12   From the left to the right: The first image presents an indoor scene with a situation of camouflage in depth. The second image presents the depth map which shows this camouflage and the third image shows the ground-truth. This sequence called "DCamSeq" comes from the RGB-D Dataset [44].

TABLE 1.4   Challenges and Solutions: An Overview

Challenges	Solutions
Noise image	*Clusters Models:* K-means [42], Codebook [163] [264], BSC [342] *Features:* Blocks [123] [240], Clusters [30]
Camera jitter	*Statistical Models:* MOG [287] [52] [263], KDE [269] [268] [304] *Statistical Models:* SVR [323], SVDD [302] [300] *Advanced Statistical:* Student's t [223], DMM [129], VarDMM [105], APMM [109] *Advanced Statistical:* KGMM [93], KGHMM [197], ViBe [23], PBAS [138] *Fuzzy Models:* FCM [167], T2-FMOG [19]
Camera automatic adjustments	*Background Maintenance:* MOG [372] *Features:* Edges [149]
Gradual Illumination changes	*Subspace Learning:* PCA [230], ICA [313], INMF [41], IRT [187], LoPP [173] *Filter Models:* Wiener Filter [312], Kalman Filter [157], Chebychev Filter [53] *RPCA Models:* PCP [46], IRLS [126], BRPCA [94], GoDec [391] *Subspace tracking models:* GRASTA [134], pROST [128] *Discriminative and Mixed Subspace Models:* IMMC [108], PCA-ILDA [213] *Sparse Models:* SS [145], DGS [144], RDL [382], SEE [353] *Preprocessing:* Spatial Illumination Compensation [234] *Test:* Gaussianity Test [227] *Features:* Illumination Ratio [325] [226] *Features:* Edge [139], Texture (SURF [264]) *Strategies:* Multi-Layers [294], Multi-Cues [146] [229] [315]
Sudden Illumination changes	*Preprocessing:* Spatial Illumination Compensation [234] *Preprocessing:* Chromaticity Compensation [21] [20] *Preprocessing:* Intensity Compensation [249] [311] [310] [231] *Preprocessing:* Tonal Compensation [288] *Test:* Gaussianity Test [227] *Features:* Illumination Ratio [325] [226] *Strategies:* Multi-Levels [312], Multi-Layers [294]
Bootstrapping	*Initialization:* SVM [191] *Initialization:* Block-based Sum of Absolute Differences (SAD) [307] *Initialization:* Patch-based Sum of Squared Distances (SSD) [72] *Cluster Models:* K-Means [42], Codebook [163] [263], BSC [342] *Features:* Blocks [140] [7] *Strategies:* Markov Random Fields (MRF) [251], Multi-Levels [76]
Camouflage	*Features:* Disparity [119] *Features:* Depth [133] [280] [178] [177]
Foreground aperture	*Background Maintenance:* MOG [317]
Moved background objects	*Background Maintenance:* MOG [379] [308] [321] *Background Maintenance:* Codebook [166]
Inserted background objects	*Background Maintenance:* MOG [379] [308] [321] *Background Maintenance:* Codebook [166]
Dynamic backgrounds	*Statistical Models:* MOG [287] [52] [263], KDE [269] [268] [304] *Statistical Models:* SVR [323], SVDD [302] [300] *Advanced Statistical:* Student's t [223], DMM [129], VarDMM [105], APMM [109] *Advanced Statistical:* KGMM [93], KGHMM [197], ViBe [23], PBAS [138] *Fuzzy Models:* FCM [167], T2-FMOG [19] *Domain Transform Models:* Waviz [341], Wave-Back [244] *Texture:* LBP [136], STLBP [375], ELBP [326], SCS-LBP [355] *Texture:* HCS-LBP [354], SC-LBP [392], OC-LBP [183], SA-LBP [228] *Texture:* LTP [190], SILTP [380], MC-SILTP [380], SCS-LTP [204] *Histogram:* LH [387], LDH [376]), Pattern (LDP [367], SLDP [368] *Strategies:* Multi-Cues [146] [132]
Beginning moving object	*Background Maintenance:* MOG [239] [307] *Strategies:* Multi-Cues [146]
Sleeping foreground object	*Background Maintenance:* MOG [239] [307] *Strategies:* Multi-Backgrounds [274] [275]
Shadows	*Features:* Colors, Edges, Textures [246] [377] [8] [256] [10] *Strategies:* Multi-Cues [146], Markov Random Fields (MRF) [27] [293]

1.6   Traditional Background Models

The traditional background representation models can be classified in the following categories: basic models, statistical models, cluster models, neural network models and estimation models. For each category, we give a table which allows the reader a quick overview.

1.6.1 Basic Models

In this case, the background is modeled using an average [180], a median [217] or an histogram analysis over time [384]. Once the model is computed, pixels of the current image are classified as foreground by thresholding the difference between the background image and the current frame as follows:

$$d\left(I_t(x,y), B_{t-1}(x,y)\right) > T \tag{1.5}$$

Otherwise, pixels are classified as background. T is a constant threshold, $I_t(x,y)$, $B_{t-1}(x,y)$ are respectively the current image at time t and the background image at time $t-1$. $d(.,.)$ is a distance measure which is usually the absolute difference between the current and the background images.

1.6.2 Statistical Models

The statistical models offer more robustness to illumination changes and dynamic backgrounds [39] [36]. The statistical background models can be classified in the following categories [39] [36]: Gaussian models, support vector models and subspace learning models.

- **Gaussian models:** The simplest way to represent the background is to assume that the history over time of pixel's intensity values can be modeled by a Gaussian. Following this idea, Wren et al. [340] have proposed to use a single Gaussian (SG). Kim et al. [162] generalized the SG using single general Gaussian (SGG) to alleviate the constraint of a strict Gaussian. However, a unimodal model cannot handle dynamic backgrounds when there are waving trees, water rippling or moving algae. To solve this problem, the Mixture of Gaussians (MOG) or Gaussian Mixture Model (GMM) [287] has been used to model dynamic backgrounds. Many improvements of the MOG were developed to be more robust and adaptive to the critical situations and a nice survey can be found in [38]. For example, Porikli and Tuzel [243] defined each pixel as layers of 3D multivariate Gaussians. Each layer corresponds to a different appearance of the pixel. Using a Bayesian approach, the probability distributions of mean and variance are estimated instead of the mean and variance themselves. In another way, Allili et al. [9] proposed the mixture of general Gaussians (MOGG) to alleviate the constraint of a strict Gaussian. However, the MOG and MOGG present several disadvantages. For example, background having fast variations cannot be accurately modeled with just a few Gaussians (usually 3 to 5), causing problems for sensitive detection. So, a non-parametric technique [100] was developed for estimating background probabilities at each pixel from many recent samples over time using Kernel Density Estimation (KDE) but it is time consuming. Several improvements of the KDE can be found in [39] [36]. For example, Sheikh and Shah [269] [268] modeled the background using a KDE method over a joint domain-range representation of image pixels. So, multi-modal spatial uncertainties and complex dependencies between the domain and range are directly modeled. Furthermore, the background and foreground models are used competitively in a MAP-MRF decision framework.
- **Support vector models:** The second category uses more sophisticated statistical models as Support Vector Machine (SVM) [191], support vector regression (SVR) [323] and support vector data description (SVDD) [302]. First, Lin et al. [191] [192] proposed to initialize the background using a probabilistic Support Vector Machine (SVM). SVM classification is applied for all pixels of each

training frame by computing the output probabilities. Newly found pixels are evaluated and determined if they should be added to the background model. The background initialization continues until there are no more new background pixels to be considered. The features used are the optical flow value and inter-frame difference. In a similar way, Wang et al. [323] [324] used a separate SVR to model each background pixel as a function of the intensity. The background initialization is made using a batch algorithm and the background maintenance is achieved with an on-line SVR algorithm [205]. For the foreground detection, they fed its intensity value to the SVR model associated with the pixel and they thresholded the output of the SVR. In another way, Tavakkoli et al. [302] [303] proposed to label pixels in video sequences into foreground and background classes using Support Vector Data Description (SVDD). Instead of parametric and non-parametric density estimation techniques, the background model is not based on the probability function of the background or foreground. Indeed, it is an analytical description of the decision boundary between background and foreground classes. So, the model accuracy is not bounded to the accuracy of the estimated probability density functions. So, the memory requirements are less than those of non-parametric techniques as, in non-parametric density estimation methods, pixel feature vectors for all background training frames need to be stored to regenerate the probability of pixels in new frames. For the background initialization, the process is made off-line and the background maintenance is made using an on-line algorithm. For the foreground detection, pixels are compared only with the support vectors, which are practically much fewer than the number of frames in the temporal window. Furthermore, SVDD explicitly models the decision boundary of the known class and results in less parameter tuning and automatic classification. The background initialization was improved by a genetic approach [301] and the background maintenance is made by an incremental SVDD algorithm [306] [305].

- **Subspace learning models:** The third category employs subspace learning methods. Subspace learning using Principal Component Analysis (SL-PCA) [230] is applied on N images to construct a background model, which is represented by the mean image and the projection matrix comprising the first p significant eigenvectors of PCA. In this way, foreground segmentation is accomplished by computing the difference between the input image and its reconstruction. However, this model presents several limitations [35]: (1) The size of the foreground objects must be small and they should not appear in the same location during a long period in the training sequence. Some authors [352] [158] [248] alleviate partially these constraints. For example, Xu et al. [352] proposed to apply recursively an error compensation process which reduces the influence of foreground moving objects on the eigenbackground model; (2) For the background maintenance, it is computationally intensive to perform model updating using the batch mode PCA. Moreover without a mechanism of robust analysis, the outliers or foreground objects may be absorbed into the background model. Some incremental mechanisms robust to outliers have been developed in [255] [189] [283]. The incremental PCA proposed by [255] needs less computation but the background image is contaminated by the foreground object. To solve this, Li et al. [189] proposed an incremental PCA which is robust in presence of outliers. However, when keeping the background model updated incrementally, it assigned the same weights to the different frames. Thus, clean frames and frames which contain foreground objects have the same contribution. The consequence is a relative pollution of the background model. In this context, Skocaj et al. [283] used a weighted incremental and robust. The weights are different following the frame and this method achieved a better background model. However, the weights were applied to the whole frame without considering the contribution of different image parts to building the background model. To achieve a pixel-wise precision for the weights, Zhang and Zhuang [374] proposed an adaptive weighted selection for an incremental PCA. This method performs a better model by assigning a weight to each pixel at each new frame during the update. Experiments [174] show that this method achieves better results than the SL-IRPCA [189]; (3) The application of this model is mostly limited to gray-scale images and pixel-wise aspect since the integration of multi-channel data is not straightforward. It involves much higher dimensional space and causes additional difficulty to manage data in general. Recently, Han and Jain [131] proposed an efficient algorithm using a weighted incremental 2-Dimensional Principal Component Analysis. The proposed algorithm was applied to 3-channel (RGB) and 4-channel (RGB+IR) data. Results show noticeable improvements in presence of multi-modal background and shadows; (4) The representation is not multi-modal so various illumination changes cannot be handled correctly. Recently, Dong et al. [95] proposed to use multi-subspace learning to handle different illumination changes. The feature space is organized into clusters which represent the different lighting conditions. A Local Principle Component Analysis (LPCA) transformation is used to learn separately an eigen-subspace for each cluster. When a current image arrives, the algorithm selects the learned subspace which shares the nearest lighting condition. The results [95] show that the LPCA algorithm outperforms the original PCA especially under sudden illumination changes. In a similar way, Kawanishi et al. [159] [160] generated the background image which well expresses the weather and the lighting condition of the scene. This method collects a huge number of images by super long term surveillance, classifies them according to their time in the day, and applies the PCA so as to reconstruct the background image. Recently, other reconstructive subspace models than the PCA one were used. For example, Yamazaki et al. [357] and Tsai et al. [313] used an Independent Component Analysis (ICA). In another way, Bucak et al. [41] proposed an Incremental Non-negative Matrix Factorization (INMF) to reduce the dimension. In order to take into account the spatial information, Li et al. [187] used an Incremental Rank-(R1,R2,R3) Tensor (IRT). Recently, Krishna et al. [173] used the Locality Preserving Projections (LoPP) also known as Laplacian eigenmaps. LoPP is a a classical linear technique that projects the data along the directions of maximal variance and optimally preserves the neighborhood structure of the observation matrix. Furthermore, LoPP shares many of the data representation properties of non-linear techniques which is interesting to separate background and foreground. A complete survey on reconstructive subspace learning models used in background modeling and foreground detection can be found [35].

The Gaussian models and support vector models are greatly designed for dynamics backgrounds and subspace learning models for illumination changes. The statistical models are the most used due to a good compromise between their performance and their computation cost. Table 1.5 shows an overview of the statistical background modeling methods. The first column indicates the category model and the second column the name of each method. Their corresponding acronym is indicated in the first parenthesis and the number of papers counted for each method in the second parenthesis. The third column gives the name of the authors and the date of the related publication.

TABLE 1.5   Statistical Models: An Overview.

Models	Methods	Authors - Dates
Gaussian	Single Gaussian (SG) (34) Single General Gaussian (SGG) (3) Mixture of Gaussians (MOG) (243) Mixture of General Gaussians (MOGG) (3) Kernel Density Estimation (KDE) (70)	Wren et al. (1997) [340] Kim et al. (2007) [162] Stauffer et al. (1999) [287] Allili et al. (2007) [9] Elgammal et al. (2000) [100]
Support Vector	Support Vector Machine (SVM) (11) Support Vector Regression (SVR) (3) Support Vector Data Description (SVDD) (6)	Lin et al. (2002) [191] Wang et al. (2006) [323] Tavakkoli et al. (2006) [302]
Subspace Learning	Principal Components Analysis (PCA) (30) Independent Component Analysis (ICA) (9) Inc. Non Negative Matrix Factorization (INMF) (3) Inc. Rank-(R1,R2,R3) Tensor (IRT) (2) Locally Preserving Projections (LoPP) (1)	Oliver et al. (1999) [230] Yamazaki et al. (2006) [357] Bucak et al.(2007 [41]) Li et al. (2008) [187] Krishna et al. (2012) [173]

1.6.3   Cluster Models

Cluster models suppose that each pixel in the frame can be represented temporally by clusters. The clustering approaches consist of using K-mean algorithm [42], using Codebook [163] or using basic sequential clustering [342].

- **K-means models:** Butler et al. [42] proposed an algorithm that represents each pixel in the frame by a group of clusters. The background initialization is achieved off-line. The clusters are ordered according to the likelihood that they model the background and are adapted to deal with background and lighting variations. Incoming pixels are matched against the corresponding cluster group and are classified according to whether the matching cluster is considered part of the background. To improve the robustness, Duan et al. [97] proposed to use a genetic K-means algorithm. The idea is to alleviate the disadvantages of the traditional K-means algorithm which has random and locality aspects causing lack of the global optimization.

- **Codebook models:** Kim et al. [163] proposed to model the background using a codebook model. For each pixel, a codebook is constructed and consists of one or more codewords. Samples at each pixel are clustered into the set of codewords based on a color distortion metric together with brightness bounds. The number of codewords is different following the pixel's activities. The clusters represented by codewords do not necessarily correspond to single Gaussian or other parametric distributions. The background is encoded on a pixel-by-pixel basis. Detection involves testing the difference of the current image from the background model with respect to color and brightness differences. If an incoming pixel verifies that 1) the color distortion to some codeword is less than the detection threshold, and 2) its brightness lies within the brightness range of that codeword, it is classified as background. Otherwise, it is classified as foreground. Practically, the codebook contains the following information: the means for each channels in the RGB color space, the minimum and maximum brightness that the codeword accepted, the frequency with which the codeword has occurred, the maximum negative run-length (MNRL) defined as the longest interval during the training period that the codeword has not recurred, the first and last access times that the codeword has occurred. The idea is to capture structural background variation due to periodic-like motion over a long period of time under limited memory. The codebook representation is efficient in memory and speed compared with other traditional background models. This original algorithm has been improved in several ways. For example, Kim et al. [164] presented a modified algorithm which a layered modeling/detection and adaptive codebook updating.

Sigari and Fatih [277] proposed a two-layer codebook model. The first layer is the main codebook, the second one is the cache codebook, and both contain some codewords relative to a pixel. Main codebook models the current background images and cache codebook is used to model new background images during input sequence. Other improvements concern the color models instead of the cone cylinder model such as in [96] where the authors used an hybrid-cone cylinder model and in [142] where the authors used a spherical model. Other modifications concern block based approach [88] [349], hierarchical approach [124] or multi-scale approach [369].

- **Basic sequential clustering:** Another clustering approach was developed in [342] based on assumption that the background would not be the parts which appear in the sequence for a short time. Firstly, pixels intensities are classified based on an on-line clustering model. Secondly, cluster centers and appearance probabilities of each cluster are calculated. Finally, a single or multi intensity clusters with the appearance probability greater than threshold are selected as the background pixel intensity value. A modified version proposed in [343] add in the second step a merging procedure to classify classes. When a cluster deviates to another and be very close to each other, the two clusters are fused in one cluster. Xiao and Zhang [344] improved this approach with a two-threshold sequential clustering algorithm. The avoiding of quick deviation of clusters to themselves is improved by the second threshold on the creation of new cluster. Recently, Benalia and Ait-Aoudia [26] proposed to address the problem of cluster deviation without the use of margin procedure or the second threshold. The algorithm consists of saving the first value of the cluster, when it is created, in another cluster center. Then, the current cluster value is compared to its past value after every updating operation of the cluster value for deviation control. If the deviation is important and bigger than threshold, a new cluster is created from the past one and takes the weight of the current one. To optimize the used space memory, the old cluster without updating are deleted basing on the assumption that a background cluster is updated with large time frequency.

The cluster models seem well adapted to deal with dynamic backgrounds and noise from video compression. Table 1.6 shows an overview of the cluster models. The first column indicates the category model and the second column the name of each method. Their corresponding acronym is indicated in the first parenthesis and the number of papers counted for each method in the second parenthesis. The third column gives the name of the authors and the date of the related publication.

TABLE 1.6 Cluster Models: An Overview.

Categories	Methods	Authors - Dates
K-Means	K-Means (KM) (9) Genetic K-Means (GKM) (1)	Butler et al. (2005) [42] Duan et al. (2011) [97]
Codebook	Original Codebook (CB) (1) Layered Codebook (LCB) (2) Hybrid Cone Cylinder Codebook (HCB) (2) Spherical Codebook (SCB) (1) Block based Codebook (BCB) (2) Hierarchical Codebook (HCB) (3) Multi-Scale Codebook (MCB) (1)	Kim et al. (2003) [163] Kim et al. (2005) [164] Doshi and Trivedi (2006) [96] Hu et al. (2012) [142] Deng et al. (2008) [88] Guo and Hsu (2010) [124] Zaharescu and Jamieson (2011) [369]
Sequential Clustering	Basic Sequential Clustering (BSC) (2) Modified BSC (MBSC)) (2) Two-Threshold SC (TTSC) (1) Improved MBSC (IMBSC) (1)	Xiao et al. (2006) [342] Xiao and Zhang (2008) [343] Xiao and Zhang (2008) [344] Benalia and Ait-Aoudia (2012) [26]

1.6.4   Neural Network Models

In this case, the background is represented by mean of the weights of a neural network suitably trained on N clean frames. The network learns how to classify each pixel as background or foreground. Typically, the neurons of the network encode the background model of one pixel or of the whole frame. In the context of background subtraction, the application of particular neural network types has turned out to be beneficial since they allow for intrinsic parallelization. This property of neural networks allows their efficient hardware implementation. The main approaches can be classified in the following way:

- **General Regression Neural Network:** Culibrk et al. [81] [82] proposed to use a neural network (NN) architecture to form an unsupervised Bayesian classifier for background modeling and foreground detection. The constructed classifier efficiently handles the segmentation in natural-scene sequences with complex background motion and changes in illumination. The weights allow to model the background and are updated to reflect the change statistics of the background. Furthermore, this algorithm is parallelized on a sub-pixel level and designed to enable efficient hardware implementation.

- **Multivalued Neural Network:** Luque et al. [203] used a foreground detection method based on the use of a multivalued discrete neural network. The multivalued neural network model is used to detect and correct some of the deficiencies and errors off the MOG algorithm. One of the advantages of using a multivalued neural network for foreground detection is that all process units (neurons) compute the solution to the problem in parallel. Another advantage of the multivalued neural model is its ability to represent nonnumerical classes or states, which can be very useful when dealing with foreground detection problems, in which pixel states are usually defined with qualitative labels: foreground, background and shadow.

- **Competitive Neural Network:** Luque et al. [200] used an unsupervised competitive neural network (CNN) based on adaptive neighborhoods to construct the background model. The weights and the adaptive neighborhood of the neurons models the background and are updated. This algorithm is parallelized on the pixel level and designed for hardware implementation to achieve real-time processing.

- **Dipolar Competitive Neural Network:** Luque et al. [202] proposed to improve the CNN approach by using a dipolar CNN which is used to classify the pixels as background or foreground. The dipolar representation is designed to deal with the problem of estimating the directionality of data at a low computational cost. The dipolar CNN achieved better results in terms of precision rate and false positive rate, whereas the false negative rate is, at least, comparable to the CNN.

- **Self Organizing Neural Network:** Maddalena and Petrosino [206] adopted a self-organizing neural network for learning motion patterns in the HSV color space. So, the background model is constructed. This algorithm named Self-Organizing Background Subtraction (SOBS) detects the moving object by using the background model through a map of motion and stationary patterns. Furthermore, an update neural network mapping method is used to make the neural network structure much simpler and the training step much more efficient. Recently, Maddalena and Petrosino [210] improved the SOBS by introducing spatial coherence into the background update procedure. This led to the so-called SC-SOBS algorithm, that provides further robustness against false detections.

- **Growing Hierarchical Self Organizing Neural Network:** The Self Organizing Neural Network (SONN) presents some limitations related to their fixed network structure in terms of number and arrangement of neurons, which has to be defined in advance, and their lack of representation of hierarchical relations among input. To address both limitations, Palomo et al. [232] proposed a growing hierarchical neural network. This neural network model has a hierarchical structure divided into layers, where each layer is composed of different single SONNs with adaptative structure that is determined during the unsupervised learning process according to input data. Experimental results show good performance in the case of illumination changes.

- **ART-type (Adaptive Resonance Theory) Neural Network:** Such networks exhibit two different fields of neurons (without the input layer) that are bi-directional completely linked: the comparison field and the recognition field. If the output of the recognition field, which is fed back to the comparison field, is close enough to the input pattern with respect to a vigilance parameter, the network is deemed to be in resonance and the weights of the network are adjusted. Both approaches developed in [201] [54] used an ART-network to represent each pixel of a video frame. For foreground segmentation the best matching neuron determines the classification of the pixel. Chen et al. [54] applied a fuzzy ART neural net on thermal imagery, whereas the ART network of Luque et al. [201] classifies RGB sequences and incorporates additional management of neurons to avoid too large numbers of neurons by merging overlapping clusters.

The SOBS [206] and the SOBS-SC [210] are the leading methods in the baseline category in the ChangeDetection.net dataset [120]. Table 1.7 shows an overview of the neural networks models. The first column indicates the name of each method. Their corresponding acronym is indicated in the first parenthesis and the number of papers counted for each method in the second parenthesis. The second column gives the name of the authors and the date of the related publication.

TABLE 1.7 Neural Networks Models: An Overview.

Methods	Authors - Dates
General Regression Neural Network (GNN)(4)	Culibrk et al. (2006) [81]
Multivalued Neural Network (MNN) (1)	Luque et al. (2008) [203]
Competitive Neural Network (CNN) (2)	Luque et al. (2008) [200]
Dipolar Competitive Neural Network (DCNN) (1)	Luque et al. (2008) [202]
Self Organizing Neural Network (SONN) (10)	Maddalena and Petrosino (2008) [206]
Growing Hierarchical SONN (GHSONN) (1)	Palomo et al. (2009) [232]
Adaptive Resonance Theory Neural Network (ART-NN) (2)	Luque et al. [201] (2010)

1.6.5 Estimation Models

The background is estimated using a filter. Any pixel of the current image that deviates significantly from its predicted value is declared foreground. This filter may be a Wiener filter [312], a Kalman filter [157] or a Chebychev filter [53].

- **Wiener filter:** Toyama et al. [312] proposed in their algorithm called Wallflower a pixel-level algorithm which makes probabilistic predictions about what background pixel values are, expected in the next live image using a one-step Wiener prediction filter. The Wiener filter is a linear predictor based on a recent history of values. Any pixel that deviates significantly from its predicted value is declared foreground. In their implementation, they use the past 50 values to compute 30 prediction coefficients. The Wiener filter works well for periodically changing pixels, and for random changes it produces a larger value of the threshold used in the foreground detection. The main advantage of the Wiener filter is that it reduces the uncertainty in a pixels value by accounting for how it varies with time. A disadvantage occurs when a moving object corrupts the history values. To solve this, Toyama et al. [312] kept a history of predicted values for each pixel as well as the history of actual values. For each new pixel, they computed two predictions based on the actual history and the predicted history. If either prediction is within the tolerance, the pixel is considered background. To handle adaptation, the prediction coefficients are recomputed for every new frame. Furthermore, they added a frame-level algorithm to deal with global illumination changes.

- **Kalman filter:** Karmann et al. [157] proposed a background estimation algorithm based on the Kalman filter which is an optimal estimator of the state of processes which verifies: (1) they can be modeled by a linear system, (2) the measurement and the process noise are white, and have zero mean Gaussian distributions. Under these conditions, knowing the input (external controls) and the output of the system, the Kalman filter provides an optimal estimate of the state of the process, by minimizing the variance of the estimation error and constraining the average of the estimated outputs and the average of the measures to be the same. It is characterized by two main equations: the state equation and the measurement equation. In the algorithm developed by [157], the system state corresponds to the background image B_t and the measurements to the input gray levels I_t. So, the method assumes that the evolution of the background pixel intensity can be described by a finite-dimension dynamic system. The system input term is set to zero, and the temporal distributions of the background intensities are considered constant. All unexpected changes are described by random noise, which by hypothesis is a zero mean Gaussian variable. In order to prevent foreground pixels modifying the background image, a different gain factor is introduced if the innovation overcomes a given threshold. In this approach, gradual illumination changes can be captured only by the random noise term which has to vary in time according to them. In this way the zero-mean hypothesis could be not respected. Moreover, sudden illumination changes cause intensity variations that are considered as foreground pixels and cannot be correctly managed. To address the previous disadvantages, Boninsegna and Bozzoli [33] introduced a new term to model the intensity variations caused by gradual illumination changes. The random noise term is split into a slow varying component and a zero mean Gaussian component. In this way, the state of the system consists of two components: the background value B_t and its changing rate β_t whose role is to model gradual illumination variations. In order to deal with foreground objects, Boninsegna and Bozzoli [33] introduced a statistical test which takes into consideration the innovation value. If the innovation is below a given threshold they say that the measure accords with the prediction and the filter works as usual, otherwise a different updating criterion is adopted. The threshold, computed on a statistical base, is related to the sum of the estimate error variance (P) and the measurement noise variance (R). Messelodi et al. [219] argued that it is incorrect to model all variations in the background intensities as noise added to the process. Indeed, changes caused by global illumination changes are external events that are different from changes due to the presence of moving objects. So, Messelodi et al. [219] developed a module that measures such global changes, and used this information as an external input to the system. The module exploits the fact that global illumination changes affect in a consistent way all the pixels in the image while variations due to foreground objects involve only a subset of them. To improve the use of the Kalman filter, some authors [388] [366] [356] used texture features instead of intensity ones. In a similar way, Gao et al. [113] set up the Kalman model with a local-region feature rather than a pixel feature, under the assumption that pixels in a small local-region have the same illumination changes and system noise. So, their algorithm employs a recursive least square adaptive filter to estimate the non-constant statistical characteristics of the noise in the region and update the background. Wang et al. [328] proposed to use an extension of the Kalman filter to non-linear systems which is called Unscented Kalman Filter (UKF) to address more robustly dynamic backgrounds, abrupt illumination changes and camera jitter. Fan et al. [104] used both static background and dynamic background information to renew a Self-Adaptive Kalman filter (SAKF). The two renewal rates in SAKF are obtained by the cumulants of the foreground detection in object region and background region automatically. This updating method is simple, efficient and can be used in real-time detection systems.

- **Correntropy filter:** The Kalman filter gives the optimal solution to the estimation problem when all the processes are Gaussian random processes and then KF offers a sub-optimal behavior in non-Gaussian settings which is the case in some challenging situations met in video-surveillance. So, Cinar and Principe [93] proposed a Correntropy filter (CF) that extract higher order information from the sequence. The information theoretic cost function is based on the similarity measure Correntropy. The Correntropy filter copes with salt and pepper noise which is not Gaussian.

- **Chebychev filter:** Chang et al. [53] proposed to use a Chebychev filter to model the background. The idea is to slowly update the background for changes in lighting and scenery while utilizing a small memory footprint as well as having a low computational complexity. The Chebychev filter used has a pass-band frequency of 0.06 Hz and ripple of 10 for a sampling rate of 30 frames per second. The correct estimation is reached after 1250 frames. The drastic changes can be detected if there is a large discrepancy between the background estimate and the current frame that persists for several frames throughout the entire image. Only 2 frames are kept in memory to represent the filter. Furthermore, only the intensity channel is utilized to generate the background image.

The estimation models seem well adapted for gradual illumination changes. Table 1.8 shows an overview of the estimation background models. The first column indicates the category model and the second column the name of each method. Their corresponding acronym is indicated in the first parenthesis and the number of papers counted for each method in the second parenthesis. The third column gives the name of the authors and the date of the related publication.

TABLE 1.8   Estimation Models: An Overview.

Estimation models	Algorithm	Authors - Dates
Wiener filter	Wiener Filter (WF) (1)	Toyama et al. (1999) [312]
Kalman filter	Kalman Filter (KF)(17) Unscented Kalman Filter (UKF)(1) Self-adaptive Kalman Filter (SAKF) (2)	Karmann et al. (1994) [157] Wang et al. (2006) [328] Fan et al. (2008) [104]
Correntropy filter	Correntropy (CF) (1)	Cinar and Principe (2011) [69]
Chebychev filter	Chebychev filter (CF) (2)	Chang et al. (2004) [53]

1.7   Performance Evaluation

Performance evaluation contains several meanings. Performance evaluation can be required in terms of time and memory consuming or in terms of how well the algorithm detects the targets with less false alarms. To evaluate performance in the first meaning, the time and the memory used can be measured easily by instruction in line code or with complexity analysis. For the evaluation in term of detection, there are three main approaches used in the literature:

- **ROC Curves Evaluation:** This evaluation using Receiver Operation Characteristics (ROC) Curves needs ground truth based metrics computed from the true positives (TP), true negatives (TN), false positives (FP) and false negatives (FN). FP and FN refer to pixels misclassified as foreground (FP) or background (FN) while TP and TN account for accurately classified pixels respectively as foreground and background. Then, different metrics can be computed such as the detection rate, the precision and the F-Measure. Detection rate gives the percentage of corrected pixels classified as background when compared with the total number of background pixels in the ground truth:

$$DR = \frac{TP}{TP + FN} \tag{1.6}$$

  Precision gives the percentage of corrected pixels classified as background as compared with the total pixels classified as background by the method:

$$Precision = \frac{TP}{TP + FP} \tag{1.7}$$

  A good performance is obtained when the Detection Rate is high without altering the Precision. The F-Measure (or effectiveness measure) is determined as follows:

$$F = \frac{2 * DR * Precision}{DR + Precision} \tag{1.8}$$

  The F-Measure characterizes the performance of classification in Precision-Detection Rate space. The aim is to maximize F closed to one. The reader can refer for a nice example to the evaluation made by Maddelena and Petrosino [209]. Furthermore, to see the progression of the performance of each algorithm, the ROC curves [87] can be used. For that, the false positive rate (FPR) and the true positive rate (TPR) are computed as follows:

$$FPR = \frac{FP}{FP + TN} \quad ; \quad TPR = \frac{TP}{TP + FN}$$

  The FPR is the proportion of background pixels that were erroneously reported as being moving object pixels. And the TPR is the proportion of moving object pixels that were correctly classified among all positive samples.

- **Similarity Evaluation:** This evaluation was introduced in this field by Li et al. [186]. Let A be a detected region and B be the corresponding ground truth, the similarity between A and B can be defined as:

$$S(A, B) = \frac{A \cap B}{A \cup B} \tag{1.9}$$

  If A and B are the same, $S(A, B)$ approaches 1, otherwise 0; that is, A and B have the least similarity.

- **Perturbation Detection Rate (PDR) Evaluation:** This evaluation was introduced by Kim et al. [165] for measuring performance of foreground-background segmentation. It has some advantages over the ROC evaluation. Specifically, it does not require foreground targets or knowledge of foreground distributions. The idea is to measure the sensitivity of a background subtraction algorithm in detecting possible low contrast moving objects against the background as a function of contrast, also depending on how well the model captures mixed dynamic background event. The reader can refer to the work of Kim et al. [165] for a nice evaluation of the codebook against the SG, the MOG and the KDE with the PDR evaluation.

Other prospective evaluation approaches using ground truth can be found in [225] [179] [194]. Recently, San Miguel and Martinez presented an evaluation without ground truth in [221]. Finally, the most used methods are the ROC Curves evaluation and the similarity one.

1.8 Traditional Datasets

Several datasets available to evaluate and compare background subtraction algorithms have been developed in the last decade. We classified them in traditional and recent datasets. The traditional datasets provide videos with several challenges and some ground-truth images but none of them address all the challenges. On the other hand, the recent datasets provide realistic large-scale videos with accurate ground truth giving balanced coverage of the range of challenges present in the real world. All these datasets are publicly available and their links are provided on the Background Subtraction Web Site in the section Available Datasets.

1.8.1 Wallflower Dataset

The Wallflower dataset[4] was provided by Toyama et al. [312] and gives representation of real-life situations typical of scenes susceptible to meet in video surveillance. Moreover, it consists of seven video sequences, with each sequence presenting one of the difficulties a practical task is likely to encounter (i.e illumination changes, dynamic backgrounds). The size of the images is 160×120 pixels. A brief description of the Wallflower image sequences can be made as follows:

- **Moved Object (MO):** A person enters a room, makes a phone call, and leaves. The phone and the chair are left in a different position. This video contains 1747 images.
- **Time of Day (TOD):** The light in a room gradually changes from dark to bright. Then, a person enters the room and sits down. This video contains 5890 images.
- **Light Switch (LS):** A room scene begins with the lights on. Then a person enters the room and turns off the lights for a long period. Later, a person walks in the room and switches on the light. This video contains 2715 images.
- **Waving Trees (WT):** A tree is swaying and a person walks in front of the tree. This video contains 287 images.
- **Camouflage (C):** A person walks in front of a monitor, which has rolling interference bars on the screen. The bars include similar color to the person's clothing.

---

[4]http://research.microsoft.com/en-us/um/people/jckrumm/wallflower/testimages.htm

FIGURE 1.13 Wallflower dataset [312]: The first line shows the original test image of each sequence and the second line shows the corresponding ground-truth in the following order from left to right: Moved Object, Time of Day, Light Switch, Waving Trees, Camouflage, Bootstrapping and Foreground Aperture.

This video contains 353 images.

- **Bootstrapping (B)**: The image sequence shows a busy cafeteria and each frame contains people. This video contains 3055 images.
- **Foreground Aperture (FA)**: A person with a uniformly colored shirt wakes up and begins to move slowly. This video contains 2113 images.

For each sequence, the ground truth is provided for one image when the algorithm has to show its robustness to a specific change in the scene. Thus, the performance is evaluated against hand-segmented ground truth. Figure 1.13 shows an overview of this dataset. The first line shows the original test image of each sequence and the second line shows the corresponding ground-truth. This dataset is the most used because it was the first in the field. However, as it provided only one ground-truth image by sequence, its use tends to disappear for the profit of the recent datasets.

1.8.2   I2R Dataset

The I2R dataset[5] provided by Lin and Huang [186] consists of nine video sequences, each sequence presenting dynamic backgrounds, illumination changes and bootstrapping issues. The size of the images is 176*144 pixels. This dataset consists of the following sequences:

- **Curtain**: A person presents a course in a meeting room with a moving curtain. This sequence contains 23893 images.
- **Campus**: Persons walk and vehicles pass on a road in front of waving trees. This sequence contains 1439 images.
- **Lobby:** Persons walk in an office building with switching on/off lights. This sequence contains 2545 images.
- **Shopping Mall**: This image sequence shows a busy shopping center and each frame contains people. This sequence contains 1286 images.
- **Airport**: This image sequence shows a busy hall of an airport and each frame contains people. This sequence contains 3584 images.
- **Restaurant**: This sequence comes from the wallflower dataset and shows a busy cafeteria. This video contains 3055 images.
- **Water Surface**: A person arrives in front of the sea. There are many waves. This sequence contains 633 images.
- **Fountain**: Persons walk in front of a fountain. This sequence contains 1523 images.

---

[5]http://perception.i2r.a-star.edu.sg/bk_model/bk_index.html

FIGURE 1.14 I2R Dataset [186]: The first line shows the original test image of each sequence and the second line shows the corresponding ground-truth in the following order from left to right: Curtain, Campus, Fountain, Water Surface, Lobby, Shopping Mall, and Airport.

The sequences Curtain, Campus, Water Surface, and Fountain present dynamic backgrounds whereas the sequence Lobby presents sudden illumination changes and the sequences Shopping Mall, Airport and Restaurant show bootstrapping issues. For each sequence, the ground truth is provided for twenty images when algorithms have to show their robustness. Figure 1.14 shows an overview of this dataset. The first line shows the original test image of each sequence and the second line shows the corresponding ground-truth. This dataset is frequently used due to the different varieties of dynamic backgrounds. In particular, it is commonly used in the evaluation of RPCA algorithms as Candes et al. [46] used in their papers on RPCA.

1.8.3 Carnegie Mellon Dataset

The sequence dataset[6] provided by Sheikh and Shah [267] involves a camera mounted on a tall tripod. The wind caused the tripod to sway back and forth causing nominal motion in the scene. Furthermore, some illumination changes occur. The ground-truth is provided for all the frames allowing a reliable evaluation.

1.8.4 LIMU Dataset

The Laboratory of Image and Media Understanding (LIMU[7]) [367] [368] provides 5 sequences with ground truth images made every 15 frames. Furthermore, three sequences from PETS 2001 dataset are given with the ground-truth images.

1.8.5 UCSD Dataset

This dataset [52] consists of 18 video sequences from the Statistical Visual Computing Lab (SVCL[8]). The frames of each sequence are provided in JPEG format. The ground truth mask is given in the form of a 3D array variable in MATLAB®, where 1 indicates foreground and 0 indicates background. For some sequences, the number of frames of the ground truth mask is smaller than the number of frames in the sequence. But the ground truth is provided for frames starting from frame 1 of the sequence.

---

[6]http://www.cs.cmu.edu/~yaser/new_backgroundsubtraction.htm
[7]http://limu.ait.kyushu-u.ac.jp/dataset/en/
[8]http://www.svcl.ucsd.edu/projects/background_subtraction

1.8.6   SZTAKI Surveillance Dataset

This benchmark set [27] contains raw video frames and binary foreground with shadow ground-truth masks, which were used for validation in publications. From this dataset, five evaluated sequences can be downloaded: two of them (Laboratory, Highway) come from the ATON benchmark set [225], but the enclosed ground truth was generated by Benedek [27]. The three remaining sequences (Sepm, Seam, Senoon) are outdoor surveillance videos. Not all frames of the video sequence have ground truth masks. Corresponding images have the same ordinary number.

1.8.7   VSSN 2006 Dataset

This dataset consists of different categories of video classified as follows: 1) Vacillate background, gradual illumination changes and shadows, 2) Sudden changes in illumination, 3) Bootstrapping and 4) Two or more cameras with many people. In each video, the moving objects are synthetic ones in real backgrounds. So, the ground-truth images are very precise. Furthermore, a development framework with Visual C++ was proposed to permit fair evaluation.

1.8.8   OTCBVS Benchmark Datasets

This benchmark related to the conference "Object Tracking and Classification in and Beyond the Visible Spectrum" (OTCBVS[9]) contains videos and images recorded in and beyond the visible spectrum. There are sequences for person detection and face detection. Three sequences are then interesting for background subtraction: (1) Dataset 01 (OSU Thermal Pedestrian) which concerns person detection in thermal imagery, (2) Dataset 03 (OSU Color-Thermal Database) on fusion-based object detection in color and thermal imagery and (3) Dataset 05 (Terravic Motion IR Database) which focuses on detection and tracking with thermal imagery.

1.8.9   PETS Datasets

These datasets related to the conference "Performance Evaluation of Tracking and Surveillance" (PETS) consist of different datasets such as PETS 2001, PETS 2003 and PETS 2006. They are more adapted for tracking evaluation than directly for background subtraction in the sense that the ground-truth is provided as bounding boxes.

1.8.10   ATON Dataset

The ATON dataset[10] [225] contains five sequences (Highway I, Highway II, Campus, Laboratory, Intelligent Room) which present shadows in indoor and outdoor environments. The ground-truth is provided for each sequence.

In summary, the Wallflower and I2R datasets provide videos with different representative challenges liable to meet in video surveillance. The CMU dataset focuses on the camera jitter. The OTCBVS dataset focuses on infrared videos. The ATON dataset is limited

---

[9]http://www.cse.ohio-state.edu/otcbvs-bench/
[10]http://cvrr.ucsd.edu/aton/shadow/index.html

to shadows. Finally, none of these datasets is a realistic large-scale dataset with accurate ground-truth providing a balanced coverage of the range of challenges present in the real world.

1.9   Conclusion

This chapter provides an overview of the different steps and challenges met in background modeling. Then, traditional models were surveyed and classified following the models used. Moreover, measures for performance evaluation were presented to compare and evaluate algorithms. In conclusion, the Gaussian models and support vector models are greatly designed for dynamic backgrounds and subspace learning models for illumination changes. The cluster models seem well adapted to deal with dynamic backgrounds and noise from video compression. The estimation models seem well adapted for gradual illumination changes. Finally, the neural network offers a good compromise between computational time and performance as the SOBS [206] and the SOBS-SC [210]. However, none of the traditional algorithm today seem to be able to simultaneously address all the key challenges that we meet in videos. So there is a need to investigate other representation models and to have a large-scale dataset to permit a complete performance evaluation.

References

1. T. Aach and A. Condurache. Transformation of adaptive thresholds by significance invariance for change detection. *International Workshop on Statistical Signal Processing*, July 2005.
2. T. Aach, L. Dumbgen, R. Mester, and D. Toth. Bayesian illumination-invariant motion detection. *IEEE International Conference on Image Processing, ICIP 2001*, 3:640–643, October 2001.
3. T. Aach and A. Kaup. Bayesian algorithms for adaptive change detection in image sequences using Markov Random Fields. *Signal Processing Image Communication*, 7:147–160, 1995.
4. T. Aach, A. Kaup, and R. Mester. Change detection in image sequences using Gibbs random fields: a Bayesian approach. *IEEE Workshop Intelligent Signal Processing and Communications Systems*, October 1993.
5. T. Aach, A. Kaup, and R. Mester. Statistical model-based change detection in moving video. *Signal Processing*, pages 165–180, 1993.
6. T. Aach, D. Toth, and R. Mester. Motion estimation in varying illumination using a total least squares distance measure. *Picture Coding Symposium, PCS 2003*, pages 145–148, April 2003.
7. R. Abbott and L. Williams. Multiple target tracking with lazy background subtraction and connected component analysis. *Machine Vision and Applications*, 20(2):93–101, January 2009.
8. N. Al-Najdawi, H. Bez, J. Singhai, and E. Edirisinghe. A survey of cast shadow detection algorithms. *Pattern Recognition Letters*, 33(6):752–764, April 2012.
9. M. Allili, N. Bouguila, and D. Ziou. A robust video foreground segmentation by using generalized Gaussian mixture modeling. *Canadian Conference on Computer and Robot Vision, CRV 2007*, pages 503–509, 2007.
10. A. Amato, I. Huerta, M. Mozerov, F. Roca, and J. Gonzalez. Moving cast shadows detection methods for video surveillance applications. *Augmented Vision and Reality*, September 2012.
11. S. Apewokin, B. Valentine, L. Wills, S. Wills, and A. Gentile. Multimodal mean adaptive

backgrounding for embedded real-time video surveillance. *IEmbedded Computer Vision Workshop, ECVW 2007*, June 2007.

12. K. Appiah, A. Hunter, P. Dickinson, and H. Meng. Accelerated hardware video object segmentation: from foreground detection to connected components labelling. *Computer Vision and Image Understanding, CVIU 2010*, 2010.

13. C. Arth, H. Bischof, and C. Leistner. TRICam - an embedded platform for remote traffic surveillance. *CVPR Workshop on Embedded Computer Vision, CVPRW 2006*, June 2010.

14. N. Avinash, M. Shashi Kumar, and S. Sagar. Automated video surveillance for retail store statistics generation. *International Conference on Signal and Image Processing, ICSIP 2012*, pages 585–596, 2012.

15. M. Azab, H. Shedeed, and A. Hussein. A new technique for background modeling and subtraction for motion detection in real-time videos. *International Conference on Image Processing, ICIP 2010*, pages 3453–3456, September 2010.

16. F. El Baf and T. Bouwmans. Comparison of background subtraction methods for a multimedia learning space. *International Conference on Signal Processing and Multimedia, SIGMAP*, July 2007.

17. F. El Baf, T. Bouwmans, and B. Vachon. Foreground detection using the choquet integral. *International Workshop on Image Analysis for Multimedia Interactive Integral, WIAMIS 2008*, pages 187–190, May 2008.

18. F. El Baf, T. Bouwmans, and B. Vachon. A fuzzy approach for background subtraction. *International Conference on Image Processing, ICIP 2008*, pages 2648–2651, October 2008.

19. F. El Baf, T. Bouwmans, and B. Vachon. Type-2 fuzzy mixture of Gaussians model: Application to background modeling. *International Symposium on Visual Computing, ISVC 2008*, pages 772–781, December 2008.

20. M. Bales, D. Forsthoefel, B. Valentine, D. Wills, and L. Wills. BigBackground-based illumination compensation for surveillance video. *EURASIP Journal on Image and Video Processing, Hindawi Publishing Corporation*, 2011:22, 2011.

21. M. Bales, D. Forsthoefel, D. Wills, and L. Wills. Chromatic sensitivity of illumination change compensation techniques. *International Symposium on Visual Computing, ISVC 2010*, pages 211–220, December 2010.

22. D. Baltieri, R. Cucchiara, and R. Vezzani. Fast background initialization with recursive hadamard transform. *International Conference on Advanced Video and Signal Based Surveillance, AVSS 2010*, September 2010.

23. O. Barnich and M. Van Droogenbroeck. ViBe: a powerful random technique to estimate the background in video sequences. *International Conference on Acoustics, Speech, and Signal Processing, ICASSP 2009*, pages 945–948, April 2009.

24. A. Bayona, J. SanMiguel, and J. Martinez. Comparative evaluation of stationary foreground object detection algorithms based on background subtraction techniques. *International Conference on Advanced Video and Signal Based Surveillance, AVSS 2009*, pages 25–30, September 2009.

25. S. Becker, E. Candes, and M. Grant. TFOCS: flexible first-order methods for rank minimization. *Low-rank Matrix Optimization Symposium, SIAM Conference on Optimization*, 2011.

26. M. Benalia and S. Ait-Aoudia. An improved basic sequential clustering algorithm for background construction and motion detection. *International Conference on Image Analysis and Recognition, ICIAR 2012*, June 2012.

27. C. Benedek and T. Sziranyi. Markovian framework for foreground-background-shadow segmentation of real world video scenes. *Asian Conference on Computer Vision, ACCV 2006*, January 2006.

28. A. Bevilacqua, G. Capelli, L. Di Stefano, and A. Lanza. A novel approach to change detection based on a coarse-to-fine strategy. *IEEE International Conference on Image Processing, ICIP 2005*, pages 434–437, 2005.

29. A. Bevilacqua, L. Di Stefano, and A. Lanza. Coarse-to-fine strategy for robust and efficient change detectors. *IEEE International Conference on Advanced Video and Signal Based Surveillance, AVSS 2005*, September 2005.

30. H. Bhaskar, L. Mihaylova, and A. Achim. Video foreground detection based on symmetric alpha-stable mixture models. *IEEE Transactions on Circuits, Systems and Video Technology*, March 2010.

31. P. Blauensteiner and M. Kampel. Visual surveillance of an airport's apron - an overview of the AVITRACK project. *Workshop of the Austrian Association for Pattern Recognition, AAPR 2004*, pages 213–220, 2004.

32. D. Bloisi, A. Pennisi, and L. Iocchi. Background modeling in the maritime domain. *Machine Vision and Applications, Special Issue on Background Modeling*, 2013.

33. M. Boninsegna and A. Bozzoli. A tunable algorithm to update a reference image. *Signal Processing: Image Communication*, 16(4):1353–365, 2000.

34. S. Boragno, B. Boghossian, J. Black, D. Makris, and S. Velastin. A DSP-based system for the detection of vehicles parked in prohibited areas. *AVSS 2007*, pages 1–6, 2007.

35. T. Bouwmans. Subspace learning for background modeling: A survey. *Recent Patents on Computer Science*, 2(3):223–234, November 2009.

36. T. Bouwmans. Recent advanced statistical background modeling for foreground detection: A systematic survey. *Recent Patents on Computer Science*, 4(3):147–171, September 2011.

37. T. Bouwmans. Background subtraction for visual surveillance: A fuzzy approach. *Handbook on Soft Computing for Video Surveillance, Taylor and Francis Group*, 5, March 2012.

38. T. Bouwmans, F. El-Baf, and B. Vachon. Background modeling using mixture of Gaussians for foreground detection: A survey. *Recent Patents on Computer Science*, 1(3):219–237, November 2008.

39. T. Bouwmans, F. El-Baf, and B. Vachon. Statistical background modeling for foreground detection: A survey. *Handbook of Pattern Recognition and Computer Vision, World Scientific Publishing*, 4(3):181–199, November 2009.

40. S. Brutzer, B. Höferlin, and G. Heidemann. Evaluation of background subtraction techniques for video surveillance. *International Conference on Computer Vision and Pattern Recognition, CVPR 2011*, pages 1937–1944, June 2011.

41. S. Bucak, B. Gunsel, and O. Gursoy. Incremental non-negative matrix factorization for dynamic background modeling. *International Workshop on Pattern Recognition in Information Systems*, June 2007.

42. D. Butler, V. Bove, and S. Shridharan. Real time adaptive foreground/background segmentation. *EURASIP*, pages 2292–2304, 2005.

43. J. Campbell, L. Mummert, and R. Sukthankar. Video monitoring of honey bee colonies at the hive entrance. *ICPR Workshop on Visual Observation and Analysis of Animal and Insect Behavior, VAIB 2008*, December 2008.

44. M. Camplani, C. Blanco, L. Salgado, F. Jaureguizar, and N. Garcia. Advanced background modeling with RGB-D sensors through classifiers combination and interframe foreground prediction. *Machine Vision and Application, Special Issue on Background Modeling*, 2013.

45. M. Camplani and L. Salgado. Background foreground segmentation with RGB-D Kinect data: an efficient combination of classifiers. *Journal on Visual Communication and Image Representation*, 2013.

46. E. Candes, X. Li, Y. Ma, and J. Wright. Robust principal component analysis? *International Journal of ACM*, 58(3), May 2011.

47. J. Carranza, C. Theobalt, M. Magnor, and H. Seidel. Free-viewpoint video of human actors. *ACM Transactions on Graphics*, 22(3):569–577, 2003.

48. M. Casares and S. Velipasalar. Resource-efficient salient foreground detection for embedded smart cameras by tracking feedback. *International Conference on Advanced Video and Signal Based Surveillance, AVSS 2010*, September 2010.

49. M. Casares, S. Velipasalar, and A. Pinto. Light-weight salient foreground detection for embedded smart cameras. *Computer Vision and Image Understanding*, 2010.

50. V. Cevher, D. Reddy, M. Duarte, A. Sankaranarayanan, R. Chellappa, and R. Baraniuk. Background subtraction for compressed sensing camera. *European Conference on Computer Vision, ECCV 2008*, October 2008.

51. M. Chacon-Murguia and S. Gonzalez-Duarte. An adaptive neural-fuzzy approach for object detection in dynamic backgrounds for surveillance systems. *IEEE Transactions on Industrial Electronic*, pages 3286–3298, August 2012.

52. A. Chan, V. Mahadevan, and N. Vasconcelos. Generalized stauffer-grimson background subtraction for dynamic scenes. *Machine Vision and Applications*, 21(5):751–766, September 2011.

53. T. Chang, T. Ghandi, and M. Trivedi. Vision modules for a multi sensory bridge monitoring approach. *ITSC 2004*, pages 971–976, 2004.

54. B. Chen, W. Wang, and Q. Qin. Infrared target detection based on fuzzy ART neural network. *International Conference on Computational Intelligence and Natural Computing*, 2:240–243, 2010.

55. C. Chen and J. Aggarwal. An adaptive background model initialization algorithm with objects moving at different depths. *International Conference on Image Processing, ICIP 2008*, pages 2264–2267, 2008.

56. Y. Chen, C. Chen, C. Huang, and Y. Hung. Efficient hierarchical method for background subtraction. *Journal of Pattern Recognition*, 40(10):2706–2715, October 2007.

57. L. Cheng and M. Gong. Real time background subtraction from dynamics scenes. *International Conference on Computer Vision, ICCV 2009*, September 2009.

58. L. Cheng, M. Gong, D. Schuurmans, and T. Caelli. Real-time discriminative background subtraction. *IEEE Transaction on Image Processing*, 20(5):1401–1414, December 2011.

59. S. Cheng, X. Luo, and S. Bhandarkar. A multiscale parametric background model for stationary foreground object detection. *International Workshop on Motion and Video Computing, WMCV 2007*, 2007.

60. S. Cheung and C. Kamath. Robust background subtraction with foreground validation for urban traffic video. *Journal of Applied Signal Processing, EURASIP*, 2005.

61. S. Chien, S. Ma, and L. Chen. Efficient moving object segmentation algorithm using background registration technique. *IEEE Transaction on Circuits and Systems for Video Technology*, 12(7):577–586, July 2012.

62. P. Chiranjeevi and S. Sengupta. Detection of moving objects using fuzzy correlogram based background subtraction. *ICSIPA 2011*, 2011.

63. P. Chiranjeevi and S. Sengupta. New fuzzy texture features for robust detection of moving objects. *IEEE Signal Processing Letters*, 19(10):603–606, October 2012.

64. P. Chiranjeevi and S. Sengupta. Robust detection of moving objects in video sequences through rough set theory framework. *Image and Vision Computing, IVC 2012*, 2012.

65. S. Chiu, Y. Sheng, and C. Chang. Robust initial background extraction algorithm based on dynamic analysis. *International Conference on Information Science and Digital Content Technology, ICIDT 2012*, 3:678–681, 2012.

66. Y. Chu, X. Ye, J. Qian, Y. Zhang, and S. Zhang. Adaptive foreground and shadow segmentation using hidden conditional random fields. *Journal of Zhejiang University*, pages 586–592, 2007.

67. T. Chua, K. Leman, and Y. Wang. Fuzzy rule-based system for dynamic texture and color based background subtraction. *IEEE International Conference on Fuzzy Systems, FUZZ-IEEE 2012*, pages 1–7, June 2012.

68. Y. Chung, J. Wang, and S. Cheng. Progressive background image generation. *IPPR Conference on Computer Vision, Graphics and Image Processing, CVGIP 2002*, pages 858–865, 2002.

69. G. Cinar and J. Principe. Adaptive background estimation using an information theoretic cost for hidden state estimation. *International Joint Conference on Neural Networks, IJCNN 2011*, August 2011.

70. S. Cohen. Background estimation as a labeling problem. *International Conference on Computer Vision, ICCV 2005*, 2:1034–1041, October 2005.

71. R. Collins, A. Lipton, T. Kanade, H. Fujiyoshi, D. Duggin, Y. Tsin, D. Tolliver, N. Enomoto, and O. Hasegawa. A system for video surveillance and monitoring. *Technical Report CMU-RI-TR-00-12, Robotics Institute, Carnegie Mellon University*, May 2000.

72. A. Colombari and A. Fusiello. Patch-based background initialization in heavily cluttered video. *IEEE Transactions on Image Processing*, 19(4):926–933, April 2010.

73. R. Mora Colque and G. Camara-Chavez. Progressive background image generation of surveillance traffic videos based on a temporal histogram ruled by a reward/penalty function. *SIBGRAPI 2011*, 2011.

74. C. Conaire, N. O'Connor, E. Cooke, and A. Smeaton. Multispectral object segmentation and retrieval in surveillance video. *IEEE International Conference on Image Processing, ICIP 2006*, pages 2381–2384, 2006.

75. M. Cristani, M. Bicego, and V. Murino. Integrated region- and pixel-based approach to background modeling. *IEEE Workshop on Motion and Video Computing, MOTION 2002*, 2002.

76. M. Cristani, M. Bicego, and V. Murino. Multi-level background initialization using Hidden Markov Models. *ACM SIGMM Workshop on Video Surveillance, IWVS 2003*, pages 11–19, 2003.

77. M. Cristani, M. Farenzena, D. Bloisi, and V. Murino. Background subtraction for automated multisensor surveillance: A comprehensive review. *EURASIP Journal on Advances in Signal Processing*, 2010:24, 2010.

78. M. Cristani and V. Murino. A spatial sampling mechanism for effective background subtraction. *International Conference on Computer Vision Theory and Applications, VISAPP 2007*, 2:403–410, March 2007.

79. M. Cristani and V. Murino. Background subtraction with adaptive spatio-temporal neighborhood analysis. *International Conference on Computer Vision Theory and Applications, VISAPP 2008*, January 2008.

80. X. Cui, Q. Liu, S. Zhang, and D. Metaxas. Temporal spectral residual for fast salient motion detection. *Neurocomputing*, pages 24–32, 2012.

81. D. Culbrik, O. Marques, D. Socek, H. Kalva, and B. Furht. A neural network approach to Bayesian background modeling for video object segmentation. *International Conference on Computer Vision Theory and Applications, VISAPP 2006*, February 2006.

82. D. Culbrik, O. Marques, D. Socek, H. Kalva, and B. Furht. Neural network approach to background modeling for video object segmentation. *IEEE Transaction on Neural Networks*, 18(6):1614–1627, 2007.

83. D. Culibrk, V. Crnojevic, and B. Antic. Multiscale background modelling and segmen-

tation. *International Conference on Digital Signal Processing, DSP 2009*, pages 922–927, 2009.

84. S. Davarpanah, F. Khalid, and M. Golchin. A block-based multi-scale background extraction algorithm. *Journal of Computer Science*, pages 1445–1451, 2010.

85. S. Davarpanah, F. Khalid, N. Lili, S. Puteri, and M. Golchin. Using multi-scale filtering to initialize a background extraction model. *Journal of Computer Science*, pages 1077–1084, 2012.

86. C. David, V. Gui, and F. Alexa. Foreground/background segmentation with learned dictionary. *International Conference on Circuits, Systems and Signals, CSS 2009*, pages 197–201, 2009.

87. J. Davis and M. Goadrich. The relationship between precision-recall and roc curves. *International Conference on Machine Learning, ICML 2006*, pages 233–240, 2006.

88. X. Deng, J. Bu, Z. Yang, C. Chen, and Y. Liu. A block-based background model for video surveillance. *IEEE International Conference on Acoustics, Speech and Signal Processing, ICASSP 2008*, pages 1013–1016, March 2008.

89. Y. Deng, Q. Dai, R. Liu, and Z. Zhang. Low-rank structure learning via log-sum heuristic recovery. *Preprint*, 2012.

90. B. Dey and M. Kundu. Robust background subtraction for network surveillance in h.264 streaming video. *IEEE Transactions on Circuits and Systems for Video Technology*, 2013.

91. R. Diaz, S. Hallman, and C. Fowlkes. Detecting dynamic objects with multi-view background subtraction. *International Conference on Computer Vision, ICCV 2013*, 2013.

92. M. Dikmen and T. Huang. Robust estimation of foreground in surveillance videos by sparse error estimation. *International Conference on Pattern Recognition, ICPR 2008*, December 2008.

93. J. Ding, M. Li, K. Huang, and T. Tan. Modeling complex scenes for accurate moving objects segmentation. *Asian Conference on Computer Vision, ACCV 2010*, pages 82–94, 2010.

94. X. Ding, L. He, and L. Carin. Bayesian robust principal component analysis. *IEEE Transaction on Image Processing*, 2011.

95. Y. Dong and G. DeSouza. Adaptive learning of multi-subspace for foreground detection under illumination changes. *Computer Vision and Image Understanding*, 2011.

96. A. Doshi and M. Trivedi. Hybrid cone-cylinder codebook model for foreground detection with shadow and highlight suppression. *AVSS 2006*, November 2006.

97. X. Duan, G. Sun, and T. Yang. Moving target detection based on genetic k-means algorithm. *International Conference on Communication Technology Proceedings, ICCT 2011*, pages 819–822, September 2011.

98. E. Durucan and T. Ebrahimi. Robust and illumination invariant change detection based on linear dependence. *European Signal Processing Conference, EUSIPCO 2000*, pages 1141–1144, September 2000.

99. E. Durucan and T. Ebrahimi. Change detection and background extraction by linear algebra. *Special Issue of IEEE on Video Communications and Processing for Third Generation Surveillance*, October 2001.

100. A. Elgammal and L. Davis. Non-parametric model for background subtraction. *European Conference on Computer Vision, ECCV 2000*, pages 751–767, June 2000.

101. A. Elgammal, T. Parag, and A. Mittal. A framework for feature selection for background subtraction. *International Conference on Pattern Recognition, CVPR 2006*, June 2006.

102. S. Elhabian, K. El-Sayed, and S. Ahmed. Moving object detection in spatial domain

using background removal techniques - state-of-art. *Patents on Computer Science*, 1(1):32–54, January 2008.

103. A. Elqursh and A. Elgammal. Online moving camera background subtraction. *European Conference on Computer Vision, ECCV 2012*, 2012.

104. D. Fan, M. Cao, and C. Lv. An updating method of self-adaptive background for moving objects detection in video. *ICALIP 2008*, pages 1497–1501, July 2008.

105. W. Fan and N. Bouguila. Online variational learning of finite Dirichlet mixture models. *Evolving Systems*, January 2012.

106. X. Fang, H. Xiong, B. Hu, and L. Wang. A moving object detection algorithm based on color information. *International Symposium on Instrumentation Science and Technology*, pages 384–387, 2006.

107. X. Fang, W. Xiong, B. Hu, and L. Wang. A moving object detection algorithm based on color information. *International Symposium on Instrumentation Science and Technology*, 48:384–387, 2006.

108. D. Farcas, C. Marghes, and T. Bouwmans. Background subtraction via incremental maximum margin criterion: A discriminative approach. *Machine Vision and Applications*, 23(6):1083–1101, October 2012.

109. A. Faro, D. Giordano, and C. Spampinato. Adaptive background modeling integrated with luminosity sensors and occlusion processing for reliable vehicle detection. *IEEE Transactions on Intelligent Transportation Systems*, 12(4):1398–1412, December 2011.

110. E. Fauske, L. Eliassen, and R. Bakken. A comparison of learning based background subtraction techniques implemented in CUDA. *NAIS 2009*, pages 181–192, 2009.

111. E. Fernandez-Sanchez, J. Diaz, and E. Ros. Background subtraction based on color and depth using active sensors. *Sensors*, 13:8895–8915, 2013.

112. B. Gao, T. Liu, Q. Cheng, and W. Ma. A linear approximation based method for noise-robust and illumination-invariant image change detection. *PCM 2004*, 2004.

113. D. Gao, J. Zhou, and L. Xin. A novel algorithm of adaptive background estimation. *International Conference on Image Processing, ICIP 2001*, 2:395–398, October 2001.

114. L. Gao, Y. Fan, N. Chen, Y. Li, and X. Li. Moving objects detection using adaptive region-based background model in dynamic scenes. *Foundations of Intelligent Systems, Advances in Intelligent and Soft Computing*, 2012.

115. T. Gao, Z. Liu, W. Gao, and J. Zhang. A robust technique for background subtraction in traffic video. *International Conference on Neural Information Processing, ICONIP 2008*, pages 736–744, November 2008.

116. M. Genovese and E. Napoli. ASIC and FPGA implementation of the Gaussian mixture model algorithm for real-time segmentation of high definition video. *IEEE Transactions on Very Large Scale Integration (VLSI) Systems*, 2013.

117. M. Genovese, E. Napoli, D. De Caro, N. Petra, and A. Strollo. FPGA implementation of Gaussian mixture model algorithm for 47fps segmentation of 1080p video. *Journal of Electrical and Computer Engineering*, 2013.

118. M. Gong and L. Cheng. Real time foreground segmentation on GPUs using local online learning and global graph cut optimization. *International Conference on Pattern Recognition, ICPR 2008*, December 2008.

119. G. Gordon, T. Darrell, M. Harville, and J. Woodfill. Background estimation and removal based on range and color. *International Conference on Computer Vision and Pattern Recognition, CVPR 1999*, pages 459–464, June 1999.

120. N. Goyette, P. Jodoin, F. Porikli, J. Konrad, and P. Ishwar. changedetection.net: A new change detection benchmark dataset. *IEEE Workshop on Change Detection, CDW 2012 at CVPR 2012*, June 2012.

121. G. Guerra-Filho. Optical motion capture: Theory and implementation. *Journal of Theoretical and Applied Informatics (RITA), Brazilian Computing Society*, 12(2):61–89, 2005.

122. P. Guha, D. Palai, K. Venkatesh, and A. Mukerjee. A multiscale co-linearity statistic based approach to robust background modeling. *Asian Conference on Computer Vision, ACCV 2006*, pages 297–306, 2006.

123. J. Guo and C. Hsu. Cascaded background subtraction using block-based and pixel-based codebooks. *International Conference on Pattern Recognition, ICPR 2010*, August 2010.

124. J. Guo and C. Hsu. Hierarchical method for foreground detection using codebook model. *International Conference on Image Processing, ICIP 2010*, pages 3441–3444, September 2010.

125. D. Gutchess, M. Trajkovic, E. Cohen, D. Lyons, and A. Jain. A background model initialization for video surveillance. *International Conference on Computer Vision, ICCV 2001*, pages 733–740, 2001.

126. C. Guyon, T. Bouwmans, and E. Zahzah. Moving object detection via robust low rank matrix decomposition with IRLS scheme. *International Symposium on Visual Computing, ISVC 2012*, pages 665–674, July 2012.

127. D. Ha, J. Lee, and Y. Kim. Neural-edge-based vehicle detection and traffic parameter extraction. *Image and Vision Computing*, 22:899–907, May 2004.

128. C. Hage and M. Kleinsteuber. Robust PCA and subspace tracking from incomplete observations using l_0-surrogates. *Preprint*, 2012.

129. T. Haines and T. Xiang. Background subtraction with Dirichlet processes. *European Conference on Computer Vision, ECCV 2012*, October 2012.

130. B. Han and L. Davis. Density-based multi-feature background subtraction with support vector machine. *IEEE Transactions on Pattern Analysis and Machine Intelligence, PAMI 2012*, 34(5):1017–1023, May 2012.

131. B. Han and R. Jain. Real-time subspace-based background modeling using multi-channel data. *International Symposium on Visual Computing, ISVC 2007*, pages 162–172, November 2007.

132. J. Han, M. Zhang, and D. Zhang. Background modeling fusing local and global cues for temporally irregular dynamic textures. *Advanced Science Letters*, 7:58–63, 2012.

133. M. Harville, G. Gordon, and J. Woodfill. Foreground segmentation using adaptive mixture models in color and depth. *International Workshop on Detection and Recognition of Events in Video*, July 2001.

134. J. He, L. Balzano, and J. Luiz. Online robust subspace tracking from partial information. *IT 2011*, September 2011.

135. J. He, D. Zhang, L. Balzano, and T. Tao. Iterative Grassmannian optimization for robust image alignment. *Image and Vision Computing*, June 2013.

136. M. Heikkila and M. Pietikainen. A texture-based method for modeling the background and detecting moving objects. *IEEE Transactions on Pattern Analysis and Machine Intelligence, PAMI 2006*, 28(4):657–62, 2006.

137. J. Hiraiwa, E. Vargas, and S. Toral. An FPGA based embedded vision system for real-time motion segmentation. *International Conference on Systems, Signals and Image Processing, IWSSIP 2010*, pages 360–363, 2010.

138. M. Hofmann, P. Tiefenbacher, and G. Rigoll. Background segmentation with feedback: The pixel-based adaptive segmenter. *IEEE Workshop on Change Detection, CVPR 2012*, June 2012.

139. P. Holtzhausen, V. Crnojevic, and B. Herbst. An illumination invariant framework for real-time foreground detection. *Journal of Real Time Image Processing*, November 2012.

140. H. Hsiao and J. Leou. Background initialization and foreground segmentation for boot-strapping video sequences. *EURASIP Journal on Image and Video Processing*, pages 1–19, 2013.

141. D. Hsu, S. Kakade, and T. Zhang. Robust matrix decomposition with sparse corruptions. *IEEE Transactions on Information Theory*, 57(11):7221–7234, 2011.

142. H. Hu, L. Xu, and H. Zhao. A spherical codebook in yuv color space for moving object detection. *Sensor Letters*, 10(1):177–189, January 2012.

143. J. Hu and T. Su. Robust background subtraction with shadow and highlight removal for indoor surveillance. *International Journal on Advanced Signal Processing*, pages 1–14, 2007.

144. J. Huang, X. Huang, and D. Metaxas. Learning with dynamic group sparsity. *International Conference on Computer Vision, ICCV 2009*, October 2009.

145. J. Huang, T. Zhang, and D. Metaxas. Learning with structured sparsity. *International Conference on Machine Learning, ICML 2009*, 2009.

146. I. Huerta, A. Amato, X. Roca, and J. Gonzalez. Exploiting multiple cues in motion segmentation based on background subtraction. *Neurocomputing, Special issue: Behaviours in video*, pages 183–196, January 2013.

147. I. Huerta, D. Rowe, J. Gonzalez, and J. Villanueva. Efficient incorporation of motionless foreground objects for adaptive background segmentation. *Articulated Motion and Deformable Objects, AMDO 2006*, pages 424–433, 2006.

148. M. Itoh, M. Kazui, and H. Fujii. Robust object detection based on radial reach correlation and adaptive background estimation for real-time video surveillance systems. *Real Time image Processing, SPIE 2008*, 2008.

149. V. Jain, B. Kimia, and J. Mundy. Background modelling based on subpixel edges. *ICIP 2007*, 6:321–324, September 2007.

150. O. Javed, K. Shafique, and M. Shah. A hierarchical approach to robust background subtraction using color and gradient information. *IEEE Workshop on Motion and Video Computing, WMVC 2002*, December 2002.

151. S. Javed, S. Oh, and S. Jung. Foreground object detection via background modeling using histograms of oriented gradient. *International Conference On Human Computer Interaction, HCI-2013*, January 2013.

152. H. Jiang, H. Ard, and V. wall. A hardware architecture for real-time video segmentation utilizing memory reduction techniques. *IEEE Transactions on Circuits and Systems for Video Technology*, 19(2):226–236, February 2009.

153. P. Jimenez, S. Bascon, R. Pita, and H. Moreno. Background pixel classification for motion detection in video image sequences. *International Work Conference on Artificial and Natural Neural Network, IWANN 2003*, pages 718–725, 2003.

154. P. KaewTraKulPong and R. Bowden. An improved adaptive background mixture model for real-time tracking with shadow detection. *AVBS 2001*, September 2001.

155. A. Kamkar-Parsi, R. Laganier, and M. Bouchard. Multi-criteria model for robust foreground extraction. *VSSN 2005*, pages 67–70, November 2007.

156. S. Kanprachar and S. Tangkawanit. Performance of RGB and HSV color systems in object detection applications under different illumination intensities. *International Multi Conference of Engineers and Computer Scientists*, 2:1943–1948, March 2007.

157. K. Karmann and A. Von Brand. Moving object recognition using an adaptive background memory. *Time-Varying Image Processing and Moving Object Recognition, Elsevier*, 1990.

158. S. Kawabata, S. Hiura, and K. Sato. Real-time detection of anomalous objects in dynamic scene. *International Conference on Pattern Recognition, ICPR 2006*, 3:1171–1174, August 2006.

159. Y. Kawanishi, I. Mitsugami, M. Mukunoki, and M. Minoh. Background image generation keeping lighting condition of outdoor scenes. *International Conference on Security Camera Network, Privacy Protection and Community Safety, SPC 2009*, October 2009.

160. Y. Kawanishi, I. Mitsugami, M. Mukunoki, and M. Minoh. Background image generation by preserving lighting condition of outdoor scenes. *Procedia - Social and Behavioral Science*, 2(1):129–136, March 2010.

161. P. Khorrami, J. Wang, and T. Huang. Multiple animal species detection using robust principal component analysis and large displacement optical flow. *Workshop on Visual Observation and Analysis of Animal and Insect Behavior, VAIB 2012*, 2012.

162. H. Kim, R. Sakamoto, I. Kitahara, T. Toriyama, and K. Kogure. Robust foreground extraction technique using Gaussian family model and multiple thresholds. *Asian Conference on Computer Vision, ACCV 2007*, pages 758–768, November 2007.

163. K. Kim, T. Chalidabhongse, D. Harwood, and L. Davis. Background modeling and subtraction by codebook construction. *IEEE International Conference on Image Processing, ICIP 2004*, 2004.

164. K. Kim, T. Chalidabhongse, D. Harwood, and L. Davis. Real time foreground background segmentation using codebook model. *Real time Imaging*, 11(3):167–256, 2005.

165. K. Kim, T. Chalidabhongse, D. Harwood, and L. Davis. PDR: performance evaluation method for foreground-background segmentation algorithms. *EURASIP Journal on Applied Signal Processing*, 2006.

166. K. Kim, D. Harwood, and L. Davis. Background updating for visual surveillance. *International Symposium, ISVC 2005*, pages 337–346, December 2005.

167. W. Kim and C. Kim. Background subtraction for dynamic texture scenes using fuzzy color histograms. *IEEE Signal Processing Letters*, 3(19):127–130, March 2012.

168. T. Kimura, M. Ohashi, K. Crailsheim, T. Schmickl, R. Odaka, and H. Ikeno. Tracking of multiple honey bees on a flat surface. *International Conference on Emerging Trends in Engineering and Technology, ICETET 2012*, pages 36–39, November 2012.

169. B. Klare and S. Sarkar. Background subtraction in varying illuminations using an ensemble based on an enlarged feature set. *IEEE Conference on Computer Vision and Pattern Recognition, CVPR 2009*, June 2009.

170. U. Knauer, M. Himmelsbach, F. Winkler, F. Zautke, K. Bienefeld, and B. Meffert. Application of an adaptive background model for monitoring honeybees. *VIIP 2005*, 2005.

171. T. Ko, S. Soatto, and D. Estrin. Background subtraction on distributions. *European Conference on Computer Vision, ECCV 2008*, pages 222–230, October 2008.

172. T. Ko, S. Soatto, and D. Estrin. Warping background subtraction. *IEEE International Conference on Computer Vision and Pattern Recognition, CVPR 2010*, June 2010.

173. M. Krishna, V. Aradhya, M. Ravishankar, and D. Babu. LoPP: Locality preserving projections for moving object detection. *International Conference on Computer, Communication, Control and Information Technology, C3IT 2012*, 4:624–628, February 2012.

174. F. Kristensen, P. Nilsson, and V. Wall. Background segmentation beyond RGB. *ACCV 2006*, pages 602–612, 2006.

175. P. Kumar and K. Sengupta. Foreground background segmentation using temporal and spatial Markov processes. *Department of Electrical and Computer Engineering, National University of Singapore*, November 2000.

176. P. Kumar, A. Singhal, S. Mehta, and A. Mittal. Real-time moving object detection algorithm on high-resolution videos using GPUs. *Journal of Real-Time Image Processing*, January 2013.

177. B. Langmann, S. Ghobadi, K. Hartmann, and O. Loffeld. Multi-model background subtraction using Gaussian mixture models. *Symposium on Photogrammetry Computer Vision and Image Analysis, PCV 2010*, pages 61–66, 2010.

178. B. Langmann, K. Hartmann, and O. Loffeld. Depth assisted background subtraction for color capable ToF-cameras. *International Conference on Image and Video Processing and Computer Vision, IVPCV 2010*, pages 75–82, July 2012.

179. N. Lazarevic-McManus, J. Renno, D. Makris, and G. Jones. An object-based comparative methodology for motion detection based on the F-Measure. *Computer Vision and Image Understanding*, 111(1):74–85, 2008.

180. B. Lee and M. Hedley. Background estimation for video surveillance. *Image and Vision Computing New Zealand, IVCNZ*, pages 315–320, 2002.

181. D. Lee. Improved adaptive mixture learning for robust video background modeling. *APR Workshop on Machine Vision for Application, MVA 2002*, pages 443–446, December 2002.

182. S. Lee, N. Kim, I. Paek, M. Hayes, and J. Paik. Moving object detection using unstable camera for consumer surveillance systems. *International Conference on Consumer Electronics, ICCE 2013*, pages 145–146, January 2013.

183. Y. Lee, J. Jung, and I. Kweon. Hierarchical on-line boosting based background subtraction. *Workshop on Frontiers of Computer Vision, FCV 2011*, pages 1–5, February 2011.

184. A. Leykin and M. Tuceryan. Detecting shopper groups in video sequences. *International Conference on Advanced Video and Signal-Based Surveillance, AVSS 2007*, 2007.

185. J. Li, J. Wang, and W. Shen. Moving object detection in framework of compressive sampling. *Journal of Systems Engineering and Electronics*, 21(5):740–745, October 2010.

186. L. Li and W. Huang. Statistical modeling of complex background for foreground object detection. *IEEE Transaction on Image Processing*, 13(11):1459–1472, November 2004.

187. X. Li, W. Hu, Z. Zhang, and X. Zhang. Robust foreground segmentation based on two effective background models. *MIR 2008*, pages 223–228, October 2008.

188. X. Li and X. Jing. FPGA based mixture Gaussian background modeling and motion detection. *International Conference on Natural Computation, ICNC 2011*, 4:2078–2081, 2011.

189. Y. Li. On incremental and robust subspace learning. *Pattern Recognition*, 37(7):1509–1518, 2004.

190. S. Liao, G. Zhao, V. Kellokumpu, M. Pietikinen, and S. Li. Modeling pixel process with scale invariant local patterns for background subtraction in complex scenes. *International Conference on Computer Vision and Pattern Recognition, CVPR 2010*, June 2010.

191. H. Lin, T. Liu, and J. Chuang. A probabilistic SVM approach for background scene initialization. *International Conference on Image Processing, ICIP 2002*, 3:893–896, September 2002.

192. H. Lin, T. Liu, and J. Chuang. Learning a scene background model via classification. *IEEE Transactions on Signal Processing*, 57(5):1641–1654, May 2009.

193. J. Lindstrom, F. Lindgren, K. Astrom, J. Holst, and U. Holst. Background and foreground modeling using an online EM algorithm. *IEEE International Workshop on Visual Surveillance VS 2006 in conjunction with ECCV 2006*, pages 9–16,

May 2006.

194. L. Liu and N. Sang. Metrics for objective evaluation of background subtraction algorithms. *International Conference on Image and Graphics, ICIG 2011*, August 2011.

195. Y. Liu, H. Yao, W. Gao, X. Chen, and D. Zhao. Nonparametric background generation. *International Conference on Pattern Recognition, ICPR 2006*, 4:916–919, 2006.

196. Z. Liu, W. Chen, K. Huang, and T. Tan. Probabilistic framework based on KDE-GMM hybrid model for moving object segmentation in dynamic scenes. *International Workshop on Visual Surveillance, ECCV 2008*, October 2008.

197. Z. Liu, K. Huang, and T. Tan. Foreground object detection using top-down information based on EM framework. *IEEE Transactions on Image Processing*, 21(9):4204–4217, September 2012.

198. W. Long and Y. Yang. Stationary background generation: An alternative to the difference of two images. *Pattern Recognition*, 12(23):1351–1359, 1990.

199. R. Luo, L. Li, and I. Gu. Efficient adaptive background subtraction based on multiresolution background modelling and updating. *Pacific-RIM Conference on Multimedia, PCM 2007*, December 2007.

200. R. Luque, E. Dominguez, E. Palomo, and J. Munoz. A neural network approach for video object segmentation in traffic surveillance. *International Conference on Image Analysis and Recognition, ICIAR 2008*, pages 151–158, 2008.

201. R. Luque, F. Dominguez, E. Palomo, and J. Muñoz. An ART-type network approach for video object detection. *European Symposium on Artificial Neural Networks*, pages 423–428, 2010.

202. R. Luque, D. Lopez-Rodriguez, E. Dominguez, and E. Palomo. A dipolar competitive neural network for video segmentation. *Ibero-American Conference on Artificial Intelligence, IBERAMIA 2008*, pages 103–112, 2008.

203. R. Luque, D. Lopez-Rodriguez, E. Merida-Casermeiro, and E. Palomo. Video object segmentation with multivalued neural networks. *IEEE International Conference on Hybrid Intelligent Systems, HIS 2008*, pages 613–618, 2008.

204. F. Ma and N. Sang. Background subtraction based on multi-channel SILTP. *Asian Conference on Computer Vision, ACCV 2012*, November 2012.

205. J. Ma and J. Theiler. Accurate on-line support vector regression. *Neural Computation*, 15:2683–2703, 2003.

206. L. Maddalena and A. Petrosino. A self organizing approach to background subtraction for visual surveillance applications. *IEEE Transactions on Image Processing*, 17(7):1729–1736, 2008.

207. L. Maddalena and A. Petrosino. Multivalued background/foreground separation for moving object detection. *International Workshop on Fuzzy Logic and Applications, WILF 2009*, 5571:263–270, June 2009.

208. L. Maddalena and A. Petrosino. Self organizing and fuzzy modelling for parked vehicles detection. *Advanced Concepts for Intelligent Vision Systems, ACIVS 2009*, pages 422–433, 2009.

209. L. Maddalena and A. Petrosino. A fuzzy spatial coherence-based approach to background/foreground separation for moving object detection. *Neural Computing and Applications, NCA 2010*, pages 1–8, 2010.

210. L. Maddalena and A. Petrosino. The SOBS algorithm: What are the limits? *IEEE Workshop on Change Detection, CVPR 2012*, June 2012.

211. L. Maddalena and A. Petrosino. The 3dSOBS+ algorithm for moving object detection. *Computer Vision and Image Understanding, CVIU 2014*, 122(65-73), May 2014.

212. D. Magee. Tracking multiple vehicles using foreground, background and motion models. *Image and Vision Computing*, 22:143–155, 2004.

213. C. Marghes, T. Bouwmans, and R. Vasiu. Background modeling and foreground detection via a reconstructive and discriminative subspace learning approach. *International Conference on Image Processing, Computer Vision, and Pattern Recognition, IPCV 2012*, July 2012.

214. S. Mashak, B. Hosseini, and S. Abu-Bakar. Background subtraction for object detection under varying environments. *International Journal of Computer Information Systems and Industrial Management Applications*, 4:506–513, 2012.

215. S. Mashak, B. Hosseini, and S. Abu-Bakar. Real-time bird detection based on background subtraction. *World Congress on Intelligent Control and Automation, WCICA 2012*, pages 4507–4510, July 2012.

216. G. Mateos and G. Giannakis. Sparsity control for robust principal component analysis. *International Conference on Signals, Systems, and Computers*, November 2010.

217. N. McFarlane and C. Schofield. Segmentation and tracking of piglets in images. *British Machine Vision and Applications*, pages 187–193, 1995.

218. A. McIvor. Background subtraction techniques. *International Conference on Image and Vision Computing, New Zealand, IVCNZ 2000*, November 2010.

219. S. Messelodi, C. Modena, N. Segata, and M. Zanin. A Kalman filter based background updating algorithm robust to sharp illumination changes. *ICIAP 2005*, 3617:163–170, September 2005.

220. R. Mester, T. Aach, and L. Duembgen. Illumination-invariant change detection using a statistical colinearity criterion. *DAGM 2001*, pages 170–177, September 2001.

221. J. San Miguel and J. Martinez. On the evaluation of background subtraction algorithms without ground-truth. *International Conference on Advanced Video and Signal Based Surveillance, AVSS 2010*, September 2010.

222. M. Molinier, T. Hame, and H. Ahola. 3d connected components analysis for traffic monitoring in image sequences acquired from a helicopter. *Scandinavian Conference, SCIA 2005*, page 141, June 2005.

223. D. Mukherjee and J. Wu. Real-time video segmentation using Student's t mixture model. *International Conference on Ambient Systems, Networks and Technologies, ANT 2012*, pages 153–160, 2012.

224. S. Nadimi and B. Bhanu. Physics-based cooperative sensor fusion for moving object detection. *IEEE Workshop on Learning in Computer Vision and Pattern Recognition, CVPR 2004*, June 2004.

225. J. Nascimento and J. Marques. Performance evaluation of object detection algorithms for video surveillance. *IEEE Transaction on Multimedia*, pages 761–774, August 2006.

226. K. Ng and E. Delp. Background subtraction using a pixel-wise adaptive learning rate for object tracking initialization. *SPIE Conference on Visual Information Processing and Communication*, January 2011.

227. K. Ng, S. Srivastava, and E. Delp. Foreground segmentation with sudden illumination changes using a shading model and a Gaussianity test. *IEEE International Symposium on Image and Signal Processing and Analysis*, 11(3):241–248, March 2011.

228. S. Noh and M. Jeon. A new framework for background subtraction using multiple cues. *Asian Conference on Computer Vision, ACCV 2012*, November 2012.

229. S. Noh and M. Jeon. A new framework for background subtraction using multiple cues. *Asian Conference on Computer Vision, ACCV 2012*, November 2012.

230. N. Oliver, B. Rosario, and A. Pentland. A Bayesian computer vision system for modeling human interactions. *International Conference on Vision Systems, ICVS 1999*, January 1999.

231. G. Pajares, J. Ruz, and J. Manuel de la Cruz. Performance analysis of homomorphic

systems for image change detection. *IBPRIA 2005*, pages 563–570, 2005.

232. E. Palomo, E. Dominguez, R. Luque, and J. Munoz. Image hierarchical segmentation based on a GHSOM. *International Conference on Neural Information Processing, ICONIP 2009*, pages 743–750, 2009.

233. G. Park. Background initialization by spatiotemporal similarity. *Journal of Broadcast Engineering*, 12(3):289–292, 2007.

234. J. Paruchuri, S. Edwin, S. Cheung, and C. Chen. Spatially adaptive illumination modeling for background subtraction. *International Conference on Computer Vision, ICCV 2011 Workshop on Visual Surveillance*, November 2011.

235. M. Paul, W. Lin, C. Lau, and B. Lee. Video coding with dynamic background. *EURASIP Journal on Advances in Signal Processing*, 2013:11, 2013.

236. D. Peng, C. Lin, W. Sheu, and T. Tsai. Architecture design for a low-cost and low-complexity foreground object segmentation with multi-model background maintenance algorithm. *ICIP 2009*, pages 3241–3244, 2009.

237. V. Pham, P. Vo, H. Vu Thanh, and B. Le Hoai. GPU implementation of extended Gaussian mixture model for background subtraction. *IEEE International Conference on Computing and Telecommunication Technologies, RIVF 2010*, November 2010.

238. M. Piccardi. Background subtraction techniques: a review. *IEEE International Conference on Systems, Man and Cybernetics*, October 2004.

239. A. Pnevmatikakis and L. Polymenakos. 2D person tracking using Kalman filtering and adaptive background learning in a feedback loop. *CLEAR Workshop 2006*, 2006.

240. D. Pokrajac and L. Latecki. Spatiotemporal blocks-based moving objects identification and tracking. *IEEE Visual Surveillance and Performance Evaluation of Tracking and Surveillance (VS-PETS 2003)*, pages 70–77, October 2003.

241. F. Porikli. Human body tracking by adaptive background models and mean-shift analysis. *IEEE International Workshop on Performance Evaluation of Tracking and Surveillance, PETS 2003*, March 2003.

242. F. Porikli. Detection of temporarily static regions by processing video at different frame rates. *IEEE International Conference on Advanced Video and Signal based Surveillance, AVSS 2007*, 2007.

243. F. Porikli and O. Tuzel. Bayesian background modeling for foreground detection. *ACM International Workshop on Video Surveillance and Sensor Networks, VSSN 2005*, pages 55–58, November 2005.

244. F. Porikli and C. Wren. Change detection by frequency decomposition: Wave-back. *International Workshop on Image Analysis for Multimedia Interactive Services, WIAMIS 2005*, April 2005.

245. F. Porikli and Z. Yin. Temporally static region detection in multi-camera systems. *PETS 2007*, October 2007.

246. A. Prati, I. Mikic, M. Trivedi, and R. Cucchiara. Detecting moving shadows: Algorithms and evaluation. *IEEE Transactions on Pattern Analysis and Machine Intelligence*, 25(4):918–923, July 2003.

247. B. Qin, J. Wang, J. Gao, T. Pang, and F. Su. A traffic video background extraction algorithm based on image content sensitivity. *CSI 2010*, pages 603–610, 2010.

248. C. Quivy and I. Kumazawa. Background images generation based on the nelder-mead simplex algorithm using the eigenbackground model. *International Conference on Image Analysis and Recognition, ICIAR 2011*, pages 21–29, June 2011.

249. R. Radke, S. Andra, O. Al-Kofahi, and B. Roysam. Image change detection algorithms: A systematic survey. *IEEE Transactions on Image Processing*, 14(3):294–307, March 2005.

250. V. Reddy, C. Sanderson, and B. Lovell. Robust foreground object segmentation via

adaptive region-based background modelling. *International Conference on Pattern Recognition, ICPR 2010*, August 2010.

251. V. Reddy, C. Sanderson, A. Sanin, and B. Lovell. MRF-based background initialisation for improved foreground detection in cluttered surveillance videos. *Asian Conference on Computer Vision, ACCV 2010*, November 2010.

252. H. Ribeiro and A. Gonzaga. Hand image segmentation in video sequence by GMM: a comparative analysis. *XIX Brazilian Symposium on Computer Graphics and Image Processing, SIBGRAPI 2006*, pages 357–364, 2006.

253. R. Rodriguez-Gomez, E. Fernandez-Sanchez, J. Diaz, and E. Ros. Codebook hardware implementation on FPGA for background subtraction. *Journal of Real-Time Image Processing*, 2012.

254. P. Rosin and E. Ioannidis. Evaluation of global image thresholding for change detection. *Pattern Recognition Letters*, 24:2345–2356, October 2003.

255. J. Rymel, J. Renno, D. Greenhill, J. Orwell, and G. Jones. Adaptive eigen-backgrounds for object detection. *International Conference on Image Processing, ICIP 2004*, pages 1847–1850, October 2004.

256. A. Sanin, C. Sanderson, and B. Lovell. Shadow detection: A survey and comparative evaluation of recent methods. *Pattern Recognition*, 45(4):1684–1689, April 2012.

257. Y. Satoh, S. Kaneko, and S. Igarashi. Robust object detection and segmentation by peripheral increment sign correlation image. *Systems and Computers in Japan*, 35(9):70–80, 2004.

258. Y. Satoh and K. Sakaue. Robust background subtraction based on bi-polar radial reach correlation. *IEEE International Conference on Computers, Communications, Control and Power Engineering, TENCON 2005*, 2005.

259. Y. Satoh, H. Tanahashi, Y. Niwa, S. Kaneko, and K. Yamamoto. Robust object detection and segmentation based on radial reach correlation. *IAPR Workshop on Machine Vision Applications, MVA 2002*, pages 512–517, 2002.

260. K. Schindler and H. Wang. Smooth foreground-background segmentation for video processing. *Asian Conference on Computer Vision, ACCV 2006*, pages 581–590, January 2006.

261. F. Seidel, C. Hage, and M. Kleinsteuber. pROST - a smoothed l_p-norm robust online subspace tracking method for realtime background subtraction in video. *Machine Vision and Applications, Special Issue on Background Modeling*, 2013.

262. N. Setiawan, S. Hong, J. Kim, and C. Lee. Gaussian mixture model in improved IHLS color space for human silhouette extraction. *International Conference on Artificial Reality and Telexistence, ICAT 2006*, pages 732–741, 2006.

263. M. Shah, J. Deng, and B. Woodford. Localized adaptive learning of mixture of Gaussians models for background extraction. *International Conference of Image and Vision Computing New Zealand, ICVNZ 2010*, November 2010.

264. M. Shah, J. Deng, and B. Woodford. Enhanced codebook model for real-time background subtraction. *International Conference on Neural Information Processing, ICONIP 2011*, November 2011.

265. M. Shakeri, H. Deldari, H. Foroughi, A. Saberi, and A. Naseri. A novel fuzzy background subtraction method based on cellular automata for urban traffic applications. *International Conference on Signal Processing, ICSP 2008*, pages 899–902, October 2008.

266. Y. Sheikh, O. Javed, and T. Kanade. Background subtraction for freely moving cameras. *IEEE International Conference on Computer Vision, ICCV 2009*, pages 1219–1225, October 2009.

267. Y. Sheikh and M. Shah. Bayesian modeling of dynamic scenes for object detection. *IEEE Transactions on Pattern Analysis and Machine Intelligence*, 27:1778–

1792, 2005.

268. Y. Sheikh and M. Shah. Bayesian modeling of dynamic scenes for object detection. *IEEE Transactions on Pattern Analysis and Machine Intelligence, PAMI 2005*, 27(11):1778–1792, November 2005.

269. Y. Sheikh and M. Shah. Bayesian object detection in dynamic scenes. *IEEE Conference on Computer Vision and Pattern Recognition, CVPR 2005*, June 2005.

270. Y. Shen, W. Hu, Mi. Yang, J. Liu, C. Chou, and B. Wei. Efficient background subtraction for tracking in embedded camera networks. *International Conference on Information Processing in Sensor Networks, IPSN 2012*, April 2012.

271. A. Shimada and R. Taniguchi. Object detection based on Gaussian mixture predictive background model under varying illumination. *International Workshop on Computer Vision, MIRU 2008*, July 2008.

272. A. Shimada and R. Taniguchi. Hybrid background model using spatial-temporal lbp. *IEEE International Conference on Advanced Video and Signal based Surveillance, AVSS 2009*, September 2009.

273. A. Shimada and R. Taniguchi. Object detection based on fast and low-memory hybrid background model. *IEEJ Transactions on Electronics, Information and Systems*, 129-C(5):846–852, May 2009.

274. A. Shimada, S. Yoshinaga, and R. Taniguchi. Adaptive background modeling for paused object regions. *International Workshop on Visual Surveillance, VS 2010*, November 2010.

275. A. Shimada, S. Yoshinaga, and R. Taniguchi. Maintenance of blind background model for robust object detection. *IPSJ Transactions on Computer Vision and Applications*, 3:148–159, 2011.

276. M. Sigari. Fuzzy background modeling/subtraction and its application in vehicle detection. *World Congress on Engineering and Computer Science, WCECS 2008*, October 2008.

277. M. Sigari and M. Fathy. Real-time background modeling/subtraction using two-layer codebook model. *International Multiconference on Engineering and Computer Science, IMECS 2008*, March 2008.

278. M. Sigari, N. Mozayani, and H. Pourreza. Fuzzy running average and fuzzy background subtraction: Concepts and application. *International Journal of Computer Science and Network Security*, 8(2):138–143, 2008.

279. J. Silveira, C. Jung, and S. Musse. Background subtraction and shadow detection in grayscale video sequences. *Brazilian Symposium on Computer Graphics and Image Processing, SIBGRAPI 2005*, pages 189–196, 2005.

280. D. Silvestre. Video surveillance using a time-of-light camera. *Master Thesis, Informatics and Mathematical Modelling, University of Denmark*, 2007.

281. M. Singh, V. Parameswaran, and V. Ramesh. Order consistent change detection via fast statistical significance testing. *IEEE Computer Vision and Pattern Recognition Conference, CVPR 2008*, June 2008.

282. M. Sivabalakrishnan and D. Manjula. Adaptive background subtraction in dynamic environments using fuzzy logic. *International Journal of Image Processing*, 4(1), 2010.

283. D. Skocaj and A. Leonardis. Weighted and robust incremental method for subspace learning. *International Conference on Computer Vision, ICCV 2003*, pages 1494–1501, 2003.

284. M. Smids. Background subtraction for urban traffic monitoring using webcams. *Master Thesis, Univ. Amsterdam*, December 2006.

285. C. Spampinato, Y. Burger, G. Nadarajan, and R. Fisher. Detecting, tracking and counting fish in low quality unconstrained underwater videos. *VISAPP 2008*, pages

514–519, 2008.

286. C. Spampinato, S. Palazzo, and I. Kavasidis. A texton-based kernel density estimation approach for background modeling under extreme conditions. *Computer Vision and Image Understanding, CVIU 2014*, 122:74–83, 2014.

287. C. Stauffer and E. Grimson. Adaptive background mixture models for real-time tracking. *IEEE Conference on Computer Vision and Pattern Recognition, CVPR 1999*, pages 246–252, 1999.

288. L. Di Stefano, F. Tombari, S. Mattoccia, and E. De Lisi. Robust and accurate change detection under sudden illumination variations. *International Workshop on Multi-dimensional and Multi-view Image Processing, ACCV 2007*, pages 103–109, 2007.

289. E. Stringa. Morphological change detection algorithms for surveillance applications. *British Machine Vision Conference, BMVC 2000*, September 2000.

290. B. Subudhi, S. Ghosh, and A. Ghosh. Change detection for moving object segmentation with robust background construction under Wronskian framework. *Machine Vision and Applications, MVA 2013*, January 2013.

291. Y. Sugaya and K. Kanatani. Extracting moving objects from a moving camera video sequence. *Symposium on Sensing via Image Information, SSII2004*, pages 279–284, 2004.

292. J. Paik T. Kim, S. Lee. Evolutionary algorithm-based background generation for robust object detection. *ICIC 2006*, pages 542–552, 2006.

293. S. Lai T. Su, Y. Chen. Over-segmentation based background modeling and foreground detection with shadow removal by using hierarchical MRFs. *Asian Conference on Computer Vision, ACCV 2010*, November 2010.

294. K. Takahara, T. Tori, and T. Zin. Making background subtraction robust to various illumination changes. *International Journal of Computer Science and Network Security*, 11(3):241–248, March 2011.

295. T. Tanaka, A. Shimada, D. Arita, and R. Taniguchi. Object detection under varying illumination based on adaptive background modeling considering spatial locality. *International Workshop on Computer Vision, MIRU 2008*, July 2008.

296. T. Tanaka, A. Shimada, D. Arita, and R. Taniguchi. Object segmentation under varying illumination based on combinational background modeling. *Joint Workshop on Machine Perception and Robotics, MPR 2008*, 2008.

297. T. Tanaka, A. Shimada, D. Arita, and R. Taniguchi. Object detection under varying illumination based on adaptive background modeling considering spatial locality. *PSVIT 2009*, pages 645–656, January 2009.

298. A. Taneja, L. Ballan, and M. Pollefeys. Modeling dynamic scenes recorded with freely. *Asian Conference on Computer Vision, ACCV 2010*, 2010.

299. G. Tang and A. Nehorai. Robust principal component analysis based on low-rank and block-sparse matrix decomposition. *CISS 2011*, 2011.

300. A. Tavakkoli. Novelty detection: An approach to foreground detection in videos. *Pattern Recognition, INTECH*, 7, 2010.

301. A. Tavakkoli, A. Ambardekar, M. Nicolescu, and S. Louis. A genetic approach to training support vector data descriptors for background modeling in video data. *International Symposium on Visual Computing, ISVC 2007*, November 2007.

302. A. Tavakkoli, M. Nicolescu, and G. Bebis. Novelty detection approach for foreground region detection in videos with quasi-stationary backgrounds. *International Symposium on Visual Computing, ISVC 2006*, pages 40–49, November 2006.

303. A. Tavakkoli, M. Nicolescu, G. Bebis, and M. Nicolescu. A support vector data description approach for background modeling in videos with quasi-stationary backgrounds. *International Journal of Artificial Intelligence Tools*, 17(4):635–658,

2008.

304. A. Tavakkoli, M. Nicolescu, G. Bebis, and M. Nicolescu. Non-parametric statistical background modeling for efficient foreground region detection. *Machine Vision and Applications*, 20(6):395–409, 2009.

305. A. Tavakkoli, M. Nicolescu, M. Nicolescu, and G. Bebis. Efficient background modeling through incremental support vector data description. *ICPR 2008*, December 2008.

306. A. Tavakkoli, M. Nicolescu, M. Nicolescu, and G. Bebis. Incremental SVDD training: Improving efficiency of background modeling in videos. *International Conference on Signal and Image Processing*, 2008.

307. L. Taycher, J. Fisher, and T. Darrell. Incorporating object tracking feedback into background maintenance framework. *IEEE Workshop on Motion and Video Computing, WMVC 2005*, 2:120–125, 2005.

308. L. Teixeira, J. Cardoso, and L. Corte-Real. Object segmentation using background modelling and cascaded change detection. *Journal of Multimedia*, 2(5):55–65, 2007.

309. H. Tezuka and T. Nishitani. A precise and stable foreground segmentation using fine-to-coarse approach in transform domain. *International Conference on Image Processing, ICIP 2008*, pages 2732–2735, October 2008.

310. D. Toth, T. Aach, and V. Metzler. Bayesian spatio-temporal motion detection under varying illumination. *European Signal Processing Conference, EUSIPCO 2000*, pages 2081–2084, 2000.

311. D. Toth, T. Aach, and V. Metzler. Illumination-invariant change detection. *IEEE Southwest Symposium on Image Analysis and Interpretation*, pages 3–7, April 2000.

312. K. Toyama, J. Krumm, B. Brumiit, and B. Meyers. Wallflower: Principles and practice of background maintenance. *International Conference on Computer Vision*, pages 255–261, September 1999.

313. D. Tsai and C. Lai. Independent component analysis-based background subtraction for indoor surveillance. *IEEE Transactions on Image Processing, IP 2009*, 8(1):158–167, January 2009.

314. T. Tsai, D. Peng, C. Lin, and W. Sheu. A low cost foreground object detection architecture design with multi-model background maintenance algorithm. *VLSI 2008*, 2008.

315. L. Unzueta, M. Nieto, A. Cortes, J. Barandiaran, O. Otaegui, and P. Sanchez. Adaptive multi-cue background subtraction for robust vehicle counting and classification. *IEEE Transactions on Intelligent Transportation Systems*, 2011.

316. L. Unzueta, M. Nieto, A. Cortes, J. Barandiaran, O. Otaegui, and P. Sanchez. Adaptive multi-cue background subtraction for robust vehicle counting and classification. *IEEE Transaction on Intelligent Transportation Systems*, 13(2):527–540, June 2012.

317. A. Utasi and L. Czuni. Reducing the foreground aperture problem in mixture of Gaussians based motion detection. *EURASIP Conference on Speech and Image Processing, Multimedia Communications and Services, EC-SIPMCS 2007*, 2, 2007.

318. A. Vacavant, T. Chateau, A. Wilhelm, and L. Lequievre. A benchmark dataset for foreground/background extraction. *International Workshop on Background Models Challenge, ACCV 2012*, November 2012.

319. J. Varona, J. Gonzalez, I. Rius, and J. Villanueva. Importance of detection for video surveillance applications. *Optical Engineering*, pages 1–9, 2008.

320. H. Wang and D. Suter. A re-evaluation of Mixture-of-Gaussian background modeling. *International Conference on Acoustics, Speech, and Signal Processing, ICASSP 2005*, pages 1017–1020, March 2005.

321. H. Wang and D. Suter. A re-evaluation of Mixture-of-Gaussian background modeling. *IEEE Internationm Conference on Acoustics, Speech, and Signal Processing, ICASSP 2005*, pages 1017–1020, March 2005.

322. H. Wang and D. Suter. A novel robust statistical method for background initialization and visual surveillance. *Asian Conference on Computer Vision, ACCV 2006*, pages 328–337, January 2006.

323. J. Wang, G. Bebis, and R. Miller. Robust video-based surveillance by integrating target detection with tracking. *IEEE Workshop on Object Tracking and Classification Beyond the Visible Spectrum in conjunction with CVPR 2006*, June 2006.

324. J. Wang, G. Bebis, M. Nicolescu, M. Nicolescu, and R. Miller. Improving target detection by coupling it with tracking. *Machine Vision and Application*, pages 1–19, 2008.

325. J. Wang and Y. Yagi. Efficient background subtraction under abrupt illumination variations. *Asian Conference on Computer Vision, ACCV 2012*, November 2012.

326. L. Wang and C. Pan. Fast and effective background subtraction based on ELBP. *International Conference on Acoustics, Speech, and Signal Processing, ICASSP 2010*, March 2010.

327. L. Wang and C. Pan. Effective multi-resolution background subtraction. *International Conference on Acoustics, Speech, and Signal Processing, ICASSP 2011*, May 2011.

328. T. Wang, G. Chen, and H. Zhou. A novel background modelling approach for accurate and real-time motion segmentation. *International Conference on Signal Processing, ICSP 2006*, 2, 2006.

329. W. Wang, D. Chen, W. Gao, and J. Yang. Modeling background from compressed video. *International Conference on Computer Vision, ICCV 2005*, 2005.

330. W. Wang and R. Wu. Fusion of luma and chroma GMMs for HMM-based object detection. *First Pacific Rim Symposium on Advances in Image and Video Technology, PSIVT 2006*, pages 573–581, December 2006.

331. X. Wang, F. Liu, and Z. Ye. Background modeling in compressed sensing scheme. *ESEP 2011*, pages 4776–4783, December 2011.

332. Y. Wang. Real-time moving vehicle detection with cast shadow removal in video based on conditional random field. *IEEE Transaction on Circuits Systems Video Technologies*, 19:437–441, 2009.

333. Y. Wang and H. Chen. The design of background subtraction on reconfigurable hardware. *International Conference on Intelligent Information Hiding and Multimedia Signal Processing*, pages 182–185, 2012.

334. Y. Wang, K. Loe, T. Tan, and J. Wu. A dynamic Hidden Markov Random field model for foreground and shadow segmentation. *Workshops on Application of Computer Vision, WACV 2005*, 1:474–480, January 2005.

335. Y. Wang, K. Loe, and J. Wu. A dynamic conditional random field model for foreground and shadow segmentation. *IEEE Transactions on Pattern Analysis and Machine Intelligence*, 28(2):279–289, February 2006.

336. A. Waters, A. Sankaranarayanan, and R. Baraniuk. SpaRCS: recovering low-rank and sparse matrices from compressive measurements. *Neural Information Processing Systems, NIPS 2011*, December 2011.

337. B. White and M. Shah. Automatically tuning background subtraction parameters using particle swarm optimization. *IEEE International Conference on Multimedia and Expo, ICME 2007*, pages 1826–1829, 2007.

338. B. Wohlberg, R. Chartrand, and J. Theiler. Local principal component pursuit for nonlinear datasets. *International Conference on Acoustics, Speech, and Signal Processing, ICASSP 2012*, March 2012.

339. M. Wojcikowski, R. Zaglewsk, and B. Pankiewicz. FPGA-based real-time implementa-

tion of detection algorithm for automatic traffic surveillance sensor network. *Journal of Signal Processing Systems*, December 2010.

340. C. Wren and A. Azarbayejani. Pfinder : Real-time tracking of the human body. *IEEE Transactions on Pattern Analysis and Machine Intelligence*, 19(7):780–785, July 1997.

341. C. Wren and F. Porikli. Waviz: Spectral similarity for object detection. *IEEE International Workshop on Performance Evaluation of Tracking and Surveillance, PETS 2005*, January 2005.

342. M. Xiao, C. Han, and X. Kang. A background reconstruction for dynamic scenes. *International Conference on Information Fusion, ICIF 2006*, pages 1–6, July 2006.

343. M. Xiao and L. Zhang. A background reconstruction algorithm based on modified basic sequential clustering. *International Colloquium on Computing, Communication, Control, and Management, CCCM 2008*, 1:47–51, August 2008.

344. M. Xiao and L. Zhang. A background reconstruction algorithm based on two-threshold sequential clustering. *International Colloquium on Computing, Communication, Control, and Management, CCCM 2008*, 1:389–393, August 2008.

345. B. Xie, V. Ramesh, and T. Boult. Sudden illumination change detection using order consistency. *Image and Vision Computing*, 22(2):117–125, February 2004.

346. L. Xiong, X. Chen, and J. Schneider. Direct robust matrix factorization for anomaly detection. *International Conference on Data Mining, ICDM 2011*, 2011.

347. H. Xu, C. Caramanis, and S. Sanghavi. Robust PCA via outlier pursuit. *NIPS 2010*, 2010.

348. J. Xu, V. Ithapu, L. Mukherjee, J. Rehg, and V. Singhy. GOSUS: grassmannian online subspace updates with structured-sparsity. *International Conference on Computer Vision, ICCV 2013*, December 2013.

349. J. Xu, N. Jiang, and S. Goto. Block-based codebook model with oriented-gradient feature for real-time foreground detection. *IEEE International Workshop on Multimedia Signal Processing*, November 2011.

350. M. Xu and T. Ellis. Illumination-invariant motion detection using color mixture models. *British Machine Vision Conference, BMVC 2001*, pages 163–172, September 2001.

351. Y. Xu and L. Li. Moving object segmentation by pursuing local spatio-temporal manifolds. *Technical Report, Sun Yat-Sen University*, 2013.

352. Z. Xu, I. Gu, and P. Shi. Recursive error-compensated dynamic eigenbackground learning and adaptive background subtraction in video. *Optical Engineering*, 47(5), May 2008.

353. G. Xue, L. Song, J. Sun, and M. Wu. Foreground estimation based on robust linear regression model. *International Conference on Image Processing, ICIP 2011*, pages 3330–3333, September 2011.

354. G. Xue, L. Song, J. Sun, and M. Wu. Hybrid center-symmetric local pattern for dynamic background subtraction. *International Conference in Multimedia and Exposition, ICME 2011*, July 2011.

355. G. Xue, J. Sun, and L. Song. Dynamic background subtraction based on spatial extended center-symmetric local binary pattern. *International Conference in Multimedia and Exposition, ICME 2010*, July 2010.

356. A. Yamamoto and Y. Iwai. Real-time object detection with adaptive background model and margined sign correlation. *Asian Conference on Computer Vision, ACCV 2009*, September 2009.

357. M. Yamazaki, G. Xu, and Y. Chen. Detection of moving objects by independent component analysis. *Asian Conference on Computer Vision, ACCV 2006*, pages

467–478, 2006.

358. H. Yang, Y. Tan, J. Tian, and J. Liu. Accurate dynamic scene model for moving object detection. *International Conference on Image Processing, ICIP 2007*, pages 157–160, 2007.

359. N. Yang, T. Yao, J. Wang, and D. Yeung. A probabilistic approach to robust matrix factorization. *European Conference on Computer Vision, ECCV 2012*, pages 126–139, 2012.

360. N. Yang and D. Yeung. Bayesian robust matrix factorization for image and video processing. *International Conference on Computer Vision, ICCV 2013*, 2013.

361. S. Yang and C. Hsu. Background modeling from GMM likelihood combined with spatial and color coherency. *ICIP 2006*, pages 2801–2804, 2006.

362. T. Yang, S. Li, Q. Pan, and J. Li. Real-time and accurate segmentation of moving objects in dynamic scene. *ACM International Workshop on Video Surveillance and Sensor Networks, VSSN 2004*, October 2004.

363. B. Yeo, W. Lim, H. Lim, and W. Wong. Extended fuzzy background modeling for moving vehicle detection using infrared vision. *IEICE Electronics Express*, pages 340–345, 2011.

364. K. Yokoi. Illumination-robust change detection using texture based featuress. *Machine Vision Applications, MVA 2007*, pages 487–491, May 2007.

365. K. Yokoi. Probabilistic BPRRC: robust change detection against illumination changes and background movements. *Machine Vision Applications, MVA 2009*, pages 148–151, May 2009.

366. H. Yoshimura, Y. Iwai, and M. Yachida. Object detection with adaptive background model and margined sign cross correlation. *International Conference on Pattern Recognition, ICPR 2006*, 3:19–23, 2006.

367. S. Yoshinaga, A. Shimada, H. Nagahara, and R. Taniguchi. Object detection using local difference patterns. *Asian Conference on Computer Vision, ACCV 2010*, pages 216–227, November 2010.

368. S. Yoshinaga, A. Shimada, H. Nagahara, and R. Taniguchi. Background model based on statistical local difference pattern. *International Workshop on Background Models Challenge, ACCV 2012*, November 2012.

369. A. Zaharescu and M. Jamieson. Multi-scale multi-feature codebook-based background subtraction. *IEEE International Conference on Computer Vision Workshops, ICCV 2011*, pages 1753–1760, November 2011.

370. Q. Zang and R. Klette. Evaluation of an adaptive composite Gaussian model in video surveillance. *CITR Technical Report 114*, August 2002.

371. Q. Zang and R. Klette. Robust background subtraction and maintenance. *International Conference on Pattern Recognition, ICPR 2004*, pages 90–93, 2004.

372. H. Zen and S. Lai. Adaptive foreground object extraction for real-time video surveillance with lighting variations. *ICASSP 2007*, 1:1201–1204, 2007.

373. H. Zhang and D. Xu. Fusing color and texture features for background model. *International Conference on Fuzzy Systems and Knowledge Discovery, FSKD*, 4223(7):887–893, September 2006.

374. J. Zhang and Y. Zhuang. Adaptive weight selection for incremental eigen-background modeling. *International Conference in Media and Expo, ICME 2007*, pages 1494–1501, July 2007.

375. S. Zhang, H. Yao, and S. Liu. Dynamic background and subtraction using spatiotemporal local binary patterns. *IEEE International Conference on Image Processing, ICIP 2008*, pages 1556–1559, October 2008.

376. S. Zhang, H. Yao, and S. Liu. Dynamic background subtraction based on local dependency histogram. *International Journal of Pattern Recognition and Artificial*

Intelligence, IJPRAI 2009, 2009.

377. W. Zhang, J. Wu, and X. Fang. Moving cast shadow detection. *INTECH*, page 546, June 2007.

378. Y. Zhang, Z. Liang, Z. Hou, H. Wang, and M. Tan. An adaptive mixture Gaussian background model with online background reconstruction and adjustable foreground mergence time for motion segmentation. *ICIT 2005*, pages 23–27, December 2005.

379. Y. Zhang, Z. Liang, Z. Hou, H. Wang, and M. Tan. An adaptive mixture Gaussian background model with online background reconstruction and adjustable foreground mergence time for motion segmentation. *ICIT 2005*, pages 23–27, December 2005.

380. Z. Zhang, C. Wang, B. Xiao, S. Liu, and W. Zhou. Multi-scale fusion of texture and color for background modeling. *International Conference on Advanced Video and Signal-Based Surveillance, AVSS 2012*, September 2012.

381. Z. Zhang, C. Wang, B. Xiao, S. Liu, and W. Zhou. Multi-scale fusion of texture and color for background modeling. *AVSS 2012*, pages 154–159, 2012.

382. C. Zhao, X. Wang, and W. Cham. Background subtraction via robust dictionary learning. *EURASIP Journal on Image and Video Processing, IVP 2011*, January 2011.

383. M. Zhao, J. Bu, and C. Chen. Robust background subtraction in HSV color space. *SPIE Multimedia Systems and Applications*, 4861:325–332, July 2002.

384. J. Zheng, Y. Wang, N. Nihan, and E. Hallenbeck. Extracting roadway background image: A mode based approach. *Journal of Transportation Research Report*, 1944:82–88, 2006.

385. B. Zhong, S. Liu, H. Yao, and B. Zhang. Multi-resolution background subtraction for dynamic scenes. *International Conference on Image Processing, ICIP 2009*, pages 3193–3196, November 2009.

386. B. Zhong, H. Yao, S. Shan, X. Chen, and W. Gao. Hierarchical background subtraction using local pixel clustering. *IEEE International Conference on Pattern Recognition, ICPR 2008*, December 2008.

387. B. Zhong, H. Yao, and X. Yuan. Local histogram of figure/ground segmentations for dynamic background subtraction. *EURASIP Journal on Advances in Signal Processing*, 2010:14, 2010.

388. J. Zhong and S. Sclaroff. Segmenting foreground objects from a dynamic textured background via a robust Kalman filter. *International Conference on Computer Vision, ICCV 2003*, pages 44–50, 2003.

389. Q. Zhong, L. Dai, Y. Song, and R. Wang. A hierarchical motion detection algorithm with the fusion of the two types of motion information. *Pattern Recognition*, 18(5):552–557, 2005.

390. D. Zhou and H. Zhang. Modified GMM background modeling and optical flow for detection of moving objects. *IEEE International Conference on Systems, Man and Cybernetics, SMC 2005*, pages 2224–2229, October 2005.

391. T. Zhou and D. Tao. GoDec: randomized low-rank and sparse matrix decomposition in noisy case. *International Conference on Machine Learning, ICML 2011*, 2011.

392. W. Zhou, Y. Liu, W. Zhang, L. Zhuang, and N. Yu. Dynamic background subtraction using spatial-color binary patterns. *International Conference on Image and Graphics, ICIG 2011*, August 2011.

393. X. Zhou, C. Yang, and W. Yu. Moving object detection by detecting contiguous outliers in the low-rank representation. *IEEE Transactions on Pattern Analysis and Machine Intelligence*, 35:597–610, 2013.

394. Z. Zhou, X. Li, J. Wright, E. Candes, and Y. Ma. Stable principal component pursuit. *IEEE ISIT Proceedings*, pages 1518–1522, June 2010.

2

Recent Approaches in Background Modeling for Static Cameras

Thierry Bouwmans
Lab. MIA, Univ. La Rochelle, France

2.1 Introduction

In Chapter 1, we provided a survey of the traditional background models for video-surveillance and how to evaluate them using measures evaluation and datasets. However, no traditional algorithm today seems to be able to simultaneously address all the key challenges that we meet in videos. This is due to two main reasons:

1. Lack of scientific progress, that is, there are mainly improvements of the state-of-the-art methods such as Mixture of Gaussians, and other representation models are sometimes investigated insufficiently.

2. The absence of a single realistic large-scale dataset with accurate ground-truth providing a balanced coverage of the range of challenges present in the real world.

These two problems have been recently addressed by the application of other mathematical tools and, the creation of large datasets such as ChangeDetection.net [75], SABS [24] and BMC [182]. In this Chapter 2, we investigate the recent advances proposed in the literature.

The rest of this Chapter 2 is organized as follows: In Section 2.2, we investigate the recent background models. In Section 2.3, we present some strategies to improve the robustness of the background model. In 2.4, real-time implementations are investigated. Then, we present recent resources, datasets and codes publicly available to test and evaluate algorithms. Finally, we provide a conclusion and perspectives.

2.2   Recent Background Models

The recent background representation models can be classified in the following categories: advanced statistical background models, fuzzy background models, discriminative subspace learning models, RPCA models, sparse models and transform domain models. For each category, we give a table which allows the reader a quick overview. The references corresponding to the number of papers found for each model are available at the Background Subtraction Web Site.

2.2.1   Advanced Statistical Models

Advanced statistical background models have been recently developed and can be classified as follows:

- **Mixture models:** In this category, the authors used another distribution than the Gaussian one used in the GMM listed in Section 1.6.2. Some authors used the Student-t Mixture Model [135] [78] or the Dirichlet Mixture Model [93] [86]. Student's t-mixture model (STMM) has proven to be very robust against noises due to its more heavily-tailed nature compared to Gaussian mixture model [135] but STMM has not been applied previously to video processing, because EM algorithm cannot be directly applied to the process. The huge increase in complexity would prevent a real-time implementation. Thus, Mukherjee et al. [135] proposed a new real-time recursive filter approach to update the parameters of the distribution effectively. So, the method allows to model the background and then to separate the foreground with high accuracy in the case of slow foreground objects and dynamic backgrounds. In another way, He et al. [93] used a Dirichlet mixture model, which constantly adapts both the parameters and the number of components of the mixture to the scene in a block-based method but the potential advantages were not completely exploited and the algorithm was computationally expensive. Recently, Haines and Xiang [86] improved the DMM to estimate a per-pixel background distribution by a probabilistic regularization. The key improvements consist of the inference for the per-pixel mode count, such that it accurately models dynamic backgrounds, and of the maintenance scheme which updates continuously the background. In a similar way, Fan and Bouguila [61] model and update the background using Dirichlet Mixture Model (DMM) which allows to deal with non Gaussian processes that are met in real videos. They first proposed a batch algorithm and then they adopted a variational Bayes framework in an on-line manner to update the background one at a time. So, all the parameters and the model complexity of the Dirichlet mixture model are estimated simultaneously in a closed form. The DMM are more robust in dynamic background than the GMM. Recently, Faro et al. [63] used an Adaptive Poisson Mixture Model (APMM) for vehicle detection. The luminosity information side channel permits to effectively handle rapid changes in illumination. A novel algorithm for detecting and removing partial and full occlusions among blobs was also designed. The APMM outperformed state-of-the-art methods in terms of both vehicle detection and processing time.

- **Hybrid models:** Ding et al. [56] used a mixture of nonparametric regional model (KDE) and parametric pixel-wise model (GMM) to approximate the background color distribution. The foreground color distribution is learned from neighboring pixels of the previous frame. The locality distributions of background and foreground are approximated with the nonparametric model (KDE). The temporal coherence is modeled with a Markov chain. So, color, locality, temporal coherence and spatial consistency are fused together in the same framework. The models of color, locality and temporal coherence are learned on-line from complex dynamic backgrounds. In the same idea, Liu et al. [124] [125] used a KDE-GMM hybrid model. Under this probabilistic framework, this method deal with foreground detection and shadow removal simultaneously by constructing probability density functions (PDFs) of moving objects and non-moving objects. Here, these PDFs are constructed based on KDE-GMM hybrid model (KGHM) which has advantages of both KDE and GMM. This KGHM models the spatial dependencies of neighboring pixel colors to deal with highly dynamic backgrounds.

- **Nonparametric models:** This group of models follows a nonparametric background modeling paradigm. Barnich et al. [13] proposed a samples based algorithm called Visual Background Extractor (ViBe) [13] that builds the background model by aggregating previously observed values for each pixel location. The key innovation of ViBe is a random selection policy that ensures a smooth exponentially decaying lifespan for the sample values that constitute the pixel models. The second innovation concerned the post-processing to give spatial consistency by using a fast and spatial information propagation method that randomly diffuses pixel values across neighboring pixels. The third innovation is related to the background initialization which is instantaneous and allows the algorithm to start from the second frame of the sequence. Although ViBe gives acceptable detection results in many scenarios, it is problematic with challenging scenarios such as darker background, shadows, and frequent background change. Another approach developed by Hofmann et al. [95] and called Pixel-Based Adaptive Segmenter (PBAS) models the background by a history of recently observed pixel values. PBAS consists of several components. As a central component, the decision block decides for or against foreground based on the current image and a background. This decision is based on the per-pixel threshold. Furthermore, the background model is updated over time in order to deal with gradual background changes. This update depends on a per-pixel learning parameter. The key innovation in PBAS approach is that both of these two per-pixel thresholds change the estimate of the background dynamics. The foreground decision depends on a decision threshold. PBAS outperforms most state-of-the-art methods.

- **Multi-kernels models:** Recently, Molina-Giraldo et al. [134] proposed a Weighted Gaussian Video Segmentation (WGKVS) which employs multiple kernel representations and incorporates different color representations.

The first column indicates the category model and the second column the name of each method.The third column gives the name of the authors and the date of the related publication.

TABLE 2.1 Advanced Statistical Models: An Overview.

Models	Methods	Authors - Dates
Mixture	Student-t Mixture Models (STMM) (2) Dirichlet Process Gaussian Mixture Model (DP-GMM) (2) Variational Dirichlet Mixture Model (varDMM) (1) Adaptive Poisson Mixture Model (APMM) (1)	Mukherjee et al. (2012) [135] He et al. (2012) [93] Fan and Bouguila (2012) [61] Faro et al. (2011) [63]
Hybrid	KDE-GMM (KGMM) (1) KDE-GMM Hybrid Model (KGHM) (2)	Ding et al. (2010) [56] Liu et al. (2008) [124]
Non Parametric	Video Background Extractor (ViBe) (6) Pixel-Based Adaptive Segmenter (PBAS) (1)	Barnich et al. (2009) [13] Hofmann et al. (2012) [95]
Multi-Kernels	Weighted Gaussian Video Segmentation (WGKVS) (2)	Molina-Giraldo et al. (2013) [134]

2.2.2   Fuzzy Models

All the critical situations generate imprecisions and uncertainties in the whole process of background subtraction. Therefore, some authors have recently introduced fuzzy concepts in the different steps of background subtraction to take into account these imprecisions and uncertainties. Different fuzzy methods have been developed as classified in the recent survey [21]:

- **Fuzzy background modeling:** The main challenge addressed here consists of modeling multi-modal backgrounds. The algorithm usually used is the Gaussian Mixture Model [169] to deal with this challenge but the parameters are determined using a training sequence which contains insufficient or noisy data. So, the parameters are not well determined. In this context, Type-2 Fuzzy Mixture of Gaussians (T2F-MOG) cite591 are used to model uncertainties when dynamic backgrounds occur. El Baf et al. proposed two algorithms, that is, one for the uncertainty over the mean and one for the uncertainty over the variance, called T2-FMOG-UM and T2-FMOG-UV, respectively. The T2-FMOG-UM and T2-FMOG-UV are more robust than the crisp MOG [169]. Practically, T2-MOG-UM is more robust than T2-FMOG-UV. Indeed, only the means are estimated and tracked correctly over time in the MOG maintenance. The variance and the weights are unstable and unreliable, so generating less robustness for the T2-FMOG-UV. In another way, Kim and Kim [109] adopted a fuzzy c-means clustering model that uses fuzzy color histogram as feature. This model allows to attenuate color variations generated by background motions while still highlighting foreground objects, and obtains better results in dynamic backgrounds than the MOG [169].

- **Fuzzy foreground detection:** In this case, a saturating linear function can be used to avoid crisp decision in the classification of the pixels as background or foreground. The background model can be unimodal, such as the running average in [154,160], or multimodal, such as the background modeling with confidence measure proposed in [149]. Another approach consists in aggregating different features such as color and texture features. As seen previously, the choice of the feature is essential and using more than one feature permits to be more robust to illumination changes and shadows. In this context, Zhang and Xu [206] have used texture and color features to compute similarity measures between current and background pixels. Then, these similarity measures are aggregated by applying the Sugeno integral. The assumption made by the authors reflects that the scale is ordinal. The moving objects are detected by thresholding the results of the Sugeno integral. Recently, El Baf et al. [9] have shown that the scale is continuum in the foreground detection. Therefore, they used the same features with the Choquet integral instead of the Sugeno integral. Ding et al. [58] used the Choquet integral too but they change the similarity measures. Recently, Azab et al. [7] have aggregated three features, i.e color, edge and texture. Fuzzy foreground detection is more robust to illumination changes and shadows than crisp foreground detection.

- **Fuzzy background maintenance:** The idea is to update the background following the membership of the pixel at the class background or foreground. The membership comes from the fuzzy foreground detection and can be introduced in the maintenance scheme in two ways. The first way [129] consists in adapting in a fuzzy manner the learning rate following the classification of the pixel. For the second way, the maintenance rule becomes a fuzzy combination of two crisp rules [10] [107]. The fuzzy adaptive background maintenance allows one to deal robustly with illumination changes and shadows.

- **Fuzzy Features:** First, Chiranjeevi and Sengupta [39] proposed a fuzzy feature, called fuzzy correlogram, which is obtained by applying fuzzy c-means algorithm on correlogram. The fuzzy correlogram is computed at each pixel in the background image and the current image. The distance between the two correlograms is obtained by a modified version of K-L divergence distance. If the distance is less than the threshold, then it implies that the current fuzzy correlogram is matched with the background model and the pixel is labeled as background otherwise foreground. Then, the background image is updated with the current fuzzy correlogram by simple adaptive filtering. In [42], Chiranjeevi and Sengupta proposed a multi-channel correlogram to exploit full color information and the inter-pixel relations on the same color planes and across the planes. They derived a feature, called multi-channel kernel fuzzy correlogram, composed by applying a fuzzy membership transformation over multi-channel correlogram. Multi-channel kernel fuzzy correlogram is less sensitivity to noise. This approach handles multimodal distributions without using multiple models per pixel unlike traditional approaches. In a similar way, Chiranjeevi and Sengupta [40] applied a membership transformation on a co-occurrence vector to derive a fuzzy transformed co-occurrence vector with shared membership values in a reduced dimensionality vector space. Fuzzy statistical texture features (FST), derived from this fuzzy transformed co-occurrence vector, are combined with the intensity feature using the Choquet integral. The FST features allow to deal better with dynamic backgrounds than the crisp statistical texture features. Recently, Chiranjeevi and Sengupta [41] adopted a fuzzy rough-set theoretic measures to embed the spatial similarity around a neighborhood as a model for the pixel. First, they extended the basic histon concept to a 3D histon one, which considers the intensities across the color planes in a combined manner, instead of considering independent color planes. Then, they incorporated fuzziness into the 3D HRI measure. The foreground detection is based on the Bhattacharyya distance between the 3D fuzzy histon model and the corresponding measure in the current image. The background maintenance is made using a selective update scheme.

- **Fuzzy post-processing:** It can be applied on the results. For example, fuzzy inference can be used between the previous and the current foreground masks to improve the detection of the moving objects, as developed recently by Sivabalakrishnan and Manjula [163].

Fuzzy models are robust in case of illumination changes and dynamic backgrounds. Table 2.2 shows an overview of the fuzzy models. The first column indicates the background subtraction steps and the second column the name of each method. Their corresponding acronym is indicated in the first parenthesis and the number of papers counted for each method in the second parenthesis. The third column gives the name of the authors and the date of their first publication that use the corresponding fuzzy concept.

TABLE 2.2 Fuzzy Models: An Overview.

Background Subtraction	Algorithm	Authors - Dates
Background Modeling	Fuzzy C-means Clustering (FCM) (1) Type-2 Fuzzy MOG (T2-FMOG) (3)	Kim and Kim (2012) [109] El Baf et al. (2008) [11]
Foreground Detection	Saturating Linear Function (SLF) (5) Sugeno Integral (SI) (1) Choquet Integral (CI) (5)	Sigari et al. (2008) [160] Zhang and Xu (2006) [206] El Baf et al. (2008) [9]
Background Maintenance	Fuzzy Learning Rate (FLR) (7) Fuzzy Maintenance Rule (FMR) (2)	Maddalena et al. (2009) [129] El Baf et al. (2008) [10]
Features	Fuzzy Correlogram (FC) (1) Multi-channel Kernel FC (MKFC) (1) Fuzzy Stat. Texture Features (FST) (1) Fuzzy 3D Histon (F3DH) (1)	Chiranjeevi and Sengupta (2011) [39] Chiranjeevi and Sengupta (2013) [42] Chiranjeevi and Sengupta (2012) [40] Chiranjeevi and Sengupta (2012) [41]
Post-Processing	Fuzzy Inference (FI) (5)	Sivabalakrishnan et al. (2010) [163]

2.2.3   Discriminative and Mixed Subspace Learning Models

In the literature, only reconstructive subspace learning models such as PCA and NMF have attracted attention in the context of background modeling and foreground detection. However, subspace learning methods can be classified into two main categories: reconstructive or discriminative methods [165]. With the reconstructive representations, the model strives to be as informative as possible in terms of well approximating the original data [166]. Their main goal is encompassing the variability of the training data gathered, meaning these representations are not task-dependent. On the other hand, the discriminative methods provide a supervised reconstruction of the data. These methods are task-dependent, but are also spatially and computationally far more efficient and they will often give better classification results when compared to the reconstructive methods [166]. Practically, reconstructive subspace learning models give more effort to construct a robust background model in an unsupervised manner rather than providing a good classification in the foreground detection. Furthermore, they assume that the foreground has a low contribution in the training step, even though this assumption can only be verified when the moving objects in question are either small or far enough away from the camera. In the end, the only advantage in modeling the background with a reconstructive subspace learning is the lack of supervision required. On the other hand, discriminative subspace learning models allow us a robust supervised initialization of the background and a robust classification of pixels as background or foreground. So, some authors achieved background and foreground separation using discriminative or mixed subspace models:

- **Discriminative subspace models:** Farcas et al. [62] proposed to use a discriminative and supervised approach. This approach is based on an incremental discriminative subspace learning algorithm, called Incremental Maximum Margin Criterion (IMMC) [199]. It derives the on-line adaptive supervised subspace from sequential data samples and incrementally updates the eigenvectors of the criterion matrix. IMMC does not need to reconstruct the criterion matrix when it receives a new sample, thus the computation is very fast. This method outperforms the reconstructive ones for the foreground detection but the main drawback is that the ground truth images are needed for the background initialization.

- **Mixed subspace models:** Recently, Marghes et al. [132] used a mixed method that combines a reconstructive method (PCA) with a discriminative one (LDA) [181] to model robustly the background. The objective is firstly to enable a robust model of the background and secondly a robust classification of pixels as background or foreground. So, Marghes et al. [132] used the PCA to model the primary distribution of the pixel values among multiple images, and regards the primary distribution as the knowledge of the background. Then, this method assumes that the low-rank principal vectors of the image space contain discriminative information. Then, Marghes et al. [132] applied on them the LDA for background/foreground classification. Experiments show that the mixed model is more robust than the reconstructive methods (PCA, ICA, INMF and IRT) and the discriminative one (IMMC).

Discriminative and mixed subspace models offer a nice framework for background modeling and foreground detection. Furthermore, there are several algorithms that can be evaluated for this field such as Linear Discriminant Analysis (LDA) [173] and Canonical Correlation Analysis (CCA) [89]. For example, LDA exists in several incremental versions as ILDA using fixed point method [32] or sufficient spanning set approximations [108]. In the same way, Partial Least Squares (PLS) methods [148] give a good perspective to model and update robustly the background.

2.2.4   Robust PCA Models

Recent research on subspace estimation by sparse representation and rank minimization shows a nice framework to separate moving objects from the background. Robust Principal Component Analysis (RPCA) solved via Principal Component Pursuit (PCP) [28] decomposes a data matrix A in two components such that $A = L + S$, where L is a low-rank matrix and S is a sparse noise matrix. Practically, A contains the training sequence. So, the background sequence is then modeled by the low-rank subspace (L) that can gradually change over time, while the moving foreground objects constitute the correlated sparse outliers (S). Robust PCA models can be classified in the following categories:

- **RPCA via Principal Component Pursuit:** The first work on RPCA-PCP developed by Candes et al. [28] proposed a convex optimization to address the robust PCA problem. Under minimal assumptions, this approach called Principal Component Pursuit (PCP) perfectly recovers the low-rank and the sparse matrices. The background sequence is then modeled by a low-rank subspace that can gradually change over time, while the moving foreground objects constitute the correlated sparse outliers. So, Candes et al. [28] showed visual results on foreground detection that demonstrated encouraging performance but PCP presents several limitations for foreground detection. The first limitation is that it required algorithms to be solved that are computational expensive. The second limitation is that PCP is a batch method that stacked a number of training frames in the input observation matrix. In real-time application such as foreground detection, it would be more useful to estimate the low-rank matrix and the sparse matrix in an incremental way quickly when a new frame comes rather than in a batch way. The third limitation is that the spatial and temporal features are lost as each frame is considered as a column vector. The fourth limitation is that PCP imposed the low-rank component being exactly low-rank and the sparse component being exactly sparse but the observations such as in video surveillance are often corrupted by noise affecting every entry of the data matrix. The fifth limitation is that PCP assumes that all entries of the matrix to be recovered are exactly known via the observation and that the distribution of corruption should be sparse and random enough without noise. These assumptions are rarely verified in the case of real applications because (1) only a fraction of entries of the matrix can be observed in some environments, (2) the observation can be corrupted by both impulsive and Gaussian noise (3) the outliers i.e moving objects are spatially localized. Many efforts have been recently concentrated to develop low-computational algorithms for solving PCP (Accelerated Proximal Gradient (APG) [121], Augmented Lagrange Multiplier (ALM) [122], Alternating Direction Method (ADM) [204]), to develop incremental algorithms of PCP to update the low-rank and sparse matrix when a new data arrives [145] and real-time implementations [3]. Moreover, other efforts have addressed problems that appear specifically in real application such as: (1) Presence of noise, (2) Quantization of the pixels, (3) Spatial constraints of the foreground pixels and (4) Local variation in the background. To address (1), Zhou et al. [214] proposed a stable PCP (SPCP) that guarantees stable and accurate recovery in the presence of entry-wise noise. Becker et al. [14] proposed an inequality constrained version of PCP to take into account the quantization error of the pixel's value (2). To address (3), Tang and Nehorai [174] proposed a Block-based PCP (BPCP) method via a decomposition that enforces the low-rankness of one part and the block sparsity of the other part. Wohlberg et al. [190] used a decomposition corresponding to a more general underlying model consisting of a union of low-dimensional subspaces for local variation in the background (4). Furthermore, RPCA-PCP can been extended to the measurement domain, rather than the pixel domain, for use in conjunction with compressive sensing [189].

- **RPCA via Outlier Pursuit:** Xu et al. [195] proposed a robust PCA via Outlier Pursuit to obtain a robust decomposition when the outliers corrupted entire columns, that is every entry is corrupted in some columns. Moreover, Xu et al. [195] proposed a stable OP (SOP) that guarantees stable and accurate recovery in the presence of entry-wise noise.

- **RPCA via Sparsity Control:** Mateos and Giannakis [133] proposed a robust PCA where a tunable parameter controls the sparsity of the estimated matrix, and the number of outliers as a by product.

- **RPCA via Sparse Corruptions:** Even if the matrix A is exactly the sum of a sparse matrix S and a low-rank matrix L, it may be impossible to identify these components from the sum. For example, the sparse matrix S may be low-rank, or the low-rank matrix L may be sparse. To address this issue, Hsu et al. [96] imposed conditions on the sparse and low-rank components in order to guarantee their identifiability from A.

- **RPCA via Log-sum heuristic Recovery:** When the matrix has high intrinsic rank structure or the corrupted errors become dense, the convex approaches may not achieve good performances. Then, Deng et al. [53] used the log-sum heuristic recovery to learn the low-rank structure.

- **RPCA via Iteratively Reweighted Least Squares:** Guyon et al. [83] [82] proposed to solve the RPCA problem by using an Iteratively Reweighted Least Squares (IRLS) alternating scheme for matrix low rank decomposition. Furthermore, spatial constraint are added in the minimization process to take into account the spatial connexity of pixels. The advantage of IRLS over the classical solvers is its fast convergence and its low computational cost. Recently, Guyon et al. [81] improved this scheme by addressing in the minimization the temporal sparseness of moving objects.

- **Bayesian RPCA:** Ding et al. [57] proposed a Bayesian framework which infers an approximate representation for the noise statistics while simultaneously inferring the low-rank and sparse components. Furthermore, Markov dependency is introduced spatially and temporarily between consecutive rows or columns corresponding to image frames. This method has been improved in a variational Bayesian framework [8].

- **Approximated RPCA:** Zhou and Tao [212] proposed an approximated low-rank and sparse matrix decomposition. This method called Go Decomposition (GoDec) produces an approximated decomposition of the data matrix whose RPCA exact decomposition does not exist due to the additive noise, the predefined rank on the low-rank matrix and the predefined cardinality of the sparse matrix. GoDec is significantly accelerated by using bilateral random projection. Furthermore, Zhou and Tao [212] proposed a Semi-Soft GoDec which adopts soft thresholding to the entries of S, instead of GoDec which imposes hard thresholding to both the singular values of the low-rank part L and the entries of the sparse part S.

These recent advances in RPCA are fundamental for background modeling and foreground detection. However, no RPCA algorithm today seems to emerge and to be able to simultaneously address all the key challenges that accompany real-world videos. This is due, in part, to the absence of a rigorous quantitative evaluation as the authors mainly present visual results. Some recent quantitative evaluations using the performance metrics have been made but they are limited to one algorithm. For example, Wang et al. [187] studied only RPCA-PCP solved by APG [121]. Xue et al. [198] evaluated the RPCA-PCP solved by Inexact ALM [122] and Guyon et al. [80] adapted the RPCA-PCP with Linearized Alternating Direction Method with Adaptive Penalty [123]. Guyon et al. [79] showed that BPCP [174] gives better performance than PCP. In a recent comparative analysis and evaluation, Guyon et al. [84] compared with RPCA-PCP solved via Exact ALM [122], RPCA-PCP solved via

Inexact ALM [122], QPCP [14] and BRPCA [57] with the Wallflower dataset [177], the I2R dataset [118] and Shah dataset [155]. Experimental results show that BRPCA that address spatial and temporal constraints outperforms the other methods. In the same way, Rueda et al. [150] compare RPCA-PCP solved Exact ALM [122], BRPCA [57] and GoDec [212]. The authors also concluded that the BRPCA offers the best results in dynamic and static scenes by exploiting the existing correlation between frames of the video sequence using Markov dependencies.

Table 2.3 shows an overview of the RPCA models. The first column indicates the category model and the second column the name of each method. Their corresponding acronym is indicated in the first parenthesis and the number of papers counted for each method in the second parenthesis. The third column gives the name of the authors and the date of the related publication.

TABLE 2.3   Robust PCA (RPCA) Models: An Overview.

RPCA Models	Algorithm	Authors - Dates
RPCA	Principal Components Pursuit (PCP) (19) Stable PCP (SPCP) (3) Quantized PCP (QPCP) (1) Block-based PCP (BPCP) (1) Local PCP (LPCP) (1)	Candes et al. (2011) [28] Zhou et al. (2010) [214] Becker et al. (2011) [14] Tang and Nehorai (2011) [174] Wohlberg et al. (2012) [190]
RPCA-OP	Outlier Pursuit (OP)(1) Stable OP (SOP) (1)	Xu et al. (2010) [195] Xu et al. (2010) [195]
RPCA-SpaCtrl	Sparsity Control (SpaCtrl) (2)	Mateos and Giannakis (2010) [133]
RPCA-SpaCorr	Sparse Corruptions (SpaCorr) (1)	Hsu et al. (2011) [96]
RPCA-LHR	Log-sum heuristic Recovery (LHR)(1)	Deng et al. (2012) [53]
RPCA-IRLS	Iteratively Reweighted Least Squares (IRLS)(3)	Guyon et al. (2012) [83]
Bayesian RPCA	Bayesian RPCA (BRPCA) (1) Variational Bayesian RPCA (VBRPCA) (1)	Ding et al. (2011) [57] Babacan et al. (2011) [8]
Approximated RPCA	GoDec (1) Semi-soft GoDec (1)	Zhou and Tao (2011) [212] Zhou and Tao (2011) [212]

2.2.5   Subspace Tracking

He et al. [91] proposed an incremental gradient descent on the Grassmannian, the manifold of all d-dimensional subspaces for fixed d. This algorithm called Grassmannian Robust Adaptive Subspace Tracking Algorithm (GRASTA) uses a robust l_1-norm cost function in order to estimate and track non-stationary subspaces when the streaming data vectors, that are image frames in foreground detection, are corrupted with outliers, that are foreground objects. this algorithm allows to separate background and foreground on-line. GRASTA shows high-quality visual separation of foreground from background. Recently, He et al. [92] proposed t-GRASTA (transformed-GRASTA) which iteratively performs incremental gradient descent constrained to the Grassmannian manifold of subspaces in order to simultaneously estimate a decomposition of a collection of images into a low-rank subspace, a sparse part of occlusions and foreground objects, and a transformation such as rotation or translation of the image. t-GRASTA is four faster than state-of-the-art algorithms, has half the memory requirement, and can achieve alignment in the case of camera jitter.

Although the l_1-norm in GRASTA leads to favorably conditioned optimization problems it is well known that penalizing with non-convex l_0-surrogates allows reconstruction even in the case when l_1-based methods fail. Therefore, Hage and Kleinsteuber [85] proposed an improved GRASTA using l_0-surrogates solving it by a Conjugate Gradient method. Seidel et al. [153] recently improved this model by approaching the problem with a smoothed l_p-

norm in their algorithm called pROST (l_p-norm Robust Subspace Tracking). Experimental results [153] show that pROST outperforms GRASTA in the case of multi-modal backgrounds.

Recently, Xu et al. [196] developed a Grassmannian Online Subspace Updates with Structured-sparsity (GOSUS), which exploits a meaningful structured sparsity term to significantly improve the accuracy of online subspace updates. Their solution is based on Alternating Direction Method of Multipliers (ADMM), where most key steps in the update procedure reduce to simple matrix operations yielding real-time performance.

Table 2.4 shows an overview of the subspace tracking models. The first column indicates the name of each method. Their corresponding acronym is indicated in the first parenthesis and the number of papers counted for each method in the second parenthesis. The second column gives the name of the authors and the date of the related publication.

TABLE 2.4   Subspace Tracking Models: An Overview.

Methods	Authors - Dates
Grassm. Robust Adaptive Subspace Tracking Algorithm (GRASTA)(2)	He et al. (2011) [91]
transformed-GRASTA (t-GRASTA)(1)	He et al. (2013) [92]
l_p-norm Robust Online Subspace Tracking (pROST)(2)	Hage and Kleinsteuber (2012) [85]
Grassm. Online Subspace Updates with Structured-sparsity (GOSUS)(1)	Xu et al. (2013) [196]

2.2.6   Low Rank Minimization

Low Rank Minimization (LRM) methods are extremely useful in many data mining tasks, yet their performances are often degraded by outliers. However, recent advances in LRM permit a robust matrix factorization algorithm that is insensitive to outliers. Robust LRM is then formulated as a matrix approximation problem with constraints on the rank of the matrix and the cardinality of the outlier set. In addition, structural knowledge about the outliers can be incorporated to find outliers more effectively. The main approaches are the following ones:

- **Contiguous Outliers Detection:** Zhou et al. [213] proposed a formulation of outlier detection in the low-rank representation, in which the outlier support and the low-rank matrix are estimated. This method is called Detecting Contiguous Outlier detection in the Low-rank Representation (DECOLOR). The decomposition involves the same model than SPCP. The objective function is non-convex and it includes both continuous and discrete variable. Zhou et al. [213] adopted an alternating algorithm that separates the energy minimization into two steps. *B*-step is a convex optimization problem and *F*-step is a combinatorial optimization problem.

- **Direct Robust Matrix Factorization:** Xiong et al. [194] proposed a direct robust matrix factorization (DRMF) assuming that a small portion of the matrix *A* has been corrupted by some arbitrary outliers. The aim is to get a reliable estimation of the true low-rank structure of this matrix and to identify the outliers. To achieve this, the ouliers are excluded from the model estimation. The decomposition involves the same model than PCP. Comparing DRMF to the conventional LRM, the difference is that the outliers *S* can be excluded from the low-rank approximation, as long as the number of outliers is not too large, that is, *S* is sufficiently sparse. By excluding the outliers from the low-rank approximation, Xiong et al. [194] ensured the reliability of the estimated low-rank structure. Computation is accelerated using a partial SVD algorithm.

- **Direct Robust Matrix Factorization-Row:** Xiong et al. [194] proposed an extension of DRMF to deal with the presence of outliers in entire columns. This method is called DRMF-Row (DRMF-R). Instead of counting the number of outlier entries, the number of of outliers patterns is counted using the structured $l_{2,0}$-norm.

- **Probabilistic Robust Matrix Factorization:** Wang et al. [201] proposed a Probabilistic Robust Matrix Factorization (PRMF) which is formulated with a Laplace error and a Gaussian prior which correspond to an l_1 loss and an l_2 regularizer, respectively. For model learning, a parallelizable expectation-maximization (EM) algorithm is developed. Furthermore, an online extension of the algorithm for sequential data is provided to offer further scalability. PRMF is comparable to other state-of-the-art robust matrix factorization methods in terms of accuracy and outperforms them particularly for large data matrices.

- **Bayesian Robust Matrix Factorization:** The outliers which correspond to moving objects in the foreground usually form groups with high within-group spatial or temporal proximity. However, PRMF treats each pixel independently with no clustering effect. To address this problem, Wang and Yeung [202] proposed a full Bayesian robust matrix factorization (BRMF). For the generative process, the model parameters have conjugate priors and the likelihood or noise model takes the form of a Laplace mixture. For Bayesian inference, an efficient sampling algorithm is used by exploiting a hierarchical view of the Laplace distribution. Finally, the BMRF is extended by assuming that the outliers form clusters which correspond to moving objects in the foreground. This extension is obtained via placing a first-order Markov random field (MRF) and is called Markov BRMF (MBRMF).

Table 2.5 shows an overview of the low-rank minimization models. The first column indicates the name of each method. Their corresponding acronym is indicated in the first parenthesis and the number of papers counted for each method in the second parenthesis. The second column gives the name of the authors and the date of the related publication.

TABLE 2.5   Low Rank Minimization Models: An Overview.

Methods	Authors - Dates
Contiguous Outliers Detection(DECOLOR)(1)	Zhou et al. (2013) [213]
Direct Robust Matrix Factorization (DRMF) (1)	Xiong et al. (2011) [194]
Direct Robust Matrix Factorization-Row (DRMF-R) (1)	Xiong et al. (2011) [194]
Probabilistic Robust Matrix Factorization (PRMF) (1)	Wang et al. (2012) [201]
Bayesian Robust Matrix Factorization (BRMF)(1)	Wang and Yeung (2013) [202]

2.2.7   Sparse Models

The sparse models can be classified in the following categories: structure sparsity models [99], dynamic group sparsity models [176] and dictionary models [178] [98] [52] [164]. In compressive sensing, the sparsity assumption is made on the observation data. In the other models, it is the foreground objects that are supposed to be sparse.

- **Compressive sensing models:** Compressive sensing measurements for an image are obtained by its K-sparse representation. Then, this sparse representation can be done on each frame of a video. In this context, Cevher et al. [31] considered the background subtraction as a sparse approximation problem and provided different solutions based on convex optimization [33] and total variation [27]. So, the background is learned and adapted in a low dimensional compressed representation, which is sufficient to determine spatial innovations. This representation is adapted overtime to be robust against variations such as illumination changes. Foreground objects are directly detected using the compressive samples. In the same idea, Li et al. [117] used a running average on compressive sensing measurements. In another way, Wang et al. [185] used a background modeling scheme, in which background evaluation is performed on the measurement vectors directly before reconstruction. The estimated background measurement vector is constructed through average, running average, median and block-based selective method respectively. Then, the estimated background image is reconstructed using the background model measurements through the gradient projection for sparse reconstruction (GPSR) algorithm [67]. Experimental results show similar performance than the spatial domain features.

- **Structured Sparsity:** Huang et al. [99] proposed to achieve background and foreground separation using Structured Sparsity (SS), which is a natural extension of the standard sparsity concept in statistical learning and compressive sensing and it can be written as follows:

$$\hat{\beta}_{L_0} = \arg\ min_{\beta \in \mathbb{R}^p}\ \hat{Q}(\beta)\ \ \text{subject to}\ \ ||\beta||_0 \leq \epsilon \tag{2.1}$$

where $Q(\beta) = ||X\beta - y||_2^2$. $X \in \mathbb{R}^{n \times p}$ is a matrix which contains a fixed set of p basis vector $x_1, ..., x_p$ where $x_i \in \mathbb{R}^n$ with n is the sample size. y is the random observation matrix, that depends on an underlying coefficient vector $\bar{\beta} \in \mathbb{R}^p$. Practically, y is the matrix which contains random training frames. The structured sparsity problem consists of the problem of estimating $\bar{\beta}$ under the assumption that the coefficient $\bar{\beta}$ is sparse. This optimization problem is NP-hard and can be approximated by a convex relaxation of L_0 regularization to L_1 regularization such as Lasso [176]. Another algorithm is the Orthogonal Matching Pursuit (OMP) [178]. These algorithms only address the sparsity but, in practice, the structure of $\hat{\beta}$ is known in addition to sparsity. Furthermore, algorithms such as Lasso and OMP do not correctly handle overlapping groups, in that overlapping components are over-counted. To address this issue, Huang et al. [99] developed an algorithm called structured OMP (StructOMP). Experimental results show that StructOMP outperforms Lasso and OMP in the case of background and foreground separation.

- **Dynamic Group Sparsity:** Huang et al. [98] used a learning formulation called Dynamic Group Sparsity (DGS). The idea is that in sparse data the nonzero coefficients are often not random but tend to be cluster such as in foreground detection. According to compressive sensing, a sparse signal $x \in \mathbb{R}^n$ can be recovered from the linear random projections:

$$y = \Phi x + e \tag{2.2}$$

where $y \in \mathbb{R}^m$ is the measurement vector, $\Phi \in \mathbb{R}^{m \times n}$ is the random projection matrix where m is very small against n, and e is the measurement noise. This problem can be written as follows:

$$x_0 = \arg\ min||x||_0\ \ \text{subject to}\ \ ||y - \Phi x||^2 \leq \epsilon \tag{2.3}$$

where ϵ is the noise level. This problem is NP hard. Practically, efficient algorithms were developed to approximate the sparsest solution but all of these algorithms do not consider sparse data priors other than sparsity. However, the nonzero sparse coefficients are often not randomly distributed but group-clustered such as in foreground detection. So, Huang et al. [98] developed a dynamic group sparsity recovery algorithm which assumes that the dynamic group clustering sparse signals consist of a union of subspaces. The group clustering characteristic implies that, if a point is in the union of subspaces, its neighboring points would also be in this union of subspaces with higher probability, and vice versa. The DGS algorithm efficiently obtain stable sparse recovery with far fewer measurements than the previous algorithms. However, DGS assumes that the sparsity number of the sparse data is known before recovery but the exact sparsity numbers are not known and tend to be dynamic group sparse. Then, Huang et al. [98] developed an adaptive DGS algorithm (AdaDGS) which incorporates an adaptive sparsity scheme. Experimental results show that AdaDGS can handle well dynamic backgrounds.

- **Dictionary learning:** David et al. [52] used a sparse representation over a learned dictionary. In the case of a static background, a current frame is decomposed as follows:

$$Y_t = B + F_t \tag{2.4}$$

where I_t is the current frame at time t, B is the static background image and F_t is the foreground image. The basic assumption of the sparse representation is that the current image can be represented as linear combination of vectors form a dictionary. In order to recover the background image, the following type of functional is minimized:

$$||B - I_t||_2^2 + ||D\alpha - B||_2^2 + ||\alpha||_0 \tag{2.5}$$

The first term ensures that the recovered image is similar to the observed version and no main structural changes are present. The second term ensures that the background image is an approximated linear combination over a dictionary D and coefficients α. Finally, the third term forces sparsity of the coefficients, as $||.||_0$ counts the null coefficients. This means that the background image is represented with the smallest possible number of vectors from the dictionary. As similarity between the current image and the background image can not be expected in practice, the corresponding term is dropped from the functional, obtaining, thus, the following expression to minimize:

$$\hat{B} = \arg \ min_{|}|D\alpha - B||_2^2 + ||\alpha||_0 \tag{2.6}$$

The initialization begins with a known or at least estimated version of the background image. David et al. [52] used a patch-wise average of recent frames as an initial background estimate. The success of this approach depends on the capacity to robustly train the dictionary and find suitable sets of coefficients for estimating the background. In training the dictionary, David et al. [52] exploited the redundancy of consecutive frames by training an initial global dictionary by a k-means classifier. The set of coefficients are obtained by applying a basis matching pursuit. Sivalingram et al. [164] estimated the sparse foreground during the training and performance phases by formulating it as a Lasso [176] problem, while the dictionary update step in the training phase is motivated from the K-SVD algorithm [1]. This method works well in the presence of foreground in the training frames, and also gives the foreground masks for the training frames as a by-product of the batch training phase. In a similar way, Huang et al. [100] initialized the data dictionary with compressive sensing measurement values and sparse basis. Then, they trained and updated it through the K-SVD which can get the sparsest representations. At the same time, the correlation between the dictionaries are taken into account to reduce the dictionary redundancy. Then, the background/foreground separation is achieved through the robust principal component pursuit (PCP) so that the image is consistent with the low-rank part of the background regions and the sparse part of the foreground regions. In order for initialization to work with corrupted training samples, Zhao et al. [208] proposed a Robust Dictionary Learning (RDL) approach, which automatically prunes unwanted foreground objects out in the learning stage. Furthermore, they model the foreground detection problem as a l_1-measured and l_1-regularized optimization. So, the global optimal solution of which can be efficiently found. For the background maintenance, a robust dictionary update scheme was specifically developed to achieve better results than the K-SVD algorithm. Indeed, K-SVD produces inaccurate dictionary atom (ghosting effect) at regions where outliers (foreground objects) are present, while this method generates a correct update completely free from outliers. Furthermore, experimental results show that RDL handle both sudden and gradual background changes better than the MOG [169] and the AdaDGS [98]. Lu et al. [126] proposed a new online framework enabling the use of l_1 sparse data fitting term in robust dictionary learning, notably enhancing the usability and practicality of this technique. Their results do not contain the "ghost" artifacts as those produced by batch dictionary learning. In another way, Zhou [211] considered a nonparametric Bayesian dictionary learning for sparse image representation using the beta process and the dependent hierarchical beta processing. The results show more robustness than the RPCA [27]. Sang et al. [151] used a dictionary for each position of the video by using the temporal-spatial information of the local region. Once the dictionaries are built, the information of the corresponding positions in the next frame are expressed by the dictionaries. The words of each dictionary are divided into two classes: One class was used to describe the background, while the other was employed to express the foreground. The corresponding position was judged to be the background or the foreground according to the word that had made the best contribution to the corresponding pixel. The dictionary was updated after the judgment. The proposed method was performed on public videos against three state-of-the-art algorithms. The experimental results show the robustness of the proposed method.

- **Sparse error estimation:** Static background and dynamic foreground can be considered as samples of signals that vary slowly in time with sparse corruption due to foreground objects. Following this idea, Dikmen et al. [54] achieved the background subtraction as a signal estimation problem, where the error sparsity is enforced through minimization of the l_1 norm of the difference between the current frame and estimated background subspace, as an approximation to the underlying l_0 norm minimization structure. The minimization problem is solved by a conjugate gradient [13] for memory efficiency. So, background subtraction is then solved as a sparse error recovery problem. Then, different base construction techniques have been compared and discussed in [55]. However, the sparse assumption on the total error in dynamic backgrounds may be inaccurate which degrades the detection performance. To address this problem, Xue et al. [197] proposed an approach that detects foreground objects based on robust linear regression model (RLM). Foreground objects are considered as outliers and the observation error is composed of foreground outlier and background noise. In order to reliably estimate the background, the outlier and noise are removed. Thus, the foreground detection task is converted into outlier estimation problem. Based on the observation that foreground outlier is sparse and background noise is dispersed, an objective function simultaneously estimates the coefficients and sparse foreground outlier. Xue et al. [197] then transformed the function to fit the problem by only estimating the foreground outlier. Experimental results outperform the MOG [169], KDE [2] and BS [54] in the case of dynamic backgrounds.

Table 2.6 shows an overview of the sparse models. The first column indicates the category model and the second column the name of each method. Their corresponding acronym is indicated in the first parenthesis and the number of papers counted for each method in the second parenthesis. The third column gives the name of the authors and the date of the related publication.

TABLE 2.6   Sparse Models: An Overview.

Sparse Models	Algorithm	Authors - Dates
Compressive Sensing	Compressive Sensing(CS) (3)	Cevher et al. (2008) [31]
Structured Sparsity	Structured Sparsity(SS) (3)	Huang et al. (2009) [99]
Dynamic Group Sparsity	Dynamic Group Sparsity (DGS)(2) Adaptive Dynamic Group Sparsity (AdaDGS)(2)	Huang et al. (2009) [98] Huang et al. (2009) [98]
Dictionary Learning	Dictionary Learning (DL) (2) Online Dictionary Learning (ODL) (3) Robust Dictionary Learning (RDL) (1) Bayesian Dictionary Learning (RDL) (1)	David et al. (2009) [52] Sivalingram et al. (2011) [164] Zhao et al. (2011) [208] Zhou (2013) [211]
Sparse error estimation	Sparse Error Estimation (SEE) (5)	Dikmen et al. (2008) [54]

2.2.8   Transform Domain Models

Fast Fourier Transform (FFT)

Wren and Porikli [193] estimated the background model that captures spectral signatures of multi-modal backgrounds by using FFT. Those signatures are then used to detect changes in the scene that are inconsistent with these signatures. Results show robustness to low-contrast foreground objects in dynamic scenes. This method is called Waviz.

Discrete Cosinus Transform (DCT)

Porikli and Wren [143] developed an algorithm called as Wave-Back that generated a representation of the background using the frequency decompositions of the pixel's history. The Discrete Cosine Transform(DCT) coefficients are computed for the background and the current images. Then, the coefficients of the current image are compared to the background coefficients to obtain a distance map for the image. Then, the distance maps are fused in the same temporal window of the DCT to improve the robustness against noise. Finally, the distance maps are thresholded to achieve foreground detection. This algorithm is efficient in the case of waving trees. Another approach developed by Zhu et al. [215] used two features: the DC parameters and the low frequency AC one, each of which focuses on intensity and texture information, respectively. The AC feature parameter consists of the sum of the low frequency coefficients. So, its distribution is assumed to be a single Gaussian which caused false decision under dynamic backgrounds. In another way, Wang et al. [184] utilized only the information from DCT coefficients at block level to construct background models at pixel level. They implemented the running average, the median and the MOG in the DCT domain. Evaluation results show that these algorithms have much lower computational complexity in the DCT domain than in the spatial domain with the same accuracy.

Walsh Transform (WT)

Tezuka and Nishitani [175] modeled the background using the MOG applied on multiple block sizes by using Walsh transform (WT) . Feature parameters of WT applied to the MOG are determined by using the vertical, horizontal and diagonal direction coefficients, having strong spatial correlation among them. Then, four neighboring WTs are merged into a WT of four times wider block without using the inverse transform. The WT spectral nature reduces the computational steps. Furthermore, Tezuka and Nishitani [175] developed a Selective Fast Walsh Transform (SFWT).

Wavelet Transform (WT)

- **Marr Wavelet:** Gao et al. [70] proposed a background model based on Marr wavelet kernel and used a feature based on binary discrete wavelet transforms to achieve foreground detection. The background model keeps a sample of intensity values for each pixel in the image and uses this sample to estimate the probability density function of the pixel intensity. The density function is estimated using a Marr wavelet kernel density estimation technique. Practically, Gao et al. [70] used difference of Gaussians (DOG) to approximate Marr wavelet. The background is initialized with the first frame. For the foreground detection, the background and current images are transformed in the Binary Discrete Wavelet domain and then the difference is performed in each sub-band. Experiments show that these methods outperform the MOG in traffic surveillance scenes, even though the objects are similar to the background.

- **Dyadic Wavelet:** Guang [76] proposed to use the Dyadic Wavelet (DW) to detect foreground objects. The difference between the background and the current images is decomposed into multi-scale wavelet components. The feature are the HSV components. The value component is used to achieve foreground detection and the saturation component is used to suppress moving shadow. However, dyadic wavelet transformation is achieved through tensor product which depends on the horizontal and vertical direction of the image signals. So, it will cause the cut-off of horizontal and vertical direction.

- **Orthogonal non-separable Wavelet:** To avoid the cut-off effect in DW, Gao et al. [68] used orthogonal non-separable wavelet transformation of the training frames and extracted the approximate information to reconstruct the background. Non-separable wavelet transformation is the real multi-dimensional wavelet transformation and processes signal image as a block rather than separate rows and columns. For the background maintenance, a running average scheme is used if the background has a gradual change, else if the background has a sudden change, the background is replaced by the current frame.

- **Daubechies complex wavelet:** Jalal and Singh [102] used the Daubechies complex wavelet transform because it is approximately shift-invariant and has better directionality information with respect to DWT [70]. The background model is the median which is updated over time. For the foreground detection, the threshold is a pixel-wise one and is also updated over time following the activity of the pixel.

Hadamard Transform (HT)

Baltieri et al. [12] proposed a fast background initialization method made at block level in a non-recursive way to obtain the best background model using the minimum number of frames as possible. For this, each frame is split into blocks, producing a history of blocks and searching among them for the most reliable ones. In this last phase, the method works at a super-block level evaluating and comparing the spectral signatures of each block component. These spectral signatures are obtained with the Hadamard Transform which is faster than DCT.

Table 2.7 shows an overview of the transform domain models. The first column indicates the category model and the second column the name of each method. Their corresponding acronym is indicated in the first parenthesis and the number of papers counted for each method in the second parenthesis. The third column gives the name of the authors and the date of the related publication.

TABLE 2.7 Transform Domain Models: An Overview.

Transform Domain Models	Algorithm	Authors - Dates
Fast Fourier	Fast Fourier Transform (FFT) (1)	Wren and Porikli (2005) [193]
Discrete Cosinus	Discrete Cosinus Transform (DCT) (3)	Porikli and Wren (2005) [143]
Walsh	Walsh Transform (WT) (3)	Tezuka et al. (2008) [175]
Wavelet	Marr Wavelet (MT) (4) Dyadic Wavelet (DT) (4) Orthogonal non-separable Wavelet (OT) (1) Binary Discrete WT (BDWT) (4) Daubechies Complex WT (DCWT) (3)	Gao et al. (2008) [70] Guan (2008) [76] Gao et al. (2008) [68] Gao et al. (2008) [70] Jalal et al. (2011) [102]
Hadamard	Hadamard Transform (HT) (1)	Baltieri et al. (2010) [12]

2.3 Strategies

Several strategies can be used to improve the robustness of a background model to the critical situations. It is important to note that these strategies can be generally applied with any background models. These strategies can be classified as follows:

- **Multi-Features:** Some authors used multiple features to improve the robustness of their background model. As developed in Section 1.3.6, the idea is to add other features to the color one. As gradients of image are relatively less sensitive to changes in illumination, Javed et al. [103] added it to the color feature to obtain quasi illumination invariant foreground detection. But the most common approach is to add texture features to be more robust to illumination changes and shadows as in [44] [45] [206] [9]. To deal with the camouflage, some authors [74] [90] used stereo features (Disparity, depth). In another way, Azab et al. [7] aggregated three features, i.e color, edge and texture using the Choquet integral to deal with illumination changes and dynamic backgrounds. Klare and Sakar [110] proposed an algorithm that incorporates multiple instantiations of the MOG algorithm with color, edge and texture (Haar features). The fusion method used is the average rule. More recently, Han and Davis [87] performed background subtraction using a Support Vector Machine over background likelihood vectors for a set of features which consist of the color, gradient and Haar-like features.

- **Multi-Scales:** This strategy consists of detecting both in the pixel and block scales to combine their respective advantages (See Section 1.3.5). The features used at the pixel and block scales may be different, that is color for pixel and texture for block [207]. Multi-scales strategy have been used for background initialization [51] [50], stationary object detection [37] and dynamic backgrounds [77]. Several details on multi-scales schemes can be found in [49].

- **Multi-Levels:** The idea is to deal with the different challenges at different levels [103] [205]. The region level is generally used to take into account the spatial connexity of the foreground objects. On the other hand, the frame level is used to deal with sudden illumination changes and generally if the amount of pixels detected is over 50 of the frame the background is replaced by the current image. Then, some authors used two levels (pixel and region, pixel and frame levels) or three levels (pixel, region and frame levels).

- **Multi-Resolutions:** The aim is to detect foreground areas in the higher levels and then to achieve progressively a per-pixel foreground detection [127] [209] [183]. The areas detected as background at the higher level are then no tested in the lower level. This strategy reduces the computational time. This strategy is also called hierarchical [34] [210] or coarse-to-fine strategy [16] [15] [69].

- **Multi-Backgrounds:** This strategy uses two background models to handle gradual and sudden illumination changes [97] [141]. The first one is the short-term model which represents the change in background over a short time. Pixels belonging to background are used to update the short-term model. It is not desired when the foreground object becomes a part of background. The second one is the long-term model which represents the change in background over a long time. Each pixel will be used to update the long-term model. It is not desired when foreground objects move slowly or when some stationary variations happened in background, but it overcomes the drawback of the short-term model. Then, the two models are combined to get a better background model [97]. A nice example in [141] [144] used two backgrounds to detect temporarily static regions in video surveillance scene. All these works used the same model for the short-term and long term models but it would be optimal to use a different model for each term. For example, Shimada and Taniguchi [158] used an STLBP based histogram and the MOG for the short-term and long-term models, respectively. Tanaka et al. [172] [171] [170] employed a RRC based SG and the KDE. In the same idea, Shimada and Taniguchi [157] [159] used a predictive model and the MOG.

- **Multi-Layers:** This strategy is generally used for dynamic backgrounds. For example, Yang et al. [200] used a two-layer Gaussian mixture model. The first layer permits to deal with gradually changing pixels, specially. The second layer focuses on those pixels changing significantly and irregularly. This method represents dynamic scenes more accurately and effectively. Another approach developed by Porikli and Tuzel [142] models each pixel as layers of 3D multivariate Gaussians. Each layer corresponds to a different appearance of the pixel in the RGB space. The number of layers required to represent a pixel is not known beforehand so background is initialized with more layers than needed (3 to five layers). Using confidence scores, significant layers are retained for each pixel and then a robust multi-modal representation is provided.

- **Multi-Classifiers:** This strategy consists of using different classifiers in the foreground detection and then to combine or cascade them to strengthen the robustness to the critical situations. For example, Reddy et al. [146] used three classifiers: 1) a probability measurement according to a multivariate Gaussian model, 2) an illumination invariant similarity measurement and 3) a temporal correlation check where decisions from the previous frame are taken into account. On average, this method achieves 36 more accurate foreground masks than the MOG [169].

- **Multi-Cues:** The aim is to address the different challenges to offer robust foreground detection. A representative work is the one developed by Huerta et al. [101]. This method takes into account the inaccuracies due to different cues such as color, edge and intensity in hybrid architecture. The color and edge models solve global and local illumination changes but also the camouflage in intensity. In addition, local information is also used to solve the camouflage in chroma. The intensity cue is applied when color and edge cues are not available. Additionally, temporal difference scheme is used to segment motion where the three cues cannot be reliably computed, for example in the background regions not visible during the initialization period. This approach is extended to deal with ghost detection too. This method obtains very accurate and robust detection in different complex indoor and outdoor scenes. Other multi-cues scheme can be found in [137] [180] [88].

- **Multi-Criteria:** This approach consists of aggregating the classification results of several foreground detection techniques. For example, Kamkar-Parsi et al. [106] aggregates three detections based on UV color deviations, probabilistic gradient information and vector deviations, respectively. The idea is to produce a single decision that is robust to more challenges.

- **Map Selection:** The idea is to use a different model following the activity of each area in the image [105]. For example, Liang et al. [120] defined two areas: uni-modal and multi-modal ones. In the uni-modal area, the SG model [192] is used. On the other hand, the KDE model [59] is used in the multi-modal area. This strategy reduces the computational time.

- **Markov Random Fields:** This strategy consists of adding temporal and spatial information by using Markov Random Fields [113] [152]. For example, Kumar and Sengupta [113] proposed to enforce in the MOG model the temporal contiguity in pixel classification by the use of simple Markov sequence models and to capture the spatial contiguity by MRF models. Results show that false positive and false negative detections are less than in the MOG [169].

Table 2.8 shows an overview of the different strategies. The first column indicates the strategy and the second column the characteristics of each method. Their corresponding acronym is indicated in the first parenthesis and the number of papers counted for each method in the second parenthesis. The third column gives the name of the authors and the date of the related publication.

TABLE 2.8   Strategies: An Overview.

Strategies	Algorithm	Authors - Dates
Multi-Features	Color, Edge (2) Color, Texture (19) Color, Disparity (2) Color, Depth (3) Color, Edge, Texture (1) Color, Gradient, Haar-like features (2)	Javed et al. (2002) [103] Chua et al. (2012) [44] Gordon et al. (1999) [74] Harville et al. (2001) [90] Azab et al. (2010) [7] Klare and Sakar (2009) [110]
Multi-Scales	Pixel and Block Scales (5)	Zhang et al. (2012) [207]
Multi-Levels	Pixel, Region and Frame Levels (5) Pixel and Region Levels (4) Pixel and Frame Levels (1)	Toyama et al. (1999) [177] Cristani et al. (2002) [47] Yang et al. (2004) [203]
Multi-Resolutions	Multi-Resolutions (3) Hierarchical (7) Coarse-to-Fine (3)	Luo et al. (2007) [127] Chen et al. (2007) [34] Bevilacqua et al. (2005) [16]
Multi-Backgrounds	Same Background Models (9) Different Background Models (9)	Hu and Su (2007) [97] Shimada and Taniguchi (2009) [158]
Multi-Layers	Two layers (9) Muliple layers (7)	Yang et al. (2007) [200] Porikli and Tuzel (2005) [142]
Multi-Classifiers	Three Classifiers (3)	Reddy et al. (2010) [146]
Multi-Cues	Three Cues (3) Two Cues (2)	Huerta et al. (2013) [43] Han et al. (2012) [88]
Multi-Criteria	Three Criteria (1)	Kamkar-Parsi et al. (2005) [106]
Map Selection	Two Areas (3)	Jimenez et al. (2003) [105]
Markov Random Fields	Markov Random Fields(4) Dynamic Hidden Markov Random Field (4) Hidden Conditional Random Fields (1)	Kumar and Sengupta (2000) [113] Wang et al. (2005) [188] Chu et al. (2007) [43]

2.4   Real-Time Implementations

Background subtraction is generally made in the context of real-time application on common PC or on smart embedded cameras. So, the computation time and memory requirement need to be reduced as possible. Real-time implementations focus mainly on the following issues:

- **GPU Implementation:** One way to achieve real-time implementation on a common PC is to use the GPU. For example, Fauske et al. [64] proposed a real-time implementation using the language CUDA for the SG, the codebook and the SOBS. Most of the investigations in the literature concern the MOG [140] [114]. Other real-time implementation can be found for the SVM [73] [35] [36].

- **Embedded Implementation:** Cost-sensitive embedded platforms with real-time performance and efficiency demands to optimize algorithms to accomplish background subtraction. Apewokin et al. [4] introduced an adaptive background modeling technique, called multi-modal mean (MM), which balances accuracy, performance, and efficiency to meet embedded system requirements. This algorithm delivers comparable accuracy of the MOG with a 6× improvement in execution time and an 18 reduction in required storage on an eBox-2300 embedded platform. For smart embedded cameras, Casares et al. [29] [30] developed a light-weight and efficient background subtraction algorithm that is robust in the presence of illumination changes and dynamic backgrounds. Furthermore, the memory requirement for each pixel is very small, and the algorithm provides very reliable results with gray-level images. This method selectively updates the background model with an automatically adaptive rate, thus can adapt to rapid changes. Results demonstrated the success of this light-weight salient foreground detection method. Another implementation for embedded cameras developed by Shen et al. [156] introduced compressive sensing to the MOG. This method decreases the computation significantly with a factor of 7 in a DSP setting but remain comparably accurate.

- **Dedicated Architecture:** An efficient way to achieve real-time background subtraction is to develop a dedicated architecture using DSP, VLSI or FPGA. A Digital Signal Processor (DSP) is a specialized microprocessor with an architecture optimized for the operational needs of digital signal processing. Very-Large-Scale Integration (VLSI) is the process of creating integrated circuits by combining thousands of transistors into a single chip. A Field-Programmable Gate Array (FPGA) is an integrated circuit designed to be configured by a designer after manufacturing. DSP architectures have been developed for the SG [6] and the MOG [18]. Dedicated VLSI architectures can be found in [179] [139]. FPGA architectures have been designed for the SG [94] [191] [186], the MOG [5] [104] [119] [72] and the codebook [147].

Table 2.9 shows an overview of the different real-time implementations. The first column indicates the name of the real-time implementation and the second column the implemented model. Their corresponding acronym is indicated in the first parenthesis and the number of papers counted for each method in the second parenthesis. The third column gives the name of the authors and the date of the related publication.

TABLE 2.9 Real-Time Implementations: An Overview.

RT Implementations	Algorithm	Authors - Dates
GPU	Single Gaussian (1) Mixture of Gaussians (4) SVM (3) Codebook (1) SOBS (1)	Fauske et al. (2009) [64] Pham et al. (2010) [140] Gong and Chen (2008) [73] Fauske et al. (2009) [64] Fauske et al. (2009) [64]
Embedded	Multimodal Mean (10) Light-Weight Salient FD (4) CS-MOG (2)	Apewokin et al. (2007) [4] Casares and Velipasalar (2010) [29] Shen et al. (2012) [156]
Dedicated Architecture	Digital Signal Processor (DSP) (4) Very-Large-Scale Integration (VLSI) (3) Field-Programmable Gate Array (FPGA) (29)	Arth et al. (2006) [6] Tsai et al. (2008) [179] Hiraiwa et al. (2010) [94]

2.5 Sensors

Sensors are essential to acquire the video sequence and they determine the availability of some features such as infrared or depth features. The different common sensors are the following ones:

- **Cameras:** Cameras in the visible spectrum are naturally the most used and they can acquire in gray [161] or color scales. Webcams can be also used with less definition [167] [111] [112]. On the other hand, HD cameras are beginning to be employed as their prices decrease [71]. In each case, specific algorithms have been developed.

- **Infrared Cameras (IR):** It is a sensor that forms an image using infrared radiation, similar to a common camera that forms an image using visible light. The interest is to handle adverse illumination conditions such as sudden illumination changes or lack of illumination in a scene, and to avoid camouflage in color. For example, Nadimi and Bhanu [136] used the IR feature in addition to the RGB ones by using two different cameras (IR, visible). This method maintains high detection rates under a variety of critical illumination conditions and visible camera failure. Another example can be found in [46].

- **Depth Cameras :** Traditional stereo systems can provide the color and depth information [74] [90]. However, it needs two cameras. On the other hand, Time of Flight Cameras (ToF) which is a range imaging camera system obtains distance based on the known speed of light, measuring the time-of-flight of a light signal between the camera and the object for each point of the image. Apart from their advantages of high frame rates and their ability to capture the scene all at once, ToF based cameras have generally the disadvantages of a low resolution. Nevertheless, the 2D/3D cameras provide the depth feature compared to ordinary video which makes it possible to overcome the case of camouflage in color [162] [116] [115]. Recently, low cost RGB-D cameras such as the Microsoft's Kinect or the Asus's Xtion Pro are completely changing the computer vision world, as they are successfully used in several applications and research areas. In this context, Camplani et al. [26] [25] used a Microsoft Kinect for objects detection through foreground/background segmentation. Another background subtraction based on RGB-D camera was developed by Fernandez-Sanchez et al. [65] [66].

2.6 Background Subtraction Web Site

This website contains a full list of the references in the field, links to available datasets and codes. In each case, the list is regularly updated and classified following the background models as in this Chapter. An overview of the content of the Background Subtraction Web Site[11] is given at the home page.

In addition to the sections which concern the steps and issues of background subtraction, this website gives references and links to surveys, traditional and recent datasets, available implementations, journals, conferences, workshops and websites as follows:

- **Surveys:** The surveys are classified as traditional surveys [38], recent surveys [60] [48], specific surveys on statistical models [23] [20], mixture of Gaussians models [22], subspace learning models [19] and fuzzy models [21].

---

[11]http://sites.google.com/site/backgroundsubtraction/Home

- **Datasets:** Several datasets have been developed in the last decade. They can be classified in traditional and recent datasets. The traditional datasets provide videos with several challenges and some ground-truth images but none of them address all the challenges. On the other hand, the recent datasets provide realistic large-scale videos with accurate ground truth giving a balanced coverage of the range of challenges present in the real-world. The most common traditional dataset are the following ones: Wallflower dataset [177], I2R dataset [118] and Carnegie Mellon dataset [155]. In addition, the conference "Performance Evaluation of Tracking and Surveillance" (PETS) offers different datasets such as PETS 2001, PETS 2003 and PETS 2006. They are more adapted for tracking evaluation than directly for background subtraction in the sense that the ground-truth is provided as bounding boxes. The recent datasets are the ChangeDetection.net [75], the BMC dataset and the SABS dataset [24]. All these datasets are publicly available and their links are provided on the Background Subtraction Web Site in the section Available Datasets[12]. In Section 1.8 and Section 2.7, we describe briefly the Wallflower dataset [177], the I2R dataset [118], the ChangeDetection.net dataset [75], the BMC dataset [182], and the SABS dataset [24].

- **Available implementations:** Some authors provide their codes on a web page. Furthermore, some libraries are available such as the BGS Library [168]. All the links are provided in the section Available implementations.

- **Journals, conferences, workshops:** This page lists journals, conferences and workshops where recent advances on background subtraction are mainly published. Relevant papers can be found in international journals such as Pattern Analysis and Machine Intelligence (PAMI), Image Processing (IP), Pattern Recognition (PR), Pattern Recognition Letter (PRL), Image and Vision Computing (IVC), Computer Vision and Image Understanding (CVIU), and Machine Vision and Applications (MVA). Furthermore, the reader can find relevant advances in the special issue "Background Modeling for Foreground Detection in real-world dynamic scenes" in Machine Vision and Applications, and in the special issue "Background Models Comparison" in Computer Vision and Image Understanding. For the conference, relevant publications are made in the well-known conferences such as the International Conference on Computer Vision and Pattern Recognition (CVPR), International Conference on Image Processing (ICIP), International Conference on Computer Vision (ICCV), International Conference on Pattern Recognition (ICPR), Asian Conference on Computer Vision (ACCV), European Conference on Computer Vision (ECCV) and the International Symposium on Visual Computing (ISVC). Furthermore, there is a conference dedicated to video surveillance, that is the International Conference on Advanced Video and Signal based Surveillance (AVSS). For the workshop, one of them focuses on visual surveillance, that is the International Workshop on Visual Surveillance in conjunction with ICCV (VS). For animal surveillance, there is the ICPR Workshop on Visual Observation and Analysis of Animal and Insect Behavior (VAIB). Moreover, two workshops in 2012 focus only on the evaluation of background subtraction algorithms on complete datasets, that are the following ones: the IEEE International Workshop on Change Detection (CDW) in conjunction with CVPR and the International Workshop on Background Models Challenge (BMC) in conjunction with ACCV.

- **Web Site:** Some authors provide their publications, results and codes on a web page. All the links are provided in the section Web Site.

---

[12]http://sites.google.com/site/backgroundsubtraction/test-sequences

2.7   Recent Datasets

The recent datasets provide realistic large-scale videos with accurate ground truth giving a balanced coverage of the range of challenges present in the real world. All these datasets are publicly available and their links are provided in the footnotes.

2.7.1   ChangeDetection.net Dataset

The ChangeDetection.net dataset[13] [75] is a realistic, large-scale video dataset for benchmarking background subtraction methods. It consists of nearly 90,000 frames in 31 video sequences representing 6 categories selected to cover a wide range of challenges in 2 modalities (color and thermal IR). A characteristic of this dataset is that each frame is rigorously annotated for ground-truth foreground, background, and shadow area boundaries. This enables objective and precise quantitative comparison and ranking of change detection algorithms. The following categories are covered by this dataset:

- **Baseline:** This category contains four videos, two indoor and two outdoor. These videos represent a mixture of mild challenges typical of the next 4 categories. Some videos have subtle background motion, others have isolated shadows, some have an abandoned object and others have pedestrians that stop for a short while and then move away.

- **Dynamic Background:** There are six videos in this category depicting outdoor scenes with dynamic backgrounds. Two videos represent boats on water, two videos show cars passing next to a fountain, and the last two depict pedestrians, cars and trucks passing in front of a tree shaken by the wind.

- **Camera Jitter:** This category contains one indoor and three outdoor videos captured by unstable cameras. The jitter magnitude varies from one video to another.

- **Shadows:** This category consists of two indoor and four outdoor videos exhibiting strong as well as faint shadows. Some shadows are fairly narrow while others occupy most of the scene. Also, some shadows are cast by moving objects while others are cast by trees and buildings.

- **Intermittent Object Motion:** This category contains six videos with scenarios known for causing "ghosting" artifacts in the detected motion, i.e., objects move, then stop for a short while, after which they start moving again. Some videos include still objects that suddenly start moving, e.g., a parked vehicle driving away, and also abandoned objects. This category is intended for testing how various algorithms adapt to background changes.

- **Thermal:** In this category, five videos (three outdoor and two indoor) have been captured by far-infrared cameras. These videos contain typical thermal artifacts such as heat stamps (e.g., bright spots left on a seat after a person gets up and leaves), heat reflection on floors and windows, and camouflage effects, when a moving object has the same temperature as the surrounding regions.

Figure 2.1 shows a representative original image of a sequence from each category and the second line shows the corresponding ground-truth. Moreover, quality metrics are computable thanks to free software. The results of the first workshop CDW 2012 are available at the related website.

---

[13]http://www.changedetection.net/

FIGURE 2.1 ChangeDetection.et dataset [75]: The first line shows an original image of a sequence from each category and the second line shows the corresponding ground-truth in the following order from left to right: Baseline, Dynamic Background, Camera Jitter, Shadows and Intermittent Object Motion.

2.7.2 SABS Dataset

The SABS (Stuttgart Artificial Background Subtraction) dataset[14] [24] represents an artificial dataset for pixel-wise evaluation of background models. Synthetic image data generated by modern ray tracing makes realistic high quality ground-truth data available. The dataset consists of video sequences for nine different challenges of background subtraction for video surveillance. These sequences are further split into training and test data. For every frame of each test sequence ground-truth annotation is provided as color-coded foreground masks. This way, several foreground objects can be distinguished and the ground-truth annotation could also be used for tracking evaluation. The dataset contains additional shadow annotation that represents for each pixel the absolute luminance distance between the frame with and without foreground objects. The sequences have a resolution of 800×600 pixels and are captured from a fixed viewpoint.

The following scenarios are covered by the dataset:

- **Basic:** This is a basic surveillance scenario combining a multitude of challenges for general performance overview.

- **Dynamic Background:** Uninteresting background movements are considered using a detail of the Basic sequence, containing moving tree branches and changing traffic light.

- **Bootstrapping:** This scenario contains no training phase, thus subtraction starts after the first frame.

- **Darkening:** Gradual scene change is simulated by decreasing the illumination constantly. Thus, background and foreground darkens and their contrast decreases.

- **Light Switch:** Once-off changes are simulated by switching off the light of the shop (frame 901) and switching it on again (frame 1101).

- **Noisy Night:** Basic sequence at night, with increased sensor noise accounting for high gain level and low background/foreground contrast resulting in more camouflage.

- **Shadow:** We use a detail of the street region to measure shadow pixels classified as foreground.

- **Camouflage:** Detail of the street region is used, too. We compare performance between a sequence with persons wearing dark clothes and gray cars, and a sequence containing colored foreground objects significantly differing from the background.

- **Video Compression:** Basic sequence compressed with different bit rates by a standard codec often used in video surveillance (H.264, $40 - 640$ kbits/s, 30 frames per second). Five video sequences with different compression rates are therefore provided.

---

[14]http://www.vis.uni-stuttgart.de/index.php?id=sabs

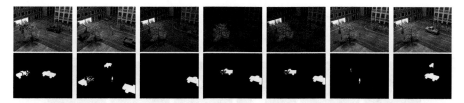

FIGURE 2.2   SABS dataset [24]: The first line shows an original image of some sequences and the second line shows the corresponding ground-truth in the following order from left to right: Basic, Bootstrapping, Darkening, Light Switch, Noisy Night, Camouflage and Video Compression.

Each scenario includes a sequence with 800 frames for initial background training (with exception of the bootstrapping scenario) and a sequence of 600 frames for validation (except for the *Darkening* and *Bootstrap* scenarios that both contain 1400 frames) with full ground truth annotation. The dataset is encoded frame-by-frame as PNG (Portable Network Graphics) images. For download, it was split and compressed to several RAR archives.

Figure 2.2 shows a representative original image of a sequence from each category and the second line shows the corresponding ground-truth. Furthermore, several quality metrics such as Precision-Recall and F-Measure can be made using a provided free software. The comparison made in [24] are dynamically available at the related website.

2.7.3   BMC Dataset

The BMC (Background Models Challenge) dataset[15] consists of both synthetic and real videos to permit a rigorous comparison of background subtraction techniques for the corresponding workshop organized within Asian Conference in Computer Vision (ACCV). This dataset [182] consists of the following sequences:

- **Synthetic Sequences:** A set of 20 urban video sequences rendered with the SiVIC simulator. With this tool, the associate ground truth was rendered frame by frame for each video at 25 fps. Several complex scenarios are simulated such as fog, sun and acquisition noise for two environments (a rotary and a street). A first part of 10 synthetic videos are devoted to the learning phase, while 10 others are used for the evaluation.

- **Real Sequences:** The BMC dataset also contains 9 real videos acquired from static cameras in video-surveillance contexts for evaluation. This real dataset has been built in order test the algorithms reliability during time and in difficult situations such as outdoor scenes. So, real long videos about one hour and up to four hours are available, and they may present long time change in luminosity with small density of objects in time compared to previous synthetic ones. Moreover, this real dataset allows to test the influence of some difficulties encountered during the object extraction phase, as the presence of vegetation, cast shadows or sudden light changes in the scene.

Figure 2.3 shows a representative original image of a sequence from each category and the second line shows the corresponding ground-truth. Furthermore, several quality metrics are computable thanks to a free software named BMC Wizard. The results of the first workshop BMC 2012 are available at the related website.

---

[15]http://bmc.univ-bpclermont.fr

FIGURE 2.3   BMC dataset [182]: The first line shows an original image of some sequences and the second line shows the corresponding ground-truth in the following order from left to right: Sequence Street, Sequence Rotary, Sequence Video 001, Sequence Video 003, Sequence Video 008 and Sequence Video 009.

FIGURE 2.4   MAR dataset [17]: The first line shows an original image of some sequences and the second line shows the corresponding ground-truth.

2.7.4   MAR Dataset

Maritime Activity Recognition (MAR[16]) [17] is a dataset containing data coming from different video sources (fixed and Pan-Tilt-Zoom cameras) and from different scenarios. There are 10 videos from fixed cameras with ground-truth images and 15 form PTZ cameras. The aim of this dataset is to provide a set of videos that can be used to develop intelligent surveillance system for the maritime environment. Figure 2.4 shows representative original images of some sequences and their corresponding ground-truth.

2.7.5   RGB-D Object Detection Dataset

This dataset [25] available from the UPM[17] provides five sequences of indoor environments, acquired with the Microsoft Kinect RGB-D cameras. Each sequence contain different challenges such as cast shadows, color and depth camouflage. For each sequence a hand-labeled ground truth is provided. A brief description of this dataset can be made as follows:

- **General:** This sequence acquired at a frame rate of 30 fps can be used to test the overall performance in case of complex scenarios taking into account all the possible error contributions in the scene: shadows of the moving objects, color camouflage, noisy depth data and interaction between foreground objects and the background. It is composed of 300 frames and the corresponding ground truth is composed of 39 frames spanning 115 frames of the sequence where the moving object is present. One every three frames has been labeled.

- **Shadow:** It permits to analyze the impact of shadows by considering both color and depth data. It is composed by 250 frames and the corresponding ground truth is composed of 25 frames spanning 120 frames of the sequence where the moving object

---

[16]http://labrococo.dis.uniroma1.it/MAR/
[17]http://www.gti.ssr.upm.es/~mac/

FIGURE 2.5 RGB-D dataset [25]: The first line shows an original image of some sequences and the second line shows the corresponding ground-truth in the following order from left to right: General, Shadows, Depth Camouflage and Color Camouflage.

is present. One every five frames has been labeled.

- **Depth Camouflage:** It can be used to analyze the performance when depth camouflage situation occurs when both color and depth data are available. It contains 670 frames and the ground truth is composed of 102 frames that cover 400 frames of the sequence where the moving object is present, in this case one every four frames has been labeled.

- **Color Camouflage:** It is to test the performance when the color camouflage problem occurs when both data are available. It contains 360 frames and the ground truth is composed of 45 frames that cover 240 frames of the sequence that are the one where the moving object is present. In this case one every six frames have been labeled.

- **Move Background:** It helps to analyze the impact of the moved background object problem when depth and color data are available. It contains 250 frames. The aim of this sequence is to highlight the impact of the moved background object problems on the algorithms' performance. In the scene there are present two static bags on the floor, that are rapidly removed from their position.

It is worth noting that due to the limitation of depth sensors, these sequences represent indoor environment. For sequences 3 and 4, the groundtruth is available for the entire scene and also considering only those regions in the images where each particular type of problem is present. This procedure guarantees that other sources of errors do not bias the algorithms performance comparison with respect to the considered error factor. Figure 2.5 shows representative original images of some sequences and their corresponding ground-truth.

2.7.6 CITIC RGB-D Dataset

This dataset [65] available from the CITIC[18] contains sequences recorded with rectified stereo cameras, and some frames have been hand-segmented to provide ground-truth information.

- **Suitcase:** A person enters the scene with a suitcase and leaves it on the floor. The main difficulty of the sequence is the low lighting and color saturation, as well as the similar color between the suitcase and the floor.

- **Crossing:** Two people walk in and out of the camera field. The dark floor complicates the detection by color when they get near the camera, while the range is less useful when they get near the wall.

---

[18]http://atcproyectos.ugr.es/mvision/

FIGURE 2.6 CITIC dataset [65]: The first line shows an original image of some sequences and the second line shows the corresponding ground-truth in the following order from left to right: Stereo cameras sequences (LCDScreen, Labdoor), Kinect sequences (Chairbox, Wall, Shelves and Hallway).

- **LCDScreen:** A person walks into a lab and deposits a black box in front of a black LCD screen. In addition, there are flickering lights in the ceiling.

- **LabDoor:** A person walks in and out of the camera field, projecting shadows on background objects. In addition, there are occlusions due to background objects, flickering lights in the ceiling and sudden illumination changes.

Furthermore, four sequences recorded with the Microsoft Kinect sensor (both RGB images and depth) are provided with hand-segmented ground-truth for some frames. The sequences are the following:

- **ChairBox:** A person enters the field of view and leaves a box on a chair. There are flickering lights, as well as areas where depth cannot be obtained by infrared active sensors.

- **Wall:** A flat object (paper sheet) appears close to a wall, creating shadows and highlighted regions. The main difficulties are the similarity of depth between foreground and background and the change of lighting.

- **Shelves:** A person enters the scene and puts two objects on shelves. There are changes of exposure, as well as difficult depth estimation.

- **Hallway:** This sequence presents some reflections, complicated lighting, objects similar to the background and sudden illumination changes.

Figure 2.6 shows representative original images of some sequences and their corresponding ground-truth.

In summary, there are two recent large-scale datasets: 1) the ChangeDetection dataset which contains visible and infrared videos and the BMC dataset which contains synthetic and real videos. The SABS dataset concerns synthetic videos with different scenarios. On the other hand, the MAR dataset concerns maritime scenes. Finally, two datasets permit the evaluation when the depth information is used as feature.

2.8   Background Subtraction (BGS) Libraries

There are several libraries available for background subtraction algorithms. The first one is under the library OpenCV, but there are firstly two algorithms: the MOG [169] and the foreground detection method developed by Li et al. [118]. Then, Parks and Fels [138] evaluated background subtraction algorithms using Visual Microsoft C and OpenCV but their code cannot be directly integrated under OpenCV. More recently, Laurence Bender developed a first library using OpenCV at the Laboratory of Arte Electronico e Inteligencia Artificial of the Universidad Nacional de Tres de Febrero. This library called Scene is an open source multiplatform computer vision framework that performs background subtraction using two traditional algorithms (SG [192], MOG [169]) and three recent algorithms based on fuzzy classification rules and neural networks (Fuzzy Gaussian [160], SOBS [128] and SOBS-SC [130]). It was mainly designed as a toolkit for the development of interactive art projects that investigate dynamics of complex environments.

In 2012, Andrews Sobral [168] developed the BGS Library which provides a C++ framework to perform background subtraction. The code works either on Windows or on Linux and requires OpenCV version 2.0 or superior. Currently, the library offers more than 29 background subtraction algorithms. A large number of algorithms were provided by several authors which guarantee their accuracy. The source code is available under GNU GPL v3 license, the library is free and open source. Any user can download the latest project source code using a SVN client. In Windows, a demo project for Visual Studio 2010 is provided. An executable version of BGSLibrary is available for Windows 32 bits and 64 bits, and all required DLL's are included in package. For Linux users, a Makefile can be used to compile all files and generate an executable example. In the BGSLibrary website[19], the users can download also the BGSLibraryGUI developed in Java. It is a friendly Graphical User Interface to configure and run BGSLibrary. The algorithms are classified according to their similarities. Finally, this library is the first large-scale library which permits a fair comparison.

2.9 Conclusion

This chapter provides an overview of recent models for background modeling models with several promising directions. Then, strategies and real-time implementations were investigated. Finally, resources such as datasets and libraries are presented. Algorithms that seem to be able to simultaneously address all the key challenges that accompany real-world videos are the ViBe [13], PBAS [95], SOBS [128], SOBS-SC [130] and 3dSOBS+ [131] . Their key advantages consist of that they use robust update models to deal with illumination changes and dynamic backgrounds. In our opinion, future developments may concern recent advances on RPCA and sparse models which show a great potential but they actually need investigation to be implemented in both incremental and real-time ways. Other investigations may concern fuzzy models which offer a nice potential too and feature selection which seems to be an open problem in this field.

2.10 Acknowledgements

The author would like to thank F. Porikli (MERL, ChangeDetection.net), B. Hörferlin (Univ. of Stuggart, SABS dataset), A. Vacavant (Univ. of Auvergne, BMC dataset), M. Camplani (Univ. of Madrid, RBD Object Detection dataset) and E. Fernandez-Sanchez (Univ, of Granada, CITIC RGB-D Dataset) for providing the brief description of their datasets.

References

1. M. Aharon, M. Elad, and A. Bruckstein. The K-SVD: An algorithm for designing of overcomplete dictionaries for sparse representation. *IEEE Transactions on Signal Processing,*

---

[19]http://code.google.com/p/bgslibrary/

24(11):4311–4322, 2006.

2. M. Allili, N. Bouguila, and D. Ziou. A robust video foreground segmentation by using generalized Gaussian mixture modeling. *Canadian Conference on Computer and Robot Vision, CRV 2007*, pages 503–509, 2007.

3. M. Anderson, G. Ballard, J. Demme, and K. Keutzer. Communication-avoiding QR decomposition for GPUs. *Technical Report, ECCS*, 2010.

4. S. Apewokin, B. Valentine, L. Wills, S. Wills, and A. Gentile. Multimodal mean adaptive backgrounding for embedded real-time video surveillance. *IEmbedded Computer Vision Workshop, ECVW 2007*, June 2007.

5. K. Appiah, A. Hunter, P. Dickinson, and H. Meng. Accelerated hardware video object segmentation: from foreground detection to connected components labelling. *Computer Vision and Image Understanding, CVIU 2010*, 2010.

6. C. Arth, H. Bischof, and C. Leistner. TRICam - an embedded platform for remote traffic surveillance. *CVPR Workshop on Embedded Computer Vision, CVPRW 2006*, June 2010.

7. M. Azab, H. Shedeed, and A. Hussein. A new technique for background modeling and subtraction for motion detection in real-time videos. *International Conference on Image Processing, ICIP 2010*, pages 3453–3456, September 2010.

8. S. Babacan, M. Luessi, R. Molina, and A. Katsaggelos. Sparse Bayesian methods for low-rank matrix estimation. *IEEE Transactions on Signal Processing*, 2011.

9. F. El Baf, T. Bouwmans, and B. Vachon. Foreground detection using the choquet integral. *International Workshop on Image Analysis for Multimedia Interactive Integral, WIAMIS 2008*, pages 187–190, May 2008.

10. F. El Baf, T. Bouwmans, and B. Vachon. A fuzzy approach for background subtraction. *International Conference on Image Processing, ICIP 2008*, pages 2648–2651, October 2008.

11. F. El Baf, T. Bouwmans, and B. Vachon. Type-2 fuzzy mixture of Gaussians model: Application to background modeling. *International Symposium on Visual Computing, ISVC 2008*, pages 772–781, December 2008.

12. D. Baltieri, R. Cucchiara, and R. Vezzani. Fast background initialization with recursive hadamard transform. *International Conference on Advanced Video and Signal Based Surveillance, AVSS 2010*, September 2010.

13. O. Barnich and M. Van Droogenbroeck. ViBe: a powerful random technique to estimate the background in video sequences. *International Conference on Acoustics, Speech, and Signal Processing, ICASSP 2009*, pages 945–948, April 2009.

14. S. Becker, E. Candes, and M. Grant. TFOCS: flexible first-order methods for rank minimization. *Low-rank Matrix Optimization Symposium, SIAM Conference on Optimization*, 2011.

15. A. Bevilacqua, G. Capelli, L. Di Stefano, and A. Lanza. A novel approach to change detection based on a coarse-to-fine strategy. *IEEE International Conference on Image Processing, ICIP 2005*, pages 434–437, 2005.

16. A. Bevilacqua, L. Di Stefano, and A. Lanza. Coarse-to-fine strategy for robust and efficient change detectors. *IEEE International Conference on Advanced Video and Signal Based Surveillance, AVSS 2005*, September 2005.

17. D. Bloisi, A. Pennisi, and L. Iocchi. Background modeling in the maritime domain. *Machine Vision and Applications, Special Issue on Background Modeling*, 2013.

18. S. Boragno, B. Boghossian, J. Black, D. Makris, and S. Velastin. A DSP-based system for the detection of vehicles parked in prohibited areas. *AVSS 2007*, pages 1–6, 2007.

19. T. Bouwmans. Subspace learning for background modeling: A survey. *Recent Patents on Computer Science*, 2(3):223–234, November 2009.

20. T. Bouwmans. Recent advanced statistical background modeling for foreground detection:

A systematic survey. *Recent Patents on Computer Science*, 4(3):147–171, September 2011.

21. T. Bouwmans. Background subtraction for visual surveillance: A fuzzy approach. *Handbook on Soft Computing for Video Surveillance, Taylor and Francis Group*, 5, March 2012.

22. T. Bouwmans, F. El-Baf, and B. Vachon. Background modeling using mixture of Gaussians for foreground detection: A survey. *Recent Patents on Computer Science*, 1(3):219–237, November 2008.

23. T. Bouwmans, F. El-Baf, and B. Vachon. Statistical background modeling for foreground detection: A survey. *Handbook of Pattern Recognition and Computer Vision, World Scientific Publishing*, 4(3):181–199, November 2009.

24. S. Brutzer, B. Höferlin, and G. Heidemann. Evaluation of background subtraction techniques for video surveillance. *International Conference on Computer Vision and Pattern Recognition, CVPR 2011*, pages 1937–1944, June 2011.

25. M. Camplani, C. Blanco, L. Salgado, F. Jaureguizar, and N. Garcia. Advanced background modeling with RGB-D sensors through classifiers combination and inter-frame foreground prediction. *Machine Vision and Application, Special Issue on Background Modeling*, 2013.

26. M. Camplani and L. Salgado. Background foreground segmentation with RGB-D Kinect data: an efficient combination of classifiers. *Journal on Visual Communication and Image Representation*, 2013.

27. E. Candes. Compressive sampling. *International Congress of Mathematicians*, 1998.

28. E. Candes, X. Li, Y. Ma, and J. Wright. Robust principal component analysis? *International Journal of ACM*, 58(3), May 2011.

29. M. Casares and S. Velipasalar. Resource-efficient salient foreground detection for embedded smart cameras by tracking feedback. *International Conference on Advanced Video and Signal Based Surveillance, AVSS 2010*, September 2010.

30. M. Casares, S. Velipasalar, and A. Pinto. Light-weight salient foreground detection for embedded smart cameras. *Computer Vision and Image Understanding*, 2010.

31. V. Cevher, D. Reddy, M. Duarte, A. Sankaranarayanan, R. Chellappa, and R. Baraniuk. Background subtraction for compressed sensing camera. *European Conference on Computer Vision, ECCV 2008*, October 2008.

32. D. Chen and L. Zhang. An incremental linear discriminant analysis using fixed point method. *ISSN 2006*, 3971:1334–1339, 2006.

33. S. Chen, D. Donoho, and M. Saunders. Atomic decomposition by basis pursuit. *Journal on Scientific Computing*, 1998.

34. Y. Chen, C. Chen, C. Huang, and Y. Hung. Efficient hierarchical method for background subtraction. *Journal of Pattern Recognition*, 40(10):2706–2715, October 2007.

35. L. Cheng and M. Gong. Real time background subtraction from dynamics scenes. *International Conference on Computer Vision, ICCV 2009*, September 2009.

36. L. Cheng, M. Gong, D. Schuurmans, and T. Caelli. Real-time discriminative background subtraction. *IEEE Transaction on Image Processing*, 20(5):1401–1414, December 2011.

37. S. Cheng, X. Luo, and S. Bhandarkar. A multiscale parametric background model for stationary foreground object detection. *International Workshop on Motion and Video Computing, WMCV 2007*, 2007.

38. S. Cheung and C. Kamath. Robust background subtraction with foreground validation for urban traffic video. *Journal of Applied Signal Processing, EURASIP*, 2005.

39. P. Chiranjeevi and S. Sengupta. Detection of moving objects using fuzzy correlogram based background subtraction. *ICSIPA 2011*, 2011.

40. P. Chiranjeevi and S. Sengupta. New fuzzy texture features for robust detection of moving objects. *IEEE Signal Processing Letters*, 19(10):603–606, October 2012.

41. P. Chiranjeevi and S. Sengupta. Robust detection of moving objects in video sequences

through rough set theory framework. *Image and Vision Computing, IVC 2012*, 2012.

42. P. Chiranjeevi and S. Sengupta. Detection of moving objects using multi-channel kernel fuzzy correlogram based background subtraction. *IEEE Transactions on Systems, Man, and Cybernetics*, October 2013.

43. Y. Chu, X. Ye, J. Qian, Y. Zhang, and S. Zhang. Adaptive foreground and shadow segmentation using hidden conditional random fields. *Journal of Zhejiang University*, pages 586–592, 2007.

44. T. Chua, K. Leman, and Y. Wang. Fuzzy rule-based system for dynamic texture and color based background subtraction. *IEEE International Conference on Fuzzy Systems, FUZZ-IEEE 2012*, pages 1–7, June 2012.

45. T. Chua, K. Leman, and Y. Wang. Fuzzy rule-based system for dynamic texture and color based background subtraction. *IEEE International Conference on Fuzzy Systems, FUZZ-IEEE 2012*, pages 1–7, June 2012.

46. C. Conaire, N. O'Connor, E. Cooke, and A. Smeaton. Multispectral object segmentation and retrieval in surveillance video. *IEEE International Conference on Image Processing, ICIP 2006*, pages 2381–2384, 2006.

47. M. Cristani, M. Bicego, and V. Murino. Integrated region- and pixel-based approach to background modeling. *IEEE Workshop on Motion and Video Computing, MOTION 2002*, 2002.

48. M. Cristani, M. Farenzena, D. Bloisi, and V. Murino. Background subtraction for automated multisensor surveillance: A comprehensive review. *EURASIP Journal on Advances in Signal Processing*, 2010:24, 2010.

49. D. Culibrk, V. Crnojevic, and B. Antic. Multiscale background modelling and segmentation. *International Conference on Digital Signal Processing, DSP 2009*, pages 922–927, 2009.

50. S. Davarpanah, F. Khalid, and M. Golchin. A block-based multi-scale background extraction algorithm. *Journal of Computer Science*, pages 1445–1451, 2010.

51. S. Davarpanah, F. Khalid, N. Lili, S. Puteri, and M. Golchin. Using multi-scale filtering to initialize a background extraction model. *Journal of Computer Science*, pages 1077–1084, 2012.

52. C. David, V. Gui, and F. Alexa. Foreground/background segmentation with learned dictionary. *International Conference on Circuits, Systems and Signals, CSS 2009*, pages 197–201, 2009.

53. Y. Deng, Q. Dai, R. Liu, and Z. Zhang. Low-rank structure learning via log-sum heuristic recovery. *Preprint*, 2012.

54. M. Dikmen and T. Huang. Robust estimation of foreground in surveillance videos by sparse error estimation. *International Conference on Pattern Recognition, ICPR 2008*, December 2008.

55. M. Dikmen, S. Tsai, and T. Huang. Base selection in estimating sparse foreground in video. *International Conference on Image Processing, ICIP 2009*, November 2009.

56. J. Ding, M. Li, K. Huang, and T. Tan. Modeling complex scenes for accurate moving objects segmentation. *Asian Conference on Computer Vision, ACCV 2010*, pages 82–94, 2010.

57. X. Ding, L. He, and L. Carin. Bayesian robust principal component analysis. *IEEE Transaction on Image Processing*, 2011.

58. Y. Ding, W. Li, T. Fan, and H. Yang. Robust moving object detection under complex background. *Computer Science and Information Systems*, 7(1), February 2010.

59. A. Elgammal and L. Davis. Non-parametric model for background subtraction. *European Conference on Computer Vision, ECCV 2000*, pages 751–767, June 2000.

60. S. Elhabian, K. El-Sayed, and S. Ahmed. Moving object detection in spatial domain using background removal techniques - state-of-art. *Patents on Computer Science*, 1(1):32–

54, January 2008.

61. W. Fan and N. Bouguila. Online variational learning of finite Dirichlet mixture models. *Evolving Systems*, January 2012.

62. D. Farcas, C. Marghes, and T. Bouwmans. Background subtraction via incremental maximum margin criterion: A discriminative approach. *Machine Vision and Applications*, 23(6):1083–1101, October 2012.

63. A. Faro, D. Giordano, and C. Spampinato. Adaptive background modeling integrated with luminosity sensors and occlusion processing for reliable vehicle detection. *IEEE Transactions on Intelligent Transportation Systems*, 12(4):1398–1412, December 2011.

64. E. Fauske, L. Eliassen, and R. Bakken. A comparison of learning based background subtraction techniques implemented in CUDA. *NAIS 2009*, pages 181–192, 2009.

65. E. Fernandez-Sanchez, J. Diaz, and E. Ros. Background subtraction based on color and depth using active sensors. *Sensors*, 13:8895–8915, 2013.

66. E. Fernandez-Sanchez, L. Rubio, J. Diaz, and E. Ros. Background subtraction model based on color and depth cues. *Machine Vision and Application, Special Issue on Background Modeling*, 2013.

67. M. Figueiredo, R. Nowak, and S. Wright. Gradient projection for sparse reconstruction: Application to compressed sensing and other inverse problems. *IEEE Journal on Selected Topics in Signal Processing*, 1:586–597, 2007.

68. D. Gao, Z. Jiang, and M. Ye. A new approach of dynamic background modeling for surveillance information. *International Conference on Computer Science and Software Engineering, CSSE 2008*, 1:850–855, 2008.

69. L. Gao, Y. Fan, N. Chen, Y. Li, and X. Li. Moving objects detection using adaptive region-based background model in dynamic scenes. *Foundations of Intelligent Systems, Advances in Intelligent and Soft Computing*, 2012.

70. T. Gao, Z. Liu, W. Gao, and J. Zhang. A robust technique for background subtraction in traffic video. *International Conference on Neural Information Processing, ICONIP 2008*, pages 736–744, November 2008.

71. M. Genovese and E. Napoli. ASIC and FPGA implementation of the Gaussian mixture model algorithm for real-time segmentation of high definition video. *IEEE Transactions on Very Large Scale Integration (VLSI) Systems*, 2013.

72. M. Genovese, E. Napoli, D. De Caro, N. Petra, and A. Strollo. FPGA implementation of Gaussian mixture model algorithm for 47fps segmentation of 1080p video. *Journal of Electrical and Computer Engineering*, 2013.

73. M. Gong and L. Cheng. Real time foreground segmentation on GPUs using local online learning and global graph cut optimization. *International Conference on Pattern Recognition, ICPR 2008*, December 2008.

74. G. Gordon, T. Darrell, M. Harville, and J. Woodfill. Background estimation and removal based on range and color. *International Conference on Computer Vision and Pattern Recognition, CVPR 1999*, pages 459–464, June 1999.

75. N. Goyette, P. Jodoin, F. Porikli, J. Konrad, and P. Ishwar. changedetection.net: A new change detection benchmark dataset. *IEEE Workshop on Change Detection, CDW 2012 at CVPR 2012*, June 2012.

76. Y. Guan. Wavelet multi-scale transform based foreground segmentation and shadow elimination. *Open Signal Processing Journal*, 1:1–6, 2008.

77. P. Guha, D. Palai, K. Venkatesh, and A. Mukerjee. A multiscale co-linearity statistic based approach to robust background modeling. *Asian Conference on Computer Vision, ACCV 2006*, pages 297–306, 2006.

78. L. Guo and M. Du. Student's t-distribution mixture background model for efficient object detection. *IEEE International Conference on Signal Processing, Communication and Computing, ICSPCC 2012*, pages 410–414, August 2012.

79. C. Guyon, T. Bouwmans, and E. Zahzah. Foreground detection based on low-rank and block-sparse matrix decomposition. *IEEE International Conference on Image Processing, ICIP 2012*, September 2012.

80. C. Guyon, T. Bouwmans, and E. Zahzah. Foreground detection by robust PCA solved via a linearized alternating direction method. *International Conference on Image Analysis and Recognition, ICIAR 2012*, June 2012.

81. C. Guyon, T. Bouwmans, and E. Zahzah. Foreground detection via robust low rank matrix decomposition including spatio-temporal constraint. *International Workshop on Background Model Challenges, ACCV 2012*, November 2012.

82. C. Guyon, T. Bouwmans, and E. Zahzah. Foreground detection via robust low rank matrix factorization including spatial constraint with iterative reweighted regression. *International Conference on Pattern Recognition, ICPR 2012*, November 2012.

83. C. Guyon, T. Bouwmans, and E. Zahzah. Moving object detection via robust low rank matrix decomposition with IRLS scheme. *International Symposium on Visual Computing, ISVC 2012*, pages 665–674, July 2012.

84. C. Guyon, T. Bouwmans, and E. Zahzah. Robust principal component analysis for background subtraction: Systematic evaluation and comparative analysis. *INTECH, Principal Component Analysis, Book 1, Chapter 12*, pages 223–238, March 2012.

85. C. Hage and M. Kleinsteuber. Robust PCA and subspace tracking from incomplete observations using l_0-surrogates. *Preprint*, 2012.

86. T. Haines and T. Xiang. Background subtraction with Dirichlet processes. *European Conference on Computer Vision, ECCV 2012*, October 2012.

87. B. Han and L. Davis. Density-based multi-feature background subtraction with support vector machine. *IEEE Transactions on Pattern Analysis and Machine Intelligence, PAMI 2012*, 34(5):1017–1023, May 2012.

88. J. Han, M. Zhang, and D. Zhang. Background modeling fusing local and global cues for temporally irregular dynamic textures. *Advanced Science Letters*, 7:58–63, 2012.

89. D. Hardoon, S. Szedmak, and S. Shawe. Canonical correlation analysis: An overview with application to learning methods. *Neural Computation*, 16:2639–2664, 2006.

90. M. Harville, G. Gordon, and J. Woodfill. Foreground segmentation using adaptive mixture models in color and depth. *International Workshop on Detection and Recognition of Events in Video*, July 2001.

91. J. He, L. Balzano, and J. Luiz. Online robust subspace tracking from partial information. *IT 2011*, September 2011.

92. J. He, D. Zhang, L. Balzano, and T. Tao. Iterative Grassmannian optimization for robust image alignment. *Image and Vision Computing*, June 2013.

93. Y. He, D. Wang, and M. Zhu. Background subtraction based on nonparametric Bayesian estimation. *International Conference Digital Image Processing*, July 2011.

94. J. Hiraiwa, E. Vargas, and S. Toral. An FPGA based embedded vision system for real-time motion segmentation. *International Conference on Systems, Signals and Image Processing, IWSSIP 2010*, pages 360–363, 2010.

95. M. Hofmann, P. Tiefenbacher, and G. Rigoll. Background segmentation with feedback: The pixel-based adaptive segmenter. *IEEE Workshop on Change Detection, CVPR 2012*, June 2012.

96. D. Hsu, S. Kakade, and T. Zhang. Robust matrix decomposition with sparse corruptions. *IEEE Transactions on Information Theory*, 57(11):7221–7234, 2011.

97. J. Hu and T. Su. Robust background subtraction with shadow and highlight removal for indoor surveillance. *International Journal on Advanced Signal Processing*, pages 1–14, 2007.

98. J. Huang, X. Huang, and D. Metaxas. Learning with dynamic group sparsity. *International Conference on Computer Vision, ICCV 2009*, October 2009.

99. J. Huang, T. Zhang, and D. Metaxas. Learning with structured sparsity. *International Conference on Machine Learning, ICML 2009*, 2009.

100. X. Huang, F. Wu, and P. Huang. Moving-object detection based on sparse representation and dictionary learning. *AASRI Conference on Computational Intelligence and Bioinformatics*, 1:492–497, 2012.

101. I. Huerta, A. Amato, X. Roca, and J. Gonzalez. Exploiting multiple cues in motion segmentation based on background subtraction. *Neurocomputing, Special issue: Behaviours in video*, pages 183–196, January 2013.

102. A. Jalal and V. Singh. A robust background subtraction approach based on daubechies complex wavelet transform. *Advances in Computing and Communication, ACC 2011*, pages 516–524, 2011.

103. O. Javed, K. Shafique, and M. Shah. A hierarchical approach to robust background subtraction using color and gradient information. *IEEE Workshop on Motion and Video Computing, WMVC 2002*, December 2002.

104. H. Jiang, H. Ard, and V. wall. A hardware architecture for real-time video segmentation utilizing memory reduction techniques. *IEEE Transactions on Circuits and Systems for Video Technology*, 19(2):226–236, February 2009.

105. P. Jimenez, S. Bascon, R. Pita, and H. Moreno. Background pixel classification for motion detection in video image sequences. *International Work Conference on Artificial and Natural Neural Network, IWANN 2003*, pages 718–725, 2003.

106. A. Kamkar-Parsi, R. Laganier, and M. Bouchard. Multi-criteria model for robust foreground extraction. *VSSN 2005*, pages 67–70, November 2007.

107. H. Kashani, S. Seyedin, and H. Yazdi. A novel approach in video scene background estimation. *International Journal of Computer Theory and Engineering*, 2(2):274–282, April 2010.

108. T. Kim, S. Wong, B. Stenger, J. Kittler, and R. Cipolla. Incremental linear discriminant analysis using sufficient spanning set approximations. *CVPR*, pages 1–8, June 2007.

109. W. Kim and C. Kim. Background subtraction for dynamic texture scenes using fuzzy color histograms. *IEEE Signal Processing Letters*, 3(19):127–130, March 2012.

110. B. Klare and S. Sarkar. Background subtraction in varying illuminations using an ensemble based on an enlarged feature set. *IEEE Conference on Computer Vision and Pattern Recognition, CVPR 2009*, June 2009.

111. T. Ko, S. Soatto, and D. Estrin. Background subtraction on distributions. *European Conference on Computer Vision, ECCV 2008*, pages 222–230, October 2008.

112. T. Ko, S. Soatto, and D. Estrin. Warping background subtraction. *IEEE International Conference on Computer Vision and Pattern Recognition, CVPR 2010*, June 2010.

113. P. Kumar and K. Sengupta. Foreground background segmentation using temporal and spatial Markov processes. *Department of Electrical and Computer Engineering, National University of Singapore*, November 2000.

114. P. Kumar, A. Singhal, S. Mehta, and A. Mittal. Real-time moving object detection algorithm on high-resolution videos using GPUs. *Journal of Real-Time Image Processing*, January 2013.

115. B. Langmann, S. Ghobadi, K. Hartmann, and O. Loffeld. Multi-model background subtraction using Gaussian mixture models. *Symposium on Photogrammetry Computer Vision and Image Analysis, PCV 2010*, pages 61–66, 2010.

116. B. Langmann, K. Hartmann, and O. Loffeld. Depth assisted background subtraction for color capable ToF-cameras. *International Conference on Image and Video Processing and Computer Vision, IVPCV 2010*, pages 75–82, July 2012.

117. J. Li, J. Wang, and W. Shen. Moving object detection in framework of compressive sampling. *Journal of Systems Engineering and Electronics*, 21(5):740–745, October 2010.

118. L. Li and W. Huang. Statistical modeling of complex background for foreground object

detection. *IEEE Transaction on Image Processing*, 13(11):1459–1472, November 2004.

119. X. Li and X. Jing. FPGA based mixture Gaussian background modeling and motion detection. *International Conference on Natural Computation, ICNC 2011*, 4:2078–2081, 2011.

120. Y. Liang, Z. Wang, X. Xu, and X. Cao. Background pixel classification for motion segmentation using mean shift algorithm. *ICLMC 2007*, pages 1693–1698, 2007.

121. Z. Lin, M. Chen, L. Wu, and Y. Ma. The augmented Lagrange multiplier method for exact recovery of corrupted low-rank matrices. *UIUC Technical Report*, November 2009.

122. Z. Lin, A. Ganesh, J. Wright, L. Wu, M. Chen, and Y. Ma. Fast convex optimization algorithms for exact recovery of a corrupted low-rank matrix. *UIUC Technical Report*, August 2009.

123. Z. Lin, R. Liu, and Z. Su. Linearized alternating direction method with adaptive penalty for low-rank representation. *NIPS 2011*, December 2011.

124. Z. Liu, W. Chen, K. Huang, and T. Tan. Probabilistic framework based on KDE-GMM hybrid model for moving object segmentation in dynamic scenes. *International Workshop on Visual Surveillance, ECCV 2008*, October 2008.

125. Z. Liu, K. Huang, and T. Tan. Foreground object detection using top-down information based on EM framework. *IEEE Transactions on Image Processing*, 21(9):4204–4217, September 2012.

126. C. Lu, J. Shi, and J. Jia. Online robust dictionary learning. *EURASIP Journal on Image and Video Processing, IVP 2011*, January 2011.

127. R. Luo, L. Li, and I. Gu. Efficient adaptive background subtraction based on multi-resolution background modelling and updating. *Pacific-RIM Conference on Multimedia, PCM 2007*, December 2007.

128. L. Maddalena and A. Petrosino. A self organizing approach to background subtraction for visual surveillance applications. *IEEE Transactions on Image Processing*, 17(7):1729–1736, 2008.

129. L. Maddalena and A. Petrosino. Multivalued background/foreground separation for moving object detection. *International Workshop on Fuzzy Logic and Applications, WILF 2009*, 5571:263–270, June 2009.

130. L. Maddalena and A. Petrosino. The SOBS algorithm: What are the limits? *IEEE Workshop on Change Detection, CVPR 2012*, June 2012.

131. L. Maddalena and A. Petrosino. The 3dSOBS+ algorithm for moving object detection. *Computer Vision and Image Understanding, CVIU 2014*, 122(65-73), May 2014.

132. C. Marghes, T. Bouwmans, and R. Vasiu. Background modeling and foreground detection via a reconstructive and discriminative subspace learning approach. *International Conference on Image Processing, Computer Vision, and Pattern Recognition, IPCV 2012*, July 2012.

133. G. Mateos and G. Giannakis. Sparsity control for robust principal component analysis. *International Conference on Signals, Systems, and Computers*, November 2010.

134. S. Molina-Giraldo, J. Carvajal-Gonzalez, A. Alvarez-Meza, and C. Castellanos-Dominguez. Video segmentation based on multi-kernel learning and feature relevance analysis for object classification. *International Conference on Pattern Recognition Applications and Methods, ICPRAM 2013*, February 2013.

135. D. Mukherjee and J. Wu. Real-time video segmentation using Student's t mixture model. *International Conference on Ambient Systems, Networks and Technologies, ANT 2012*, pages 153–160, 2012.

136. S. Nadimi and B. Bhanu. Physics-based cooperative sensor fusion for moving object detection. *IEEE Workshop on Learning in Computer Vision and Pattern Recognition, CVPR 2004*, June 2004.

137. S. Noh and M. Jeon. A new framework for background subtraction using multiple cues. *Asian Conference on Computer Vision, ACCV 2012*, November 2012.

138. D. Parks and S. Fels. Evaluation of background subtraction algorithms with post-processing. *IEEE International Conference on Advanced Video and Signal-based Surveillance, AVSS 2008*, 2008.

139. D. Peng, C. Lin, W. Sheu, and T. Tsai. Architecture design for a low-cost and low-complexity foreground object segmentation with multi-model background maintenance algorithm. *ICIP 2009*, pages 3241–3244, 2009.

140. V. Pham, P. Vo, H. Vu Thanh, and B. Le Hoai. GPU implementation of extended Gaussian mixture model for background subtraction. *IEEE International Conference on Computing and Telecommunication Technologies, RIVF 2010*, November 2010.

141. F. Porikli. Detection of temporarily static regions by processing video at different frame rates. *IEEE International Conference on Advanced Video and Signal based Surveillance, AVSS 2007*, 2007.

142. F. Porikli and O. Tuzel. Bayesian background modeling for foreground detection. *ACM International Workshop on Video Surveillance and Sensor Networks, VSSN 2005*, pages 55–58, November 2005.

143. F. Porikli and C. Wren. Change detection by frequency decomposition: Wave-back. *International Workshop on Image Analysis for Multimedia Interactive Services, WIAMIS 2005*, April 2005.

144. F. Porikli and Z. Yin. Temporally static region detection in multi-camera systems. *PETS 2007*, October 2007.

145. C. Qiu and N. Vaswani. Real-time robust principal components pursuit. *International Conference on Communication Control and Computing*, 2010.

146. V. Reddy, C. Sanderson, and B. Lovell. Robust foreground object segmentation via adaptive region-based background modelling. *International Conference on Pattern Recognition, ICPR 2010*, August 2010.

147. R. Rodriguez-Gomez, E. Fernandez-Sanchez, J. Diaz, and E. Ros. Codebook hardware implementation on FPGA for background subtraction. *Journal of Real-Time Image Processing*, 2012.

148. R. Rosipal and N. Kramer. Overview and recent advances in partial least squares. *SLSFS*, 3940:34–35, 2005.

149. J. Rossel-Ortega, G. Andrieu, A. Rodas-Jorda, and V. Atienza-Vanacloig. A combined self-configuring method for object tracking in colour video. *International Conference on Computer Vision, ICPR 2010*, August 2010.

150. H. Rueda, L. Polania, and K. Barner. Robust tracking and anomaly detection in video surveillance sequences. *SPIE Airborne Intelligence, Surveillance, Reconnaissance, ISR 2012, Systems and Applications*, May 2012.

151. N. Sang, T. Zhang, B. Li, and X. Wu. Dictionary-based background subtraction. *Journal of Huazhong University of Science and Technology*, 41(9):28–31, September 2013.

152. K. Schindler and H. Wang. Smooth foreground-background segmentation for video processing. *Asian Conference on Computer Vision, ACCV 2006*, pages 581–590, January 2006.

153. F. Seidel, C. Hage, and M. Kleinsteuber. pROST - a smoothed l_p-norm robust online subspace tracking method for realtime background subtraction in video. *Machine Vision and Applications, Special Issue on Background Modeling*, 2013.

154. M. Shakeri, H. Deldari, H. Foroughi, A. Saberi, and A. Naseri. A novel fuzzy background subtraction method based on cellular automata for urban traffic applications. *International Conference on Signal Processing, ICSP 2008*, pages 899–902, October 2008.

155. Y. Sheikh and M. Shah. Bayesian modeling of dynamic scenes for object detection. *IEEE Transactions on Pattern Analysis and Machine Intelligence*, 27:1778–1792, 2005.

156. Y. Shen, W. Hu, Mi. Yang, J. Liu, C. Chou, and B. Wei. Efficient background subtraction for tracking in embedded camera networks. *International Conference on Information Processing in Sensor Networks, IPSN 2012*, April 2012.

157. A. Shimada and R. Taniguchi. Object detection based on Gaussian mixture predictive background model under varying illumination. *International Workshop on Computer Vision, MIRU 2008*, July 2008.

158. A. Shimada and R. Taniguchi. Hybrid background model using spatial-temporal lbp. *IEEE International Conference on Advanced Video and Signal based Surveillance, AVSS 2009*, September 2009.

159. A. Shimada and R. Taniguchi. Object detection based on fast and low-memory hybrid background model. *IEEJ Transactions on Electronics, Information and Systems*, 129-C(5):846–852, May 2009.

160. M. Sigari, N. Mozayani, and H. Pourreza. Fuzzy running average and fuzzy background subtraction: Concepts and application. *International Journal of Computer Science and Network Security*, 8(2):138–143, 2008.

161. J. Silveira, C. Jung, and S. Musse. Background subtraction and shadow detection in grayscale video sequences. *Brazilian Symposium on Computer Graphics and Image Processing, SIBGRAPI 2005*, pages 189–196, 2005.

162. D. Silvestre. Video surveillance using a time-of-light camera. *Master Thesis, Informatics and Mathematical Modelling, University of Denmark*, 2007.

163. M. Sivabalakrishnan and D. Manjula. Adaptive background subtraction in dynamic environments using fuzzy logic. *International Journal of Image Processing*, 4(1), 2010.

164. R. Sivalingam, A. De Souza, V. Morellas, N. Papanikolopoulo, M. Bazakos, and R. Miezianko. Dictionary learning for robust background modeling. *IEEE International Conference on Robotics and Automation, ICRA 2011*, May 2011.

165. D. Skocaj and A. Leonardis. Canonical correlation analysis for appearance-based orientation and self-estimation and self-localization. *CogVis Meeting*, January 2004.

166. D. Skocaj, A. Leonardis, M. Uray, and H. Bischof. Why to combine reconstructive and discriminative information for incremental subspace learning. *Computer Vision Winter Workshop, Czech Society for Cybernetics and Informatics*, February 2006.

167. M. Smids. Background subtraction for urban traffic monitoring using webcams. *Master Thesis, Univ. Amsterdam*, December 2006.

168. A. Sobral, L. Oliveira, L. Schnitman, and F. de Souza. Highway traffic congestion classification using holistic properties. *IASTED International Conference on Signal Processing, Pattern Recognition and Applications, SPPRA2013*, February 2013.

169. C. Stauffer and E. Grimson. Adaptive background mixture models for real-time tracking. *IEEE Conference on Computer Vision and Pattern Recognition, CVPR 1999*, pages 246–252, 1999.

170. T. Tanaka, A. Shimada, D. Arita, and R. Taniguchi. Object detection under varying illumination based on adaptive background modeling considering spatial locality. *International Workshop on Computer Vision, MIRU 2008*, July 2008.

171. T. Tanaka, A. Shimada, D. Arita, and R. Taniguchi. Object segmentation under varying illumination based on combinational background modeling. *Joint Workshop on Machine Perception and Robotics, MPR 2008*, 2008.

172. T. Tanaka, A. Shimada, D. Arita, and R. Taniguchi. Object detection under varying illumination based on adaptive background modeling considering spatial locality. *PSVIT 2009*, pages 645–656, January 2009.

173. F. Tang and H. Tao. Fast linear discriminant analysis using binary bases. *International Conference on Pattern Recognition, ICPR 2006*, 2, 2006.

174. G. Tang and A. Nehorai. Robust principal component analysis based on low-rank and block-sparse matrix decomposition. *CISS 2011*, 2011.

175. H. Tezuka and T. Nishitani. A precise and stable foreground segmentation using fine-to-coarse approach in transform domain. *International Conference on Image Processing, ICIP 2008*, pages 2732–2735, October 2008.

176. R. Tibshirani. Regression shrinkage and selection via the Lasso. *Journal of the Royal Statistical Society*, 58:267–288, 1996.

177. K. Toyama, J. Krumm, B. Brumiit, and B. Meyers. Wallflower: Principles and practice of background maintenance. *International Conference on Computer Vision*, pages 255–261, September 1999.

178. J. Tropp and A. Gilbert. Signal recovery from random measurements via orthogonal matching pursuit. *IEEE Transactions on Information Theory*, 53:4655–4666, 2007.

179. T. Tsai, D. Peng, C. Lin, and W. Sheu. A low cost foreground object detection architecture design with multi-model background maintenance algorithm. *VLSI 2008*, 2008.

180. L. Unzueta, M. Nieto, A. Cortes, J. Barandiaran, O. Otaegui, and P. Sanchez. Adaptive multi-cue background subtraction for robust vehicle counting and classification. *IEEE Transactions on Intelligent Transportation Systems*, 2011.

181. M. Uray, D. Skocaj, P. Roth, H. Bischof, and A. Leonardis. Incremental LDA learning by combining reconstructive and discriminative approaches. *BMVC 2007*, pages 272–281, 2007.

182. A. Vacavant, T. Chateau, A. Wilhelm, and L. Lequievre. A benchmark dataset for foreground/background extraction. *International Workshop on Background Models Challenge, ACCV 2012*, November 2012.

183. L. Wang and C. Pan. Effective multi-resolution background subtraction. *International Conference on Acoustics, Speech, and Signal Processing, ICASSP 2011*, May 2011.

184. W. Wang, D. Chen, W. Gao, and J. Yang. Modeling background from compressed video. *International Conference on Computer Vision, ICCV 2005*, 2005.

185. X. Wang, F. Liu, and Z. Ye. Background modeling in compressed sensing scheme. *ESEP 2011*, pages 4776–4783, December 2011.

186. Y. Wang and H. Chen. The design of background subtraction on reconfigurable hardware. *International Conference on Intelligent Information Hiding and Multimedia Signal Processing*, pages 182–185, 2012.

187. Y. Wang, Y. Liu, and L. Wu. Study on background modeling method based on robust principal component analysis. *Annual Conference on Electrical and Control Engineering, ICECE 2011*, pages 6787–6790, September 2011.

188. Y. Wang, K. Loe, T. Tan, and J. Wu. A dynamic Hidden Markov Random field model for foreground and shadow segmentation. *Workshops on Application of Computer Vision, WACV 2005*, 1:474–480, January 2005.

189. A. Waters, A. Sankaranarayanan, and R. Baraniuk. SpaRCS: recovering low-rank and sparse matrices from compressive measurements. *Neural Information Processing Systems, NIPS 2011*, December 2011.

190. B. Wohlberg, R. Chartrand, and J. Theiler. Local principal component pursuit for nonlinear datasets. *International Conference on Acoustics, Speech, and Signal Processing, ICASSP 2012*, March 2012.

191. M. Wojcikowski, R. Zaglewsk, and B. Pankiewicz. FPGA-based real-time implementation of detection algorithm for automatic traffic surveillance sensor network. *Journal of Signal Processing Systems*, December 2010.

192. C. Wren and A. Azarbayejani. Pfinder : Real-time tracking of the human body. *IEEE Transactions on Pattern Analysis and Machine Intelligence*, 19(7):780–785, July 1997.

193. C. Wren and F. Porikli. Waviz: Spectral similarity for object detection. *IEEE International Workshop on Performance Evaluation of Tracking and Surveillance, PETS 2005*, January 2005.

194. L. Xiong, X. Chen, and J. Schneider. Direct robust matrix factorization for anomaly detection. *International Conference on Data Mining, ICDM 2011*, 2011.

195. H. Xu, C. Caramanis, and S. Sanghavi. Robust PCA via outlier pursuit. *NIPS 2010*, 2010.

196. J. Xu, V. Ithapu, L. Mukherjee, J. Rehg, and V. Singhy. GOSUS: grassmannian online

subspace updates with structured-sparsity. *International Conference on Computer Vision, ICCV 2013*, December 2013.

197. G. Xue, L. Song, J. Sun, and M. Wu. Foreground estimation based on robust linear regression model. *International Conference on Image Processing, ICIP 2011*, pages 3330–3333, September 2011.

198. Y. Xue, X. Gu, and X. Cao. Motion saliency detection using low-rank and sparse decomposition. *International Conference on Acoustics, Speech, and Signal Processing, ICASSP 2012*, March 2012.

199. J. Yan, B. Zhang, S. Yan, Q. Yang, H. Li, Z. Chen, W. Xi, W. Fan, W. Ma, and Q. Cheng. IMMC: incremental maximum margin criterion. *KDD 2004*, pages 725–730, August 2004.

200. H. Yang, Y. Tan, J. Tian, and J. Liu. Accurate dynamic scene model for moving object detection. *International Conference on Image Processing, ICIP 2007*, pages 157–160, 2007.

201. N. Yang, T. Yao, J. Wang, and D. Yeung. A probabilistic approach to robust matrix factorization. *European Conference on Computer Vision, ECCV 2012*, pages 126–139, 2012.

202. N. Yang and D. Yeung. Bayesian robust matrix factorization for image and video processing. *International Conference on Computer Vision, ICCV 2013*, 2013.

203. T. Yang, S. Li, Q. Pan, and J. Li. Real-time and accurate segmentation of moving objects in dynamic scene. *ACM International Workshop on Video Surveillance and Sensor Networks, VSSN 2004*, October 2004.

204. X. Yuan and J. Yang. Sparse and low-rank matrix decomposition via alternating direction methods. *Optimization Online*, November 2009.

205. Q. Zang and R. Klette. Robust background subtraction and maintenance. *International Conference on Pattern Recognition, ICPR 2004*, pages 90–93, 2004.

206. H. Zhang and D. Xu. Fusing color and texture features for background model. *International Conference on Fuzzy Systems and Knowledge Discovery, FSKD*, 4223(7):887–893, September 2006.

207. Z. Zhang, C. Wang, B. Xiao, S. Liu, and W. Zhou. Multi-scale fusion of texture and color for background modeling. *AVSS 2012*, pages 154–159, 2012.

208. C. Zhao, X. Wang, and W. Cham. Background subtraction via robust dictionary learning. *EURASIP Journal on Image and Video Processing, IVP 2011*, January 2011.

209. B. Zhong, S. Liu, H. Yao, and B. Zhang. Multi-resolution background subtraction for dynamic scenes. *International Conference on Image Processing, ICIP 2009*, pages 3193–3196, November 2009.

210. B. Zhong, H. Yao, S. Shan, X. Chen, and W. Gao. Hierarchical background subtraction using local pixel clustering. *IEEE International Conference on Pattern Recognition, ICPR 2008*, December 2008.

211. M. Zhou. Nonparametric Bayesian dictionary learning and count and mixture modeling. *PhD thesis*, 2013.

212. T. Zhou and D. Tao. GoDec: randomized low-rank and sparse matrix decomposition in noisy case. *International Conference on Machine Learning, ICML 2011*, 2011.

213. X. Zhou, C. Yang, and W. Yu. Moving object detection by detecting contiguous outliers in the low-rank representation. *IEEE Transactions on Pattern Analysis and Machine Intelligence*, 35:597–610, 2013.

214. Z. Zhou, X. Li, J. Wright, E. Candes, and Y. Ma. Stable principal component pursuit. *IEEE ISIT Proceedings*, pages 1518–1522, June 2010.

215. J. Zhu, S. Schwartz, and B. Liu. A transform domain approach to real-time foreground segmentation in video sequences. *ICASSP 2005*, 2005.

3

Background Model Initialization for Static Cameras

Lucia Maddalena
National Research Council, Naples, Italy

Alfredo Petrosino
University of Naples Parthenope, Naples, Italy

3.1 Introduction

The background modeling process is basically characterized by three fundamental tasks which are the followings: 1) model representation, that describes the kind of model used to represent the background; 2) model initialization, that regards the initialization of this model; and 3) model adaptation, that concerns the mechanism used for adapting the model to the background changes. In the last two decades, several methods have been proposed in order to address these tasks, as witnessed by several surveys [10, 18, 36, 40, 41] and by many other Chapters in this Book. Most of these methods focus on the representation and the adaption issues, whereas limited attention is given to the model initialization. In the following, we will focus our attention on existing approaches that specifically address background model initialization, referring the interested reader to the above mentioned literature for other background modeling issues.

Given an image sequence, the availability of an initial background model that describes the scene without foreground objects is the prerequisite, or at least can be of help, for many applications, including:

- **Video surveillance**, where an accurate estimate of the background is essential for high quality moving object detection based on background subtraction [34];

- **Video segmentation**, where the background provides rich information to extract foreground objects [11];

- **Video compression**, where the estimated background represents redundant information that can be suppressed [39];

- **Video inpainting** (or *video completion*), whose techniques try to fill-in user defined spatio-temporal holes in a video sequence using information extracted in the existent

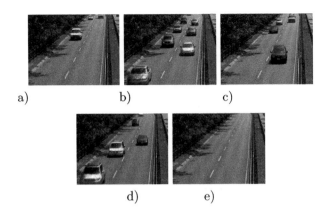

FIGURE 3.1   The background initialization problem: From(a) to (d) example frames from a bootstrap sequence where the background is always occluded by foreground objects; (e) estimated background.

spatio-temporal volume, according to consistency criteria evaluated both in time and space [14];

- **Privacy protection for videos**, whose main aim is to avoid the infringement on the privacy right of people taken in the many videos uploaded to video sharing services, such as YouTube, that may contain privacy sensitive information of the people [37];

- **Computational photography**, where the user wants to obtain a clean background plate from a set of input images containing cluttering foreground objects [1, 23].

In order to train a proper initial background model, often the assumption is made that one or more of the initial sequence frames are *empty*, that is, there are no *foreground objects* (also referred to as *occluders* [14], *clutter* [15], or *outliers* [47]). However, in some uncontrolled environments, such as train stations, airports, or motorways, directly obtaining an empty background is almost impossible. Furthermore, issues such as strong illumination changes can render the existing background model ineffective, thereby enforcing the computation of a new background model [42]. In all such cases, it may be necessary to build the background model using a sequence which does contain foreground objects.

The afforded general problem of *background initialization*, also known as *bootstrapping* [46], *background estimation* [20], *background reconstruction* [3], or *initial background extraction* [11], can be stated as follows:

> *Given a set of images taken with a stationary camera, in which the background is occluded by any number of foreground objects, the aim is to output a single model describing the background of the scene with no foreground objects.*

Depending on the application, the set of images can consist of a subset of sequence frames adopted for background training (e.g., for video surveillance), a set of non-time sequence photographs (e.g., for computational photography), or the entire available sequence. In the following, it will be generally referred to as the *bootstrap sequence*.

An exemplification of the background initialization problem is provided in Figure 3.1 for sequence *highway* of the Change Detection Challenge dataset [22].

In the following, we provide a summary of the classifications adopted in the literature (Section 3.2). Next, we give hints on frequently adopted approaches to the problem, grouped by the common schemes they rely on (Sections 3.3 through 3.6). Some considerations are drawn in Section 3.7, while concluding remarks are given in 3.8.

3.2   Classifications

Background initialization approaches have been classified according to different criteria, including the data-abstraction level used (as in [4, 14, 42]), the number of sequence frames simultaneously taken into account (as in [4]), or their selectivity (as in [6]).

According to the data-abstraction level used, the approaches can be classified as:

- **Pixel-level** methods, also referred to as *temporal* methods [6], that mainly exploit the pixel intensity temporal constancy, assuming that the time series of observations is independent at each pixel. Usually, these methods are very fast, but, neglecting spatial relations, they cannot handle background maintenance issues that require higher-level analysis, such as local or global sudden illumination changes.
- **Region-level** methods, also referred to as *spatio-temporal* methods [6], that take advantage of inter-pixel relations, subdividing the images into regions or refining the low-level result obtained at the pixel level. Usually, these methods provide robust background estimates, but are characterized by a high computational cost.
- **Hybrid** methods, that combine pixel- and region-level analyses, looking for an optimal compromise between efficiency and accuracy.

According to the number of sequence frames simultaneously taken into account, background initialization methods can also be classified as:

- **Recursive** methods, also called *sequential* methods [42], that analyze each new sequence frame and use it to iteratively estimate and update a single background model. These methods are generally computationally efficient and have minimal memory requirements, avoiding the need to store all the sequence frames. However, model update could easily include into the background slowly moving or stationary foreground objects.
- **Non-recursive** methods, that store a buffer of previous sequence frames and estimate a background model based on the statistical properties of these frames. Having explicit access to the most recent video frames, these methods can model aspects of the data which cannot be analyzed with recursive techniques. However, memory requirements can be significant.

According to selectivity, background initialization methods can be further classified as:

- **Blind** methods, that generate a background model for each pixel by means of temporal statistics computed using the whole time series of the pixel intensities. These background statistics may be "dirty", by retaining information not just about the background process, but also about some possible foreground processes due to the moving objects covering the pixel along the bootstrap sequence.
- **Selective** methods, that try to isolate the background process for each pixel by the possible foreground processes, thus computing "clean" background statistics.

Clearly, these classifications can also be combined, as in [6], where selective methods are further sub-classified as temporal or spatio-temporal, according to the data-abstraction level adopted.

Although so many classifications can be considered, in the following we will briefly describe some of the existing approaches to background initialization by grouping them according to common schemes they rely on. The correspondence of the described methods to the above classes will be provided later (see Section 3.7).

3.3   Methods Based on Temporal Statistics

Often, when dealing with background modeling, the specific problem of background initialization is neglected, by merging it into the estimation and update steps: starting from an unreliable background model, errors are subsequently identified and corrected by analyzing the extracted foreground objects. For these approaches, initializing a background model is viewed more or less as part of the process for background maintenance.

As an example, using the well-known Mixture of Gaussians (MoG) representation proposed by Stauffer and Grimson (1999) [45], the background is modeled probabilistically at each pixel location by fitting a MoG to the pixel values observed in a recent temporal window. In this case, moving objects can to some extent be accepted during initialization, since each foreground object will be represented by its own distribution, which is likely to have a low weight. However, this erroneous distribution is likely to produce false positives in the classification process [36]. Indeed, if a foreground object persists on the same location sufficiently long, the background model will be overfitted to this object, absorbing it into the background. Also Elgammal et al. (2000) [17] model the background in a similar way, but using a more flexible, nonparametric probabilistic model (KDE). Although it is capable of modeling the distribution of background pixels more precisely, it still has the same problem as [45].

The problem could be reduced by updating the background model in a selective way, that is, only using image elements that are not classified as foreground. As an example, in [34] the Self-Organizing Background Subtraction (SOBS) algorithm has been proposed, which implements an approach to moving object detection based on the neural background model automatically generated by a self-organizing method without prior knowledge about the involved patterns. Based on the learned background model, the SOBS algorithm can detect motion and selectively update the background model. The neural network initial learning provides an initial estimate of the background, obtained through the same selective update procedure, but using a higher foreground/background segmentation threshold, that allows to limit selectivity in the update of the background model during the calibration phase, as well as a monotonically decreasing temporal learning rate, that ensures the neural network convergence. Although moving objects are learned during the background training phase, and thus are included into the initial background model, the subsequent update of the background model during the online phase allows one to reduce their contribution to the model itself, as long as they are not anymore present in the background.

Coming back to methods specific for background initialization, one of the first methods adopted is the one proposed by Shio and Sklansky (1991) in [44], where the temporal mode is voted as the background value for each pixel. The background image is obtained by extracting the most frequent value of image intensity at each pixel, implicitly assuming that the background value will be more frequent than any other possible foreground value.

Very frequently, applying a temporal median filter on pixels at each location across all the frames has been preferred [19, 21, 32, 35]. The temporal median avoids blending pixel values, but the background is correctly estimated only if it is exposed for more than 50% of the time. Moreover, from the computational point of view, the median must be computed at each time step, and all the frames must be stored in memory.

Dawson-Howe (1996) [16] proposed a similar technique, named Dynamic Background Subtraction (DBS), using a fixed window of three frames to recursively update the first sequence image taken as background, avoiding the rather expensive cost of computing the median.

A fast estimate of the stationary background is achieved with the Scoreboard algorithm by Lai and Yung (1998) in [29]. The idea of the algorithm is to estimate the background either by the running-mode or running-average method. The decision of using which method is based on the intensity differences of the pixels between the current frame and the previously estimated background, kept in a scoreboard. If the difference is low, then the running-average method is used, as it can accurately handle small variations, with low computational complexity. Otherwise, the running-mode method is used, which provides higher accuracy for wide variations, at the price of higher computational complexity. By selecting the method in this way, the speed of the algorithm is greatly improved and accuracy is not severely affected.

In the VSAM system [13], Collins et al. (2000) incrementally compute running mean and variance for each pixel over the bootstrap sequence. When the variance drops below a predefined threshold, the pixel is considered stable and the mean is voted as the background value. The method is quite efficient, but stationary foreground objects can be easily included into the background model.

The temporal median is also exploited in the W4 system proposed by Haritaoglu et al. (2000) in [25], presenting a two-stage algorithm to generate the background model. The first stage extracts a temporary background by means of a median filter applied to a bootstrap sequence of several seconds. The second stage uses that background for detecting reliable background regions where to extract the clean statistics to be used for generating the final background model.

A selective spatio-temporal method is presented by Bevilacqua et al. (2002) in [5] as a single-stage approach based on the Bayes theory. It estimates, for every pixel, the intensity value to which that pixel has the maximum posterior probability restricted to non-moving pixels, estimated through temporal frame difference. In case of slow moving objects, this method can include into the background model the foreground pixel intensities, thus requiring a long bootstrap sequence.

The method by Bevilacqua et al. (2005) in [6] works with pixel-wise temporal statistics and consists of three subsequent stages. The first two stages try to isolate for each pixel the stationary background process from the possible foreground processes due to the moving objects covering the pixel, voting the background process temporal median as the good background value for stationary pixels. The third stage completes the background generation by means of a nonparametric statistical model of the temporal camera noise, inferred from the statistics computed in the previous stage. However, the method assumes that lighting changes and background motion are negligible, that is not always the case.

The approximated median filter adopted by Roth et al. (2005) in [43] is a computationally efficient method, that computes the median by incrementing the current median estimate by one if the input pixel value is larger than the previous estimate and by decreasing it by one if smaller. This way, only one reference image must be stored in memory.

More recently, Alexa (2007) [2] adapted the mean shift algorithm to obtain robust estimates of the mode at every pixel location. In order to achieve spatial coherence, a metric which penalizes pixels differing from their neighbors is used. This metric is also used for a non-local means reconstruction of image areas where local mean shift fails to converge, i.e., that have no consistent explanation in the source images.

In [3], Amri et al. (2010) propose an iterative approach for sequences from moving cameras, based on the combination of temporal median and spatial segmentation. After aligning some key-frames of the bootstrap sequence onto a reference plane in order to compensate the camera motion, an initial estimate of the background is obtained through temporal median, to be iteratively refined in order to eliminate moving object residuals. At each iteration, coarse moving object masks are estimated by comparing each motion-compensated key-frame with the corresponding part in the input background image. These masks are then refined by spatial segmentation, while exploiting the semantic information offered by region maps. The refined background estimate is achieved by temporal median filtering restricted only to those regions that do not belong to the foreground regions.

3.4   Methods Based on Subintervals of Stable Intensity

Several approaches to background initialization share an analogous two-stage structure, relying on the assumption that a background value always has the longest stable value:

1. First, pixels (or image blocks) are divided into temporal subintervals with similar values ("stable intervals");

2. Then, the best subinterval belonging to the background is chosen, according to some criterion.

Long and Yang (1990) [33] propose the Adaptive Smoothness Detector (ASD) method that finds for each pixel all intervals of stable intensity in the bootstrap sequence and then heuristically chooses the value of the longest stable interval to most likely represent the background. This approach is effective, but it requires a quite long batch processing of the bootstrap sequence. Moreover, it performs well when all foreground objects are in motion throughout the sequence, but, as observed in [24], many pixels are incorrectly classified in sequences where foreground objects are stationary for a long period of time.

The Local Image Flow (LIF) algorithm presented by Gutchess et al. (2001) in [24] is a hybrid approach performing a two-stage batch processing of the bootstrap sequence. The first stage works in the time domain and generates multiple hypotheses of the background value at each pixel by locating all the stable intervals, in a way similar to that of [33]. The second stage exploits spatio-temporal information for choosing the time interval most likely representing the background process. The likelihood of each hypothesis is evaluated using optical flow information from the neighborhood around the pixel, and the most likely hypothesis is chosen to represent the background. While motion information potentially adds valuable information, most optical flow computation methods themselves are computationally complex and very sensitive to noise, and the quality of the derived background model critically depends on the accuracy of the pixel-wise optical flow estimations.

In [31], Lipton and Haering (2002) present a two-stage algorithm called ComMode (Competitive Mode Estimation). The first stage detects the stable intervals for each pixel through a region growing algorithm in the time domain, using as seeds the local minima of intensity variance in a temporal window. The second stage exploits spatial information in the neighborhood of each pixel in order to choose the best time interval. To this end, a competitive spatial propagation of the clusters detected in the first stage is performed until stability is reached. Finally, for each pixel the temporal mean of the chosen cluster is voted as the background value. The approach is effective, but it requires a long batch processing of the bootstrap sequence.

The region-level method proposed by Farin et al. (2003) in [20] performs an initial rough segmentation of the bootstrap sequence into foreground and background regions, in order to exclude the foreground objects from the background estimation. To this end, each frame is divided into blocks, the temporal sum of absolute differences (SAD) of the co-located blocks is calculated, and a block similarity matrix is formed. The matrix elements that correspond to small SAD values are considered as stationary elements, while high SAD values correspond to non-stationary elements. Background periods in each block are obtained by searching for the subset of frames for which the sum of stationary matrix elements is minimized and that of non-stationary matrix elements is maximized. The stable interval of a block is used to initialize the stable interval of the adjacent blocks, thereby exploiting the spatio-temporal continuity of the sequence. The background image is then synthesized using the median algorithm only on the blocks classified as background. The algorithm works well in most scenarios. However, the spatial correlation of a given block with its neighboring blocks already filled by background is not exploited, which can result in estimation errors if the objects are quasi-stationary for extended periods.

For the purpose of background initialization, Kim et al. (2005) [28] quantize the temporal values of each pixel into distinct bins, called codewords, and use a temporal filter to eliminate any codeword that has not recurred for a long period of time[20]. Indeed, for each codeword, their method keeps a record of the maximum time interval during which it has not recurred. If this time period is greater than half the number of frames in the sequence, the corresponding codeword is discarded as foreground pixel.

Wang and Suter (2006) [47] employ a two-staged approach that can tolerate over 50% of noise in the data, including foreground pixels. The first stage, similar to that of [24,33], locates all non-overlapping stable subsequences of pixel values. The second stage consists of choosing the most reliable subsequence, which is more likely to arise from the background, where the reliability definition is motivated by RANSAC. Mean values in the selected subsequence provide the estimated background model.

In [8], Chen and Aggarwal (2008) propose an algorithm that begins with locating stable intervals of each pixel. Optical flow is then computed between every pair of successive frames. The likelihood of a pixel being background or foreground is decided by the relative coordinates of optical flow vector vertices in its neighborhood. The effect of multiple foreground object depths, where objects closer to the camera may dominate the value of background likelihood, is equalized based on the local optical flow density. Finally, the value of each background pixel is estimated as the median pixel value of the stable interval having highest average background likelihood.

The system recently proposed by Chiu et al. (2010) [11] estimates the background for the purpose of object segmentation. Pixels obtained from each location along its time axis are clustered according to their intensity variations. The pixel corresponding to the cluster having the maximum probability and greater than a time-varying threshold is extracted as background pixel. The main advantages are the low computational complexity and the automatic adjustment of the number of frames needed to compute the background according to speed and size of moving objects. However, this approach relies on the hypothesis that, for any given pixel, the probability of the background color is higher than that of foreground colors, i.e., the same hypothesis adopted for the temporal mode, and it may fail when the hypothesis is not verified.

3.5 Methods Based on Iterative Model Completion

Several recent methods are based on a two-step approach that iteratively construct the background model, articulated in:

1. Selection of static image regions adopted as reference;
2. iterative model completion.

Colombari et al. (2005) [14] consider the background initialization process as an instance of video inpainting, aiming at eliminating from the sequence all the foreground objects (considered as holes to be filled in) using the remaining visual information to estimate the statistics of the entire background scene. The method is based on the hypotheses that a portion of the background, called *sure background,* can be identified with high certainty by using only per-pixel reasoning and that the remaining scene background can be generated using exemplars of the sure background. The initialization of the iterative algorithm provides a partition of the scene into a sure background zone, whose pixels are modeled by Gaussian distributions, and a confusion zone, where foreground and background pixels are mixed

---

[20]http://www.umiacs.umd.edu/~knkim/

together. Then, at each iteration, a patch-level background expansion looks for background patches in the confusion volume, and a pixel-level background expansion computes per-pixel statistics exploiting previously obtained per-patch statistics. These steps are iterated until the confusion zone is empty or it is not possible to proceed.

The overall idea of Lin et al. (2009) [30], is to identify background blocks from each image frame through online classification, and to iteratively integrate these background blocks into a complete model. They train a collection of image blocks of the background extracted from different sequences, and formulate a decision function that provides, for each incoming new frame, probabilities for each image block being part of the background. Based on the classification result, exploiting inter-frame difference and optical flow features, a bottom-up process handles block updating, in either a gradual or an abrupt fashion, for capturing the background variations and scene changes, respectively. A top-down process validates the inter-block consistency for all the updated background blocks. The adoption of learning approaches to identify background blocks not only provides a convenient way of defining some preferred background types from image examples, but also avoids manually setting of discriminating parameters, which are learned from the training data. Moreover, the progressive estimation scheme and the adoption of fast classifiers help achieving computational efficiency.

In [4], Baltieri et al. (2010) propose a background initialization method working at block level in a non-recursive way, specially conceived for achieving the best background model using the minimum number of frames as possible. They firstly split each frame into blocks and fix non variable blocks as background. Then, the holes left by moving objects are filled in by blocks whose frequency spectrum is coherent with the already fixed neighbors, where block similarity is based on the Hadamard Transform. Finally, the spatial coherence along the block borders is also checked, leading to block rejection if it is not verified.

Colombari and Fusiello (2010) [15] propose the Patch-based Background Initialization (PBI) technique for background initialization that exploits both spatial and temporal consistency of the static background. First, the bootstrap sequence is subdivided into patches that are clustered along the temporal line, in order to narrow down the number of background candidates. Then, the background is grown incrementally by selecting at each step the patch that provides the best continuation of the current background, according to the continuity measure implemented by the spatial graph-cuts segmentation. Thus, the spatial correlations that naturally exist within small regions of the background image are taken into account during the estimation process. The method can cope with serious occlusions caused by moving objects, can deal with any number of frames greater or equal than two, and always recovers the background when the assumptions are satisfied. However, it can have problems with blending of the foreground and background due to slow moving or quasi-stationary objects, and it is unlikely to achieve real-time performance due to its complexity.

Reddy et al. (2011) [42] propose a block-level, recursive technique[21]. For each block location of the image sequence, a representative set is maintained which contains distinct blocks obtained along its temporal line. The background estimation is recast as a labeling problem in a Markov Random Field (MRF) framework, where the clique potentials are computed based on the combined frequency response of the candidate block and its neighborhood. It is assumed that the most appropriate block results in the smoothest response, indirectly enforcing the spatial continuity of structures within a scene. The method is robust to noise, can handle stationary foreground objects, and has low computational complexity, but it cannot handle dynamic backgrounds, such as waving trees.

---

[21]http://arma.sourceforge.net/background_est/

In [26], a block representation approach is proposed by Hsiao and Leou (2013), based on motion estimation and the computation of the correlation coefficient. Each block of the current frame is classified into one of four categories: background, still object, illumination change, and moving object. This classification is exploited for the background updating phase, where static blocks are selected as reference and the remaining blocks are suitably used for the iterative completion of the background model.

All the approaches based on the selection of static regions as reference and the iterative model completion tend to be degraded by error propagation if some blocks in a video frame are erroneously estimated.

3.6 Background Initialization as an Optimal Labeling Problem

A different approach to background estimation consists of formulating the problem as an optimal labeling problem. For each image region, the label provides the number of the bootstrap sequence frame in which that region contains exclusively the scene background. The background model is then constructed by assembling, for each region, the background information contained in the selected frame.

Going more into detail, for each pixel p of the background to be estimated, a label l is assigned by determining an input sequence frame I_l where the scene background in that pixel is visible. In order to select the best input sequence frame I_l, a cost is assigned to each possible labeling, and the final labeling is chosen as the one that minimizes the cost. The cost function generally comprises two terms: a *data term*, that defines the cost of assigning a label to a single pixel and indicates how well a pixel satisfies the model assumptions, and an *interaction term*, that denotes the cost of assigning labels to pairs of neighboring pixels and enforces visual smoothness in the resulting background.

Once an optimal labeling is obtained (e.g., through Loopy Belief Propagation (LBP) [49] or graph cuts [7]), two different ways to estimate the background can be considered. The more frequently adopted basic method, also referred to as *image domain composition* [48], simply puts the selected pixels together; i.e., if label l has been computed for pixel p, the value $I_l(p)$ of the l-th frame in pixel p is copied into the pixel p of the output background frame. The more sophisticated method, named *gradient domain composition* [1], collects the gradients of every pixel, instead of the pixel values, from the original frames and puts them together as a composite gradient field. The background image, with gradient field closest to the composite one, is reconstructed by solving a discretized Poisson equation. The choice of the composition method depends on the application. For surveillance applications, such as background subtraction, the background recovered with image domain composition, which is computationally much cheaper, is usually satisfactory enough. For applications where high visual quality is a concern, such as computational photography, gradient domain composition is a better choice [48].

Besides the choice of the composition method, the methods following the optimal labeling approach basically differ in the choice of the energy function to be minimized and of the algorithm adopted for its minimization.

In [1], Agarwala et al. (2004) propose a unified energy minimization-based framework for interactive image composition, under which various image editing tasks can be done by choosing appropriate energy functions[22]. The cost function, minimized through graph

---

[22]http://grail.cs.washington.edu/projects/photomontage/

cuts, consists of an interaction term that penalizes perceivable seams in the composite image, and a data term that reflects various objectives of different image editing tasks. For background estimation, the data terms adopted for achieving visual smoothness are the maximum likelihood and the minimum contrast image objectives. As observed in [48], the choice of the parameter weighting the contribution of the data and the interaction terms is rather hard.

The energy function adopted by Cohen (2005) [12], minimized through graph cuts, consists of a data term that accounts for pixel color stationarity and motion boundary consistency, and an interaction term that looks for spatial consistency in the neighborhood. The method can cope with foreground objects that are not always in motion or that have large textureless areas, even if the background is visible for only a small fraction of time. Moreover, it has also been extended to the case of moving cameras, by aligning the input frames before computing the labeling.

In [23], Granados et al. (2008) present an automatic algorithm for estimating a scene background from a set of non-time sequence photographs featuring several moving transient objects. The cost function includes a *data term* that combines a measure for the likelihood of each input pixel being the background with a motion boundary consistency term, and an *interaction term* that penalizes intensity differences along labeling discontinuities. The objective function is further weighted by an entropy measure that indicates how reliable the background likelihood is, and the algorithm is prevented from cutting through objects by explicitly enforcing the output to be locally consistent with the input data. The cost function is minimized through graph cuts and remaining gradient mismatches in the result are removed using gradient domain fusion.

In [27], Kim and Hong (2008) consider the estimation of a *practical background*, i.e., a background that is as complete and free of moving objects as possible, but where moving objects that eventually cover the background in all the sequence frames are guaranteed to be completely included into the result. The afforded problem is considered as the problem of deciding the boundary between the image patches to be included into the background and, as such, is closely related to mosaic blending from sequences taken by moving cameras. Formulated as a labeling problem over a patch-based MRF, the energy function, minimized through graph cuts, includes a patch similarity-based data term and an interaction term that encourages smoothness between adjacent patches, at the same time reducing noise-like patterns and simplifying the shape of the boundaries between patches.

In the approach proposed by Xu and Huang (2008) [48], the interaction term entails visual smoothness of the estimated background, thus looking for a piecewise smooth image composed of pixels from input frames. The data term, speeding up the energy minimization process, exploits the assumed stationarity of background values, favoring pixels to be labeled as background if other pixels in the same timeline have similar color. The background can be estimated if each background pixel is revealed at least once, and the method does not absorb foreground objects into the background even if they stay still for a long time. Moreover, no motion information needs to be known or estimated from the input frames; therefore, the method also applies to input frames sampled at large time interval, or even frames with no explicit temporal correlation. The minimization problem is solved with LBP, with consequent high computational complexity. Moreover, the assumption that the background is stationary excludes all frequent cases where part of the background is undergoing local motion.

In the approach proposed by Chen et al. (2010) [9], the cost function, minimized by applying graph cuts optimization, includes a data term that is composed of a stationary pixel color term and a predicted term for unstable pixels obtained using an image inpainting technique, and a smoothness term that guarantees that the output is visually smooth. The method achieves robust results on image sequences with both similar illumination conditions and different exposure settings. However, since image inpainting is an essential step, there must be some background regions that are never occluded throughout the input image sequence.

In [37], Nakashima et al. (2011) achieve background estimation for videos taken with mobile video cameras through an optimal labeling approach similar to [27], but operating only on the most representative frames extracted by clustering. The resulting estimated background is employed for the automatic generation of privacy-protected videos. Indeed, intended human objects, that are essential for the camera person's capture intention, substitute the covered background pixels, leading to videos that are free of unintended human objects.

Park and Byun (2013) [38] view background adaptation as a unified problem with background initialization and stationary object detection. The problem is formulated as an energy minimization problem in a dynamic MRF, finding the optimal background as a set of labels. Constraining the connections among the sites with spatio-temporal reliabilities, they robustly handle object-wise changes and efficiently minimize the energy terms with a coordinate descent method.

As observed in [30], even though methods based on optimal labeling are capable of deriving a background model even for dynamic scenes, they are often non-recursive, in that they need to process a video sequence in batch.

3.7 Considerations

A classification of the described methods for background initialization, according to the recurrent schemes underneath them highlighted in the previous sections and to the generally adopted criteria described in Section 3.2, is reported in Table 3.1. Besides any consideration that can be drawn, the applicability of the methods to the specific problem under consideration is clearly dictated by the different assumptions they rely on, that concern

- **Background occlusions:** A frequently adopted assumption (e.g., [4, 6, 9, 15, 24, 47]) is that in each pixel (or small region) the background is revealed for at least a short time interval of the sequence. This assumption is necessary if we want to use only observed values to fill the background at each location, as opposed, for example, to video inpainting [14], where "plausible" values are used for filling in holes.

  The previous hypothesis is often made stronger (e.g., [19, 21, 32, 35]), requiring that the background is revealed for at least 50% of the entire bootstrap sequence length. Indeed, in this case a quite accurate background can be obtained as the temporal median of each pixel color distribution, because this robust statistical method can tolerate up to only 50% outliers.

  There are also methods (e.g., [9]) based on the assumption that some background regions are never occluded throughout the bootstrap sequence, that is not always the case.

- **Foreground motion:** Some methods require that foreground objects are always in motion [33], or can remain stationary only for a short time interval in the bootstrap sequence; however, the interval should be no longer than the interval from the revealed static background [47]. As observed in [15, 24], all the techniques based on temporal stability are doomed to fail in presence of very persistent clutter, i.e, foreground objects that stand still for a considerable portion of time, eventually becoming part of the background.

- **Background motion:** Some methods (e.g., [24, 42, 47, 48]) assume that the background is approximately stationary. In principle, this assumption prevents parts of the background to be moving, because this would require discrimination between different types

TABLE 3.1 Classifications of the methods briefly described in Sections 3.3 through 3.6.

	pixel-level	region-level	hybrid	recursive	non-recursive	blind	selective
Temporal Statistics							
Mode (1991) [44]	X				X	X	
Median (1995) [21]	X				X	X	
DBS (1996) [16]	X			X			X
Scoreboard (1998) [29]	X			X		X	
MoG (1999) [45]	X			X		X	
VSAM (2000) [13]	X			X			X
KDE (2000) [17]	X			X		X	
W4 (2000) [25]	X						X
Bevilacqua (2002) [5]	X			X			X
Bevilacqua (2005) [6]	X						X
Roth (2005) [43]	X			X		X	
Alexa (2007) [2]			X		X		X
SOBS (2008) [34]			X	X			X
Amri (2010) [3]			X		X		X
Subintervals of Stable Intensity							
ASD (1990) [33]	X				X	X	
LIF (2001) [24]			X		X	X	
ComMode (2002) [31]			X		X	X	
Farin (2003) [20]		X			X	X	
Codebook (2005) [28]	X				X	X	
Wang (2006) [47]	X				X	X	
Chen (2008) [8]	X				X	X	
Chiu (2010) [11]	X				X	X	
Model Completion							
Colombari (2005) [14]			X		X		X
Lin (2009) [30]		X		X			X
Baltieri (2010) [4]		X			X		X
PBI (2010) [15]		X			X		X
Reddy (2011) [42]		X		X			X
Hsiao (2013) [26]		X		X			X
Optimal Labeling							
Agarwala (2004) [1]			X		X		X
Cohen (2005) [12]			X		X		X
Granados (2008) [23]			X		X		X
Kim(2008) [27]		X			X		X
Xu(2008) [48]			X		X		X
Chen(2010) [9]			X		X		X
Nakashima (201) [37]			X		X		X
Park (2013 [38]			X	X			X

of motion; motion of the foreground objects would have to be distinguishable from that of the background, and task-specific assumptions would be necessary to distinguish them. In practice, it prevents also suitable handling of dynamic backgrounds showing small variations caused by, e.g., waving trees.

A less restrictive version of this assumption requires that the background be globally stationary, i.e., only small local motion may occur [15], thus enabling to properly handle the so-called *waving trees* background maintenance problem [46].

- **Further restrictions on background:** Some methods (e.g., [15]) assume that foreground objects introduce a color discontinuity with the background, thus excluding camouflage problems.

Some others assume that the background contains less occluding boundaries than an image involving foreground objects, i.e., the background should be visually smoother than the foreground. This assumption is common for methods that afford the problem as an optimal labeling problem (see Section 3.6).

Further assumptions require that sequence frames are obtained under similar illumination conditions (e.g., [9, 23]). This is particularly true in the context of background initialization for non-time sequence images, where the temporal order of input frames is not taken into account, and thus gradual illumination variations cannot be temporally compensated.

- **Batch vs. on-line computation:** While some methods are explicitly designed to be fast enough to perform on-line computation (e.g., [4]), some other methods assume batch processing of the bootstrap sequence (e.g., [15, 24], as well as all non-recursive methods), clearly preventing real-time processing. However, batch algorithms make use of information from the entire sequence and output a single decision, while on-line algorithms must perform sequential decision-making using information from past frames only. Depending on the application, the penalty of a small processing delay at system startup caused by an initialization procedure can be outweighed by the resulting performance improvement.

Other considerations can be evidenced from Table 3.1. Most of the recent methods are region-level or hybrid, aiming to achieve higher robustness through exploitation of spatial relations too, at the price of higher time complexity. Moreover, most of thems are non-recursive, aiming to exploit the information provided by past frames, at the price of higher memory requirements. Finally, most recent methods adopt selective approaches, explicitly trying to avoid using foreground information for the construction of the background model. Therefore, it clearly appears that the main trend in background initialization goes towards non-recursive, selective, region-level/hybrid methods, as long as sufficient computational resources are available.

3.8 Conclusion

In this chapter, we focused our attention on the background initialization problem that is at the basis of many applications, ranging from video surveillance to computational photography. We gave hints on frequently adopted approaches to the problem, trying to highlight the common schemes they rely on, as well as their pro's and con's. Several considerations have been reported, going through possible classifications, working assumptions, and open issues.

Besides the reported considerations, it should be observed that there is no common agreement on the definition of what a background model should include. Namely, should stopped objects be included into the background model or not? Some methods favor the most continuous background, thus preventing the inclusion of stopped objects into the background model; others incorporate into the background model moving objects that become stationary over a certain period of time, since this yields an initial background model accommodating the most recent statistics about the background scene, e.g., a parking car or an occluded area. The dilemma is still open.

References

1. A. Agarwala, M. Dontcheva, M. Agrawala, S. Drucker, A. Colburn, B. Curless, D. Salesin, and M. Cohen. Interactive digital photomontage. *ACM Trans. Graph.*, 23(3):294–302, August 2004.
2. M. Alexa. Extracting the essence from sets of images. In Douglas W. Cunningham, Gary W. Meyer, Lszl Neumann, Alan Dunning, and Raquel Paricio, editors, *Computational Aesthetics 2007: Eurographics Workshop on Computational Aesthetics in Graphics, Visualization and Imaging*, pages 113–120. Eurographics Association, 2007.
3. S. Amri, W. Barhoumi, and E. Zagrouba. Unsupervised background reconstruction based on iterative median blending and spatial segmentation. In *Imaging Systems and Techniques (IST), 2010 IEEE International Conference on*, pages 411–416, 2010.

4. D. Baltieri, R. Vezzani, and R. Cucchiara. Fast background initialization with recursive Hadamard transform. In *IEEE International Conference on Advanced Video and Signal Based Surveillance, AVSS 2010*, pages 165 –171, Sept. 2010.

5. A. Bevilacqua. A novel background initialization method in visual surveillance. In *MVA*, pages 614–617, 2002.

6. A. Bevilacqua, L. Di Stefano, and A. Lanza. An effective multi-stage background generation algorithm. In *IEEE Conf. on Advanced Video and Signal Based Surveillance (AVSS)*, pages 388–393, September 2005.

7. Y. Boykov, O. Veksler, and R. Zabih. Fast approximate energy minimization via graph cuts. *IEEE Transactions on Pattern Analysis and Machine Intelligence*, 23(11):1222–1239, 2001.

8. C. Chen and J. Aggarwal. An adaptive background model initialization algorithm with objects moving at different depths. In *IEEE International Conference on Image Processing, ICIP 2008*, pages 2664–2667, 2008.

9. X. Chen, Y. Shen, and Y. Yang. Background estimation using graph cuts and inpainting. In *Proceedings of Graphics Interface 2010*, GI '10, pages 97–103, Toronto, Ont., Canada, Canada, 2010. Canadian Information Processing Society.

10. S. Cheung and K. Chandrika. Robust techniques for background subtraction in urban traffic video. In *Proc. Visual Communications and Image Processing*, volume 5308, pages 881–892. SPIE, 2004.

11. C. Chiu, M. Ku, and L. Liang. A robust object segmentation system using a probability-based background extraction algorithm. *IEEE Transactions on Circuits and Systems for Video Technology*, 20(4):518–528, Apr. 2010.

12. S. Cohen. Background estimation as a labeling problem. In *IEEE International Conference on Computer Vision, ICCV 2005*, volume 2, pages 1034–1041, October 2005.

13. R. Collins, A. Lipton, T. Kanade, H. Fujiyoshi, D. Duggins, Y. Tsin, D. Tolliver, N. Enomoto, O. Hasegawa, P. Burt, and L. Wixson. A system for video surveillance and monitoring. Technical Report CMU-RI-TR-00-12, Carnegie Mellon University, Pittsburgh, PA, 2000.

14. A. Colombari, M. Cristani, V. Murino, and A. Fusiello. Exemplar-based background model initialization. In *ACM International Workshop on Video Surveillance & Sensor Networks*, VSSN 2005, pages 29–36, New York, NY, USA, 2005. ACM.

15. A. Colombari and A. Fusiello. Patch-based background initialization in heavily cluttered video. *IEEE Transactions on Image Processing*, 19(4):926 –933, Apr. 2010.

16. K. Dawson-Howe. Active surveillance using dynamic background subtraction. *Trinity College Dublin*, 1996.

17. A. Elgammal, D. Harwood, and L. Davis. Non-parametric model for background subtraction. *Proc. ECCV*, pages 751–767, 2000.

18. S. Elhabian, K. El Sayed, and S. Ahmed. Moving object detection in spatial domain using background removal techniques: State-of-art. *Recent Patents on Computer Science*, 1(1):32–54, Jan. 2008.

19. H. Eng, K. Toh, A. Kam, J. Wang, and W. Yau. An automatic drowning detection surveillance system for challenging outdoor pool environments. In *IEEE International Conference on Computer Vision, ICCV 2003*, volume 1, pages 532–539, October 2003.

20. D. Farin, P. de With, and K. Effelsberg. Robust background estimation for complex video sequences. In *IEEE International Conference on Image Processing, ICIP 2003*, volume 1, pages I–145–8, Sept. 2003.

21. B. Gloyer, K. Aghajan, K. Siu, and T. Kailath. Video-based freeway-monitoring system using recursive vehicle tracking. *SPIE 2421, Image and Video Processing*, pages 173–180, March 1995.

22. N. Goyette, P. Jodoin, F. Porikli, J. Konrad, and P. Ishwar. Changedetection.net: A new change detection benchmark dataset. In *Computer Vision and Pattern Recognition*

Workshops (CVPRW), 2012 IEEE Computer Society Conference on, pages 1–8, 2012.

23. M. Granados, H. Seidel, and H. Lensch. Background estimation from non-time sequence images. In *Graphics Interface*, pages 33–40, 2008.

24. D. Gutchess, M. Trajkovics, E. Cohen-Solal, D. Lyons, and A.K. Jain. A background model initialization algorithm for video surveillance. In *Computer Vision, 2001. ICCV 2001. Proc. Eighth IEEE Int. Conf. on*, volume 1, pages 733–740, 2001.

25. I. Haritaoglu, D. Harwood, and L. Davis. W4: Real-time surveillance of people and their activities. *IEEE Transactions on Pattern Analysis and Machine Intelligence*, 22:809–830, 2000.

26. H. Hsiao and J. Leou. Background initialization and foreground segmentation for bootstrapping video sequences. *EURASIP Journal on Image and Video Processing*, 2013(1):12, 2013.

27. D. Kim and K. Hong. Practical background estimation for mosaic blending with patch-based Markov Random Fields. *Pattern Recognition*, 41(7):2145–2155, 2008.

28. K. Kim, T. Chalidabhongse, D. Harwood, and L. Davis. Real-time foreground-background segmentation using codebook model. *Real-Time Imaging*, 11(3):172–185, 2005.

29. A. Lai and N. Yung. A fast and accurate scoreboard algorithm for estimating stationary backgrounds in an image sequence. In *IEEE International Symposium on Circuits and Systems, ISCAS 1998*, volume 4, pages 241–244, 1998.

30. H. Lin, T. Liu, and J. Chuang. Learning a scene background model via classification. *Signal Processing, IEEE Transactions on*, 57(5):1641 –1654, May 2009.

31. A. Lipton and N. Haering. ComMode: an algorithm for video background modeling and object segmentation. In *International Conference on Control, Automation, Robotics and Vision, ICARCV 2002*, volume 3, pages 1603–1608, December 2002.

32. B. Lo and S. Velastin. Automatic congestion detection system for underground platforms. In *International Symposium on Intelligent Multimedia, Video and Speech Processing*, pages 158–161, 2001.

33. W. Long and Y. Yang. Stationary background generation: An alternative to the difference of two images. *Pattern Recognition*, 23(12):1351 – 1359, 1990.

34. L. Maddalena and A. Petrosino. A self-organizing approach to background subtraction for visual surveillance applications. *IEEE Transactions on Image Processing*, 17(7):1168–1177, July 2008.

35. M. Massey and W. Bender. Salient stills: Process and practice. *IBM Systems Journal*, 35(3.4):557–573, 1996.

36. T. Moeslund, A. Hilton, and V. Krüger. A survey of advances in vision-based human motion capture and analysis. *Comput. Vis. Image Underst.*, 104(2):90–126, November 2006.

37. Y. Nakashima, N. Babaguchi, and J. Fan. Automatic generation of privacy-protected videos using background estimation. In *IEEE International Conference on Multimedia and Expo, ICME 2011*, pages 1–6, July 2011.

38. D. Park and H. Byun. A unified approach to background adaptation and initialization in public scenes. *Pattern Recognition*, 46(7):1985 – 1997, 2013.

39. M. Paul. Efficient video coding using optimal compression plane and background modelling. *Image Processing, IET*, 6(9):1311–1318, 2012.

40. M. Piccardi. Background subtraction techniques: a review. *Proc. IEEE SMC*, 4:3099–3104, October 2004.

41. R. Radke, S. Andra, O. Al-Kofahi, and B. Roysam. Image change detection algorithms: A systematic survey. *IEEE Transactions on Image Processing*, 14:294–307, 2005.

42. V. Reddy, C. Sanderson, and C. Lovell. A low-complexity algorithm for static background estimation from cluttered image sequences in surveillance contexts. *J. Image Video Process.*, 2011:1:1–1:14, January 2011.

43. P. Roth, H. Bischof, D. Skočaj, and A. Leonardis. Object detection with bootstrapped learning. In Allan Hanbury and Horst Bischof, editors, *Proc. 10th Computer Vison Winter Workshop*, pages 33–42, 2005.

44. A. Shio and J. Sklansky. Segmentation of people in motion. In *IEEE Workshop on Visual Motion, WVM 1991*, pages 325 –332, October 1991.

45. C. Stauffer and W.E.L. Grimson. Adaptive background mixture models for real-time tracking. In *Proc. CVPR*, volume 2, page 252, 1999.

46. K. Toyama, J. Krumm, B. Brumitt, and B. Meyers. Wallflower: principles and practice of background maintenance. In *Proc. ICCV*, volume 1, pages 255–261, 1999.

47. H. Wang and D. Suter. A novel robust statistical method for background initialization and visual surveillance. In *Asian Conference on Computer Vision, ACCV 2006*, pages 328–337, Berlin, Heidelberg, 2006. Springer-Verlag.

48. X. Xu and T. Huang. A loopy belief propagation approach for robust background estimation. In *IEEE International Conference on Computer Vision and Pattern Recognition, CVPR 2008*, pages 1–7, June 2008.

49. J. Yedidia, W. Freeman, and Y. Weiss. Understanding belief propagation and its generalizations. Technical Report TR-2001-22, Mitsubishi Electric Research Laboratories, 2001.

4

Background Subtraction for Moving Cameras

Ahmed Elgammal
Department of Computer Science, Rutgers University, USA

Ali Elqursh
Department of Computer Science, Rutgers University, USA

4.1 Introduction

The detection and segmentation of objects of interest, such as humans, vehicles, etc., is the first stage in automated visual event detection applications. It is always desirable to achieve a very high accuracy in the detection of targets with the lowest possible false alarm rates. We can characterize the domains of videos into four classes, depending on camera motion (static vs mobile) and the processing requirement for decision-making (offline vs online). Figure 4.1 shows these four categories with an example of each. Most surveillance videos are captured using static cameras. Whether processing is needed online or offline for this type of videos, traditional background subtraction techniques provide effective solutions for the detection and segmentation of targets in these scenarios.

However, most of todays videos are captured from moving platforms (camera phones, cameras mounted on ground vehicles or drones, mounted on robots, soldiers, etc.). In many applications online real-time processing is needed to support decision-making. In other applications, the videos are available offline, e.g. YouTube videos, and offline processing can be afforded. Traditional background subtraction algorithms are not applicable for the moving-camera case. There have been some extensions of background subtraction algorithms to deal with mobile cameras where cameras motion is known or scene geometry is restricted, such as stationary mounted pan-tilt-zoom cameras, or videos captured from high altitudes (e.g. drones). Such extensions utilize motion compensation to generalize background subtraction to such videos. We review these extensions in Section 4.3. Unfortunately, there is no effective solution for the case of videos captured from a freely moving camera.

In the last decade, many object-specific foreground detection algorithms have been introduced. Detecting objects of interest can be achieved through designing and learning specialized detectors using large training data, e.g., pedestrian detector, face detector, vehicle detector, bicycle detector, etc. Given a trained classifier for the object, the occurrences

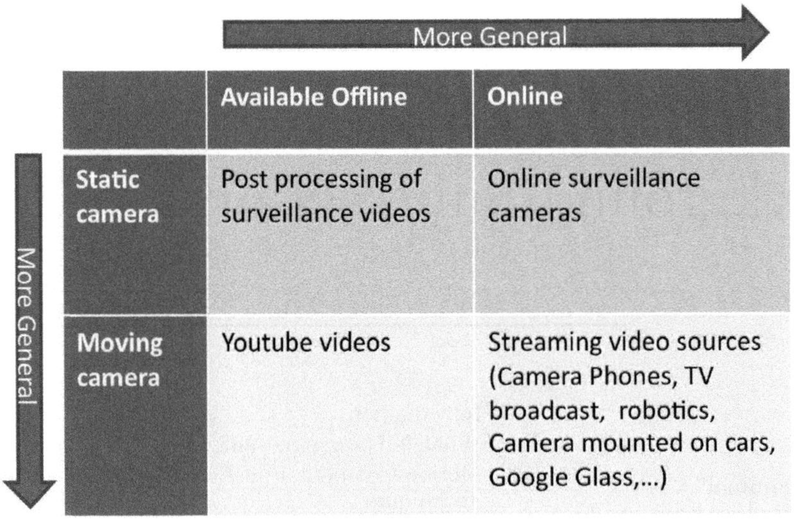

	Available Offline	**Online**
Static camera	Post processing of surveillance videos	Online surveillance cameras
Moving camera	Youtube videos	Streaming video sources (Camera Phones, TV broadcast, robotics, Camera mounted on cars, Google Glass,...)

More General →

↓ *More General*

FIGURE 4.1 Classification of the different types of videos based on the camera motion and processing needs.

of that object in the image are found by exhaustively scanning the image and applying the detector at each location and scale. This concept has been successfully used in detecting pedestrians and vehicles in images [5, 15, 37]. Also this concept is widely used for face detection [48]. These algorithms work best when the object has limited appearance variability, for example these algorithms work best for pedestrian detector, however it is very hard to train these detectors for various human poses, variations in shape, and appearance that are expected to appear in unconstrained videos. Also, these algorithms cannot scale to detect a large number of objects of interest with desired accuracy/false alarm rates. Alternatively, feature-based/part-based detectors can accommodate large variability of object poses and within-class variability. In such approaches candidate object parts are detected and geometric constraints are used on these parts to detect objects. For example pictorial structure-based models can deal with large within-class variability in human poses [10, 11].

For the case of static-camera videos, an object-specific detector is far from matching the performance of a typical background subtraction algorithm, in terms of false positives, false negatives, and moreover, they do not provide exact segmentation of objects but rather a bounding box(s) around the object and its parts. To the contrary, for the case of mobile-camera videos, object-specific foreground detection algorithms are directly applicable, as they can be applied to every frame in the video, since they do not assume stationary camera or scene. However, despite providing an applicable solution, there are three fundamental limitations for these algorithms that make them not effective in dealing with videos.

First, these algorithms are designed to detect objects from images, and hence do not utilize any temporal constraints in the video. Second, even in images, the performance of these algorithms are far from being readily useful in real-world applications. Results of the Pascal Challenge 2011 showed that the best-achieved precision for detecting people and cars is 51.6% and 54.5% respectively. The performance degrades significantly if the targets are smaller in size or occluded. A recent evaluation [22] on a challenging dataset showed that state-of-the-art face detection algorithms could only achieve 70% accuracy with more than 1 FPPI (False Positive Per Image). A recent evaluation [7] on state-of-the-art pedestrian detection algorithms found that such algorithms have a recall of no more than 60% for un-occluded pedestrians with 80 pixel height at a 1 FPPI (False Positive Per Image). The performance of state-of-the-art and widely used deformable part models [10] is 33% and 43% (precision) for detecting cars and people respectively, evaluated on the PASCAL VOC 2008 dataset. Thirdly, scaling these algorithms for a large number of classes is challenging, since they require training for each object category. At run time, a detector for each object class needs to be used. Recently, Dean et al. [6] show how part-based models can be scaled up for detecting 100K classes efficiently, however the accuracy reported on PASCAL VOC 2007 dataset is limited to 24% precision.

Video segmentation generalizes the concept of image segmentation, and attempts to group pixels into spatiotemporal regions that exhibit coherence in both appearance and motion. Several existing approaches for video segmentation are formulated as clustering of pixels from all frames or frames of a sliding window. First, multidimensional features are extracted representing photometric and motion properties. Then clustering is used to gather evidence of pixel groupings in the feature space. Such clustering can be done using Gaussian mixture models [16], mean-shift [32, 51], and spectral clustering [13]. However, it is unclear how to use the resulting segmented regions for detecting objects of interest in video. In practice, one can identify objects that move coherently but are composed of multiple regions with different appearances. Similarly, articulated objects may be formed of multiple regions of different motions with common appearance. Typically, a video segmentation technique produces an over-segmentation of such objects.

Whereas objects can be composed of different parts with different appearances, they typically move as a coherent region. Therefore, one can argue, that by segmenting differently-moving objects in the scene one can easily detect objects of interest in the scene. Traditionally, object detection in video sequence has been formulated as detecting outliers to dominant apparent image motion between two frames, e.g. [20, 21, 36]; such approaches assume that objects are relatively small compared to the scene. However, there exist several challenges with formulating the detection problem as a motion segmentation problem. The observed image motion is the result of the perspective projection of the 3D motion. Thus, the straightforward approach of segmenting based on translational motion over two frames typically leads to poor results. On the other hand, models using richer models, such as affine camera motion, can only segment a very small subset of pixels; namely those that can be tracked through the entire frame sequence. Furthermore, many methods make the assumption that objects are rigid, and as a consequence over-segment articulated and non-rigid objects. For all those reasons, motion segmentation is still an active area of research.

The above discussion highlights the fact that, despite that there are applicable solutions for detecting objects in videos, these solutions are far from being effective. The goal of the research in the area of background subtraction from moving camera is to develop algorithms that can be as effective as traditional background subtraction but applicable to moving-camera videos.

In this chapter we review traditional and recent work in this area. We start in Section 4.2 by describing the difficulties in defining what is the background in the case of a mobile camera. In Section 4.3 we discuss traditional extensions to the background subtraction concept to the case of a moving camera, with restrictions on the camera motion or scene geometry. In Section 4.4 we present a short review on motion segmentation approaches, which is a fundamental building block to recent approaches for moving-camera background subtraction. We follow that by a short review on layered-motion segmentation approaches, in Section 4.5. In Section 4.6 we describe some recent algorithms for background subtraction for freely moving cameras, which are based on motion segmentation.

4.2   What Is Foreground and What Is Background

In the stationary camera case, the definition of foreground and background is clear; a foreground object is anything that does not look like the stationary background, in terms of appearance or local motion. Therefore, it suffices to model the statistical distribution of the scene background features, and detecting the foreground is mainly a process of detecting outliers to that distribution. In the freely moving camera case, the definition of foreground and background might be ambiguous.

The problem of figure/ground separation is well studied in the psychology of vision. Results of such studies have highly influenced the field of perceptual grouping in computer vision. In the case of a single image, the definition of figure/ground can be quite ambiguous. Among the most important factors psychology studies pointed out, to determine figure/-ground are: Surroundness (a region that is completely surrounded by another); Size (smaller regions are perceived as figure); Orientation (vertical and horizontal regions perceived more frequently as figure); Contrast (region with greatest contrast to its surrounding is perceived as figure); Symmetry, Convexity, and Parallelism of the contour. Such low level cues have been widely used in figure/ground segmentation literature and in salient region detection.

Higher-level cues are quite important as well; figure/ground processing is highly influenced by knowledge of specific object shapes. Peterson et al. [38] experiments clearly gave evidence to the role of familiar shapes in figure/ground processing. However, it is not clear whether figure/ground organization precedes object recognition and later recognition process gives feedback to the figure/ground process, or there exists prefigural recognition process before any figure/ground decision. In both cases it is clear that figure/ground separation is a process that is intertwined with recognition. This issue received attention in the computer vision community as well through many recent papers that uses shape priors in segmentation.

Dynamic figure/ground processing seems to be far less ambiguous compared to single image figure/ground processing. It is clear from psychological experiments that common fate (the tendency to group tokens that move with the same direction and speed) plays a significant role in perceptual grouping and figure/ground separation in dynamic scenes. The concept of common fate can be generalized to include any kind of rigid motion and also configural [25] and biological motion [26].

A successful Moving-camera background subtraction should utilize and integrate various perceptual cues to segment the foreground including:

Motion discontinuity: objects and image regions that are moving independently from the rest of the scene.

Depth discontinuity: objects that are closer to the camera than the rest of the scene.

Appearance discontinuity: salient objects or image regions that look different from the surrounding.

Familiar shape: image regions that look like familiar shapes; i.e., object of interest such as people, vehicle, and animals.

We argue that none of these cues by itself is enough to solve the problem, and therefore integrating these four cues, besides high-level reasoning about the scene semantics is necessary. We believe that the decision of foreground/background separation is not only low-level discontinuity detection, but also a high-level semantic decision. In what follows we justify our argument.

There is a large literature on motion segmentation, which exploits motion discontinuity, however, these approaches do not necessarily aim at modeling scene background and segmenting the foreground layers. Fundamentally 3D motion segmentation by itself is not enough to separate the foreground from the background in case both of them constitutes a rigid or close to rigid motion, e.g., a car parked in the street will have the same 3D motion as the rest of the scene. Apparent 2D motion (optical flow) is also not enough to properly segment objects; for example an articulated object will exhibit different apparent motion at different parts.

Similarly depth discontinuity by itself, is not enough since objects of interest can be at a distance from the camera with no much depth difference than the background. There has been a lot of interest recently in saliency detection from images as a pre-step for object categorization. However, most of these works use static images and do not exploit rich video information. The performance of such low-level saliency detection is far from being reliable in detecting foreground objects accurately. Similarly, as we discussed earlier, the state-of-the-art object detection is far from being able to segment foreground objects with the desired accuracy.

This highlights the need for an integrated approach for the problem that combines low-level vision cues, namely: motion discontinuity, depth discontinuity, and appearance saliency with high-level object and scene accumulated and learned models.

4.3 Motion-Compensation Based Background Subtraction Techniques

Motion-compensation based approaches estimate the motion or the geometric transformation between consecutive frames, or between each frame and an extended background map. These methods typically use a Homography or a 2D affine transform to compensate for the motion. The extended background map is a representation of the scene background, and any of the traditional methods such as a Gaussian model or a Mixture of Gaussian can be used to statistically model the pixel process in this representation. Once a transformation is recovered between each frame and the background map, an object can be detected. However, the success of these approaches depends on the accuracy of the motion estimation, which depends on how the assumed motion model is suitable to the actual scene. The accuracy of the motion estimation also depends on the existence of foreground objects (independent motion) which do not follow the motion model assumption. Robust estimation of the motion parameter is typically needed to tolerate these two problems. However, even with robust estimation, residual error exists, and approaches for dealing with these errors have to take place.

Many surveillance applications involve monitoring planer scenes, for example a parking lot or an aerial view. The geometric transformation between images of a given plane, under the pinhole camera model, can be characterized by a Homography, a transformation between two projective planes which preserves lines. A Homography between two images involves eight parameters, and therefore can be estimated from four point matches. Least-squares can be used to estimate Homographies from many matches. Typically RANSAC [12] and its variants, such as MLESAC [46] are used to achieve robust estimation using many matches. Estimating a Homography between two images means that we can directly map the coordinate of one image to another, and hence an image mosaic can be established. Therefore, if the camera is overlooking a planer scene, an image mosaic of that scene can be established and every frame can be mapped to that mosaic via the computed Homography.

If the camera motion is pure rotation, or pure zoom (with no translation), the geometric transformation between images can also be modeled by a Homography, without the need for the planer scene assumption. In particular, this is very relevant for surveillance applications since pan-tilt-zoom (PTZ) cameras are typically used. Physically, panning and tilting of a

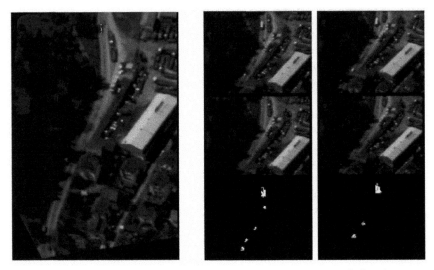

FIGURE 4.2   Example results from the approach of Mittal and Huttenlocher [33]. Left: The background mosaic from an aerial video of 150 frames. Right top: Frame number 42 and 57; middle: synthesized frames with the moving objects removed; bottom: the corresponding moving objects detected. (IEEE ©2000)

PTZ camera are not pure rotations since the rotation of the camera head is not around the center of projection. Therefore, a small translation is involved in the camera motion and hence a parallax effect.

Some of the early papers that aimed at detecting foreground objects from PTZ cameras used motion compensation between consecutive frames and then image differencing, i.e., these approaches do not maintain a background model. For example, Murray and Basu [34] computed the coordinate transformation between two images analytically as a function of the pan and tilt angles, assuming they are known. To compensate for the error in coordinate estimation, resulting from the small translation, adaptive morphology is used, where the size of the morphology filter is derived as a function of the noise level. The assumption of knowing the pan and tilt angle is reasonable in active vision application and in online PTZ surveillance systems. Araki et al. [1] used Least Median of Squares to estimate an affine motion model between consecutive frames, and after compensating for the motion, image difference was used to detect the foreground each frame and a background model.

There have been several approaches for building a background model from a panning camera based on building an image mosaic and the use of a traditional Mixture of Gaussian background model, e.g. [33, 40]. Mittal and Huttenlocher [33] used image registration to align each new image to a panoramic background image. The background image is the mean of the highest weighted Gaussian from a mixture of Gaussian model used at each pixel. The registered image is then used to update the background model. Image registration is achieved on two steps based on tracked features: first, an affine transformation is estimated using robust least squares, and it is then used as an initial solution to estimate a projective transformation through nonlinear optimization. Figure 4.2 illustrates an example of this approach. The figure shows the mosaic built from a sequence of 150 frames from an aerial video and detection results for two frames. Hayman and Eklundh [18] also built a mixture of Gaussian mosaic background model. However, they use MLESAC [46] to estimate the pan and tilt angle between a given image and the mosaic; assuming the calibration parameters are known. Alternatively, Wada and Matsuyama [49] proposed to use a virtual sphere, denoted by "appearance sphere" to map images taken by a PTZ camera, assuming pure rotation. Practically, they proposed a representation of the scene background as a finite set of images on a virtual polyhedral appearance model. Jin et al. [24] modeled the scene as a set of planer regions where each background pixel is assumed to belong to one of these regions. Homographies were used to rectify each region to its corresponding planer representation in the model.

Several approaches have been proposed to deal with the residual errors resulting from the parallax effect due to the small translation involved in the pan and tilt motion, as well as other sources of errors, such as sub-pixel camera motion, etc. Basically, most of these approaches boil down to pooling a background representation from an "uncertainity" region around each target pixel in the background mosaic, with different statistical strategies. Ren et al. [39] uses motion compensation to predict the position of each pixel in a background map, and model the uncertainty of that prediction by a spatial Gaussian distribution. This results in an uncertainty region for each pixel in the background map. Traditional single Gaussian model is used at each pixel in the background map. Hayman and Eklundh [18] showed, that given a Gaussian mixture component at each pixel in an uncertainty region, the mean and variance of the observation likelihood follows a so-called Mixel Distribution [29], where its mean and variance can be computed efficiently by convolution with separable filters over the background mosaic.

4.4 Motion Segmentation: A Review

Motion segmentation refers to the problem of segmenting a video based on motion. It comes in different flavors depending on what is being segmented and what cues are used to achieve the segmentation. First, there are approaches that segment a set of extracted point trajectories using motion information alone. We refer to these as sparse motion segmentation algorithms since they only use the motion information at sparse locations to segment them (Not to be confused with algorithms that use sparsity). Second, there are algorithms that segment the entire set of pixels based on motion information in addition to static cues such as color or texture to produce the segmentation. Finally, there are hybrid algorithms that either use sparse motion segmentation with static cues or dense motion segmentation using only motion.

Existing work in Motion Segmentation has largely assumed that cameras are affine. This assumption leads to the following formulation for the motion segmentation problem. Let $\mathbf{X} = [X_1 X_2 \ldots X_N]$ be a $4 \times N$ matrix representing a set of N points in 3D using a homogeneous coordinate system. Similarly, let $\mathbf{M} = [M_1 \ldots M_F]^T$ be the $2F \times 4$ matrix formed by vertically concatenating F camera matrices, corresponding to F video frames. Such model is only valid if the object on which the 3D points lie is rigid, where the motion of the object can be represented as motion of the camera with respect to a stationary object. Under the assumption of affine cameras, such camera matrices can be specified by the first two rows of the projection matrix. Taking the product of the two matrices $\mathbf{X}$ and $\mathbf{M}$, we get a matrix $\mathbf{W}$ where each column represent the projected locations of a single 3D point over F frames. The columns of $\mathbf{W}$ represent the trajectories in 2D image space, and can be written as

$$\mathbf{W} = \mathbf{MX}$$

$$
\begin{bmatrix}
x_1^1 & \cdots & x_1^N \\
y_1^1 & & y_1^N \\
\vdots & & \vdots \\
x_F^1 & & x_F^N \\
y_F^1 & & y_F^N
\end{bmatrix}
=
\begin{bmatrix}
M_1 \\
M_2 \\
\vdots \\
M_F
\end{bmatrix}
\begin{bmatrix} X_1 X_2 \ldots X_N \end{bmatrix}
$$

Since the rank of a product of two matrices is bounded by the smallest dimension, the rank of the matrix of trajectories $\mathbf{W}$ is at most 4. It follows that Trajectories generated from a single rigidly moving object lies on a subspace of dimension 4. In general when there are multiple moving objects, the columns of the trajectory matrix $\mathbf{W}$ will be sampled from multiple subspaces. The motion segmentation problem can be formulated as that of subspace separation, i.e. assign each trajectory to one of the subspaces and estimate the subspaces.

Approaches to motion segmentation (and similarly subspace separation) can be roughly divided into four categories: statistical, factorization-based, algebraic, and spectral clustering. Statistical methods alternate between assigning points to subspaces and re-estimating the subspaces. For example, in [17] the Expectation-Maximization (EM) algorithm was used to tackle the clustering problem. Robust statistical methods, such as RANSAC [12], repeatedly fits an affine subspace to randomly sampled trajectories and measures the consensus with the remaining trajectories. The trajectories belonging to the subspace with the largest number of inliers are then removed and the procedure is repeated.

Factorization-based methods such as [19, 28, 44] attempt to directly factorize a matrix of trajectories. These methods work well when the motions are independent. However, it is frequently the case that multiple rigid motions are dependent, such as in articulated motion. This has motivated the development of algorithms that handle dependent motion. Algebraic methods, such as GPCA [47] are generic subspace separation algorithms. They do not put assumptions on the relative orientation and dimensionality of motion subspaces. However, their complexity grows exponentially with the number of motions and the dimensionality of the ambient space.

Spectral clustering-based methods [4, 35, 54], use local information around the trajectories to compute a similarity matrix. It then uses spectral clustering to cluster the trajectories into different subspaces. One such example is the approach by Yan et al. [54], where neighbors around each trajectory are used to fit a subspace. An affinity matrix is then built by measuring the angles between subspaces. Spectral clustering is then used to cluster the trajectories. Similarly, sparse subspace clustering [8] builds an affinity matrix by representing each trajectory as a sparse combination of all other trajectories and then applies spectral clustering on the resulting affinity matrix. Spectral clustering methods represent the state-of-the-art in motion segmentation. We believe this can be explained because the trajectories do not exactly form a linear subspace. Instead, such trajectories fall on a non-linear manifold.

With the realization of accurate trackers for dense long term trajectories such as [41, 43] there has been great interest in exploiting dense long term trajectories in motion segmentation. In particular, Brox et al. [4] achieves motion segmentation by creating an affinity matrix capturing similarity in translational motion across all pairs of trajectories. Spectral clustering is then used to over-segment the set of trajectories. A final grouping step then achieves motion segmentation. More recently, Fragkiadaki et al. [14] proposes a two step process that first uses trajectory saliency to segment foreground trajectories. This is followed by a two-stage spectral clustering of an affinity matrix computed over figure trajectories. The success of such approaches can be attributed in part to the large number of trajectories available. Such trajectories help capture the manifold structure empirically in the spectral clustering framework.

4.5   Layered-Motion Segmentation

Layered-motion segmentation refers to approaches that model the scene as a set of moving layers [50]. Layers have a depth ordering, which together with mattes and motion parameters model how an image is generated. Let n_l denote the number of layers, the ith layer can be characterized by an appearance A_i and optionally a matte L_i. To generate a frame at time t, the appearance of the i-th layer is first transformed according to the transformation parameters M_i^t using $f(A_i, L_i, M_i^t)$. The transformed appearances are then overlaid in the order specified by the depth of each layer d_i. Different variations are possible using different parameterizations of the variables or by including additional variables that describe, for example, lighting effects at each frame.

It is obvious that if one knows the assignment of pixels to different segments, it is trivial to estimate the appearance and transformation parameters. Similarly, if one knows the appearance and transformation parameters it is easy to assign pixels to segments. Since initially we know neither of them, this is an instance of a chicken-and-egg problem.

Wang et al. [50] used an iterative method to achieve layered motion segmentation. For every pair of frames they initialize using a set of motion models computed from square patches from the optical flow. Next, they iterate between assigning pixels to motion models and refining the motion models. This process is repeated until the number of pixel reassignments is less than a threshold or a maximum number of iterations is reached. Using the support maps for each segment, all pixels belonging to one segment are wrapped into a reference frame and combined using a median filter. A final counting step is used to establish depth ordering based on the assumption that occluded pixels appear in fewer frames.

Originally used for optical flow [23], mixture models were also used to probabilistically model the image generation process. Weiss et al. [52] used Expectation Maximization (EM) algorithm to update the model parameters. In the E-step, the system updates the conditional expectation of L_i given the fixed parameters. This is done by computing the residual of observed measurement and the predicted appearance given the motion parameters M_i^t. In the M-step, the model parameters A_i, M_i^t are updated based on these "soft" assignments. To incorporate spatial constraints in the mixture model, two methods were proposed. In the first, the images are pre-segmented based on static cues, and assignment in the E-step is based on the residual of all the pixels in each segment. In the second, a MRF prior is imposed on the labels L_i. The motion model of the ith layer is represented by the six parameters of an affine transformation. Although the approach reasons about occlusions by assigning pixels to the correct model, it does not infer the depth ordering of different layers. Torr et al. [45] extend the approach by modeling the layers as planes in 3D and integrating priors in a Bayesian framework. Both approaches rely on a key frame for estimation, and therefore can handle a few number of frames. Flexible sprites [27] are another variation of layered motion segmentation where transformations are restricted to a discrete set of predetermined transformations. A Gaussian model for pixel-wise appearance and mattes is used. A variational optimization is then used to learn the parameters of the probabilistic model. Most of these approaches either use expectation-maximization or variational inference to learn the model parameters.

Wills et al. [53], noted the importance of spatial continuity prior for learning layered models. Given an initial estimate, they learn the shape of the regions using the α-expansion algorithm [3], which guarantees a strong local minima. However, their method does not deal with multiple frames. Kumar et al. [30] propose a method that models spatial continuity, while representing each layer as composed of a set of segments. Each segment is allowed to transform separately thus enabling the method to handle nonrigid objects by segmenting them into multiple segments distributed over multiple layers. This flexibility comes at the expense of a less semantically meaningful representation.

A common theme of most layered models is the assumption that the video is available before an hand. Such assumption prevents the use of such approaches for processing videos from streaming sources. In addition, typically the computational complexity increases exponentially with the length of the videos.

4.6   Motion-Segmentation Based Background Subtraction Approaches

Segmenting independent motions has been one of the traditional approaches for detection objects in video sequences. Most of the traditional approaches, e.g. [20, 21, 36], fits a motion model between two frames, e.g., affine motion, and independently moving objects are detected as outliers to the model fitting. However, these techniques do not maintain an appearance model of the scene background. In this section we focus on recent techniques [9, 31, 42] that utilize motion segmentation to separate independent moving objects, and maintain a representation of the scene background. Interestingly these recent techniques share a common feature, which is maintaining representations for both the background and the foreground.

Sheikh et al. [42] use orthographic motion segmentation over a sliding window to segment a set of trajectories. This is followed by sparse per-frame appearance modeling to densely segment images. The process starts with extracting point trajectories from a sequence of frames. These 2D trajectories are generated from various objects in the scene. Let $\mathbf{W}$ be the matrix formed by the set of all extracted trajectories. Assuming an affine camera, trajectories belonging to the stationary background will lie in a subspace of dimension three. The next step is to identify which of these trajectories belong to the background subspace. Without additional assumptions this problem is ill-posed, as there may exist other moving objects in the scene that also lie on a subspace of dimension 3. To tackle this problem, it is assumed that the background subspace will contain the largest number of trajectories. This is reasonable only when objects in the scene do not span large areas of the image.

RANSAC [12] is used to extract the background subspace. RANSAC works in iterations, with each iteration consisting of a sampling step and a fitting step. In the sampling step, a set of three trajectories w_i, w_j, and w_k are sampled. The fitting step then measures the consensus, i.e. how many other trajectories fits the subspace identified by the sampled trajectories. Given the assumption that the background subspace will contain the largest number of trajectories, it is expected that when our sampled set of trajectories belongs to the background we get the largest consensus. The projection error on the subspace is used to measure the consensus. Given the RANSAC sample as a matrix of trajectories $\mathbf{W_S} = \begin{bmatrix} w_i & w_j & w_k \end{bmatrix}$, the projection matrix is constructed by

$$P = \mathbf{W}_S(\mathbf{W}_S^T \mathbf{W}_S)^{-1}\mathbf{W}_S^T.$$

The projection error can be written as $E_P(w_i) = \| \mathbf{P}w_i - w_i \|_2$. By comparing this error to a fixed threshold τ, the number of inlier trajectories can be determined. If we have low consensus, the process is repeated by sampling a new set of trajectories. Figure 4.3 illustrates an example of trajectory bases learned from a 30 frame time window, and examples of the foreground and background trajectories.

Motion segmentation using RANSAC into foreground and background trajectories gives only a sparse segmentation of the pixels in the frame. To achieve background subtraction, background and foreground appearance models are built using the sparse set of pixels belonging to the trajectories in the current frame. The algorithm labels the pixels in the current frame by maximizing the posterior of labeling given the set of sparsely labeled pixels. Let $\mathcal{L} = [l_1 \ldots l_N]$ denote the labels assigned to each pixel, and $x = [x_1 \ldots x_N]$ denote the spatial location and color of the pixels, the goal is to estimate

$$\mathcal{L}^* = arg \max_{\mathcal{L}} p(\mathcal{L}|x),$$

$$p(\mathcal{L}|x) \propto p(\mathcal{L}) \prod_{i=1}^{N} p(x_i|l_i)$$

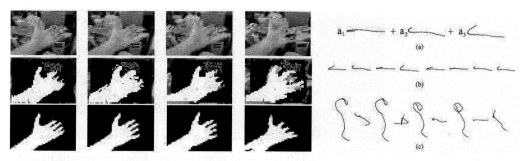

FIGURE 4.3 Left: Example results from Sheikh et al. [42] Right: (a) The background trajectory bases of a 30 frame window. (b) background trajectories lie in this subspace, (c) foreground trajectories do not lie in this subspace. (IEEE ©2009)

where the posterior is further factored using Bayes Theorem in prior and likelihood terms. The likelihood $p(x|\mathcal{L})$ is computed as:

$$p(x_i|l_i) = p(x_i|\mathcal{B})^{(l_i-1)}p(x_i|\mathcal{F})^{l_i},$$

where $p(x_i|\mathcal{B})$ and $p(x_i|\mathcal{F})$ are the likelihood of a pixel given the background foreground models respectively. Kernel Density Estimation (KDE) was then used to model the appearance of the background and foreground from the sparse features' color and spatial locations, (see Chapter 5 for details about KDE). Finally a Markov Random Field was used to achieve the final labeling and enforce smoothness over the labels.

Figure 4.3 shows example segmentation in a short sequence. More results can be seen in Figure 4.6. This method suffers from three drawbacks. The use of orthographic motion segmentation based on a sliding window means that motion information outside that window is lost. The fact that affine motion segmentation can only be applied on complete trajectories raises a trade-off. On one hand increasing the window size captures more history, but at the cost of less trajectories. On the other hand with a small window size more trajectories can be used at the cost of loss of motion history. In addition, the appearance models are built using sparse color information at the trajectory locations in the current frame and discarded away after the frame is segmented. The sparse appearance modeling fails to capture object boundaries when appearance information is ambiguous. Due to the dependence on long term trajectories only, regions with no trajectories may be disregarded as background altogether. Finally, the method fails if the assumption of orthographic projection is not satisfied.

Maintaining the background and foreground appearance representation at the level of sparse features in [42] results in arbitrary segmentation boundaries at the dense segmentation, at areas where there are not many tracked features, e.g. see Figure 4.6. In contrast, recent approaches like [31] and [9] maintain a dense appearance representation at the block level in [31] and at the pixel level in [9]. Both approaches uses a Baysian filtering framework, where the appearance representation from the previous frame is considered as a prior model, which is propagated to the current frame using a motion model to obtain the appearance posterior. We explain the details of these two approaches next.

Kwak et al. [31] proposed a method that maintains block based appearance models for the background and foreground within a Bayesian filtering framework. Optical flow is computed and is used to robustly estimate motion by nonparametric belief propagation in a Markov Random Field. To update the appearance models, the method iterates between estimating the motion of the blocks and inferring the labels of the pixels. Once converged, the appearance models are used as priors for the next frame and the process continues. Due to the iterative nature of the approach, the method is susceptible to reach a local

FIGURE 4.4   Elqursh and Elgammal [9] Left: Graphical model. Center: Automatic segmentation obtained on long sequence with fast articulated motion. The method is able to segment the tennis ball even when no trajectories exist on it. Right: One sample from the background pixel based appearance model learned automatically. (©(2012) Springer)

minimum. Although motion information between successive frames is estimated, it is only used to estimate the labels in the current frame and does not carry on to future frames. In contrast, we maintain motion information via long term trajectory and maintain a belief over different labeling in a Bayesian filtering framework.

One way to avoid the trade-off caused by the affine motion segmentation model, as in [42], is to attempt to use an alternative method for motion segmentation. Elqursh and Elgammal [9] proposed to use nonlinear manifold embedding to achieve online motion segmentation into foreground and background clusters. The approach moves away from traditional affine factorization and formulate the motion segmentation as a manifold separation problem. This approach is online, with constant computation per frame, and does not depend on a sliding window, in contrast long-term trajectory informations are utilized. The approach can be seen as a layered motion segmentation one, however, in contrast to the methods described in Section 4.5, which are mainly offline, this approach is online.

Moreover, in contrast to the previous methods, where the motion segmentation output was used directly to model the appearance, the result of the motion segmentation is used to build dense motion models $\mathcal{M} = (\mathbf{M}_f, \mathbf{M}_b)$ for the foreground and background layers respectively. These motion models, together with the input frame I_t are used to update pixel-level appearance models $\mathcal{A} = \{A_b, A_f\}$ computed from the previous frame in a Bayesian tracking framework. The process is then repeated for each incoming frame. Figure 4.4 illustrates the graphical model for the approach, and an example of the foreground segmentation and background model learned. Algorithm 4.1 summarizes the algorithm.

To perform the motion segmentation, for all pairs of trajectories a distance matrix is computed $\mathbf{D}_t$ and an affinity matrix $\mathbf{W}_t$ is formed. The distance between trajectories does not require that the trajectories should have the same length. A multi-step approach is then used to achieve motion segmentation. First, Laplacian Eigenmaps [2] is applied to compute a low-dimensional representation of the trajectories. Then a Gaussian Mixture Model (GMM) is used to cluster the set of trajectories. To initialize the clustering the labels of the trajectories from the previous frame $\mathbf{c}_{t-1}$ are used. Figure 4.5 shows an example of such a clustering in the embedding coordinates. A final grouping step based on several cues is used to merge the clusters into exactly two sets; namely a foreground manifold and background manifold.

Output of the motion segmentation is used to compute the motion models. Each motion model $\mathbf{M}$ is a grid of motion vectors $\mathbf{m}^i = [u_i v_i]$, where u_i, and v_i denote the motion of pixel i in the x and y directions respectively. The motion segmentation result gives motion information and layer assignments at a subset of the pixels (albeit some noise in the motion due to tracking). To propagate this information over the entire pixels and layers, Gaussian

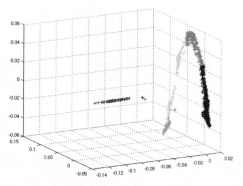

FIGURE 4.5 (See color insert.) Example trajectories in the image and their corresponding embedding coordinates, from Elqursh and Elgammal [9]. Each cluster is marked with a different color. Black dots are short trajectories not yet added to the embedding. The foreground cluster is marked with two red concentric circles around its trajectories. (©(2012) Springer)

belief propagation (GaBP) is used. The result is the marginal distribution of the motion at each pixel and is represented by a mean and variance.

Assuming that the previous appearance models are provided. The approach then computes the predicted appearance models $\mathcal{A}_t = \{A_{b,t}, A_{f,t}\}$ and labels $L_t = \{l_t^i\}$. The models are again represented as a grid of pixel-wise representation a_b^i, a_f^i, and l^i where a_b^i, a_f^i, l^i denote the color in background, color in foreground, and label of pixel i. To simplify the notation we drop the subscript denoting the layer assignment when the equation applies to all layers. Prediction then follows

$$p(a_t^i) = \sum_{m_t^i \in \mathbb{N}^2} p(m_t^i)p(a_{t-1}^j), \tag{4.1}$$

$$p(l_t^i = f) = \sum_{m_{f,t}^i \in \mathbb{N}^2} p(m_f^i)p(l_{t-1}^j = f), \tag{4.2}$$

where j is the pixel m_f^i units away from pixel i. The distribution $p(a^i)$ is approximated using kernel density estimation as a set of N_{KDE} color samples. In the update step the new frame I_t is incorporated. Given the predicted label prior and appearance models a MAP estimate of the labels is inferred using the equation

$$p(l_t^i = k|I_t^i) \propto \int_{a_k^i} p(I_t^i|l_t^i, a_{k,t}^i)p(a_{k,t}^i)\, \mathbf{d}a_{k,t}^i\, p(l_t^i) \tag{4.3}$$

Next the inferred labels are used to update the corresponding appearance model. This is done by adding the observed pixel color I_t^i to the set KDE samples of pixel i in the foreground layer if $p(l_t^i = f|I_t^i) > \frac{1}{2}$ and to the background otherwise.

Figure 4.6 shows qualitative comparisons between [9] and [42] on three sequences. Notice that the dense segmentation of [42] erroneously extends around the foreground in areas where there are no tracked features, this is due to modeling the appearance at the sparse tracked features, as described above. Also notice that when the orthographic camera assumption in [42] fails, as in the third case, the resulting segmentation is totally erroneous. In contrast the approach in [9] does not have such assumption and accurately detect and segment targets in these situations.

Algorithm 4.1 - BSMC using nonlinear manifold segmentation and appearance modeling

Input: Images $I_1, I_2, \ldots$ and trajectories $W_1, W_2, \ldots$, Output pixel labeling at $\mathbf{L}_1, \mathbf{L}_2, \ldots$

Algorithm:

1. for $t = 1$ to K

   - Perform initialization, compute $(\mathbf{c}_K, \mathbf{D}_K, \mathcal{A}_K, \mathbf{L}_K)$

2. If frame t exists, i.e. received $\mathbf{I}_t$

   (a) Update motion segmentation

      i. $\mathbf{D}_t = \max (\mathbf{D_{t-1}}, \mathbf{D_{t-1:t}})$, $\mathbf{W_t} = exp(-\mathbf{D}_t/\lambda)$

      ii. Compute embedding coordinates $\mathbf{X}_t = laplacian - eigenmaps(\mathbf{W}_t)$

      iii. Apply GMM using $\mathbf{c}_{t-1}$ for initialization.

      iv. Label each cluster foreground/background using several cues.

   (b) Estimate dense motion model $\mathbf{M}_f$, and $\mathbf{M}_b$.

   (c) Update appearance models $\mathcal{A}_t = (\mathbf{A}_f, \mathbf{A}_b)$

      i. Predict the appearance $\mathcal{A}_{t|t-1}$ and labels $\mathbf{L}_{t|t-1}$ given $\mathbf{M_b}, \mathbf{M_f}$

      ii. Compute the updated appearance models $\mathcal{A}_{t|t}$ and $\mathbf{L}_{t|t}$ given the observed image $\mathbf{I}_t$

   (d) $t = t + 1$

FIGURE 4.6 Comparative results on three sequences. For each sequence, top: Elqursh and Elgammal [9], bottom Sheikh et al. [42]. For visualization purposes, background regions are darkened while foreground regions maintain their colors (Best viewed in color). (ⓒ(2012) Springer)

References

1. S. Araki, T. Matsuoka, N. Yokoya, and H. Takemura. Real-time tracking of multiple moving object contours in a moving camera image sequence. *IEICE TRANSACTIONS on Information and Systems*, 83(7):1583–1591, 2000.

2. M. Belkin and P. Niyogi. Laplacian Eigenmaps for Dimensionality Reduction and Data Representation. *Neural Computation*, 1396:1373–1396, 2003.

3. Y. Boykov, O. Veksler, and R. Zabih. Fast Approximate Energy Minimization via Graph Cuts. In *IEEE International Conference on Computer Vision*, pages 377–384 vol.1. Ieee, 1999.

4. T. Brox and J. Malik. Object Segmentation by Long Term Analysis of Point Trajectories. In *European Conference on Computer Vision*, pages 282–295, 2010.

5. N. Dalal and B. Triggs. Histograms of oriented gradients for human detection. In *Computer Vision and Pattern Recognition, 2005. CVPR 2005. IEEE Computer Society Conference on*, volume 1, pages 886–893. IEEE, 2005.

6. T. Dean, M. Ruzon, M. Segal, J. Shlens, S. Vijayanarasimhan, and J. Yagnik. Fast, accurate detection of 100,000 object classes on a single machine. In *IEEE Conference on Computer Vision and Pattern Recognition*, 2013.

7. P. Doll, C. Wojek, B. Schiele, and P. Perona. Pedestrian Detection : A Benchmark. In *Computer Vision and Pattern Recognition*, 2009.

8. E. Elhamifar and R. Vidal. Sparse subspace clustering. In *Computer Vision and Pattern Recognition*, pages 2790–2797. IEEE, June 2009.

9. A. Elqursh and A. Elgammal. Online moving camera background subtraction. In *Computer Vision, ECCV 2012*, pages 228–241. Springer, 2012.

10. P. Felzenszwalb, R. Girshick, D. McAllester, and D. Ramanan. Object detection with discriminatively trained part-based models. *IEEE Transactions on Pattern Analysis and Machine Intelligence*, 32(9):1627–1645, 2010.

11. P. Felzenszwalb and D. Huttenlocher. Pictorial structures for object recognition. *International Journal of Computer Vision*, 61(1):55–79, 2005.

12. M. Fischler and R. Bolles. Random sample consensus: a paradigm for model fitting with applications to image analysis and automated cartography. *Commun. ACM*, 24(6):381–395, June 1981.

13. C. Fowlkes, S. Belongie, F. Chung, and J. Malik. Spectral grouping using the Nystrom method. *IEEE Transactions on Pattern Analaysis and Machine Intelligence*, 2004.

14. K. Fragkiadaki and J. Shi. Detection Free Tracking: Exploiting Motion and Topology for Segmenting and Tracking under Entanglement. In *Computer Vision and Pattern Recognition*, pages 2073–2080. IEEE, June 2011.

15. D. Gavrila. Pedestrian detection from a moving vehicle. In *Computer VisionECCV 2000*, pages 37–49. Springer, 2000.

16. H. Greenspan, J. Goldberg, and A. Mayer. A Probabilistic Framework for Spatio-Temporal Video Representation & Indexing. In *European Conference on Computer Vision*, 2002.

17. A. Gruber and Y. Weiss. Multibody factorization with uncertainty and missing data using the EM algorithm. In *Computer Vision and Pattern Recognition*, volume 1, pages 707–714. IEEE, 2004.

18. E. Hayman and J. Eklundh. Statistical background subtraction for a mobile observer. In *ICCV 2003*, pages 67–74, 2003.

19. N. Ichimura. Motion Segmentation Based on Factorization Method and Discriminant Criterion. In *International Conference on Computer Vision*, volume 1, 1999.

20. M. Irani and P. Anandan. A unified approach to moving object detection in 2D and 3D scenes. *IEEE Transactions on Pattern Analysis and Machine Intelligence*, 20(6):577–589, 1998.

21. M. Irani, B. Rousso, and S. Peleg. Computing occluding and transparent motions. *International Journal of Computer Vision*, 12(1):5–16, February 1994.

22. V. Jain and E. Learned-Miller. Online Domain Adaptation of a Pre-Trained Cascade of Classifiers. In *Computer Vision and Pattern Recognition*, pages 577–584. IEEE, June 2011.

23. A. Jepson and M. Black. Mixture Models for Optical Flow Computation. In *Computer Vision and Pattern Recognition*, pages 760–761. IEEE Comput. Soc. Press, 1993.

24. Y. Jin, L. Tao, H. Di, N. Rao, and G. Xu. Background modeling from a free-moving camera by multi-layer homography algorithm. In *Image Processing, 2008. ICIP 2008. 15th IEEE International Conference on*, pages 1572–1575. IEEE, 2008.

25. G. Johansson. *Configurations in event perception*. PhD thesis, University of Oxford, 1950.

26. G. Johansson. Visual perception of biological motion and a model for its analysis. *Perception & psychophysics*, 14(2):201–211, 1973.

27. N. Jojic and B. Frey. Learning Flexible Sprites in Video Layers. In *Computer Vision and Pattern Recognition*, pages I-199–I-206. IEEE Comput. Soc, 2001.

28. K. Kanatani. Motion Segmentation by Subspace Separation and Model Selection. In *International Conference on Computer Vision*, volume 2, pages 586–591. IEEE, 2001.

29. A. Kitamoto. The moments of the mixel distribution and its application to statistical image classification. In *Advances in Pattern Recognition*, pages 521–531. Springer, 2000.

30. M. Kumar, P. Torr, and A. Zisserman. Learning Layered Motion Segmentations of Video. *International Conference on Computer Vision*, 2005.

31. S. Kwak, T. Lim, W. Nam, B. Han, and J. Han. Generalized Background Subtraction Based on Hybrid Inference by Belief Propagation and Bayesian Filtering. In *International Conference on Computer Vision*, 2011.

32. D. De Menthon and R. Megret. Spatio-temporal segmentation of video by hierarchical mean shift analysis. In *Workshop Statistical Methods in Video Processing ECCV*, 2002.

33. A. Mittal and D. Huttenlocher. Scene modeling for wide area surveillance and image synthesis. In *CVPR*, 2000.

34. D. Murray and A. Basu. Motion tracking with an active camera. *Pattern Analysis and Machine Intelligence, IEEE Transactions on*, 16(5):449–459, 1994.

35. P. Ochs and T. Broxs. Higher order motion models and spectral clustering. In *Computer Vision and Pattern Recognition*, pages 614–621. IEEE, June 2012.

36. J. Odobez and P. Bouthemy. Separation of moving regions from background in an image sequence acquired with a mobile camera. In *Video Data Compression for Multimedia Computing*, pages 283–311. Springer, 1997.

37. M. Oren, C. Papageorgiou, P. Sinha, E. Osuna, and T. Poggio. Pedestrian detection using wavelet templates. In *Computer Vision and Pattern Recognition, 1997. Proceedings, 1997 IEEE Computer Society Conference on*, pages 193–199. IEEE, 1997.

38. M. Peterson and B. Gibson. The initial identification of figure-ground relationships: Contributions from shape recognition processes. *Bulletin of the Psychonomic Society*, 29(3):199–202, 1991.

39. Y. Ren, C. Chua, and Y. Ho. Statistical background modeling for non-stationary camera. *Pattern Recognition Letters*, 24(1-3):183–196, 2003.

40. S. Rowe and A. Blake. Statistical mosaics for tracking. *Image and Vision Computing*, 14(8):549–564, 1996.

41. P. Sand and S. Teller. Particle Video: Long-Range Motion Estimation Using Point Trajectories. *International Journal of Computer Vision*, 80(1):72–91, May 2008.

42. Y. Sheikh, O. Javed, and T. Kanade. Background Subtraction for Freely Moving Cameras. In *International Conference on Computer Vision, ICC 2009*, 2009.

43. N. Sundaram, T. Brox, and K. Keutzer. Dense Point Trajectories by GPU-Accelerated Large Displacement Optical Flow. In *European Conference on Computer Vision*, pages

438–451, 2010.

44. C. Tomasi and T. Kanade. Shape and Motion from Image Streams under Orthography: a Factorization Method. *International Journal of Computer Vision*, 9(2):137–154, November 1992.

45. H. Torr, R. Szeliski, and P. Anandan. An Integrated Bayesian Approach to Layer Extraction from Image Sequences. *IEEE Transactions on Pattern Analysis and Machine Intelligence*, 23(3):297–303, 2001.

46. P. Torr and A. Zisserman. Mlesac: A new robust estimator with application to estimating image geometry. *Computer Vision and Image Understanding*, 78(1):138–156, 2000.

47. R. Vidal and R. Hartley. Motion segmentation with missing data using powerfactorization and GPCA. In *Computer Vision and Pattern Recognition*, volume 2, pages 310–316, 2004.

48. P. Viola and M. Jones. Robust real-time face detection. *International journal of computer vision*, 57(2):137–154, 2004.

49. T. Wada and T. Matsuyama. Appearance sphere: Background model for pan-tilt-zoom camera. In *13th International Conference on Pattern Recognition*, 1996.

50. J. Wang and E. Adelson. Representing Moving Images with Layers. *IEEE Transactions on Image Processing*, 2(5), 1994.

51. J. Wang, B. Thiesson, Y. Xu, and M. Cohen. Image and Video Segmentation by Anisotropic Kernel Mean Shift. In *European Conference on Computer Vision*, 2004.

52. Y. Weiss and E. Adelson. A unified mixture framework for motion segmentation : incorporating spatial coherence and estimating the number of models. In *Computer Vision and Pattern Recognition*, 1996.

53. J. Wills, S. Agarwal, and S. Belongie. What Went Where. In *Computer Vision and Pattern Recognition*, 2003.

54. J. Yan and M. Pollefeys. A General Framework for Motion Segmentation: Independent, Articulated, Rigid, Non-rigid, Degenerate and Non-degenerate. In *European Conference on Computer Vision*, 2006.

II

Traditional and Recent Models

5

Statistical Models for Background Subtraction

Ahmed Elgammal
Department of Computer Science, Rutgers University, USA

5.1 Introduction

In visual surveillance applications stationary cameras or pan-tilt-zoom (PTZ) cameras are used to monitor activities at outdoor or indoor sites. Since the cameras are stationary, the detection of moving objects can be achieved by comparing each new frame with a representation of the scene background. This process is called *background subtraction* and the scene representation is called the *background model*. The scene here is assumed to be stationary or quasi stationary.

The concept of background modeling is rooted to the early days of photography in the first half of the 19th century, where it was typical to expose a film for an extended period of time, leading to the capture of the scene background without fast moving objects. The use of background subtraction process to detect moving objects was originated in image analysis, and emanated from the concept of change detection, a process in which two images of the same scene taken at different time instances are compared, for example in Landsat Imagery, *e.g.* [10, 28]. Background subtraction nowadays is a widely used concept to detect moving objects in videos taken from a static camera, and even extends to videos taken from moving cameras (Chapter 4). In the last two decades several algorithms have been developed for background subtraction and were used in various important applications such as visual surveillance, sports video analysis, motion capture, *etc.*

Typically, the background subtraction process forms the first stage in automated visual surveillance systems as well as other applications such as motion capture, sport analysis, *etc.* Results from this process are used for further processing, such as tracking targets and understanding events. One main advantage of target detection using background subtraction is that the outcome is an accurate segmentation of the foreground regions from the scene background. For human subjects, the process gives accurate silhouettes of the human body, which can be further used for tracking, fitting body limbs, pose and posture estimation, *etc.* This is in contrast to classifier-based object-based detectors, which mainly decides whether a bounding box or a region in the image contains the object of interest or not, *e.g.* pedestrian detectors [6].

The concept of background subtraction has been widely used since the early human motion analysis systems such as Pfinder [54], W4 [22], *etc.* Efficient and more sophisticated background subtraction algorithms that can address challenging situations have been developed since then. The success of these algorithms lead to the growth of automated visual surveillance industry as well as many commercial applications, for example sports monitoring. Unlike earlier background subtraction algorithms while the cameras and the scenes are assumed to be stationary, many approaches have been proposed to overcome these limitations, for example dealing with quasi-stationary scenes and moving cameras. We will discuss such approaches later in this chapter. In this chapter we review the concept and the practice in background subtraction. We discuss several basic statistical background subtraction models, including parametric Gaussian models and nonparametric models. We discuss the issue of shadow suppression, which is essential for human motion analysis applications. We also discuss approaches and tradeoffs for background maintenance. We also point out many of the recent developments in background subtraction paradigms, which are details in the other chapters of this book.

The organization of this chapter is as follows. Section 5.2 discusses some of the challenges in building a background model for detection. Section 5.3 discusses some of the basic and widely used background modeling techniques. Section 5.4 discusses how to deal with color information to avoid detecting shadows. Section 5.5 discusses the tradeoffs and challenges in updating background models.

5.2 Challenges in Scene Modeling

In any indoor or outdoor scene there are changes that occur over time. It is important for any background model to be able to tolerate these changes, either by being invariant to them or by adapting to them. These changes can be local, affecting only parts of the background, or global affecting the entire background. The study of these changes is essential to understand the motivations behind different background subtraction techniques. Toyama *et al.* [53] identified a list of ten challenges that a background model has to overcome, and denoted them by: *Moved objects, Time of day, Light switch, Waving trees, Camouflage, Bootstrapping, Foreground aperture, Sleeping person, Walking person, Shadows.* We can classify the possible changes in a scene background according to their source:

Illumination changes:

- Gradual change in illumination, as might occur in outdoor scenes due to the change in the relative location of the sun during the day.

- Sudden change in illumination as might occur in an indoor environment by switching the lights on or off, or in an outdoor environment, *e.g,* a change between cloudy and sunny conditions.

- Shadows cast on the background by objects in the background itself (*e.g,* buildings and trees) or by moving foreground objects, *i.e.*, moving shadows.

Motion changes:

- Global image motion due to small camera displacements. Despite the assumption that cameras are stationary, small camera displacements are common in outdoor situations due to wind load or other sources of motion which cause global motion in the images.

- Motion in parts of the background. For example, tree branches moving with the wind, or rippling water.

Structural Changes:

- These are changes introduced to the background, including any change in the geometry or the appearance of the background of the scene introduced by targets. Such changes typically occur when something relatively permanent is introduced into the scene background. For example, if somebody moves (introduces) something from (to) the background, or if a car is parked in the scene or moves out of the scene, or if a person stays stationary in the scene for an extended period, *etc.* Toyama *et al.* [53] denoted these situations by "Moved Objects", "Sleeping Person" and "Walking Person" scenarios.

A central issue in building a representation for the scene background is the choice of the statistical model that explains the observation at a given pixel or region in the scene. The choice of the proper model depends on the type of changes expected in the scene background. Such a choice highly affects the accuracy of the detection. Section 5.3 discusses some of the statistical models that are widely used in background modeling context.

Another fundamental issue in building a background representation is what features to use for this representation or, in other words, what to model in the background. In the literature a variety of features have been used for background modeling including pixel-based features (pixel intensity, edges, disparity) and region-based features (*e.g*, image blocks). The choice of the features affects how the background model will tolerate the changes in the scene and the granularity of the detected foreground objects.

Beyond choosing the statistical model and suitable features, maintaining the background representation is another challenging issue that we will discuss in Section 5.5.

5.3 Statistical Scene Modeling

In this section we will discuss some of the existing and widely used statistical background modeling approaches. For each model we will discuss how the model is initialized and how it is maintained. For simplicity of the discussion we will use pixel intensity as the observation. Instead, color or any other features can be used.

5.3.1 Probabilistic Background Modeling

The process of background/foreground segmentation can be formulated as follows: At the pixel level, the observed intensity is assumed to be a random variable that takes a value based on whether the pixel belongs to the background or the foreground. Given the intensity observed at a pixel at time t, denoted by x_t, we need to classify that pixel to either the background $\mathcal{B}$ or foreground classes $\mathcal{F}$. This is a two-class classification problem. In a Bayesian setting, we can decide whether that pixel belongs to the background or the foreground based on the ratio between the posterior probabilities $p(\mathcal{B}|x_t)/p(\mathcal{F}|x_t)$. Using Bayes' rule, we can compute the posterior probability as

$$p(\mathcal{B}|x_t) = \frac{p(x_t|\mathcal{B}) \cdot p(\mathcal{B})}{p(x_t)}, \ p(\mathcal{F}|x_t) = \frac{p(x_t|\mathcal{F}) \cdot p(\mathcal{F})}{p(x_t)},$$

where $p(x_t|\mathcal{B})$ and $p(x_t|\mathcal{F})$ are the likelihoods of the observation x_t given the background and foreground models respectively. The terms $p(\mathcal{B})$ and $p(\mathcal{F})$ are the prior beliefs that the pixel belongs to the background or the foreground. If we assume equal priors then the ratio between the posterior reduces to the ratio between the likelihood, which is typically denoted by the likelihood ratio

$$\frac{p(x_t|\mathcal{B})}{p(x_t|\mathcal{F})}.$$

Therefore, statistical background modeling aims at providing estimates for the likelihood of the observation gives a model for the background and a model for the foreground. However, since the intensity of a foreground pixel can arbitrarily take any value, unless some further information about the foreground is available, we can just assume that the foreground distribution is uniform. Therefore, the problem reduces to a one-class classification problem, and a decision can be made by comparing the observation likelihood of the background model to a threshold, *i.e.*,

$$x_t \text{ belongs to } \mathcal{B} \quad \text{if } p(x_t|\mathcal{B}) \geq \text{ threshold} \qquad (5.1)$$

$$x_t \text{ belongs to } \mathcal{F} \qquad\qquad \text{otherwise} \qquad\qquad (5.2)$$

Therefore, any background subtraction approach needs a statistical model to estimate the likelihood of the observation given the background class, *i.e.*, $p(x_t|\mathcal{B})$, which can be achieved if a history of background observations are available at that pixel. If the history observations are not purely coming from the background, *i.e.*, foreground objects are present in the scene, the problem becomes more challenging. Mainly background models differ in the way this likelihood is estimated based on the underlying statistical model that is used. In what follows we will discuss parametric methods that assume a Gaussian model or a mixture of Gaussian model for the pixel process. We also will discuss non-paramteric methods that do not assume a specific model for the pixel process. Coupled with the choice of the statistical model is the mechanism that the model will adapt to the scene changes. Any background subtraction approach also has to provide a mechanism to choose the threshold used in the Background/Foreground decision.

5.3.2   Parametric Background Models

A Single Gaussian Background Model

Pixel intensity is the most commonly used feature in background modeling. In a static scene, a simple noise model that can be used for the pixel process is the independent stationary additive Gaussian noise model [14]. According to that model, the noise distribution at a given pixel is a zero mean Gaussian distribution $\mathcal{N}(0, \sigma^2)$. It follows that the observed intensity at that pixel is a random variable with a Gaussian distribution, *i.e.*,

$$P(x_t|\mathcal{B}) \approx \mathcal{N}(\mu, \sigma^2) = \frac{1}{\sigma\sqrt{2\pi}} e^{\frac{(x_t - \mu)^2}{2\sigma^2}}.$$

This Gaussian distribution model for the intensity value of a pixel is the underlying model for many background subtraction techniques and widely known as a single Gaussian background model.

Learning the background model, reduces to estimating the sample mean and variance from history observations at each pixel. The background subtraction process in this case is a classifier that decides whether a new observation at that pixel comes from the learned background distribution. Assuming the foreground distribution is uniform, this amounts to putting a threshold on the tail of the Gaussian likelihood, *i.e.*, the classification rule reduces to marking a pixel as foreground if

$$\|x_t - \hat{\mu}\| > \text{threshold},$$

where the threshold is typically set to $k\hat{\sigma}$. Here $\hat{\mu}$ and $\hat{\sigma}$ are the estimated mean and standard deviation and k is a free parameter. The standard deviation σ can be even assumed to be the same for all pixels. So, literally, this simple model reduces to subtracting a background image B from each new frame I_t and checking the difference against a threshold. In such case the background image B is the mean of the history background frames.

For the case of color images, as well as where the observation at the pixel is a high-dimensional vector, $x_t \in \mathbb{R}^d$, a multivariate Gaussian density is used, *i.e.*,

$$P(x_t|\mathcal{B}) \approx \mathcal{N}(\mu, \Sigma) = \frac{1}{\sqrt{2\pi|\Sigma|}} e^{\frac{-1}{2}(x_t-\mu)^{\mathsf{T}}\Sigma^{-1}(x_t-\mu)},$$

where $\mu \in \mathbb{R}^d$ is the mean, and $\Sigma \in \mathbb{R}^{d \times d}$ is the covariance matrix of the distribution. Typically, the color channels are assumed to be independent. Therefore, the covariance is a diagonal matrix, which reduces a multivariate Gaussian to a product of single Gaussians, one for each color channel. More discussion about dealing with color will be presented in Section 5.4.

This basic single Gaussian model can be made adaptive to slow changes in the scene (for example, gradual illumination changes) by recursively updating the mean with each new frame to maintain a background image

$$B_t = \frac{t-1}{t}B_{t-1} + \frac{1}{t}I_t,$$

where $t \geq 1$. Here, B_t denotes the background image computed up to frame t. Obviously this update mechanism does not forget the history and, therefore, the effect of new images on the model tends to zero. This is not suitable when the goal is to adapt the model to illumination changes. Instead the mean and variance can be computed over a sliding window of time. However, a more practical and efficient solution is to recursively update the model via temporal blending, also known as exponential forgetting, *i.e.*

$$B_t = \alpha I_t + (1 - \alpha)B_{t-1}. \tag{5.3}$$

The parameter α controls the speed of forgetting old background information. This update equation is a low-pass filter with a gain factor α that effectively separates the slow temporal process (background) from the fast process (moving objects). Notice that the computed background image is no longer the sample mean over the history but captures the central tendency over a window in time [17].

This basic adaptive model seems to be a direct extension to earlier work on change detection between two images. One of the earliest papers that suggested this model with a full justification of the background and foreground distributions is the paper by Donohoe *et al.* [8]. Karmann *et al.* [30, 31] used a similar recursive update model without explicit assumption about the background process. Koller *et al.* [34] used a similar model for traffic monitoring. Similar model was also used in early people tracking systems such as the Pfinder [54].

A Mixture Gaussian Background Model

Typically, in outdoor environments with moving trees and bushes, the scene background is not completely static. For example, one pixel can be the image of the sky in one frame, a tree leaf in another frame, a tree branch in a third frame and some mixture subsequently. In each of these situations the pixel will have a different intensity (color). Therefore, a single Gaussian assumption for the probability density function of the pixel intensity will not hold. Instead, a generalization based on a Mixture of Gaussians (MoG) has been proposed in [15,

18,19,46] to model such variations. This model was introduced by Friedman and Russell [15], where a mixture of three Gaussian distributions was used to model the pixel value for traffic surveillance applications. The pixel intensity was modeled as a weighted mixture of three Gaussian distributions corresponding to road, shadow and vehicle distribution. Fitting a mixture of Gaussian (MoG) model can be achieved using the Expectation Maximization (EM) algorithm [7]. However this is impractical for a real time background subtraction application. An incremental EM algorithm [41] was used to learn and update the parameters of the model.

Stauffer and Grimson [18, 19] proposed a generalization to the previous approach. The intensity of a pixel is modeled by a mixture of K Gaussian distributions (K is a small number from 3 to 5). The mixture is weighted by the frequency with which each of the Gaussians explains the background. The likelihood that a certain pixel has intensity x_t at time t is estimated as

$$p(x_t|\mathcal{B}) = \sum_{i=1}^{K} w_{i,t} G(x_t; \mu_{i,t}, \Sigma_{i,t}),\qquad(5.4)$$

where $G(\cdot; \mu_{i,t}, \Sigma_{i,t})$ is a Gaussian density function with a mean $\mu_{i,t}$ and covariance $\Sigma_{i,t} = \sigma_{i,t}^2 \mathbf{I}$. $w_{i,t}$ is the weight for the i-th Gaussian components. The subscript t indicates that the weight, the mean, and the covariance of each component are updated at each time step.

The parameters of the distributions are updated recursively using online approximation of the EM algorithm. The mixture is weighted by the frequency with which each of the Gaussians explains the background, *i.e.*, a new pixel value is checked against the existing K Gaussians and when a match is found the weight for that distribution is updated as follows

$$w_{i,t} = (1 - \alpha)w_{i,t-1} + \alpha M(i,t),$$

where $M(i,t)$ is an indicator variable which is 1 if the i-th component is matched, 0 otherwise. The parameter of the matched distributions are updated as follows

$$\mu_t = (1 - \rho)\mu_{t-1} + \rho x_t,$$

$$\sigma_t^2 = (1 - \rho)\sigma_{t-1}^2 + \rho(x_t - \mu_t)^T(x_t - \mu_t).$$

The parameters α and ρ are two learning rates. The K distributions are ordered based on w_j/σ_j^2 and the first B distributions are used as a model of the background of the scene where B is estimated as

$$B = \arg\min_b \left(\sum_{j=1}^{b} w_j > T \right).\qquad(5.5)$$

The threshold T is the fraction of the total weight given to the background model. Background subtraction is performed by marking any pixel that is more than 2.5 standard deviations away from any of the B distributions as a foreground pixel.

The MoG background model was shown to perform very well in indoor and outdoor situations. Many variations have been suggested to the Stauffer and Grimson's model [18], *e.g.* [23, 29, 39], *etc.* The model also was used with different feature spaces and/or with a subspace representations. Gao *et al.* [17] studied the statistical error characteristic of MoG background models.

5.3.3   Non-parametric Background Models

In outdoor scenes, typically there are wide range of variations, which can be very fast. Outdoor scenes usually contain dynamic areas such as waving trees and bushes, rippling water, ocean waves. Such fast variations are part of the scene background. Modeling such dynamic areas requires a more flexible representation of the background probability distribution at each pixel. This motivates the use of a non-parametric density estimator for background modeling [13].

A particular nonparametric technique that estimates the underlying density and is quite general is the Kernel Density Estimation (KDE) technique [9, 47]. Given a sample $S = \{x_i\}_{i=1..N}$ from a distribution with density function $p(x)$, an estimate $\hat{p}(x)$ of the density at x can be calculated using

$$\hat{p}(x) = \frac{1}{N} \sum_{i=1}^{N} K_\sigma(x - x_i), \tag{5.6}$$

where K_σ is a kernel function (sometimes called a "window" function) with a bandwidth (scale) σ such that $K_\sigma(t) = \frac{1}{\sigma}K(\frac{t}{\sigma})$. The kernel function K should satisfy $K(t) \geq 0$ and $\int K(t)dt = 1$. Kernel density estimators asymptotically converge to any density function with sufficient samples [9, 47]. In fact, all other nonparametric density estimation methods, e.g., histograms, can be shown to be asymptotically kernel methods [47]. This property makes these techniques quite general and applicable to many vision problems where the underlying density is not known [3, 11]. We can avoid having to store the complete data set by weighting a subset of the samples as

$$\hat{p}(x) = \sum_{x_i \in B} \alpha_i K_\sigma(x - x_i),$$

where α_i are weighting coefficients that sum up to one and B is a sample subset. A good discussion of KDE techniques can be found in [47].

A variety of kernel functions with different properties have been used in the literature of nonparametric estimation. Typically kernel functions are symmetric and unimodal functions that fall off to zero rapidly away from the center, *i.e.*, the kernel function should have finite local support and points beyond certain window will have no contribution. The Gaussian function is typically used as a kernel for its continuity, differentiability and locality properties although it violates the finite support criterion [11]. Note that choosing the Gaussian as a kernel function is different from fitting the distribution to a mixture of Gaussian model. Here, the Gaussian is only used as a function to weight the data points. Unlike parametric fitting of a mixture of Gaussians, kernel density estimation is a more general approach that does not assume any specific shape for the density function, and does not require model parameter estimation.

Elgammal *et al.* [13] introduced a background modeling approach based on kernel density estimation. Let $x_1, x_2, ..., x_N$ be a sample of intensity values for a pixel. Given this sample, we can obtain an estimate of the probability density function of the pixel intensity at any intensity value using kernel density estimation using Eq. 5.6. This estimate can be generalized to use color features or other high-dimensional features by using kernel products as

$$p(x_t|\mathcal{B}) = \frac{1}{N} \sum_{i=1}^{N} \prod_{j=1}^{d} K_{\sigma_j}(x_t^j - x_i^j), \tag{5.7}$$

where x_t^j is the j-th dimension of the feature at time t and K_{σ_j} is a kernel function with bandwidth σ_j in the jth color space dimension. If the Gaussian function is used as a kernel, then the density can be estimated as

$$p(x_t|\mathcal{B}) = \frac{1}{N} \sum_{i=1}^{N} \prod_{j=1}^{d} \frac{1}{\sqrt{2\pi\sigma_j^2}} e^{-\frac{1}{2} \frac{(x_{t_j} - x_{i_j})^2}{\sigma_j^2}}. \tag{5.8}$$

Using this probability estimate, the pixel is considered a foreground pixel if $p(x_t|\mathcal{B}) < th$, where the threshold th is a global threshold over all the image, which can be adjusted to achieve a desired percentage of false positives. The estimate in Eq. 5.7 is based on the most recent N samples used in the computation. Therefore, adaptation of the model can be achieved simply by adding new samples and ignoring older samples [13], *i.e.*, using a sliding window over time.

Practically, the probability estimation in Eq. 5.8 can be calculated in a very fast way using precalculated lookup tables for the kernel function values given the intensity value difference, $(x_t - x_i)$, and the kernel function bandwidth. Moreover, a partial evaluation of the sum in equation 5.8 is usually sufficient to surpass the threshold at most image pixels, since most of the image is typically from the background. This allows a real time implementation of the approach.

This nonparametric technique for background subtraction was introduced in [13] and has been tested for a wide variety of challenging background subtraction problems in a variety of setups and was found to be robust and adaptive. We refer the reader to [13] for details about the approach; such as model adaptation and false detection suppression. Figures 5.1 shows two detection results for targets in a wooded area where tree branches move heavily and the target is highly occluded. Figure 5.2-top shows the detection results using an omni-directional camera. The targets are camouflaged and walking through the woods. Figure 5.2-bottom shows the detection result for a rainy day where the background model adapts to account for different rain and lighting conditions.

One major issue that needs to be addressed when using kernel density estimation technique is the choice of suitable kernel bandwidth (scale). Theoretically, as the number of samples increases the bandwidth should decrease. Practically, since only a finite number of samples are used and the computation must be performed in real time, the choice of suitable bandwidth is essential. Too small bandwidth will lead to a ragged density estimate, while a too wide bandwidth will lead to an over-smoothed density estimate [9,11]. Since the expected variations in pixel intensity over time are different from one location to another in the image, a different kernel bandwidth is used for each pixel. Also, a different kernel bandwidth is used for each color channel.

In [13] a procedure was proposed for estimating the kernel bandwidth for each pixel as a function of the median of absolute differences between consecutive frames. To estimate the kernel bandwidth σ_j^2 for the jth color channel for a given pixel the median absolute deviation for consecutive intensity values of the pixel is computed over the sample. That is, the median, m, of $| x_i - x_{i+1} |$ for each consecutive pair (x_i, x_{i+1}) in the sample is calculated independently for each color channel. The motivation behind the use of median of absolute deviation is that pixel intensities over time are expected to have jumps because different objects (e.g., sky, branch, leaf and mixtures when an edge passes through the pixel) are projected onto the same pixel at different times. Since we are measuring deviations between two consecutive intensity values, the pair (x_i, x_{i+1}) usually comes from the same local-in-time distribution, and only a few pairs are expected to come from cross distributions (intensity jumps). The median is a robust estimate and should not be affected by few jumps. If this local-in-time distribution is assumed to be a Gaussian $N(\mu, \sigma^2)$, then the distribution for the deviation $(x_i - x_{i+1})$ is also Gaussian $N(0, 2\sigma^2)$. Since this distribution is symmetric, the median of the absolute deviations, m, is equivalent to the quarter percentile of the deviation distribution. That is,

$$Pr(N(0, 2\sigma^2) > m) = 0.25 \,,$$

and therefore the standard deviation of the first distribution can be estimated as

$$\sigma = \frac{m}{0.68\sqrt{2}} \,.$$

Since the deviations are integer gray scale (color) values, linear interpolation can be used to obtain more accurate median values. In [40] an adaptive approach for estimation of kernel bandwidth was proposed. Parag *et al.* [42] proposed an approach using boosting to evaluate different kernel bandwidth choices for bandwidth selection.

KDE-Background Practice and Other Nonparametric models:

One of the drawbacks of the KDE background model is the requirement to store a large number of history samples for each pixel. In KDE literature many approaches were proposed to avoid storing a large number of samples. Within the context of background modeling, Piccardi and Jan [44] proposed an efficient mean shift approach to estimate the modes of a pixel's history PDF then a few number of Gaussians were used to model the PDF. Mean shift is a nonparametric iterative mode seeking procedure [2, 4, 16]. With the same goal of reducing memory requirement, Han *et al.* [21] proposed a sequential kernel density estimation approach where variable bandwidth mean shift was used to detect the density modes. Unlike mixture of Gaussian methods where the number of Gaussian is fixed, a technique such as [21, 44] can adaptively estimate a variable number of modes to represent the density, therefore keeping the flexibility of a nonparametric model while achieving the efficiency of a parametric model.

Efficient implementation of KDE can be achieved through building look-up tables for the kernel function values, which facilitate real time performance. The Fast Gauss Transform has been proposed for efficient computation of KDE [12], however, the Fast Gauss Transform is only justifiable with a large number of samples required for the density estimation as well as the need for estimation at many pixels in batches. For example, Fast Gauss implementation was effectively used in a layered background representation [43].

Many variations have been suggested to the basic nonparametric KDE background model. In practice nonparametric KDE has been used at the pixel level, as well as at the region level, and in a domain-range representation to model a scene background. For example, in [43] a layered representation was used to model the scene background where the distribution of each layer is modeled using KDE. Such layered representation facilitates detecting the foreground under static or dynamic background and in the presence of nominal camera motion. In [49] KDE was used in a joint domain-range representation of image pixel (r,g,b,x,y), which exploits the spatial correlation between neighboring pixels. Parag *et al.* [42] proposed an approach for feature selection for the KDE framework where boosting based ensemble learning was used to combine different features. The approach also can be used to evaluate different kernels bandwidth choices for bandwidth selection. Recently, Sheikh *et al.* [48] used a KDE approach in a joint domain-range representation within a foreground/background segmentation framework from a freely moving camera as discussed in Chapter 4.

5.3.4 Other Statistical Models

The statistical models for background subtraction that are described in this section are the basis for many other algorithms in the literature. In [53], linear prediction using the Wiener filter is used to predict pixel intensity given a recent history of values. The prediction coefficients are recomputed each frame from the sample covariance to achieve adaptivity. Linear prediction using the Kalman Filter was also used in [30, 31, 34], such models assume a single Gaussian distribution of the pixel process.

Another approach to model a wide range of variations in the pixel intensity is to represent these variations as discrete states corresponding to modes of the environment, e.g., lights on/off, cloudy/sunny. Hidden Markov Models (HMM) have been used for this purpose in [45, 52]. In [45], a three-state HMM was used to model the intensity of a pixel for

FIGURE 5.1   Example background subtraction detection results: Left: original frames, Right: detection results.

traffic monitoring application where the three states correspond to the background, shadow, and foreground. The use of HMMs imposes a temporal continuity constraint on the pixel intensity, *i.e.*, if the pixel is detected as a part of the foreground then it is expected to remain part of the foreground for a period of time before switching back to be part of the background. In [52], the topology of the HMM representing global image intensity is learned while learning the background. At each global intensity state the pixel intensity is modeled using a single Gaussian. It was shown that the model is able to learn simple scenarios like switching the lights on-off.

Background subtraction techniques can successfully deal with quasi-moving backgrounds, *e.g.* scenes with dynamic textures. The nonparametric model using Kernel Density Estimation (KDE), described above, has very good performance in scenes with dynamic backgrounds, such as outdoor scenes with trees in the background. Several approaches were developed to address such dynamic scenes. In [50] an Auto Regressive Moving Average Model (ARMA) model was proposed for modeling dynamic textures. ARMA is a first order linear prediction model. In [57] an ARMA model was used for background modeling of scenes with dynamic texture where a robust Kalman filter was used to update the model. In [40] a combination of optical flow and appearance features was used within an adaptive kernel density estimation framework to deal with dynamic scenes.

In [36] a biologically inspired nonparametric background subtraction approach was proposed where a self-organizing artificial neural network model was used to model the pixel process. Each pixel is modeled with a sample arranged in a shared 2D grid of nodes, where each node is represented with a weight vector with the same dimensionality as the input observation. An incoming pixel observation is mapped to the node whose weights are most similar to the input where a threshold function is used to decide background/foreground. The weights of each node are updated at each new frame using a recursive filter similar to Eq. 5.3. An interesting feature of this approach is that the shared 2D grid of nodes allows the spatial relationships between pixels to be taken into account at both the detection and update phases.

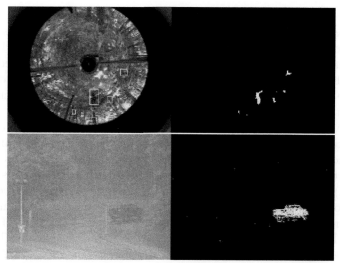

FIGURE 5.2   Top: Detection of camouflaged targets from an omni-directional camera. Bottom: Detection result for a rainy day.

5.3.5   Features for Background Modeling

Intensity has been the most commonly used feature for modeling the background. Alternatively, edge features have also been used to model the background. The use of edge features to model the background is motivated by the desire to have a representation of the scene background that is invariant to illumination changes, as discussed in Section 5.4. In [55] foreground edges are detected by comparing the edges in each new frame with an edge map of the background which is called the background "primal sketch". The major drawback of using edge features to model the background is that it would only be possible to detect edges of foreground objects instead of the dense connected regions that result from pixel intensity based approaches. Fusion of intensity and edge information was used in [26, 27, 38, 56]. Among many other features studied, optical flow was used in [40] to help capture background dynamics. A general framework for feature selection based on boosting for background modeling was proposed in [42].

Besides pixel-based approaches, block-based approaches have also been used for modeling the background. Block matching has been extensively used for change detection between consecutive frames. In [25] each image block is fit to a second order bivariate polynomial and the remaining variations are assumed to be noise. A statistical likelihood test is then used to detect blocks with significant change. In [37] each block was represented with its median template over the background learning period and its block standard deviation. Subsequently, at each new frame, each block is correlated with its corresponding template and blocks with too much deviation relative to the measured standard deviation are considered to be foreground. The major drawback with block-based approaches is that the detection unit is a whole image block and therefore they are only suitable for coarse detection.

5.4   Shadow Suppression

A background subtraction process on gray scale images, or on color images without carefully selecting the color space, is bound to detect the shadows of moving objects along with the objects themselves. While shadows of static objects can typically be adapted in the background process, shadows cast by a moving object, *i.e.*, dynamic shadows, constitute a severe challenge for foreground segmentation. Since the goal of background subtraction is to obtain accurate segmentation of moving foreground regions for further processing, it is highly desirable to detect such foreground regions without cast shadows attached to them. This is particularly important for human motion analysis since shadows attached to silhouettes would cause problems in fitting body limbs and estimating body poses, consider the example shown in Figure 5.3. Therefore, extensive research has addressed the detection/supression of moving (dynamic) shadows.

Avoiding the detection of shadows or suppressing the detected shadows can be achieved in color sequences by understanding how shadows affect color images. This is also useful to achieve a background model that is invariant to illumination changes. Cast shadows have a dark part (umbra) where a light source is totally occluded, and a soft transitional part (penumbra) where light is partially occluded [51]. In visual surveillance scenarios, the penumbra shadows are common since diffused and indirect lights are common in indoor and outdoor scenes. Penumbra shadows can be characterized by low value of intensity while preserving the chromaticity of the background, *i.e.*, achromatic shadows. Therefore, most research on detecting shadows have focused on achromatic shadows [5, 13, 24].

Let us consider the RGB color space, which is a typical output of a color camera. The brightness of a pixel is a linear combination or the RGB channels, here denoted by I

$$I = w_r R + w_g G + w_b B.$$

When an object casts a shadow on a pixel, less light reaches that pixel and the pixel seems darker. Therefore, a shadow casted on a pixel can be characterized by a change in brightness of that pixel such that:

$$\tilde{I} = \alpha I$$

where $\tilde{I}$ is the pixel's new brightness. Similar effect happens under certain changes in illumination, e.g., turning on/off the lights. Here $\alpha < 1$ for the case of shadow, which means the pixel is darker under shadow, while $\alpha > 1$ for the case of highlights, the pixel seems brighter. A change in the brightness of a pixel will affect all the three color channels R, G, and B. Therefore any background model based on the RGB space, and of course gray scale imagery, is bound to detect moving shadows as foreground regions.

So, which color spaces are invariant or less sensitive to shadows and highlights? For simplicity, let us assume that the effect of the change in a pixel brightness is the same in the three channels. Therefore, the observed colors are $\alpha R, \alpha G, \alpha B$. Any chromaticity measure of a pixel where the effect of the α factor is cancelled, is in fact invariant to shadows and highlights. For example, in [13] chromaticity coordinates based on normalized RGB were used for modeling the background. Given three color variables, R, G and B, the chromaticity coordinates are defined as [35]:

$$r = \frac{R}{R+G+B}, g = \frac{G}{R+G+B}, b = \frac{B}{R+G+B} \tag{5.9}$$

Obviously only two coordinates are enough to represent the chromaticity since $r + g + b = 1$. The above equation describes a central projection to the plane $R + G + B = 1$. This is analogous to the transformation used to obtain CIE xy chromaticity space from CIE XYZ color space. The CIE XYZ color space is a linear transformation to the RGB space [1]. The chromaticity space defined by the variable r,g is therefore analogous to the CIE xy chromaticity space. It can be easily seen that the chromaticity variables r, g, b are invariant to shadows and highlights (according to our assumption) since the α factor does not have an effect on them. Figure 5.3 shows the results of detection using both (R, G, B) space and (r, g) space. The figure shows that using the chromaticity coordinates allows detection of the target without detecting its shadow.

Some other color spaces also have chromaticity variables that are invariant to shadows and highlights in the same way. For example, the reader can verify that the Hue and Saturation variables in the HSV color space are invariant to the α factor, and thus insensitive to shadows and highlights, while the Value variable, which represents the brightness is variant to them. Therefore, the HSV color space has been used in some background subtraction algorithms that suppress shadows, *e.g.* [5]. Similarly, HSL, CIE xy spaces have the same property. On the other hand color spaces such as YUV, YIQ, YCbCr are not invariant to shadows and highlights since they are just linear transformations from the RGB space.

Although using chromaticity coordinates helps in the suppression of shadows, they have the disadvantage of losing lightness information. Lightness is related to the differences in whiteness, blackness and grayness between different objects [20]. For example, consider the case where the target wears a white shirt and walks against a gray background. In this case there is no color information. Since both white and gray have the same chromaticity coordinates, the target will not be detected using only chromaticity variables. In fact in the r, g space all the gray line (R=G=B) projects to the point $(1/3, 1/3)$ in the space, similarly for CIE xy. Therefore, there is no escape of using a brightness variable. In [13] a third "lightness" variable $s = R + G + B$ was used besides r, g. While the chromaticity variable r, g are not expected to change under shadow, s is expected to change within limits which corresponds to the expected shadows and highlights in the scene.

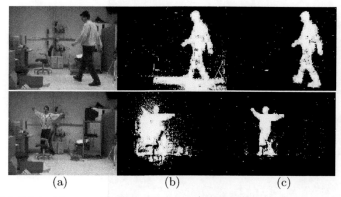

(a)   (b)   (c)

FIGURE 5.3 (a) Original frames, (b) Detection using (R, G, B) color space, (c) detection using chromaticity coordinates (r, g) and the lightness variable s. (©(2010) Springer)

Most approaches for shadow suppression rely on the above reasoning of separating the chromaticity distortion from brightness distortion where each of these distortions are treated differently, *e.g.* [5, 13, 24, 26, 32]. In Horprasert *et al.* [24] both brightness and color distortions are defined using a chromatic cylinder model. By projecting an observed pixel color to the vector defined by that pixel's background value in the RGB color space (chromaticity line), the color distortion is defined as the orthogonal distance, while the projection defines the brightness distortion. Here a single Gaussian background model is assumed. These two measures were used to classify an observation to either background, foreground, shadows or highlights. Figure 5.4 shows an example of using that model. Notice that the orthogonal distance between an observed pixel's RGB color and a chromaticity line is affected by brightness of that pixel, while the distance measured in the r-g space (or xy space) corresponds to the angles between the observed color vector and the chromaticity line, *i.e.*, the r-g space used in [13] is a projection of a chromatic cone. In [26] a chromatic and brightness distortion model is used similar to [24, 32], however using a chromatic cone instead of a chromatic cylinder distortion model.

Another class of algorithms for shadow suppression are approaches that depend on im-

FIGURE 5.4   An illustration showing the detection using the model of Horprasert *et al.* [24]. At the middle of the sequence, half of the fluorescence lamps are turned off. The result shows that the system still detects the moving object successfully.

age gradient to model the scene background. The idea is that texture information in the background will be consistent under shadow, hence using the image gradient as a feature will be invariant to cast shadows, except at the shadow boundary. These approaches utilize a background edge or gradient model besides the chromaticity model to detect shadows, *e.g.* [26,27,38,56]. In [26] a multistage approach was proposed to detect chromatic shadows. In the first stage potential shadow regions are detected by fusing color (using the invariant chromaticity cone model described above) and gradient information. In the second stage pixels in these regions are classified using different cues including spatial and temporal analysis of chrominance, brightness, and texture distortion; and a measure of diffused sky lighting denoted by a "bluish effect". The approach can successfully detect chromatic shadows.

5.5 Tradeoffs in Background Maintenance

As discussed in Section 5.1 there are different changes that can occur in a scene background, which can be classified to: Illumination changes, Motion Changes, Structural Changes. The goal of background maintenance is to be able to cope with these changes and keep an updated version of the scene background model. In parametric background models, recursive update in the form of Eq. 5.3 (or some variant of it) is typically used for background maintenance, *e.g.* [18,31,34]. In nonparametric models, the sample of each pixel history is updated continuously to achieve adaptability [13,40]. These recursive updates along with careful choice of the color space are typically enough to deal with both the illumination changes and motion changes previously described.

The most challenging case is where changes are introduced to the background (objects moved in or from the background) denoted here by "Structural Changes". For example, if a vehicle came and parked in the scene. A background process should detect such a car but should also adapt it to the background model in order to be able to detect other targets that might pass in front of it. Similarly if a vehicle that was already part of the scene moved out, a false detection "hole" will appear in the scene where that vehicle was parked. There are many examples similar to these scenarios. Toyama *et al.* [53] denoted these situations "sleeping person" and "walking person" scenarios.

Here we point out two interwound tradeoffs that associate with maintaining any background model

Background update rate: The speed or the frequency in which a background model gets updated highly influence the performance of the process. In most parametric models, the learning rate α in Eq. 5.3 controls the speed in which the model adapts to changes. In nonparametric models, the frequency in which new samples are added to the model has the same effect. Fast model update makes the model able to rapidly adapt to scene changes such as fast illumination changes, which leads to high sensitivity in foreground/background classification. However, the model can also adapt to targets in the scene if the update is done blindly in all pixels or errors occurs in masking out foreground regions. Slow update is safer to avoid integrating any transient changes to the model. However, the classifier will lose its sensitivity in case of fast scene changes.

Selective vs. Blind update: Given a new pixel observation, there are two alternative mechanisms to update a background model: 1) Selective Update: update the model only if the pixel is classified as a background sample. 2) Blind Update: just update the model regardless of the classification outcome. Selective update is commonly used by masking out foreground-classified pixels from the update since updating the model with foreground information would lead to increased false negative,*e.g*, holes in the detected targets. The problem with selective update is that any incorrect detection decision will result in persistent incorrect detection later, which is a deadlock situation, as denoted by Karmann *et al.* [31]. For example, if a tree branch is displaced and stayed fixed in the new location for a long time, it would be continually detected. This is what leads to the "Sleeping/Walking person" problems as denoted in [53].

Blind update does not suffer from this deadlock situation since it does not involve any update decisions; it allows intensity values that do not belong to the background to be added to the model. This might lead to more false negatives as targets erroneously become part of the model. This effect can be reduced if the update rate is slow.

The interwound effects of these two tradeoffs is shown in Table 5.1. Most background models chose a selective update approach and tried to avoid the effects of detection errors by using a slow update rate. However, this is bound to deadlocks. In [13] the use of a combination of two models was proposed: a short-term model (selective and fast) and a long-term model (blind and slow). This combination tries to achieve high sensitivity and, at the same time, avoid deadlocks.

TABLE 5.1   Tradeoffs in Background Maintenance

	Fast Update	Slow Update
Selective Update	Highest sensitivity Adapts to fast illumination changes bound to Deadlocks	Less sensitivity bound to Deadlocks
Blind Update	Adapts to targets (more False Negatives) No deadlocks	Slow adaptation No deadlocks

Several approaches have been proposed for dealing with specific scenarios with structural changes. The main problem is that dealing with such changes requires a higher level of reasoning about what are the objects causing such structural changes (vehicle, person, animal) and what should be done with them, which mostly depends on the application. Such high level of reasoning is typically beyond the design goal of the background process, which is mainly a low level process that knows only about pixels' appearance.

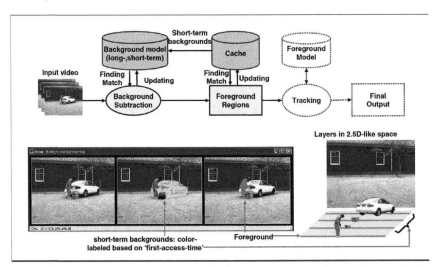

FIGURE 5.5   An overview of Kim et al. approach [33] with short-term background layers: the fore- ground and the short-term backgrounds can be interpreted in a different temporal order. (©(2005) Springer)

The idea of using multiple background models was further developed by Kim *et al.* in [33] to address scene structure changes in an elegant way. In that work, a layered background model was used where a long term background model is used besides several multiple short term background models that capture temporary changes in the background. An object that comes to the scene and stops is represented by a short term background (layer). Therefore, if a second object passes in front of the stopped object, it will also be detected and represented as a layer as well. Figure 5.5 shows an overview of the approach and detection results.

References

1. W. Burger and M. Burge. *Digital Image Processing, an Algorithmic Introduction Using Java.* Springer, 2008.
2. Y. Cheng. Mean shift, mode seeking, and clustering. *IEEE Transaction on Pattern Analysis and Machine Intelligence,* 17(8):790–799, Aug 1995.
3. D. Comaniciu. *Nonparametric Robust Methods For Computer Vision.* PhD thesis, Rutgers, The State University of New Jersey, January 2000.
4. D. Comaniciu and P. Meer. Mean shift analysis and applications. In *IEEE International Conference on Computer Vision, ICCV 1999,* volume 2, pages 1197–1203, September 1999.
5. R. Cucchiara, C. Grana, M. Piccardi, and A. Prati. Detecting moving objects, ghosts, and shadows in video streams. *IEEE Transactions on Pattern Analysis and Machine Intelligence,* 25:1337–1342, 2003.
6. N. Dalal and B. Triggs. Histograms of oriented gradients for human detection. In *International Conference on Pattern Recognition, CVPR 2005,* 2005.
7. A. Dempster, N. Laird, and D. Rubin. Maximum likelihood from incomplete data via the EM algorithm. *Journal of the Royal Statistical Society,* 39:1–38, 1977.
8. G. W. Donohoe, Don R. Hush, and Nasir Ahmed. Change detection for target detection and classification in video sequences. In *ICASSP,* 1988.
9. R. Duda, D. Stork, and P. Hart. *Pattern Classification.* Wiley, John & Sons,, 2000.
10. H. Eghbali. K-S test for detecting changes from landsat imagery data. *SMC,* 9(1):17–23, 1979.
11. A. Elgammal. *Efficient Kernel Density Estimation for Realtime Computer Vision.* PhD thesis, University of Maryland, 2002.
12. A. Elgammal, R. Duraiswami, and L. Davis. Efficient non-parametric adaptive color modeling using fast gauss transform. In *IEEE Conference on Computer Vision and Pattern Recognition,* December 2001.
13. A. Elgammal, D. Harwood, and L. Davis. Nonparametric background model for background subtraction. In *European Conference of Computer Vision, ECCV 2000,* 2000.
14. D. Forsyth and J. Ponce. *Computer Vision a Modern Approach.* Prentice Hall, 2002.
15. N. Friedman and S. Russell. Image segmentation in video sequences: A probabilistic approach. In *Uncertainty in Artificial Intelligence,* 1997.
16. K. Fukunaga and L. Hostetler. The estimation of the gradient of a density function, with application in pattern recognition. *IEEE Transaction on Information Theory,* 21:32–40, 1975.
17. X. Gao and T. Boult. Error analysis of background adaption. In *IEEE Conference on Computer Vision and Pattern Recognition,* 2000.
18. W. Grimson and C. Stauffer. Adaptive background mixture models for real-time tracking. In *IEEE Conference on Computer Vision and Pattern Recognition,* 1999.
19. W. Grimson, C. Stauffer, and R. Romano. Using adaptive tracking to classify and monitor activities in a site. In *IEEE Conference on Computer Vision and Pattern Recognition,* 1998.

20. E. Hall. *Computer Image Processing and Recognition.* Academic Press, 1979.

21. B. Han, D. Comaniciu, and L. Davis. Sequential kernel density approximation through mode propagation: Applications to background modeling. In *ACCV 2004*, 2004.

22. I. Haritaoglu, D. Harwood, and L. Davis. W4: Who? when? where? what? A real time system for detecting and tracking people. In *International Conference on Face and Gesture Recognition*, 1998.

23. M. Harville. A framework for high-level feedback to adaptive, per-pixel, mixture-of-gaussian background models. In *ECCV 2006*, pages 543–560, 2002.

24. T. Horprasert, D. Harwood, and L. Davis. A statistical approach for real-time robust background subtraction and shadow detection. In *IEEE Frame-Rate Applications Workshop*, 1999.

25. Y. Hsu, H. Nagel, and G. Rekers. New likelihood test methods for change detection in image sequences. *Computer Vision and Image Processing*, 26:73–106, 1984.

26. I. Huerta, M. Holte, T. Moeslund, and J. Gonzalez. Detection and removal of chromatic moving shadows in surveillance scenarios. *ICCV 2009*, pages 1499–1506, 2009.

27. S. Jabri, Z. Duric, H. Wechsler, and A. Rosenfeld. Detection and location of people in video images using adaptive fusion of color and edge information. In *International Conference of Pattern Recognition*, 2000.

28. R.C. Jain and H.H. Nagel. On the analysis of accumulative difference pictures from image sequences of real world scenes. *PAMI*, 1(2):206–213, April 1979.

29. O. Javed, K. Shafique, and M. Shah. A hierarchical approach to robust background subtraction using color and gradient information. In *IEEE Workshop on Motion and Video Computing*, pages 22–27, 2002.

30. K. Karmann, A. Brandt, and R. Gerl. Moving object segmentation based on adaptive reference images. In *Signal Processing V: Theories and Application.* Elsevier Science Publishers B.V., 1990.

31. K. Karmann and A. von Brandt. Moving object recognition using and adaptive background memory. In *Time-Varying Image Processing and Moving Object Recognition.* Elsevier Science Publishers B.V., 1990.

32. K. Kim, T. Chalidabhongse, D. Harwood, and L. Davis. Background modeling and subtraction by codebook construction. In *International Conference on Image Processing, ICIP 2004*, pages 3061–3064, 2004.

33. K. Kim, D. Harwood, and L. Davis. Background updating for visual surveillance. In *International Symposium on Visual Computing, ISVC 2005*, pages 1–337, 2005.

34. D. Koller, J. Weber, T. Huang, J. Malik, G. Ogasawara, B. Rao, and S. Russell. Towards robust automatic traffic scene analyis in real-time. In *International Conference of Pattern Recognition*, 1994.

35. M. Levine. *Vision in Man and Machine.* McGraw-Hill Book Company, 1985.

36. L. Maddalena and A. Petrosino. A self-organizing approach to background subtraction for visual surveillance applications. *IEEE Transactions on Image Processing*, 17(7):1168–1177, July 2008.

37. T. Matsuyama, T. Ohya, and H. Habe. Background subtraction for nonstationary scenes. In *4th Asian Conference on Computer Vision*, 2000.

38. S. Mckenna, S. Jabri, Z. Duric, H. Wechsler, and A. Rosenfeld. Tracking groups of people. *Computer Vision and Image Understanding*, 80:42–56, 2000.

39. A. Mittal and D. Huttenlocher. Scene modeling for wide area surveillance and image synthesis. In *CVPR*, 2000.

40. A. Mittal and N. Paragios. Motion-based background subtraction using adaptive kernel density estimation. In *CVPR*, pages 302–309, 2004.

41. R. Neal and G. Hinton. A new view of the EM algorithm that justifies incremental and other variants. In *Learning in Graphical Models*, pages 355–368. Kluwer Academic

Publishers, 1993.

42. T. Parag, A. Elgammal, and A. Mittal. A framework for feature selection for background subtraction. In *IEEE Computer Society Conference on Computer Vision and Pattern Recognition, CVPR 2006*, June 2006.

43. K. Patwardhan, G. Sapiro, and V. Morellas. Robust foreground detection in video using pixel layers. *IEEE Transactions on Pattern Analysis and Machine Intelligence*, 30:746–751, April 2008.

44. M. Piccardi and T. Jan. Mean-shift background image modelling. In *ICIP 2004*, pages V: 3399–3402, 2004.

45. J. Rittscher, J. Kato, S. Joga, and A. Blake. A probabilistic background model for tracking. In *6th European Conference on Computer Vision*, 2000.

46. S. Rowe and A. Blake. Statistical mosaics for tracking. *Image and Vision Computing*, 14(8):549–564, 1996.

47. D. Scott. *Mulivariate Density Estimation*. Wiley-Interscience, 1992.

48. Y. Sheikh, O. Javed, and T. Kanade. Background subtraction for freely moving cameras. In *ICCV 2009*, pages 1219–1225, 2009.

49. Y. Sheikh and M. Shah. Bayesian modeling of dynamic scenes for object detection. *PAMI*, 27:1778–1792, 2005.

50. S. Soatto, G. Doretto, and Y. Wu. Dynamic textures. In *Proceedings of the International Conference on Computer Vision*, volume 2, pages 439–446, 2001.

51. J. Stauder, R. Mech, and J. Ostermann. Detection of moving cast shadows for object segmentation. *IEEE Transactions on Multimedia*, 1:65–76, 1999.

52. B. Stenger, V. Ramesh, N. Paragios, F. Coetzee, and J. Bouhman. Topology free hidden markov models: Application to background modeling. In *IEEE International Conference on Computer Vision*, 2001.

53. K. Toyama, J. Krumm, B. Brumitt, and B. Meyers. Wallflower: Principles and practice of background maintenance. In *IEEE International Conference on Computer Vision*, 1999.

54. C. Wren, A. Azarbayejani, T. Darrell, and A. Pentland. Pfinder: Real-time tracking of human body. *IEEE Transaction on Pattern Analysis and Machine Intelligence*, 1997.

55. Y. Yang and M. Levine. The background primal sketch: An approach for tracking moving objects. *Machine Vision and Applications*, 5:17–34, 1992.

56. W. Zhang, X. Fang, and X. Yang. Moving cast shadows detection based on ratio edge. In *International Conference on Pattern Recognition, ICPR 2006*, 2006.

57. J. Zhong and S. Sclaroff. Segmenting foreground objects from a dynamic textured background via a robust Kalman filter. In *IEEE International Conference on Computer Vision, ICCV 2003*, page 44, Washington, DC, USA, 2003. IEEE Computer Society.

6

Non-parametric Background Segmentation with Feedback and Dynamic Controllers

Philipp Tiefenbacher

Institute for Human-Machine Communication, Technische Univ. Muenchen, Germany

Martin Hofmann

Institute for Human-Machine Communication, Technische Univ. Muenchen, Germany

Gerhard Rigoll

Institute for Human-Machine Communication, Technische Univ. Muenchen, Germany

6.1 Introduction

The number of video cameras in our environment is increasing. On the one hand the cameras are used in automated surveillance scenarios to protect the people against the threat of potential trespassers or to solve already committed crimes faster. On the other hand, video cameras are used more often as input devices for communication with intelligent machines. Whether for communication with machines, enhanced control of devices by gestures or for consumer electronics in the form of an innovative control of games, it is beneficial or even necessary to split up the images to areas of movement and to static regions. Thus in most cases, the obtained images are processed with the objective to detect movement in the images.

In the future, it is expected that video cameras are used more often because of the desire for faster and more intuitive interaction with machines as well as the striving of the people for security in an increasingly urbanized world. The basis for such applications is the detection of regions in the video, which successively change from frame to frame. So generally these are areas where movement occurs. This important pre-processing step can be described as a detection step, which differentiates moving foreground objects from a primarily static background. This is often stated as background subtraction or foreground segmentation.

In the simplest case, the static background model consists of one frame, which gets compared to the current frame. The regions of the current frame with high deviation to the background model are labelled as foreground. This simplistic method might work in certain specialized scenarios. However, for the most part a perfectly rigid background frame is not available at all. Background has subtle motion, light is gradually changing or new objects are emerging or vanishing from the scene. Thus, to model the background a multitude of more sophisticated methods have been developed in the recent past.

This chapter presents a detailed analysis and improvements of a method in the field of non-parametric foreground segmentation. More specifically, the Pixel-Based Adaptive Segmenter (PBAS), which outperformed several state of the art methods on the Change Detection challenge [7], is further explained and scrutinized. The PBAS approach follows a non-parametric background modelling paradigm. Therefore, the background is modelled by a history of recently observed pixel values. This avoids time consuming and error prone estimation of parameters for a background model, making PBAS fast and robust. A central idea of the method is the dynamic adaptation of the two most important parameters, namely (1) the decision threshold, to make the foreground decision, and (2) the learning rate, which controls the update of the background model. In previous non-parametric methods, these two parameters have been set to fixed values and are applied to all pixel locations. In PBAS both of these parameters are controlled dynamically at runtime and for each pixel and channel separately.

Furthermore, both controllers are steered by a new estimate of the background dynamics. Applying this new measurement for background dynamics leads to a sensitive foreground detection in regions of low dynamics (e.g. a wall or a street), but is also able to provide satisfactory results in high dynamic areas like waving trees or sparkling water. We show that this kind of dynamic adaptation proves highly beneficial and extremely important for good classification performance.

A thorough evaluation of the involved parameters and settings gives insights to the way the controllers work and on how to optimize the method for specific use cases and application scenarios. Overall, the results show state-of-the-art performance on the Change Detection database [7]. Furthermore, due to the non-parametric nature of the algorithm, the method runs extremely fast and efficiently and is able to make use of multi-core CPUs. The remainder of this chapter is organized as follows: First (in Section 6.2), we give an overview of related background subtraction algorithms, indicating the similarities and differences to our approach.

Subsequently, we recap the general principles of PBAS as it was presented in [8]. Afterwards, we go into more detail regarding the processing details in Section 6.4. Next in Section 6.6, the Change Detection dataset is envisaged. This collection of databases features a wide range of scenarios of potentially relevant applications. This makes the database an ideal test bed for competitive performance evaluation. Also, the available performance metrics and the characteristics of the ground truth are presented. Then a profound evaluation is carried out.

Thus, we expose the dynamics of the PBAS on the Change Detection dataset [7]. Hereby, we concentrate on detailed examination of the system dynamics in Section 6.7. The following Section 6.8 focuses on the evaluation of the impact of single parameter value changes. As the PBAS uses two features for the classification, the influence of the new magnitude based feature is elaborated in Section 6.8.1. Afterwards, the effects on the categories of the Change Detection challenge through different settings in a pre- and post-processing step are shown in Section 6.8.2 and in Section 6.8.3. In conclusion, the results gained due to the deep examination of the PBAS dynamics are summarized and we present optimized settings for the PBAS as well as for the pre- and post-processing steps. Here, we concentrate on finding ideal parameter sets of the PBAS for each category of the Change Detection database. Consequently, an optimization for each category (like *Camera Jitter, Intermittent Object Motion* etc.) is realized in Section 6.9. Lastly, a conclusion and outlook is given in Section 6.10.

6.2   Related Work

Over the recent past, a multitude of algorithms and methods for background modelling have been developed. Mentioning all of them would go beyond the scope of this chapter. Excellent survey papers can be found in [11], [12] and [4]. However, we want to give a short overview of different background modelling approaches and state to which type our method belongs.

One of the most prominent categories is statistical background modelling, where the background is modelled as a mixture of weighted Gaussian distributions [15] or as a kernel density estimation [6]. In [15], pixels which are detected as background are used to improve the Gaussian mixtures by an iterative update rule.

Furthermore, background clustering methods compare input pixel values to clusters and classify the pixels according to the label of the nearest cluster [5], [9]. Another idea is to detect foreground with the use of well elaborated filters like the Wiener [17] filter or lately the Chebyshev [10] filter. Therefore, the filters constitute an estimation of the background. If new pixel values differ from the assumed values of the filters, the pixels are classified as foreground.

More recently, attempts have been made to use fuzzy integrals to fuse features for a background model [19]. This concept has been enhanced in [2] due to the use of the Choquet integral by also adding an adaptive way for the maintenance of the background model. The maintaining is achieved by taking the fuzzy membership values into account and increasing the update rate more significantly, if the pixel is clearly classified as background and more slowly if the pixel is classified to foreground.

Our method can be referred as an advanced statistical background modelling technique as we develop new measurements for describing the background statistics and additionally employ a non-parametric method. Like in SACON (SAmple CONsensus) [18] the background model is defined by a non-parametric method. Each pixel in the background model is defined by a history of the N most recent image values at each pixel. Here, N ranges from 20 to 200 with maximum performance reported at $N = 60$. The history of background images is filled using a first-in first-out strategy.

By contrast to this in-order filling, in ViBe [3], which is also a non-parametric method, the N background values are updated by a random scheme. More over updated pixels can "diffuse" their current pixel value into neighbouring pixels. This is achieved by using another random selection method. The non-parametric methods have in general lower memory complexity as statistical background methods.

Our PBAS method can also be categorized as a non-parametric method, since we also use a history of N image values as the background model as in SACON. We use a similar random update rule as in ViBe. However, in ViBe the randomness parameters as well as the decision threshold are fixed for all pixels. In contrast, we do not treat these values as parameters, but instead as adaptive state variables, which can dynamically change over time for each pixel separately.

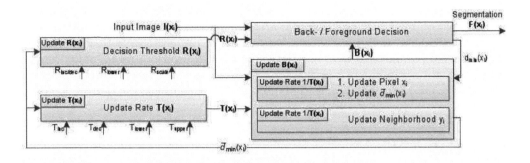

FIGURE 6.1   Overview of the Pixel-Based Adaptive Segmenter.

6.3   Non-parametric Background Modeling

This section describes the concept of the Pixel-Based Adaptive Segmenter. The PBAS is a foreground segmenter following a non-parametric paradigm. So a background model for every pixel x_i exists. The background model ideally states the real background. In this case, the background model consists of recently observed feature values, which are not part of moving objects. In Section 6.4.2 we go more into detail regarding the used feature values. Our method consists of several components which are depicted as a state machine in Figure 6.1. As a central component, the decision block decides for or against foreground based on the current image and a background model $\boldsymbol{B}(x_i)$. This decision is based on the per-pixel threshold $R(x_i)$. Moreover, the background model has to be updated over time in order to allow for gradual background changes. In our model, this update depends on a per-pixel learning parameter $T(x_i)$.

Now, the essential and novel idea of our PBAS approach is that both of these two per-pixel thresholds dynamically change based on an estimate of the background dynamics. In the following, we first describe the decision process and the update of the background. Then, we detail our dynamic update method of both the decision threshold $R(x_i)$ and the learning parameter $T(x_i)$.

6.3.1   Segmentation Decision

The goal of every background segmentation method is to come to a binary decision, whether a pixel belongs to the foreground or to the background. This decision process takes the input image and compares it in some way to a model of the background. In our case, the background model $\boldsymbol{B}(x_i)$ is defined by an array of N recently observed pixel features:

$$\boldsymbol{B}(x_i) = \{B_1(x_i), \ldots, B_k(x_i), \ldots, B_N(x_i)\} \tag{6.1}$$

A pixel x_i is decided to belong to the background, if its pixel feature $I(x_i)$ is closer than a certain decision threshold $R(x_i)$ to at least $\#_{min}$ of the N background values. Thus, the foreground segmentation mask is calculated as

$$F(x_i) = \begin{cases} 1 & \#\{\text{dist}(I(x_i), B_k(x_i)) < R(x_i)\} < \#_{min} \\ 0 & \text{else} \end{cases} \tag{6.2}$$

Here, $F = 1$ implies foreground. It can thus be seen, that the decision making involves two parameters: (1) The distance threshold $R(x_i)$, which is defined for each pixel separately and which can change dynamically; and (2) the minimum number $\#_{min}$, which is a fixed global parameter.

6.3.2 Update of Background Model

Updating the background model **B** is essential in order to account for changes in the background, such as lighting changes, shadows and moving background objects such as trees. Since foreground regions cannot be used for updating, the background model is only updated for those pixels that are currently background (i.e. $F(x_i) = 0$). Updating means that for a certain index $k \in 1 \ldots N$ (chosen uniformly at random), the corresponding feature of the background model $B_k(x_i)$ is replaced by the current pixel feature $I(x_i)$. This allows the current pixel feature to be "learned" into the background model. Tavakkoli *et al.* propose a traditional way for maintaining the background model in [16]. Here, due to the use of the Adaptive Kernel Density Estimation (AKDE), a history of old pixel values defines the probability of each pixel being background. Whereas the oldest pixel values are dismissed for new ones and the update of the history happens every 100 to 150 frames. In the classification step the probability of being background for each pixel is calculated and then compared to the estimated threshold.

However, also the update rate of the PBAS is adaptive and is performed with a probability of $P(x_i) = 1/T(x_i)$. Otherwise no update is carried out at all. Therefore, the parameter $T(x_i)$ defines the inverse update rate. The lower $T(x_i)$, the more likely a pixel will be updated.

We also update (with probability $1/T(x_i)$) a randomly chosen neighboring pixel $y_i \in \mathcal{N}(x_i)$. Thus, the background model $B_l(y_i)$ at this neighboring pixel is replaced by its current feature value $I(y_i)$. Here, not necessarily the same background component B_k is updated. Hence, another random process selects the background component B_l for updating the neighboring pixel. This is contrary to the approach in [3] where $B_k(y_i)$ is replaced by the pixel feature $I(x_i)$ of the current pixel (called "diffusion" in their approach). In general, a pixel x_i is only updated, if it is classified as background. However, a neighboring pixel y_i, which might be foreground, can be updated as well. This means that certain foreground pixels at the boundary will gradually be included into the background model. With this method, every foreground object will be "eaten-up" from the outside after a certain time, depending on the inverse update parameter $T(x_i)$. The advantage of this property is that erroneous foreground objects will quickly vanish as shown in Figure 6.2. Obviously this will also include slowly moving foreground objects into the background. Therefore, in Section 6.3.4 we present a dynamic adaptation of the update parameter $T(x_i)$, such that big objects are only "eaten-up" a little bit, while small erroneous blobs are "eaten" completely. We call this property the implicit erosion effect.

6.3.3 Dynamic Decision Threshold $R(x_i)$

In a video sequence, there can be areas with high background dynamics (i.e. water, trees in the wind, etc.) and areas with little to no changes (i.e. a wall). Ideally, for highly dynamic areas, the threshold $R(x_i)$ should be increased as to not include objects to the foreground. For static regions, $R(x_i)$ should be low, such that small deviations lead to a decision for foreground. Thus, the threshold $R(x_i)$ needs to be able to automatically adapt accordingly. To allow for these changes, there needs to be a measure of background dynamics, which is done as follows:

First of all, besides saving an array of recently observed feature values in the background model $\mathbf{B}(x_i)$, we also create an array $\mathbf{D}(x_i) = \{D_1(x_i), \ldots, D_N(x_i)\}$ of minimal decision distances. Whenever an update of $B_k(x_i)$ is carried out, the currently observed minimal distance $d_{min}(x_i) = \min_k \mathrm{dist}(I(x_i), B_k(x_i))$ is written to this array: $D_k(x_i) \leftarrow d_{min}(x_i)$. Thus, a history of minimal decision distances is created. The average of these values $\bar{d}_{min}(x_i) = 1/N \sum_k D_k(x_i)$ is a measure of the background dynamics. So only those pixels

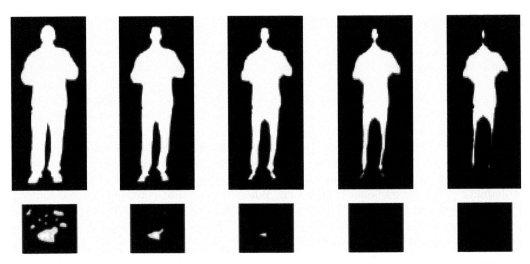

FIGURE 6.2 The "erosion effect" depicted for a big static object and small static object. The big object is slowly learned into the background, whereas the small object (i.e. noise) is quickly merged into background.

which are classified as background lead to an update of $\bar{d}_{min}(x_i)$. This ensures that the parameter $\bar{d}_{min}(x_i)$ describes the quality of the background model. Here, the current background pixel has the distance zero to an ideal background model. From another point of view, $\bar{d}_{min}(x_i)$ contains the dynamics of the background, because in the case of an almost static background the update rate of the background model is fast enough to handle minimal changes in the background. But in high dynamic areas the background is changing more rapidly, so when a pixel x_i is determined as background, it is more likely that the distance to the background model **B** is higher, inducting a higher $\bar{d}_{min}(x_i)$.

For example, assuming a completely static background, $\bar{d}_{min}(x_i)$ will be zero. For a more dynamic background, there will always be a (small) deviation of the currently observed feature to the previously seen ones, and thus $\bar{d}_{min}(x_i)$ will be higher.

With this estimate of the background dynamics, the decision threshold can be dynamically adapted, therefore we first verify, if the current $R(x_i)$ value should be increased or decreased with:

$$\Delta R(x_i) = \begin{cases} -1, & \text{if } R(x_i) > \bar{d}_{min}(x_i) \cdot R_{scale} \\ +1, & \text{else} \end{cases} \tag{6.3}$$

Then, the $R(x_i)$ value is changed relatively to its current amount:

$$R(x_i) = R(x_i) \cdot (1 + \Delta R(x_i) \cdot R_{inc/dec}) \tag{6.4}$$

Here, $R_{inc/dec}, R_{scale}$ are fixed parameters. This can be seen as a dynamic controller for the state variable $R(x_i)$. For a constant $\bar{d}_{min}(x_i)$, the decision threshold $R(x_i)$ approaches the product of $\bar{d}_{min}(x_i) \cdot R_{scale}$. Thus, a (sudden) increase in background dynamics leads to a (slow) increase of $R(x_i)$ towards a higher decision threshold $R(x_i)$.

Controlling the update rate in the way presented above leads to robust handling of varying amounts of background dynamics. An example picture with high dynamic background showing the spatial distribution of the state variable $R(x_i)$ is depicted in Figure 6.3. In this, brighter pixel values indicate a higher value for $R(x_i)$.

a) Image $I(x_i)$ b) Groundtruth

c) Segmentation   d) $R(x_i)$ e) $T(x_i)$

FIGURE 6.3 Example frame 2049 of the video "autumn" in the *Dynamic Background* category. The threshold of $R(x_i)$ is high in the regions with high dynamic background, as in this case the regions with waving branches. The inverse update parameter $T(x_i)$ is higher (whiter) in the regions with lower update probability.

6.3.4   Dynamic Inverse Update Probability $T(x_i)$

As mentioned in Section 6.3.2, independent of the foreground state $F(x_i)$, eventually, every object will be merged into the background depending on the inverse update parameter $T(x_i)$. A problem of this is that wrongly classified foreground is only slowly learned into the background and thus remains foreground. To alleviate the problem, the idea is to introduce a (second) dynamic controller for $T(x_i)$, such that the probability of background learning is (slowly) increased when the pixel is background and (slowly) decreased when the pixel is foreground. It can be assumed that pixels are mostly wrongly classified as foreground in areas of high dynamic background. Thus, the dynamic estimator $\bar{d}_{min}(x_i)$, which exactly represents this property is used for adjusting the rate of controlling of $T(x_i)$. We define the dynamic controller for $T(x_i)$ as follows:

$$T(x_i) = \begin{cases} T(x_i) + \frac{T_{inc}}{\bar{d}_{min}(x_i)+\Delta}, & \text{if } F(x_i) = 1 \\ T(x_i) - \frac{T_{dec}}{\bar{d}_{min}(x_i)+\Delta}, & \text{if } F(x_i) = 0 \end{cases} \tag{6.5}$$

Here, T_{inc}, T_{dec} are fixed parameters. There are different parameters for the two cases; because we assume that most of the time the pixels are background, which is true in most cases. Choosing independent parameters for background and foreground therefore leads to a balanced regulation of $T(x_i)$. The Δ is a small offset to guarantee that a division by zero is impossible. Furthermore, we define an upper and lower bound

$$T_{lower} < T(x_i) < T_{upper} \tag{6.6}$$

such that values cannot go out of a specified bound. The above controller ensures that in case of highly dynamic background (i.e. high $\bar{d}_{min}(x_i)$), the learning parameter $T(x_i)$ stays constant or only slightly changes.

In this case of highly dynamic background, erroneously detected foreground will not remain for long, because the inverse update probability $T(x_i)$ does not reach the T_{upper} value too fast. Generally, the update probability keeps the same and so ensures that once false detected foreground is learned into the background model, if it is classified as background again. In the other ideal case of a fully static background, a classification as foreground is

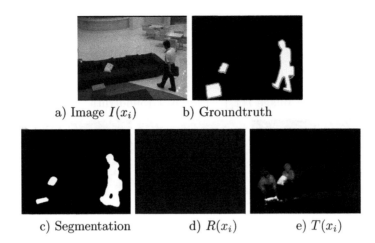

a) Image $I(x_i)$ b) Groundtruth

c) Segmentation   d) $R(x_i)$ e) $T(x_i)$

FIGURE 6.4   Example frame 1150 of the video "sofa" in the *Intermittent Object Motion* category. Higher values of $R(x_i)$ correspond to brighter pixel values. In the case of static background, like in the sofa scenario, $R(x_i)$ is low, so a very sensitive foreground detection is available. The inverse update parameter $T(x_i)$ is higher in the regions of static background and more frequent foreground detections. White areas imply less updates, thus static foreground objects remain longer in the foreground.

quite solid, hence $T(x_i)$ rapidly increases, validating the background model, in the way that it retains less updates.

Figure 6.4 depicts the inverse update parameter $T(x_i)$ in case of a slowly moving, but significantly large foreground object. For the pixels corresponding to the person, a high $T(x_i)$ indicates low update probability. Assume a background pixel just outside of the person's silhouette. This pixel can update a foreground silhouette boundary pixel by means of the neighbor update rule. Thus, a boundary pixel can become background and therefore the silhouette is "eaten-up" from the outside. The low update probability at this pixel, however, will avoid further shrinking of the silhouette, such that the majority of the silhouette is still detected as foreground.

6.4   Processing Details

In this section we go into more detail about the precise processing steps of the PBAS. In Section 6.4.1 we explain the procedure to initialize the background model **B**. Afterwards, we describe the exact steps for the calculation of the before mentioned pixel features $I(x_i)$. As pre- and post-processing is crucial to obtain the best classification performance, we shortly refer to the applied pre- and post-processing steps in Section 6.4.3 and Section 6.4.4.

6.4.1   Initialization

Due to the random background update process, for each run, multiple random variables need to be calculated. Particularly, there are the following random variables: If pixel x_i is denoted as background, one random variable is needed for determining if a background update should be performed. The neighbouring pixel update procedure contains another random variable. Thus, the update of the pixel x_i is independent of an update of the neighboring pixel. For the update procedure itself a randomly chosen background component $B_k(x_i)$ of the background model $\mathbf{B}(x_i)$ is selected. The chosen component $B_k(x_i)$ of the background model $\mathbf{B}(x_i)$ receives the current calculated features. Furthermore, the update of the minimal distance array is also utilized here. So a further random variable is necessary. In the case of an update of the neighbouring pixel, three random numbers are needed. One random number for choosing another background component B_l and the other two for selecting x- and a y-coordinate. This can be very time consuming. Hence, there exist a maximum necessary number of seven random numbers per run. So to improve performance, the random arrays are initialized at the beginning. At runtime just one random variable is calculated selecting the index of the previously initialized random arrays. Therefore, calculation time can be reduced to the cost of memory usage. But we do not need strong random numbers as in cryptography, thus a vector length of 10.000 initialized random numbers per case is completely sufficient. For the initialization of the background model the first N frames of the video are used as background model, not taking into account if a pixel x_i is classified as background or foreground. In the first iteration the array for the dynamic threshold $R(x_i)$ obtains the initialization value R_{lower}, which is also the lower boundary of $R(x_i)$. Additionally, the initialization value T_{init} is assigned to the dynamic inverse update probability $T(x_i)$.

6.4.2   Input Features

For each color channel, in addition to the pixel value, we also use gradient magnitudes. Thus, the input $I(x_i) = \{I^v(x_i), I^m(x_i)\}$ consists of the pixel value $I^v(x_i)$ itself and the gradient magnitude $I^m(x_i)$ at the pixel. Consequently, each element of the background history $B_k(x_i) = \{B_k^v(x_i), B_k^m(x_i)\}$ (in Equation 6.1) also consists of two corresponding entries. The gradient values are calculated through the convolution of the source image with a 3×3 Sobel operator. The definitions of the horizontal S_x and vertical S_y derivative approximations with Sobel filters are

$$\mathbf{I}_x = \begin{bmatrix} -1 & 0 & +1 \\ -2 & 0 & +2 \\ -1 & 0 & +1 \end{bmatrix} * \mathbf{I} \quad \text{and} \quad \mathbf{I}_y = \begin{bmatrix} -1 & -2 & -1 \\ 0 & 0 & 0 \\ +1 & +2 & +1 \end{bmatrix} * \mathbf{I}. \tag{6.7}$$

Finally, to obtain the gradient magnitude, we calculate

$$\mathbf{I^m} = \sqrt{\mathbf{I}_x{}^2 + \mathbf{I}_y{}^2} \tag{6.8}$$

For the distance calculation in Equation 6.2, we use the following equation:

$$\text{dist}(I(x_i), B_k(x_i)) = \frac{\alpha}{I_{mean}^m} \cdot |I^m(x_i) - B_k^m(x_i)| + |I^v(x_i) - B_k^v(x_i)| \tag{6.9}$$

Here, I_{mean}^m is the mean value of all former calculated distances between gradient magnitude values, which have also been classified as foreground. This value is ascertained due to a summarization of all gradient magnitude distances, which are greater than $R(x_i)$. We get

$$I_{sum}^m = I_{sum}^m + \text{dist}(I^m(x_i), B_k^m(x_i)), \quad \text{if } R(x_i) < \text{dist}(I(x_i), B_k(x_i)). \tag{6.10}$$

Based on this result, I_{mean}^m is defined as

$$I_{mean}^m = \frac{I_{sum}^m}{n_{er}}. \tag{6.11}$$

Here, n_{er} holds the total number of exceedances of the calculated distance over $R(x_i)$. The variables I_{sum}^m and n_{er} are set to zero before every new frame. Thus, the fraction $\frac{\alpha}{I^m}$ weighs the importance of pixel values against the gradient magnitude.

6.4.3   Pre-Processing

A common step to reduce noise in images is to convolve the image with a 2-dimensional Gaussian filter, which suppresses high frequencies and therefore acts as a low-pass. The filter is given in Equation 6.12. The x and y variables hold the distance to the origin and σ is the standard deviation of the Gaussian distribution. By convolving the image with the Gaussian filter, the pixel values are weighted depending on the distance to the origin. Here the origin has the highest weighting, while the weighting decreases with increasing distance. Leading to the effect, that distant pixels only slightly influence the value of the origin pixel.

$$F_{Gaussian}(x, y) = \frac{1}{2\pi\sigma^2} e^{-\frac{x^2+y^2}{2\sigma^2}} \tag{6.12}$$

We have to define two parameters, a number for the size of the two-dimensional kernel and the standard deviation σ. Increasing the filter size of the Gaussian leads to a stronger blurring, whereas a bigger σ leads to a greater decay of the weighting distribution according to the distance to the origin. In Section 6.8.2, the effect of different filter sizes and different σ are evaluated on different scenarios of foreground / background classification.

6.4.4   Post-Processing

As a result of our classification we obtain an image classified into two categories. Generally, a binary image is used for easy distinction through human perception. Thus, the image is split into zero values (black) for background and to one (white) for foreground. Hence, the post-processing should preserve this binary property. As the necessary time for processing a single image is important, we decided not to use more enhanced methods like blob detecting or Connected Component Analysis [1], which commonly need more time.

The PBAS algorithm makes only pixel by pixel decisions. Thus, it can happen that a single pixel is classified wrongly, whereas the other pixels around the wrong pixel are classified correctly. This mostly happens in the case of dynamic backgrounds. In image processing this phenomenon that a single pixel strongly differs from the other pixels around, is called "salt and pepper noise". A common way to reduce this kind of noise is to apply a median filter or a morphological filter. This filter fits to the binary problem in the sense that no average value of a series of input values is calculated. Hence, the result of the post-processing step always fits to the binary space. In Section 6.8.3 different sizes of a median filter are applied to a variety of scenarios.

6.5   Evaluation on Change Detection Challenge

In the previous sections, the algorithm PBAS was extensively described. In the following sections the PBAS is evaluated. Firstly, we examine the behavior of the individual state variables under different conditions in Section 6.7. Additionally, we evaluate our approach on the database provided for the Change Detection Challenge [7]. This database features 31 videos from six categories including scenarios with indoor views, outdoor views, shadows, camera jitter, small objects, dynamic background and thermal images. Human-annotated ground truth is available for all scenarios and is used for performance evaluation. Thus, exhaustive competitive comparison of methods is possible on this database. The goal of the Change Detection challenge [7] is to allow for competitive performance evaluation of recent and future background modeling methods.

6.6   The Change Detection Dataset

The Change Detection dataset [7] consists of a large video dataset with human annotated ground truth, therefore 31 videos divided into six categories are present. The categories are featuring real world videos of a broad field of applications, starting with the *Baseline* category including videos of traffic monitoring or the well-known PETS dataset [14]. The *Camera Jitter* category further includes surveillance videos of parking spots with jitter due to wind. The videos of the dynamic category mainly contain outdoor scenes with waving trees or water. As an important part of public surveillance, *Intermittent Object Motion* includes scenes with left behind luggage or cars fitting into a parking spot or moving off after hours of parking.

The category *Shadow* contains images with hard and soft shadow of objects as well as fast gradual light changes. Lastly, *Thermal* includes videos of moving and still humans captured by a thermal camera. A complete overview of the total number of pictures is depicted in Table 6.1. The numbers in the column "rated pictures" show the actual number of rated images for the evaluation. This number is smaller than the "total number of pictures", because the first frames of each video are used for creating a satisfactory background model and therefore are excluded of the rating. Thus, it is not possible to assess how long the different background modelling algorithms need to create a stable background model.

Category	Videos	Total number of pictures	Rated pictures
Baseline	4	6049	4413
Camera Jitter	4	6420	3134
Dynamic Background	6	18871	13276
Intermittent Object Motion	6	18650	12111
Shadow	6	16949	14105
Thermal	5	21100	18055
Total	31	88039	65094

TABLE 6.1   Overview of all categories of the Change Detection dataset.

6.6.1   Annotation

The following section briefly describes the annotation applied at the Change Detection dataset. Figure 6.6 shows a ground truth image of the dataset, the Figure 6.5 besides presents the five possible labels a pixel can be divided to. On the one hand, there are labels for background and foreground, which are used for a determination of the recognition ability. Hence, in Section 6.6.2 the related metrics are specified. Additionally, a label for the shadow of the moving objects is available and is named as "hard shadow".

Furthermore, it is considered that different degrees of camera motion blur (*unknown motion*) occur. In these regions, no unique assignment, whether foreground or background is present, is possible. Therefore, these pixels are not rated. Finally, the label "outside region of interest" hides the regions in the image, that are not to be evaluated. This is advisable, because in practice, it is difficult to produce results based on real videos, which just belong to a certain category. Each category of the Change Detection datasets holds a distinct property, but in some regions of a video unfavored actions may occur. This may be subtle dynamic noise in the *Shadow* category. Hence, these areas are labelled as "outside region of interest".

Pixel value	Label
0	Background
50	Hard shadow
85	Outside region of interest
170	Unknown motion
255	Foreground

FIGURE 6.5 Possible labels of the ground truth.

FIGURE 6.6 Ground truth of an image of the "bungalows" video.

6.6.2 Performance Metrics

In order to extensively compare the algorithms, appropriate metrics are used to evaluate different quality characteristics. Therefore, a total of seven different metrics have been defined, which are based on a pixel-wise evaluation. Thereby, each classified pixel is summed up, based on the ground truth, into one of four different parameters. Hereby, TP stands for the *true positives* and holds the absolute number of correctly classified foreground pixels. TN (*true negatives*) contains the absolute number of correctly recognized background points. The misinterpreted picture elements are divided into $FN = number\ of\ false\ negatives$ and $FP = number\ of\ false\ positives$. According to that, the metrics are defined in the following way:

- False Positive Rate (FPR) $= \dfrac{FP}{FP+TN}$

- False Negative Rate (FNR) $= \dfrac{FN}{TP+FN}$

- Percentage of Wrong Classifications (PWC) $= 100 \cdot \dfrac{FN+FP}{TP+FN+FP+TN}$

- Specificity (Sp) $= \dfrac{TN}{TN+FP}$

- Recall (Re) $= \dfrac{TP}{TP+FN}$

- Precision (Pr) $= \dfrac{TP}{TP+FP}$

- F1-Measure (F1) $= 2 \cdot \dfrac{Re \cdot Pr}{Re+Pr}$

The sum of all pixels of a single category is used to calculate the above metrics. Except for the *Overall* category, where the mean is computed based on the results of each category. To easily rank the algorithms by their performance, two relative measures were introduced at the Change Detection challenge. The metric *average ranking* is a per-category-wise metric, hence this metric is calculated for each category. Thus, each result of one of the seven metrics is ranked according to the results of the other algorithms. So, there are seven rankings in sum for each algorithm. These rankings are averaged to conclude the *average ranking* in a specific category.

The second relative measure is the *average ranking across categories*, which only exists for the *Overall* category. For this purpose, the results of the *average ranking* measure, which have to be computed for each category (*Baseline, Camera Jitter, Dynamic Background, Intermittent Object Motion, Shadow, Thermal*) first, are averaged. The result is the *average ranking across categories* measure.

6.7   System Behavior

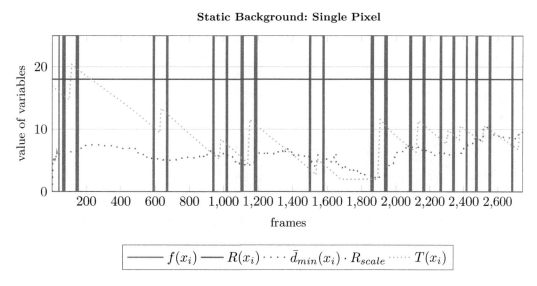

FIGURE 6.7   (See color insert.) Inspection of a pixel ($x = 280$, $y = 120$) in the video *sofa* of the Change Detection competition.

Until now, the function of the PBAS was described only theoretically, so now we want to show how the system behaves under real conditions and whether it differs from the theoretical considerations. For this purpose, a single pixel is observed over a whole video sequence and corresponding key values are stored. Hence, three pixels are selected by hand, which are subjected to three different, relevant and practical everyday foreground and background dynamics. First, a pixel is selected which has completely static background and is properly set as foreground due to a moving object, which appears from time to time. Secondly, we observe the system variables of a pixel, which is part of an intermittent foreground object for a long time. At the end, we illustrate how the system maintains the background model **B** in the case of high *Dynamic Background*.

The graphs contain two different system variables, introduced in Sections 6.3.3 and 6.3.4. Thus, $T(x_i)$ represents the inverse update rate, $R(x_i)$ the threshold for the decision as foreground and $\bar{d}_{min}(x_i)$ represents the average distance to the background model. $f(x_i)$ includes the result of the segmentation. Accordingly, by detection as foreground $f(x_i)$ is filled and otherwise empty. The average distance to the background model $\bar{d}_{min}(x_i)$ is scaled with the factor R_{scale}, in order to demonstrate the system behavior more clearly. The first case is shown in Figure 6.7, which includes the system variables of the entire sequence of the video *sofa* for the selected pixel. $T(x_i)$ always increases greatly if x_i is classified as foreground and decreases slightly when the pixel is background again. Furthermore, $T(x_i)$ changes strongly since the slope is inverse to $\bar{d}_{min}(x_i)$, which contains rather low values for the whole video. The threshold value $R(x_i)$ remains unmodified with the initialization value of 18, because the average background distance multiplied by R_{scale} never exceeds the start value. In the case of a static background, this is a perfect behavior, since $R(x_i)$ should only raise if the background dynamics are high. Thus, to prevent false positive detections, the decision threshold is increased. As a first conclusion, we can derive that $\bar{d}_{min}(x_i)$, as theoretically considered, can be used as a metric for background dynamics.

In Figure 6.8, a pixel part of an intermittent moving object is depicted. Again, we see that $T(x_i)$ immediately increases sharply when the image element is labelled as foreground. $T(x_i)$ reaches the upper threshold T_{upper} in approximately ~ 500 frames. This corresponds to a capture length of roughly $\sim 16.6\,s$ at a frame rate of 30 frames/sec. The value is kept until the pixel is re-classified as a background pixel at frame ~ 1675. The drop is smaller

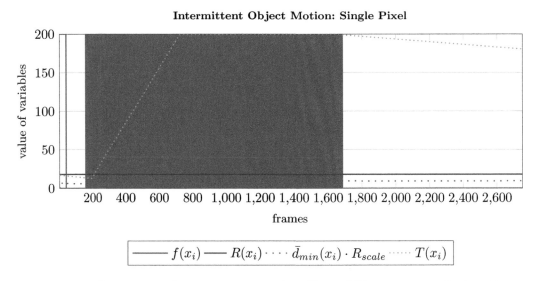

FIGURE 6.8 (See color insert.) Inspection of a pixel ($x = 60, y = 200$) in the video *sofa* of the Change Detection competition.

due to a small value for the decreasing parameter T_{dec} of the inverse update rate $T(x_i)$. Hence, for a reduction of 20 points 1100 frames are necessary, leading to around ~ 10000 pictures ($333\,s$) until the initial value of 18 is reached. Nevertheless, these conclusions are only valid for the existing average distance to the background model $\bar{d}_{min}(x_i)$, which is low in this scenario. Consequently, greater values of $\bar{d}_{min}(x_i)$ are corresponding to an even smaller change or a slower system. This is related due to the inverse relationship between $\bar{d}_{min}(x_i)$ and the inverse update rate $T(x_i)$ as shown in Section 6.3.4 with Equation 6.5. Furthermore, the pixel is correctly labelled as foreground for 1400 frames and declared as part of the background in the correct frame. So the approach to reduce the refresh rate of the foreground pixel, to further recognize the pixel as foreground, is in this case reliable.

One drawback of the algorithm occurs in the case of a background pixel, which is erroneously detected as foreground. Here, the probability of an update of the background model decreases also. Thus, the pixel is kept in the foreground. This behavior appears mostly in *Dynamic Background* scenarios, where most probably background pixels are considered as foreground.

Thus, the last graph in Figure 6.9 depicts a pixel of the video *canoe* of the Change Detection dataset [7]. The pixel is part of waving water in the video and therefore can be stated as dynamic background. According to the ground truth, the pixel should be classified as background for the whole sequence. At first, consider the beginning of the video. The selected pixel is assigned the value of the initialization and then regulated virtue of the dynamics in the video. Note that already the first 40 frames are enough to adjust the state variable $R(x_i)$ properly. Only in the frames before, the pixel is erroneously detected as foreground. $T(x_i)$ decreases during 1100 processed images only by the value of 5, which is a consequence of a high average distance to the background model $\bar{d}_{min}(x_i)$. Thus, as a further effect, the short false recognition as foreground at frame 1020 results in no detectable change of the inverse update rate $T(x_i)$. This is advantageous, would $T(x_i)$ change more rapidly, as in the cases before, the updates of the background would be less frequent. Thereby, the wrong background model remains and the pixel stays as part of the foreground.

Finally, it can be acknowledged that the the use of the average distance to the background model $\bar{d}_{min}(x_i)$ as a metric for background dynamics and detection uncertainty is valid. In addition, the calculation of $T(x_i)$ based on $\bar{d}_{min}(x_i)$ offers the advantage that, the system acts suitable in the case of a intermittent moving foreground objects, as well as in presence of dynamic background.

6.8   General Parameter Settings

Our method consists of few parameters, which have to be tuned for optimal performance. Since we evaluate on the database of the Change Detection Challenge, a multitude of possible scenarios are covered. We seek one unique optimal set of parameters which gives the best performance on the complete dataset. Nevertheless, for certain applications, parameters could be fine tuned for a specific need. Overall, the 9 parameters, detailed below, were tuned on the Change Detection database and should give a quick overview of possible ranges for each parameter.

(a) **N = 35:** N is the number of components of the background model. In Figure 6.10, it can be seen that less advanced scenarios such as *Baseline*, *Shadow* and *Thermal* require as little as 10 background components. Thus, with $N = 10$ the *Overall PWC* is already

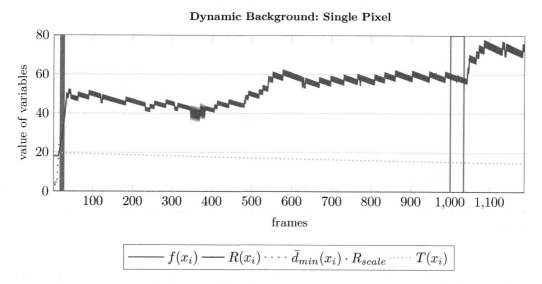

FIGURE 6.9   (See color insert.) Inspection of a pixel ($x = 40$, $y = 200$) in the video *canoe* of the Change Detection competition.

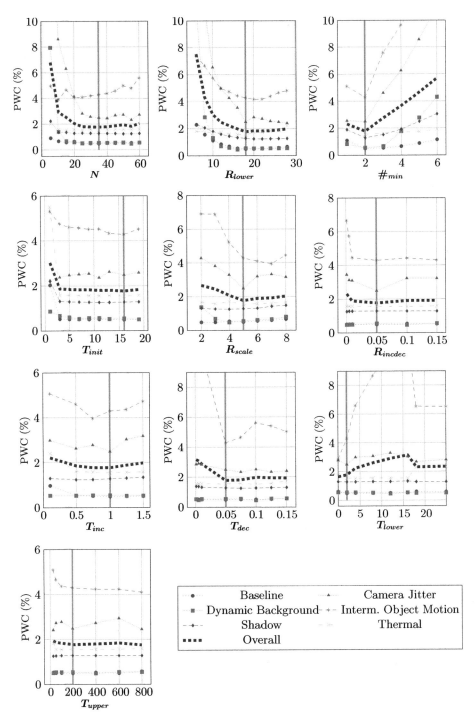

FIGURE 6.10   (See color insert.) Impact on the PWC metric for different values of the parameters of the PBAS.

at 2.5 %. The *Dynamic Background* category is rather stable for more than 15 background models. Additionally, only the failure of the *Camera Jitter* category decreases with higher N. Furthermore, the only PWC, which increases for higher N, is the PWC of the *Inter-*

mittent Object Motion category. A reason could be that, with greater N, the time until foreground objects are learned into the background rises. Finally note that, increasing N further increases memory and computational complexity.

(b) $R_{lower} = 18$: Lower bound and initial value of the decision threshold. Setting it too low leads to ghosts and false positives (thus, low precision), too high will lead to misses (and thus low recall). Setting R_{lower} smaller than 10 is not recommendable for any category, while in the range between $10 - 15$ the PWC decreases slightly in all categories, except *Camera Jitter* and *Intermittent Object Motion*, where the decrease is greater. A value between $18-22$ for the R_{lower} parameter performs well in every category, a further increase may result in noticeable misses, mostly in the *Intermittent Object Motion* category.

(c) $\#_{min} = 2$: The number of components that have to be smaller than $R(x_i)$ in order to set the pixel to be background. An optimum for all categories is found at $\#_{min} = 2$. The same optimal value has been found in [3] and in [18]. A value of $\#_{min} = 1$ leads to a higher PWC in every category, because the decision for a background pixel is too sensitive (low recall). In the opposite case of a higher $\#_{min}$, the precision is lower leading also to a higher PWC. The highest impact can be noted for the categories *Camera Jitter, Intermittent Object Motion* and *Dynamic Background*, the lowest for *Thermal* and *Baseline*. Hence, in these cases the background models are fitting to the current background the most.

(d) $T_{init} = 16$: The initial value of the $T(x_i)$ state variable. This value does not alter the system behavior significantly as long as T_{init} is bigger than T_{lower}. If T_{init} is smaller than T_{lower} no update of the inverse update rate $T(x_i)$ is performed and $T(x_i)$ has a constant value of T_{init}. The salient point at three represents this transition. To neglect the setting of another parameter, T_{init} can be set to the same value as R_{lower}.

(e) $R_{scale} = 5$: Scaling factor in the controller for the decision threshold $R(x_i)$. This controls the equilibrium value $\bar{d}_{min}(x_i) \cdot R_{scale}$ and thus low values lead to low precision, while high values lead to low recall. The slope of the *Overall PWC* metric is at five, while only the *Intermittent Object Motion* category benefits by higher values. Again the categories *Camera Jitter* and *Intermittent Object Motion* are sensitive to this parameter. *Baseline, Shadow* and *Thermal* are almost independent from this value, because these videos have no dynamic in the background. Therefore, a dynamic adjustment is not profitable. In the case of *Dynamic Background* a noticeable increase of the PWC is identifiable at $R_{scale} = 2$, higher values than two lead already to adequate results.

(f) $R_{inc/dec} = 0.05$: The rate, at which the decision threshold $R(x_i)$ is regulated by the controller. Recognition rate is not very sensitive to this value as long as R_{incdec} is bigger than 0.01. The categories *Camera Jitter* and *Intermittent Object Motion* perform worst for low regulation.

(g) $T_{inc} = 1.0$: If x_i is foreground, T_{inc} is the rate at which $T(x_i)$ is increased. Because a priori, foreground is less likely than background, a higher adaptation for foreground is necessary to remain small standing foreground objects. Otherwise, these objects are completely eroded due to the background updates. A very small value has the most impact on *Baseline, Thermal* and on *Intermittent Object Motion* dataset yielding to a higher PWC. Whereas higher values for T_{inc} increase the PWC in the categories *Camera Jitter* and *Intermittent Object Motion*. A reason for the observed influence of T_{inc} in the *Camera Jitter* category is that, wrong classifications of foreground stay foreground due to high T_{inc}. A second reason for the influence in the *Camera Jitter* category is that, the static objects are learned into the background. Then these objects start to move and therefore, the old static positions of the objects are wrongly classified as foreground. This leads, in the case of a high T_{inc}, to a long lasting wrong classification, because the probability of a background update is quite low after a few classifications as background.

(h) $T_{dec} = 0.05$: If x_i is background, T_{dec} is the rate at which $T(x_i)$ is decreased (i.e the rate at which the probability of background update is increased). A very significant change can be observed for the *Intermittent Object Motion*, using small values for T_{dec}, the *PWC* increases over 10 %. The reason is the same as the one mentioned for T_{inc}. Once static objects move again, the old still position of the objects are generally classified as foreground. These areas need to be learned into the background.

After an update of the background model of the foreground pixel due to the neighbour updating process, the former foreground pixel with probably a high $T(x_i)$ is now classified as background. But if T_{dec} is too small, $T(x_i)$ decreases very slowly, thus the probability of an update of a neighbour pixel remains quite low. So the implicit erosion effect executes very slowly. Finally, *Camera Jitter* is also slightly effected by this property.

(i) $\mathbf{T_{lower} = 2}$: Lower bound of $T(x_i)$. In the tested range quite constant. In the case of *Intermittent Object Motion* an increase of *PWC* for high values is observed. The T_{lower} boundary has a salient point (especially for *Intermittent Object Motion*) after the value 16, because the T_{init} parameter is initialized with exact the same value. Thus, the $T(x_i)$ value is always outside the boundary of Equation 6.6. This leads to a constant $T(x_i)$ parameter with the value of T_{init}.

(j) $\mathbf{T_{upper} = 200}$: The upper bound may not be chosen too low. Values around 200 lead to good results. This concludes that a minimum update probability of maximum $\approx 1\%$ is required. The update probability is attained by taking the pixel update and the neighbour pixel update probability to account. For the neighbouring update process, we only consider pixels with a distance of one to the current pixel. Hence, if all eight surrounding pixels are classified as background, there is the chance of an update of $8 \cdot 1/8 \cdot 1/200 = 1/200$.

6.8.1   Influence of Gradient Magnitudes

In this section the effect of different weighting factors between the two features are evaluated. As mentioned in Section 6.4.2, the features are on the one hand the colour values and on the other hand the gradient magnitudes of the image. Therefore, we change the parameter α, which alters the weighting of the gradient features. For these tests the colour features are fixed to a weighting factor of one. The resulting graph is shown in Figure 6.11. By giving higher weight to the gradient component, the overall *PWC* metric is slightly decreasing until an α of ~ 10, afterwards the *PWC* is increasing. For *Intermittent Object Motion* α should at least be five, higher values have no major effect on *PWC*. In the *Camera Jitter* and *Dynamic Background* categories the *PWC* successively increases for higher α. Here, a smaller α should be selected. The *PWC* for *Thermal* improves until an α of 10. In the *Baseline* and *Shadow* category a change can hardly be recognized. The lowest *Overall PWC* is obtained for an α between $5 - 10$ with $PWC \sim 1.7\%$.

Now, we want to state if the additional use of gradient magnitudes lead to a lower *PWC* and hence to a performance gain. For that reason, we evaluated in a second step the performance of the PBAS by using only the colour features with a constant scale of one, but with different values of R_{lower}. Our guess is, if the additional use of gradient magnitudes does not lead to any gain in detection quality, we should observe the same results as with the use of the gradient features. At least with one specifically tuned R_{lower}. The corresponding figure is depicted on the right side of Figure 6.11.

Primarily, the lowest *Overall PWC* is around 1.9 %. Thus, an extending of the features by gradient magnitudes enables the PBAS to find additional foreground image elements, which are missing in the case of a pure consideration of the colour values. A category wise comparison shows that the *Intermittent Object Motion* benefits the most of the additional features, followed by the *Thermal* and *Baseline* categories.

Furthermore, a more general foreground detection could be achieved. In the single feature approach, using only the color values, the lowest *PWC* of the *Thermal* category is received at $R_{lower} = 13$. Whereas the *Camera Jitter* has approximately 0.8 % higher failure rate

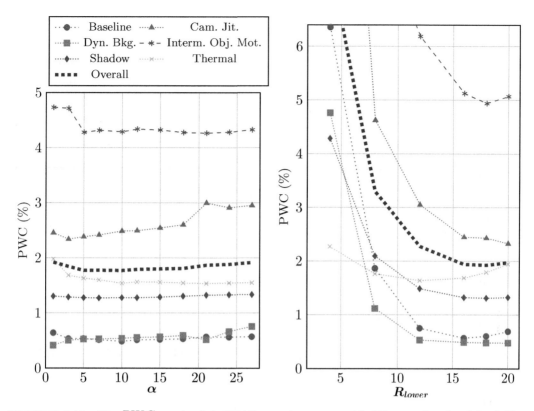

FIGURE 6.11 The PWC metric of the PBAS on each category with different values for alpha (left). On the right side, the PWC metric on each category with an alpha of zero. Thus, in the graph on the right side no magnitude values are used for classification.

than in the best case at $R_{lower} = 20$. This is also true for the categories *Baseline* and *Intermittent Object Motion*, but the differences are smaller. On the contrary, including the gradient magnitudes show that, the optimum α between $5 - 10$ is valid for all categories.

6.8.2 Pre-Processing with Gaussian Filter

In this section the influence on the PWC metric of a Gaussian filter, used in a pre-processing step is evaluated. Therefore, different values for the Gaussian filter size and the standard deviation σ are selected and the corresponding results are depicted in Figure 6.12. Here, each plot shows a specific category with three different filter sizes (3×3, 5×5 and 7×7) for the Gaussian filter at the horizontal axis. For each filter size, six different σ are tested. The dashed black line in each plot represents the PWC value with no pre-filtering at all. It is important to note that these results are received without any post-processing step (like median filtering). Hence, these figures illustrate solely the influence of the pre-processing step.

In the *Baseline* category the smoothing leads to a small gain, if the filter size and particularly the standard deviation σ is small. For instance, a filter with the dimension 7×7 only improves the PWC with a $\sigma = 0.5$. Otherwise the results are worse than without a pre-filter. Almost the same characteristics are seen at the categories *Shadow*, *Thermal* and *Overall*. In the case of *Intermittent Object Motion*, no improvements via Gaussian pre-filtering are noticeable. The results are even worse, consequently no pre-filtering should be preferred.

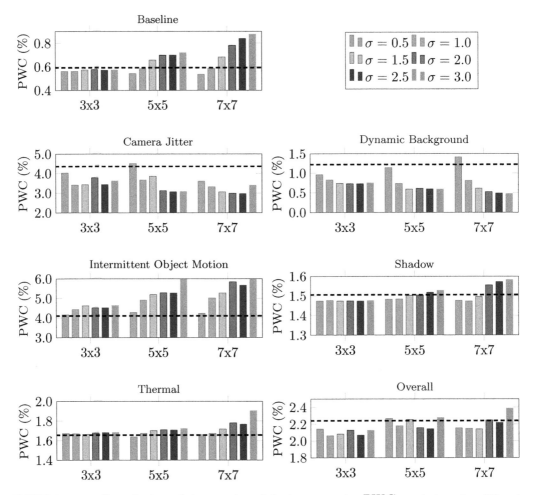

FIGURE 6.12 (See color insert.) An overview of the impact on the PWC metric by using different filter sizes and standard deviations σ in a Gaussian pre-filter step.

Then again, the impact is different in the categories *Camera Jitter* and *Dynamic Background*. The PWC metric can be decreased due to the use of a Gaussian filter. Also greater filter sizes are leading to better results. But the results are not surprising, in the scenarios with static background an additional smoothing led to more bloated boundaries of the foreground silhouette. Consequently, the PWC metric is worse. In the other case, blurring a noisy background can remove false detections verifiable through our results. The less sharp foreground boundaries seem to be neglectable in this case.

6.8.3   Post-Processing with Median Filter

Our proposed method is a pixel by pixel method, which means that the segmentation decision is made independently for each pixel. The resulting output can thus benefit from spatial smoothing, which is done using simple median filtering. Figure 6.13 depicts results for different median filter sizes applied to each category. Therefore, the single results attained without any post-processing are labelled in Figure 6.13 as *no*, the other labels represent different filter sizes for the median.

The PWC is lower for high median filter sizes through all categories. The only exception

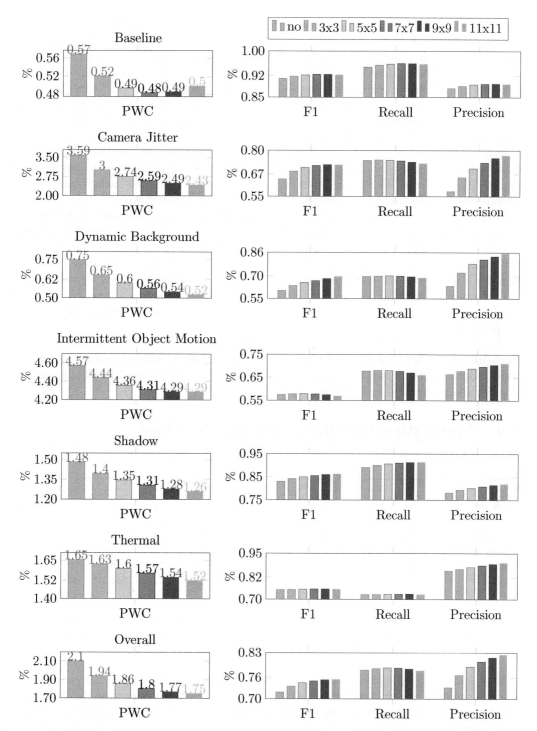

FIGURE 6.13 (See color insert.) Results of each category for PWC, Recall, Precision and $F1$ measure with different sizes of median filter for post-processing.

is the *Baseline* category, where the smallest PWC is obtained with a 9×9 median filter. Higher dimensions of the median filter have mentionable positive influence on the recall metric only in two scenarios, the *Baseline* with around 1 % and the *Shadow* with 2.5 %. On the downside, the recall minimally decreases for the categories *Camera Jitter*, *Dynamic Background* and *Intermittent Object Motion* for greater filter sizes. Thus, a maximum decay of 2.5 % at the *Intermittent Object Motion* category and 2 % at the *Camera Jitter* category are obtained, when comparing no post-processing with the case of a 11×11 median filter. Median filtering is most advantageous for increasing the precision of the segmentation. Hence, it is beneficial for every category, with the lowest improvement in the *Baseline* category with around 2 %. But here the precision of close to 90 % without any filtering has already a high base value, which make it harder to improve it even further. The highest improvements are received in the *Camera Jitter* and *Dynamic Background* categories with around 20 % increase. These categories have also the most alterable background. Hence, background modelling is rather complicated. Bringing back to mind, that we are conducting just a pixel by pixel image to background model comparison, without any further cooperation between pixel neighbourhoods. It is clear that using a median filter for reducing heavily background noise is advantageous. It can be seen that strong median filtering leads to a higher $F1$, because the positive effect on the precision metric is higher than the partly negative progression of the recall metric. Once again, the *Camera Jitter* and *Dynamic Background* category are most affected by the median filtering. In the case of the *Thermal* and the *Intermittent Object Motion* categories, the median filter has only a small effect according to the $F1$ metric.

Finally, it is important to note that while median filtering reduces noise, it also smoothes the blobs and leads to less sharp boundaries. Depending on the scenario, strong median filtering might not be appropriate. Thus, we decided to use a 9×9 median filter for all our experiments in the Change Detection challenge.

6.9   Application Specific Parameter Settings

Parameters	N	R_{init}	$\#_{min}$	T_{init}	R_{scale}	R_{incdec}	T_{inc}	T_{dec}	T_{lower}	T_{upper}
Baseline	35	18	2	16	5	0.05	1	0.05	2	200
Camera Jitter	35	**20**	2	16	**4**	0.05	**0.5**	0.05	**0**	**800**
Dyn. Background	**55**	18	2	**2**	5	0.05	1	**0.075**	2	**25**
Inter. Obj. Mot.	35	18	2	16	5	0.05	1	0.05	2	200
Shadow	**50**	**20**	2	**18**	5	0.05	**0.75**	**0.075**	**10**	200
Thermal	35	**16**	2	**25**	5	0.05	1	0.05	**25**	**600**
Overall	35	18	2	16	5	0.05	1	0.05	2	200

TABLE 6.2   Selected parameter values for the PBAS algorithm to receive the best results for the specific categories of the Change Detection challenge. The bold numbers in the table indicate a parameter modification in relation to the parameters of the Change Detection challenge.

The PBAS has some parameters to tune, which can change the control loop and thereby the segmentation results significantly. Hence, we optimize the parameters of the PBAS taking the former evaluation results to account. In Table 6.2 the best parameters for each categories are shown, whereas Table 6.3 displays the properties of the pre- and post-processing step. Lastly, Table 6.4 shows the gain in certain categories compared to our results at the Change Detection challenge [7]. For the *Baseline* category the standard parameters of the Change Detection challenge are used. Here, an optimization is only achievable by choosing different parameters for the Gaussian pre-filter. So, the PWC decreases from 0.486 % to 0.435 % while the $F1$ metric marginal changes from 92.42 % to 93.02 %.

Having videos with jitter like in the *Camera Jitter* category, a reduced blurring, a higher value for the threshold R_{lower}, a greater filter size for the median filter in the post-processing step and most important half the scale of the T_{inc} lead to an increased $F1$. Thus, the $F1$ changes from formerly 72.2 % to 74.32 % and the PWC decreases from 2.49 % to 2.11 %. The biggest improvement due to the category wise optimization is obtained for the *Dynamic Background* category. Therefore, the N is increased to 55, which also extends the memory usage. In a second step, the dynamic inverse update probability $T(x_i)$ is adjusted. $T(x_i)$ initializes with a low value of two and also T_{upper} has just a value of 25. Consequently, there is always a good chance of a background update. The parameters of the decision threshold $R(x_i)$ remain the same. Last but not least, a strong Gaussian blurring and a big filter size for the median post-processing improve the results excellently. This is depicted in Table 6.4. The *Intermittent Object Motion* category can be separated into two scenarios. The first sce-

Category	Pre-processing		Post-processing
	Filter size	σ	Filter size
Baseline	7×7	0.5	7×7
Camera Jitter	3×3	0.5	11×11
Dyn. Background	7×7	2.5	13×13
Inter. Obj. Motion	7×7	0.5	7×7
Shadow	3×3	3.0	9×9
Thermal	3×3	0.5	11×11
Overall	7×7	0.5	9×9

TABLE 6.3 Shows the best parameters for the depicted categories. On the one hand the best parameters for the pre-processing step with the formerly mentioned Gaussian smoothing filter are presented. On the other hand the optimal properties for the post-processing step using a median filter are shown.

nario holds a long lasting still object, which is already there in the first frame of the video and then suddenly starts to move. The second scenario handles an object, which is moving into the scene and then intermits at a certain position and keeps standing there for a few hundred frames.

It is important to note that from the view of optimizing the background model, these scenarios are contrary. In the first case, the background model should be updated more frequently, even when the background is almost static. In the second case, foreground objects which are still, should remain in the foreground as long and complete as possible. For that reason, an optimization of the update procedure is almost impossible. Nevertheless, through the use of different parameter values in the pre-filter step and a smaller median filter for post-processing, the $F1$ and also the PWC could be improved. The obtained results at the Change Detection challenge for the *Shadow* category seem to be well optimized already. So, no change in the parameter settings is favourable. For superior results in the *Thermal* category mainly the inverse update threshold $T(x_i)$ needs to be altered. An improvement, compared to the results of the Change Detection challenge, of approximately 5 % in $F1$ measure is recognizable.

Lastly, we tried to improve the *Overall* performance. In this case, like in the official challenge, only one parameter set is used for all categories. Due to the optimization of the parameters for the *Overall* category at the Change Detection challenge, no better parameters could be spotted, but a different pre-filter is advantageous.

Scenarios		Recall (%)	Specificity (%)	PWC (%)	Precision (%)	F1 (%)
	old	95.94	99.70	0.4858	89.41	92.42
Baseline		↑ +0.01%	↑ +0.04%	↓ -0.05%	↑ +1.09%	↑ +0.60%
	new	95.95	99.74	0.4351	90.50	93.02
	old	73.73	98.38	2.488	75.86	72.20
Camera Jitter		↓ -1.13%	↑ +0.48%	↓ -0.38%	↑ +4.08%	↑ +2.12%
	new	72.59	98.86	2.105	79.95	74.32
	old	69.55	99.89	0.5394	83.26	68.29
Dynamic Background		↑ +8.52%	↑ +0.04%	↓ -0.23%	↑ +5.13%	↑ +12.88%
	new	78.07	99.93	0.3127	88.39	81.12
	old	67.00	97.51	4.287	70.45	57.45
Inter. Object Motion		↑ +2.47%	↑ +0.31%	↓ -0.41%	↑ +2.27%	↑ +5.34%
	new	69.47	97.82	3.879	72.72	62.80
	old	91.33	99.04	1.275	81.43	85.97
Shadow		↓ -3.16%	↑ +0.22%	↓ -0.07%	↑ +4.80%	↑ +1.13%
	new	88.17	99.26	1.204	86.22	87.10
	old	72.83	99.34	1.540	89.22	75.56
Thermal		↑ -4.91%	↓ -0.07%	↓ -0.07%	↓ +1.52%	↑ +5.48%
	new	77.74	99.26	1.471	87.70	81.04
	old	78.40	98.98	1.769	81.60	75.32
Overall		↓ -0.90%	↑ +0.15%	↓ -0.08%	↑ +1.53%	↑ +1.02%
	new	77.50	99.13	1.688	83.13	76.34

TABLE 6.4 Results of the PBAS on all six scenarios and *Overall*. The rows titled as "old" contain the results of the Change Detection challenge. These results can also be verified at the official homepage of the challenge. The rows named as "new" represent the improvements obtained due to category-wise optimization. The line in-between indicates the absolute difference between the "old" and the "new" results. Whereas the numbers in green color symbolize an improvement and the red color a worsening.

6.10   Conclusion

We have presented an advanced statistical method for detecting foreground. We started with the basic idea of just using a simple parameterless background model and improved through two controllers with feedback loops to adapt the parameters to the current video dynamics by changing the decision threshold and the background update parameter. Furthermore, we wanted to verify and understand the system dynamics more deeply. Therefore, we observed carefully selected pixels for three contrary scenarios. In those cases we could prove that the actual made assumptions in the theoretical part are true. Then, we stated that the use of the gradient magnitudes as a feature for the classification, in addition to the color values, improve the detection performance. Afterwards, an exhaustive evaluation of the PBAS algorithm has been performed. For that reason, also the influence of basic filters for pre- and post-processing have been tested in conjunction with the PBAS. These evaluations have been carried out on well sorted categories of common video scenes by using the Change Detection dataset. Thus finally, it was possible to present application specific recommendations for pre- and post-processing as well as for the PBAS parameters.

Future work could focus on explicit modelling of shadow as well as to create a framework with an additional detection of objects. Having an accurate recognition of static and moving objects, the parameters of the PBAS could be adjusted for dynamic background cases by on the other site handling stopped moving objects equally well. Additionally, an integration of probabilistic Superpixel Markov Random Fields [13] could also improve the actual results.

The current implementation of the PBAS utilizes the same number of cores as number of channels in the image. For future implementations a way to make the processing time almost independent from the number of cores is beneficial. This includes a different way of the calculation of the distance between background component and current image. Thus, for processing a multi-channel image in one step, the use of the Euclidean distance instead of the absolute distance would be a straight forward approach.

Furthermore, the procedure to initialize the background model is not sophisticated, as we just save the first N frames as the individual background components. Here, an approach which considers the classification of the pixels already in the initialization phase of the background model could improve the results.

The PBAS consists of a few parameters. Although we show in this chapter recommendable parameter sets for a range of applications, it would be advantageous to reduce them or develop a way to train them automatically. ViBe and PBAS prove that random update rules lead to great results in a variety of applications. Nevertheless, an approach which does not completely neglect the estimated properties of the neighbor pixels could increase the performance even more. The neighborhood of the pixel can be involved in the classification step as well as for the background update rule. Here, also a dynamic pixel by pixel re-definition of the neighborhood size could be included in the controlling loop.

Lastly, the maximum size of the background model is fixed to N for the whole image. An intelligent way to enhance the background model in areas of high background dynamics and decrease the number of background components in static regions of the image could improve the flexibility of the algorithm. Therefore, a broader range of scenarios would be processable with the same parameter set.

References

1. R. Abbott and L. Williams. Multiple target tracking with lazy background subtraction and connected components analysis. *Machine Vision and Application*, 20(2):93–101, 2009.

2. F. El Baf, T. Bouwmans, and B. Vachon. A fuzzy approach for background subtraction. In *IEEE International Conference Image Processing, ICIP 2008*, pages 2648–2651, 2008.

3. O. Barnich and M. Van Droogenbroeck. Vibe: A universal background subtraction algorithm for video sequences. *IEEE Transactions on Image Processing*, 20(6):1709 –1724, June 2011.

4. T. Bouwmans. Recent advanced statistical background modeling for foreground detection: A systematic survey. *Recent Patents on Computer Science*, 4(3), 2011.

5. D. Butler, V. Bove, and S. Sridharan. Real-time adaptive foreground/background segmentation. *EURASIP J. Appl. Signal Process.*, 2005:2292–2304, January 2005.

6. A. Elgammal, D. Harwood, and L. Davis. Non-parametric model for background subtraction. In *European Conference on Computer Vision, ECCV 2000*, pages 751–767, London, UK, UK, 2000. Springer-Verlag.

7. N. Goyette, P. Jodoin, F. Porikli, J. Konrad, and P. Ishwar. Changedetection.net: A new change detection benchmark dataset. In *Computer Vision and Pattern Recognition Workshops (CVPRW), 2012 IEEE Computer Society Conference on*, 2012. `http://www.changedetection.net`.

8. M. Hofmann, P. Tiefenbacher, and G. Rigoll. Background segmentation with feedback: The pixel-based adaptive segmenter. *IEEE Workshop on Change Detection*, 6(7):8, 2012.

9. K. Kim, T. Chalidabhongse, D. Harwood, and L. Davis. Real-time foreground / background segmentation using codebook model. *Real-Time Imaging*, 11(3):172–185, 2005. Special Issue on Video Object Processing.

10. A. Morde, X. Ma, and S. Guler. Learning a background model for change detection. In *Computer Vision and Pattern Recognition Workshops, CVPRW 2012*, pages 15–20. IEEE, June 2012.

11. M. Piccardi. Background subtraction techniques: a review. In *IEEE International Confer-*

      ence Systems, Man and Cybernetics, SMC 2004, volume 4, pages 3099–3104, October 2004.

12. R. Radke, S. Andra, O. Al-Kofahi, and B. Roysam. Image change detection algorithms: a systematic survey. *IEEE Transactions on Image Processing*, 14(3):294 –307, March 2005.

13. A. Schick, M. Bauml, and R. Stiefelhagen. Improving Foreground Segmentations with Probabilistic Superpixel Markov Random Fields. In *Change Detection CVPR*. IEEE Workshop, 2012.

14. K. Smith, P. Quelhas, D. Gatica-Perez, K. Smith, P. Quelhas, and D. Gatica-Perez. Detecting abandoned luggage items in a public space. In *IEEE Performance Evaluation of Tracking and Surveillance Workshop (PETS)*, 2006.

15. C. Stauffer and W. Grimson. Adaptive background mixture models for real-time tracking. In *IEEE International Conference on Computer Vision and Pattern Recognition, CVPR 1999*, volume 2, pages 637–663, 1999.

16. A. Tavakkoli, M. Nicolescu, G. Bebis, and M. N. Nicolescu. Non-parametric statistical background modeling for efficient foreground region detection. *Machine Vision and Applications*, 20(6):395–409, 2009.

17. K. Toyama, J. Krumm, B. Brumitt, and B. Meyers. Wallflower: principles and practice of background maintenance. In *IEEE International Conference on Computer Vision, ICCV 1999*, volume 1, pages 255–261, Los Alamitos, CA, USA, 1999. IEEE.

18. H. Wang and D. Suter. A consensus-based method for tracking: Modelling background scenario and foreground appearance. *Pattern Recognition*, 40(3):1091–1105, 2007.

19. H. Zhang and D. Xu. Fusing color and gradient features for background model. In *International Conference on Signal Processing*, volume 2, 2006.

7

ViBe: A Disruptive Method for Background Subtraction

Marc Van Droogenbroeck
University of Liège, Belgium

Olivier Barnich
EVS Broadcast, Belgium

7.1 Introduction

The proliferation of video surveillance cameras, which account for the vast majority of cameras worldwide, has resulted in the need to find methods and algorithms for dealing with the huge amount of information that is gathered every second. This encompasses processing tasks, such as raising an alarm or detouring moving objects, as well as some semantic tasks like event monitoring, trajectory or flow analysis, counting people, etc.

Many families of tools related to motion detection in videos are described in the literature (see [21] for tools related to the visual analysis of humans). Some of them track objects, frame by frame, by following features in the video stream. Others operate by comparing the current frame with a static background frame, pixel by pixel. This is the basis of *background subtraction*, whose principle is very popular for fixed cameras. The purpose of background subtraction is therefore formulated as the process to extract moving objects by detecting zones where a significant change occurs. Moving objects are referred to as the *foreground*, while static elements of the scene are part of the *background*. Practice however shows that this distinction is not always obvious. For example, a static object that starts moving, such as a parked car, creates both a hole in the background and an object in the foreground. The hole, named *ghost*, is usually wrongly assigned to the foreground despite that it should be discarded as there is no motion associated to it. Another definition of the background considers that background pixels should correspond to values visible most of the time, or in statistical terms, whose probabilities are the highest. But this poses problems when the background is only visible for short periods of time, like a road with heavy traffic. The diversity of scenes and specific scenarios explains why countless papers have been devoted to background subtraction, as well as additional functionalities such as the detection of shadows or the robustness to camera jitter, etc.

In this chapter, we elaborate on a background subtraction technique named ViBe, that has been described in three previous papers [1,2,37], and three granted patents [34–36]. After some general considerations, we describe the principles of ViBe in Section 7.2. We review the innovations introduced with ViBe: a background model, an update mechanism composed of random substitutions and spatial diffusion, and a fast initialization technique. We also discuss some enhancements that broaden up the possibilities for improving background subtraction techniques, like the distinction between the segmentation map and the updating mask, or a controlled diffusion mechanism. Section 7.3 discusses the computational cost of ViBe. In particular, we introduce the notion of *background subtraction complexity factor* to express the speed of the algorithm, and show that ViBe has a low complexity factor. Section 7.4 concludes the chapter.

7.1.1　What Are We Looking For?

The problem tackled by background subtraction involves the comparison of the last frame of a video stream with a reference background frame or model, free of moving objects. This comparison, called *foreground detection*, is a classification process that divides the image into two complementary sets of pixels (there is no notion of object at this stage): (1) the *foreground*, and (2) the *background*. This binary classification process is illustrated in Figure 7.1.

FIGURE 7.1　The image of a road with moving cars (from the baseline/highway directory of the "Change Detection" dataset [10]) and the corresponding binary segmentation map (produced by ViBe). Foreground pixels are drawn in white.

It results in a binary output map, called *segmentation map* hereafter. For humans, it might seem obvious to delineate moving objects, because humans incorporate knowledge from the semantic level (they know what a car is, and understand that there are shadows). However, such knowledge is not always available for computers to segment the image. Therefore segmentation is prone to classification errors. These errors might disturb the interpretation, but as long as their number is small, this is acceptable. In other words, there is no need for a perfect segmentation map. In addition, as stated in [27], it is almost impossible to specify a gold-standard definition of what a background subtraction technique should detect as a foreground region, as the definition of foreground objects relates to the application level. Therefore, initiatives such as the i-LIDS evaluation strategy [13] are directed more towards the definition of scenarios than the performance of background subtraction.

7.1.2 Short Review of Background Subtraction Techniques

Many background subtraction techniques have been proposed, and several surveys are devoted to this topic (see for example [3, 5, 9]). Background subtraction requires to define an underlying background model and an update strategy to accommodate for background changes over time. One common approach to background modeling assumes that background pixel values (called background samples) observed at a given location are generated by a random variable, and therefore fit a given probability density function. Then it is sufficient to estimate the parameters of the density function to be able to determine if a new sample belongs to the same distribution. For example, one can assume that the probability density function is a gaussian and estimate its two parameters (mean and variance) adaptively [39]. A simpler version of it considers that the mean which, for each pixel, is stored in a memory and considered as the background model, can be estimated recursively by

$$B_t = \alpha B_{t-1} + (1 - \alpha) I_t, \tag{7.1}$$

where B_t and I_t are the background model and current pixel values at time t respectively, and α is a constant ($\alpha = 0.05$ is a typical value). We name this filter the *exponential filter* because B_t adapts to I_t exponentially. With this filter, the decision criterion is simple: if I_t is close to B_{t-1}, that is if the difference between them is lower than a fixed threshold, $|I_t - B_{t-1}| \leq T$, then I_t belongs to the background. This filter is one of the simplest algorithms for background subtraction, except the algorithm that uses a fixed background. There is one comparison and decision per pixel, and the background model is updated once per processed frame. To evaluate the computational speed of ViBe, we use this filter as one reference in Section 7.3.

The exponential filter is simple and fast. When it comes to computational speed and embedded processing, methods based on $\Sigma - \Delta$ (sigma-delta) motion detection filters [18,20] are popular too. As in the case of analog-to-digital converters, a $\Sigma - \Delta$ motion detection filter consists of a simple non-linear recursive approximation of the background image, which is based on comparisons and on an elementary increment/decrement (usually -1, 0, and 1 are the only possible updating values). The $\Sigma - \Delta$ motion detection filter is therefore well suited to many embedded systems that lack a floating point unit.

Unimodal techniques can lead to satisfactory results in controlled environments while remaining fast, easy to implement, and simple. However, more sophisticated methods are necessary when dealing with videos captured in complex environments where moving background, camera egomotion, and high sensor noise are encountered [3]. Over the years, increasingly complex pixel-level algorithms have been proposed. Among these, by far the most popular is the Gaussian Mixture Model (GMM). First presented in [32], this model consists of modeling the distribution of the values observed over time at each pixel by a weighted mixture of gaussians. Since its introduction, the model has gained vastly in popularity among the computer vision community (see the excellent extended review by Bouwmans [4]), and it is still raising a lot of interest as authors continue to revisit the method.

One of the downsides of the GMM algorithm resides in its strong assumptions that the background is more frequently visible than the foreground and that its variance is significantly lower. None of this is valid for every time window. Furthermore, if high- and low-frequency changes are present in the background, its sensitivity cannot be accurately tuned and the model may adapt to the targets themselves or miss the detection of some high speed targets. Also, the estimation of the parameters of the model (especially the variance) can become problematic in real-world noisy environments. This sometimes leaves one with no other practical choice than to use a fixed variance. Finally, it should be noted that the statistical relevance of a gaussian model is debatable as some authors claim that natural images exhibit non-gaussian statistics [31].

Because the determination of parameters can be problematic, and in order to avoid the difficult question of finding an appropriate shape for the probability density function, some authors have turned their attention to non-parametric methods to model background distributions. One of the strengths of non-parametric kernel density estimation methods is their ability to circumvent a part of the delicate parameter estimation step due to the fact that they rely on sets of pixel values observed in the past to build their pixel models. For each pixel, these methods build a histogram of background values by accumulating a set of real values sampled from the pixel's recent history. These methods then estimate the probability density function with this histogram to determine whether or not a pixel value of the current frame belongs to the background. Non-parametric kernel density estimation methods can provide fast responses to high-frequency events in the background by directly including newly observed values in the pixel model. However, the ability of these methods to successfully handle concomitant events evolving at various speeds is questionable since they update their pixel models in a first-in first-out manner. This has led some authors to represent background values with two series of values or models: a short term model and a long term model [8]. While this can be a convenient solution for some situations, it leaves open the question of how to determine the proper time interval. In practical terms, handling two models increases the difficulty of fine-tuning the values of the underlying parameters. ViBe incorporates a smoother lifespan policy for the sampled values. More importantly, it makes no assumption on the obsolescence of samples in the model. This is explained in Section 7.2.

The background subtraction method that is the closest to ViBe was proposed in [38]. Wang and Suter base their technique on the notion of "consensus". They keep the 100 last observed background values for each pixel and classify a new value as background if it matches most of the values stored in the pixel's model. ViBe has a similar approach except that the amount of stored values is limited to 20, thanks to a clever selection policy.

While mixtures of gaussians and non-parametric models are background subtraction families with the most members, other techniques exist. In [15], each pixel is represented by a codebook, which is a compressed form of background model for a long image sequence. Codebooks are believed to be able to capture background motion over a long period of time with a limited amount of memory.

A different approach to background subtraction considers that it is not important to tune a background subtraction, but instead that the results of the classification step can be refined. A two-level mechanism based on a classifier is introduced in [17]. A classifier first determines whether an image block belongs to the background. Appropriate blockwise updates of the background image are then carried out in the second stage, depending on the results of the classification. Schick et al. [28] use a similar idea and define a superpixel Markow random field to post-process the segmentation map. They show that their technique improves the performance of background subtraction most of the time; only results of background subtraction techniques that already produce accurate segmentations are not improved.

This raises the general question of applying a post-processing filter to the segmentation map. Parks et al. [25] consider several post-processing techniques that can be used to improve upon the segmentation maps: noise removal, blob processing (morphological closing is used to fill internal holes and small gaps whereas area thresholding is used to remove small blobs), etc. Their results indicate that the performance is improved for square kernel morphological filter as long as the size is not too large. The same yields for area thresholding where a size threshold well below the size of the smallest object of interest (e.g. 25%) is recommended. The post-processing is also interesting from a practical perspective. When it is applied, the segmentation step becomes less sensitive to noise, and subsequently to the exact values of the method's parameters. To produce the right-hand side image of

Figure 7.1, we have applied a 5×5 close/open filter and a median filter on the segmentation map. For ViBe, Kryjak and Gorgon [16] propose to use a 7×7 median filter to post-process the segmentation map, and two counters per pixel, related to the consecutive classification as foreground or background, to filter out oscillating pixels.

7.1.3 Evaluation

The difficulty of assessing background subtraction algorithms originates from the lack of a standardized evaluation framework; some frameworks have been proposed by various authors but mainly with the aim of pointing out the advantages of their own method. The "Change Detection" (CD) initiative is a laudable initiative to help comparing algorithms. The CD dataset contains 31 video sequences panning a large variety of scenarios and includes groundtruth maps. The videos are grouped in 6 categories: baseline, dynamic background, camera jitter, intermittent object motion, shadow, and thermal. The CD web site (`http://www.changedetection.net`) computes several metrics for each video separately.

These metrics rely on the assumption that the segmentation process is similar to that of a binary classifier, and they involve the following quantities: the number of True Positives (TP), which counts the number of correctly detected foreground pixels; the number of False Positives (FP), which counts the number of background pixels incorrectly classified as foreground; the number of True Negatives (TN), which counts the number of correctly classified background pixels; and the number of False Negatives (FN), which accounts for the number of foreground pixels incorrectly classified as background. As stated by Goyette et al. [10], finding the right metric to accurately measure the ability of a method to detect motion or change without producing excessive false positives and false negatives is not trivial. For instance, the *recall* metric favors methods with a low False Negative Rate. On the contrary, the *specificity* metric favors methods with a low False Positive Rate. Having the entire precision-recall tradeoff curve or the ROC curve would be ideal, but not all methods have the flexibility to sweep through the complete gamut of tradeoffs. In addition, one cannot, in general, rank-order methods based on a curve.

One of the encountered difficulties is that there are four measurable quantities (TP, TN, FP, FN) for a binary classifier, and a unique criterion based on them will not reflect the trade-offs when parameter values are to be selected for the classifier. This is the reason why the characterization by multiple criteria helps determine the optimal parameter values. Another difficult question is the comparison of techniques. Again, there is a difference in optimizing a technique and the willingness to compare techniques. In this chapter, we do not enter the discussion of comparing ViBe to other background subtraction techniques (see [10] for a comparative study of techniques, including two variants of ViBe: PBAS and ViBe+). We prefer to concentrate on aspects proper to ViBe and to discuss some choices. Because ViBe has a conservative updating policy, it tends to minimize the amount of False Positives. As a consequence, precision is high. But simultaneously, ViBe has an inherent mechanism to incorporate ghosts and static objects into the background. For some applications, this results in an increase of the number of False Negatives. If foreground objects only represent a very small part of the image, which is common in video-surveillance, the amount of True Positives is small, and then the recall is low. Therefore, we select the Percentage of Bad Classifications (PBC), that is a combination of the four categories, as a compromise and as the evaluation criteria. It is defined as

$$PBC = 100 \times \frac{FN + FP}{TP + FN + FP + TN}. \tag{7.2}$$

Note the PBC assesses the quality of the segmentation map. For some applications however, the quality of segmentation is not the main concern. For example, the exact shape of the segmentation map is not relevant if the role of the background subtraction is to raise an alarm in the presence of motion in the scene. Then a criterion based on the rate of False Negatives might be more appropriate.

7.2   Description of ViBe

The background/foreground classification problem that is handled by background subtraction that requires defining a model and a decision criterion. In addition, for real time processing of continuous video streams, pixel-based background subtraction techniques compensate for the lack of spatial consistency by a constant updating of their model parameters.

In this section, we first explain the rationale behind the mechanisms of ViBe. Then, we describe the model (and some extensions), which is intrinsically related to the classification procedure, an updating strategy, and initialization techniques.

7.2.1   Rationale

ViBe is a technique that collects background samples to build background models, and introduces new updating mechanisms. The design of ViBe was motivated by the following rules:

- Many efficient vision-based techniques that classify objects, including the successful pose recognition algorithm of the Kinect [29], operate at the pixel level. Information at the pixel level should be preferred to aggregated features, because the segmentation process is local and dealing with pixels directly broadens up possibilities to optimize an implementation for many hardware architectures.

- We prefer the approach that collects samples from the past for each pixel, rather than building a statistical model. There are two reasons for that. Firstly, because sample values have been observed in the past, their statistical significance is higher than values that have never been measured. Secondly, the choice of a model introduces a bias towards that model. For example, if one assumes that the probability distribution function of a pixel is gaussian, then the method tries to determine its mean and variance. It does not test the relevance of the model itself, even if the distribution appears to be bi-modal.

- The number of collected samples should be kept relatively small. Taking larger numbers both increases the computational load and memory usage. In ViBe, we keep 20 background samples for each pixel. This might seem to be a large figure, but remember that a typical video framerate is 25 or 30 Hz; keeping 20 background samples therefore represents a memory of less than 1 second.

- There should be no "planned obsolescence" notion for samples. The commonly (and exclusively) adopted substitution rule replaces oldest samples first or reduces their associated weight. This assumes that old samples are less relevant than newer samples. But this assumption is questionable, as long as the model only contains background samples and is not corrupted by foreground samples. In ViBe, old and recent values are considered equally when they are replaced.

- A mechanism to ensure spatial consistency should be foreseen. Pixel-based techniques tend to ignore neighboring pixel values in their model. This allows for a sharp detection capability, but provides no warranty that neighboring decisions are consistent. ViBe proposes a new mechanism that inserts background values in the models of neighboring pixels. Once the updating policy decides to replace a value of the model with the current value of the pixel, it also inserts that value in the model of one of the neighboring pixels.

- The decision process to determine if a pixel belongs to the background should be kept simple because, for pixel-based classification approaches, the computational load is directly proportional to the decision process. As shown in Section 7.3, the overall computational cost of ViBe is about that of 6 comparisons per pixel, which is very low compared to other background subtraction strategies.

In the next section, we present the details of ViBe. There are differences with the original version of ViBe (as described in [2]) that we point out when appropriate.

7.2.2 Pixel Classification

Classical approaches to background subtraction and most mainstream techniques rely on a probability density function (pdf) or statistical parameters of the underlying background generation process. But the question of their statistical significance is rarely discussed, if not simply ignored. In fact, the choice of a particular form for the pdf inevitably introduces a bias towards that pdf, when in fact there is no imperative to compute the pdf as long as the goal of reaching a relevant background segmentation is achieved. An alternative is to consider that one should enhance statistical significance over time, and one way to proceed is to build a model with real observed pixel values. The underlying assumption is that this makes more sense from a stochastic point of view, as already observed values should have a higher probability of being observed again than would values not yet encountered.

If we see the problem of background subtraction as a classification problem, we want to classify a new pixel value with respect to previously observed samples by comparing them. A major, and important difference in comparison with existing algorithms, is that when a new value is compared to background samples, it should be close to only a few sample values to be classified as background, instead of the majority of all sample values. Some authors believe that comparing a value with some sample values is equivalent to a non-parametric model that sums up probabilities of uniform kernel functions centered on the sample values. That is an acceptable statement for the model of [38], that considers 100 values and require a close match with at least 90 values, because the model then behaves as a pdf whose values are compared to the new value to determine if the probability is larger than 0.9. However, this reasoning is not valid for ViBe, because with the standard set of values, we only require 2 matches out of 20 possible matches, which would correspond to a probability threshold of 0.1. One can hardly claim that such a low probability is statistically significant, and consequently that the model is that of a pdf. Of course, if one trusts the values of the model, it is crucial to select background pixel samples carefully. The insertion of pixels in the background therefore needs to be conservative, in the sense that only *sure* background pixels should populate the background models.

To be more formal, let us denote by $v(p)$ the value of a feature, for example the RGB components, taken by the pixel located at p in the image, and by v_i a background feature with an index i. Each background pixel at p is then modeled by a collection of N background samples

$$\mathcal{M}(p) = \{v_1, v_2, \ldots, v_N\}, \tag{7.3}$$

taken in previous frames.

In order to classify a new pixel feature value $v(p)$, we compare it to all the values contained in the model $\mathcal{M}(p)$ with respect to a distance function $d()$ and a distance threshold T. The classification procedure is detailed in Algorithm 7.1.

Algorithm 7.1 - Pixel Classification Algorithm of ViBe

```
1   int width;                          // width of the image
2   int height;                         // height of the image
3   byte image[width*height];           // current image
4   byte segmentationMap[width*height]; // classification result
5
6   int numberOfSamples = 20;      // number of samples per pixel
7   int requiredMatches = 2;       // #_min
8   int distanceThreshold = 20;
9
10  byte samples[width*height][numberOfSamples]; // background model
11
12  for (int p = 0; p < width*height; p++) {
13    int count=0, index=0, distance=0;
14
15    // counts the matches and stops when requiredMatches are found
16    while ( (count < requiredMatches) && (index < numberOfSamples) ) {
17      distance = getDistanceBetween(image[p], samples[p][index]);
18      if (distance < distanceThreshold) {
19        count++; }
20      index++;
21    }
22
23    // pixel classification
24    if (count < requiredMatches)
25      segmentationMap[p] = FOREGROUND_LABEL;
26    else
27      segmentationMap[p] = BACKGROUND_LABEL;
28  }
```

A typical choice for $d()$ is the Euclidean distance, but any metric that measures a match between two values is usable. Benezeth et al. [3] compared several metrics for RGB images. They conclude that four of the six metrics they tested, including the common Euclidean distance, globally produce the same classification results; only the simplest zero and first order distances are less precise. In many video-surveillance applications, $v(p)$ is either the luminance or the RGB components. For the luminance, the distance $d()$ is reduced to the absolute difference between the pixel value and a sample value. As shown in the algorithm, it is important to note that, for each pixel, the classification process stops when $\#_{min}$ matches have been found; to the contrary of non-parametric methods, there is no need to compare $v(p)$ to every sample. This speeds up the classification process considerably. Discussions about the computational load of ViBe are further given in Section 7.3.

7.2.3 Updating

The classification step of our algorithm compares the current pixel feature value at time t, $v^t(p)$, directly to the samples contained in the background model $\mathcal{M}^{t-1}(p)$, built at time $t-1$. Consequently, questions regarding *which* samples have to be memorized by the model and for *how long* are essential.

Many practical situations, like the response to sudden lighting changes, the presence of new static objects in the scene, or changing backgrounds, can only be addressed correctly if the updating process incorporates mechanisms capable to cope with dynamic changes in the scene. The commonly adopted approach to updating discards and replaces old values after a number of frames or after a given period of time (typically about a few seconds); this can be seen as a sort of *planned obsolescence*. Although universally applied, there is no evidence that this strategy is optimal. In fact, ViBe proposes other strategies, based on random substitutions, that when combined, improve the performance over time.

General discussions on an updating mechanism

Many updating strategies have been proposed [4] and, to some extent, each background subtraction technique has its own updating scheme. Updating strategies are either intrinsically related to the model, or they define methods for adapting parameters over time. Considering the model, there are two major updating techniques. Parks et al. [25] distinguish between *recursive* techniques, which maintain a single background model that is updated with each new video frame, and *non-recursive* techniques which maintain a buffer of n previous video frames and estimate a background model based solely on the statistical properties of these frames.

Another typology relates to the segmentation results. *Unconditional* updating or *blind* updating considers that, for every pixel, the background model is updated, whereas in *conditional* updating, also called *selective* or *conservative* updating, only background pixels are updated. Both updating schemes are used in practice. Conservative updating leads to sharp foreground objects, as the background model will not become polluted with foreground pixel information. However the major inconvenience of that approach is that false positives (pixels incorrectly classified as foreground values) will continually be misclassified as the background model will never adapt to it. Wang et al. [38] propose to operate at the blob level and define a mechanism to incorporate pixels in the background after a given period of time. To the contrary, blind updating tends to remove static objects and requires additional care to keep static objects in the foreground.

Other methods, like kernel-based pdf estimation techniques, have a softer approach to updating. They are able to smooth the appearance of a new value by giving it a weight prior to inclusion. For example, Porikli et al. [26] define a GMM method and a Bayesian updating mechanism, to achieve accurate adaptation of the models.

With ViBe, we developed a new conservative updating method that incorporates three important mechanisms: (1) a memoryless update policy, which ensures a smooth decaying lifespan for the samples stored in the background pixel models, (2) a random time subsampling to extend the time windows covered by the background pixel models, and (3) a mechanism that propagates background pixel samples spatially to ensure spatial consistency and to allow the adaptation of the background pixel models that are masked by the foreground. These components are discussed in detail in [2]. We review their key features in the next three subsections, and provide some additional comments.

A memoryless updating policy

Many sample-based methods use first-in first-out policies to update their models. In order to deal properly with wide ranges of events in the scene background, Wang et al. [38] propose the inclusion of large numbers of samples in pixel models. Others authors [8, 40] mix two temporal sub-models to handle both fast and slow changes. These approaches are effective, but the management of multiple temporal windows is complex. In addition, they might not be sufficient for high or variable framerates.

From both practical and theoretical perspectives, we believe that it is more appropriate for background values to fade away smoothly. In terms of a non-parametric model such as ViBe, this means that the probability for a value to be replaced over time should decay progressively. Therefore, we define a policy that replaces sample values chosen randomly, according to a uniform probability density function. This guarantees a monotonic decay of the probability for a sample value to remain inside the set of samples, as established in [2]. There are two remarkable consequences to this updating policy: (1) there is no underlying notion of time window in the collection of samples, and (2) the results of the background subtraction are not deterministic anymore. In other words, if the same video sequence is processed several times, all the results will be slightly, but not significantly, different. Remember that this approach is necessarily combined with a conservative updating policy, so that foreground values should never be absorbed into the background models. The conservative updating policy is necessary for the stability of the process to avoid models diverginng over time.

The complete updating algorithm of ViBe is given in Algorithm 7.2. The core of the proposed updating policy is described at lines 18-19: one sample of the model is selected randomly and replaced by the current value. Note that this process is applicable to a scalar value, but to an RGB color vector or even more complex feature vectors as well. In fact, except for the distance calculations between values or feature vectors and their corresponding elements in the background model, all principles of ViBe are transposable as such to any feature vector.

A random updating strategy contradicts the belief that a background subtraction technique should be entirely deterministic and predictable. From our perspective, there is no reason to prefer an updating policy that would replace the oldest sample first as it reduces the temporal window dramatically. In addition, it could happen that a dynamic background has a cycle that is longer than the temporal window. The model would then not be able to characterize that background. Likewise, dynamic backgrounds with zones of different time frequencies will be impossible to handle. With our updating policy, the past is not "ordered"; one could say that the past has no effect on the future. This property is called the memoryless property (see [24]). We believe that many background subtraction techniques could benefit from this updating policy.

Until now, we have considered pixels individually; this is part of the design of pixel based methods. However, the random updating policy has introduced a spatial inhomogeneity.

Algorithm 7.2 - Updating Algorithm of ViBe

```
1   int width;                    // width of the image
2   int height;                   // height of the image
3   byte image[width*height];     // current image
4   byte updatingMask[*height];   // updating mask (1==updating allowed)
5
6   int subsamplingFactor = 16;   // amount of random subsampling
7
8   int numberOfSamples = 20;     // number of samples per pixel
9   byte samples[width*height][numberOfSamples]; // background model
10
11  for (int p = 0; p < width*height; p++)
12    if (updatingMask[p] == 1) { // updating is allowed
13
14      // eventually updates the model of p (in-place updating)
15      int randomNumber = getRandomIntegerIn(1, subsamplingFactor);
16      if (randomNumber == 1) { // random subsampling
17        // randomly selects a sample in the model to be replaced
18        int randomNumber = getRandomIntegerIn(0, numberOfSamples - 1);
19        samples[p][randomNumber] = image[p];
20      }
21
22      // eventually diffuses in a neighboring model (spatial diffusion)
23      int randomNumber = getRandomIntegerIn(1, subsamplingFactor);
24      if (randomNumber == 1) { // random subsampling
25        // chooses a neighboring pixel randomly
26        q = getPixelLocationFromTheNeighborhoodOf(p);
27        // diffuses the current value in the model of q
28        // (uncomment the following check to inhibit diffusion in the
                 foreground)
29        // if (updatingMask[q] == 1) {
30          int randomNumber = getRandomIntegerIn(0, numberOfSamples - 1);
31          samples[q][randomNumber] = image[p];
32        // }
33      }
34
35    } // end of "if"
36  } // end of "for"
```

Indeed, the index of the sample values that is replaced depends on the result of a random test and, therefore, differs from one location to another one. This is a second important element of the design of ViBe: we should consider the spatial neighborhood of a pixel, and not only the pixel itself. Two mechanisms are used for that: (1) ensure that all the pixels are processed differently, and (2) diffuse information locally. The random selection of an index, that defines which values should be replaced, is a first method to process pixels differently. In the following, we introduce two additional mechanisms: *time subsampling*, to increase the spatial inhomogeneity, and a *spatial diffusion* mechanism. We also show that the diffusion mechanism is an essential part of ViBe.

Time subsampling

With the random index selection of samples described in the previous section, we have suppressed an explicit reference to time in the model. The use of a random replacement policy allows the sample collection to cover a large, theoretically infinite, time window with a limited number of samples. In order to extend the size of the time window even more, we resort to *random time subsampling*. The main idea is that background is a slow varying information, except for the cases of a sudden global illumination change or scene cuts that need a proper handling at the frame level. Therefore, it is not necessary to update each background model for each new frame. By making the background updates less frequent, we artificially extend the expected lifespan of the background samples and ensure a spatial inhomogeneity because the decision to update or not is pixel dependent. As in the presence of periodic or pseudo-periodic background motions, the use of fixed subsampling intervals might prevent the background model from properly adapting to these motions, we prefer to use a *random* subsampling policy. As shown at lines 15-16 of Algorithm 7.2, a random process determines if a background pixel will be inserted in the corresponding pixel model.

In most cases, we adopted a time subsampling factor, denoted ϕ, of 16: a background pixel value has 1 chance in 16 of being selected to update its pixel model. For some specific scenarios, one may wish to tune this parameter to adjust the length of the time window covered by the pixel model. A smaller subsampling factor, 5 for example, is to be preferred during the first frames of a video sequence, when there is a lot of motion in the image, or if the camera is shaking. This allows for a faster updating of the background model.

Spatial diffusion to ensure spatial consistency

One of the problems with a conservative updating policy is that foreground zones hide background pixels so that there is no way to uncover the background, and subsequently to update it. A popular way to counter this drawback is what the authors of the W^4 algorithm [11] call a *"detection support map"* which counts the number of consecutive times that a pixel has been classified as foreground. If this number reaches a given threshold for a particular pixel location, the current pixel value at that location is inserted into the background model. A variant includes, in the background, groups of connected foreground pixels that have been found static for a long time, as in [7]. Some authors, like those of [38], use a combination of a pixel-level and an object-level background update.

For ViBe, we did not want to define a detection support map that would add parameters, whose determination would be application dependent, and increase the computational complexity, neither to include a mechanism that would be defined at the object level.

We believe that a progressive inclusion of foreground samples in the background models is more appropriate in general. We assume that neighboring background pixels share a similar temporal distribution and consider that a new background sample of a pixel should also update the models of neighboring pixels. According to this policy, background models hidden by the foreground will be updated with background samples *from neighboring pixel locations* from time to time. This allows a spatial diffusion of information regarding the background evolution that relies on samples *exclusively* classified as background. Our background model is thus able to adapt to a changing illumination and to structural evolutions (added or removed background objects) while relying on a *strict* conservative updating scheme. Lines 22 to 31 of Algorithm 7.2 detail the spatial diffusion mechanism. A spatial neighborhood of a pixel p is defined, typically a 4- or 8-connected neighborhood, and one location q in the neighborhood is chosen randomly. Then, the value $v(p)$ is inserted in the model of q. Since pixel models contain many samples, irrelevant information that could accidentally be inserted into a neighboring model does not affect the accuracy of the detection. Furthermore, the erroneous diffusion of irrelevant information is blocked by the need to match an observed value before it can propagate further. This natural containment inhibits the diffusion of errors. Note that in [12], the authors use the spatial updating mechanism of ViBe except that they insert $v(q)$, instead of $v(p)$, in the model of q. Also note that, for reasons similar to those exposed earlier, the spatial propagation method is also random. A random policy is convenient in the absence of prior knowledge of the scene, and to avoid a bias towards a scene model that would unnecessarily decrease the model flexibility.

It is worth mentioning that the original diffusion mechanism of ViBe (which uses the segmentation map as the updating mask) is not limited to the inside of background blobs. At the borders, it could happen that a background value is propagated into the model of a neighboring foreground pixel. If needed, background samples propagation into pixel models of the foreground may be prevented in at least two ways. The first way prevents sample diffusion across the border of the updating mask by uncommenting lines 29 and 32 of Algorithm 7.2. Another technique removes the inner border of the segmentation map when it serves as the updating mask, e.g. by using a morphologically eroded segmentation map as the updating mask.

An interesting question for ViBe relates to the respective effects of *in-place substitution* (a value replaces one value of its own model) and *spatial diffusion* (a value replaces one value of a neighboring model). To evaluate the impact of these strategies, we made several tests where we modified the proportion of in-place substitution and spatial diffusion. In addition, we compared the original diffusion mechanism of ViBe and a strategy that does not allow to modify models of foreground pixels; this variant, called *intra diffusion*, is obtained by uncommenting lines 29 and 32 of Algorithm 7.2 if the updating mask is equal to the background/foreground segmentation map. Table 7.1 shows the percentage of bad classification (PBCs) obtained for different scenarios; these results are averages for the 31 sequences of the Change Detection dataset.

	100%	90%	80%	70%	60%	50%	40%	30%	20%	10%	0%
intra diffusion	3.79	3.74	3.55	3.49	3.40	3.39	3.32	3.27	3.25	3.24	3.20
inter diffusion	2.99	2.32	2.19	2.21	2.27	2.32	2.33	2.36	2.40	2.40	2.41

TABLE 7.1 Percentage of bad classification (PBC) for different updating strategies. The top row indicates the percentage of in-place substitution, with respect to the percentage of spatial diffusion. A 100% rate thus means that no values are propagated into the model of a neighboring pixel. To the contrary, in the 0% scenario, every pixel model update results from spatial diffusion. *Intra diffusion* is defined as a diffusion process that forbids the diffusion of background samples into models located in the foreground. In the original version of ViBe, diffusion is always allowed, even if that means that the value of a background pixel is put into the model of a foreground pixel, as could happen for a pixel located at the border of the background mask. This corresponds to the bottom row, named *inter diffusion*.

One can observe that, for the intra diffusion scenario, the PBC decreases as the percentage of spatial diffusion increases. By extension, we would recommend, for this dataset, to always diffuse a value and never use it to update a pixel's own model. This observation might also be applicable to other background subtraction techniques. The effects of intra diffusion are mainly to maintain a larger local variance in the model, and to mimic small camera displacements, except that all these displacements differ locally.

The last row of Table 7.1 (inter diffusion) provides the percentages of bad classification for the original diffusion mechanism proposed in ViBe. It consists to diffuse background values into the neighborhood, regardless of the classification result of neighboring pixels. It appears that the original spatial diffusion process is always preferable to a diffusion process restricted to the background mask (intra diffusion). In other words, it is a good idea to allow background values to cross the frontiers of background blobs. The experiments also show that the original spatial diffusion mechanism of ViBe performs better when in-place substitution and spatial diffusion are balanced; a 50%-50% proportion, as designed intuitively in [2], is close to being optimal.

Why it is useful to define an updating mask that is different from the segmentation map

The purpose of a background subtraction technique is to produce a binary map with background and foreground pixels. Most of the time, it is the segmentation results that users are looking for. In a conservative approach, the segmentation map is used to determine which values are allowed to enter the background model. In other words, the segmentation map plays the role of an updating mask. But this is not a requirement. In fact, the unique constraint is that foreground values should never be used to update the model. In [37], we proposed to produce the updating mask by processing the segmentation map. For example, small holes in the foreground are filled to avoid the appearance of background seeds inside of background objects. Also, we remove very small foreground objects, such as isolated pixels, in order to include them in the background model as fast as possible. Another possibility consists to slow down the spatial diffusion across background borders. An idea to achieve this consists of allowing the diffusion only when the gradient value is lower than a given threshold. This prevents, or at least decreases, the pollution of background models covered by foreground objects (such as abandoned objects), while it does not affect the capability of ViBe to incorporate ghosts smoothly. Note that reducing diffusion enhances the risk that an object is never incorporated into the background.

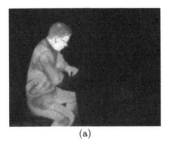 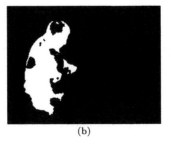 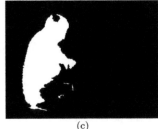

         (a)   (b)   (c)

FIGURE 7.2 Comparison of the effects of modifying the updating mask: (a) infrared input image, (b) segmentation map of the original version of ViBe, (c) segmentation map obtained when the updating mask inhibits samples diffusion across contrasted background borders (this inhibition mechanism is described in [37]).

These discussions clearly show that the purpose of differentiating the segmentation map and the updating mask is to induce a behavior for the background subtraction technique that is adapted to the application needs, while keeping the principles of ViBe. For example, an abandoned object might need to be left in the foreground for a certain amount of time. After that time, the application layer can decide to switch the label from foreground to background in the updating mask, in order to include the abandoned object into the background. In other words, the distinction between the updating mask and the segmentation map introduces some flexibility and permits to incorporate high-level information (such as objects blobs, regions of interest, time notions, etc.) into any low-level background subtraction technique.

Initialization

Many popular techniques described in the literature, such as [8, 15], need a sequence of several dozens of frames to initialize their models. From a statistical point of view, it is logical to wait for a long period of time to estimate the temporal distribution of the background pixels. But one may wish to segment the foreground of a sequence that is even shorter than the typical initialization sequence required by some background subtraction algorithms, or even to be operational from the second frame on. Furthermore, many applications require the ability to provide a mechanism to refresh or re-initialize the background model in circumstances such as in the presence of sudden lighting changes, that cannot be handled properly by the regular update mechanism of the algorithm.

A convenient solution provides a technique that will initialize the background model from a single frame. Given such a technique, the response to sudden illumination changes is straightforward: the existing background model is discarded and a new model is initialized instantaneously. Furthermore, being able to provide a reliable foreground segmentation as early on as the second frame of a sequence has obvious benefits for short sequences in video-surveillance or for devices that embed a motion detection algorithm.

In the original version of ViBe [2], we used the same assumption as the authors of [14], which is that neighboring pixels share a similar temporal distribution, to populate the pixel models with values found in the spatial neighborhood of each pixel. From our experiments, selecting samples randomly in the 8-connected neighborhood of each pixel has proved to be satisfactory. As no temporal information is available from the first frame, there seems to be no other alternative than to take values from the spatial neighborhood. However, other strategies have been tested as well. The first alternative initializes $N-2$ samples of the models with values that are uniformly distributed over the possible range, and 2 samples with the current pixel value. This accounts for a clean background while providing the capability to accommodate to some motion in the next frames. In other terms, this ensures to have a background that fills the entire image and a non zero variance. If one is ready to wait for a few frames before segmenting the image, then it is also possible to simply put random values in the model. We then recommend to temporarily decrease the time subsampling factor ϕ to speed up the incorporation of background values into the model.

All these strategies have proved to be successful, and they rapidly converge to identical segmentation results when dealing with long sequences. For short time periods, the presence of a moving object in the first frame introduces an artifact commonly called a *ghost*. According to [30], a ghost is a set of connected points, detected as in motion but not corresponding to any real moving object. In this particular case, the ghost is caused by the unfortunate initialization of pixel models with samples taken from the moving object. In subsequent frames, the object moves and uncovers the real background, which will be learned progressively through the regular model updating process, making the ghost fade over time. It appears that the propagation mechanism is capable to overcome ghost effects. In fact, as shown in [2], ghosts are absorbed rapidly compared to static objects that remain visible for a longer period of time.

Other issues relate to sudden natural illumination changes, scene cuts, or changes caused by the Automatic Gain Control (AGC) of cameras that artificially improve their dynamic range to produce usable images. In all these cases, changes are not due to the motion of objects, and the amount of foreground pixels indicates that there is a major change in the scene. It also happens that these changes are temporary and that the scene rapidly returns to its previous state. This is typical for IP cameras whose gain changes when an object approaches the sensor and returns to its previous state when the object leaves the scene. This situation also occurs on a pixel basis when an object stays at the same location for a short period of time, and then moves again.

The solution that we propose to cope with these problems consists to partly re-initialize the model of each pixel with the new value. More precisely, we only replace $\#_{min}$ samples of each model; the $N - \#_{min}$ other samples are left unchanged. With this mechanism, the background model is re-initialized, while it memorizes the previous background model as well. It is then possible to return to the previous scene and to get a meaningful segmentation instantaneously. Theoretically, we could even store $N/\#_{min}$ background models at a given time. This technique is similar to that of GMM based models when they would keep a few of the significant gaussians to memorize a previous model.

7.2.4　Variants of ViBe

Several variants of ViBe have been proposed recently. In [6], it was suggested to use a threshold in relation to the samples in the model for a better handling of camouflaged foreground. This adaptation was done by Hofmann et al. [12]. Their proposed background subtraction technique uses all the principles of ViBe, and add learning heuristics that adapt the model pixel classification and updating dynamics to the scene content. They also change the diffusion mechanism to guarantee that a model is never updated with a value of a neighboring pixel, by updating a neighboring model with the current value for that same location. However, previously in this section, we have discussed the diffusion mechanism and clearly showed that the original diffusion mechanism is preferable. Mould and Havlicek also propose variants of ViBe that adapt themselves to the scene content [22, 23]. Their major modification applies a learning algorithm that integrates new information into the models by replacing the most outlying values with respect to the current sample collections.

While it is obvious that adapting fixed parameters to the image content should increase the performance, one can observe that a learning phase is then needed to train the parameters of the heuristics of these authors. To some extent, they have changed the dynamic behavior of ViBe by replacing fixed parameters by heuristics which have their own fixed parameters. This might be an appropriate strategy to optimize ViBe to deal with specific scene contents, but the drawbacks are that it introduces a bias towards some heuristics, and that it reduces the framerate dramatically.

In [19], Manzanera describes a background subtraction technique that extends the modeling capabilities of ViBe. The feature vectors are enriched by local jets and an appropriate distance is defined. Manzanera also proposes to regroup the samples in clusters to speed up the comparison step. These extensions do not modify the dynamic behavior of ViBe, but Manzanera shows that its method improves the performance. Enriching the feature vector slightly increases the performance, but it also increases the computation times, more or less significantly, depending on the complexity of the additional features and that of the adapted distance.

Another technique, inspired by ViBe, is described by Sun et al. [33]. They propose to model the background of a pixel not only with samples gathered from its own past, but also in the history of its spatial neighborhood.

Note that, while this Chapter focuses on the implementation of ViBe on CPUs, small modifications might be necessary for porting ViBe to GPUs, or to FPGA platforms [16]. However, the fundamental principles of ViBe are unaltered.

7.3 Complexity Analysis and Computational Speed of ViBe

Except for papers dedicated to hardware implementation, authors of background subtraction techniques tend to prove that their technique runs in real time, for small or medium sized images. It is then almost impossible to compare techniques, because many factors differ from one implementation to another (CPU, programming language, compiler, etc.). In this section, we analyze the complexity of the ViBe algorithm in terms of the number of operations, and then compare execution times of several algorithms. We also define the notion of *background subtraction complexity factor* that expresses the practical complexity of an algorithm.

To express the complexity of ViBe, we have to count the number of operations for processing an image. There are basically four types of operations involved: distance computation, distance thresholding, counter checks, and memory substitutions. More precisely, the number of operations is as follows for ViBe:

- Segmentation/classification step.
  Remember that we compare a new pixel value to background samples to find $\#_{min}$ matches ($\#_{min} = 2$ in all our experiments). Once these matches have been found, we step over to the next pixel and ignore the remaining background samples of a model; this is different from techniques that requires a 90% match, for example. Operations involved during the segmentation step are:

  - comparison of the current pixel value with the values of the background model. Most of the time, the $\#_{min}$ first values of the model of a background pixel are close to the new pixel value. The algorithm then reaches its minimal number of comparisons. For foreground pixels however, the number of comparisons can be as large as the number of samples in the model, N. Therefore, if $\#F$ denotes the proportion of foreground pixels, the minimal number of comparisons per pixel N_{comp} is, on average,

    $$N_{\text{comp}} = \#_{min} + (N - \#_{min})\#F. \qquad (7.4)$$

    Each of these comparisons involves one distance calculation and one thresholding operation on the distance. For a typical number of foreground pixels in the video, this accounts for 2.8 comparisons on average per pixel: $N_{\text{comp}} = 2.8$. This is an experimentally estimated average value as computed on the Change Detection dataset (see last row, column (e) of Table 7.2).

  - comparisons of the counter to check if there are at least 2 matching values in the model. We need to start comparing the counter value only after the comparison between the current pixel value and the second value of the background model. Therefore, we have $N_{\text{comp}} - 1 = 1.8$ comparisons.

- Updating step:

  - 1 pixel substitution per 16 background pixels (the update factor, ϕ, is equal to 16). Because we have to choose the value to substitute and access the appropriate memory block in the model, we perform an addition on memory addresses. Then we perform a similar operation, for a pixel in the neighborhood (first we locate which pixel in the neighborhood to select, then which value to substitute).
    In total, we evaluate the cost of the update step as 3 additions on memory addresses per 16 background pixels.

- Summary (average per pixel, assuming that $\#_{min} = 2$ and $N_{\text{comp}} = 2.8$, that is that most pixels belong to the background):

  - 2.8 distance computations and 2.8 thresholding operations. In the simple case of a distance that is the absolute difference between two grayscale values, this accounts for 5.6 comparisons.

  - 1.8 counter checking operations. Note that a test such as *count* < 2 is equivalent to *count* $- 2 < 0$, that is to a subtraction and sign test. If we ignore the sign test, one comparison corresponds to one subtraction in terms of computation times.

  - $\frac{3}{16}$ addition on memory addresses.

Profiling tests show that the computation time of the updating step is negligible, on a CPU, compared to that of the segmentation step. If for the segmentation step, we approximate the complexity of a distance computation to that of a comparison, and if we assume that the image is a grayscale image, the number of operations is $3N_{\text{comp}} - 1$ per pixel; this results in 7.4 comparisons per pixel. In order to verify this number, we have measured the computation times for different algorithms. Table 7.2 shows measures obtained for each sequence of the Change Detection dataset.

Column (a) provides the computation times of the simple background subtraction technique with a static background, whose corresponding code is given in Algorithm 7.3. To avoid any bias related to input and output handling, 1000 images were first decoded, converted to grayscale, and then stored into memory. The computation times therefore only relate to the operations necessary for background subtraction.

Algorithm 7.3 - Background Subtraction with a Static Background (This corresponds to the minimum number of operations for any pixel based background subtraction. Here, the code does not include any updating step, neither the initialization of the static background model.)

```
1   int width;                            // width of the image
2   int height;                           // height of the image
3   byte image[width*height];             // current image
4   byte segmentationMap[width*height];   // classification result
5
6   int distanceThreshold = 20;
7
8   byte sample[width*height];            // static background model
9
10  ...
11  for (int p = 0; p < width*height; p++) {
12    if (getDistanceBetween(image[p], sample[p]) < distanceThreshold)
13      segmentationMap[p] = BACKGROUND_LABEL;
14    else
15      segmentationMap[p] = FOREGROUND_LABEL;
16  }
17  ...
```

Values of column (a) constitute the absolute lower bound on the computation time for any background subtraction technique, because one needs at least to compare the current pixel value to that of a reference, and allocate the result in the segmentation map (this last

Sequences	(a) static backgr.	(b) ViBe↓	(c) expon. filter	(d) ViBe	(e) N_{comp}	(f) theoretical C_F	(g) real C_F
highway	241	367	897	1583	3.1	4.1	6.6
office	277	416	1014	1598	2.7	3.6	5.8
pedestrians	242	381	981	1528	2.2	2.8	6.3
PETS2006	1068	1734	4619	8148	2.4	3.0	7.6
badminton	1070	1641	4017	9414	3.6	4.8	8.8
boulevard	322	457	1040	2253	3.1	4.2	7.0
sidewalk	326	462	1050	2404	3.4	4.6	7.4
traffic	281	407	944	2324	4.4	6.0	8.3
boats	298	420	951	1658	2.4	3.1	5.6
canoe	336	459	990	2184	3.0	4.0	6.5
fall	1697	2247	4628	10357	3.6	4.9	6.1
fountain01	434	633	1492	2544	2.4	3.0	5.9
fountain02	408	608	1466	2215	2.1	2.6	5.4
overpass	306	430	962	1434	2.2	2.7	4.7
abandonedBox	437	637	1496	2694	2.9	3.8	6.2
parking	218	344	876	1304	2.1	2.76	6.0
sofa	206	331	865	1378	2.4	3.0	6.7
streetLight	298	424	957	1559	2.8	3.7	5.2
tramstop	400	601	1460	2436	3.1	4.1	6.1
winterDriveway	277	402	934	1418	2.2	2.9	5.1
backdoor	260	385	917	1460	2.4	3.2	5.6
bungalows	254	395	993	1802	3.2	4.3	7.1
busStation	257	398	995	1565	2.4	3.2	6.1
copyMachine	931	1487	3881	7354	3.0	3.9	7.9
cubicle	244	381	967	1418	2.2	2.8	5.8
peopleInShade	256	407	1049	1835	2.9	3.9	7.2
corridor	198	324	857	1436	2.6	3.4	7.2
diningRoom	194	319	852	1353	2.6	3.4	7.0
lakeSide	193	317	851	1256	2.0	2.5	6.5
library	192	317	851	1373	4.0	5.5	7.1
park	263	431	1134	1779	2.2	2.8	6.8
Mean	**399**	**599**	**1451**	**2679**	**2.8**	**3.7**	**6.5**

TABLE 7.2 Computation times, in milliseconds and for 1000 images (converted to grayscale), of several algorithms for each sequence of the Change Detection dataset, on a Intel(R) Core(TM)2 Duo CPU T7300 @2.00GHz. The considered algorithms are: (a) background subtraction with a static background (1 distance comparison and 1 distance thresholding operation), (b) downscaled version of ViBe (1 value in the model, 1 comparison), (c) exponential filter ($\alpha = 0.05$), and (d) ViBe. The average number of distance computations per pixel is given, for ViBe, in column (e). Considering that all images were first converted to grayscale, (f) is the theoretical background subtraction complexity factor and (g) is the measured background subtraction complexity factor.

step is generally negligible with respect to that of distance computations or comparisons). In our implementation, we have converted the input video stream to grayscale to achieve the lowest computation times, because the distance calculation is then equivalent to the absolute difference between the two values. For more complex distances, the distance calculation will require more processing power than distance thresholding. We note that, on an Intel(R) Core(TM)2 Duo CPU T7300 @2.00GHz, the average computation time per image is 0.4 ms.

In [2], we proposed a downscaled version of ViBe, named ViBe↓ hereafter and in Table 7.2, that uses only one sample for each background model and, consequently, proceeds to only one comparison per pixel. Computation times of that downscaled version of ViBe are given in column (b) of Table 7.2. On average, the computation is 200 ms slower for 1000 images, that is 0.2 ms per image. Because the only difference with the technique with a static background is the updating step, we can estimate the cost of the updating to half that of a distance computation and thresholding. Therefore, ViBe↓ is probably the fastest background subtraction technique that includes an updating mechanism. Results of ViBe↓ are shown in column (b) of Figure 7.3.

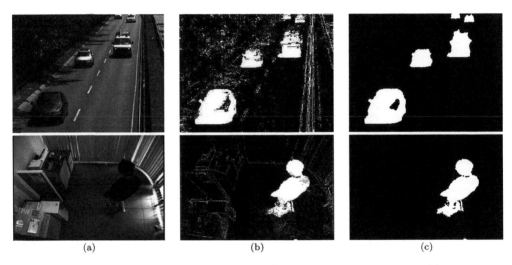

(a)   (b)   (c)

FIGURE 7.3 Segmentation maps obtained with ViBe↓, a downscaled version of ViBe: (a) original image, (b) unfiltered segmentation map of ViBe↓, and (c) same as (b) after an area opening and a close filter.

Columns (c) and (d) of Table 7.2 provide the computation times for the exponential filter method and ViBe, respectively. One can see that the computational cost of ViBe is less than twice that of the exponential filter, and about 6.5 times that of the simplest background subtraction technique.

In further tests, we explored the average number of distance computations per pixel of ViBe. This number is given, for each video sequence of the Change Detection dataset, in column (e) of Table 7.2. Despite that the model size is 20, we have an average number of distance computations of 2.76. This means that, on average, ViBe only has to perform about less than 1 distance computation in addition to the $\#_{min} = 2$ unavoidable distance computations. While this figure is interesting for the optimization of parameters, it does not translate directly in terms of computation time because many other operations are involved in a software implementation. While these additional operations, such as loops unrolling, are necessary, they are often neglected in complexity analyses. In order not to underestimate the computation times and the complexity, we define a new ratio, named the *background subtraction complexity factor*, as follows:

$$C_F = \frac{\text{measured computation time}}{\text{measured computation for one distance measure and check}}. \tag{7.5}$$

It corresponds to the ratio between the measured computation time and the time needed to perform the simplest background subtraction technique (with a static background). For the denominator, we calculate the time of the algorithm such as described in Algorithm 7.3. Because this simple algorithm has a fixed number of operations per pixel and all the operations are known, it can be used as a yardstick to measure the complexity of another algorithm. Assuming the specific case of a grayscale video stream, the algorithm requires to compute one distance and one thresholding per pixel. If we approximate these operations as two comparisons per pixel, we have a theoretical estimate of the denominator of Equation 7.5. We can also estimate the numerator as follows. Column (e) of Table 7.2 corresponds to the measured number of distance computations per pixel N_{comp}. As indicated previously, we have $3N_{\text{comp}} - 1$ operations per pixel. The complexity factor can thus be estimated as $(3N_{\text{comp}} - 1)/2$. The theoretical estimates of the complexity factor for each video sequence are given in column (f) of Table 7.2. When we compare them to the real complexity factors (see column (g)), based on experimental measures, we see that the theoretical complexity factor is often underestimated. This is not surprising as we have neglected some operations such as loops, updating, memory reallocation, etc.

7.4 Conclusion

ViBe is a pixel based background subtraction technique that innovated by introducing several new mechanisms: segmentation by comparing a pixel value to a small number of previously collected samples, memoryless updating policy, and spatial diffusion. This chapter elaborates on the underlying ideas that motivated us to build ViBe. One of the main keys is the absence of any notion of time when building and updating the background models. In addition, ViBe introduces a spatial diffusion mechanism that consists to modify the model of a neighboring pixel while updating the model of a pixel. We have showed that the use of spatial diffusion always increases the performance and that, in particular for the Change Detection dataset, it is even preferable to diffuse a value in the neighborhood of a pixel rather than to use it to update its own model. Furthermore, crossing background borders in the segmentation map to adapt models of foreground pixels is also beneficial for the suppression of ghosts or static objects. The complexity of ViBe is also discussed in this chapter. We introduce the notion of complexity factor that compares the time needed by a background subtraction technique to the computational cost of the simplest background subtraction algorithm. The measured complexity factor of ViBe is about 6.5, on average, on the Change Detection dataset. The complexity factor even drops to 1.5 for a downscaled version of ViBe, which still produces acceptable results!

References

1. O. Barnich and M. Van Droogenbroeck. ViBe: a powerful random technique to estimate the background in video sequences. In *IEEE International Conference on Acoustics, Speech and Signal Processing, ICASSP*, pages 945–948, April 2009.
2. O. Barnich and M. Van Droogenbroeck. ViBe: A universal background subtraction algorithm for video sequences. *IEEE Transactions on Image Processing*, 20(6):1709–1724, June 2011.
3. Y. Benezeth, P. Jodoin, B. Emile, H. Laurent, and C. Rosenberger. Review and evaluation

of commonly-implemented background subtraction algorithms. In *IEEE International Conference on Pattern Recognition, ICPR*, pages 1–4, December 2008.

4. T. Bouwmans. Recent advanced statistical background modeling for foreground detection - a systematic survey. *Recent Patents on Computer Science*, 4(3):147–176, 2011.

5. T. Bouwmans, F. El Baf, and B. Vachon. Statistical background modeling for foreground detection: A survey. In *Handbook of Pattern Recognition and Computer Vision (volume 4)*, chapter 3, pages 181–199. World Scientific Publishing, January 2010.

6. S. Brutzer, B. Höferlin, and G. Heidemann. Evaluation of background subtraction techniques for video surveillance. In *IEEE International Conference on Computer Vision and Pattern Recognition, CVPR*, pages 1937–1944, Colorado Spring, USA, June 2011.

7. R. Cucchiara, C. Grana, M. Piccardi, and A. Prati. Detecting moving objects, ghosts, and shadows in video streams. *IEEE Transactions on Pattern Analysis and Machine Intelligence*, 25(10):1337–1342, October 2003.

8. A. Elgammal, D. Harwood, and L. Davis. Non-parametric model for background subtraction. In *European Conference on Computer Vision, ECCV 2000*, volume 1843 of *Lecture Notes in Computer Science*, pages 751–767, London, UK, June 2000.

9. S. Elhabian, K. El-Sayed, and S. Ahmed. Moving object detection in spatial domain using background removal techniques – State-of-art. *Recent Patents on Computer Science*, 1:32–54, January 2008.

10. N. Goyette, P.-M. Jodoin, F. Porikli, J. Konrad, and P. Ishwar. changedetection.net: A new change detection benchmark dataset. In *Change Detection Workshop, CDW 2012*, Providence, Rhode Island, June 2012.

11. I. Haritaoglu, D. Harwood, and L. Davis. W^4: Real-time surveillance of people and their activities. *IEEE Transactions on Pattern Analysis and Machine Intelligence*, 22(8):809–830, August 2000.

12. M. Hofmann, P. Tiefenbacher, and G. Rigoll. Background segmentation with feedback: The pixel-based adaptive segmenter. In *Change Detection Workshop, CDW 2012*, Providence, Rhode Island, June 2012.

13. Home Office, Scientific Development Branch. *i-LIDS User Guide*, 2011.

14. P.-M. Jodoin, M. Mignotte, and J. Konrad. Statistical background subtraction using spatial cues. *IEEE Transactions on Circuits and Systems for Video Technology*, 17(12):1758–1763, December 2007.

15. K. Kim, T. Chalidabhongse, D. Harwood, and L. Davis. Real-time foreground-background segmentation using codebook model. *Real-Time Imaging*, 11(3):172–185, June 2005.

16. T. Kryjak and M. Gorgon. Real-time implementation of the vibe foreground object segmentation algorithm. In *Proceedings of the Federated Conference on Computer Science and Information Systems, FedCSIS*, pages 591–596, Krakow, Poland, September 2013.

17. H.-H. Lin, T.-L. Liu, and J.-C. Chuang. Learning a scene background model via classification. *IEEE Signal Processing Magazine*, 57(5):1641–1654, May 2009.

18. A. Manzanera. Σ-Δ background subtraction and the Zipf law. In *Progress in Pattern Recognition, Image Analysis and Applications*, volume 4756 of *Lecture Notes in Computer Science*, pages 42–51. Springer, November 2007.

19. A. Manzanera. Local jet feature space framework for image processing and representation. In *International Conference on Signal Image Technology & Internet-Based Systems*, pages 261–268, Dijon, France, November 2011.

20. A. Manzanera and J. Richefeu. A new motion detection algorithm based on Σ-Δ background estimation. *Pattern Recognition Letters*, 28(3):320–328, February 2007.

21. T. Moeslund, A. Hilton, V. Krüger, and L. Sigal. *Visual Analysis of Humans: Looking at People*. Springer, 2011.

22. N. Mould and J. Havlicek. A conservative scene model update policy. In *Southwest Symposium on Image Analysis and Interpretation*, Santa Fee, New Mexico, USA, April

2012.

23. N. Mould and J. Havlicek. A stochastic learning algorithm for pixel-level background models. In *IEEE International Conference on Image Processing, ICIP*, pages 1233–1236, September 2012.

24. A. Papoulis. *Probability, random variables, and stochastic processes.* McGraw-Hill, 1984.

25. D. Parks and S. Fels. Evaluation of background subtraction algorithms with post-processing. In *IEEE International Confersence on Advanced Video and Signal Based Surveillance*, pages 192–199, Santa Fe (New Mexico, USA), September 2008.

26. F. Porikli and O. Tuzel. Bayesian background modeling for foreground detection. In *ACM international workshop on Video surveillance & sensor networks*, pages 55–58, Singapore, November 2005.

27. R. Radke, S. Andra, O. Al-Kofahi, and B. Roysam. Image change detection algorithms: A systematic survey. *IEEE Transactions on Image Processing*, 14(3):294–307, March 2005.

28. A. Schick, M. Bauml, and R. Stiefelhagen. Improving foreground segmentation with probabistic superpixel Markov random fields. In *Change Detection Workshop, CDW 2012*, Providence, Rhode Island, June 2012.

29. J. Shotton, A. Fitzgibbon, M. Cook, T. Sharp, M. Finocchio, R. Moore, A. Kipman, and A. Blake. Real-time human pose recognition in parts from single depth images. In *IEEE International Conference on Computer Vision and Pattern Recognition, CVPR*, Colorado Springs, June 2011.

30. B. Shoushtarian and H. Bez. A practical adaptive approach for dynamic background subtraction using an invariant colour model and object tracking. *Pattern Recognition Letters*, 26(1):5–26, January 2005.

31. A. Srivastava, A. Lee, E. Simoncelli, and S. Zhu. On advances in statistical modeling of natural images. *Journal of Mathematical Imaging and Vision*, 18(1):17–33, January 2003.

32. C. Stauffer and E. Grimson. Adaptive background mixture models for real-time tracking. In *IEEE International Conference on Computer Vision and Pattern Recognition, CVPR*, volume 2, pages 246–252, Ft. Collins, USA, June 1999.

33. L. Sun, Q. De Neyer, and C. De Vleeschouwer. Multimode spatiotemporal background modeling for complex scenes. In *European Signal Processing Conference, EUSIPCO*, pages 165–169, Bucharest, Romania, August 2012.

34. M. Van Droogenbroeck and O. Barnich. Visual background extractor. European Patent Office, EP 2_015_252_B1, February 2010.

35. M. Van Droogenbroeck and O. Barnich. Visual background extractor. Japan Patent Office, JP 2011 4699564 B2, June 2011.

36. M. Van Droogenbroeck and O. Barnich. Visual background extractor. United States Patent and Trademark Office, US 8,009,918 B2, 18 pages, August 2011.

37. M. Van Droogenbroeck and O. Paquot. Background subtraction: Experiments and improvements for ViBe. In *Change Detection Workshop, CDW 2012*, pages 32–37, Providence, Rhode Island, June 2012.

38. H. Wang and D. Suter. A consensus-based method for tracking: Modelling background scenario and foreground appearance. *Pattern Recognition*, 40(3):1091–1105, March 2007.

39. C. Wren, A. Azarbayejani, T. Darrell, and A. Pentland. Pfinder: Real-time tracking of the human body. *IEEE Transactions on Pattern Analysis and Machine Intelligence*, 19(7):780–785, July 1997.

40. Z. Zivkovic and F. van der Heijden. Efficient adaptive density estimation per image pixel for the task of background subtraction. *Pattern Recognition Letters*, 27(7):773–780, May 2006.

8

Online Learning by Stochastic Approximation for Background Modeling

Ezequiel López-Rubio
University of Málaga, Spain

Rafael M. Luque-Baena
University of Málaga, Spain

8.1 Introduction

The task of obtaining a reliable model of the background of a scene is hampered by many undesirable effects such that sensor noise, camouflage, cast shadows and others. Moreover, the model must be built in real time, as new frames are captured. Stochastic approximation is a methodology to estimate unknown parameters of a dynamic system, i.e. one which evolves through time, under the presence of noise. Consequently, it is suitable for the background modeling section of computer vision systems. Stochastic approximation is based on the online update of the parameter estimates. The updates are done in a way that tend to remove the noise in the observations on the long run. The methods that we consider here are based in the Robbins-Monro stochastic approximation algorithm [10, 20, 21, 29, 31].

The earliest methods for foreground detection were based on frame differences [9, 35]. These approaches lack a probabilistic model, so it is difficult to quantify the uncertainties associated to this application. In particular, it is advisable that the background detection algorithm outputs the probability that a pixel belongs to the background, rather than a binary background/foreground flag. A probabilistic output allows more robust processing in the subsequent stages of the system, since a quantitative measure of the reliability of the results is made available. Unlike the above mentioned methods, stochastic approximation can be employed to estimate the parameters of a probabilistic model.

Many well known approaches use probabilistic models of the observed color values at each pixel. Hence, they do not suffer from the disadvantages explained before. The typical architecture consists of several multivariate Gaussian components, where one or more of them are associated with the background and the rest are associated with the foreground [19, 36, 44]. This poses the fundamental problem whether multivariate Gaussians are adequate to model both background and foreground. For static backgrounds, a multivariate Gaussian can be a good approximation to reality, since the physical noise processes relevant in these situations are well modeled by Gaussians [25]. However, the foreground cannot be modeled adequately by Gaussians in any way, since the color of incoming objects can be completely different from that of previous objects. Hence, the approaches based on Gaussian components only have serious limitations. This is closely related to the kind of algorithms which are used to update the parameters of the Gaussians. Typically an online version of the Expectation Maximization method for mixtures of Gaussians is used. The use of non Gaussian mixture components is difficult since Expectation Maximization update equations are heavily dependent on the kind of mixture component used, and the equations for non Gaussian cases are computationally demanding. On the other hand, stochastic approximation does not assume a particular family of probability densities. Consequently, uniform distributions can be used to model the foreground objects, which removes these inconveniences.

This freedom with respect to the form of the probability density function is also a characteristic of the methods based on Kernel Density Estimation (KDE), which are non-parametric [14, 15]. However, KDE tends to be slow when compared to all the strategies considered before, since it needs to keep track of a large number of previously observed samples. By contrast, stochastic approximation is faster because it only considers the current sample to modify the parameter estimations, i.e. it is inherently online.

Now it is time to consider stochastic approximation in more detail. Let Θ be a parameter to be estimated such that

$$\zeta(\Theta) = 0 \qquad (8.1)$$

where ζ is a function whose values cannot be obtained directly due to the noise in the observations. What we have is a random variable z which is a noisy estimate of ζ:

$$E[z(\Theta) \mid \Theta] = \zeta(\Theta) \qquad (8.2)$$

That is, we assume that the noise process has zero mean. The Robbins-Monro algorithm implements a discrete time update of the estimation of Θ at time instant $n + 1$:

$$\Theta(n+1) = \Theta(n) + \epsilon(n) z(\Theta(n)) \qquad (8.3)$$

where $\epsilon(n)$ is a suitable *step size*. If the step size is too small, the convergence to the true value of Θ is small. On the other hand, if ϵ is too large, the approximation is too sensitive to noise. Hence, a fundamental balance between these two tendencies must be found.

Now, let us focus on the estimation of the expectation $E[S]$ of certain random variable S from its samples s [24]. In this case we may take

$$\zeta(\Theta) = E[S] - \Theta \qquad (8.4)$$

$$z(\Theta) = s - \Theta \qquad (8.5)$$

which satisfies the condition (8.2). Hence equation (8.3) reads

$$\Theta(n+1) = \Theta(n) + \epsilon(n)(s_n - \Theta(n)) \qquad (8.6)$$

and $\Theta(n)$ is a running estimation of $E[S]$. If we assume that the data come from a mixture probability density, the conditional expectation of a function $\varphi(\mathbf{t})$ given a mixture component i can be estimated. As explained before, in the background modeling context one or more mixture components are associated with the background and the rest are associated with the foreground objects. First we set:

$$S = P(i \mid \mathbf{t}) \varphi(\mathbf{t}) \tag{8.7}$$

$$s_n = P(i \mid \mathbf{t}_n) \varphi(\mathbf{t}_n) \tag{8.8}$$

Hence from (8.4) we obtain:

$$\zeta(\mathbf{\Theta}) = E[P(i \mid \mathbf{t}) \varphi(\mathbf{t})] - \mathbf{\Theta} \tag{8.9}$$

Now the discrete time update rule (8.6) reads:

$$\mathbf{\Theta}(n+1) = \mathbf{\Theta}(n) + \epsilon(n)(P(i \mid \mathbf{t}_n) \varphi(\mathbf{t}_n) - \mathbf{\Theta}(n)) \tag{8.10}$$

This allows to approximate the conditional expectation as follows:

$$E[\varphi(\mathbf{t}) \mid i] \approx \frac{E[P(i \mid \mathbf{t}) \varphi(\mathbf{t})]}{E[P(i \mid \mathbf{t})]} \tag{8.11}$$

where it is assumed that the estimated probability density $p(\mathbf{t})$ is a good approximator of the true input density $\bar{p}(\mathbf{t})$:

$$\bar{p}(\mathbf{t}) \approx p(\mathbf{t}) \tag{8.12}$$

Also, please note that the a priori probabilities of the mixture components π_i are estimated by setting $\varphi(\mathbf{t}) = 1$ in (8.7).

Due to its robustness and online learning properties, stochastic approximation has been employed for several computer vision tasks. Moving object tracking is a hard problem given the large uncertainties provoked by the lack of precision in the location of the objects and the errors in the associations of foreground blobs from one frame to another. This means that the techniques considered here are suitable for this issue [43]. In particular, models of deformable objects can be built with the help of them [1,2]. Stereo reconstruction from monocular video sequences poses similar challenges associated with occlusions, lighting changes, camera errors, etc. In this context stochastic approximation is able to cope with incomplete information [8]. Finally, classical mixtures of Gaussians have been adapted with the help of this methodology [7], although models with Gaussian components only have serious limitations, as detailed before and in Chapter 12 of this book.

8.2   Mixture of Distributions

Many background modeling approaches are based on mixtures of multivariate Gaussians with diagonal covariance matrices. In this sense, Stauffer et al. [12,37] present a well-known theoretical framework for the upgrade background which apply this strategy and use an expectation-maximization (EM) algorithm to update the model. Since it is not always necessary to use a fixed number of Gaussians to represent the background, in [45] the number of Gaussians is estimated online, reducing slightly the time complexity of the algorithm. Additionally, in order to determine as fast as possible whether a pixel value t belongs to the background distribution, a background match is defined as a pixel value within $\mathcal{N}$ standard deviations of the associated distribution. This framework yields good results, but complex backgrounds are not adequately captured, and postprocessing techniques are needed.

In this section, the use of mixtures of uniform distributions and multivariate Gaussians with full covariance matrices is proposed. These mixtures are able to cope with both dynamic backgrounds and complex patterns of foreground objects. The stochastic approximation framework is used as a learning algorithm, which has a very reduced computational complexity. Hence, it is suited for real time applications. Moreover, this framework has an intrinsic tendency to assign more relevance to the newer data with respect to the older ones, which reinforces its suitability for this task. As we will see, it is also a robust strategy which needs a minimal number of tuneable parameters. The following sections provide the theoretical methodology of this approach, while the experimental results of its performance are shown in Section 8.4.

8.2.1   Model Definition

We propose to use the stochastic approximation framework to train a mixture which models the distribution of pixel values $\mathbf{t}\left(\mathbf{x}\right) = \left(t_1\left(\mathbf{x}\right), t_2\left(\mathbf{x}\right), t_3\left(\mathbf{x}\right)\right)$ at position $\mathbf{x} = \left(x_1, x_2\right)$. There is one Gaussian component for the background and one uniform component for the foreground. This leads to the following probabilistic model for the distribution of pixel values at any given position, where we drop the position indices $\mathbf{x}$ for the sake of simplicity:

$$p\left(\mathbf{t}\right) = \pi_{Back}p\left(\mathbf{t}|Back\right) + \pi_{Fore}p\left(\mathbf{t}|Fore\right) =$$

$$\pi_{Back}G\left(\mathbf{t}|\boldsymbol{\mu}_{Back}, \mathbf{C}_{Back} + \boldsymbol{\Psi}\right) + \pi_{Fore}U\left(\mathbf{t}\right) \tag{8.13}$$

We have:

$$G\left(\mathbf{t}|\boldsymbol{\mu}, \boldsymbol{\Sigma}\right) = \left(2\pi\right)^{-D/2}\det\left(\boldsymbol{\Sigma}\right)^{-1/2}\exp\left(\left(\mathbf{t} - \boldsymbol{\mu}\right)^T\boldsymbol{\Sigma}^{-1}\left(\mathbf{t} - \boldsymbol{\mu}\right)\right) \tag{8.14}$$

$$U\left(\mathbf{t}\right) = \begin{cases} 1/Vol\left(\mathcal{S}\right) & \text{iff } \mathbf{t} \in \mathcal{S} \\ 0 & \text{iff } \mathbf{t} \notin \mathcal{S} \end{cases} \tag{8.15}$$

$$\boldsymbol{\mu}_{Back} = E\left[\mathbf{t}|Back\right] \tag{8.16}$$

$$\boldsymbol{\mu}_{Fore} = E\left[\mathbf{t}|Fore\right] \tag{8.17}$$

$$\mathbf{C}_{Back} = E\left[\left(\mathbf{t} - \boldsymbol{\mu}_{Back}\right)\left(\mathbf{t} - \boldsymbol{\mu}_{Back}\right)^T|Back\right] \tag{8.18}$$

$$\forall i \in \{Back, Fore\}, \; R_{ni} = P\left(i|\mathbf{t}_n\right) = \frac{\pi_i p\left(\mathbf{t}_n|i\right)}{\pi_{Back}p\left(\mathbf{t}_n|Back\right) + \pi_{Fore}p\left(\mathbf{t}_n|Fore\right)} \tag{8.19}$$

where $\mathcal{S}$ is the support of the uniform pdf, $Vol\left(\mathcal{S}\right)$ is the D-dimensional volume of $\mathcal{S}$, and $\boldsymbol{\Psi}$ is a constant diagonal matrix which accounts for the quantization noise due to lossy compression (see Section 2.3). In background modeling, we typically have $D=3$ for tristimulus pixels, and $\mathcal{S}$ spans all the color range:

$$\mathcal{S} = \{\left(t_1, t_2, t_3\right)|t_1, t_2, t_3 \in [0, v]\} \tag{8.20}$$

with $Vol\left(\mathcal{S}\right) = v^3$. In most cases the pixels are given in RGB, but $\mathbf{t}$ could also be expressed in any other color space, such as L*u*v*. On the other hand, if we have color values with 8 bit precision then $v=255$. It must be highlighted that $\boldsymbol{\mu}_{Fore}$ is not used to compute the probability densities in any way. It is only needed when a sudden background change is detected, as explained in Section 8.2.3.

The Gaussian component $G\left(\mathbf{t}|\boldsymbol{\mu},\boldsymbol{\Sigma}\right)$ is able to capture a wide range of dynamic backgrounds because $\mathbf{C}_{Back}$ (and hence $\mathbf{C}_{Back} + \boldsymbol{\Psi}$) are not restricted to be diagonal matrices. On the other hand, the uniform component $U\left(\mathbf{t}\right)$ is able to model foreground objects of any color equally well.

8.2.2 Learning

When applied to the probabilistic model given in the previous subsection, the Robbins-Monro stochastic approximation algorithm yields the following update equations at time step n for an observed pixel value:

$$\forall i \in \{Back, Fore\}\,, \ \pi_i\left(n\right) = \left(1 - \epsilon\right)\pi_i\left(n-1\right) + \epsilon R_{ni} \tag{8.21}$$

$$\forall i \in \{Back, Fore\}\,, \ \mathbf{m}_i\left(n\right) = \left(1 - \epsilon\right)\mathbf{m}_i\left(n-1\right) + \epsilon R_{ni}\mathbf{t}_n \tag{8.22}$$

$$\forall i \in \{Back, Fore\}\,, \ \boldsymbol{\mu}_i\left(n\right) = \frac{\mathbf{m}_i\left(n\right)}{\pi_i\left(n\right)} \tag{8.23}$$

$$\mathbf{M}_{Back}\left(n\right) = \left(1 - \epsilon\right)\mathbf{M}_{Back}\left(n-1\right) + \epsilon R_{n,Back}\left(\mathbf{t}_n - \boldsymbol{\mu}_{Back}\left(n\right)\right)\left(\mathbf{t}_n - \boldsymbol{\mu}_{Back}\left(n\right)\right)^T \tag{8.24}$$

$$\mathbf{C}_{Back}\left(n\right) = \frac{\mathbf{M}_{Back}\left(n\right)}{\pi_{Back}\left(n\right)} \tag{8.25}$$

where ϵ is a constant step size, $\epsilon \approx 0.01$. The step size is constant (and not decaying) because the system is time varying, as considered in Section 3.2 of [20]. Please note that $\mathbf{m}_i$ and $\mathbf{M}_{Back}$ are auxiliary variables required to update the model parameters π_i, $\boldsymbol{\mu}_i$ and $\mathbf{C}_{Back}$.

The diagonal noise matrix $\boldsymbol{\Psi}$ models the effects of the quantization and the compression on the pixels. This is estimated separately because these effects appear and disappear suddenly from one frame to the next, so that they cannot be captured adequately by the stochastic approximation learning algorithm. Furthermore, we assume that these effects are approximately equal for all the pixels, so that an offline estimation of a global and constant value of $\boldsymbol{\Psi}$ is carried out. The matrix is given by

$$\boldsymbol{\Psi} = \begin{pmatrix} \sigma_R^2 & 0 & 0 \\ 0 & \sigma_G^2 & 0 \\ 0 & 0 & \sigma_B^2 \end{pmatrix} \tag{8.26}$$

8.2.3 Sudden Background Change Detection

Sometimes an object integrates into the background, for example when a car is parked. This produces a sudden change in the background distributions of many pixels, i.e. those which are inside the integrating object. These changes must be dealt with separately [22]. As pointed out by Kushner & Yin in Section 3.2 of [20], stochastic approximation of time varying systems is only advisable when the rate of change of the estimated parameters is slow. Hence, these sudden changes must be tackled with a different procedure which detects these situations that the stochastic approximation framework is unable to cope with.

The detection procedure needs a previous offline study. Let χ be the ground truth for a particular pixel:

$$\chi = \begin{cases} 1 & \text{iff the pixel belongs to the foreground} \\ 0 & \text{iff the pixel belongs to the background} \end{cases} \tag{8.27}$$

Then, let $\widetilde{\chi}$ be the estimation of χ carried out by the basic algorithm, i.e. without sudden background change detection:

$$\widetilde{\chi} = \begin{cases} 1 & \text{iff the pixel is classified as foreground by the basic algorithm} \\ 0 & \text{iff the pixel is classified as background by the basic algorithm} \end{cases} \tag{8.28}$$

The basic algorithm works correctly if and only if $\widetilde{\chi} = \chi$. We define three disjoint random events: undetected background change (*Change*), correct foreground detection (*Good*) and other situations that we are not interested in (*Other*):

$$Change \equiv (\widetilde{\chi} = 1) \wedge (\chi = 0) \tag{8.29}$$

$$Good \equiv (\widetilde{\chi} = 1) \wedge (\chi = 1) \tag{8.30}$$

$$Other \equiv \widetilde{\chi} = 0 \tag{8.31}$$

At frame n, the number of frames $z(n)$ that the pixel has been continuously classified as foreground is defined as follows:

$$z(n) = \max \{ N \mid \widetilde{\chi}(n) = \widetilde{\chi}(n-1) = ... = \widetilde{\chi}(n-N) = 1 \} \tag{8.32}$$

Please note that $z(n)$ is readily computed by a running counter which is incremented each frame that $\widetilde{\chi} = 1$ and reset to zero each time that $\widetilde{\chi} = 0$. The offline task consists in estimating the probabilities $P(Good \mid z, \widetilde{\chi} = 1)$ and $P(Change \mid z, \widetilde{\chi} = 1)$. This is accomplished by counting the number of times that *Change* and *Good* are verified for a certain value of z, where we take as input samples the values of $z(n)$ for all frames n and all pixels such that $\widetilde{\chi} = 1$. These pixel counts are noisy functions of z that are smoothed by kernel regression before estimating the probabilities.

Then we compute a threshold Z:

$$Z = \min \left\{ z \mid P(Change | z, \widetilde{\chi} = 1) > \frac{1}{2} \right\} \tag{8.33}$$

The computation of Z finishes the online study. When the online system is operating, in each pixel we maintain a running counter of the number of frames z. If we have at frame n that a pixel is classified as foreground and $z(n) \geq Z$, then that pixel must be reset. The pixel reset procedure involves setting $\mathbf{C}_{Back} = \mathbf{0}$. Additionally, we swap $\boldsymbol{\mu}_{Back}$ with $\boldsymbol{\mu}_{Fore}$, and $\mathbf{m}_{Back}$ with $\mathbf{m}_{Fore}$.

8.3   Probabilistic Self-Organizing Maps

In this section, a probabilistic model which includes a self-organizing map is proposed to carry out the background modelling. Many efforts have been devoted to probabilistic neural networks [16,28], and in particular self organization is pervasive in many computational intelligence applications [6], but its use for object detection in video sequences is scarce. One of the rare exceptions is the Self-Organizing Background Subtraction (SOBS) method [27,34]. However, unlike our approach, the SOBS is unable to adapt its color similarity measure to the characteristics of the input video, it is unable to yield the probability that a pixel belongs to the background (because it is not a probabilistic model), and it does not provide a model of the foreground objects. Hence, it is completely different from our proposal. The vector quantization based proposal by Doulamis et al. [11] is more related to SOBS, since it uses non probabilistic unsupervised learning like SOBS.

On the other hand, this model is able to coordinate several Gaussians in a computational map to yield a faithful representation of the background, so that each Gaussian adapts its color similarity measure to the local characteristics of a region of the input distribution. A uniform mixture component is then used for the foreground; this mixture component avoids excessive absorption of slowly moving foreground pixels by the background model, so that the trade off problems of Stauffer and Grimson and related approaches are overcome. Additionally, a feedback procedure is integrated into the learning scheme, so that the spatial coherence among nearby pixels is used to improve the segmentation performance.

Next we present our probabilistic model for the foreground object detection problem. First we define the model (Section 8.3.1), then we explain how to adapt its parameters (Section 8.3.2), and finally we outline a method to avoid suboptimal solutions which are found in the learning process (Section 8.3.3).

8.3.1 Model Definition

The model accepts as inputs (training samples) the color values of the pixels of the successive frames of a video. Its goal is to represent the distribution of the input samples faithfully at the same time that it is able to distinguish foreground pixels from background pixels. We can use the RGB, L*a*b* (CIELAB) or any other color space [5], but in all cases the input space dimension is $D = 3$. The likelihood of the observed data (pixel color value) $\mathbf{t}$ at pixel position $\mathbf{x}$ is modeled by a mixture probability density function with two mixture components (*Back* for the background and *Fore* for the foreground):

$$p_{\mathbf{x}}(\mathbf{t}) = \pi_{Back,\mathbf{x}} p_{\mathbf{x}}(\mathbf{t} \mid Back) + \pi_{Fore,\mathbf{x}} p_{\mathbf{x}}(\mathbf{t} \mid Fore) \tag{8.34}$$

$$\pi_{Back,\mathbf{x}} + \pi_{Fore,\mathbf{x}} = 1 \tag{8.35}$$

As seen from (8.34), there is a probabilistic mixture per pixel position $\mathbf{x}$, so that the particular features of the pixels are adequately accounted for. Since moving objects can have any color, the foreground pixels are best modeled by the uniform distribution over the entire color space:

$$p_{\mathbf{x}}(\mathbf{t} \mid Fore) = U(\mathbf{t}) \tag{8.36}$$

$$U(\mathbf{t}) = \begin{cases} 1/Vol(\mathcal{S}) & \text{iff } \mathbf{t} \in \mathcal{S} \\ 0 & \text{iff } \mathbf{t} \notin \mathcal{S} \end{cases} \tag{8.37}$$

where $\mathcal{S}$ is the support of the uniform pdf and $Vol(\mathcal{S})$ is the D-dimensional volume of $\mathcal{S}$. The uniform pdf is a neutral choice which does not favor any color over another. If we had prior information about the pdf of the colors of the foreground objects, we should change the uniform pdf by the supplied pdf. In most practical applications the pixels are given in the RGB color space with 8-bit precision values. In this case we have:

$$\mathcal{S} = \{(t_R, t_G, t_B) \mid t_R, t_G, t_B \in [0, 255]\} \tag{8.38}$$

$$Vol(\mathcal{S}) = 255^3 \tag{8.39}$$

The background pixel values at a given position $\mathbf{x}$ form one or more clusters depending on the complexity of the scene. We propose to approximate this background distribution by means of a probabilistic self-organizing map:

$$p_{\mathbf{x}}\left(\mathbf{t} \mid Back\right) = \frac{1}{H} \sum_{i=1}^{H} p_{\mathbf{x}}\left(\mathbf{t} \mid i\right) \tag{8.40}$$

where H is the number of mixture components (units) of the self-organizing map, and the prior probabilities or mixing proportions are assumed to be equal. A *topological distance* $d\left(i,j\right)$ must be defined between every pair of units $\left(i,j\right)$ of the map. Here the standard choices have proven to be effective in experiments: a rectangular lattice to place the units, with a Euclidean topological distance:

$$d\left(i,j\right) = \|\mathbf{r}_i - \mathbf{r}_j\| \tag{8.41}$$

where $\mathbf{r}_i$ and $\mathbf{r}_j$ are the positions of units i and j in the rectangular lattice, respectively.

As we have one self-organizing map for each position $\mathbf{x}$ in the image, it is of paramount importance that the units have a simple structure so that the computational needs are kept small. To this end we associate a spherical Gaussian probability density to each mixture component $i \in \{1, ..., H\}$ of the map [3, 17, 18]:

$$p_{\mathbf{x}}\left(\mathbf{t} \mid i\right) = \left(2\pi\right)^{-D/2} \left(\sigma_{i,\mathbf{x}}^2\right)^{-D/2}$$
$$\exp\left(-\frac{1}{2\sigma_{i,\mathbf{x}}^2}\left(\mathbf{t} - \boldsymbol{\mu}_{i,\mathbf{x}}\right)\left(\mathbf{t} - \boldsymbol{\mu}_{i,\mathbf{x}}\right)^T\right) \tag{8.42}$$

where $\boldsymbol{\mu}_{i,\mathbf{x}}$ and $\sigma_{i,\mathbf{x}}^2$ are the mean vector and the variance of mixture component i, respectively:

$$\boldsymbol{\mu}_{i,\mathbf{x}} = E\left[\mathbf{t} \mid i, \mathbf{x}\right] \tag{8.43}$$

$$\sigma_{i,\mathbf{x}}^2 = E\left[\frac{1}{D}\left(\mathbf{t} - \boldsymbol{\mu}_{i,\mathbf{x}}\right)^T\left(\mathbf{t} - \boldsymbol{\mu}_{i,\mathbf{x}}\right) \mid i, \mathbf{x}\right] \tag{8.44}$$

The overall goal of the model is to distinguish the foreground objects from the background reliably. A Bayesian estimation of the probability that the observed data (pixel color value) $\mathbf{t}$ is foreground is given by

$$P_{Fore,\mathbf{x}}\left(\mathbf{t}\right) =$$
$$\frac{\pi_{Fore,\mathbf{x}} p_{\mathbf{x}}\left(\mathbf{t} \mid Fore\right)}{\pi_{Back,\mathbf{x}} p_{\mathbf{x}}\left(\mathbf{t} \mid Back\right) + \pi_{Fore,\mathbf{x}} p_{\mathbf{x}}\left(\mathbf{t} \mid Fore\right)} \tag{8.45}$$

and the corresponding probability of the background is

$$P_{Back,\mathbf{x}}\left(\mathbf{t}\right) = 1 - P_{Fore,\mathbf{x}}\left(\mathbf{t}\right) \tag{8.46}$$

However, $P_{Fore,\mathbf{x}}\left(\mathbf{t}\right)$ and $P_{Back,\mathbf{x}}\left(\mathbf{t}\right)$ are prone to noise due to isolated pixels that change their color randomly. We propose to use the information from the 8-neighbors of a given pixel $\mathbf{x}$ to remove this noise. However, we cannot take into account all the 8-neighbors equally, since many neighboring pixels are not related. For example, this can happen at the border of a road: the pixels off the road (where no vehicles occur) are almost independent from those inside the road (where vehicles pass frequently) despite their proximity. Consequently, we need a quantitative measure of the correlation of pairs of pixels. Our selected measure is Pearson's correlation [30] between the random variables $P_{Fore,\mathbf{x}}$ and $P_{Fore,\mathbf{y}}$ corresponding to each pair of 8-neighboring pixels $\mathbf{x}$ and $\mathbf{y}$:

$$\rho_{\mathbf{x},\mathbf{y}} = \frac{\phi_{\mathbf{x},\mathbf{y}}}{\sqrt{\nu_{\mathbf{x}}}\sqrt{\nu_{\mathbf{y}}}} \tag{8.47}$$

$$\phi_{\mathbf{x},\mathbf{y}} = \text{cov}\left(P_{Fore,\mathbf{x}}, P_{Fore,\mathbf{y}}\right) =$$
$$E\left[\left(P_{Fore,\mathbf{x}} - E\left[P_{Fore,\mathbf{x}}\right]\right)\left(P_{Fore,\mathbf{y}} - E\left[P_{Fore,\mathbf{y}}\right]\right)\right] \tag{8.48}$$

$$\nu_{\mathbf{x}} = \text{var}\left(P_{Fore,\mathbf{x}}\right) = E\left[\left(P_{Fore,\mathbf{x}} - E\left[P_{Fore,\mathbf{x}}\right]\right)^2\right] \tag{8.49}$$

$$\nu_{\mathbf{y}} = \text{var}\left(P_{Fore,\mathbf{y}}\right) = E\left[\left(P_{Fore,\mathbf{y}} - E\left[P_{Fore,\mathbf{y}}\right]\right)^2\right] \tag{8.50}$$

where we have:

$$E\left[P_{Fore,\mathbf{x}}\right] = \pi_{Fore,\mathbf{x}} \tag{8.51}$$

$$E\left[P_{Fore,\mathbf{y}}\right] = \pi_{Fore,\mathbf{y}} \tag{8.52}$$

Please note that the properties of Pearson's correlation ensure that:

$$\rho_{\mathbf{x},\mathbf{y}} \in [-1, 1] \tag{8.53}$$

$$\rho_{\mathbf{x},\mathbf{y}} = \rho_{\mathbf{y},\mathbf{x}} \tag{8.54}$$

where the last equation saves one half of the computations.

The value of $\rho_{\mathbf{x},\mathbf{y}}$ is large and positive if and only if pixels $\mathbf{x}$ and $\mathbf{y}$ are usually assigned to the same class, i.e. either both are background or both are foreground. On the other hand, if both pixels are independent, then we have $\rho_{\mathbf{x},\mathbf{y}} = 0$; please remember that independence implies uncorrelatedness but not vice versa. Negative correlations are expected to be uncommon, and theoretically correspond to pairs of pixels which are usually assigned to opposite classes. However, in practice negative correlations are mostly due to noise in the estimations or the input data.

The correlations $\rho_{\mathbf{x},\mathbf{y}}$ allow us to obtain a noise-removed version of $P_{Fore,\mathbf{x}}\left(\mathbf{t}\right)$ by combining it with the information from the 8-neighbors of $\mathbf{x}$:

$$\widetilde{P}_{Fore,\mathbf{x}}\left(\mathbf{t}\right) = \text{trunc}\left(\frac{1}{9}\sum_{\mathbf{y}\in Neigh(\mathbf{x})}\rho_{\mathbf{x},\mathbf{y}}P_{Fore,\mathbf{y}}\left(\mathbf{t}\right)\right) \tag{8.55}$$

where $Neigh\left(\mathbf{x}\right)$ contains the pixel $\mathbf{x}$ and its 8-neighbors, and

$$\rho_{\mathbf{x},\mathbf{x}} = 1 \tag{8.56}$$

$$\text{trunc}\left(z\right) = \begin{cases} z & \text{iff } z \geq 0 \\ 0 & \text{otherwise} \end{cases} \tag{8.57}$$

$$\widetilde{P}_{Fore,\mathbf{x}}\left(\mathbf{t}\right) \in [0, 1] \tag{8.58}$$

The trunc function is used in (8.55) to deactivate the parameter learning by setting $\widetilde{P}_{Fore,\mathbf{x}}\left(\mathbf{t}\right) = 0$ when an excess of noisy negative correlations is present; please see the update equations in Subsection 8.3.2. It has been found in experiments that the argument of the trunc function in equation (8.55) is negative in less than 0.5% of the cases, because $\rho_{\mathbf{x},\mathbf{y}}$ is nearly always positive. Nevertheless, the performance is significantly increased by selectively deactivating learning with this strategy.

In the following subsection we explain how the model we have just defined can be trained. This is done by means of stochastic approximation [20, 21], which is a standard method to develop online learning algorithms for probabilistic self-organizing maps [17, 23, 26, 42]. The proposed learning method is specifically tailored to yield a very low computational complexity $O(HD)$, which is linear both in the number of neurons H and in the input space dimension D, like the SOBS approach [27].

8.3.2   Learning

The model parameters are learned by means of Robbins-Monro stochastic approximation algorithm [20, 21, 29], where $\widetilde{P}_{Fore,\mathbf{x}}(\mathbf{t}_{n,\mathbf{x}})$ is our best estimation of the a posteriori probability that the input sample $\mathbf{t}_{n,\mathbf{x}}$ at pixel position $\mathbf{x}$ and time instant n belongs to the foreground:

$$P_{\mathbf{x}}(Fore \mid \mathbf{t}_{n,\mathbf{x}}) = \widetilde{P}_{Fore,\mathbf{x}}(\mathbf{t}_{n,\mathbf{x}}) \tag{8.59}$$

$$P_{\mathbf{x}}(Back \mid \mathbf{t}_{n,\mathbf{x}}) =$$

$$\widetilde{P}_{Back,\mathbf{x}}(\mathbf{t}_{n,\mathbf{x}}) = 1 - P_{\mathbf{x}}(Fore \mid \mathbf{t}_{n,\mathbf{x}}) \tag{8.60}$$

The adaptation is done in online mode. That is, at time instant n a new input frame (image) is presented to the system, and the model parameters for every pixel position $\mathbf{x}$ are modified according to the pixel color $\mathbf{t}_{n,\mathbf{x}}$ at that position. The corresponding update equations are as follows:

$$\pi_{Fore,\mathbf{x}}(n) = (1 - \epsilon(n))\,\pi_{Fore,\mathbf{x}}(n-1) +$$
$$\epsilon(n)\,\widetilde{P}_{Fore,\mathbf{x}}(\mathbf{t}_{n,\mathbf{x}}) \tag{8.61}$$

$$\pi_{Back,\mathbf{x}}(n) = 1 - \pi_{Fore,\mathbf{x}}(n) \tag{8.62}$$

$$\nu_{\mathbf{x}}(n) = (1 - \epsilon(n))\,\nu_{\mathbf{x}}(n-1) +$$
$$\epsilon(n)\left(\widetilde{P}_{Fore,\mathbf{x}}(\mathbf{t}_{n,\mathbf{x}}) - \pi_{Fore,\mathbf{x}}(n)\right)^2 \tag{8.63}$$

$$\phi_{\mathbf{x},\mathbf{y}}(n) = (1 - \epsilon(n))\,\phi_{\mathbf{x},\mathbf{y}}(n-1) +$$
$$\epsilon(n)\left(\widetilde{P}_{Fore,\mathbf{x}}(\mathbf{t}_{n,\mathbf{x}}) - \pi_{Fore,\mathbf{x}}(n)\right)$$
$$\left(\widetilde{P}_{Fore,\mathbf{y}}(\mathbf{t}_{n,\mathbf{y}}) - \pi_{Fore,\mathbf{y}}(n)\right) \tag{8.64}$$

$$\rho_{\mathbf{x},\mathbf{y}}(n) = (1 - \epsilon(n))\,\rho_{\mathbf{x},\mathbf{y}}(n-1) +$$
$$\epsilon(n)\,\frac{\phi_{\mathbf{x},\mathbf{y}}(n)}{\sqrt{\nu_{\mathbf{x}}(n)}\sqrt{\nu_{\mathbf{y}}(n)}} \tag{8.65}$$

$$\boldsymbol{\mu}_{i,\mathbf{x}}(n) = (1 - \xi_{\mathbf{x}}(i,n))\,\boldsymbol{\mu}_{i,\mathbf{x}}(n-1) + \xi_{\mathbf{x}}(i,n)\,\mathbf{t}_{n,\mathbf{x}} \tag{8.66}$$

$$\sigma_{i,\mathbf{x}}^2(n) = (1 - \xi_{\mathbf{x}}(i,n))\,\sigma_{i,\mathbf{x}}^2(n-1) +$$
$$\xi_{\mathbf{x}}(i,n)\,\frac{1}{D}\,(\mathbf{t}_{n,\mathbf{x}} - \boldsymbol{\mu}_{i,\mathbf{x}}(n))^T\,(\mathbf{t}_{n,\mathbf{x}} - \boldsymbol{\mu}_{i,\mathbf{x}}(n)) \tag{8.67}$$

$$\xi_{\mathbf{x}}(i,n) = \epsilon(n)\,\frac{\Lambda(i, Winner(\mathbf{x},n))}{\pi_{Back,\mathbf{x}}(n)}\,\widetilde{P}_{Back,\mathbf{x}}(\mathbf{t}_{n,\mathbf{x}}) \tag{8.68}$$

where we use a Gaussian *neighborhood function* Λ (not to be confused with the probability density function $p_{\mathbf{x}}(\mathbf{t} \mid i)$ of the mixture components), which varies with the time step n and depends on a decaying *neighborhood radius* $\Delta(n)$ and the topological distance $d(i,j)$ to the winner:

$$\Lambda(i, Winner(\mathbf{x},n)) =$$
$$\exp\left(-\left(\frac{d(i, Winner(\mathbf{x},n))}{\Delta(n)}\right)^2\right) \tag{8.69}$$

$$\Delta(n+1) \leq \Delta(n) \tag{8.70}$$

Please note that we use $\widetilde{R}_{Fore,\mathbf{x}}$ and $\widetilde{R}_{Back,\mathbf{x}}$ instead of $R_{Fore,\mathbf{x}}$ and $R_{Back,\mathbf{x}}$ in equations (8.61), (8.63), (8.64) and (8.68) because the two former values are more reliable than the latter, i.e. the former values have much less noise. Moreover, this is a mechanism to feed back the foreground detection information of neighboring pixels to the learning procedure. Consequently, the models of nearby pixels cooperate to yield a coordinated, coherent result.

8.3.3 Suboptimal Solutions Avoidance

As in any other application, the probabilistic mixture learning method we have just outlined can fall into a suboptimal solution, i.e. one which does not represent the input distribution faithfully. In our case we have identified two problems that arise in practice:

1. The variance $\sigma_{i,\mathbf{x}}^2$ of a unit diminishes too much. This makes the unit excessively specialized in a tiny cluster of input samples, rendering it useless.

2. The variance $\sigma_{i,\mathbf{x}}^2$ of a unit grows excessively. As a consequence, the unit starts to model all the input samples, no matter whether they belong to the foreground or the background, and the detection performance falls abruptly.

We have devised a procedure to tackle both problems at a time. We define two thresholds, σ_{max}^2 and σ_{min}^2, and we require that each unit i satisfies the following:

$$\sigma_{min}^2 \leq \sigma_{i,\mathbf{x}}^2 \leq \sigma_{max}^2 \tag{8.71}$$

This is enforced by setting $\sigma_{i,\mathbf{x}}^2$ to σ_{max}^2 or σ_{min}^2 every time that the update equation (8.67) produces a new value that does not fulfill (8.71). That way, the two mentioned problems are removed. As an additional advantage, we have found that it is not necessary to tune the thresholds σ_{max}^2 and σ_{min}^2, since a fixed value works smoothly for all the tested videos.

8.4 Experimental Results

In this section a principled evaluation methodology is developed in order to compare the two previous approaches, namely, the mixture of Gaussians with an stochastic approximation learning (AE)[23] described in Section 8.2 and the probabilistic self-organising model (SOM)[24] described in Section 8.3. The experimental design is exposed in Section 8.4.1, where the methods and sequences to compare are selected and the performance measures are discussed. The qualitative and quantitative results are evaluated in Sections 8.4.2 and 8.4.3.

8.4.1 Experimental Design

Methods

Considering that the proposed methods are included in the so-called pixel-level methods since they analyze the scene at a low level pixel by pixel, the comparison is carried out with

---

[23] Further information about the source code and test videos is available in http://www.lcc.uma.es/~ezeqlr/backsa/backsa.html

[24] The source code and a demo of our proposal are available at http://www.lcc.uma.es/ezeqlr/fsom/fsom.html

TABLE 8.1 Parameter settings tested during evaluation of the object detection algorithms. The combination of all the values of the parameters generate a set of configurations for each method.

Algorithm	Test Parameters
PF [41]	Learning rate, $\alpha = \{1e^{-4}, 5e^{-4}, 0.001, 0.005, 0.01, 0.05, 0.1\}$ Threshold for Gaussian tiles in standard deviations, $\mathcal{N} = \{2, 2.5, 3, 5\}$
$AGMM$ [45]	Learning rate, $\alpha = \{1e^{-4}, 5e^{-4}, 0.001, 0.005, 0.01, 0.05, 0.1\}$ No. of Gaussian mixture components, K$= \{3, 5, 7\}$ Threshold for Gaussian tiles in standard deviations, $\mathcal{N} = \{2, 2.5, 3, 5\}$ Weight threshold, $\mathcal{T} = \{0.6, 0.8\}$
FGD [22]	Learning rate, $\alpha_1 = \{0.01, 0.05, 0.1\}$ $\alpha_2 = \{1e^{-4}, 5e^{-4}, 0.001, 0.005, 0.01, 0.05, 0.1\}$ $\alpha_3 = \{0.01, 0.05, 0.1\}$ Similarity threshold, $\delta = \{2, 2.5, 3\}$
$SOBS$ [27]	Number of sequence frames for calibration, K$= \{100, 200, 300\}$ Calibration distance threshold, $\epsilon_1 = \{0.1, 0.2\}$ Online distance threshold, $\epsilon_2 = \{0.001, 0.005, 0.01\}$ Learning factor, $c_2 = \{0.005, 0.01, 0.05\}$
AE	Step size, $\varepsilon_0 = \{1e^{-4}, 5e^{-4}, 0.001, 0.005, 0.01, 0.05, 0.1\}$
SOM	Learning Rate, $\alpha = \{0.001, 0.005, 0.01, 0.05\}$ Step size, $\varepsilon_0 = \{0.001, 0.005, 0.01, 0.05\}$ Number of neurons, $H = \{6, 12, 18\}$

regard to several techniques of the same class. The *Pfinder* method (PF) proposed in [41], the adaptive Mixture of Gaussians approach $AGMM$, which is developed in [45], the Li et al. approach [22] called FGD, and the background model based on self-organizing maps called Self-organizing Background Subtraction ($SOBS$) by Maddalena and Petrosino [27,34]. This latter technique is included to compare the SOM method with similar alternatives.

Although our AE model defines several parameters, we have empirically found that a fixed value for some of them is suitable for all the studied scenes. Specifically, the number of frames used in the initialization phase is always assigned to $K = 200$, and the global smoothing parameter is held fixed to $h = 2$. Therefore, in this model only the step size parameter ϵ_0, which influences the learning process, must be tuned for each scene. According to the SOM approach, the model has some parameters whose variation does not significantly influence the results. Thus, we have empirically selected a one-dimensional line topology, the radius of neighborhood at the convergence point is assigned to 0.1, and the boundaries of the variance are within the range $\sigma^2_{min} = 9$ and $\sigma^2_{max} = 25$. The learning rate associated to the map (λ), the step size (ε_0), which handles how the neurons are updated with the input data, and the number of neurons of the linear topology H are considered as the tunable parameters.

The values of the tunable parameters both the comparison methods as the two presented approaches can be found in Table 8.1.

Our SOM approach and the FGD technique apply an implicit post-processing analysis to improve the segmentation results and take into account the spatial information of each pixel. With the aim of performing a fair comparison, morphological operators are used to enhance the output mask in three of the remaining methods ($AGMM$, PF and $SOBS$). Pearson correlation, which is used in SOM to reinforce the model (Section 8.3.1), is considered as well as a filter-noise method and it is used to improve the output mask of the AE strategy. All the techniques incorporate a filter of spurious objects which removes that ones whose area is less than ten pixels.

Sequences

A representative set of twelve outdoor and indoor video sequences have been used to assess the analyzed methods, which come together with the ground truth information (GT) to validate the effectiveness of an approach from a quantitative viewpoint. The GT consists of a set of manually-segmented images where the sets of pixels belonging to the foreground and the background are defined. At present, some public repositories comprise valuable and useful sequences to compare and contrast object detection methods:

- The repository created by the International Conference VSSN 2006 on the occasion of an algorithm competition in Foreground/Background Segmentation contains some interesting sequences. These sequences are considered as synthetic ones since they have a real background together with artificial 3D objects moving along the scene. The most useful feature of these datasets is that we can get the GT information of all the analyzed frames. Specifically, the sequences named Video2 (*V2*) and Video4 (*V4*), with 749 and 819 frames respectively, have been selected for our comparison. Their frame size is 384 x 240.

- A set of complex sequences has been developed by Li et al. and is available in their website[25]. These sequences try to show most of the common and difficult situations which a robust and reliable object detection technique has to face. Some of them deal with background variability in outdoor scenes (campus scene, *CAM*, 1439 frames); to ignore the stationary objects with repetitive movements in indoor scenes (meeting room, *MR*, and subway station, *SS*, with 2964 and 3417 frames respectively), or to handle both gradual and sudden illumination changes, which are caused, for instance, along the day in an outdoor scene (water surface, *WS*, 633 frames) or when a light is switched off in an indoor one, respectively. Additionally, the small scrolling of fixed cameras caused by weather effects (a fountain in a public square, *FT*, 523 frames) can reduce significantly the quality of the results. Each sequence has a frame size of 320 x 240.

- Three videos from the IPPR (Image Processing and Pattern Recognition) contest held in Taiwan in 2006 are also taken, whose names are IPPRData1, IPPRData2 and IPPRData3, with manually generated ground-truth images[26]. Two of them correspond to indoor sequences in which a corridor is observed from a different point of view, whereas IPPRData3 is the result of recording a highway from above. The frame size is the same as the previous datasets (320x240) and their number of frames is 300.

- Another very popular repository called CAVIAR[27] has been also used to get some sequences. Its only drawback is that the ground truth information is mainly related to the tracking phase and it does not include the ideal segmentation of a set of frames. For this reason we have selected the only sequence with this kind of ground truth information, where people are walking on a corridor in a leisure center (*OS*) and whose frame size is 384 x 288 and 1377 frames.

- Finally, one of the sequences included in [33] and set as a resource in the Sheikh's web page[28] is evaluated. With 500 frames and a frame size of 360 x 240, it shows a railroad crossing without barriers in which both people and vehicles are crossing.

---

[25] http://perception.i2r.a-star.edu.sg/bk_model/bk_index.html
[26] http://media.ee.ntu.edu.tw/Archer_contest/
[27] http://homepages.inf.ed.ac.uk/rbf/CAVIARDATA1/
[28] http://www.cs.cmu.edu/~yaser/GroundtruthSeq.rar

Performance Measures

Many algorithms have been proposed for objective evaluation of motion detection methods. Since successful tracking relies heavily on accurate object detection, the evaluation of object detection algorithms within a surveillance systems plays an important role in overall performance analysis of the whole system. Generally, the selection of a set of evaluation metrics depends largely on the kind of applied segmentation. Provided that the ground truth is supplied, evaluation techniques based on comparison with that ideal output can be further classified according to the type of metrics they propose. Typically, pixel-based metrics [4, 13, 32] consist of a combination of *true positives* (tp), *false positives* (fp), *true negatives* (tn) and *false negatives* (fn), computed over each frame and whose sum corresponds to the image size. It should be noted that *fp* and *fn* refer to pixels misclassified as foreground (*fp*) or background (*fn*), while *tp* and *tn* account for accurately classified pixels.

In this paper a similarity standard measure based on the previous pixel-based metrics is applied. It was defined as a metric to check the MPEG Segmentation Quality [40] and used as the starting point of other developed metrics [39]. It has also been used to compare different object detection methods [22, 38]. This quantitative measure named *spatial accuracy* is defined as follows:

$$AC = \frac{\text{card}\,(A \cap B)}{\text{card}\,(A \cup B)} \tag{8.72}$$

where "card" stands for the number of elements of a set, A is the set of all pixels which belong to the foreground, and B is the set of all pixels which are classified as foreground by the analyzed method:

$$A = \{\mathbf{x} \mid \chi(\mathbf{x}) = 1\} \qquad B = \{\mathbf{x} \mid \widetilde{\chi}(\mathbf{x}) = 1\} \tag{8.73}$$

In order to take into account some information about false negatives and false positives, the proportions of these metrics are also defined and normalized using the foreground mask:

$$FN = \frac{\text{card}\,(A \cap \overline{B})}{\text{card}\,(A \cup B)} \qquad FP = \frac{\text{card}\,(\overline{A} \cap B)}{\text{card}\,(A \cup B)} \tag{8.74}$$

From set theory we know that:

$$AC,\ FN,\ FP \in [0, 1] \qquad AC + FN + FP = 1 \tag{8.75}$$

The optimal performance would be achieved for $AC = 1,\ FN = 0,\ FP = 0$. On the other hand, the worst possible performance corresponds to $AC = 0,\ FN + FP = 1$. Overall evaluation metrics are found as their average over the entire video sequence. It should be noted that the AC measure is faithful provided that there are foreground objects in the evaluated frame. Otherwise, the most probable configuration is $FP = 1,\ FN = 0,\ AC = 0$ because of the more than likely occurrence of false positives after the detection. Therefore, this configuration does not truly reflect the effectiveness of the methods since it is only based on the detected foreground and not in the background. Moreover, in outdoor scenes with great background variability, and where moving objects are quite small compared to the image size, the results of the assessment will significantly decrease due to this foreground importance. To handle this problem, only frames containing foreground objects are evaluated with the aim of not affecting the overall accuracy measure adversely.

As an alternative measure which tries to avoid the previous drawback, we use two information retrieval measurements, recall and precision, to quantify how well each algorithm matches the ground-truth. Like the previous accuracy measure, the new ones have also

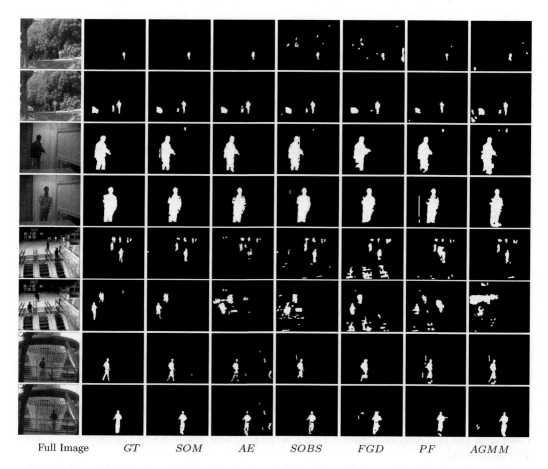

| Full Image | *GT* | *SOM* | *AE* | *SOBS* | *FGD* | *PF* | *AGMM* |

FIGURE 8.1 Experimental results on scenes with high variability on the background. An outdoor campus scene (*CAM*) with plentiful vegetation moving continuously corresponds to the first two rows with frames 1392 and 1831. A meeting room environment (*MR*) with wavering curtains is shown in the third and fourth rows with frames 2847 and 3893. The last rows represent the analysis of the frames 3078 and 3279 of a subway station environment (*SS*) with moving escalators and a waterwise scene showing a fountain (*FT*) with frames 1422 and 1440 respectively.

been rather applied in vision for comparing the performance of motion detection algorithms. Both are considered as a standard within this scope. Therefore, the precision (*PR*) and recall (*RC*) measures can be computed as follows (higher is better):

$$PR = \frac{\mathrm{card}\,(A \cap B)}{\mathrm{card}\,(B)} \qquad RC = \frac{\mathrm{card}\,(A \cap B)}{\mathrm{card}\,(A)} \qquad PR,\ RC \in [0,1] \qquad (8.76)$$

RC is defined as the percentage of the correctly detected pixels to the real moving objects whereas *PR* is defined as the percentage of the detected pixels to all detected moving object pixels. High recall means less segmentation misses while high precision means less false segmentation. Often there is a trade-off between precision and recall, i.e. we can make one of them grow as desired at the expense of diminishing the other.

8.4.2　Qualitative Results

The quality of the segmentation results is shown for each method in Figures 8.1, and 8.2. Each row exhibits a qualitative comparison among the best configurations of all the methods, in order to get a visual idea of which is the best performing method. All the rows correspond to some frames from the selected datasets. The first column displays some frames from the different benchmark sequences whereas in the second column the manually-segmented foreground mask (GroundTruth) is depicted. From the third column to the eighth one, the segmentation results of the following methods are showed: the probabilistic self-organizing map approach (SOM), the mixture of distributions with stochastic approximation (AE), the Maddalena and Petrosino technique ($SOBS$), the Li et al. approach (FGD), the single Gaussian distribution model (PF) and the adaptive mixture of Gaussians ($AGMM$), respectively.

Several visual conclusions can be extracted from observing the images. In most cases, the SOM algorithm achieves better results than the other alternatives, both in indoor and outdoor scenes. In particular, the detection of foreground objects is more reliable and effective with regard to the real ones in this approach, as we can observe in the campus (CAM) and subway station (SS) scenes (the two first and the fifth and sixth rows in Figure 8.1). Unlike the other methods, repetitive movements of background objects are also implicitly avoided in our model, especially in the SS sequence, where the escalators are explicitly considered as foreground in the rest of the methods. The feedback process of the SOM approach achieves to improve the quality of the segmentation progressively, removing spurious pixels and little objects, and dealing with slight camera movements which can appear in outdoor scenes, caused by hard winds or vibrations in the camera stand. This effect is particularly annoying after applying pixel level methods. In spite of morphological operations correct it slightly, the results are not so good as in our method, as we can observe in the last rows of Figure 8.1 (FT sequence) and first rows of Figure 8.1 ($IP1$ and $IP2$ sequences).

The results for the AE model provide lights and shadows. On one hand, the quality of the segmentation for the sequences CAM, MR, $V2$, $V4$ and WS is considerably good, equaling or even overcoming the results of the SOM method. However, for sequences $IP1$, $IP2$ and $IP3$ both incomplete foreground objects as spurious noise are generated, blurring the identification of what is actually movement ($IP2$ sequence). Furthermore, the method is very sensitive to any vibration of the camera, something that is clearly observable by analyzing the last two rows of Figure 8.1 (sequence FT). In this case, the use of Pearson correlation as postprocessing technique does not help to totally remove spurious noise generated by the method.

The problem of *camouflage*, in which there is a significant similarity between the pixel colors from the background and foreground and causing the appearance of holes in the foreground mask, is also presented in several sequences as $IP3$, OC or WS. This effect is partially corrected in the two approaches, completing the man's feet in the WS scene, the car in the railway sequence (LC) and the woman's body in the corridor sequence (OC), which is not handled in some of the competing techniques such as $AGMM$, FGD and PF. In stationary scenes like $V2$ (seventh row in Figure 8.1), there is not much difference between the segmentation quality in all the assessed techniques as we can observe. However, a non desirable effect is produced after applying the FGD method in sequences with quick movements on the foreground, which leaves a trace of the motion objects in previous frames. This is particularly evident in sequences MR, $IP1$, $V2$ and $V4$.

These figures have provided a qualitative assessment of the selected methods performance. In order to show more clearly the improvements achieved by our SOM and AE approaches, the corresponding quantitative evaluations are listed in Yables 8.2 and 8.3,

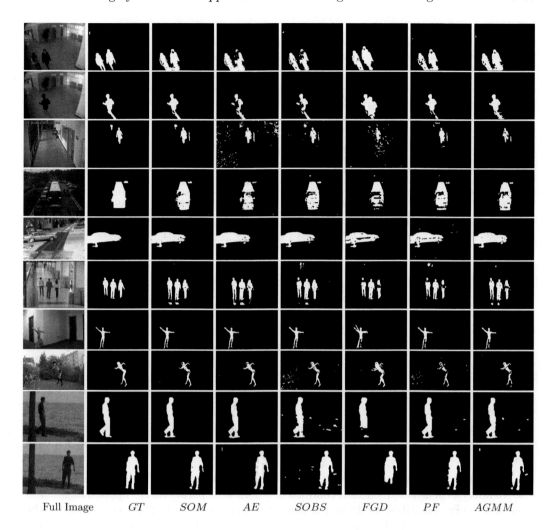

| Full Image | GT | SOM | AE | SOBS | FGD | PF | AGMM |

FIGURE 8.2 Another set of experimental results. Two corridor indoor sequences (*IP1* and *IP2*) can be observed in the first, second and third rows (frames 129, 159 and 285). The fourth row represents a highway (*IP3*), sight from above with frame 294. From the fifth row it is possible to observe level crossing environments (*LC*, frame 413), scenes with corridors in leisure centers (*OC*), sixth row (frame 390), synthetic sequences which model 3D objects over real backgrounds (*V2* and *V4*), corresponding to the seventh and eighth lines, with frames 555 and 794, and seaside scenes, displayed in the last rows with frames 1499 and 1624 respectively.

where a comparison is provided for each sequence, using a set of performance evaluation measures. This point is explained in the following subsection.

8.4.3 Quantitative Results

A comparison between the studied methods for each sequence is displayed in Table 8.2 in which the accuracy (*AC*) previously computed in the equation (8.72) is used to measure the results of each method performance. According to the results, both our *SOM* and *AE* approaches are the best for four of the twelve test sequences. On average, the *SOM*

TABLE 8.2 Quantitative evaluation using the accuracy measure (equation 8.72) and comparison results over the test sequences. The results show the mean and standard deviation after analyzing each dataset, where the best performance for each sequence is highlighted in bold. The SOM approach is the best method in seven of twelve sequences and on average over all the scenes analyzed.

	AE	SOM	SOBS	FGD	PF	AGMM
Campus (*CAM*)	**0.772±0.070**	0.724±0.088	0.639±0.164	0.640±0.155	0.488±0.138	0.645±0.139
Meeting Room (*MR*)	0.810±0.058	0.811±0.071	**0.851±0.056**	0.829±0.062	0.688±0.110	0.749±0.065
Subway Station (*SS*)	0.461±0.135	**0.632±0.108**	0.545±0.120	0.371±0.152	0.421±0.146	0.524±0.153
Fountain (*FT*)	0.543±0.087	0.706±0.075	0.690±0.053	0.559±0.112	0.656±0.078	**0.718±0.067**
IPPRData1 (*IP1*)	0.671±0.167	**0.740±0.107**	0.727±0.107	0.639±0.092	0.688±0.104	0.715±0.100
IPPRData2 (*IP2*)	0.519±0.131	**0.760±0.081**	0.732±0.086	0.485±0.176	0.696±0.095	0.635±0.133
IPPRData3 (*IP3*)	**0.639±0.166**	0.637±0.224	0.597±0.252	0.516±0.298	0.583±0.222	0.620±0.188
Level Crossing (*LC*)	0.900±0.030	0.910±0.030	**0.910±0.026**	0.779±0.053	0.812±0.055	0.885±0.032
Corridor (*OC*)	0.716±0.039	0.739±0.054	**0.767±0.036**	0.755±0.026	0.721±0.036	0.719±0.049
Video2 (*V2*)	0.918±0.020	**0.923±0.022**	0.902±0.021	0.745±0.061	0.890±0.029	0.896±0.031
Video4 (*V4*)	**0.760±0.114**	0.748±0.104	0.596±0.098	0.694±0.078	0.557±0.068	0.704±0.092
WaterSurface (*WS*)	**0.909±0.024**	0.891±0.027	0.858±0.027	0.769±0.064	0.858±0.031	0.832±0.036
Average	0.718±0.156	**0.769±0.083**	0.735±0.087	0.648±0.111	0.672±0.093	0.720±0.090

method is selected as the best of all the proposed alternatives whereas the *AE* method falls to the fourth place. By analyzing the *AE* results in deeper detail, we can observe that this drop in performance is because the accuracy is rather poor in comparison with the other methods for the sequences *SS*, *FT* and *IP2*. Nevertheless, for the rest of sequences (*CAM*, *MR*, *LC*, *OC*, *V2*, *V4*, and *WS*) the performance is almost always above its main competitor, *AGMM*, because of their similar characteristics. This implies that *AE* could be quite sensitive to flickering camera scenes (*FT*), or background elements in motion (*SS*), and the attenuating effect of the postprocessing does not positively influence in a improvement of the performance.

The results of the best performing configuration are shown for each sequence in Table 8.3, including three different quality measures and the parameter configuration for each method.

The influence of the postprocessing phase could be significant and is also studied in Table 8.4. An analysis of nine sequences is conducted, where we separate all the applied algorithms in a basic stage identified as *B* and a postprocessing stage, which is recognised by the initials *PP*. The second phase involves the execution of 1-step of morphological operators (erosion and dilation) for *FGD*, *PF* and *AGMM*, and the Pearson correlation for our approaches *SOM* and *AE*. *SOBS* already includes intrinsically neighborhood information as post-processing method. A removal filter for spurious objects (with less than ten pixels) is applied in the postprocessing phase for all methods. In particular, although the *SOM* approach works as a whole, it is divided in three steps in order to analyze them separately: a basic stage which includes the probabilistic self organizing model, the postprocessing stage to exploit the neighborhood of the pixels (Pearson correlation) and a feedback process (identified as F) which uses the enhanced foreground mask to update the model, exclusive only for this method.

By observing the information provides in Table 8.4, it is possible to highlight several ideas. Postprocessing techniques (both Pearson correlation as morphological operations) significantly improve the output of the object detection methods for scenes with a lot of variability in the background, which implies many false positives in the mask, as happen in the sequences Campus (*CAM*) and Video4 (*V4*) due to the existence of abundant vegetation, and in the scenes Fountain (*FT*) and WaterSurface (*WS*) by the movement of water. On the other hand, in stable scenes where the background is fairly constant with little movement, such as occurs in Level Crossing (*LC*), Corridor (*OC*) or Video2 (*V2*) scenes, the postprocessing phase produces a reduced improvement and an increase in the processing time of the algorithm.

TABLE 8.3   Best configuration for each method over all the analyzed sequences. The sequence names are shown in the first column whereas the second column one indicates the applied methods. The accuracy, precision and recall measures (third, fourth and fifth column) are listed for each method in the second column. The test parameter configuration is in the last column. The row highlights in bold correspond to the best option for that sequence between all the alternatives.

Sequence	Method	Accuracy	Precision	Recall	Parameters
	AE	**0.772±0.070**	0.85±0.09	0.90±0.06	ε_0=0.005
	SOM	0.724±0.088	0.87±0.10	0.82±0.09	H=12, α=0.01, ε_0=0.01
	SOBS	0.638±0.164	0.75±0.20	0.82±0.09	$\mathcal{K}$=100, ε_1=0.2, ε_2=0.01, c_2=0.05
	FGD	0.640±0.155	0.76±0.19	0.83±0.11	α_1=0.01, α_2=0.001, α_3=0.05, δ=3
	PF	0.488±0.138	0.86±0.14	0.55±0.17	$\mathcal{N}$=5, α=0.005
	AGMM	0.645±0.139	0.72±0.17	0.87±0.07	$\mathcal{N}$=2, α=0.01, $\mathcal{K}$=7, $\mathcal{T}$=0.8
	AE	0.810±0.058	0.97±0.03	0.83±0.06	ε_0=0.0005
	SOM	0.811±0.071	0.98±0.02	0.83±0.07	H=6, α=0.005, ε_0=0.005
	SOBS	**0.851±0.056**	0.94±0.05	0.90±0.04	$\mathcal{K}$=200, ε_1=0.2, ε_2=0.005, c_2=0.05
	FGD	0.829±0.062	0.96±0.03	0.86±0.06	α_1=0.01, α_2=0.001, α_3=0.05, δ=2
	PF	0.688±0.110	0.85±0.07	0.77±0.09	$\mathcal{N}$=3, α=0.001
	AGMM	0.749±0.065	0.87±0.04	0.84±0.07	$\mathcal{N}$=3, α=0.001, $\mathcal{K}$=5, $\mathcal{T}$=0.6
	AE	0.461±0.135	0.69±0.17	0.60±0.17	ε_0=0.05
	SOM	**0.632±0.108**	0.83±0.11	0.73±0.12	H=6, α=0.005, ε_0=0.001
	SOBS	0.545±0.120	0.64±0.15	0.79±0.10	$\mathcal{K}$=300, ε_1=0.2, ε_2=0.01, c_2=0.01
	FGD	0.371±0.152	0.48±0.23	0.69±0.09	α_1=0.01, α_2=0.01, α_3=0.05, δ=2
	PF	0.421±0.146	0.67±0.20	0.54±0.16	$\mathcal{N}$=5, α=0.005
	AGMM	0.524±0.153	0.66±0.21	0.74±0.11	$\mathcal{N}$=2.5, α=0.001, $\mathcal{K}$=5, $\mathcal{T}$=0.8
	AE	0.543±0.087	0.70±0.05	0.71±0.12	ε_0=0.001
	SOM	0.706±0.075	0.86±0.08	0.80±0.07	H=6, α=0.01, ε_0=0.001
	SOBS	0.690±0.053	0.92±0.04	0.74±0.06	$\mathcal{K}$=300, ε_1=0.1, ε_2=0.005, c_2=0.05
	FGD	0.559±0.112	0.70±0.09	0.73±0.15	α_1=0.1, α_2=0.01, α_3=0.1, δ=2.5
	PF	0.656±0.078	0.79±0.09	0.80±0.07	$\mathcal{N}$=2, α=0.01
	AGMM	**0.718±0.066**	0.78±0.07	0.90±0.06	$\mathcal{N}$=2, α=0.005, $\mathcal{K}$=5, $\mathcal{T}$=0.8
	AE	0.671±0.167	0.91±0.07	0.73±0.19	ε_0=0.01
	SOM	**0.739±0.107**	0.91±0.06	0.80±0.12	H=18, α=0.005, ε_0=0.001
	SOBS	0.727±0.107	0.88±0.07	0.80±0.11	$\mathcal{K}$=100, ε_1=0.1, ε_2=0.005, c_2=0.05
	FGD	0.639±0.091	0.70±0.11	0.88±0.07	α_1=0.01, α_2=0.01, α_3=0.1, δ=3
	PF	0.688±0.104	0.89±0.07	0.76±0.12	$\mathcal{N}$=3, α=0.01
	AGMM	0.715±0.100	0.85±0.07	0.82±0.11	$\mathcal{N}$=3, α=0.01, $\mathcal{K}$=5, $\mathcal{T}$=0.6
	AE	0.519±0.131	0.63±0.18	0.77±0.09	ε_0=0.01
	SOM	**0.760±0.081**	0.90±0.05	0.83±0.09	H=6, α=0.01, ε_0=0.05
	SOBS	0.732±0.086	0.90±0.05	0.80±0.09	$\mathcal{K}$=100, ε_1=0.2, ε_2=0.01, c_2=0.05
	FGD	0.485±0.176	0.80±0.13	0.55±0.19	α_1=0.01, α_2=0.01, α_3=0.1, δ=3
	PF	0.696±0.095	0.87±0.07	0.78±0.08	$\mathcal{N}$=3, α=0.005
	AGMM	0.635±0.133	0.91±0.10	0.67±0.14	$\mathcal{N}$=5, α=0.005, $\mathcal{K}$=3, $\mathcal{T}$=0.8
	AE	**0.639±0.166**	0.75±0.15	0.82±0.15	ε_0=0.05
	SOM	0.637±0.224	0.72±0.19	0.81±0.20	H=18, α=0.05, ε_0=0.001
	SOBS	0.597±0.251	0.69±0.24	0.75±0.22	$\mathcal{K}$=100, ε_1=0.2, ε_2=0.01, c_2=0.05
	FGD	0.516±0.298	0.63±0.33	0.64±0.32	α_1=0.01, α_2=0.001, α_3=0.1, δ=2
	PF	0.583±0.222	0.63±0.23	0.85±0.13	$\mathcal{N}$=3, α=0.001
	AGMM	0.620±0.188	0.76±0.17	0.76±0.17	$\mathcal{N}$=5, α=0.005, $\mathcal{K}$=3, $\mathcal{T}$=0.8
	AE	0.900±0.030	0.91±0.03	0.98±0.01	ε_0=0.01
	SOM	0.910±0.030	0.94±0.02	0.96±0.02	H=12, α=0.01, ε_0=0.001
	SOBS	**0.910±0.026**	0.95±0.02	0.96±0.02	$\mathcal{K}$=300, ε_1=0.2, ε_2=0.005, c_2=0.01
	FGD	0.779±0.053	0.93±0.04	0.83±0.05	α_1=0.01, α_2=0.001, α_3=0.1, δ=2
	PF	0.812±0.055	0.90±0.06	0.89±0.02	$\mathcal{N}$=2.5, α=0.005
	AGMM	0.885±0.031	0.92±0.03	0.96±0.02	$\mathcal{N}$=3, α=0.005, $\mathcal{K}$=7, $\mathcal{T}$=0.6
	AE	0.716±0.039	0.76±0.02	0.92±0.04	ε_0=0.01
	SOM	0.739±0.054	0.79±0.03	0.91±0.04	H=6, α=0.05, ε_0=0.01
	SOBS	**0.767±0.036**	0.83±0.01	0.91±0.05	$\mathcal{K}$=100, ε_1=0.2, ε_2=0.01, c_2=0.05
	FGD	0.755±0.026	0.92±0.01	0.80±0.03	α_1=0.01, α_2=0.001, α_3=0.05, δ=2
	PF	0.721±0.036	0.89±0.02	0.79±0.03	$\mathcal{N}$=5, α=0.005
	AGMM	0.719±0.049	0.81±0.03	0.86±0.04	$\mathcal{N}$=5, α=0.005, $\mathcal{K}$=3, $\mathcal{T}$=0.6

TABLE 8.3   (continued)

Sequence	Method	Accuracy	Precision	Recall	Parameters
	AE	0.918±0.020	0.94±0.01	0.98±0.02	ε_0=0.0001
	SOM	**0.923±0.022**	0.95±0.01	0.97±0.02	H=12, α=0.005, ε_0=0.001
	SOBS	0.902±0.021	0.93±0.01	0.97±0.02	K=100, ε_1=0.1, ε_2=0.005, c_2=0.01
	FGD	0.745±0.061	0.83±0.07	0.88±0.06	α_1=0.01, α_2=0.001, α_3=0.01, δ=2
	PF	0.890±0.029	0.94±0.02	0.94±0.04	$\mathcal{N}$=2.5, α=0.001
	AGMM	0.896±0.031	0.93±0.01	0.96±0.04	$\mathcal{N}$=3, α=0.001, $\mathcal{K}$=7, $\mathcal{T}$=0.8
	AE	**0.760±0.114**	0.83±0.13	0.90±0.03	ε_0=0.005
	SOM	0.748±0.104	0.83±0.14	0.89±0.04	H=18, α=0.01, ε_0=0.05
	SOBS	0.596±0.098	0.64±0.12	0.89±0.03	K=100, ε_1=0.2, ε_2=0.01, c_2=0.01
	FGD	0.693±0.078	0.81±0.09	0.83±0.06	α_1=0.01, α_2=0.001, α_3=0.01, δ=3
	PF	0.557±0.068	0.80±0.11	0.65±0.05	$\mathcal{N}$=5, α=0.001
	AGMM	0.704±0.092	0.83±0.12	0.83±0.06	$\mathcal{N}$=2.5, α=0.005, $\mathcal{K}$=5, $\mathcal{T}$=0.8
	AE	**0.909±0.024**	0.97±0.02	0.94±0.02	ε_0=0.0001
	SOM	0.891± 0.027	0.97±0.01	0.91±0.02	H=6, α=0.001, ε_0=0.001
	SOBS	0.858±0.027	0.89±0.03	0.96±0.01	K=200, ε_1=0.1, ε_2=0.001, c_2=0.05
	FGD	0.769±0.063	0.96±0.04	0.79±0.05	α_1=0.01, α_2=0.001, α_3=0.05, δ=2
	PF	0.858±0.031	0.91±0.04	0.94±0.02	$\mathcal{N}$=3, α=0.001
	AGMM	0.832±0.036	0.98±0.01	0.85±0.04	$\mathcal{N}$=5, α=0.001, $\mathcal{K}$=3, $\mathcal{T}$=0.6

Nevertheless, it should be noted that this effect also depends on the quality of the output of the basic method (B). For instance, the *PF* method applied over the *LC* scene provides an accuracy of 0.535 without postprocessing (B strategy), whereas if morphological operators are subsequently applied (strategy B+PP) the accuracy tends to 0.812. The same effect happens for the scene *V4*, improving the accuracy from 0.374 (B) to 0.557 (B+PP). In italics it is possible to find the results of the best basic method in terms of accuracy for all the sequences. Thus, our *AE* approach is quite robust for most of them (together with

TABLE 8.4   Comparison of the segmentation methods studied using various strategies (second column): the basic approach (B), a subsequent post-processing step (B+PP) and a feedback process in the update of the model, only applied to the *SOM* method. In bold the best results for each scene are shown, while the italic style highlights the best basic approach (B).

Sequence	Strategy	AE	SOM	SOBS	FGD	PF	AGMM
CAM	B	*0.594±0.168*	0.280±0.177	0.559±0.161	0.502±0.219	0.177±0.112	0.261±0.134
	B+PP	**0.772±0.070**	0.513±0.182	0.639±0.164	0.640±0.155	0.488±0.138	0.645±0.139
	B+PP+F	-	0.724±0.088	-	-	-	-
MR	B	0.775±0.060	0.779±0.075	*0.843±0.058*	0.825±0.053	0.655±0.077	0.712±0.085
	B+PP	0.810±0.058	0.776±0.074	**0.851±0.056**	0.829±0.062	0.688±0.110	0.749±0.065
	B+PP+F	-	0.811±0.071	-	-	-	-
SS	B	0.458±0.124	0.462±0.094	*0.506±0.118*	0.278±0.123	0.314±0.102	0.398 ±0.091
	B+PP	0.461±0.135	0.567±0.128	0.545±0.120	0.371±0.152	0.421±0.146	0.524±0.153
	B+PP+F	-	**0.632±0.108**	-	-	-	-
FT	B	0.407±0.150	0.417±0.156	*0.664±0.042*	0.463±0.161	0.561±0.081	0.592±0.085
	B+PP	0.543±0.087	0.506±0.101	0.690±0.053	0.559±0.112	0.656±0.078	**0.718±0.067**
	B+PP+F	-	0.706±0.075	-	-	-	-
LC	B	0.879±0.035	0.772±0.116	*0.901±0.028*	0.778±0.049	0.535±0.149	0.750±0.076
	B+PP	0.900±0.030	0.886±0.034	**0.910±0.026**	0.779±0.053	0.812±0.055	0.885±0.032
	B+PP+F	-	0.910±0.030	-	-	-	-
OC	B	0.709±0.037	0.657±0.034	*0.761±0.037*	0.752±0.025	0.708±0.034	0.702±0.047
	B+PP	0.716±0.039	0.691±0.041	**0.767±0.036**	0.755±0.026	0.721±0.036	0.719±0.049
	B+PP+F	-	0.739±0.054	-	-	-	-
V2	B	0.920±0.019	*0.920±0.018*	0.903±0.020	0.757±0.059	0.862±0.031	0.907±0.024
	B+PP	0.918±0.020	0.918±0.021	0.902±0.021	0.745±0.061	0.890±0.029	0.896±0.031
	B+PP+F	-	**0.923±0.022**	-	-	-	-
V4	B	*0.682±0.111*	0.453±0.070	0.561±0.089	0.664±0.093	0.374±0.053	0.487±0.068
	B+PP	**0.760±0.114**	0.707±0.103	0.596±0.098	0.694±0.078	0.557±0.068	0.704±0.092
	B+PP+F	-	0.748±0.104	-	-	-	-
WS	B	*0.870±0.023*	0.821±0.021	0.852±0.034	0.783±0.059	0.771±0.029	0.799±0.026
	B+PP	**0.909±0.024**	0.820±0.049	0.858±0.027	0.769±0.064	0.858±0.031	0.832±0.036
	B+PP+F	-	0.891±0.027	-	-	-	-

SOBS and *FGD*), although when it provides discrete results (*FT* or *SS*) the postprocessing stage (B+PP) does not help in excess.

The influence of the postprocessing over the *FGD* technique is quite reduced and even inadvisable for some scenes, probably because the method itself includes a filter that performs these functions in an intrinsic way. The use of the feedback strategy (F) in our *SOM* method is crucial to improve the results, being even more influential than the postprocessing. Its performance makes the neural model competitive with regard to other alternatives, becoming the best object detection method on average.

8.5   Conclusion

Stochastic approximation is an estimation methodology which can be used for online learning of the parameters of a background model. Its inherent robustness and low computational complexity are major reasons to consider it as a valuable alternative to other approaches. Two specific approaches have been presented. One of them models the background with a multivariate Gaussian, while the other uses a probabilistic self-organizing map. Both of them associate a uniform distribution to the foreground so as to cope with any incoming objects adequately. An experimental design has been developed, which includes a set of relevant quantitative performance. Experimental results have been reported which show that stochastic approximation methods perform well both in qualitative and quantitative terms when compared with a selection of state of the art algorithms. However, more research is needed to improve this methodology, since early appearing foreground objects can degrade the quality of the initial background model, and no specific provision is made for commonly encountered difficulties such as illumination changes. Future works include the incorporation of illumination change and shadow detection mechanisms, and the use of color spaces other than RGB. Moreover, GPU implementations should be considered to further reduce the computation time of the proposals.

References

1. S. Allassonnire and E. Kuhn. Stochastic algorithm for Bayesian mixture effect template estimation. *ESAIM - Probability and Statistics*, 14(6):382–408, 2010.

2. S. Allassonnire, E. Kuhn, and A. Trouv. Construction of Bayesian deformable models via a stochastic approximation algorithm: A convergence study. *Bernoulli*, 16(3):641–678, 2010.

3. C. Bishop and M. Svenson. The generative topographic mapping. *Neural Computation*, 10(1):215–234, 1998.

4. J. Black, T. Ellis, and P. Rosin. A novel method for video tracking performance evaluation. In *IEEE International Workshop on Visual Surveillance and Performance Evaluation of Tracking and Surveillance (VS-PETS)*, pages 125–132, 2003.

5. D. Brainard. *The Science of Color*, chapter Color Appearance and Color Difference Specification, pages 191–216. Elsevier, 2003.

6. Y. Chen and K. Young. An som-based algorithm for optimization with dynamic weight updating. *International Journal of Neural Systems*, 17(3):171–181, 2007.

7. J. Cheng, J. Yang, Y. Zhou, and Y. Cui. Flexible background mixture models for foreground segmentation. *Image and Vision Computing*, 24(5):473–482, 2006.

8. A. Chowdhury and R. Chellappa. Stochastic approximation and rate-distortion analysis for robust structure and motion estimation. *International Journal of Computer Vision*, 55(1):27–53, 2003.

9. R. Cucchiara, M. Piccardi, and P. Mello. Image analysis and rule-based reasoning for a

     traffic monitoring system. *IEEE Transactions on Intelligent Transportation Systems*, 1(2):119–130, June 2000.

10. B. Delyon, M. Lavielle, and E. Moulines. Convergence of a stochastic approximation version of the EM algorithm. *Annals of Statistics*, 27(1):94–128, 1999.

11. A. Doulamis. Dynamic tracking re-adjustment: a method for automatic tracking recovery in complex visual environments. *Multimedia Tools and Applications*, 50(1):49–73, 2010.

12. W.E.L. Grimson, C. Stauffer, R. Romano, and L. Lee. Using adaptive tracking to classify and monitor activities in a site. In *IEEE Conference on Computer Vision and Pattern Recognition, CVPR*, pages 22–29, Jun 1998.

13. D. Hall, J. Nascimento, P. Ribeiro, E. Andrade, and P. Moreno. Comparison of target detection algorithms using adaptive background models. In *IEEE International Workshop on Visual Surveillance and Performance Evaluation of Tracking and Surveillance (VS-PETS)*, pages 113–120, Oct. 2005.

14. B. Han, D. Comaniciu, Y. Zhu, and L. Davis. Sequential kernel density approximation and its application to real-time visual tracking. *IEEE Transactions on Pattern Analysis and Machine Intelligence*, 30(7):1186–1197, 2008.

15. B. Han and L. Davis. Density-based multifeature background subtraction with support vector machine. *IEEE Transactions on Pattern Analysis and Machine Intelligence*, 34(5):1017–1023, 2012.

16. A. Hojjat and P. Ashif. A probabilistic neural network for earthquake magnitude prediction. *Neural Networks*, 22(7):1018–1024, 2009.

17. M. Van Hulle. Kernel-based topographic map formation by local density modeling. *Neural Computation*, 14(7):1561–1573, 2002.

18. M. Van Hulle. Maximum likelihood topographic map formation. *Neural Computation*, 17(3):503–513, 2005.

19. P. KaewTrakulPong and R. Bowden. A real time adaptive visual surveillance system for tracking low-resolution colour targets in dynamically changing scenes. *Image and Vision Computing*, 21(9):913–929, September 2003.

20. H. Kushner and G. Yin. *Stochastic approximation and Recursive Algorithms and Applications.* Springer-Verlag, New York, NY, USA, 2003.

21. T. Lai. Stochastic approximation. *Annals of Statistics*, 31(2):391–406, 2003.

22. L. Li, W. Huang, I.Gu, and Q. Tian. Statistical modeling of complex backgrounds for foreground object detection. *IEEE Transactions on Image Processing*, 13(11):1459–1472, Nov. 2004.

23. E. López-Rubio. Multivariate Student-t self-organizing maps. *Neural Networks*, 22(10):1432–1447, 2009.

24. E. López-Rubio. Probabilistic self-organizing maps for continuous data. *IEEE Transactions on Neural Networks*, 21(10):1543–1554, 2010.

25. E. López-Rubio. Restoration of images corrupted by Gaussian and uniform impulsive noise. *Pattern Recognition*, 43(5):1835–1846, 2010.

26. E. López-Rubio, J. M. Ortiz-de-Lazcano-Lobato, and D. López-Rodríguez. Probabilistic PCA self-organizing maps. *IEEE Transactions on Neural Networks*, 20(9):1474–1489, 2009.

27. L. Maddalena and A. Petrosino. A self-organizing approach to background subtraction for visual surveillance applications. *IEEE Transactions on Image Processing*, 17(7):1168–1177, 2008.

28. A. Mehran and A. Hojjat. Enhanced probabilistic neural network with local decision circles: A robust classifier. *Integrated Computer-Aided Engineering*, 17(3):197–210, August 2010.

29. H. Robbins and S. Monro. A stochastic approximation method. *Annals of Mathematical Statistics*, 22(3):400–407, 1951.

30. J. Rodgers and W. Nicewander. Thirteen ways to look at the correlation coefficient. *The*

American Statistician, 42(1):59–66, 1988.

31. M. Sato and S. Ishii. On-line EM algorithm for the normalized Gaussian network. *Neural Computation*, 12(2):407–432, 2000.

32. T. Schlogl, C. Beleznai, M. Winter, and H. H. Bischof. Performance evaluation metrics for motion detection and tracking. In *International Conference on Pattern Recognition, ICPR 2004*, volume 4, pages 519–522, Aug. 2004.

33. Y. Sheikh and M. Shah. Bayesian modelling of dynamic scenes for object detection. *IEEE Transactions on Pattern Analysis and Machine Intelligence*, 27(11):1778–1792, 2005.

34. Y. Singh, P. Gupta, and V. Yadav. Implementation of a self-organizing approach to background subtraction for visual surveillance applications. *International Journal of Computer Science and Network Security*, 10(3):136–143, 2010.

35. P. Spagnolo, T. Orazio, M. Leo, and A. Distante. Moving object segmentation by background subtraction and temporal analysis. *Image and Vision Computing*, 24(5):411–423, 2006.

36. C. Stauffer and W. Grimson. Adaptive background mixture models for real-time tracking. In *IEEE Computer Society Conference on Computer Vision and Pattern Recognition*, pages 246–252, 1999.

37. C. Stauffer and W. Grimson. Learning patterns of activity using real-time tracking. *IEEE Transactions on Pattern Analysis and Machine Intelligence*, 22(8):747–757, 2000.

38. K. Toyama, J. Krumm, B. Brumitt, and B. Meyers. Wallflower: Principles and practice of background maintenance. In *IEEE International Conference on Computer Vision, ICCV*, pages 255–261, 1999.

39. P. Villegas and X. Marichal. Perceptually-weighted evaluation criteria for segmentation masks in video sequences. *IEEE Transactions on Image Processing*, 13(8):1092–1103, Aug. 2004.

40. N. Wollborn and R. Mech. Procedure for objective evaluation of vop generation algorithm. *MPEG Committee, Document ISO/IEC/JTC1/SC29/WG11 M270*, 1997.

41. C.R. Wren, A. Azarbayejani, T. Darrell, and A. Pentl. Pfinder: Real-time tracking of the human body. *IEEE Transactions on Pattern Analysis and Machine Intelligence*, 19(7):780–785, 1997.

42. H. Yin and N. Allinson. Self-organizing mixture networks for probability density estimation. *IEEE Transactions on Neural Networks*, 12(2):405–411, Mar 2001.

43. X. Zhou, Y. Lu, J. Lu, and J. Zhou. Abrupt motion tracking via intensively adaptive Markov-chain Monte Carlo sampling. *IEEE Transactions on Image Processing*, 21(2):789–801, 2012.

44. Z. Zivkovic and F. Van der Heijden. Recursive unsupervised learning of finite mixture models. *IEEE Transactions on Pattern Analysis and Machine Intelligence*, 26(5):651–656, 2004.

45. Z. Zivkovic and F. van der Heijden. Efficient adaptive density estimation per image pixel for the task of background subtraction. *Pattern Recognition Letters*, 27(7):773–780, May 2006.

9

Sparsity Driven Background Modeling and Foreground Detection

Junzhou Huang
University of Texas, Arlington, USA

Chen Chen
University of Texas, Arlington, USA

Xinyi Cui
Facebook, USA

9.1 Introduction

Background subtraction is an important pre-processing step in video monitoring applications. It aims to extract foreground objects (*e.g.* humans, cars, text etc.) in videos from static and moving cameras for further processing (object recognition etc.).

A robust background subtraction algorithm should be able to handle lighting changes, repetitive motions from clutter and long-term scene changes [39]. The earliest background subtraction methods use frame difference to detect foreground [19]. Many subsequent approaches have been proposed to model the uncertainty in background appearance. The Mixture of Gaussians (MoG) background model assumes the color evolution of each pixel can be modeled as a MoG and are widely used on realistic scenes [35]. Elgammal et al. [12] proposed a non-parametric model for the background under similar computational constraints as the MoG. Spatial constraints are also incorporated into their model. Sheikh and Shah consider both temporal and spatial constraints in a Bayesian framework [33], which results in good foreground segmentations even when the background is dynamic. The model in [25] also uses a similar scheme. All these methods only implicitly model the background dynamics. In order to better handle dynamic scenes, some recent works [26,52] explicitly model the background as dynamic textures. Most dynamic texture modeling methods are based on the Auto Regressive and Moving Average (ARMA) model, whose dynamics are driven by a linear dynamic system (LDS). While this linear model can handle background dynamics with certain stationarity, it will cause over-fitting for more complex scenes.

Recently, Huang et al. presented a new method for modeling foreground in dynamic scenes using dynamic group sparsity [16]. It is assumed that the pixels belong to the foreground should be sparse, and more importantly, cluster together. They extend the CS theory to efficiently handle data with both sparsity and dynamic group clustering priors. A dynamic group sparsity recovery algorithm is then proposed based on the extended CS theory. This algorithm iteratively prunes the signal estimations according to both sparsity and group clustering priors. The group clustering trend implies that, if a pixel lives in the foreground, its neighboring pixels would also tend to live in this foreground with higher probability, and vice versa. By enforcing these constraints, the degrees of freedom of the sparse solutions have been significantly reduced to a narrower union of subspaces. It leads to several advantages: 1) accelerating the result pruning process; 2) decreasing the minimal number of necessary measurements; and 3) improving robustness to noise and preventing the recovered foreground from having artifacts. These advantages enable the algorithm to efficiently obtain stable background subtraction even when the background is dynamic.

So far, all the methods mentioned assume the cameras are fixed and stationary. In recent years, videos from moving cameras have also been studied [32] because the number of moving cameras (such as smart phones and digital videos) increased significantly. The research for moving cameras has recently attracted people's attention. Motion segmentation approaches [29,47] segment point trajectories based on subspace analysis. These algorithms provide interesting analysis on sparse trajectories, though do not output a binary mask as many background subtraction methods do. Another popular way to handle camera motion is to have strong priors of the scene, *e.g.*, approximating background by a 2D plane or assuming that camera center does not translate [14,30], assuming a dominant plane [48], etc. In [22], a method is proposed to use belief propagation and Bayesian filtering to handle moving cameras. However, handling diverse types of videos robustly is still a challenging problem.

For sparsity based methods, Cui et al. propose a unified framework for background subtraction [8], which can robustly deal with videos from stationary or moving cameras with various number of rigid/non-rigid objects. This method is based on two sparsity constraints applied on foreground and background levels, *i.e.*, low rank [7] and group sparsity constraints [49]. It is inspired by recently proposed sparsity theories [6, 34]. There are two "sparsity" observations behind this method. *First*, when the scene in a video does not have any foreground moving objects, video motion has a low rank constraint for orthographic cameras [21]. Thus the motion of background points forms a low rank matrix. *Second*, foreground moving objects usually occupy a small portion of the scene. In addition, when a foreground object is projected to pixels on multiple frames, these pixels are not randomly distributed. They tend to group together as a continuous trajectory. Thus these foreground trajectories usually satisfy the group sparsity constraint.

These two observations provide important information to differentiate independent objects from the scene. Based on them, the video background subtraction problem is formulated as a matrix decomposition problem. First, the video motion is represented as a matrix on *trajectory* level (*i.e.* each row in the motion matrix is a trajectory of a point). Then it is decomposed into a background matrix and a foreground matrix, where the background matrix is low rank, and the foreground matrix is group sparse. This low rank constraint is able to automatically model background from both stationary and moving cameras, and the group sparsity constraint improves the robustness to noise.

The trajectories recognized by the above model can be further used to label a frame into foreground and background at the pixel level. Motion segments on a video sequence are generated using fairly standard techniques. Then the color and motion information gathered from the trajectories is employed to classify the motion segments as foreground or background. Cui et. al.'s approach is validated on various types of data, *i.e.*, synthetic data, real-world video sequences recorded by stationary cameras or moving cameras and/or nonrigid foreground objects. Extensive experiments demonstrate this method compared favorably to the recent state-of-the-art methods.

The remainder of the chapter is organized as follows. Section 9.2 briefly reviews the sparsity theory. In Section 9.3, we review the dynamic group sparsity model for background subtraction. In Section 9.4, we review the background subtraction using low rank and group sparsity constraints, which could be used for both stationary and moving cameras. We conclude this chapter in Section 9.5.

9.2 Sparsity Techniques

Suppose I_t is a frame in a video at time t, in ideal conditions, it can be represented as the combination of background and an object in foreground:

$$I_t = b + f \tag{9.1}$$

where $b \in \mathbb{R}^{n \times 1}$ denotes the background and $f \in \mathbb{R}^{n \times 1}$ denotes the foreground. An example is shown in Figure 9.1. We could observe that the foreground f is sparse, with most of it values are zeros. In this section, we review the sparsity techniques and the benefit when apply them to background subtraction.

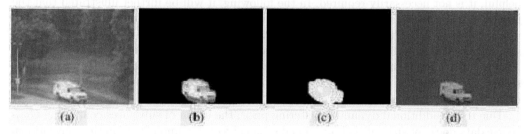

FIGURE 9.1 An example of background subtraction: (a) one video frame, (b) the foreground image, (c) the foreground mask and (d) the background subtracted image with AdaDGS [16].

9.2.1 Standard Sparsity

Sparsity based methods have been very popular since the emerging of compressive sensing (CS) [4, 11]. In CS, the capture of a sparse signal and compression are integrated into a single process. A signal $x \in \mathbb{R}^n$ is k-sparse if only $k \ll n$ entries of x are nonzero. We call the set of indices corresponding to the nonzero entries the support of x and denote it by $supp(x)$ [1,18]. Thus, we do not capture a sparse signal x directly but rather capture $m < n$ linear measurements $y = \Phi x$ based on a measurement matrix $\Phi \in \mathbb{R}^{m \times n}$. Suppose the set of k-sparse signal x lives in the union Ω_k of k-dimensional subspaces, the union Ω_k thus includes C_n^k subspaces. To stably recover the k-sparse signal x from m measurements, the measurement matrix Φ is required to satisfy the Restricted Isometry Property (RIP) [4]. CS result shows that, if $m = \mathcal{O}(k + k \log(n/k))$, then measurement matrix Φ whose entries are i.i.d. Gaussian, Bernoulli (± 1) can satisfy the RIP with high probability.

9.2.2 Group Sparsity

In CS, the assumption cares the sparsity number k of the data, while positions of such k non-zeros entries can be random. There is no prior information utilized other than sparsity. Background subtracted images are typical structured sparsity data in static video surveillance applications. They generally correspond to the foreground objects of interest. Unlike the whole scene, these images are not only spatially sparse but also inclined to cluster into groups, which correspond to different foreground objects. The success of sparse recovery in compressive sensing motivates us to further improve background substraction by utilizing such prior information. Compared with that of standard sparse data, the locations of non-zeros entries in grouped sparse data should be more fixed. Intuitively, the measurement number bound may be further reduced or the solution will be more accurate with the number of measurements.

Similar to the definition of k-sparse data, we can define dynamic group sparse (DGS) data as follow:

Definition:($G_{k,q}$-**sparse data**) *A data $x \in \mathbb{R}^n$ is defined as the dynamic group sparse data ($G_{k,q}$-sparse data) if it can be well approximated using $k \ll n$ nonzero coefficients under some linear transforms and these k nonzero coefficients are clustered into $q \in \{1, \cdots, k\}$ groups.*

From this definition, we can know that $G_{k,q}$-sparse data only requires that the nonzero coefficients in the sparse data have the group clustering trend and does not require to know any information about the group size and location. Although the information may be helpful, it is not necessarily required. In the following, it will be further illustrated that the group number q is also not necessary to be known. The group structures can be dynamic and unknown. We can find that nonzero coefficients are not randomly distributed but clustered spatially in the background subtracted image (Figure 9.1 (b)) and the foreground mask (Figure 9.1 (c)). More specially for color images, the R, G and B channels of the background subtracted image share a common support set although the nonzero coefficients are spatially clustered in each channel respectively.

Due to the additional dynamic clustering prior, the union of subspaces containing $G_{k,q}$-sparse data does not span all k-dimensional subspaces of the union Ω_k as in the conventional CS [5, 9, 11, 28, 41]. The former is far narrower than the latter in most cases. The dynamic group clustering prior significantly reduces the degrees of freedom of the sparse signal since it only permits certain combinations of its support set rather than all random combinations. This will make it possible for us to decrease the minimal measurement number m for stable recovery.

Structured sparsity theories [1, 18] show that the number of measurements required for robustly recovering dynamic group sparsity data is $m = \mathcal{O}(k + q \log(n/q))$, which is a significant improvement over the $m = \mathcal{O}(k + k \log(n/k))$ that would be required by conventional CS recovery algorithms [5, 11, 28, 41]. While the group number q is smaller, more improvements can be obtained. While q is far smaller than k and k is close to $log(n)$, we can get $m = \mathcal{O}(k)$. Note that, this is a sufficient condition. If we know more priors about group settings, we can further reduce this bound.

9.3   Sparsity Driven Background Modeling and Foreground Detection for Stationary Cameras

The inspiration for the AdaDGS background subtraction came from the success in online DT video registration based on the sparse representation constancy assumption (SRCA) [15]. The SRCA states that a new coming video frame should be represented as a linear combination of as few preceding image frames as possible. As a matter of fact, the traditional brightness constancy assumption seeks that the current video frame can be best represented by a single preceding frame, while the SRCA seeks that the current frame can be best sparsely represented by all preceding image frames. Thus, the former can be thought as a special case of SRCA.

9.3.1 Group Sparsity Based Model

Suppose a video sequence consists of frames $I_1, ..., I_n \in \mathbb{R}^m$. Without loss of generality, we can assume that background subtraction has already been performed on the first t frames. Let $A = [I_1, ..., I_t] \in \mathbb{R}^{m \times t}$. Denote the background image and the background subtracted image by b and f, respectively, for I_{t+1}. From the introduction in a previous section, we know that f is dynamic group sparse data with unknown sparsity number k_f and group structure. According to SRCA, we have $b = Ax$, where $x \in \mathbb{R}^t$ should be k_x-sparse vector and $k_x << t$. Let $\Phi = [A, I] \in \mathbb{R}^{m \times (t+m)}$, where $I \in \mathbb{R}^{m \times m}$ is an identity matrix. Then, we have:

$$I_{t+1} = Ax + f = [A, I] \begin{bmatrix} x \\ f \end{bmatrix} = \Phi z \tag{9.2}$$

where $z \in \mathbb{R}^{t+m}$ is the DGS data with unknown sparsity $k_x + k_f$. Background subtraction is thus formulated as the following sparse recovery problem:

$$(x_0, f_0) = argmin\|z\|_0, \quad \|I_{t+1} - \Phi z\|^2 < \varepsilon \tag{9.3}$$

9.3.2 Algorithm

Although problem (9.3) can be solved by conventional methods for standard sparsity, the benefit of group sparsity equips us to propose a new recovery algorithm for dynamic group sparse data z, namely dynamic group sparsity (DGS) recovery algorithm. We demonstrate how to seamlessly integrate the dynamic group clustering prior into previous framework [9, 28]. The algorithm includes five main steps in each iteration: 1) pruning the residue estimation; 2) merging the support sets; 3) estimating the signal by least square; 4) pruning the signal estimation and 5) updating the signal/residue estimation and support set. One can observe that it is similar to that of SP/CoSaMP algorithms [9, 28]. The difference only exists in the pruning process in step 1 and step 4. The modification is simple. We prune the estimation in the step 1 and step 4 using DGS approximation pruning rather than k-sparse approximation, as we only need to search over subspaces of $\mathcal{A}_{k,q}$ instead of C_n^k subspaces of Ω_k. It directly leads to fewer measurement requirement for stable data recovery.

The DGS pruning algorithm is described in algorithm 9.1. There exist two prior-dependent parameters J_y and J_b. J_y is the number of tasks if the problem can be represented as a multi-task CS problem [20]. J_b is the block size if the interested problem can be modelled as a block sparsity problem [37, 42, 43, 49]. Their default values are set as 1, which is the case of traditional sparse recovery in compressive sensing. Moreover, there are two important user-tuning parameters, the weight w of neighbors and the neighbor number τ of each element in sparse data. In practice, it is very straightforward to adjust them since they have the physical meanings. The first one controls the balance between the sparsity prior and the group clustering prior. While w is smaller/bigger, it means that the degree of dynamic group clustering is lower/higher in the sparse signal. Generally, they are set as $0.5's$ if there is not more knowledge about that in practice. The parameter τ controls the number of neighbors that can be affected by each element in sparse data. Generally, it is good enough to set it as 2, 4 and 6 for 1D, 2D and 3D data respectively.

Up to now, we assume that we know the sparsity number k of the sparse data before recovery. However, it is not always true in practical applications. For example, we do not

---

Algorithm 9.1 - DGS Approximation Pruning

Input: $x \in \mathbb{R}^n$ {estimations}; k {the sparsity number}; J_y {task number}; J_b {block size}; $N_x \in \mathbb{R}^{n \times \tau}$ {values of x's neighbors}; $w \in \mathbb{R}^{n \times \tau}$ {weights for neighbors}; τ {neighbor number}

$J_x = J_y J_b$; $x \in \mathbb{R}^n$ is shaped to $x \in \mathbb{R}^{\frac{n}{J_x} \times J_x}$

$N_x \in \mathbb{R}^{n \times \tau}$ is shape to $N_x \in \mathbb{R}^{\frac{n}{J_x} \times J_x \times \tau}$;

for all $i = 1, ..., \frac{n}{J_x}$ **do**

   Combing each entry with its neighbors

$$z(i) = \sum_{j=1}^{J_x} x^2(i,j) + \sum_{j=1}^{J_x} \sum_{t=1}^{\tau} w^2(i,t) N_x^2(i,j,t)$$

end for

$\Omega \in \mathbb{R}^{\frac{n}{J_x} \times 1}$ is set as indices corresponding to the largest k/J_x entries of z

for all $j = 1, ..., J_x$ **do**

   for all $i = 1, ..., \frac{k}{J_x}$ **do**

      Obtain the final list
      $\Gamma((j-1)\frac{k}{J_x} + i) = (j-1)\frac{k}{J_x} + \Omega(i)$

   end for

end for

Output: $supp(x,k) \leftarrow \Gamma$

---

know the exact sparsity numbers of the background subtracted images although we know they tend to be dynamic group sparse. Motivated by the idea in [10], a new recovery algorithm called AdaDGS is developed by incorporating an adaptive sparsity scheme into the above DGS recovery algorithm.

Suppose the range of the sparsity number is known to be $[k_{min}, k_{max}]$. We can set the step size of sparsity number as $\triangle k$. The whole recovery process is divided into several stages, each of which includes several iterations. Thus, there are two loops in AdaDGS recovery algorithm. The sparsity number is initialized as k_{min} before iterations. During each stage (inner loop), we iteratively optimize sparse data with the fixed sparsity number k_{curr} until the halting condition within the stage is true (for example, the residue norm is not decreasing). We then switch to the next stage after adding $\triangle k$ into the current sparsity number k_{curr} (outer loop). The whole iterative process will stop whenever the halting condition is satisfied. For practical applications, there is a trade-off between the sparsity step size $\triangle k$ and the recovery performance. Smaller step sizes require more iterations and bigger step size may cause inaccuracy. The sparsity range depends on the applications. Generally, it can be set as $[1, n/3]$, where n is the dimension of the sparse data. Algorithm 9.2 describes the AdaDGS recovery algorithm.

With the AdaDGS recovery algorithm, problem (9.3) can be efficiently solved. Similar ideas are used for face recognition robust to occlusion [46]. It is worth mentioning that the coefficients in w corresponding to the x part are randomly sparse while those corresponding to f are dynamic group sparse. During the DGS approximation pruning, we thus can set those coefficients in weight w for the x-related part as zeros and those for f as nonzeros. Since we do not know the sparsity number k_x and k_f, we can set sparsity ranges for them respectively and run the AdaDGS recovery algorithm until the halting condition is true. Then, we can obtain the optimized background subtracted image f and background image $b = Ax$. For long video sequences, it is impractical to build a model matrix $A = [I_1, ..., I_t] \in$

Algorithm 9.2 - AdaDGS Recovery [16]

1: **Input:** $\Phi \in \mathbb{R}^{m \times n}${sample matrix}; $y \in \mathbb{R}^m${sample vector};$[k_{min}, k_{max}]$ {sparsity range}; $\triangle k$ {sparsity step size}

2: Initialization: residue $y_r = y$; $\Gamma = supp(x) = \varnothing$; sparse data $x = 0$; sparsity number $k = k_{min}$

3: **repeat**

4:     Perform DGS recovery algorithm with sparsity number k to obtain x and the residue

5:     **if** halting criterion false **then**

6:         Update Γ, y_r and $k = k + \triangle k$

7:     **end if**

8: **until** halting criterion true

9: **Output:** $x = \Phi_\Gamma^\dagger y$

$\mathbb{R}^{m \times t}$, where t denotes the last frame number. In order to cope with this case, we can set a time window width parameter τ. We then build the model matrix, $A = [I_{t-\tau+1}, ..., I_t] \in \mathbb{R}^{m \times (t-\tau)}$, for the $(t+1)$ frame, which can avoid the memory requirement blast for a long video sequence. The complete algorithm for AdaDGS based background subtraction is summarized in Algorithm 9.3. As we know, the sparsity number must be provided in most of current recovery algorithms, which make them impractical for this problem. In contrast, the AdaDGS can apply well to this task since it not only can automatically learn the sparsity number and group structures but also is a fast enough greedy algorithm.

Algorithm 9.3 - AdaDGS Background Subtraction [16]

1: **Input:** The video sequence $I_1, ..., I_n$, the number t which means $1^{st} \sim t^{th}$ have been performed background subtraction, the time window width $\tau \leq t$

2: **for all** $j = t + 1, ..., n$ **do**

3:     Set $A = [I_{j-\tau}, ..., I_{j-1}]$ and form $\Phi = [A, I]$

4:     Set $y = I_j$ and the sparsity ranges/step-sizes

5:     $(x_0, f_0) = AdaDGS(\Phi, y)$

6: **end for**

7: **Output:** Background subtracted images

9.3.3   Evaluations

The experiment is designed to validate the advantage of the AdaDGS model. The AdaDGS algorithm is tested on Zhong's dataset [52]. The background subtracted images can be directly obtained with the AdaDGS. The corresponding binary mask of these images are obtained with the simple threshold. The Zhong's results with robust Kalman model are also shown for comparisons. Figure 9.2 shows the results. Note that all results with AdaDGS are not post-processed with morphological operations and the results are directly the solutions of the optimization problem in Equation 9.3. It is clear that our AdaDGS produces clean background subtracted images, which shows the advantages of the DGS model. Figure 9.3 and Figure 9.4 show the background subtraction results on two other videos [12, 26]. Results of AdaDGS without postprocessing can compete with others with postprocessing. The results show the AdaDGS model can handle well highly dynamic scenes by exploiting the effective sparsity optimization scheme.

    All experimental results show the AdaDGS algorithm gains marked improvement over

FIGURE 9.2   Results on the Zhong's dataset (a) original frame, (b) background subtracted image with AdaDGS [16], (c) the binary mask with AdaDGS [16] and d) with robust Kalman model [52].

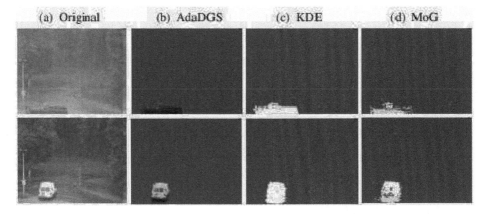

FIGURE 9.3   Results on the Elgammal's dataset. (a) original frame, (b) with AdaDGS [16], (c) with KDE model [12] (d) with MoG [35].

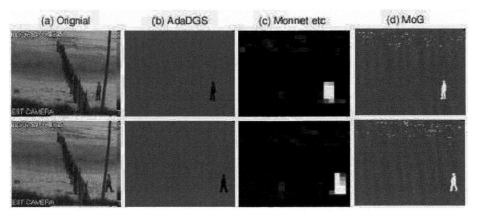

FIGURE 9.4   Results on Monnet's dataset. (a) original frame, (b) with AdaDGS [16], (c) with Monnet's method [26] and (d) with MoG [35].

previous algorithms when DGS priors are available. From a practical perspective, the DGS/AdaDGS can recover DGS data with higher accuracy and lower computational complexity from fewer measurements. From a theoretical point of view, structured sparsity theories [1, 18] offer a stronger guarantee for DGS/AdaDGS to achieve stable recovery. Moreover, we review a generalized framework for priors-driven background substraction algorithm. Group structure and sparsity number are not must-knows for this algorithm, which make it flexible and applicable in many practical applications.

9.4 Sparsity Driven Background Modeling and Foreground Detection for Moving Cameras

Besides the topic of background substraction for stationary cameras in the previous section, the research for moving cameras has recently attracted people's attention. Motion segmentation approaches [29, 47] segment point trajectories based on subspace analysis. These algorithms provide interesting analysis on sparse trajectories, though do not output a binary mask as many background subtraction methods do. Another popular way to handle camera motion is to have strong priors of the scene, *e.g.*, approximating background by a 2D plane or assuming that camera center does not translate [14, 30], assuming a dominant plane [48], etc. Kwak et al. [22] proposed a method to use belief propagation and Bayesian filtering to handle moving cameras. Different from their work, long-term trajectories we introduce here encode more information for background subtraction. Recently, [32] has proposed to build a background model using RANSAC to estimate the background trajectory basis. This approach assumes that the background motion spans a three dimensional subspace. Then sets of three trajectories are randomly selected to construct the background motion space until a consensus set is discovered, by measuring the projection error on the subspace spanned by the trajectory set. However, RANSAC based methods are generally sensitive to parameter selection, which makes it less robust when handling different videos.

Group sparsity [16,17] and low rank constraint [7] have also been applied to background subtraction problem. However, these methods only focus on stationary camera and their constraints are at spatial pixel level. Different from these works, we review a LRGS (low rank and group sparsity) method [8] which is based on constraints in temporal domain and analyzing the trajectory properties.

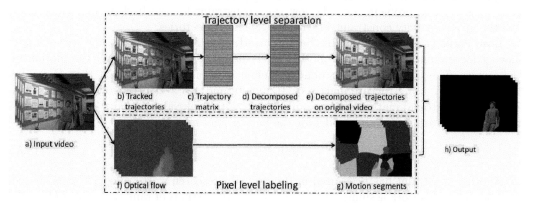

FIGURE 9.5 (See color insert.) The framework. LRGS method takes a raw video sequence as input, and produces a binary labeling as the output. Two major steps are trajectory level separation and pixel level labeling.

This background subtraction algorithm takes a raw video sequence as input, and generates a binary labeling at the pixel level. Fig. 9.5 shows the framework. It has two major steps: trajectory level separation and pixel level labeling. In the first step, a dense set of points is tracked over all frames. An off-the-shelf dense point tracker [38] is used to produce the trajectories. With the dense point trajectories, a low rank and group sparsity based model is proposed to decompose trajectories into foreground and background. In the second step, motion segments are generated using optical flow [13] and graph cuts [2]. Then the color and motion information gathered from the recognized trajectories builds statistics to label motion segments as foreground or background.

9.4.1   Low Rank and Group Sparsity Based Model

Different from the dynamic group sparsity method in the previous section, the LRGS method for moving cameras does not process the image pixels directly. Background substraction are performed on the decomposition of the tracked trajectories, and then pixel level labeling is based on the separated trajectories. Tracking algorithms used in this framework are not of discussion scope in this chapter. Interested readers may refer to [36,38,45] for more details.

Notations: Given a video sequence, K points are tracked over l frames. Each trajectory is represented as $p_i = [x_{1i}, y_{1i}, x_{2i}, y_{2i}, ...x_{li}, y_{li}] \in \mathbb{R}^{1 \times 2l}$, where x and y denote the 2D coordinates in each frame. The collection of K trajectories is represented as a $K \times 2l$ matrix, $\phi = [p_1^T, p_2^T, ..., p_l^T]^T$, $\phi \in \mathbb{R}^{K \times 2l}$.

In a video with moving foreground objects, a subset of K trajectories comes from the foreground, and the rest belongs to the background. The goal is to decompose tracked K trajectories into two parts: m background trajectories and n foreground trajectories. If we already know exactly which trajectories belong to the background, then foreground objects can be easily obtained by subtracting them from K trajectories, and vice versa. In other words, ϕ can be decomposed as:

$$\phi = B + F, \tag{9.4}$$

where $B \in \mathbb{R}^{K \times 2l}$ and $F \in \mathbb{R}^{K \times 2l}$ denote matrices of background and foreground trajectories, respectively. In the ideal case, the decomposed foreground matrix F consists of n rows of foreground trajectories and m rows of flat zeros, while B has m rows of background trajectories and n rows of zeros.

Eq. 9.4 is a severely under-constrained problem. It is difficult to find B and F without any prior information. In the LRGS method [8], two effective priors are incorporated to robustly solve this problem, *i.e.*, the low rank constraint for the background trajectories and the group sparsity constraint for the foreground trajectories.

Low rank constraint for the background. In a 3D structured scene without any moving foreground objects, video motion solely depends on the scene and the motion of the camera. The LRGS [8] background modeling is inspired from the fact that B can be factored as a $K \times 3$ structure matrix of 3D points and a $3 \times 2l$ orthogonal matrix [21]. Thus the background matrix is a low rank matrix with rank value at most 3. This leads us to build a low rank constraint model for the background matrix B:

$$rank(B) \leq 3, \tag{9.5}$$

Another constraint has been used in the previous research work using RANSAC based method [32]. This work assumes that the background matrix is of rank three: $rank(B) = 3$. This is a very strict constraint for the problem. We refer the above two types of constraints as the General Rank model (GR) and the Fixed Rank model (FR). The GR model in LRGS [8] method is more general and handles more situations. A rank-3 matrix models 3D scenes under moving cameras; a rank-2 matrix models a 2D scene or 3D scene under stationary cameras; a rank-1 matrix is a degenerated case when a scene only has one point. The usage of GR model allows us to develop a unified framework to handle both stationary cameras and moving cameras. The experiment section (Section 9.4.5) provides more analysis on the effectiveness of the GR model when handling diverse types of videos.

Group sparsity constraint for the foreground. Foreground moving objects, in general, occupy a small portion of the scene. This observation motivates us to use another important prior, *i.e.*, the number of foreground trajectories should be smaller than a certain ratio of all trajectories: $m \leq \alpha K$, where α controls the sparsity of foreground trajectories.

Another important observation is that each row in ϕ represents one trajectory. Thus the entries in ϕ are not randomly distributed. They are spatially clustered within each row. If one entry of the ith row ϕ_i belongs to the foreground, the whole ϕ_i is also in the foreground. This observation makes the foreground trajectory matrix F satisfy the group sparsity constraint:

$$\|F\|_{2,0} \leq \alpha k, \tag{9.6}$$

where $\| \cdot \|_{2,0}$ is the mixture of both L_2 and L_0 norm. The L_2 norm constraint is applied to each group separately (*i.e.*, each row of F). It ensures that all elements in the same row are either zero or nonzero at the same time. The L_0 norm constraint is applied to count the nonzero groups/rows of F. It guarantees that only a sparse number of rows are nonzero. Thus this group sparsity constraint not only ensures that the foreground objects are spatially sparse, but also guarantees that each trajectory is treated as one unit. Group sparsity is a powerful tool in computer vision problems [17, 50]. The standard sparsity constraint, L_0 norm, (we refer this as *Std. sparse* method) has been intensively studied in recent years. However, it does not work well for this problem compared to the group sparsity one. *Std. sparse* method treats each element of F independently. It does not consider any neighborhood information. Thus it is possible that points from the same trajectory are classified into two classes. In the experiment section (Section 9.4.4), we discuss the advantage of group sparsity constraint over sparsity constraint through synthetic data analysis, and also show that this constraint improves the robustness of LRGS model [8].

Based on the low rank and group sparsity constraints, we formulate our objective function as:

$$\left(\hat{B}, \hat{F}\right) = \underset{B,F}{\arg\min} \left(\| \phi - B - F \|_F^2\right),$$

$$\text{s.t. } rank(B) \leq 3, \| F \|_{2,0} < \alpha k, \tag{9.7}$$

where $\| \cdot \|_F$ is the Frobenius norm. This model leads to a good separation of foreground and background trajectories. Fig. 9.6 illustrates the model.

Eq. 9.7 only has one parameter α, which controls the sparsity of the foreground trajectories. In general, user-tuning parameter is a key issue for a good model. It is preferable that the parameters are easy to tune and not sensitive to different datasets. LRGS is relatively insensitive to parameter selection. Using $\alpha = 0.3$ works for all tested videos.

Low rank constraint is a powerful method in computer vision and machine learning area [24]. Low rank constraints and Robust PCA have been recently used to solve vision problems [7, 27, 51], including background subtraction at the pixel level [7]. It assumes that the stationary scenes satisfy a low rank constraint. However, this assumption does not hold when the camera moves. Furthermore, that formulation does not consider any group information, which is an important constraint to make sure neighbor elements are considered together.

9.4.2 Optimization Framework

This subsection discusses how to effectively solve Eq. 9.7. The first challenge is that it is not a convex problem, because of the nonconvexity of the low rank constraint and the group

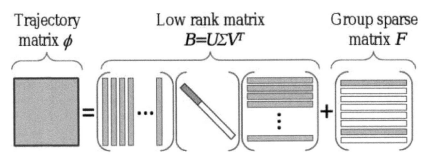

Trajectory matrix ϕ Low rank matrix $B = U\Sigma V^T$ Group sparse matrix F

FIGURE 9.6   Illustration of our model. Trajectory matrix ϕ is decomposed into a background matrix B and a foreground matrix F. B is a low rank matrix, which only has a few nonzero eigenvalues (*i.e.* the diagonal elements of Σ in SVD); F is a group sparse matrix. Elements in one row belongs to the same group (either foreground or background), since they lie on one trajectory. The foreground rows are sparse compared to all rows. White color denotes zero values, while blue color denotes nonzero values.

sparsity constraint. Furthermore, we also need to simultaneously recover matrix B and F, which is generally a Chicken-and-Egg problem.

Alternating optimization and greedy methods are employed to solve this problem. We can first focus on the fixed rank problem (*i.e.*, rank equals to 3), and then will discuss how to deal with the more general constraint of $rank \leq 3$.

Eq. 9.7 is divided into two subproblems with unknown B or F, and solved by using two steps iteratively:

Step 1: Fix B, and update F. The subproblem is:

$$\left(\hat{F}\right) = \arg\min_{F} \left(\| \ \phi' - F \ \|_F^2\right), \ \text{s.t.} \ \|F\|_{2,0} < \alpha k, \tag{9.8}$$

where $\phi' = \phi - B$.

Step 2: Fix F, and update B. The subproblem is:

$$\left(\hat{B}\right) = \arg\min_{B} \left(\| \ \phi'' - B \ \|_F^2\right), \ \text{s.t.} \ rank(B) = 3, \tag{9.9}$$

where $\phi'' = \phi - F$.

To initialize this optimization framework, we can simply choose $B_{init} = \phi$, and $F_{init} = 0$. Greedy methods are used to solve both subproblems. To solve Eq. 9.8, we compute $\|F_i\|_2, i \in 1, 2, ..., k$, which represents the L_2 norm of each row. Then the αk rows with largest values are preserved, while the rest rows are set to zero. This is the estimated F in the first step. In the second step, ϕ'' is computed as per newly-updated F. To solve Eq. 9.9. Singular value decomposition (SVD) is applied on ϕ''. Then three eigenvectors with largest eigenvalues are used to reconstruct B. Two steps are alternatively employed until a stable solution of $\hat{B}$ is found. Then $\hat{F}$ is computed as $\phi - \hat{B}$. The reason of updating $\hat{F}$ after all iterations is that the greedy method of solving Eq. 9.8 discovers exact αk number of foreground trajectories, which may not be the real foreground number. On the contrary, B can be always well estimated, since a subset of unknown number of background trajectories is able to have a good estimation of background subspace. Thus we finalize $\hat{F}$ by $\phi - \hat{B}$. Since the whole framework is based on greedy algorithms, it does not guarantee a global minimum. In the experiments, however, it is able to generate reliable and stable results. The above-mentioned method solves the fixed rank problem, but the rank value in the background problem usually cannot be pre-determined. To handle this undetermined rank issue, a multiple rank iteration method is used. First, B and F are initialized as $B_{init}^{(0)} = \phi$

and $F_{init}^{(0)} = \mathbf{0}^{k \times 2l}$. Then the fixed rank optimization procedure is performed on each specific rank starting from 1 to 3. The output of the current fixed rank procedure is fed to the next rank as its initialization. We obtain the final result $B^{(3)}$ and $F^{(3)}$ in the rank-3 iteration. Given a data matrix of $K \times 2L$ with K trajectories over L frames, the major calculation is $O(KL^2)$ for SVD on each iteration. Convergence of each fixed rank problem is achieved 6.7 iterations on average. The overall time complexity is $O(KL^2)$.

To explain why this framework works for the general rank problem, we discuss two examples. First, if the rank of B is 3 (*i.e.*, moving cameras), then this framework discovers an optimal solution in the third iteration, *i.e.*, using rank-3 model. The reason is that the first two iterations, *i.e.* the rank-1 and rank-2 models, cannot find the correct solution as they are using the wrong rank constraints. Second, if the rank of the matrix is 2 (*i.e.*, stationary cameras), then this framework obtains stable solution in the second iteration. This solution will not be affected in the rank-3 iteration. The reason is that the greedy method is used to solve Eq. 9.9. When selecting the eigenvectors with three largest eigenvalues, one of them is simply flat zero. Thus B does not change, and the solution is the same in this iteration. Note that low rank problems can also be solved using convex relaxation on the constraint problem [7]. However, the greedy method on unconstrained problem is better than convex relaxation in this application. Convex relaxation is not able to make use of the specific rank value constraint (≤ 3 in our case). The convex relaxation uses λ to implicitly constrain the rank level, which is hard to constrain a matrix to be lower than a specific rank value.

9.4.3 Pixel Level Labeling

The labeled trajectories from the previous step are then used to label each frame at the pixel level (*i.e.* return a binary mask for a frame). In this step, each frame is treated as an individual labeling task. First, the optical flow [13] is calculated between two consecutive frames. Then motion segments are computed using graph cuts [2] on optical flow. After collecting the motion segments, we want to label each motion segment s as f or b, where f and b denotes the label of foreground and background. There are two steps to label the segments. First, segments with high confidence belonging to f and b are selected. Second, a statistical model is built based on those segments. This model is used to label segments with low confidence. The confidence of a segment is determined by the number of labeled f and b trajectories. The low confidence segments are those ambiguous areas. To predict the labels of these low confidence segments, a statistical model is built for f and b based on high confidence ones. First, 20% pixels are uniformly sampled on segments with high confidence. Each sampled pixel w is represented by color in hue-saturation space (h, s), optical flow (u, v) and position on the frame (x, y). The reason we use sampled points to build the model instead of the original trajectories is that the sparse trajectories may not cover enough information (*i.e.*, color and positions) on the motion unit. Uniform sampling covering the whole segment can build a richer model. The segments are then evaluated using a kernel density function: $P(s_i|c) = \frac{1}{N \cdot |s_i|} \sum_{i=1}^{N} \sum_{j \in s_i} \kappa(e_j - w_i), c \in \{f, b\}$, where $\kappa(\cdot)$ is the Normal kernel, N is the total number of sampled pixels, and $|s_i|$ is the pixel number of s_i. For every pixel j lying on s_i, e_j denotes the vector containing color, optical flow and position.

To evaluate the performance of the LRGS algorithm, experiments on different data sources are conducted: synthetic data, real-world videos from both moving and stationary cameras. The performance is evaluated by F-Measure, the harmonic mean of recall and precision. This is a standard measurement for background subtraction [3, 23]: $F = 2 \cdot recall \cdot precision/(recall + precision)$. The comparisons and analysis mainly focus on the first step: trajectory level labeling.

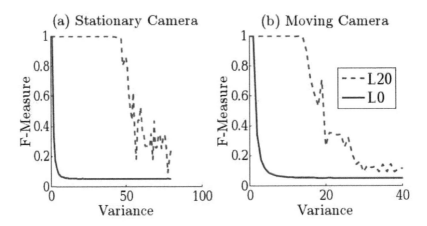

FIGURE 9.7   Comparison between group sparsity constraint ($L_{2,0}$ norm) and *Std. sparse* constraint. The left is synthetic data simulated with a stationary camera, and the right is simulated with a moving camera.

9.4.4   Evaluations on Synthetic Data

Experimental settings. A grid of background points and four shapes are generated to simulate objects moving under camera movements. The homogeneous representation of each point is $(X, Y, Z, 1)$ as its position in the 3D world. Then the projected points (x, y) are obtained in a 2D image by $(x, y, 1)^T = C \cdot (X, Y, Z, 1)^T$, where C is a 3×4 camera projection matrix. The depth value Z in the 3D world for foreground shapes and background grid is 10 ± 5 and 20 ± 10, respectively. The foreground shapes move and the background grid stays still. Changing the camera projection matrix C simulates the camera movement and generates projected images.

Group sparsity constraint versus standard sparsity constraint. The performance between the group sparsity constraint ($L_{2,0}$ norm) and *Std. sparse* constraint (L_0 norm) is compared. The sparsity constraint aims to find a sparse set of nonzero elements, which is $\|F\|_0 < \alpha K \times 2l$ in this problem. $K \times 2l$ denotes the total number of nonzero elements. It is equivalent to the total number of nonzero elements in the group sparsity constraint. Note that the formulation with this *Std. sparse* method is similar to Robust PCA method [7]. The difference is that LRGS use it at the trajectory level instead of pixel level.

Two sets of data are generated to evaluate the performance. One is simulated with a stationary camera, and the other is from a moving one. Foreground keeps moving in the whole video sequence. Random noise with variance v is added to both the foreground moving trajectories and camera projection matrix C. The performance is shown in Fig. 9.7. In the noiseless case (*i.e.* $v = 0$), the motion pattern from the foreground is distinct from the background in the whole video sequence. Thus each element on the foreground trajectories is different from the background element. Sparsity constraint produces the same perfect result as group sparsity constraint. When v goes up, the distinction of elements between foreground and background goes down. Thus some elements from the foreground may be recognized as background. On the contrary, the group sparsity constraint connects the elements in the neighboring frames. It treats the elements on one trajectory as one unit. Even some elements on these trajectories are similar to the background, the distinction along the whole trajectory is still large from the background. As shown in Fig. 9.7, using the group sparsity constraint is more robust than using the sparsity constraint when variance increases.

9.4.5   Evaluations on Videos

Experimental settings. The LRGS algorithm is then tested on publicly available videos from various sources. One video source is provided by Sand and Teller [31] (refer to as ST sequences). ST sequences are recorded with hand held cameras, both indoors and outdoors, containing a variety of non-rigidly deforming objects (hands, faces and bodies). They are high resolution images with large frame-to-frame motion and significant parallax. Another source of videos is provided from Hopkins 155 dataset [40], which has two or three motions from indoor and outdoor scenes. These sequences contain degenerate and non-degenerate motions, independent and partially dependent motions, articulated and non-rigid motions. The algorithm is also tested on a typical video for stationary camera: "*truck*". The trajectories in these sequences were created using an off-the-shelf dense particle tracker [38]. The pixels are manually labeled into foreground/background (binary maps). If a trajectory falls into the foreground area, it is considered as foreground, and vice versa.

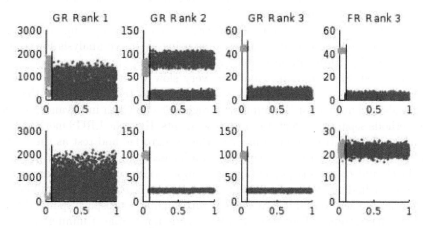

FIGURE 9.8   (See color insert.) $\|\hat{F}_i\|_2$ distribution of the GR and the FR model. Green means foreground and blue means background. Separation means good result. Top row is from moving camera, and bottom row is from stationary camera. The four columns are GR-1, GR-2, GR-3 and FR.

Handling stationary and moving cameras. We first demonstrate that our approach handles both stationary cameras and moving cameras automatically in a unified framework, by using the General Rank constraint (GR) instead of the Fixed Rank constraint (FR). We use two videos to show the difference. One is "*VHand*" from a moving camera ($rank(B) = 3$), and the other is "*truck*" captured by stationary camera ($rank(B) = 2$). We use the distribution of L_2 norms of estimated foreground trajectories (*i.e.*, $\|\hat{F}_i\|_2, i \in 1, 2, ..., k$) to show how well background and foreground is separated in the LRGS model. For a good separation result, F should be well estimated. Thus $\|\hat{F}_i\|_2$ is large for foreground trajectories and small for background ones. In other words, its distribution has an obvious difference between the foreground region and the background region (see examples in Fig. 9.8).

GR-$i, i \in 1, 2, 3$ is used to denote the optimization iteration on each rank value. $\|\hat{F}_i\|_2$ of each specific rank iteration is plotted in Fig. 9.8. The GR method works for both cases. When the rank of B is 3 (the first row of Fig. 9.8), the FR model also finds a good solution, since rank-3 perfectly fits the FR model. However, the FR constraint fails when the rank of B is 2, where the distribution of $\|\hat{F}_i\|_2$ between B and F are mixed together. On the other hand, GR-2 can handle this well, since the data perfectly fits the constraint. On GR-3 stage, it uses the result from GR-2 as the initialization, thus the result on GR-3 still holds. The figure shows that the distribution of $\|\hat{F}_i\|_2$ from the two parts has been clearly separated in the third column of the bottom row. This experiment demonstrates that the

GR model can handle more situations than the FR model. Since in real applications it is hard to know the specific rank value in advance, the GR model provides a more flexible way to find the right solution.

TABLE 9.1   Quantitative evaluation at trajectory level labeling.

Sequence	RANSAC-b	GPCA	LSA	RANSAC-m	Std.sparse	LRGS
VPerson	0.786	0.648	0.912	0.656	0.616	**0.981**
VHand	0.952	0.932	0.909	0.930	0.132	**0.987**
VCars	0.867	0.316	0.145	0.276	0.706	**0.993**
VPerson	0.786	0.648	0.912	0.656	0.616	**0.981**
cars2	0.750	0.773	0.568	0.958	0.625	**0.976**
cars5	0.985	0.376	0.054	0.637	0.779	**0.990**
people1	0.932	0.564	0.087	0.743	0.662	**0.955**
truck	0.351	0.368	0.140	0.363	0.794	**0.975**

Performance evaluation on trajectory labeling. We compare LRGS method [8] with four state-of-the-art algorithms: RANSAC-based background subtraction (referred as *RANSAC-b* here) [32], Generalized GPCA (*GPCA*) [44], Local Subspace Affinity (*LSA*) [47] and motion segmentation using RANSAC (*RANSAC-m*) [40]. *GPCA*, *LSA* and *RANSAC-m* are motion segmentation algorithms using subspace analysis for trajectories. The code of these algorithms are available online. When testing these methods, the same trajectories are used. Since *LSA* method runs very slow when using trajectories of more than 5000, we randomly sample 5000 trajectories for each test video. The three motion segmentation algorithms ask for the number of regions to be given in advance. Motion segmentation methods separate trajectories into n segments. Here the LRGS method treats the segment with the largest trajectory number as the background and rest as the foreground. For *RANSAC-b* method, two major parameters influence the performance: projection error threshold th and consensus percentage p. Inappropriate selection of parameters may result in failure of finding the correct result. In addition, as *RANSAC-b* randomly selects three trajectories in each round, it may end up with finding a subspace spanned by part of foreground and background. The result it generates is not stable. Running the algorithm multiple times may give different separation of background and foreground, which is undesirable. In order to have a fair comparison with it, we grid search the best parameter set over all test videos and report the performance under the optimal parameters.

The quantitative and qualitative results on the trajectory level separation is shown in Fig. 9.9 and Table 9.1, respectively. LRGS method [8] works well for the test videos. Take "*cars5*" for example. *GPCA* and *LSA* misclassify some trajectories. *RANSAC-b* randomly selects three trajectories to build the scene motion. On this frame, the three random trajectories all lie in the middle region. The background model built from these 3 trajectories do not cover the left and right region of the scene, thus the left and right regions are misclassified as foreground. *RANSAC-m* produces similar behavior to *RANSAC-b*. *Std. sparse* method does not have any group constraint in the consecutive frames, thus some trajectories are classified as foreground in one frame, and classified as background in the next frame. Note that the quantitative results are obtained by averaging on all frames over 50 iterations. Fig. 9.9 only shows performance on one frame, which may not reflect the overall performance shown in Table 9.1.

Performance evaluation at pixel level labeling. We also evaluate the performance at the pixel level using four methods: *RANSAC-b* [32], *MoG* [36], *Std. sparse* method and the LRGS method [8]. *GPCA*, *LSA*, *RANSAC-m* are not evaluated in this part, since these three algorithms do not provide pixel level labeling. Due to space limitations, one typical video "*VPerson*" is shown here in Fig. 9.10 and the quantitative evaluation using F-measure is shown in Table 9.2.

MoG does not perform well, because a statistical model of *MoG* is built on pixels of fixed positions over multiple frames, but background objects do not stay in the fixed

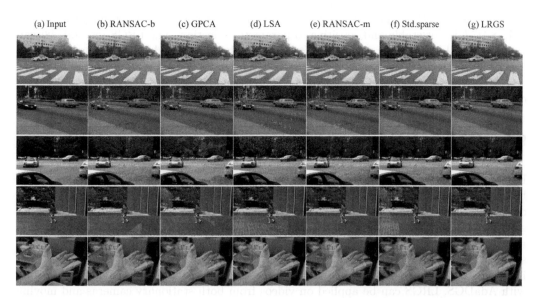

FIGURE 9.9 (See color insert.) Results at trajectory level labeling. Green: foreground; purple: background. From top to bottom, five videos are "*VCars*", "*cars2*", "*cars5*", "*people1*" and "*VHand*".

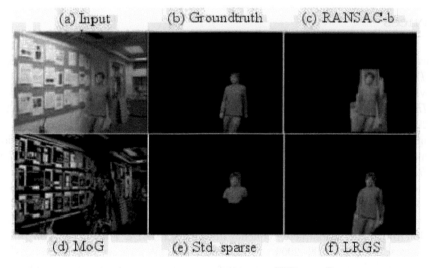

FIGURE 9.10 Quantitative results at pixel level labeling on "*VPerson*" sequence.

TABLE 9.2 Quantitative evaluation at pixel level labeling.

	RANSAC-b	MoG	Std.sparse	LRGS
VPerson	0.868	0.132	0.504	**0.892**

positions under moving cameras. *RANSAC-b* can accurately label pixels if the trajectories are well classified. However, it is also possible to build a wrong background subspace because *RANSAC-b* is not stable and sensitive to parameter selection. The LRGS method [8] can robustly handle these diverse types of data, due to the generalized low rank constraint and group sparsity constraint. One limitation of LRGS method [8] is that it classifies the shadow as part of the foreground (*e.g.*, on the left side of the person in "*VPerson*" video). This could be further refined by using shadow detection/removal techniques [23].

9.5   Conclusion

In this chapter, we have reviewed sparsity based methods AdaDGS [16] and LRGS [8] for background substraction. The AdaDGS method is motivated by the group sparsity property of the pixel values on foreground images, that they are not only sparse but also clustered into groups. The LRGS method is developed by decomposing the motion trajectory matrix into a low rank one and a group sparsity one. The group sparsity constraint for AdaDGS is at spatial pixel level while LRGS is at trajectory level. In addition, the group setting in LRGS is known and has been utilized, which is unknown but not required in AdaDGS. Compared with AdaDGS, LRGS can be applied on videos from both stationary cameras and moving cameras. However, this method depends on trajectory-tracking technique, which is also an active research area in computer vision. When the tracking technique fails, the LRGS method may not work well. Benefit from group sparsity property, these two algorithms outperform all the other methods that are compared in this chapter.

References

1. R. Baraniuk, V. Cevher, M. Duarte, and C. Hegde. Model-based compressive sensing. *IEEE Transactions on Information Theory*, 56(4):1982–2001, 2010.
2. Y. Boykov, O. Veksler, and R. Zabih. Fast approximate energy minimization via graph cuts. *IEEE Transactions on Pattern Analysis and Machine Intelligence*, 23(11):1222–1239, 2001.
3. S. Brutzer, B. Hoferlin, and G. Heidemann. Evaluation of background subtraction techniques for video surveillance. In *IEEE Conference on Computer Vision and Pattern Recognition, CVPR 2011*, pages 1937–1944, 2011.
4. E. Candès. Compressive sampling. In *International Congress of Mathematicians*, pages 1433–1452, 2006.
5. E. Candès, J. Romberg, and T. Tao. Robust uncertainty principles: Exact signal reconstruction from highly incomplete frequency information. *IEEE Transactions on Information Theory*, 52(2):489–509, 2006.
6. E. Candes and T. Tao. Near-optimal signal recovery from random projections: Universal encoding strategies? *IEEE Transactions on Information Theory*, 52(12):5406–5425, 2006.
7. J. Candès, X. Li, Y. Ma, and J. Wright. Robust principal component analysis? *arXiv preprint arXiv:0912.3599*, 2009.
8. X. Cui, J. Huang, S. Zhang, and D. Metaxas. Background subtraction using low rank and group sparsity constraints. *European Conference on Computer Vision, ECCV 2012*, pages 612–625, 2012.
9. W. Dai, O. Milenkovic, and O. Olgica. Subspace pursuit for compressive sensing: Closing the gap between performance and complexity. Technical report, DTIC Document, 2008.
10. T. Do, L. Gan, N. Nguyen, and T. Tran. Sparsity adaptive matching pursuit algorithm

for practical compressed sensing. In *Proceedings of Asilomar Conference on Signals, Systems and Computers*, pages 581–587, 2008.

11. D. Donoho. Compressed sensing. *IEEE Transactions on Information Theory*, 52(4):1289–1306, 2006.

12. A. Elgammal, D. Harwood, and L. Davis. Non-parametric model for background subtraction. *European Conference on Computer Vision, ECCV 2000*, pages 751–767, 2000.

13. C. Liu et al. *Beyond pixels: exploring new representations and applications for motion analysis.* PhD thesis, Massachusetts Institute of Technology, 2009.

14. E. Hayman and J. Eklundh. Statistical background subtraction for a mobile observer. In *IEEE International Conference on Computer Vision, ICCV 2003*, pages 67–74, 2003.

15. J. Huang, X. Huang, and D. Metaxas. Simultaneous image transformation and sparse representation recovery. In *IEEE Conference on Computer Vision and Pattern Recognition, CVPR 2008*, pages 1–8, 2008.

16. J. Huang, X. Huang, and D. Metaxas. Learning with dynamic group sparsity. In *IEEE International Conference on Computer Vision, ICCV 2009*, pages 64–71, 2009.

17. J. Huang and T. Zhang. The benefit of group sparsity. *The Annals of Statistics*, 38(4):1978–2004, 2010.

18. J. Huang, T. Zhang, and D. Metaxas. Learning with structured sparsity. *Journal of Machine Learning Research*, 12:3371–3412, 2011.

19. R. Jain and H. Nagel. On the analysis of accumulative difference pictures from image sequences of real world scenes. *IEEE Transactions on Pattern Analysis and Machine Intelligence*, 1(2):206–214, 1979.

20. S. Ji, D. Dunson, and L. Carin. Multitask compressive sensing. *IEEE Transactions on Signal Processing*, 57(1):92–106, 2009.

21. T. Kanade. Shape and motion from image streams under orthography: a factorization method. *International Journal of Computer Vision*, 9(2):137–154, 1992.

22. S. Kwak, T. Lim, W. Nam, B. Han, and J. Han. Generalized background subtraction based on hybrid inference by belief propagation and bayesian filtering. In *IEEE International Conference on Computer Vision, ICCV 2011*, pages 2174–2181, 2011.

23. S. Liao, G. Zhao, V. Kellokumpu, M. Pietikainen, and S. Li. Modeling pixel process with scale invariant local patterns for background subtraction in complex scenes. In *IEEE Conference on Computer Vision and Pattern Recognition, CVPR 2010*, pages 1301–1306, 2010.

24. G. Liu, Z. Lin, and Y. Yu. Robust subspace segmentation by low-rank representation. In *International Conference on Machine Learning, ICML 2010*, volume 3, 2010.

25. A. Mittal and N. Paragios. Motion-based background subtraction using adaptive kernel density estimation. *IEEE Conference on Computer Vision and Pattern Recognition, CVPR 2004*, 2:II–302, 2004.

26. A. Monnet, A. Mittal, N. Paragios, and V. Ramesh. Background modeling and subtraction of dynamic scenes. In *International Conference on Computer Vision, ICCV 2003*, pages 1305–1312, 2003.

27. Y. Mu, J. Dong, X. Yuan, and S. Yan. Accelerated low-rank visual recovery by random projection. In *IEEE Conference on Computer Vision and Pattern Recognition, CVPR 2011*, pages 2609–2616, 2011.

28. D. Needell and J. Tropp. CoSaMP: iterative signal recovery from incomplete and inaccurate samples. *Applied and Computational Harmonic Analysis*, 26(3):301–321, 2009.

29. S. Rao, R. Tron, R. Vidal, and Y. Ma. Motion segmentation in the presence of outlying, incomplete, or corrupted trajectories. *IEEE Transactions on Pattern Analysis and Machine Intelligence*, 32(10):1832–1845, 2010.

30. Y. Ren, C. Chua, and Y. Ho. Statistical background modeling for non-stationary camera. *Pattern Recognition Letters*, 24(1):183–196, 2003.

31. P. Sand and S. Teller. Particle video: Long-range motion estimation using point trajectories. In *IEEE Conference on Computer Vision and Pattern Recognition, CVPR 2006*, volume 2, pages 2195–2202, 2006.

32. Y. Sheikh, O. Javed, and T. Kanade. Background subtraction for freely moving cameras. In *IEEE International Conference on Computer Vision, ICCV 2009*, pages 1219–1225, 2009.

33. Y. Sheikh and M. Shah. Bayesian modeling of dynamic scenes for object detection. *IEEE Transactions on Pattern Analysis and Machine Intelligence*, 27(11):1778–1792, 2005.

34. J. Starck, M. Elad, and D. Donoho. Image decomposition via the combination of sparse representations and a variational approach. *IEEE Transactions on Image Processing*, 14(10):1570–1582, 2005.

35. C. Stauffer and W. Grimson. Adaptive background mixture models for real-time tracking. In *IEEE Conference on Computer Vision and Pattern Recognition, CVPR 1999*, volume 2, 1999.

36. C. Stauffer and W. Grimson. Learning patterns of activity using real-time tracking. *IEEE Transactions on Pattern Analysis and Machine Intelligence*, 22(8):747–757, 2000.

37. M. Stojnic, F. Parvaresh, and B. Hassibi. On the reconstruction of block-sparse signals with an optimal number of measurements. *IEEE Transactions on Signal Processing*, 57(8):3075–3085, 2009.

38. N. Sundaram, T. Brox, and K. Keutzer. Dense point trajectories by GPU-accelerated large displacement optical flow. *European Conference on Computer Vision, ECCV 2010*, pages 438–451, 2010.

39. B. Tamersoy. Background subtraction- lecture notes. Technical report, University of Texas at Austin, 2009.

40. R. Tron and R. Vidal. A benchmark for the comparison of 3-d motion segmentation algorithms. In *IEEE Conference on Computer Vision and Pattern Recognition, CVPR 2007*, pages 1–8, 2007.

41. J. Tropp and A. Gilbert. Signal recovery from random measurements via orthogonal matching pursuit. *IEEE Transactions on Information Theory*, 53(12):4655–4666, 2007.

42. J. Tropp, A. Gilbert, and M. Strauss. Algorithms for simultaneous sparse approximation. part i: Greedy pursuit. *Signal Processing*, 86(3):572–588, 2006.

43. E. van den Berg, M. Schmidt, M. Friedlander, and K. Murphy. Group sparsity via linear-time projection. *Dept. Comput. Sci., Univ. British Columbia, Vancouver, BC, Canada*, 2008.

44. R. Vidal, Y. Ma, and S. Sastry. Generalized principal component analysis (GPCA). *IEEE Transactions on Pattern Analysis and Machine Intelligence*, 27(12):1945–1959, 2005.

45. C. Wren, A. Azarbayejani, T. Darrell, and A. Pentland. Pfinder: Real-time tracking of the human body. *IEEE Transactions on Pattern Analysis and Machine Intelligence*, 19(7):780–785, 1997.

46. J. Wright, A. Yang, A. Ganesh, S. Sastry, S. Shankar, and Y. Ma. Robust face recognition via sparse representation. *IEEE Transactions on Pattern Analysis and Machine Intelligence*, 31(2):210–227, 2009.

47. J. Yan and M. Pollefeys. A general framework for motion segmentation: Independent, articulated, rigid, non-rigid, degenerate and non-degenerate. *European Conference on Computer Vision, ECCV 2006*, pages 94–106, 2006.

48. C. Yuan, G. Medioni, J. Kang, and I. Cohen. Detecting motion regions in the presence of a strong parallax from a moving camera by multiview geometric constraints. *IEEE Transactions on Pattern Analysis and Machine Intelligence*, 29(9):1627–1641, 2007.

49. M. Yuan and Y. Lin. Model selection and estimation in regression with grouped variables. *Journal of the Royal Statistical Society: Series B (Statistical Methodology)*, 68(1):49–67, 2005.

50. S. Zhang, J. Huang, Y. Huang, Y. Yu, H. Li, and D. Metaxas. Automatic image annotation using group sparsity. In *IEEE Conference on Computer Vision and Pattern Recognition, CVPR 2010*, pages 3312–3319, 2010.

51. Z. Zhang, X. Liang, and Y. Ma. Unwrapping low-rank textures on generalized cylindrical surfaces. In *IEEE International Conference on Computer Vision, ICCV 2011*, pages 1347–1354, 2011.

52. J. Zhong and S. Sclaroff. Segmenting foreground objects from a dynamic textured background via a robust Kalman filter. In *International Conference on Computer Vision, ICCV 2003*, pages 44–50, 2003.

10

Robust Detection of Moving Objects through Rough Set Theory Framework

Pojala Chiranjeevi
Indian Institute of Technology, Kharagpur, India

Somnath Sengupta
Indian Institute of Technology, Kharagpur, India

10.1 Introduction

Research on visual surveillance is gaining momentum due to security requirements in sensitive areas such as airports, offices, railway stations, etc. Moving object detection is a fundamental step in many visual surveillance applications like object tracking, action recognition, high level semantic description, gesture recognition, etc. Out of three major classes of moving object detection techniques, namely, image differencing, optical flow and background subtraction, the last is considered to be somewhat robust, as compared to the others.

Background subtraction approaches rely on constructing a background model, and detecting moving objects as those that deviate from the background model, under the assumption of static camera. These approaches work for slow moving objects unlike temporal differencing, and are less noisy and simpler than optical flow. But these are sensitive to background variations such as shadows, swaying vegetation, rippling water, illumination variations, etc. The background model needs to be robust to overcome such challenging situations. Though several approaches on background subtraction exist, and have been reviewed well in the literature [11] [4] [3] [13], problems are yet to be solved under challenging situations. The background subtraction approaches are broadly classified as pixel-based and region-based, depending on the methodology used to establish the background model. In pixel-based approaches [36] [37] [12] [19] [24] [1], each pixel is considered to be independent while constructing the model for it using its characteristics. The background perturbations, which affect the pixels' characteristics, will also affect their background model. Hence, these approaches are not effective for environments having fast background variations.

Region or block based approaches divide each frame into overlapped [25] [8] [10] [7] [15] [33] [44] [42] [43] [49] [20] [45] [48] [30] [46] or non-overlapped blocks [34] [2] [18] [32] [26] and calculate block level properties (such as covariance, histogram, correlation, etc..) to model the block. In either of these two approaches, since only a few pixels in the block are affected by the background disturbances, block based background modeling provides robustness against background variations. In non-overlapping block-based approaches, the modeling as well as the foreground detection is done at the block level. These approaches segment moving objects too coarsely than the overlapping block-based approaches. For overlapped block-based approaches, every pixel is modeled using the properties of the block, surrounding the pixel, and foreground detection is done at the pixel level. As these approaches perform modeling as well as detection at the pixel level, their shape accuracy is better than the non-overlapping ones, and hence overlapped block-based approaches are preferred.

Most of the overlapping block based approaches use histograms as block models in many forms - RGB histogram [25], edge histogram [25], Local Binary Pattern (LBP) histograms [15] [33] [44] [42] [43] [49] [20] [45] [48], local kernel histogram [30], local dependency histogram [46] etc., which store only the feature without the spatial information. Though LBP captures local texture information, it is very sensitive to background disturbances, as it is a relation based feature. Gradient and LBP also fail in uniform regions. Hence the performance of these histogram-based algorithms fall short of expectations in the presence of complex backgrounds having many challenging situations, such as swaying vegetation, camera jitter, rippling water, camouflage, illumination variations, etc. Moreover, these block-based approaches detect moving objects coarsely. Block size can be reduced for a better shape localization, but it may cause an increase in false positives. Furthermore, most of these approaches use multiple models per pixel to handle multi-modal distributions. But, prior estimation of number of models required is difficult in different environments. The computational complexity of the algorithm is directly proportional to the number of models. In the work reported in this chapter, we attempted to address these problems by considering a single robust spatial model per pixel in all types of environments to achieve finer extraction and robust performance in dynamic backgrounds (with less false positives) simultaneously. The single enriched model is derived based on the spatial distribution of pixels within a block. We followed an overlapping block based approach, and hence the distributions are computed on a pixel-by-pixel basis.

10.2   Proposed Approach

Being inspired by the still image segmentation performance of rough-set theoretic features used by Mushrif and Ray [29], we decided to adopt the concept of histon, proposed by Mohabey and Ray [28] [27], in rough set theoretic framework for background modeling. Histon is a recent concept to visualize the color information for evaluating the similar colored regions. For each intensity value in the base histogram, the number of pixels falling under the similar color sphere is calculated, and this value is added to the histogram value to get the histon value of that intensity. Histogram and histon distributions, when used in a combined measure, are indicative of the color as well as the spatial information. Identical distributions of both the histogram and the histon in a region suggest that the region lacks spatial homogeneity or spatial similarity. This property of the region is used to model the pixel existing in the center of the region, and is mathematically formulated by histon roughness index (HRI) after correlating the histon concepts with the rough set theory. The advantage of HRI is that it incorporates both the color and the spatial information.

In the original formulation of histon [28] [27], each color channel is considered separately, rather independently, to derive the histon for that color channel. But, having each pixel with its three color component values together and working out a 3D spatial distribution out of it, gives us more information than what the spatial distributions of three independent color planes would give.

With this in mind, an integrated 3D histon is proposed, where the histon distribution is calculated by considering the color value on three channels jointly, and its effectiveness as compared to the basic histon is shown. The 3D HRI distribution for a region, centered at a pixel, is calculated using 3D histon and 3D histogram. In 3D histon, whether a pixel is similar to its neighbors or not is determined crisply. By determining the extent of similarity using Gaussian membership function, 3D fuzzy histon is proposed, which is an extension of the basic fuzzy histon proposed in [29]. 3D Fuzzy histon is subsequently used to compute 3D Fuzzy histon roughness index (3D FHRI). Foreground detection in a video sequence is performed by evaluating the Bhattacharyya distance between the model roughness index distribution and the roughness index distribution computed in the current frame in all three types of histons - basic histon, 3D histon, and 3D fuzzy histon.

Proposed approach uses the first frame of the video for initial HRI model construction. Our proposed basic HRI or its 3D and fuzzy extensions as block model would not succeed without an efficient model updating strategy, especially in bootstrapping sequences, where the first frame without moving objects is not available in general. Because if the initial frame contains moving objects, which is unavoidable in many practical situations, the characteristics of the pixels of the moving objects is used to model the corresponding background pixels, resulting in incorrect modeling. Such incorrect modeling may at best be allowed for a few initial frames, and eventually the model should adapt to the true background. To enforce this, we have adopted a novel background updating strategy, where we not only update the model of the pixels labeled as background, but also update the model of the foreground labeled pixels with a lower learning rate. Our strategy of updating the foreground labeled pixels' model succeeds, because the foreground labeling of a pixel is usually short-lived. Fig. 10.1 illustrates the block diagram of the proposed approach.

Our approach offers the following advantages. Being region-based, it performs well in the presence of dynamic backgrounds due to the robustness of spatial background modeling, without compromising the moving objects' shape accuracy. To handle multi-modal background distributions, techniques using multiple models per pixel were reported in literature (e.g. [36] [37] [15] [47]). Our approaches perform better than these using just a single model. Furthermore, many approaches in the literature require a series of ideal background frames for model initialization, whereas our's is simply initialized using the first frame of the video. It is simple, yet robust in handling various challenges, as shown in the results. A journal version of this chapter[29] was published in [9].

The rest of the chapter is organized as follows. The roughness index formulation of the rough-set theory is presented in Section 10.3, along with the concepts of basic histon and the motivation and the formulation of integrated 3D histon. Section 10.4 presents the proposed algorithm for moving object detection. Detailed results of the proposed algorithm under different challenging situations are reported in Section 10.5 by comparing with state-of-the-art approaches. Section 10.6 summarizes the chapter.

---

[29]Reprinted from [9] ©(2012) with permission from Elsevier.

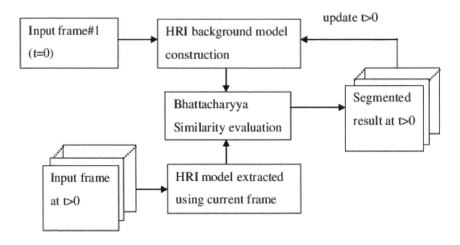

FIGURE 10.1   Block diagram of the proposed approach (©(2012) Elsevier).

10.3   Robust Detection of Moving Objects through Rough Set Theory Framework

A connection between rough set theory and histon concepts is established to use roughness index for modeling each pixel. Therefore, some basic concepts of rough set theory are presented in this section.

10.3.1   Rough Sets

Pawlak [31] introduced rough set theory, representing a new mathematical approach to vagueness and uncertainty. According to Pawlak, $I = (U, A)$ is an information system, where U is a non-empty finite set of objects and A is a non-empty finite set of attributes such that $a : U \rightarrow V_a$, where $a \in A$, V_a is the set of values that attribute a may take. The information table assigns a value $a(x)$ from V_a to each attribute a and object x in the universe U. With any $P \subseteq A$, there is an associated equivalence relation $IND(P)$, called P-indiscernibility relation.

$$IND(P) = \{(x, y) \in U^2 \mid \forall a \in P, a(x) = a(y)\}$$

$(x, y) \in IND(P)$ implies x and y are indiscernible by the attributes from P and will form one equivalence class, denoted by $[x]_P$. The partition of U containing a family of all equivalence classes of $IND(P)$ is denoted by $U|IND(P)$.

Definition 1: Rough Set

Let $X \subseteq U$ be the target set to be represented using the equivalence classes induced by the attribute subset P. In general, X cannot be expressed exactly, because the set may include and exclude objects which are indiscernible on the basis of attributes P. In that case, the target set can be expressed by P-upper and P-lower approximations. The P-lower approximation ($\underline{P}X$) is the union of all equivalence classes, which are the subsets of target set. The P-upper approximation is the union of all equivalence classes, which have non-empty intersection with the target set.

$$\underline{P}X = \bigcup\{[x]_p \in U | IND(P) : [x]_p \subseteq X\}$$

$$\bar{P}X = \bigcup\{[x]_p \in U | IND(P) : [x]_p \cap X \neq \phi\}$$

The tuple $\langle \underline{P}X, \bar{P}X \rangle$ composed of two crisp sets, one representing the lower boundary of the target set X and the other representing the upper boundary of the target set X, is called a rough set.

Definition 2: Roughness Index

The borderline region, given by the set difference $\bar{P}X - \underline{P}X$, represents the inexactness of the set X with respect to the knowledge P, consists of those objects that can neither be ruled in nor ruled out as the members of the target set X. The greater the border line region, the more the inexactness. This idea can be expressed more precisely by *roughness index measure*, defined as

$$\rho_P(X) = 1 - \frac{|\underline{P}X|}{|\bar{P}X|} \tag{10.1}$$

where, $|\underline{P}X|$ and $|\bar{P}X|$ are the cardinalities of the lower and the upper approximation sets.

The roughness index of the rough set representation of X, $\rho_P(X)$ ($0 \leq \rho_P(X) \leq 1$ for every P and $X \subseteq U$), provides a measure of how closely the rough set is approximating the target set. When the upper and the lower approximations are equal, then $\rho_P(X) = 0$, implies that the border line region of X is empty, i.e., X is crisp with respect to knowledge P, otherwise, the set X has some non empty border line region, i.e., X is rough with respect to knowledge P.

10.3.2 Concepts of Basic and 3D Histons

Definition 1: Basic Histon

Basic histon, by definition [28] [27], is a contour plotted on the top of the histograms of three primary color components of a region in a manner that the collection of all points falling under the similar color sphere of predefined radius, called the similarity threshold (S_{th}), belong to one single value. The similar color sphere is the region in RGB color space, such that all the colors falling in that region can be classified as one color. For every intensity in the base histogram, the number of pixels falling under the similar color sphere is calculated, and this value is added to the histogram value to get the histon value of that intensity.

Construction of a basic histon is given below:

The histogram count of the gth bin on c^{th} color channel ($c \in R, G, B$) of a rectangular block of $X \times Y$ pixels is given by

$$h_c(g) = \sum_{x=1}^{X} \sum_{y=1}^{Y} \delta(I(x, y, c) - g) \tag{10.2}$$

The corresponding definition of histon is given by

$$H_c(g) = \sum_{x=1}^{X} \sum_{y=1}^{Y} (1 + S(x,y))\delta(I(x,y,c) - g) \tag{10.3}$$

$$\text{for } 0 \le g \le l - 1 \text{ and } c \in \{R, G, B\}$$

where, $\delta(\cdot)$ is the digital impulse function, l is the number of quantized intensity levels, and $I(x, y, c)$ is the intensity of the pixel (x, y) on c^{th} color channel.

Whether the pixel (x, y) belongs to the similar color sphere or not is given by the similarity function $S(x, y)$ (Eq. (10.4), used in Eq. (10.3)).

$$S(x,y) = \begin{cases} 1 & \text{if } d_T(x,y) < S_{th} \\ 0 & \text{otherwise} \end{cases} \tag{10.4}$$

$$d_T(x,y) = \sum_{p=x-\lfloor R/2 \rfloor}^{x+\lfloor R/2 \rfloor} \sum_{q=y-\lfloor R/2 \rfloor}^{y+\lfloor R/2 \rfloor} \sqrt{\sum_{c \in R,G,B} (I(x,y,c) - I(p,q,c))^2} \tag{10.5}$$

where, $d_T(x, y)$ is the sum of color distances of a pixel at location (x, y) with the neighborhood pixels (p, q), existing in a region of size $R \times R$.

A simple illustrative diagram of histon for a sample image is shown in Fig. 10.2.

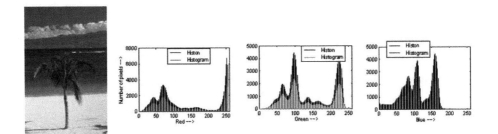

FIGURE 10.2   Histon and histogram distributions of R,G,B color components of a sample image.

Definition 2: 3D Histon

Instead of deriving histon for each color channel, as proposed in the original formulation Eq. 10.3, we propose to calculate a single 3D histon by considering three color component values jointly as bin indices, thereby achieving a complete utilization of the color information. It has the advantage because of the fact that the color distribution has much information embedded into it than the three monochrome distributions, where a very few shades of color in each image exist. Moreover, the fact that the three color channels are dependent, is being ignored, while considering the distributions on each color channel separately. This motivated us to formulate the 3D histon, which is a surface plotted on the top of the 3D histogram of a region in a manner that the collection of all points falling within the color sphere of predefined radius belongs to one single value.

3D Histogram (h) and 3D histon (H) are defined as follows:

$$h_{3D}(i,j,k) = \sum_{x=1}^{X}\sum_{y=1}^{Y} \delta(I(x,y) - (i,j,k)) \tag{10.6}$$

$$H_{3D}(i,j,k) = \sum_{x=1}^{X}\sum_{y=1}^{Y}(1 + S(x,y))\delta(I(x,y) - (i,j,k)) \tag{10.7}$$

for $0 \le i \le l_1 - 1, 0 \le j \le l_2 - 1$ and $0 \le k \le l_3 - 1$

where, $S(x,y)$ and $d_T(x,y)$ are, as defined earlier (Eq. 10.4 and Eq. 10.5), $I(x,y)$ represents the RGB value at the pixel (x,y), existing in a region, of size $X \times Y$, and (i,j,k) is the RGB bin value. l_1, l_2, l_3 are the quantized intensity levels on each color channel. We quantize each color channel to l_i levels ($l_i << 256$ and $i \in 1,2,3$) in order to reduce the size of the histogram and the histon for computational and memory efficiency.

Definition 3: 3D Fuzzy Histon

In 3D histon calculation, whether the pixel (x,y) belongs to the similar color sphere or not is decided crisply using S_{th}. Threshold applied thus is heuristic and may have considerable effects on the segmented output. Instead of adopting a crisp classification, the extent of similarity of a pixel with its neighbors is determined by Gaussian membership function (Eq. (10.8)), thereby formulating 3D fuzzy histon, which is an extension of the basic fuzzy histon, proposed by Mushrif and Ray [29]. The membership of a pixel to the similar color sphere is crisp in conventional histon, whereas in fuzzy histon, this is graded according to Gaussian membership function.

$$\tilde{S}(x,y) = e^{-\frac{1}{2}(\frac{d_T}{\sigma})^2} \tag{10.8}$$

Lower value of d_T implies higher similarity and vice versa. σ is the standard deviation of d_T. The 3D fuzzy histon $(\tilde{H})$ is obtained by applying $\tilde{S}(x,y)$ in place of $S(x,y)$ in Eq. 10.7.

Definition 4: Histon Roughness Index (HRI)

Histogram and histon can be related with the concepts of approximation space in the rough set theory [31]. Histogram gives the number of pixels having exactly the same intensity as I ($I \in (0, l-1)$), and it can be related with the lower approximation. Histon gives the number of pixels having the similar intensity as I. That is, for every intensity value in the base histogram, the number of pixels which are similar to their neighbors is calculated, and this value is added to the histogram count to obtain the histon value at that intensity. Hence, it can be related with the upper approximation. For a region centered at a pixel, similarity of both the values (histogram and histon) at each bin indicate that the region is completely non-homogeneous, rather, it can be said that no spatial correlation exists in that region. This property is mathematically formulated by Histon Roughness Index (HRI), defined in Eq. (10.9), and is utilized to model that pixel. HRI is high in homogeneous region and low in non-homogeneous region. Both the spatial similarity and the color information are embedded in HRI, whereas histogram has only color information, which is an advantage of HRI.

$$HRI = 1 - \frac{h(\bullet)}{H(\bullet)} \tag{10.9}$$

Basic Histon Roughness Index (BHRI) [29] and 3D Histon Roughness Index (3D HRI) at a bin is obtained by substituting Equations 10.2 and 10.3 and Equations 10.6 and 10.7 respectively in Equation 10.9, as follows:

$$BHRI = \rho_c(g) = 1 - \frac{h_c(g)}{H_c(g)} \tag{10.10}$$

$$3D\ HRI = \rho_{3D}(i,j,k) = 1 - \frac{h_{3D}(i,j,k)}{H_{3D}(i,j,k)} \tag{10.11}$$

To obtain 3D Fuzzy Histon Roughness Index ($\tilde{\rho}$) (3D FHRI), H in Eq. 10.11 is replaced with $\tilde{H}$.

10.4   Foreground/Background Separation Using Rough Set Measures

We have used three different roughness measures – BHRI, 3D HRI and 3D FHRI as the background models in three different versions of the same algorithm, considering basic histon, 3D histon, and 3D fuzzy histon respectively. The algorithm has three essential steps – background modeling, foreground detection, and background maintenance.

(a) Background Modeling

The histogram and the histon of a region, of size $X \times Y$, centered at a pixel, are calculated from the first frame of the video sequence. The histon roughness index (BHRI, or 3D HRI or 3D FHRI) of that region is thereby calculated using the histogram and the histon, and is used to initialize the model for that pixel.

(b) Foreground Detection

From the next frame, HRI at the corresponding pixel is calculated, similar to the procedure described in the background modeling step. The background model HRI and the current HRI distribution at a pixel are normalized along each color channel using Eq. (10.12) for the basic histon version of the algorithm.

$$\bar{\rho}_c(i) = \frac{\rho_c(i)}{\sum\limits_{i=0}^{l_c-1} \rho_c(i)} \tag{10.12}$$

$\rho_c(i)$ is the roughness index of the ith bin on c^{th} color channel.

The color channels are considered in an integrated manner for normalization in the 3D histon version (3D HRI or 3D FHRI) of the algorithm, as given below:

$$\bar{\rho}(i,j,k) = \frac{\rho(i,j,k)}{\sum\limits_{i=0}^{l_1-1}\sum\limits_{j=0}^{l_2-1}\sum\limits_{k=0}^{l_3-1} \rho(i,j,k)} \tag{10.13}$$

The similarity between the normalized background model HRI distribution and the current HRI distribution at a pixel is calculated using Bhattacharyya distance (d_B) [38], given by:

$$d_B = \sqrt{1 - q(\bar{\rho}^{cu}, \bar{\rho}^{bg})} \tag{10.14}$$

If d_B is less than distance threshold, then the spatial similarity as well as the color distributions are similar for both the background model and the current, meaning that the pixel is a background pixel, otherwise it is a foreground pixel. The superscripts bg and cu indicate the background model and the current HRI respectively.

In the basic histon version of the algorithm, the three color channels are considered to be independent as the histon is derived for each color channel separately. Hence, the overall Bhattacharyya co-efficient $(q(\bar{\rho}^{cu}, \bar{\rho}^{bg}))$ for this version of the algorithm is obtained as the product of Bhattacharyya co-efficients on each color channel.

$$q(\bar{\rho}^{cu}, \bar{\rho}^{bg}) = \sum_{i=0}^{l_1-1} \sqrt{\bar{\rho}_r^{cu}(i)\bar{\rho}_r^{bg}(i)} \times \sum_{i=0}^{l_2-1} \sqrt{\bar{\rho}_g^{cu}(i)\bar{\rho}_g^{bg}(i)} \times \sum_{i=0}^{l_3-1} \sqrt{\bar{\rho}_b^{cu}(i)\bar{\rho}_b^{bg}(i)} \quad (10.15)$$

The 3D histon version of the algorithm calculates the Bhattacharyya co-efficient as follows:

$$q(\bar{\rho}^{cu}, \bar{\rho}^{bg}) = \sum_{i=0}^{l_1-1}\sum_{j=0}^{l_2-1}\sum_{k=0}^{l_3-1} \sqrt{\bar{\rho}^{cu}(i,j,k)\bar{\rho}^{bg}(i,j,k)} \quad (10.16)$$

Foreground detection step is the same for 3D fuzzy histon except that ρ is replaced with $\tilde{\rho}$.

(c) Background Maintenance

The first frame of the video is used for background model initialization. But, in many practical situations, such as airports, shopping malls, railway stations, etc., it is difficult to obtain a proper approximation of the background in the first frame. To overcome this, a new background updating strategy is introduced that facilitates us to initialize our model with the moving objects also, described as follows. Generally, any moving object exists for a shorter duration at a pixel. That is, the probability of the moving object intensity is lower than the probability of the background intensity at a pixel. Suppose an object exists in a region in the first frame. Assume that the object left that position after n frames. Application of background subtraction algorithm will result in a ghost in that region from the $(n+1)^{th}$ frame. To alleviate this, if we update the model of a pixel in the ghost with its intensity value at that position, we will get true background approximation after a few frames, thereby facilitating the detection of the correct moving objects further. But, this approach takes a few frames for model convergence to the true background. This foreground learning rate must be low, as otherwise, the slow moving foreground objects will also be absorbed into the background model, creating false negatives. The general updating equations are described below.

$$\bar{\rho}_{x,y}^{bg}(t) = \begin{cases} (1-\alpha)\bar{\rho}_{x,y}^{bg}(t-1) + \alpha\bar{\rho}_{x,y}^{cu}(t), & \text{if } (x,y) \text{ is a background pixel} \\ (1-\beta)\bar{\rho}_{x,y}^{bg}(t-1) + \beta\bar{\rho}_{x,y}^{cu}(t), & \text{otherwise} \end{cases} \quad (10.17)$$

where, $\beta \ll \alpha, t$ is the time index.

10.5   Results and Discussions

We have tested the performance of our approaches (3D HRI and 3D FHRI) on a variety of outdoor and indoor video sequences under realistic environmental conditions like airport, campus, parking lot, office, subway station, etc., and having various challenging situations, to show their effectiveness qualitatively and quantitatively. No post processing techniques such as median filtering, morphological operations, and shadow removal algorithms are applied in any of the algorithms to maintain the fairness in comparison.

In order to reduce the size of the histon and the histogram, we quantized each color channel to a smaller number of bins ($4 \times 4 \times 8$) and maintained constant. With this minimum number of bins (128), we got optimum results. Parameters such as neighborhood size ($= 3 \times 3$), $X \times Y (= 5 \times 5)$, $\alpha\ (= 0.01)$, $\beta\ (= 0.005)$, $S_{th}\ (= 30)$, $\sigma\ (= 6)$ are retained constants for all the experiments. Values of the distance threshold (T_P) are kept between [0.65, 0.75] for all the video sequences. Distance thresholds of the other competing algorithms are also varied so that their best results are used for comparison. Quantitative evaluation of the approaches are carried out using $f_0 - measure$ (background), $f_1 - measure$ (foreground), joint $f - measure$ (f_{joint}), average classification error, and matching index (η), as defined in equations Eq. (10.18), Eq. (10.19), Eq. (10.20), Eq. (10.21) and Eq. (10.22) respectively.

$$f_0 = \frac{2P_0 R_0}{P_0 + R_0} \tag{10.18}$$

$$f_1 = \frac{2P_1 R_1}{P_1 + R_1} \tag{10.19}$$

$$\text{where, } P_0 = \frac{TN}{TN + FN}, R_0 = \frac{TN}{TN + FP}, P_1 = \frac{TP}{TP + FP} \text{ and }, R_1 = \frac{TP}{TP + FN}$$

$$f_{joint} = \frac{2f_0 f_1}{f_0 + f_1} \tag{10.20}$$

f_0, f_1 are the measures of the background and the foreground respectively, and are proposed in [16]. f_{joint} is a joint measure of the foreground and the background, and is formulated by us.

Average classification error of an algorithm for a video sequence is measured using total false positives (FP) and total false negatives (FN) as follows:

$$Average\ classification\ error = \frac{FP + FN}{n_g} \tag{10.21}$$

n_g is the number of frames in a sequence, for which the ground truth is generated.

where, TP, FP, TN, FN indicate total true positives, total false positives, total true negatives, and total false negatives respectively.

Let A be the detected foreground region for a frame, B be the corresponding ground truth and $M \times N$ be the size of the frame. Then the matching index (η) between the regions A and B is defined as

$$\eta = \frac{\sum_{i=1}^{M} \sum_{j=1}^{N} A_{i,j} \equiv B_{i,j}}{M \times N} \tag{10.22}$$

where $\equiv$ is the EX-NOR operation and $M \times N$ is the size of the image.

Matching index is the extent of matching between the segmented result and the ground truth. When the detected region and the ground truth are exactly the same, the matching index is equal to 1, otherwise $0 \leq \eta < 1$. On an average, we had generated 50 ground truths for each sequence used in quantitative evaluation. The next three subsections indicate the different comparisons adopted by the proposed approaches.

10.5.1 Comparison between BHRI, 3D HRI and 3D FHRI

Fig. 10.3 depicts the performance comparison between BHRI, 3D HRI, and 3D FHRI, under different challenging situations. As we notice, in case of swaying vegetation (EE, WT, CA), a significant improvement in the segmented result can be observed for 3D HRI and 3D FHRI over BHRI, where many noisy background pixels are detected in clusters. In foreground aperture case (WT), where the challenge is to segment a large homogeneous colored object, our 3D approaches extract the complete object than the BHRI approach, where many false negatives are detected. Even in low contrast scenario (the CA sequence), where the foreground object is hardly visible, BHRI fails to detect the foreground object, whereas 3D HRI and 3D FHRI are so sensitive that these are able to extract the smaller object against dynamic backgrounds.

In FO sequence, where a combination of two challenging situations – swaying vegetation and sprinkling water occurs, our 3D approaches mitigate both the challenges in a better manner than BHRI. In WK sequence, our 3D approaches extract the moving objects finely than BHRI, by withstanding the local intensity variations, as noted from Fig. 10.3. During the existence of shadows in IND sequence, one can notice that the BHRI approach is susceptible to shadows, whereas the performance of 3D models is visibly better than the BHRI model. The overall better performance of 3D approaches compared to BHRI is attributed to the complete usage of color information in the formulation of 3D HRI and 3D FHRI.

In addition, we notice a better performance of 3D FHRI over 3D HRI in all the challenging situations considered in Fig. 10.3 due to the incorporation of fuzziness in the formulation of 3D FHRI. Fig. 10.4 depicts the quantitative evaluation done using the matching index (Eq. (10.22)) and the average classification error (Eq. (10.21)), where 3D FHRI shows the highest matching index and the least average error, followed by 3D HRI, when compared to BHRI, for all the video sequences considered.

10.5.2 Comparison of Proposed Approaches with the CB [19] over a Sequence of Frames

We compared the performance of our 3D approaches with the CodeBook (CB) approach on a sequence of frames taken from the CA sequence, as shown in Fig. 10.5, to show how their performance varies in a sequence. We observe that our approaches detect the vehicle perfectly than the CB approach despite the camouflage of its window with the background as well as the background swaying trees, with 3D FHRI outperforming 3D HRI significantly.

10.5.3 Comparison of 3D HRI and 3D FHRI with The State-of-Art Approaches

We further compared our approaches with four others – improved MOG (IMOG) [37] (this is an improvement over Stauffer and Grimson's MOG [36]), CB, SOBS [24], and LBP histogram (LBPH) [15]. MOG and CB are the standard algorithms, while the other two are the recent ones. CB, IMOG, and SOBS are the pixel level models, whereas LBPH is the block level model, constructed using the texture histogram. Our approaches also use block

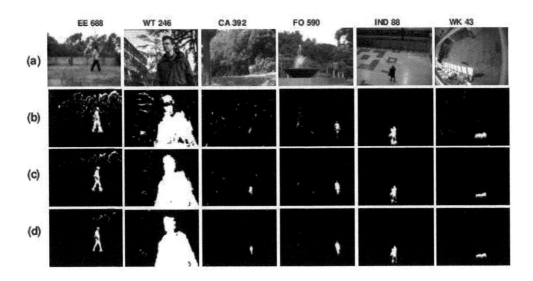

FIGURE 10.3   Qualitative performance comparison between three types of HRI's for various challenging situations: (a) original frames, (b) BHRI, (c) 3D HRI, and (d) 3D FHRI (©(2012) Elsevier).

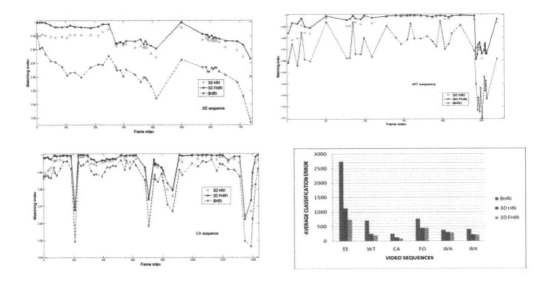

FIGURE 10.4   Quantitative performance comparison of BHRI, 3D HRI, and 3D FHRI (©(2012) Elsevier).

level models, constructed using the HRI distributions.

The sequences considered in Fig. 10.6 and 10.7 exhibit the challenging situations, such as swaying vegetation (WT, EE, CA, NN), foreground aperture (WT), low contrast (CA), and shape localization of smaller objects (CA, NN). In swaying vegetation sequences, we observe that our approaches capture less of background movements and segment moving objects better than the other competing ones. In the foreground aperture case also, our approaches are quite successful by generating the least false negatives, when compared to

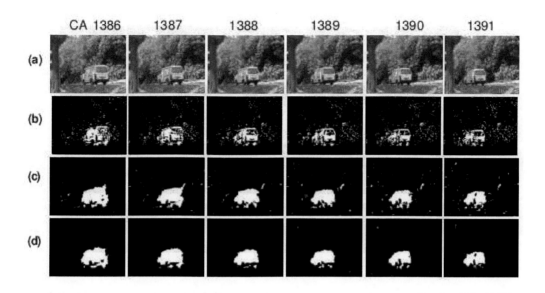

FIGURE 10.5 Qualitative performance comparison of 3D approaches with the codebook approach on a sequence of frames from CA sequence: (a) original frames, (b) CB, (c) 3D HRI, and (d) 3D FHRI (©(2012) Elsevier).

the others, especially MOG. In CA sequence, our approaches are able to detect very small moving objects, having similar chrominance features to the background, against dynamic backgrounds, better than the others. Figure. 10.8 shows our results under multiple challenging situations - the FO sequence has sprinkling water as well as swaying vegetation. The WA sequence is combined with water ripples and camouflage (leg portion) challenges. The success of our algorithms over the others is evident in both the sequences. In CA sequence (Fig. 10.7), the car window is camouflaged with the background, so also the leg region in WA sequence and the shirt region in EE sequence. Our approaches perform better than the pixel based models, where holes are detected in those regions for MOG, SOBS and CB. In dynamic backgrounds, the overall performance of our approaches is better than the LBPH, both being the block level models.

Few more challenging situations are considered in Fig. 10.9 and 10.10. Against the local intensity variations (WK) and the camera jitter (TF), the pixel based models are not that effective, whereas the performance of ours is encouraging. In mild shadows cases (MSA, IND), we may notice that IMOG and LBPH detect shadows as the foreground, whereas ours make the correct labeling in these cases. CB and SOBS also perform well in case of mild shadows. Though both ours and LBPH are block-wise models, ours achieve good shape localization than LBPH, at par with pixel based models, in all the sequences considered in Fig. 10.6 to Fig. 10.10. Furthermore, for LBPH algorithm, one can notice false negatives inside the detected foreground objects in most of the video sequences considered, due to the textural similarities between the foreground and the background, resulting in similar LBP values in both. In all the challenging situations (Fig. 10.6 to 10.10), as explained, 3D FHRI outperforms 3D HRI due to the incorporation of fuzziness.

Quantitative evaluation was also done using matching index, average classification error and f-measures. Figure. 10.11 shows frame index verses matching index for WT, CA, EE sequences. It also shows average classification error for different video sequences. In all

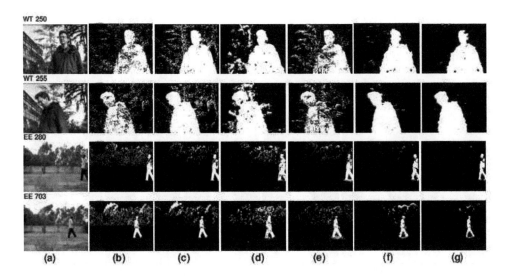

FIGURE 10.6 Qualitative performance comparison between all the competing approaches for WT and EE sequences: (a) original frames, (b) CB, (c) SOBS, (d) LBPH, (e) IMOG, (f) 3D HRI, and (g) 3D FHRI.

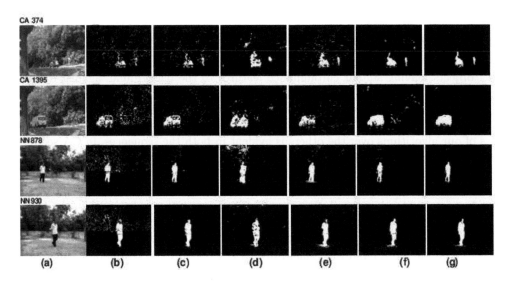

FIGURE 10.7 Qualitative performance comparison between all the competing approaches for CA and NN sequences: (a) original frames, (b) CB, (c) SOBS, (d) LBPH, (e) IMOG, (f) 3D HRI, and (g) 3D FHRI.

these plots, our proposed approaches consistently offer higher matching index and lesser average error than the others, with 3D FHRI outperforming 3D HRI. Finally, $f - measure$ are calculated for all the competing approaches, as shown in Table. 10.1. Our approaches show higher $f_0 - measure$ values than the other competing ones for all the video sequences. With respect to $f_1 - measure$, our approaches show higher values for most of the video sequences. In a few isolated cases, our approaches are marginally lower than the best performance among the competing ones. On an average (AVG) over all the sequences, our approaches dominate the rest. From $f_{joint} - measure$, our approaches outperform in all the video sequences except FO sequence, where SOBS performs the best. In FO sequence,

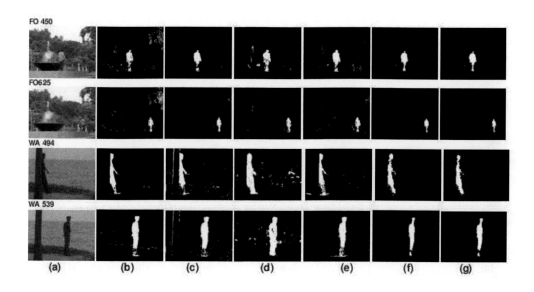

FIGURE 10.8   Qualitative performance comparison between all the competing approaches for FO and WA sequences: (a) original frames, (b) CB, (c) SOBS, (d) LBPH, (e) IMOG, (f) 3D HRI, and (g) 3D FHRI.

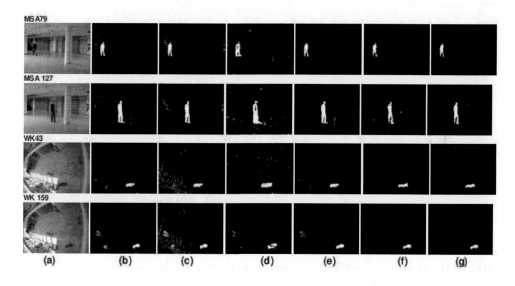

FIGURE 10.9   Qualitative performance comparison between all the competing approaches for MSA and WK sequences: (a) original frames, (b) CB, (c) SOBS, (d) LBPH, (e) IMOG, (f) 3D HRI, and (g) 3D FHRI.

our qualitative performance is not significantly worse than the SOBS approach, as shown in Fig. 10.8. Among our approaches, on an average (AVG), 3D FHRI dominates 3D HRI in all the three $f - measures$ across all the video sequences.

In addition to the above experiments, some other canonical challenges considered in Fig. 10.12 are explained as follows. The first frame of each video sequence is also included in the figure to demonstrate the challenge existing in that sequence. It may be noted that our approaches extract the moving objects in a robust way, despite significant illumination

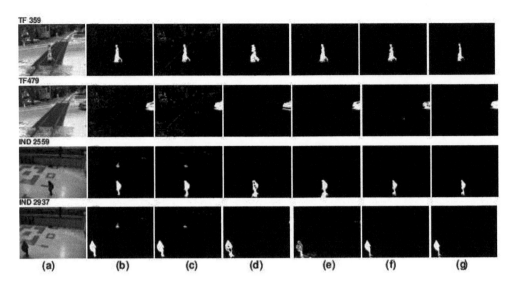

FIGURE 10.10   Qualitative performance comparison between all the competing approaches for TF and IND sequences: (a) original frames, (b) CB, (c) SOBS, (d) LBPH, (e) IMOG, (f) 3D HRI, and (g) 3D FHRI.

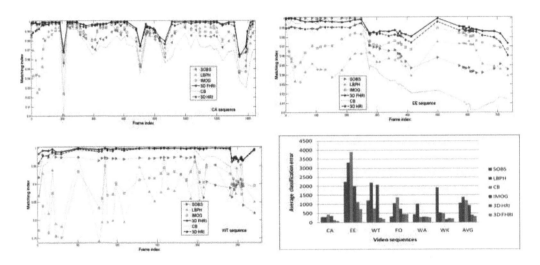

FIGURE 10.11   Quantitative performance comparison between all the competing approaches using matching index and average classification error.

variations (TD, CU) (happened with respect to the first frame) and escalator motion (ES). Even if the first frame contains moving objects (for example, BT, AP and ES), our models converge to the true background fast, thereby extracting the correct moving objects. But our models take few initial frames for convergence. In the PE sequence, a car is newly introduced in the background. But our approaches are capable of adapting to such changes in the background characteristics, thereby extracting the moving objects only. Multiple moving objects as well as small moving objects existing in BT, AP, ES and PE sequences are extracted in an isolated way with good shape localization. Curtain movements (CU) in the indoor video sequences significantly change the appearance of the background. Even then our approaches are able to capture those changes, contributing to negligible false

TABLE 10.1 Comparison of F-measures between all the competing approaches: (a) F_0-measure, (b) F_1-measure, and (c) F_{joint}-measure

	WT	EE	CA	FO	WA	WK	AVG
SOBS	0.965	0.985	0.992	0.997	0.988	0.987	0.986
LBPH	0.935	0.977	0.992	0.993	0.972	0.996	0.978
IMOG	0.939	0.986	0.990	0.995	0.992	0.999	0.984
3D HRI	0.993	0.992	0.997	0.997	0.992	0.999	0.995
3D FHRI	0.995	0.995	0.998	0.997	0.993	0.999	0.996

	WT	EE	CA	FO	WA	WK	AVG
SOBS	0.672	0.490	0.375	0.855	0.734	0.192	0.553
LBPH	0.535	0.419	0.444	0.634	0.554	0.460	0.508
IMOG	0.503	0.535	0.334	0.722	0.777	0.749	0.603
3D HRI	0.908	0.638	0.616	0.789	0.759	0.621	0.722
3D FHRI	0.927	0.705	0.700	0.773	0.774	0.632	0.752

	WT	EE	CA	FO	WA	WK	AVG
SOBS	0.792	0.655	0.545	0.920	0.842	0.322	0.680
LBPH	0.680	0.587	0.614	0.774	0.706	0.630	0.665
IMOG	0.655	0.694	0.499	0.837	0.870	0.856	0.735
3D HRI	0.948	0.776	0.762	0.880	0.860	0.766	0.832
3D FHRI	0.959	0.826	0.823	0.870	0.870	0.774	0.854

positives. Shadows are also dealt well in AP and BT sequences using 3D HRI and 3D FHRI based models. In the toll gate motion sequence (CR), our approaches adapt well to the pole motion, thereby detecting truly moving foreground objects only.

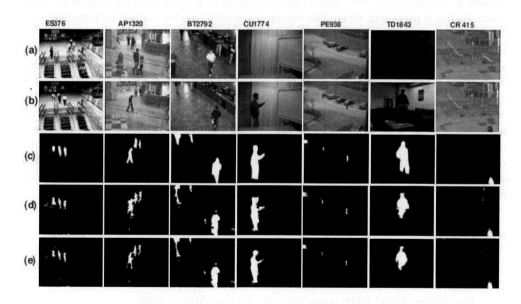

FIGURE 10.12 Qualitative evaluation of 3D approaches for the other canonical challenges: (a) first frame of the video, (b) original frames, (c) ground truth, (d) 3D HRI, and (e) 3D FHRI.

Overall, we got better results, even with less number of bins on each color channel and a region size of just 5×5 pixels. Moreover, our approaches inherit the advantages of both the pixel and the region based modelings by extracting moving objects with accurate shape in various outdoor challenging conditions.

All the experiments are carried out on a system with configuration 2.53 GHZ, 2 GB

RAM, and Intel core-two-duo processor. We have achieved a processing rate of 3fps for a frame size of 320×240 and 12 fps for a frame size of 160×120. Entire programming codes were written in C language under Microsoft Visual C++ platform. The sources of the video sequences used for bench-marking are given in Table. 10.2.

TABLE 10.2　Video sequences used in this chapter

Sequence label	Datasets	Download link	References
FO		[5]	[6]
BT,WT,TD	Wallflower	[39]	[40]
CA,ES,AP,CU,WA	I2R dataset	[21]	[22]
WK	Caviar dataset	[17]	
CR		[41]	
MSA		[23]	
IND	PETS dataset	[14]	
TF			[35]

10.6　Conclusion

Histon roughness index distribution is introduced as the background model, which captures both the spatial correlation and the color information in a single formulation. BHRI distribution is calculated for each color channel using conventional histon, which is derived for each color channel. Instead of using one histon per color channel, an integrated 3D histon is proposed to exploit the complete color information together, thereby computing 3D HRI. 3D histon is calculated by determining the similarity of a pixel with its neighbors crisply. 3D fuzzy histon is proposed, where the extent of similarity is determined by Gaussian membership function.

A novel background model updating procedure is proposed that enables our approaches to achieve robust segmentations, even if the model is initialized with moving objects. A comparison made between BHRI, 3D HRI, and 3D FHRI shows that the 3D FHRI offers the best performance, followed by 3D HRI. Effectiveness of 3D HRI and 3D FHRI in handling various challenges are illustrated in comparison with the standard approaches. Our proposed methods extract accurate shape of the moving objects, though these are region-based in nature. Our approaches are effective in multi-modal background distributions, existing in dynamic backgrounds, even without using multiple models per pixel. We initialized the background model simply using the first frame, unlike the traditional approaches which require a series of ideal background frames for model initialization.

References

1. F. El Baf, T. Bouwmans, and B. Vachon. Type-2 fuzzy mixture of Gaussians model: Application to background modeling. In *International Symposium on Advances in Visual Computing, ISVC 2008*, pages 772–781. Springer-Verlag, 2008.
2. R. Bourezak and G. Bilodeau. Iterative division and correlograms for detection and tracking of moving objects. In *Advances in Machine Vision, Image Processing, and Pattern Analysis, 2006*, volume 4153 of *LNCS*, pages 46–55. Springer Berlin, 2006.
3. T. Bouwmans. Recent advanced statistical background modeling for foreground detection: A systematic survey. *Recent Patents on Computer Science*, 4(3):147–176, November 2011.
4. T. Bouwmans, F. El Baf, and B. Vachon. Background Modeling using Mixture of Gaussians for Foreground Detection - A Survey. *Recent Patents on Computer Science*, 1(3):219–237, November 2008.

5. Y. Chen. test videos. Website. `http://imp.iis.sinica.edu.tw/ytchen/testvideos.rar`.

6. Y. Chen, C. Chen, C. Huang, and Y. Hung. Efficient hierarchical method for background subtraction. *Pattern Recognition*, 40(10):2706–2715, 2007.

7. P. Chiranjeevi and S. Sengupta. Detection of moving objects using fuzzy correlogram based background subtraction. In *IEEE International Conference Signal and Image Processing Applications, 2011*, pages 255–259, 2011.

8. P. Chiranjeevi and S. Sengupta. Moving object detection in the presence of dynamic backgrounds using intensity and textural features. *J. Electron. Imaging (SPIE)*, 20:043009, 2011.

9. P. Chiranjeevi and S. Sengupta. Robust detection of moving objects in video sequences through rough set theory framework. *Image and Vision Computing*, 30(11):829–842, 2012.

10. P. Chiranjeevi and S. Sengupta. Spatially correlated background subtraction, based on adaptive background maintenance. *J. Vis. Commun. Image R.*, 23(6):948–957, 2012.

11. M. Cristani, M. Farenzena, D. Bloisi, and V. Murino. Background subtraction for automated multisensor surveillance: A comprehensive review. *EURASIP J. Adv. Sig. Proc.*, 2010, 2010.

12. A. Elgammal, D. Harwood, and L. Davis. Non-parametric model for background subtraction. In *European Conference on Computer Vision, ECCV 2000*, volume 1843 of *Lecture Notes in Computer Science*, pages 751–767. Springer, 2000.

13. S. Elhabian, K. El-Sayed, and S. Ahmed. Moving object detection in spatial domain using background removal techniques - state-of-art. *Recent Patents on Computer Science*, 1(1):32–54, 2008.

14. J. Ferryman. PETS data set. Website. `http://www.cvg.rdg.ac.uk/slides/pets.html`.

15. M. Heikkila and M. Pietikainen. A texture-based method for modeling the background and detecting moving objects. *IEEE Transactions on Pattern Analysis and Machine Intelligence*, 28(4):657–662, April 2006.

16. S. Herrero and J. Bescos. Background subtraction techniques: Systematic evaluation and comparative analysis. In *Advanced Concepts for Intelligent Vision Systems, 2009*, volume 5807 of *LNCS*, pages 33–42. Springer Berlin, 2009.

17. INRIA. CAVIAR data set. Website. `http://homepages.inf.ed.ac.uk/rbf/CAVIAR/`.

18. D. Jang, X. Jin, Y. Choi, and T. Kim. Background subtraction based on local orientation histogram. In *Computer-Human Interaction*, volume 5068 of *LNCS*, pages 222–231. Springer Berlin, 2008.

19. K. Kim, T. Chalidabhongse, D. Harwood, and L. Davis. Real-time foreground-background segmentation using codebook model. *Real-Time Imaging*, 11(3):172–185, 2005.

20. Y. Lee, J. Jung, and I. Kweon. Hierarchical on-line boosting based background subtraction. In *Korea-Japan Joint Workshop on Frontiers of Computer Vision, FCV 2011*, pages 1–5, 2011.

21. L. Li. I2R data set. Website. `http://perception.i2r.a-star.edu.sg/bk_model/bk_index.html`.

22. L. Li, W. Huang, I. Gu, and Q. Tian. Statistical modeling of complex backgrounds for foreground object detection. *IEEE Transactions on Image Processing*, 13(11):1459–1472, November 2004.

23. L. Maddalena. Moving object detection sequences. Website. `http://cvprlab.uniparthenope.it`.

24. L. Maddalena and A. Petrosino. A self-organizing approach to background subtraction for visual surveillance applications. *IEEE Transactions on Image Processing*, 17(7):1168–1177, July 2008.

25. M. Mason and Z. Duric. Using histograms to detect and track objects in color video. In

Applied Imagery Pattern Recognition Workshop, 2001, pages 154–159, 2001.

26. T. Matsuyama, T. Ohya, and H. Habe. Background subtraction for non-stationary scenes. In *Asian Conference on Computer Vision, ACCV 2000,* pages 662–667, 2000.

27. A. Mohabey and A. Ray. Fusion of rough set theoretic approximations and FCM for color image segmentation. In *IEEE International Conference on Systems, Man, and Cybernetics,* volume 2, pages 1529–1534, 2000.

28. A. Mohabey and A. Ray. Rough set theory based segmentation of color images. In *Int. Conf. North Amer. Fuzzy Inform. Process. Soc.,* pages 338–342, 2000.

29. M. Mushrif and A. Ray. Color image segmentation: Rough-set theoretic approach. *Pattern Recognition Letters,* 29(4):483–493, 2008.

30. P. Noriega, B. Bascle, and O. Bernier. Local kernel color histograms for background subtraction. In *VISAPP 2006,* volume 1, pages 213–219, 2006.

31. Z. Pawlak. *Rough sets: theoretical aspects of reasoning about data.* Kluwer Academic Publishers, 1991.

32. D. Russell and S. Gong. A highly efficient block-based dynamic background model. In *International Conference on Advanced Video and Signal Based Surveillance, AVSS 2005,* pages 417–422, 2005.

33. A. Satpathy, H. Eng, and X. Jiang. Difference of Gaussian edge-texture based background modeling for dynamic traffic conditions. In *International Symposium on Advances in Visual Computing, ISVC 2008,* pages 406–417. Springer-Verlag, 2008.

34. M. Seki, T. Wada, H. Fujiwara, and K. Sumi. Background subtraction based on cooccurrence of image variations. In *International Conference on Computer Vision and Pattern Recognition, CVPR 2003,* volume 2, pages 65–72, 2003.

35. Y. Sheikh and M. Shah. Bayesian modeling of dynamic scenes for object detection. *IEEE Transactions on Pattern Analysis and Machine Intelligence,* 27(11):1778–1792, November 2005.

36. C. Stauffer and W. Grimson. Adaptive background mixture models for real-time tracking. In *International Conference on Computer Vision and Pattern Recognition, CVPR 1999,* volume 2, pages 2246–2252, 1999.

37. Z. Tang and Z. Miao. Fast background subtraction and shadow elimination using improved Gaussian mixture model. In *IEEE International Workshop on Haptic, Audio and Visual Environments and Games, 2007,* pages 541–544, 2007.

38. N. Thacker, F. Aherne, and P. Rockett. The Bhattacharyya metric as an absolute similarity measure for frequency coded data. *Kybernetika,* 34(4):363–368, 1997.

39. K. Toyama. Wallflower data set. Website. `http://research.microsoft.com/en-us/um/people/jckrumm/wallflower/testimages.htm`.

40. K. Toyama, J. Krumm, B. Brumitt, and B. Meyers. Wallflower: principles and practice of background maintenance. In *IEEE International Conference on Computer Vision, ICCV 1999,* volume 1, pages 255–261, 1999.

41. M. Trivedi. Shadow detection. Website. `http://cvrr.ucsd.edu/aton/shadow/index.html`.

42. L. Wang, H. Wu, and C. Pan. Adaptive ELBP for background subtraction. In *ACCV 2010,* volume 6494 of *LNCS,* pages 560–571. Springer, 2010.

43. G. Xue, L. Song, J. Sun, and M. Wu. Hybrid center-symmetric local pattern for dynamic background subtraction. In *IEEE International Conference on Multimedia and Expo, ICME 2011,* ICME 2011, pages 1–6. IEEE Computer Society, 2011.

44. G. Xue, J. Sun, and L. Song. Dynamic background subtraction based on spatial extended center-symmetric local binary pattern. In *IEEE International Conference on Multimedia and Expo, ICME 2010,* pages 1050–1054, 2010.

45. G. Yuan, Y. Gao, D. Xu, and M. Jiang. A new background subtraction method using texture and color information. In *International Conference on Advanced Intelligent Comput-*

ing Theories and Applications: With Aspects of Artificial Intelligence, pages 541–548. Springer-Verlag, 2011.

46. S. Zhang, H. Yao, and S. Liu. Dynamic background subtraction based on local dependency histogram. *Inter. J. Pattern Recognition and Artificial Intelligence*, 23(07):1397–1419, 2009.

47. S. Zhang, H. Yao, and S. Liu. Dynamic background subtraction based on local dependency histogram. *International Journal of Pattern Recognition and Artificial Intelligence*, 23(7):1397–1419, 2009.

48. S. Zhang, H. Yao, S. Liu, X. Chen, and W. Gao. A covariance-based method for dynamic background subtraction. In *International Conference on Pattern Recognition, ICPR 2008*, pages 1–4, 2008.

49. W. Zhou, Y. Liu, W. Zhang, L. Zhuang, and N. Yu. Dynamic background subtraction using spatial-color binary patterns. In *International Conference on Image and Graphics, ICIG 2011*, pages 314–319, Washington, DC, USA, 2011.

III

Applications in Video Surveillance

11

Background Learning with Support Vectors: Efficient Foreground Detection and Tracking for Automated Visual Surveillance

Alireza Tavakkoli
Univ. of Houston-Victoria, USA

Mircea Nicolescu
CV Lab., Univ. of Nevada, Reno, USA

Junxian Wang
Microsoft Research, USA

George Bebis
CV Lab., Univ. of Nevada, Reno, USA

11.1 Introduction

With the increase in the availability and computational power of digital imaging devices, it is natural to think about integrating artificial intelligence models with processing of digital images and videos used for visual surveillance applications to improve their performance. However, for such applications to be efficient, there is a need for addressing a number of significant challenges. Accounting for the presence of regions that do not belong to objects of interest, global ambient illumination variations in the environment over long periods of time, and the real-time constraints inherent to visual surveillance applications are among such obstacles. This chapter demonstrates two main categories of mathematical and computational techniques developed to help improve the accuracy and efficiency of automated visual surveillance systems.

First, a statistical modeling approach based on non-parametric density estimation is presented with the goal of accurately detecting foreground regions in videos with quasi-stationary background. Second, we show that pixel models may alternatively be learned analytically to help with the issue of unknown probabilistic distribution of background pixels. In order to train pixel models analytically, Support Vector Machines are utilized to learn the single-class of pixel models – namely the background. This can be achieved by training Support Vector Data Descriptions (SVDD) for each pixel and by utilizing Support Vector Regression (SVR). We demonstrate the applicability of each approach in different surveillance applications - i.e. videos with non-empty backgrounds, indoor and outdoor environments, videos with sudden global illumination changes, etc.- to highlight their strengths and accuracy in delivering robust and reliable foreground regions. Such robust and accurate foreground region detection mechanisms help improve the end results of automated visual surveillance applications. The chapter concludes the contributions made within the proposed unified framework for visual surveillance and presents a road-map for future investigations.

11.2   Literature Review

Detecting foreground regions in videos is an important task in high-level video processing applications. One of the major issues in detecting foreground regions is that because of inherent changes in the background (such as fluctuations in monitors and fluorescent lights, waving flags and trees, water surfaces, etc.) the background may not be completely stationary.

In the presence of these types of backgrounds, referred to as quasi-stationary, a single background frame is not enough to accurately detect moving regions. Therefore the background of the video has to be modeled in order to detect foreground regions - e.g. newly introduced objects to the scene, while allowing for quasi-stationary backgrounds.

There is also a great amount of diversity in scenarios where the background modeling techniques are used to detect foreground regions. Applications vary from indoors scenes to outdoors, from completely stationary to dynamic backgrounds, from high quality videos to low contrast scenes and so on. Therefore, a single system that addresses all possible situations while being time and memory efficient has yet to be devised.

11.2.1   Statistical Background Modeling

In the presence of quasi-stationary backgrounds, a single background frame is not enough to accurately detect foreground regions. Pless *et al.* [16] evaluated different models for dynamic backgrounds. Depending on the complexity of the problem the background models employ expected pixel features (i.e. colors) [17], consistent motion [15], [35], or fusion of color/contrast and motion [4]. They also may employ pixel-wise information [36] or regional models of features [30]. To improve robustness to noise, spatial [14] or spatio-temporal [12] features may be used.

In [36] a single 3-D Gaussian model for each pixel is built and the mean and covariance of the model are learned in each frame. However, the system failed to label a pixel as foreground or background when it has more than one modality due to fluctuations in its values, such as in a fluctuating monitor.

A mixture of Gaussians modeling technique was proposed in [20], and [19] to address the multi-modality of the underlying background. In this technique background pixels are modeled by a mixture of Gaussians. During the training stage, parameters and weights of the Gaussians are trained and used in the background subtraction where the probability of each pixel is generated using the mixture of Gaussians. The pixel is labeled as foreground or background based on its probability.

There are several shortcomings for mixture learning methods. First, the number of Gaussians needs to be specified. Second, this method does not explicitly handle spatial dependencies. Even with the use of incremental expectation maximization, the parameter estimation and its convergence is noticeably slow where the Gaussians adapt to a new cluster.

A recursive filter formulation is proposed by Lee in [11] to speed up the convergence. However, the problem of specifying the number of Gaussians as well as the adaptation in later stages still exists. This model does not account for situations in which the number of Gaussians changes due to occlusion or uncovered parts of the background.

In [7], Elgammal et al. proposed a non-parametric kernel density estimation method (KDE) for pixel-wise background modeling without making any assumption about its probability distribution. Therefore, this method can easily deal with multi-modality in background pixel distributions without specifying the number of modes in the background. However, there are several issues to be addressed using non-parametric kernel density estimation.

These methods are memory and time consuming since for each pixel in each frame the system has to compute the average of all kernels centered at each training sample. The size of temporal window used as the background model needs to be specified. Too small a window increases speed, while it does not incorporate enough history for the pixel, resulting in a less accurate model.

In order to update the background for scene changes such as moved objects, parked vehicles or opened/closed doors, Kim et al. in [9] proposed a layered modeling technique. This technique needs an additional model called *cache* and assumes that the background modeling is performed over a long period of time. It should also be used as a post-processing stage after the background is modeled.

Recently, we investigated two statistical methods for background modeling, based on adaptive kernel density estimation (AKDE) [24], [22], and recursive modeling (RM) [25], [23]. These techniques will be further investigated and discussed in this chapter.

There is a major drawback inherent to statistical modeling methods including the AKDE and the RM techniques. The accuracy of these methods is limited to the accuracy of the estimated probability density function for the background pixels. In this paper we present a non-statistical method that addresses this difficulty.

Furthermore, there is an additional issue with all statistical foreground detection techniques including the AKDE and the RM methods. In all statistical methods the assumption is that there are two classes, namely foreground and background, and that the model is trained on background samples which are present during a short period of time (the AKDE) or from the beginning of the video (the RM). Note that until a foreground object appears in the scene, there is no information about the foreground class. This problem is addressed by using thresholds in the classification stage to label pixels as foreground or background.

11.2.2 Analytical Learning of Background Pixel Models

To overcome the aforementioned disadvantage of statistical learning tools a non-statistical background modeling technique is proposed in [26], based on Support Vector data description modeling (SVDDM). This novel technique in describing one class of known data samples, is called Support Vector Data Description Modeling [26]. The backbone of the proposed method is a theory based on describing a data set using their Support Vectors [29], [28]. The SVDDM uses Support Vectors to generate a description for the known data class; e.g. the background. These Support Vectors along with the classifier information for each pixel are stored and used in the classification stage to label pixels in new frames as foreground/background. The performance of this system is studied and its experimental results on real video sequences are compared with other existing techniques in the literature.

11.2.3   Target Detection and Tracking

Target behavior analysis depends heavily on the reliability of target detection and tracking which can provide important information about the location of targets and their temporal correspondences over time. Both target detection and tracking have been investigated widely over the last two decades with the majority of approaches employing detection alone, tracking alone, or hybrid schemes such *"detect-then-track"* where detection and tracking work sequentially and independently of each other [21].

Tracking methods can be divided into two main categories. In the first category, the state sequence of a target is iteratively predicted and updated using prior information from past measurements and likelihood information from current measurements, respectively. Various filters have been employed to predict the state sequence of a target including Kalman filters [2] and extended Kalman filters for linear predictions, as well as unscented Kalman filters [2] for non-linear predictions. The most general class of filters, however, includes particle filters [10], also called bootstrap filters [8], which are based on Monte Carlo integration methods. Methods belonging in the second category use various target characteristics, such as color or gray-level information, shape, and motion information. These methods perform tracking by building the unique correspondence relationship in the appearance of the target from frame to frame [3].

In tracking alone methods, the initial location of a target is usually specified manually. The majority of methods employing detection along with tracking use a *detect-then-track* approach where the target is detected in the first frame and then turned over to the tracker in subsequent frames. The main problem with these methods is that they aim to resolve detection and tracking sequentially and independently of each other. An important issue considered in this work is improving the performance of target detection by feeding temporal information from tracking back to the detection stage. In this context, we propose a *detect-and-track* scheme where detection and tracking are addressed simultaneously in a unified framework (i.e., detection results trigger tracking, and tracking re-enforces detection). One approach to deal with this problem is by using a Bayesian decision framework which combines prior probability information provided by tracking with likelihood information provided by frame-based detection [33]. However, the performance of target detection depends heavily on the threshold used to distinguish between foreground and background objects. Another approach is propagating the probabilities of detection parameters (e.g., at several scales and poses) over time using condensation and factored sampling [32].

11.3   Non-Parametric Statistical Estimation of Pixel Distribution for Background Modeling

In this section we present a non-parametric statistical learning approach for modeling background pixel probability distribution [27]. Our focus here is to find a common ground that would cover a general scenario for background modeling. This solution is based on a non-parametric framework. This base-line system is called Adaptive Kernel Density Estimation (AKDE) [27]. To enhance the base-line statistical modeling technique, we derive a universal modeling tool. The proposed general method is called Recursive Modeling (RM) [25]. This technique addresses the issue of robust background training in slowly changing backgrounds, non-empty backgrounds, and backgrounds with irregular global motion (e.g. hand-held cameras).

Algorithm 11.1 - The proposed AKDE modeling algorithm

```
 1: for each frame at time t do
 2:     // Training Stage
 3:     for each pixel:=uv do
 4:         Σ [u,v]←CalcCovariance(frame_t)
 5:         th[u,v]←CalcThreshold(frame_t)
 6:     end for
 7:     // Classification Stage:
 8:     for each pixel:=uv do
 9:         Median [u,v]←CalcMedian(frame_t, [u,v],[w×w])
10:         if Median[u,v]≤th[u,v] then
11:             FG_t[u,v]← 1 // Foreground Detected
12:         else
13:             FG_t[u,v]← 0 // Background Detected
14:         end if
15:     end for
16:     // Update Stage:
17:     if Size(FG)≥Size(frame) then
18:         for each pixel:=[u,v] do
19:             OldestFrame_t[u,v]←frame[u,v]
20:         end for
21:     else
22:         for each pixel:=[u,v] do
23:             OldestFrame_t[u,v|FG[u,v]==0]←frame[u,v|FG[u,v]==0]
24:         end for
25:     end if
26: end for
```

11.3.1 The AKDE Algorithm

Algorithm 11.1 shows the pseudo-code for the AKDE algorithm, consisting of three major stages: training, classification and update. In the training stage the background model is generated. In new frames, pixel model values are used to estimate the probability that the pixel belongs to the background model. Since we only have samples of the background class before any foreground object appears in the scene, there should be a mechanism to label low probability values to foreground models.

The only parameter in kernel density estimation is the kernel bandwidth. In theory, as the number of training samples grows without a bound the estimated density converges to the actual underlying density regardless of the kernel bandwidth value [6]. In the AKDE method a non-parametric model for each pixel is generated and its classifier is trained. The training stage employs the history of pixel values. The algorithm then estimates the probability of each pixel being background in new frames as the classification criterion. In the classification stage, each pixel is classified as foreground or background based on its estimated probability, computed by:

$$P_t(\mathbf{x_t}) = \frac{1}{N 2\pi |\mathbf{\Sigma}|^{1/2}} \sum_{i=1}^{N} e^{\left[-\frac{1}{2}(\mathbf{x}_t - \mathbf{x}_i)^T \mathbf{\Sigma}^{-1} (\mathbf{x}_t - \mathbf{x}_i)\right]} \tag{11.1}$$

where $\mathbf{x}_t$ is the pixel feature vector at time t and $\mathbf{x}_i$ are its values in the training sequence. $\mathbf{\Sigma}$ is a positive definite symmetric matrix which is the kernel bandwidth matrix and N is

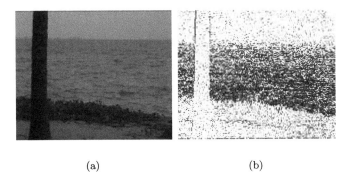

(a)   (b)

FIGURE 11.1   Adaptive threshold map: (a) An arbitrary frame. (b) Threshold map.

the number of frames used to train the background model. In order to capture dependencies between features for each pixel, Σ has to be a full (non-diagonal) matrix.

Due to limited memory and computational power, a rather short term memory of the background frames may be stored as training samples. This makes the non-parametric kernel density estimation dependent on the choice of its kernel bandwidth. In order to achieve an accurate and automatic background model, which is adaptive to the spatial information in the scene, the kernel bandwidth matrix needs to be trained.

For each pixel the training samples are vectors $\mathbf{X}_N = \{\mathbf{x}_i : i = 1 \cdots N\}$, where N is the number of training frames. The successive deviation of the above vectors is a matrix Δ_X whose columns are $\left[\mathbf{x}_i - \mathbf{x}_{i-1}\right]^T$. For each pixel, the kernel bandwidth matrix is defined such that it represents the temporal scatter of training samples [27].

Note that for pixels that change more frequently the kernel bandwidth matrix has larger elements, while for pixels that do not change much its elements are smaller. Moreover, since the kernel bandwidth matrix is computed using successive deviations, it accounts for temporal dependencies in pixel feature vectors.

To allow for the pixel probability estimation adaptation to the different amount of change, the classifier threshold values need to be trained for each pixel during the training stage. For each pixel a threshold value (th) is selected such that its classifier results in 5% false reject rate. That is, 95% of the time the pixel is correctly classified as belonging to background model [27].

This adaptive classifier threshold training can be seen in Fig. 11.1, where (a) shows an arbitrary frame of a sequence containing a water surface and (b) shows the trained threshold map for this frame. Darker pixels in Fig. 11.1(b) represent smaller threshold values and lighter pixels correspond to larger threshold values. The thresholds in areas that tend to change more (the water surface) are lower than those in areas with less amount of change (the sky).

In the classification stage, each pixel background probability in new frames is estimated using equation (11.1) to label the pixel. If we directly apply the trained threshold of each pixel to its estimated probability, due to impulse (salt and pepper) noise, isolated pixels may be erroneously classified. One of the properties of this type of noise is that, if strong noise affects a pixel, it is less likely to affect its neighbors with the same strength.

Median filtering is known to be a suitable tool to remove this type of noise. In order to remove the process noise we apply the median of estimated probabilities in a region around a pixel. After estimating the probability of each pixel in the new frame, the median of probabilities in its 8-connected neighborhood is compared with its threshold to label each pixel as background or foreground. Fig. 11.2 shows the effect of enforcing spatial consistency

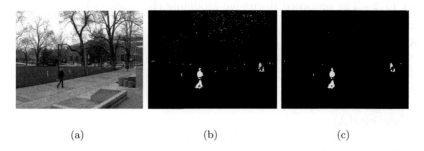

(a)   (b)   (c)

FIGURE 11.2 Original frame (a), and detected foreground regions by applying thresholds directly on the estimated probability (b), and on median of probabilities in a neighborhood (c) (©(2009) Springer).

using the median of probabilities in foreground region detection.

11.3.2 Recursive Modeling (RM)

One of the main disadvantages of the kerned density estimation approaches to background modeling is the number of background frames needed to train probabilistic models – i.e. N from equation (11.1).

The Recursive Modeling (RM) technique, in pseudo-code, is shown in Algorithm 11.2. θ_t^B is the background model and θ_t^F is the foreground model for each pixel. Let x_t be the intensity value (or the chromaticity vector) of a pixel at time t. The non-parametric estimation of the background model that accurately follows its multi-modal distribution can be reformulated in terms of recursive filtering [25]:

$$\hat{\theta}_t^B(x) = [1 - \beta_t] \cdot \theta_{t-1}^B(x) + \alpha_t \cdot H_\Delta(x - x_t) \tag{11.2}$$

where $x \in [0, 255]$ and θ_t^B is the non-normalized background pixel model at time t. The background model represents the probability of each pixel belonging to the background and must be normalized accordingly; i.e. $\sum_{x=0}^{255} \theta_t^B(x) = 1$.

$\hat{\theta}_t^B$, before normalization, is updated by the local kernel $H(\cdot)$ with bandwidth Δ centered at x_t. Parameters α_t and β_t are the learning rate and forgetting rate schedules, respectively. The kernel H should satisfy the following conditions: $\sum_x H_\Delta(x) = 1$ and $\sum_x x \times H_\Delta(x) = 0$.

Fig. 11.3 shows the process of using the proposed RM technique. The trained model (solid line) converges to the actual one (dashed line) as new samples are introduced. The actual model is the probability density function of a randomly generated sample population.

Scheduled Learning

In order to speed up the modeling convergence and recovery from stale models a schedule for learning the background model at each pixel based on the pixel's history is utilized. This schedule makes the adaptive learning process converge faster, without compromising the stability and memory requirements of the system. The learning rate changes according to the schedule $\alpha_t = \frac{1-\alpha_0}{h(t)} + \alpha_0$. In this context, α_t is the learning rate at time t and $\alpha_0 = 1/256 \times \sigma_\theta$ is a small target rate. σ_θ is the model variance. The function $h(t)$ is a monotonically increasing function $h(t) = t - t_0 + 1$. We denote the time at which a sudden global change is detected as t_0.

According to this schedule the learning occurs faster ($\alpha_t = 1$) shortly after a global illumination change is detected and decreases to converge to the target rate α_0. In section 11.6 we discuss the effect of this schedule on improving the convergence and recovery speed.

Algorithm 11.2 - The proposed Recursive Modeling (RM) algorithm

---

1: Initialization($\Delta, \alpha_0, \beta, \kappa, \text{th}$)
2: **for** each frame at time t **do**
3:    **for** each pixel:=uv **do**
4:       $x_r \leftarrow \text{frame}_t[\text{u,v}]$
5:       // **Training Stage:**
6:       $\alpha_t \leftarrow \frac{1-\alpha_0}{h_t} + \alpha_0$
7:       updateDelta(Δ)
8:       $\theta_t^B(x) \leftarrow (1-\beta_t)\theta_{t-1}^B(x) + \alpha_t \times H_\Delta(x - x_t)$
9:       **if** ($\theta_t^B(x) \leq th$) **then**
10:          $\theta_t^F(x) \leftarrow (1-\beta_t)\theta_{t-1}^F(x) + \alpha_t \times H_\Delta(x - x_t)$
11:       **end if**
12:       // **Classification Stage:**
13:       **if** ($\ln(\frac{median(\theta_t^B(x)}{\theta_t^F(x)} \leq \kappa$) **then**
14:          frame[u,v] $\leftarrow$ background
15:       **else**
16:          frame[u,v] $\leftarrow$ foreground
17:       **end if**
18:       // **Update Stage:**
19:       updateKappa(κ)
20:       updateTh(th)
21:    **end for**
22: **end for**

11.4 Support Vector Data Description Background Modeling

This section presents analytical background modeling techniques based on single class Support Vector classification techniques [26].

To overcome the main disadvantage of statistical learning tools in explicitly addressing the dependence of statistical models to probability estimation accuracy and the single-class classification problem a series of novel analytical background model learning tools are proposed in this section. In the following we present the SVDDM theory and the algorithm which detects foreground regions using this theory in detail.

11.4.1 The SVDD Theory

A normal data description is a description which gives a closed boundary around the data. A simple normal data description can be considered as a sphere with center $\mathbf{a}$ and radius $R > 0$, which encloses all of the training samples $\mathbf{x}_i$. The data description is achieved by minimizing the error function $F(R, \mathbf{a}) = R^2$ subject to $\|\mathbf{x}_i - \mathbf{a}\|^2 \leq R^2$ for every smaple.

In order to allow for outliers in the training data set, the distance of each training sample $\mathbf{x}_i$ to the center of the sphere $\mathbf{a}$ should not be strictly smaller than R^2. However, large distances should be penalized. Therefore, after introducing slack variables $\epsilon_i \geq 0$ the minimization problem becomes:

$$F(R, \mathbf{a}) = R^2 + C \sum_i \epsilon_i \tag{11.3}$$

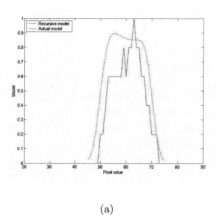
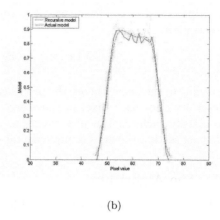

(a)   (b)

FIGURE 11.3   Recursive modeling: Model after (a) 10 frames. (b) 467 frames (©(2009) Springer).

subject to the new constraints:

$$\|\mathbf{x}_i - \mathbf{a}\|^2 \leq R^2 + \epsilon_i, \quad \forall i \tag{11.4}$$

where C controls the trade-off between the sphere volume and the description error.

The minimization problem in equation (11.3) can be solved by introducing the constraints of equation (11.4) to the error function using Lagrange multipliers:

$$L\left(R, \mathbf{a}, \alpha_i, \gamma_i, \epsilon_i\right) = R^2 + C \sum_i \epsilon_i - \sum_i \alpha_i \left[R^2 + \epsilon_i - \left(\|\mathbf{x}_i - \mathbf{a}\|^2\right)\right] - \sum_i \gamma_i \epsilon_i \tag{11.5}$$

where $\alpha_i \geq 0$ and $\gamma_i \geq 0$ are Lagrange multipliers. Optimization is achieved by minimizing for the value of L with respect to R the radius of the circle. According to our detailed explanation in [26], and after solving the minimization problems while replacing the results into equation (11.5) we have:

$$L = \sum_i \alpha_i(\mathbf{x}_i \cdot \mathbf{x}_i) - \sum_{i,j} \alpha_i \alpha_j(\mathbf{x}_i \cdot \mathbf{x}_j) \quad \forall \alpha_i : 0 \leq \alpha_i \leq C \tag{11.6}$$

A set of α_i values can be achieved using equation (11.6). If a sample $\mathbf{x}_i$ satisfies the inequality in equation (11.4) its corresponding Lagrange multiplier will be zero ($\alpha_i = 0$). For all the training samples for which the equality in equation (11.4) is satisfied the Lagrange multipliers become greater than zero ($\alpha_i > 0$).

Note that the center of data descriptor from equation (11.3), $\mathbf{a}$, is a linear combination of the training samples. Only those training samples $\mathbf{x}_i$ which satisfy (11.4) by equality are needed to generate the description since their coefficients are not zero. These samples are called *Support Vectors* of the data descriptor.

The main assumption in the above theory states that the data description is normal –i.e. the data boundary is the smallest sphere surrounding the training samples. However, this simple, normal description is not enough for more complex data which does not fit into a sphere, i.e. the description needs more complex boundaries. To achieve a more flexible description, instead of a simple dot product of the training samples ($\mathbf{x}_i \cdot \mathbf{x}_j$) in equation (11.6), we perform the dot product using a kernel function $K(\mathbf{x}_i, \mathbf{x}_j) = \Phi(\mathbf{x}_i) \cdot \Phi(\mathbf{x}_j)$. This is done by using a mapping function Φ which maps the data into another (higher dimensional) space. By performing this mapping any complicated boundary (description of data) in low dimension can be modeled by a hyper-sphere in a higher dimension. Several kernel functions have been proposed in the literature [31], among which the Gaussian kernel gives a closed data description:

$$K(\mathbf{x}_i, \mathbf{x}_j) = exp\left(-\frac{\|\mathbf{x}_i - \mathbf{x}_j\|^2}{\sigma^2}\right) \tag{11.7}$$

According to the above theory the proposed SVDDM method generates a Support Vector data description for each pixel in the scene using its history. These descriptions are then used to classify each pixel in new frames as a background or a novel/foreground pixel. In the following section the actual implementation of the system is presented.

11.4.2   The Algorithm

The methodology described above is used in our technique to build a descriptive boundary for each pixel in the background training frames in order to generate its model for the background. These boundaries are then used to classify their corresponding pixels in new frames as background or novel (foreground) pixels. There are several advantages in using the SVDD method in detecting foreground regions:

- It explicitly addresses the single-class classification problem. Existing statistical approaches try to estimate the probability of a pixel being background, and use (a set of) thresholds to classify pixels. It is impossible to have an estimate of the foreground probabilities, since there are no foreground samples in the training frames.

- The SVDD method has lower memory requirements compared to non-parametric density estimation techniques. This technique only requires a very small portion of the training samples, the *Support Vectors*, to classify new pixels.

Algorithm 11.3 shows the proposed algorithm in pseudo-code format. The only critical parameter is the number of training frames (N) that needs to be initialized. The Support Vector data description confidence parameter C is the target false reject rate of the system, which accounts for the system tolerance.

The background model in this technique is the description of the data samples (color and or intensity of pixels). The background training buffer is a First In First Out (FIFO) buffer with Round Robin replacement policy. The data description is generated in the training stage in which for each pixel Support Vectors and their Lagrange multipliers (α_i) are trained.

The Support Vectors and their corresponding Lagrange multipliers are stored as the classifier information for each pixel. This information is used for the classification step of the algorithm. The training stage can be performed off-line in cases where there are no global changes in the illumination or can be performed in parallel with the classification stage to achieve efficient foreground detection [26].

In the classification stage, for each frame its pixels are evaluated by their corresponding classifier to label them as background or foreground. To test each pixel $\mathbf{z}_t$ the distance to the center of the description hyper-sphere is calculated:

$$\|\mathbf{z}_t - \mathbf{a}\|^2 = (\mathbf{z}_t \cdot \mathbf{z}_t) - 2\sum_i \alpha_i(\mathbf{z}_t \cdot \mathbf{x}_i) + \sum_{i,j} \alpha_i\alpha_j(\mathbf{x}_i \cdot \mathbf{x}_j) \tag{11.8}$$

A pixel is classified as a background pixel if its distance to the center of the hyper-sphere is smaller or equal to the data descriptor's radius (i.e. $\|\mathbf{z}_t - \mathbf{a}\|^2 \le R^2$).

Algorithm 11.3 - The SVDDM algorithm

1: // C: Confidence, N: No. of frames, σ: Bandwidth
2: Initialization(C, N, σ)
3: **for** each pixel:=uv **do**
4:    $x_{uv} \leftarrow$ frames$_u v[1 \cdots N]$
5:    // **Training Stage:**
6:    SVDD$_{uv} \leftarrow$trainSVDD($x_{uv}[1 \cdots N]$)
7: **end for**
8: **for** each frame at time t **do**
9:   **for** each pixel:=uv **do**
10:     $x_{uv} \leftarrow$ frames$_u v[t]$
11:     // **Classification Stage:**
12:     DV$_{uv} \leftarrow$classifySVDD($x_{uv}[t]$, SVDD$_{uv}$) // DV=Description Value
13:     **if** (DV$_{uv} > 0$) **then**
14:       pixel:=uv$\leftarrow$ foreground
15:     **else**
16:       pixel:uh$\leftarrow$ background
17:     **end if**
18:     // **Update Stage:**
19:     **if** (t%10) **then**
20:       SVDD$_{uv} \leftarrow$trainSVDD($x_{uv}[t - N \cdots t]$)
21:     **end if**
22:   **end for**
23: **end for**

R^2 is the distance from the center of the hyper-sphere to its boundary which is also equivalent to the distance of each support vector to the center of the hyper-sphere:

$$R^2 = (\mathbf{x}_k \cdot \mathbf{x}_k) - 2\sum_i \alpha_i(\mathbf{x}_i \cdot \mathbf{x}_k) + \sum_{i,j} \alpha_i \alpha_j(\mathbf{x}_i \cdot \mathbf{x}_j) \tag{11.9}$$

11.5 Unified Framework for Target Detection and Tracking

Figure 11.4 illustrates the proposed framework for integrating target detection with tracking. This framework includes three main modules: (i) background modeling, (ii) target detection, and (iii) target tracking. The purpose of background modeling is to construct the intensity variation model of the pixels belonging to the background. Here, a SVR approach is exploited to fit the intensity distribution of background pixels using a Gaussian kernel. Target detection is performed by subtracting those pixels that fit the background model. Finally, the tracking module is used to establish a unique correspondence relationship among the detected targets over time.

In order to improve target-to-target correspondences over time, we calculate confidence coefficients based on shape, size, color and motion (i.e., velocity) information. Most importantly, the shape confidence coefficient computed in the tracking module is further exploited to iteratively update the threshold used in the detection stage to decide whether a pixel belongs to background or not. The detection threshold can be iteratively increased or reduced to improve detection results by considering the temporal correspondences of targets between adjacent frames. A voting-based strategy has been adopted to enhance matching results during target tracking under illumination changes and shape deformations due to

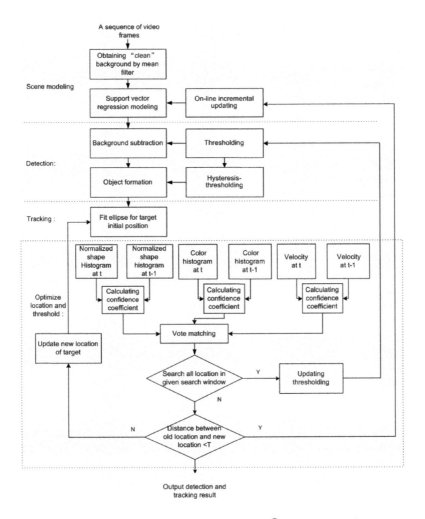

FIGURE 11.4   Framework for target detection and tracking (©(2008) Springer).

perspective projection. Specific details are provided in the following sections.

11.5.1 Background Modeling and Initialization of Target Location

In order to effectively detect the precise location of targets in a scene but also to avoid missing small targets, an accurate background model is required. Moreover, an effective way is required to incorporate background changes by updating the background model fast and effectively. The target detection state and background modeling may employ any of the statistical techniques presented in Section 11.3, the Support Vector Data Description in Section 11.4, or Support Vector Regression proposed in [18] and [13].

11.5.2 Integrating Target Detection with Tracking

When multiple targets are present, the proposed system maintains a list of targets which are actively tracked over time. The tracking is implemented through target feature matching within continuous frames. This matching can build the correspondence relationships between the previously tracked targets and each potential targets at the current frame, detected by thresholding the outputs of background models. If the matching is successful and reliable, then the target is added to the list of targets for further tracking.

Specifically, the matching procedure searches iteratively for target candidates in the current frame that have similar shape and appearance with target models defined in the previous frame. First, we compute a similarity score based on weighted normalized shape projection histograms. Then, to discriminate between targets having similar shape, we compute additional information based on target's size, color and motion and apply a voting-based strategy. Targets that have been tracked consistently over a number of frames are added to the list of targets for tracking. This list is properly maintained to include new targets and remove targets that disappear from the scene. The same procedure is also used to handle undesired merging of targets. Potential targets in the list of detected objects are tracked using shape projection histograms only. The ratio between projection histograms of candidate and model targets, called confidence coefficient, is used to localize the targets accurately as well as to define the range of detection threshold.

In the following, we describe the framework for integrating target detection with tracking. First, we discuss our target representation scheme. Then, we describe the algorithm used to predict the location of targets in subsequent frames. Finally, we present the feedback mechanism for optimizing the detection threshold.

Target Representation

Our target representation scheme is based on shape, size, color and motion information. In order to make it robust to perspective projection, scale, and rotation transformations, we employ normalized shape projection histograms.

Normalized Shape Projection Histograms: The location of a target is denoted by (x_i, y_i) and it corresponds to the location of the best-fitting ellipse. To compute the projection histograms, we project the target horizontally and vertically by counting the number of pixels in each row and each column correspondingly. To make the projection histograms invariant to target orientation, first we transform the target to a default coordinate system. This is done in two steps: (i) we find the best-fitting ellipse of the target, and (ii) we align its major and minor axes with the x- and y-axis of the default coordinate system. The main assumption here is that the targets are approximately 2-D; this is a valid assumption in our application since the depth of the targets is much smaller compared to their distance from the camera.

Weighted Shape Projection Histograms: In order to reduce the effects of background noise and image outliers, we introduce weights to improve the robustness of matching. This is done by employing an isotropic kernel function $k(\cdot)$ in a similar way as in [3]. In particular, the role of the kernel function is to assign smaller weights to pixels farther away from the center bin of the projection histogram. Then, the weighted target model histograms, denoted as H_x^T and H_y^T are calculated according to [34].

To predict the location of targets in subsequent frames, we search a window of size $W \times H$. Candidate targets are identified in this window by thresholding the outputs of the background models [34].

Predicting Target Location

To find a target location in subsequent frames, we need to define a similarity measure between the target model, computed in previous frames, and the target candidate, detected in the current frame. A Manhattan distance based measure between the corresponding weighted shape projection histograms of model and candidate targets is used.

To accurately localize a target in the search window, we minimize the objective function shown below in the case of horizontal shape projection histograms:

$$
\begin{aligned}
\Phi &= \min_{k} \sum_{m=1}^{M} [H_{x^k}^{C_k^S}(m) - H_x^T(m)] \\
&= \sum_{k} w_k \sum_{m=1}^{M} [H_{x^k}^{C_k^S}(m) - H_x^T(m)] \\
&= \sum_{k} w_k \sum_{x_i \in R} [H_{x^k}^{C_k^S}(x_i - x + M/2) \\
&\qquad - H_x^T(x_i - x + M/2)] \\
&\longrightarrow \min \text{ over } S \text{ and } x^k
\end{aligned}
\tag{11.10}
$$

where S is the threshold used to find the target candidates in the search window and w_k restricts the spatial position x^k of the target candidates around the geometric center x of the target model. $H_{x^k}^{C_k^S}(m)$ is the weighted shape projection histogram of the k-th target candidate detected using threshold S [34]. A similar calculation is applied on the vertical direction.

To perform the above minimization, an iterative scheme which gradually decreases the value of the threshold S used for target detection and changes the spatial center position of the search window is applied. The objective function is updated iteratively as follows:

$$
\begin{aligned}
\Phi(l) &= \sum_{k} w_k \sum_{m=1}^{M} [H_{x^k(l)}^{C_k^{S(l)}}(m) - H_x^T(m)] \\
&= \sum_{k} w_k \sum_{x_i \in R(l)} [H_{x^k(l)}^{C_k^{S(l)}}(x_i - x^k(l) + M/2) \\
&\qquad - H_x^T(x_i - x + M/2)]
\end{aligned}
\tag{11.11}
$$

where l corresponds to the iteration number.

Confidence Coefficient

A key issue in implementing the above idea is how to choose an appropriate function for decreasing S as well as to change the geometric center (x, y) of the candidate targets at each iteration l. For this, we use the ratio between the weighted shape projection histogram of the target model and the candidates. We refer to this ratio as the *confidence coefficient*. Its horizontal component is defined as follows:

$$
\xi_x(l) = \sum_{x_i \in R(l)} \sqrt{\frac{H_{x^k(l)}^{S(l)}[x_i - x^k(l) + M/2]}{H_x^C[x - x^k(l) + M/2]}}
\tag{11.12}
$$

where x_i is the horizontal spatial location of pixels belonging to the candidate target $R(l)$. Similar calculation applies for the vertical location. The confidence coefficient becomes a weight factor in the iterative procedure used to update the spatial location of the targets as well as to select the threshold range for target detection (see next section). Specifically, using the confidence coefficient, the center of the search window is updated as follows:

TABLE 11.1 Per-pixel memory requirements for the AKDE, the RM and the SVDDM.

Memory Req.	Intensity	Chrominance	both	asymptotic
The AKDE [24]	$N + 8$	$8N + 20$	$9N + 40$	$O(N)$
The RM [23]	1024	2048	3072	$O(1)$
The SVDDM	$[f(C, \sigma) \times 5] \geq 10$	$[f(C, \sigma) \times 8] \geq 24$	$f(C, \sigma) \geq 32$	$O(1)$

$$x^k(l) = x^k(l-1) \times \xi_x(l-1) \quad \text{and} \quad y^k(l) = y^k(l-1) \times \xi_y(l-1) \tag{11.13}$$

Adaptive Threshold Optimization

The confidence coefficient is also used to update the threshold S used in the target detection stage. Specifically, let us denote the threshold at the $l-1$ iteration as $S(l-1)$, then the threshold at the l iteration $S(l)$ is updated as follows:

$$S(l) = S(l-1) - \left[1 - \sqrt{\xi_x^2(l-1) + \xi_y^2(l-1)}\right] \tag{11.14}$$

This procedure decreases D_x and D_y while iteratively moving the spatial center of the search window closer to the geometric center of the target. The procedure terminates when the distance between the weighted shape projection histogram of target model and the target candidates is smaller than a threshold. However, when the confidence coefficient is too low, we increase the detection threshold to avoid under-segmentation which could cause differences in the shape of the targets in successive frames (see Fig. 11.10).

Tracking Multiple Targets

Using shape information alone to track multiple targets is not sufficient as it might lead to false matches. To eliminate such matches, we need to use additional information based on the target's size, color and motion. The key idea is using a voting strategy based on a majority rule – for more information please refer to [34].

It should be mentioned that the same equations used to compute the confidence coefficient in the case of shape projection histograms (i.e., Eq. (11.12)) can also be used to compute a confidence coefficient using size, color, and motion information.

11.6 Experimental Results and Comparison of Proposed Methods

In this section we compare the performance of proposed techniques using several real video sequences that pose significant challenges. In the real video experiments conducted below, $N = 300$ frames are used in the background training stage unless otherwise stated, corresponding to 10 seconds of the video's sequence. The background training buffer is a First In First Out (FIFO) buffer with Round Robin replacement policy.

11.6.1 Support Vector Modeling vs. Statistical Methods

From Table 11.1 we notice that the memory requirements of the AKDE technique presented in [24] and the SVDDM are lower than those of the RM method [23] if the number of training frames is small enough. That is, for situations when a small number of frames can cover most changes that occur in the background, by using a sliding window the AKDE method needs less memory than the RM technique. Note that SVDDM requires even less memory

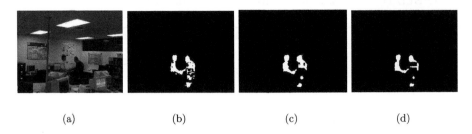

(a)   (b)   (c)   (d)

FIGURE 11.5   *Handshake* sequence (a). Detected foreground with AKDE (b), RM (c), SVDDM (d).

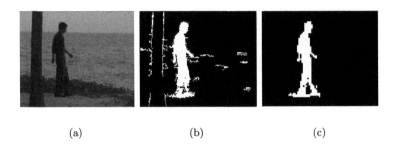

(a)   (b)   (c)

FIGURE 11.6   *Water* sequence (a), detected foreground region using AKDE (b) and SVDDM (c).

than the AKDE and like the RM its memory requirements are *independent* of the number of training samples.

Suitable Scenarios to Employ Statistical Background Modeling

For videos with rapidly changing backgrounds, the AKDE method has a better performance in terms of memory requirements and speed. Fig. 11.5 shows the detection results of the AKDE, RM and the SVDDM algorithms on the *Handshake* video sequence where the pixel values corresponding to monitors fluctuate rapidly. As it can be seen from this figure, capturing dependencies between chrominance features results in more accurate foreground regions (Fig. 11.5(b)), showing that AKDE performs better than both the RM and the SVDDM. Note that in this particular frame the color of foreground objects is very close to the background in some regions. The SVDDM technique results in very smooth and reliable foreground regions. Moreover, it uses the confidence factor C to guide the classification. This may lead to missing some parts of the foreground which are very close in color to the background.

Suitable Scenarios to Employ Support Vector Background Modeling

In videos with slowly changing or non-periodic backgrounds, the AKDE method needs more training frames to generate a good model for the background. This increases the memory requirements and drastically decreases its training and foreground detection speed. In these situations the SVDDM technique is a very good alternative, since its detection speed and memory requirements are independent of the number of training frames. Although the SVDDM like AKDE still require a large number of training frames, once the background model is trained the SVDDM only retains Support Vectors as pixel classifiers. As seen in Fig. 11.6 the SVDDM results are better than those of the AKDE.

(a) (b) (c)

FIGURE 11.7 *Mall* sequence (a), RM background model after 5 frames (b), and after 95 frames (c) (©(2009) Springer).

From this figure we can conclude that the SVDDM method has a better performance compared to the AKDE in situations where the background has a slow and irregular motion. Also in the AKDE there is a sliding window of limited size which may not cover all changes in the background resulting in an inaccurate probability density estimation but the model built in the SVDDM uses the decision boundaries of the single training class instead of bounding the training accuracy to the accuracy of the probability estimation.

Suitable Scenarios to Employ the RM Background Modeling

Fig. 11.7 shows the background model in the *Mall* video sequence in which the background is never empty. In this situation off-line methods fail unless a post-processing on the detected foreground regions is performed to generate models for uncovered parts of the background. In the RM method however, the background model is updated at every frame from the beginning of the video. When an object moves, the new pixel information is used to update the background model to the new one. Fig. 11.7(b) and (c) show the background model after 5 and 95 frames, respectively. In this scenario consistent background regions are temporarily occluded by transient moving objects. Therefore the background itself contributes more consistent information to the model. As a result, the model converges to the empty background.

In situations when the camera is not completely stationary, such as the case of a hand-held camera, the off-line methods are not suitable. In these situations there is a consistent, slow and irregular global motion in the scene, which cannot be modeled by a limited size sliding window of training frames. In such cases the RM method is highly preferable.

Fig. 11.8(a) shows an arbitrary frame of the *Room* video sequence. In this video a camera is held by hand and the scene only consists of the background. Fig. 11.8(b) compares the modeling error using different techniques. The modeling error using a constant window size in the AKDE (the dotted line) is between 20%-40%, and it does not decrease with time. This shows that the system using the AKDE method with a constant sized sliding window never converges to the actual model. The dashed line shows the modeling error using the RM method with a constant learning rate, and the solid line shows the modeling error of the RM with scheduled learning. We conclude that the model generated by the RM technique eventually converges to the actual background model and its error goes to zero.

Another important issue in the recursive learning is the convergence speed of the system (how fast the model converges to the actual background). Fig. 11.8(c)-(d) illustrates the convergence speed of the RM with scheduled learning, compared to constant learning and kernel density estimation with constant window size.

Fig. 11.8(e)-(f) shows the comparison of the recovery speed from an expired background

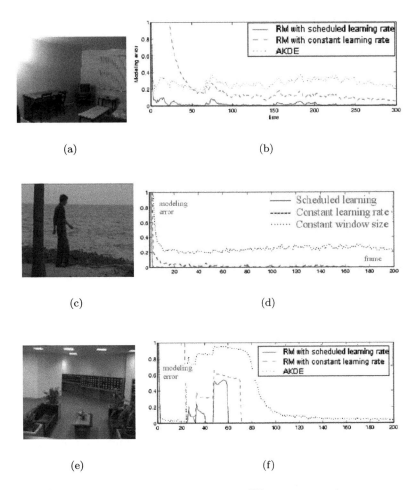

FIGURE 11.8   *Room* sequence (a), modeling error (b). *Water* sequence (c), Convergence speed (d). *Lobby* sequence (e), Recovery speed (f).

model to the new one. In Fig. 11.8(e) lights go from on to off through three global but sudden changes occurring at frames 23, 31 and 47. The scheduled learning RM method (solid curve) recovers the background model after these changes faster than non-scheduled RM and the AKDE with constant window size. The constant, large learning rate recovers more slowly (dashed curve) while the AKDE technique (dotted curve) is not able to recover even after 150 frames.

Comparison Summary

Table 11.2 summarizes this study and provides a comparison between different traditional methods for background modeling and our proposed method. The comparison includes the classification type, memory requirements, computation cost and type of parameter selection.

11.6.2   Integration of Background Modeling and Tracking

The integration framework has been evaluated by detecting vehicles and pedestrians using visible and thermal video sequences. The visible video sequence was captured at a traffic

TABLE 11.2   Comparison between the proposed methods and traditional techniques.

	SVDDM	AKDE	RM	KDE	Sp-tmp [12]	MoG [20]	Wallflower [30]
Automated	Yes	Yes	Yes	No	No	No	No
Post proc.	No	No	No	No	Yes	No	No
Classifier	SVD	Bayes	MAP	Bayes	Bayes	Bayes	K-means
Memory req.*	$O(1)$	$O(N)$	$O(1)$	$O(N)$	$O(N)$	$O(1)$	$O(N)$
Comp. cost*	$O(N)$	$O(N)$	$O(1)$	$O(N)$	$O(N)$	$O(1)$	$O(N)$

$*$: Per-pixel
N: number of training frames

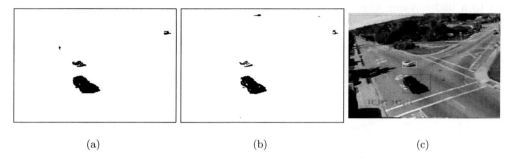

(a)   (b)   (c)

FIGURE 11.9   (See color insert.) Results using the proposed method (a) and frame-based (b). Final tracking results (c).

intersection and contains a total of two hours of video with a sampling rate 4 frames/second. The thermal video data was captured at a university campus walkway intersection over several days (morning and afternoon) using a Raytheon 300D thermal sensor core with 75mm lens mounted on an 8-story building [5].

In the following, the performance of the integration algorithm is demonstrated in terms of the following aspects: (1) detection alone, (2) integrating detection with tracking, (3) undesired merging of targets, and (4) comparisons with the state-of-the-art.

Fig. 11.9 presents comparison results between frame-based detection without feedback from tracking and the proposed method which integrates detection with tracking. Fig. 11.9 (a) and (c), show detection maps and tracking results using the proposed method. Fig. 11.9 (b) presents detection results using frame-difference and no threshold optimization. Among the results shown, it is interesting to note that the small target, labeled by a green rectangle in Fig. 11.9 (c), is very difficult to detect using frame-based detection and non-optimized thresholds as shown in Fig. 11.9 (b).

Table 11.3 shows quantitative comparisons in terms of true positives and false alarms for frame-based detection and the proposed approach. Obviously, the proposed approach has lower false alarm and higher true positive rates than frame-based detection.

TABLE 11.3   Quantitative comparisons in terms of True Positives (TP), False Alarms (FA), and Ground Truth (GT)

Data sets	Methods	Ground truth	True Positive	False Alarm
Visible video	Frame-based detection	346	296	30
	Integrating detection with tracking	346	340	5
Thermal video	Frame-based detection	371	371	35
	Integrating detection with tracking	371	371	0

Figs. 11.10 (a) and (b) show another quantitative comparison between frame-based detection and the proposed method by counting the number of pixels in two different seg-

mented regions moving away from the camera. The red curve indicates ground truth size. The green and blue curves show the performance of the proposed method and frame-based detection respectively. From the figure, the green curves are closer to the red curves, indicating that the proposed method higher accuracy compared to frame-based detection.

Fig. 11.10 (c) shows the adaptive threshold values over time for two targets with different motion characteristics (i.e., a car and a pedestrian). As it can be observed, the thresholds were iteratively decreased based on the confidence coefficient computed from the shape projection histogram matching process. To avoid under-segmentation, the threshold was reset to a higher value when the confidence coefficient fell below a certain value. Finally, Fig. 11.10 (d) demonstrates the average number of iterations for each frame.

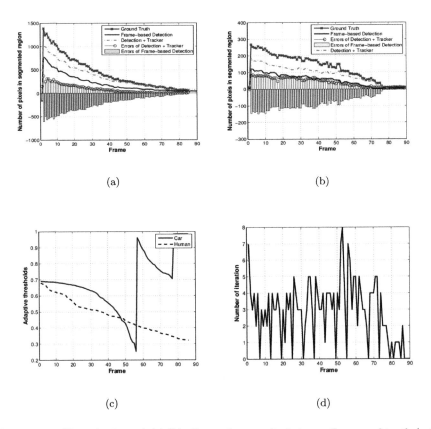

FIGURE 11.10 (See color insert.) (a),(b): Comparison results between the ground truth (red), frame-based detection (blue) and the proposed approach (green). (c): Adaptive threshold. (d): Average number of iterations (ⓒ(2008) Springer).

Tracking Merged Targets

The proposed detection and tracking approach can handle undesired target merging using the track-to-track stitching scheme reported in [1]. The voting-based matching scheme described is used to track accurately the targets when their shape is deformed due to perspective projection. Fig. 11.11 demonstrates how our proposed approach handles the undesired merging of targets. When two targets merge with each other, as shown in Fig. 11.11(b), a new track is assigned to these targets. After the undesired merge is resolved, the tracks are recovered by stitching their new tracks with previous ones [1].

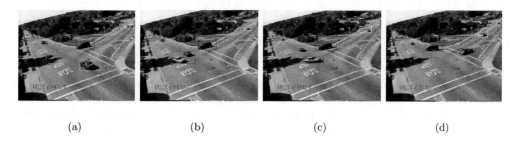

(a)	(b)	(c)	(d)

FIGURE 11.11   An example of handling undesired merging of targets. Targets merge in (b) and the issue is resolved in (c).

11.7   Conclusion

In this chapter we have presented a number of novel techniques for learning background pixel models based on both statistical and analytical tools. As statistical modeling techniques, the non-parametric density estimation and recursive modeling approaches are discussed.

The advantage of the adaptive kernel density estimation method (AKDE) over existing techniques is that instead of a global threshold for all pixels in the video scene, different and adaptive thresholds are used for each pixel. By training these thresholds the system works robustly on different video scenes without changing or tuning any parameter. Since each pixel is classified by using adaptive thresholds and exploiting its color dependency, the background model is more accurate. The modeling method (RM) updates a statistical model for background pixels on-line. This method is superior and more robust than other techniques for situations in which background changes are slow and not periodic.

To overcome the issues inherent to statistical learning models, an alternative modeling tool is proposed to label pixels in video sequences into foreground and background classes using a Support Vector data description. The advantages of training Support Vectors for background modeling include:

- The model accuracy is not bounded to the accuracy of the estimated probability density functions.

- The memory requirements are lower than those of non-parametric techniques and are independent of the number of training samples.

- Support vector data description is more suitable for novelty detection since it explicitly models the decision boundary of the known class.

Furthermore, to enhance the quality of foreground detection, a framework for improving video-based surveillance by integrating target detection with tracking is presented. From Section 11.5, on-line SVR was used to model the background and to accurately detect the initial locations of the targets. To predict the location of targets in successive frames shape projection histograms were exploited. At the same time, a confidence coefficient based on shape matching was computed to suppress false alarms. Using weights derived from the confidence coefficient of shape matching, we were able to optimize the threshold used in the target detection stage. Additional cues based on size, color, and motion were used to eliminate false positives when tracking multiple targets. Experiments show good performance, especially on small targets and targets undergoing perspective projection distortions.

References

1. A. Amer. Voting-based simultaneous tracking of multiple video objects. *IEEE Transactions on Circuits and Systems for Video Technology*, 11(15):1448–1462, 2005.
2. Y. Bar-shalom and T. Fortmann. *Tracking and Data Association*. Academic Press, 1998.
3. D. Comaniciu, V. Ramesh, and P. Meer. Kernel-based object tracking. *EEE Trans. on Pattern Analysis and Machine Intelligence*, 25(5):564–557, May 2005.
4. C. Criminisi, G. Gross, A. Blake, and V. Kolmogorov. Bilayer Segmentation of Live Video. *International Conference on Computer Vision and Pattern Recognition, CVPR 2006*, pages 17–22, June 2006.
5. J. W. Davis and M. A. Keck. A two-stage template approach to person detection in thermal imagery. *IEEE International Conference on Computer Vision and Pattern Recognition, CVPR 2005*, 2005.
6. R. O. Duda, P. E. Hart, and D. G. Stork. *Pattern Classification*. Wiley, New York, 2001.
7. A. Elgammal, R. Duraiswami, D. Harwood, and L. Davis. Background and foreground modeling using nonparametric kernel density estimation for visual surveillance. *In proceedings of the IEEE*, 90:1151–1163., 2002.
8. N. Gordon, D. Salmond, and A. Smith. A novel approach to non-linear and non-Gaussian Bayesian state estimation. *Part-F: Radar and Signal Processing.*, 140:107–113, 1993.
9. K. Kim, D. Harwood, and L. S. Davis. Background updating for visual surveillance. *International Symposium on Visual Computing, ISVC 2005*, 1:337–346, December 2005.
10. G. Kitagawa. Non-Gaussian state-space modeling of nonstationary time-series. *Journal American Statistical Association*, 82:1032–1063, 1987.
11. D. Lee. Effective Gaussian mixture learning for video background subtraction. *IEEE Transactions on Pattern Analysis and Machine Intelligence*, 27(5):827–832, May 2005.
12. L. Li, W. Huang, I.Y. Gu, and Q. Tian. Statistical modeling of complex backgrounds for foreground object detection. *IEEE Transactions on Image Processing*, 13(11):1459–1472, November 2004.
13. J. Ma and J. Theiler. Accurate on-line support vector regression. *Neural Computation*, 15:2683–2703, 2003.
14. N. Paragios and V. Ramesh. A MRF-based approach for real-time subway monitoring. *IEEE Transactions on Pattern Analysis and Machine Intelligence*, 1:1030–1040, December 2001.
15. R. Pless, T. Brodsky, and Y. Aloimonos. Detecting independent motion: The statistics of temporal continuity. *IEEE Transactions on Pattern Analysis and Machine Intelligence*, 22(8):68–73, 2000.
16. R. Pless, J. Larson, S. Siebers, and B. Westover. Evaluation of local models of dynamic backgrounds. *International Conference on Computer Vision and Pattern Recognition, CVPR 2003*, 2:73–78, June 2003.
17. Y. Sheikh and M. Shah. Bayesian object detection in dynamic scenes. *International Conference on Computer Vision and Pattern Recognition, CVPR 2005*, 1:74–79, June 2005.
18. A. Smola and B. Scholkopf. A tutorial on support vector regression. *NeuroCOLTS technical report Series NC2-TR-1998-030*, 1998.
19. C. Stauffer and W. Grimson. Adaptive background mixture models for real-time tracking. *International Conference on Computer Vision and Pattern Recognition, CVPR 1999*,

2:246–252, 1999.

20. C. Stauffer and W. Grimson. Learning patterns of activity using real-time tracking. *IEEE Transactions on Pattern Analysis and Machine Intelligence*, 22(8):747–757, August 2000.

21. Z. Sun, G. Bebis, and R. Miller. On-road vehicle detection: A review. *IEEE Transactions on Pattern Anaysis and Machine Intelligence*, 28(5):694–711, 2006.

22. A. Tavakkoli, M. Nicolescu, and G. Bebis. Automatic robust background modeling using multivariate non-parametric kernel density estimation for visual surveillance. *International Symposium on Visual Computing, ISVC 2005*, LNSC 3804:363–370, December 2005.

23. A. Tavakkoli, M. Nicolescu, and G. Bebis. An adaptive recursive learning technique for robust foreground object detection. *International Workshop on Statistical Methods in Multi-Image and Video Processing in conjunction with ECCV 2006*, May 2006.

24. A. Tavakkoli, M. Nicolescu, and G. Bebis. Automatic statistical object detection for visual surveillance. *Southwest Symposium on Image Analysis and Interpretation, SSIAI 2006*, pages 144–148, March 2006.

25. A. Tavakkoli, M. Nicolescu, and G. Bebis. Robust recursive learning for foreground region detection in videos with quasi-stationary backgrounds. *International Conference on Pattern Recognition, ICPR 2006*, August 2006.

26. A. Tavakkoli, M. Nicolescu, and G. Bebis. A support vector data description approach for background modeling in videos with quasi-stationary backgrounds. *International Journal on Artificial Intelligence Tools*, 17(4):635658, August 2008.

27. A. Tavakkoli, M. Nicolescu, and G. Bebis. Non-parametric statistical background modeling for efficient foreground region detection. *International Journal of Machine Vision and Applications*, 20(6):395–409, October 2009.

28. D. Tax and R. Duin. Support vector domain description. *Pattern Recognition Letters*, 1(20):1191–1199, 1999.

29. D. Tax and R. Duin. Support vector data description. *Machine Learning*, 54(1):45–66, 2004.

30. K. Toyama, J. Krumm, B. Brumitt, and B. Meyers. Wallflower: principles and practice of background maintenance. *International Conference on Computer Vision, ICCV 1999*, 1:255–261, September 1999.

31. V. Vapnik. *Statistical Learning Theory.* Wiley, New York, 1998.

32. R. Verma, C. Schmid, and K. Mikolajczyk. Face detection and tracking in a video by propagating detection probabilities. *IEEE Transactions on Pattern Analysis and Machine Intelligence*, 25(10):1215–1227, 2003.

33. J. Wang, H. Eng, A. Kam, and W. Yau. A framework for foreground detection in complex environments. *European Conference on Computer Vision, Workshop of Statistical Modeling for Video Processing*, pages 129–140, 2004.

34. X. Wang, G. Bebis, M. Nicolescu, M. Nicolescu, and R. Miller. Improving target detection by coupling it with tracking. *International Journal of Machine Vision and Applications*, 20(4):205–223, 2009.

35. L. Wixson. Detecting salient motion by accumulating directionary-consistent flow. *IEEE Transactions on Pattern Analysis and Machine Intelligence*, 22:774–780, August 2000.

36. C. Wren, A. Azarbayejani, T. Darrel, and A. Pentland. Pfinder: real-time tracking of human body. *IEEE Transactions on Pattern Analysis and Machine Intelligence*, 19(7):780–785, July 1997.

12

Incremental Learning of an Infinite Beta-Liouville Mixture Model for Video Background Subtraction

Wentao Fan
CIISE, Concordia University, Canada

Nizar Bouguila
CIISE, Concordia University, Canada

12.1 Introduction

Video analysis problems have been the subject of extensive research in the past [37] [34] [26]. One of these problems is video background modeling and foreground detection, which is actually the process of separating out foreground objects from the background in a sequence of video frames. It is a crucial task in computer vision and has been employed in many video applications involving video surveillance, traffic monitoring, human motion analysis and object tracking. Video background subtraction is a significant challenging problem due to the dynamic nature of video backgrounds generally characterized by changing illumination levels, temporal background clutter as often found in outdoor scenes and non-stationary background objects such as rain, snow, moving leaves and shadows cast by moving objects. In recent years, many research efforts have been devoted to the study of video background subtraction and different techniques have been deployed [2] such as high level region analysis [36], kernel density estimation [12], Markov random fields [27], and hidden Markov models [33]. All of the aforementioned approaches have their own advantages and shortcomings. A review of background subtraction techniques can be found in [28].

Mixture models are fundamental tools for data analysis and have been widely used to model video sequences in several applications such as motion segmentation, event detection, tracking and particularly background modeling and foreground detection [38] [9] [29]

[32] [25] [39] [13]. Approaches based on mixture models are considered as pixel-level evaluations, since each pixel is represented by a mixture of density functions. Compared to other techniques, these approaches are more robust and able to handle multi-modal background distributions. However, there are two main limitations regarding the majority of existing mixture modeling-based approaches: the Gaussian assumption and the difficulty of model selection (i.e. choosing the appropriate number of components). Until now, most of the background subtraction approaches based on finite mixture modeling rely on the Gaussian assumption [32] [39] due to its simplicity, which means that each pixel in a frame is represented as a mixture of Gaussian distributions. However, this assumption is not realistic in practice when the data clearly appear with a non-Gaussian structure which is the case of several video processing applications [4]. This is especially true for applications involving proportional data (or normalized histograms)[30]. Proportional data are naturally generated by many applications, such as text, image and video modeling. Recent works have shown that, in many real-world applications, other models such as the Dirichlet, generalized Dirichlet or Beta-Liouville mixtures can be better alternatives to the Gaussian for modeling proportional data [7] [8] [6]. The most noteworthy model is the mixture of Beta-Liouville which contains the Dirichlet distribution as a special case and which has a smaller number of parameters than the generalized Dirichlet. Furthermore, Beta-Liouville mixture models have shown better performance than both the Dirichlet and the generalized Dirichlet mixtures as detailed in [8]. Thus, we shall focus on this mixture model in this work.

A crucial issue regarding mixture-based video background subtraction approaches is the difficulty of determining the optimal number of mixture components. Indeed, many of the existing works and formulations consider generally finite mixture models. Thus, the number of mixture components is rather specified in advance or selected using classic selection criteria (e.g. MDL, BIC, MML, AIC, etc.). An alternative and elegant way to tackle the model selection problem is through nonparametric Bayesian techniques, such as Dirichlet process mixture models [17] [24] [18] [35], by assuming that there is an infinite number of mixture components. As a nonparametric Bayesian technique, the complexity of the Dirichlet process mixture model increases as the dataset grows. Therefore, Dirichlet process mixture models can be also considered as infinite mixture models. One of the most common choices for learning infinite mixture models is to use Markov chain Monte Carlo (MCMC) techniques [30]. Nevertheless, the use of MCMC techniques is often limited to small-scale problems in practice because of its high computational cost. A good alternative to the MCMC technique is a deterministic approximation technique known as variational inference or variational Bayes [1] [23] [3], which only requires a modest amount of computational power and has provided promising performance in many applications involving mixture models [11] [10] [15] [14] [16].

In this chapter, we attempt to propose a novel approach to video background subtraction through an incremental (or online) infinite Beta-Liouville mixture model . Incremental clustering plays an important role in many computer vision applications and has been the topic of extensive research in the past (see, for instance, [20]). It is adopted here to maintain temporal consistency which can be lost when performing subtraction in a batch mode (i.e. frame by frame). The main contribution of our work is threefold: Firstly, we extend the finite Beta-Liouville mixture model to the infinite case using Dirichlet process mixture framework with a stick-breaking construction; Secondly, an incremental variational learning algorithm is developed to learn the proposed model; Lastly, we apply the proposed model to address the problem of video background subtraction and demonstrate the advantages of our approach by comparing it with other well-defined mixture-based approaches.

---

[30]Proportional data are the data that contain two constraints: non-negativity and unit-sum.

The rest of this chapter is organized as follows: Section 12.2 presents our infinite Beta-Liouville mixture model. We describe our incremental variational Bayes framework for learning the proposed model in Section 12.3. Section 12.4 presents our experimental results for subtracting video backgrounds. Finally, conclusion is provided in Section 12.5.

12.2 Infinite Beta-Liouville Mixture Model

Finite mixture models are often adopted as effective tools to capture the multimodality of the data. In contrast to finite mixture modeling in which one major concern is the selection of the optimal number of mixture components, infinite mixture models are able to overcome this obstacle by assuming that there is an infinite number of components. In this section, we shall first briefly review the finite Beta-Liouville mixture model. Then, we extend it to the infinite case by using a Dirichlet process mixture framework with a stick-breaking representation.

12.2.1 Finite Beta-Liouville Mixture Model

Given a D-dimensional random vector $\vec{X} = (X_1, \ldots, X_D)$ which follows a Liouville distribution of the second kind with positive parameters $(\alpha_1, \ldots, \alpha_D)$ and density generator $g(\cdot)$, then the probability density function of $\vec{X}$ is given by [6]

$$p(\vec{X}|\alpha_1, \ldots, \alpha_D) = g(u) \prod_{l=1}^{D} \frac{X_l^{\alpha_l - 1}}{\Gamma(\alpha_l)} \tag{12.1}$$

where $u = \sum_{l=1}^{D} X_l < 1$ and $X_l > 0$, $l = 1, \ldots, D$. The mean and covariance of the Liouville distribution are given by

$$E(X_l) = E(u) \frac{\alpha_l}{\sum_{l=1}^{D} \alpha_l} \tag{12.2}$$

$$Cov(X_a, X_b) = \frac{\alpha_a \alpha_b}{\sum_{a=1}^{D} \alpha_a} \left(\frac{E(u^2)}{\sum_{l=1}^{D} \alpha_l + 1} - \frac{E(u)^2}{\sum_{l=1}^{D} \alpha_l} \right) \tag{12.3}$$

where $E(u)$ and $E(u^2)$ are the first and second moments of a random variable u which follows a probability density function $f(\cdot)$ namely the generating density and is related to the density generator $g(\cdot)$ as

$$g(u) = \frac{\Gamma(\sum_{l=1}^{D} \alpha_l)}{u^{\sum_{l=1}^{D} \alpha_l - 1}} f(u) \tag{12.4}$$

Accordingly, the Liouville distribution of the second kind in Eq.(12.1) can be rewritten as

$$p(\vec{X}|\alpha_1, \ldots, \alpha_D) = \frac{\Gamma(\sum_{l=1}^{D} \alpha_l)}{u^{\sum_{l=1}^{D} \alpha_l - 1}} f(u) \prod_{l=1}^{D} \frac{X_l^{\alpha_l - 1}}{\Gamma(\alpha_l)} \tag{12.5}$$

There are two noticeable and interesting properties regarding the Liouville distribution: First, it has a more general covariance structure (can be positive or negative), in contrast to the Dirichlet distribution that only allows negative covariance structure; Second, similar to the Dirichlet, the Liouville distribution is conjugate to the multinomial distribution. More interesting properties of the Liouville distribution can be viewed in [6]. Since the Beta distribution has a flexible shape and can approximate nearly any arbitrary distribution, it is a favorable choice as the generating density for u with parameters α and β

$$f(u|\alpha, \beta) = \frac{\Gamma(\alpha + \beta)}{\Gamma(\alpha)\Gamma(\beta)} u^{\alpha - 1} (1 - u)^{\beta - 1} \tag{12.6}$$

where $\alpha > 0$ and $\beta > 0$. By substituting Eq.(12.6) into Eq.(12.5), we obtain the so-called Beta-Liouville distribution [6]:

$$\mathrm{BL}(\vec{X}|\vec{\theta}) = \frac{\Gamma(\sum_{l=1}^{D}\alpha_l)\Gamma(\alpha+\beta)}{\Gamma(\alpha)\Gamma(\beta)}\prod_{l=1}^{D}\frac{X_l^{\alpha_l-1}}{\Gamma(\alpha_l)}\left(\sum_{l=1}^{D}X_l\right)^{\alpha-\sum_{l=1}^{D}\alpha_l}\left(1-\sum_{l=1}^{D}X_l\right)^{\beta-1} \tag{12.7}$$

where $\vec{\theta} = (\alpha_1,\ldots,\alpha_D,\alpha,\beta)$ are the parameters of the Beta-Liouville distribution. It is worth mentioning that Beta-Liouville distribution includes the Dirichlet as a special case. Specifically, when the density generator has a Beta distribution with parameters $\sum_{l=1}^{D}\alpha_l$ and α_{D+1}, Eq.(12.1) is reduced to the Dirichlet distribution with parameters $\alpha_1,\ldots,\alpha_{D+1}$. Assume that we have observed a set of N vectors $\mathcal{X} = \{\vec{X}_1,\ldots,\vec{X}_N\}$, where each vector $\vec{X}_i = (X_{i1},\ldots,X_{iD})$ is represented in a D-dimensional space and assumed to be generated from a finite Beta-Liouville mixture model with M components as

$$p(\vec{X}_i|\vec{\pi},\vec{\theta}) = \sum_{j=1}^{M}\pi_j\mathrm{BL}(\vec{X}_i|\theta_j) \tag{12.8}$$

where $\theta_j = (\alpha_{j1},\ldots,\alpha_{jD},\alpha_j,\beta_j)$ are the parameters of a Beta-Liouville distribution corresponding to component j. In addition, $\vec{\pi} = (\pi_1,\ldots,\pi_M)$ denotes the vector of mixing coefficients which are positive and sum to one.

12.2.2   Infinite Beta-Liouville Mixture Model via Dirichlet Process

A conventional finite mixture model can be extended to have an infinite number of components using the Dirichlet process mixture model with a stick-breaking representation [31] [22]. The Dirichlet process is a stochastic process whose sample paths are probability measures with probability one [35]. The formal definition of a Dirichlet process is the following: a random distribution G is distributed according to a Dirichlet process if its marginal distributions are Dirichlet distributed. More specifically, let ξ be a positive real number and H be a distribution over some probability space Φ, then G is a Dirichlet process with base distribution H and concentration parameter ξ, denoted as $G \sim \mathrm{DP}(\xi,H)$, if

$$(G(\phi_1),\ldots,G(\phi_t)) \sim \mathrm{Dir}(\xi H(\phi_1),\ldots,\xi H(\phi_t)) \tag{12.9}$$

where $\{\phi_1,\ldots,\phi_t\}$ is the set of the finite partitions of Φ.

A more intuitive and straightforward way to represent the Dirichlet process is through the stick-breaking construction [31]: $G \sim \mathrm{DP}(\xi,H)$ if the following conditions are satisfied

$$\lambda_j \sim \mathrm{Beta}(1,\xi)\,, \qquad \theta_j \sim H\,, \qquad \pi_j = \lambda_j\prod_{k=1}^{j-1}(1-\lambda_k)\,, \qquad G = \sum_{j=1}^{\infty}\pi_j\delta_{\theta_j} \tag{12.10}$$

where δ_{θ_j} is a probability measure concentrated at θ_j. The mixing weights π_j are obtained by recursively breaking a unit length stick into an infinite number of pieces such that the size of each successive piece is proportional to the rest of the stick. For more details about Dirichlet processes and stick-breaking construction, the reader is referred to [35].

Assuming now that the observed dataset $\mathcal{X} = (\vec{X}_1,\ldots,\vec{X}_N)$ is generated from a Beta-Liouville mixture model with a countably infinite number of components, the infinite Beta-Liouville mixture model can be defined by

$$p(\vec{X}_i|\vec{\pi},\vec{\theta}) = \sum_{j=1}^{\infty}\pi_j\mathrm{BL}(\vec{X}_i|\theta_j) \tag{12.11}$$

Next, we introduce a vector $\vec{Z} = (Z_1, \ldots, Z_N)$ as the mixture component assignment variable where each element Z_i takes an integer value j denoting the component from which $\vec{X}_i$ is drawn. The marginal distribution over $\vec{Z}$ is specified in terms of the mixing coefficients $\vec{\pi}$:

$$p(\vec{Z}|\vec{\pi}) = \prod_{i=1}^{N} \prod_{j=1}^{\infty} \pi_j^{\mathbf{1}[Z_i=j]} \tag{12.12}$$

where $\mathbf{1}[\cdot]$ is an indicator function which has the value 1 when $Z_i = j$ and 0 otherwise. Since π_j is a function of $\vec{\lambda}$ according to the stick-breaking construction as shown in Eq.(12.10), we can rewrite Eq.(12.12) as

$$p(\vec{Z}|\vec{\lambda}) = \prod_{i=1}^{N} \prod_{j=1}^{\infty} \left[\lambda_j \prod_{k=1}^{j-1} (1 - \lambda_k) \right]^{\mathbf{1}[Z_i=j]} \tag{12.13}$$

Based on the definition of stick-breaking construction in Eq.(12.10), the prior distribution of $\vec{\lambda}$ is a specific Beta distribution, in the form

$$p(\vec{\lambda}|\vec{\xi}) = \prod_{j=1}^{\infty} \mathrm{Beta}(1, \xi_j) = \prod_{j=1}^{\infty} \xi_j (1 - \lambda_j)^{\xi_j - 1} \tag{12.14}$$

Next, we need to introduce prior distributions over random variables α_l, α and β in our Bayesian model. Since α_l, α and β are positive, Gamma distribution $\mathcal{G}(\cdot)$ is adopted as the priors for these parameters:

$$p(\alpha_l|u_l, v_l) = \mathcal{G}(\alpha_l|u_l, v_l) = \frac{v_l^{u_l}}{\Gamma(u_l)} \alpha_l^{u_l - 1} e^{-v_l \alpha_l} \tag{12.15}$$

$$p(\alpha|g, h) = \mathcal{G}(\alpha|g, h) = \frac{h^g}{\Gamma(g)} \alpha^{g-1} e^{-h\alpha} \tag{12.16}$$

$$p(\beta|s, t) = \mathcal{G}(\beta|s, t) = \frac{t^s}{\Gamma(s)} \beta^{s-1} e^{-t\beta} \tag{12.17}$$

A directed graphical representation of this model is illustrated in Fig. 12.1.

12.3 Model Learning via Incremental Variational Bayes

In order to cope with sequentially arriving data, an incremental variational Bayes learning algorithm proposed by [19] is adopted in this work to learn the proposed infinite Beta-Liouville mixture model. This incremental algorithm includes two learning phases, namely the building phase and compression phase. The goal of the model building phase is to infer the current optimal mixture model, while the target of the compression phase is to determine which mixture component that groups of data points should be assigned to. In this algorithm, data points can be sequentially processed in small batches where each one may contain one or more data points.

12.3.1 Model Building Phase

Given an observed dataset $\mathcal{X}$, we define $\Omega = \{\vec{Z}, \vec{\theta}, \vec{\lambda}\}$ as the set of random variables. The main goal of variational Bayes is to find a proper approximation $q(\Omega)$ for the posterior distribution $p(\Omega|\mathcal{X})$, which is achieved by maximizing the free energy $\mathcal{F}(\mathcal{X}, q)$:

$$\mathcal{F}(\mathcal{X}, q) = \int q(\Omega) \ln[p(\mathcal{X}, \Omega)/q(\Omega)] d\Omega \tag{12.18}$$

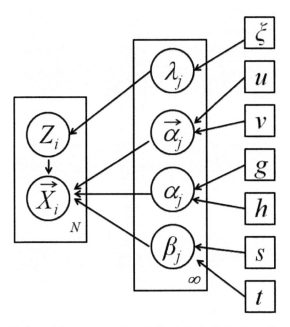

FIGURE 12.1   Graphical model representation of the infinite Beta-Liouville mixture model. Symbols in circles denote random variables; symbols in squares denote hyperparameters. Plates indicate repetition (with the number of repetitions in the lower right), and arcs describe conditional dependencies between variables.

Motivated by the approach is [5], we truncate the variational distribution $q(\Omega)$ at a value M, such that

$$\lambda_M = 1, \qquad \pi_j = 0 \quad \text{when} j > M, \qquad \sum_{j=1}^{M} \pi_j = 1 \tag{12.19}$$

where the truncation level M is a variational parameter which can be freely initialized and will be optimized automatically during the learning process [5]. Moreover, in order to achieve tractability, we assume that the approximated posterior distribution $q(\Omega)$ can be factorized into disjoint tractable factors as:

$$q(\Omega) = \left[\prod_{i=1}^{N} q(Z_i) \right] \left[\prod_{j=1}^{M} \prod_{l=1}^{D} q(\alpha_{jl}) \right] \left[\prod_{j=1}^{M} q(\alpha_j)q(\beta_j)q(\lambda_j) \right] \tag{12.20}$$

Then, we can obtain the following solutions for the variational factors by maximizing the free energy with respect to $q(\vec{Z})$, $q(\vec{\alpha}_l)$, $q(\vec{\alpha})$, $q(\vec{\beta})$ and $q(\vec{\lambda})$:

$$q(\vec{Z}) = \prod_{i=1}^{N} \prod_{j=1}^{M} r_{ij}^{\mathbb{1}[Z_i=j]}, \qquad q(\vec{\lambda}) = \prod_{j=1}^{M} \text{Beta}(\lambda_j | a_j, b_j) \tag{12.21}$$

$$q(\vec{\alpha}_l) = \prod_{j=1}^{M} \prod_{l=1}^{D} \mathcal{G}(\alpha_{jl} | u_{jl}^*, v_{jl}^*), \qquad q(\vec{\alpha}) = \prod_{j=1}^{M} \mathcal{G}(\alpha_j | g_j^*, h_j^*), \qquad q(\vec{\beta}) = \prod_{j=1}^{M} \mathcal{G}(\beta_j | s_j^*, t_j^*) \tag{12.22}$$

where the corresponding hyperparamters can be calculated as

$$r_{ij} = \frac{\exp(\widetilde{r}_{ij})}{\sum_{j=1}^{M} \exp(\widetilde{r}_{ij})} \tag{12.23}$$

$$\tilde{r}_{ij} = \exp\left[\widetilde{\mathcal{I}}_j + \widetilde{\mathcal{H}}_j + (\bar{\alpha}_j - \sum_{l=1}^{D}\bar{\alpha}_{jl})\ln(\sum_{l=1}^{D}X_{il}) + \sum_{k=1}^{j-1}\langle\ln(1-\lambda_k)\rangle + \sum_{l=1}^{D}(\bar{\alpha}_{jl}-1)\ln X_{il}\right.$$

$$\left. + (\bar{\beta}_j - 1)\ln(1 - \sum_{l=1}^{D}X_{il}) + \langle\ln\lambda_j\rangle\right] \tag{12.24}$$

$$a_j = 1 + \sum_{i=1}^{N}\langle Z_{ij}\rangle, \qquad b_j = \xi_j + \sum_{i=1}^{N}\sum_{k=j+1}^{M}\langle Z_{ik}\rangle \tag{12.25}$$

$$u_{jl}^* = u_{jl} + \sum_{i=1}^{N}\langle Z_{ij}\rangle\bar{\alpha}_{jl}\left[\Psi(\sum_{l=1}^{D}\bar{\alpha}_{jl}) + \Psi'(\sum_{l=1}^{D}\bar{\alpha}_{jl})\sum_{d\neq l}^{D}(\langle\ln\alpha_{jd}\rangle - \ln\bar{\alpha}_{jd})\bar{\alpha}_{jd} - \Psi(\bar{\alpha}_{jl})\right] \tag{12.26}$$

$$g_j^* = g_j + \sum_{i=1}^{N}\langle Z_{ij}\rangle\left[\bar{\beta}_j\Psi'(\bar{\alpha}_j+\bar{\beta}_j)(\langle\ln\beta_j\rangle - \ln\bar{\beta}_j) - \Psi(\bar{\alpha}_j) + \Psi(\bar{\alpha}_j+\bar{\beta}_j)\right]\bar{\alpha}_j \tag{12.27}$$

$$s_j^* = s_j + \sum_{i=1}^{N}\langle Z_{ij}\rangle\left[\bar{\alpha}_j\Psi'(\bar{\alpha}_j+\bar{\beta}_j)(\langle\ln\alpha_j\rangle - \ln\bar{\alpha}_j) - \Psi(\bar{\beta}_j) + \Psi(\bar{\alpha}_j+\bar{\beta}_j)\right]\bar{\beta}_j \tag{12.28}$$

$$v_{jl}^* = v_{jl} - \sum_{i=1}^{N}\langle Z_{ij}\rangle\left[\ln X_{il} - \ln(\sum_{l=1}^{D}X_{il})\right] \tag{12.29}$$

$$h_j^* = h_j - \sum_{i=1}^{N}\langle Z_{ij}\rangle\ln(\sum_{l=1}^{D}X_{il}) \tag{12.30}$$

$$t_j^* = t_j - \sum_{i=1}^{N}\langle Z_{ij}\rangle\ln(1 - \sum_{l=1}^{D}X_{il}) \tag{12.31}$$

where $\Psi(\cdot)$ is the digamma function, and $\langle\cdot\rangle$ is the expectation evaluation. $\widetilde{\mathcal{I}}_j$ and $\widetilde{\mathcal{H}}_j$ in Eq.(12.24) are the lower bounds of $\mathcal{I}_j = \langle\ln\frac{\Gamma(\sum_{l=1}^{D}\alpha_{jl})}{\prod_{l=1}^{D}\Gamma(\alpha_{jl})}\rangle$ and $\mathcal{H}_j = \langle\ln\frac{\Gamma(\alpha_j+\beta_j)}{\Gamma(\alpha_j)\Gamma(\beta_j)}\rangle$, respectively. Since these expectations are analytically intractable, we adopt the second-order Taylor series expansion to compute their lower bounds. The expected values in the above formulas are defined as

$$\bar{\alpha}_{jl} = \frac{u_{jl}^*}{v_{jl}^*}, \qquad \bar{\alpha}_j = \frac{g_j^*}{h_j^*}, \qquad \bar{\beta}_j = \frac{s_j^*}{t_j^*} \tag{12.32}$$

$$\langle Z_{ij}\rangle = r_{ij}, \qquad \langle\ln\alpha_{jl}\rangle = \Psi(u_{jl}^*) - \ln v_{jl}^* \tag{12.33}$$

$$\langle\ln\alpha_j\rangle = \Psi(g_j^*) - \ln h_j^*, \qquad \langle\ln\beta_j\rangle = \Psi(s_j^*) - \ln t_j^* \tag{12.34}$$

$$\langle\ln\lambda_j\rangle = \Psi(a_j) - \Psi(a_j+b_j), \qquad \langle\ln(1-\lambda_j)\rangle = \Psi(b_j) - \Psi(a_j+b_j) \tag{12.35}$$

After convergence, the observed data points are clustered into M groups according to corresponding responsibilities r_{ij}. Following [19], we denote these newly formed groups of data points as "clumps" which subject to the constraint that all data points $\{\vec{X}_i\}$ in the clump c share the same $q(Z_i) \equiv q(Z_c)$ which is a key factor in the following compression phase.

12.3.2 Compression Phase

The major task of the compression phase is to evaluate clumps that possibly belong to the same mixture component while taking into account future arriving data points. Assume that a set of N data points has already been observed, our target now is to make an inference of clustering at some target time T where $T \geq N$. Specifically, we estimate this future clustering process by scaling the current observed data to the target size T, which is equivalent to leveraging the variational posterior distribution of the observed data N as a predictive model of the future data [19]. Accordingly, the modified free energy for the compression phase can be calculated as the following

$$
\mathcal{F} = \sum_{j=1}^{M} \sum_{l=1}^{D} \langle \ln \frac{p(\alpha_{jl})}{q(\alpha_{jl})} \rangle + \sum_{j=1}^{M} \left[\langle \ln \frac{p(\lambda_j)}{q(\lambda_j)} \rangle + \langle \ln \frac{p(\alpha_j)}{q(\alpha_j)} \rangle + \langle \ln \frac{p(\beta_j)}{q(\beta_j)} \rangle \right]
$$

$$
+ \frac{T}{N} \sum_c |n_c| \ln \sum_{j=1}^{M} \exp(\widetilde{r}_{cj}) \tag{12.36}
$$

where $\frac{T}{N}$ is the data magnification factor and $|n_c|$ denotes the number of data points in clump c. The updating solutions of variational posteriors in the compression phase for maximizing this free energy function are given by

$$
q(Z_c) = \prod_{j=1}^{M} r_{cj}^{\mathbf{1}[Z_c=j]}, \qquad q(\vec{\lambda}) = \prod_{j=1}^{M} \mathrm{Beta}(\lambda_j | a_j, b_j) \tag{12.37}
$$

$$
q(\vec{\alpha}_l) = \prod_{j=1}^{M} \prod_{l=1}^{D} \mathcal{G}(\alpha_{jl} | u_{jl}^*, v_{jl}^*) \qquad q(\vec{\alpha}) = \prod_{j=1}^{M} \mathcal{G}(\alpha_j | g_j^*, h_j^*), \qquad q(\vec{\beta}) = \prod_{j=1}^{M} \mathcal{G}(\beta_j | s_j^*, t_j^*)
$$

$$
\tag{12.38}
$$

where the corresponding hyperparamters are calculated as

$$
r_{cj} = \frac{\exp(\widetilde{r}_{cj})}{\sum_{j=1}^{M} \exp(\widetilde{r}_{cj})} \tag{12.39}
$$

$$
\widetilde{r}_{cj} = \exp \left[\widetilde{\mathcal{I}}_j + \widetilde{\mathcal{H}}_j + (\bar{\alpha}_j - \sum_{l=1}^{D} \bar{\alpha}_{jl}) \ln(\sum_{l=1}^{D} \langle X_{cl} \rangle) + \sum_{k=1}^{j-1} \langle \ln(1-\lambda_k) \rangle + \sum_{l=1}^{D} (\bar{\alpha}_{jl} - 1) \ln \langle X_{cl} \rangle \right.
$$

$$
+ (\bar{\beta}_j - 1) \ln(1 - \sum_{l=1}^{D} \langle X_{cl} \rangle) + \langle \ln \lambda_j \rangle \Big] \tag{12.40}
$$

$$
a_j = 1 + \frac{T}{N} \sum_c |n_c| \langle Z_c = j \rangle, \qquad b_j = \zeta_j + \frac{T}{N} \sum_c |n_c| \sum_{k=j+1}^{M} \langle Z_c = k \rangle \tag{12.41}
$$

$$
u_{jl}^* = u_{jl} + \frac{T}{N} \sum_c |n_c| r_{cj} \bar{\alpha}_{jl} \left[\Psi(\sum_{l=1}^{D} \bar{\alpha}_{jl}) + \Psi'(\sum_{l=1}^{D} \bar{\alpha}_{jl}) \sum_{d \neq l}^{D} (\langle \ln \alpha_{jd} \rangle - \ln \bar{\alpha}_{jd}) \bar{\alpha}_{jd} - \Psi(\bar{\alpha}_{jl}) \right]
$$

$$
\tag{12.42}
$$

$$
g_j^* = g_j + \frac{T}{N} \sum_c |n_c| r_{cj} \left[\bar{\beta}_j \Psi'(\bar{\alpha}_j + \bar{\beta}_j)(\langle \ln \beta_j \rangle - \ln \bar{\beta}_j) - \Psi(\bar{\alpha}_j) + \Psi(\bar{\alpha}_j + \bar{\beta}_j) \right] \bar{\alpha}_j \tag{12.43}
$$

$$
s_j^* = s_j + \frac{T}{N} \sum_c |n_c| r_{cj} \left[\bar{\alpha}_j \Psi'(\bar{\alpha}_j + \bar{\beta}_j)(\langle \ln \alpha_j \rangle - \ln \bar{\alpha}_j) - \Psi(\bar{\beta}_j) + \Psi(\bar{\alpha}_j + \bar{\beta}_j) \right] \bar{\beta}_j \tag{12.44}
$$

$$v_{jl}^* = v_{jl} - \frac{T}{N} \sum_c |n_c| r_{cj} \left[\ln\langle X_{cl}\rangle - \ln(\sum_{l=1}^{D} \langle X_{cl}\rangle) \right] \tag{12.45}$$

$$h_j^* = h_j - \frac{T}{N} \sum_c |n_c| r_{cj} \ln(\sum_{l=1}^{D} \langle X_{cl}\rangle) \tag{12.46}$$

$$t_j^* = t_j - \frac{T}{N} \sum_c |n_c| r_{cj} \ln(1 - \sum_{l=1}^{D} \langle X_{cl}\rangle) \tag{12.47}$$

where $\langle X_{cl}\rangle$ represents the average over all data points contained in clump c. The first step in the compression phase is to hard assign each clump or data point to the component with the highest responsibility r_{cj} obtained from the previous model building phase as

$$I_c = \arg\max_j r_{cj} \tag{12.48}$$

where $\{I_c\}$ represents which component the clump (or data point) c belongs to in the compression phase. Then, we cycle through each component and split it into two sub-components along its principal component. This split process can be refined by updating Equations(12.37)-(12.38). After the convergence criterion is reached for refining the split, the clumps are then assigned to one of the two candidate components. Among all the potential splits, we choose the one that results in the largest change in the free energy Eq.(12.36). We iterate this split process until a stopping criterion is satisfied. Based on [19], a possible stopping criterion for the splitting process can be expressed as the limit on the amount of memory required to store the components. In our case, the component memory cost for the mixture model is $\mathcal{MC} = (D+2)N_c$, where $D+2$ is the number of parameters contained in a D-variate Beta-Liouville component, while N_c represents the number of clumps. Thus, we can define an upper limit on the component memory cost $\mathcal{C}$, and the compression phase stops when $\mathcal{MC} \geq \mathcal{C}$. As a result, the computational time and the space requirement are bounded in each learning round. After the compression phase, the currently observed data points are discarded while the resultant components are treated in the same way as data points in the next round of learning. The proposed incremental variational inference algorithm for infinite Beta-Liouville mixture model is summarized in Algorithm 1.

12.4 Video Background Subtraction

In this section, the proposed incremental infinite Beta-Liouville mixture model (denoted as *InBLM*) performance is illustrated using extensive simulations by applying it to the video background subtraction problem. Following the idea proposed by [32], we treat the problem of video background subtraction as pixel-level evaluations. Specifically, each pixel in our approach is represented by a mixture of density functions, and our goal is to make an inference whether the pixel belongs to the background or some foreground object.

12.4.1 Experimental Methodology

Suppose that we have observed a frame $\mathcal{X}$ which contains N pixels, i.e. $\mathcal{X} = \{\vec{X}_1, \ldots, \vec{X}_N\}$. Then, each pixel $\vec{X}_i$ can be modeled as a mixture of infinite Beta-Liouville distributions: $p(\vec{X}_i|\vec{\pi}, \vec{\alpha}) = \sum_{j=1}^{\infty} \pi_j \mathrm{BL}(\vec{X}_i|\vec{\alpha}_j)$, where $\vec{X}_i$ represents the RGB color (three-dimensional) or the intensity value (one-dimensional) of the pixel. The background subtraction methodology used in our work can be summarized as follows: First, the pixel values in an observed frame are normalized to the unit sum as a preprocessing step. Next, the background model is

---

Algorithm 12.1

1: Choose the initial truncation level M.

2: Initialize hyper-parameters: ξ_j, u_{jl}, v_{jl}, g_j, h_j, s_j and t_j.

3: Initialize the values of r_{ij} by K-Means algorithm.

4: **while** More data to be observed **do**

5:    Perform the model building phase through Eqs.(12.21)$\sim$(12.22).

6:    Initialize the compression phase using Eq.(12.48).

7:    **while** $\mathcal{MC} \geq \mathcal{C}$ **do**

8:       **for** $j = 1$ **to** M **do**

9:          **if** $evaluated(j) = $ **false then**

10:             Split component j and refine this split using Eqs.(12.37)$\sim$(12.38).

11:             $\Delta\mathcal{F}(j) = $ change in Eq.(12.36).

12:             $evaluated(j) = $ **true**.

13:          **end if**

14:       **end for**

15:       Split component j with the largest value of $\Delta\mathcal{F}(j)$.

16:       $M = M + 1$.

17:    **end while**

18:    Discard the currently observed data points.

19:    Save the resultant components for next learning round.

20: **end while**

---

learned using the proposed *InBLM* via incremental variational Bayes. It is worth mentioning that some of the mixture components within our mixture model are used to model the scene background while others are utilized to model the foreground objects. Thus, our final step is to determine whether $\vec{X}_i$ is a foreground or background pixel. In our work, we adopt the assumption that a mixture component is considered to belong to the background if it occurs frequently (high π_j) and does not vary much (low standard deviation σ_j). Therefore, the first B components are chosen as background components after ordering all the estimated components based on the ratio $\pi_j/\|\sigma_j\|$, where B is defined as

$$B = \arg\min_m \sum_{j=1}^{m} \pi_j > \Lambda \tag{12.49}$$

where Λ is a threshold that represents the minimum portion of the data that should be accounted for by the background, and the rest of the components are defined as foreground objects. Therefore, background subtraction can be performed for an observed frame by determining if the testing pixel $\vec{X}_i$ belongs to any of the components B.

12.4.2 Datasets

We evaluate the performance of the proposed background subtraction approach using six publicly available video datasets with different characteristics. These video sequences were used previously by [21] [36] and were selected to test the efficiency of our algorithm under diverse scenarios, such as: dynamic backgrounds, illumination changes, etc. These video sequences are described as following:

- **S1**: This video sequence shows some vehicles driving in the rain;

- **S2**: The video sequence is about a plastic drum floating on the surface of sea;

- **S3**: In this video, a person is walking on a beach;

- **S4**: This video sequence shows a person walking in front of swaying trees;

- **S5**: In this video, people come into a room with light switches on and off;

- **S6**: The video sequence consists of several minutes of an overhead view of a cafeteria.

Each individual frame in the aforementioned video sequences contains 160×120 pixels. Sample frames from these datasets are displayed in Figures 12.2 and 12.3. In our experiments, we initialize the truncation level M to 20. According to our experiments, a good choice of the initial values of the hyperparameters is: $(\xi_j, u_{jl}, v_{jl}, g_j, h_j, s_j, t_j) = (0.1, 1, 0.01, 0.1, 0.01, 0.1, 0.01)$.

FIGURE 12.2   Sample frames from datasets: **S1**, **S2** and **S3**.

FIGURE 12.3   Sample frames from datasets: **S4**, **S5** and **S6**.

12.4.3   Experimental Results

The performance of the proposed approach for video background subtraction is evaluated on pixel-level. Thus, it is straightforward to consider foreground objects detection as a binary classification of each pixel, resulting in a segmentation mask. Representative foreground masks generated by *InBLM* for each video sequence are shown in Figures 12.4 and 12.5, where the threshold Λ is set to 0.75. The ground truth frames are generated by manually

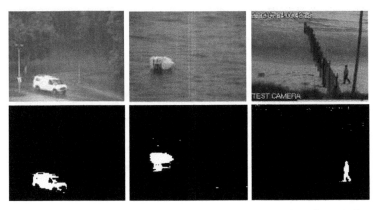

FIGURE 12.4   Foreground masks generated by *InBLM* for datasets: **S1**, **S2** and **S3**.

FIGURE 12.5   Foreground masks generated by *InBLM* for datasets: **S4**, **S5** and **S6**.

highlighting all the moving objects in sequences. As we can see in these two figures, the proposed *InBLM* is able to provide good results for all tested datasets. Although some noise can be noticed in the results of datasets **S1** ∼ **S4** due to the dynamic backgrounds, foreground objects were clearly identified by our approach. This shows the ability and flexibility of the proposed approach to handle and model dynamic backgrounds. Moreover, the appealing result of dataset **S5** demonstrates the robustness of our approach against illumination changes.

We have also evaluated quantitatively the performance for subtracting video background through *recall* and *precision* measures which are defined as the following:

$$\text{Recall} = \frac{\text{number of correctly identified foreground pixels}}{\text{number of foreground pixels in ground truth}}$$

$$\text{Precision} = \frac{\text{number of correctly identified foreground pixels}}{\text{number of foreground pixels detected}}$$

In our case, the results of recall and precision are based on the averages over all the measured frames. A trade off needs to be considered between recall and precision: an increase in recall by detecting more foreground pixels causes a decrease in precision. Therefore, we should attempt to maintain as high recall value as possible without sacrificing too much precision. For comparison, we have also applied other three mixture-based background subtraction algorithms on the same datasets: the infinite generalized Dirichlet mixture model

(*InGDM*), the infinite Dirichlet mixture model (*InDM*), and the infinite Gaussian mixture model (*InGM*).

Figures 12.6 ∼ 12.11 illustrate the comparison results in terms of precision-recall graph by varying the threshold Λ for each dataset. According to the results shown in these figures, the proposed *InBLM* outperforms the other three approaches by providing the best precision and recall result for each video sequence. These results demonstrate the better modeling capability when adopting Beta-Liouville mixtures. We may also notice that *InGDM* has the worst performance among all tested approaches. Indeed, this is as expected and confirms the fact that Gaussian mixture models are not appropriate for modeling normalized data.

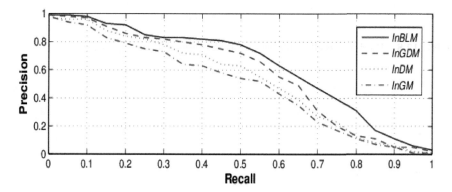

FIGURE 12.6   Precision-recall graph for dataset: **S1** using different approaches.

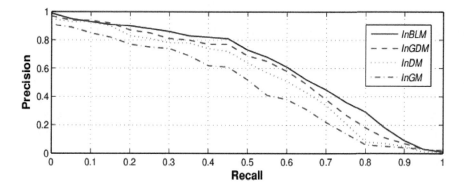

FIGURE 12.7   Precision-recall graph for dataset: **S2** using different approaches.

12.5   Conclusion

Video background subtraction is an essential task in computer vision for detecting moving objects in video sequences. In this chapter, we have sketched a statistical framework based on a Bayesian nonparametric approach to subtract video background. The proposed approach considers a mixture of Dirichlet processes with Beta-Liouville distributions, which can be viewed as an infinite Beta-Liouville mixture model (i.e. a learning machine which estimates a given probability density function as an infinite weighted sum of Beta-Liouville distributions). The proposed model is learned via variational Bayesian inference which has been

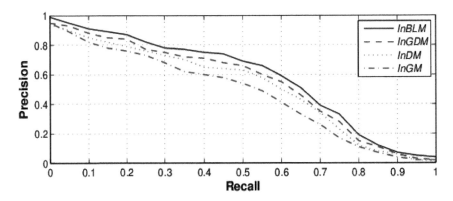

FIGURE 12.8   Precision-recall graph for dataset: **S3** using different approaches.

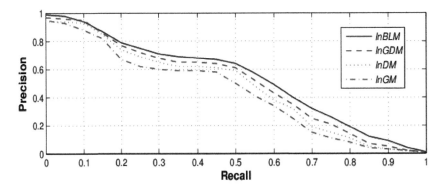

FIGURE 12.9   Precision-recall graph for dataset: **S4** using different approaches.

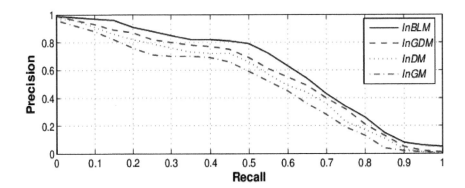

FIGURE 12.10   Precision-recall graph for dataset: **S5** using different approaches.

proposed as an efficient deterministic alternative to purely Bayesian inference. Compared to other background subtraction approaches, the proposed one has the advantages that it is more robust and adaptive to dynamic background, and it has the ability to handle multi-modal background distributions. Moreover, thanks to the nature of nonparametric Bayesian models, the determination of the correct number of components is sidestepped by assuming that there is an infinite number of components. Our results demonstrate the improved background subtraction performance by comparing our approach with other mixture-based

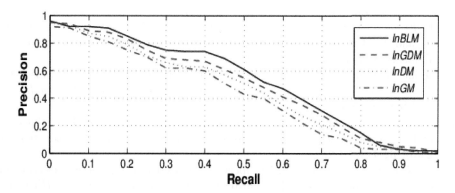

FIGURE 12.11   Precision-recall graph for dataset: **S6** using different approaches.

approaches.

References

1. H. Attias. A variational Bayes framework for graphical models. In *Advances in Neural Information Processing Systems (NIPS)*, pages 209–215, 1999.
2. A. Aumberg and D. Hogg. Learning flexible models from image sequences. In Jan-Olof Eklundh, editor, *European Conference on Computer Vision, ECCV 1994*, volume 800 of *Lecture Notes in Computer Science*, pages 297–308. Springer Berlin Heidelberg, 1994.
3. C. Bishop. *Pattern Recognition and Machine Learning*. Springer, 2006.
4. M. Black. Explaining optical flow events with parameterized spatio-temporal models. In *IEEE Computer Society Conference on Computer Vision and Pattern Recognition, CVPR 1999*, pages 326–332 Vol. 1, 1999.
5. D.M. Blei and M.I. Jordan. Variational inference for Dirichlet process mixtures. *Bayesian Analysis*, 1:121–144, 2005.
6. N. Bouguila. Hybrid generative/discriminative approaches for proportional data modeling and classification. *IEEE Transactions on Knowledge and Data Engineering*, 24(12):2184–2202, 2012.
7. N. Bouguila and D. Ziou. Unsupervised selection of a finite Dirichlet mixture model: An MML-based approach. *IEEE Transactions on Knowledge and Data Engineering*, 18(8):993–1009, 2006.
8. N. Bouguila and D. Ziou. High-dimensional unsupervised selection and estimation of a finite generalized Dirichlet mixture model based on minimum message length. *IEEE Transactions on Pattern Analysis and Machine Intelligence*, 29(10):1716–1731, 2007.
9. C. Bregler. Learning and recognizing human dynamics in video sequences. In *IEEE Computer Society Conference on Computer Vision and Pattern Recognition, CVPR 1997*, pages 568–574, 1997.
10. C. Constantinopoulos, M.K. Titsias, and A. Likas. Bayesian feature and model selection for Gaussian mixture models. *IEEE Transactions on Pattern Analysis and Machine Intelligence*, 28(6):1013 –1018, 2006.
11. A. Corduneanu and C. M. Bishop. Variational Bayesian model selection for mixture distributions. In *International Conference on Artificial Intelligence and Statistics, AISTAT 2001*, pages 27–34, 2001.

12. A. Elgammal, D. Harwood, and L. Davis. Non-parametric model for background subtraction. In *European Conference on Computer Vision, ECCV 2000*, pages 751–767, 2000.

13. W. Fan and N. Bouguila. Online variational learning of finite Dirichlet mixture models. *Evolving Systems*, 3(3):153–165, 2012.

14. W. Fan and N. Bouguila. Variational learning of a Dirichlet process of generalized Dirichlet distributions for simultaneous clustering and feature selection. *Pattern Recognition*, 46(10):2754–2769, 2013.

15. W. Fan, N. Bouguila, and D. Ziou. Variational learning for finite Dirichlet mixture models and applications. *IEEE Transactions on Neural Networks and Learning Systems*, 23(5):762–774, 2012.

16. W. Fan, N. Bouguila, and D. Ziou. Unsupervised hybrid feature extraction selection for high-dimensional non-Gaussian data clustering with variational inference. *IEEE Transactions on Knowledge and Data Engineering*, 25(7):1670–1685, 2013.

17. T. Ferguson. A Bayesian analysis of some nonparametric problems. *The Annals of Statistics*, 1(2):209–230, 1973.

18. T. Ferguson. Bayesian density estimation by mixtures of normal distributions. *Recent Advances in Statistics*, 24:287–302, 1983.

19. R. Gomes, M. Welling, and P. Perona. Incremental learning of nonparametric Bayesian mixture models. In *Proc. of IEEE Conference on Computer Vision and Pattern Recognition (CVPR)*, pages 1–8, 2008.

20. S. Guha, A. Meyerson, N. Mishra, R. Motwani, and L. O'Callaghan. Clustering data streams: Theory and practice. *IEEE Transactions on Knowledge and Data Engineering*, 15(3):515–528, 2003.

21. J. Huang, X. Huang, and D. Metaxas. Learning with dynamic group sparsity. In *IEEE International Conference on Computer Vision, ICCV 2009*, pages 64–71, 2009.

22. H. Ishwaran and L. James. Gibbs sampling methods for stick-breaking priors. *Journal of the American Statistical Association*, 96:161–173, 2001.

23. M. I. Jordan, Z. Ghahramani, T. S. Jaakkola, and L. K. Saul. An introduction to variational methods for graphical models. In *Learning in Graphical Models*, pages 105–162, 1998.

24. R. Korwar and M. Hollander. Contributions to the theory of Dirichlet processes. *The Annals of Probability*, 1:705–711, 1973.

25. D. Lee. Effective Gaussian mixture learning for video background subtraction. *IEEE Transactions on Pattern Analysis and Machine Intelligence*, 27(5):827–832, 2005.

26. T. Meier and K. Ngan. Automatic segmentation of moving objects for video object plane generation. *IEEE Transactions on Circuits and Systems for Video Technology*, 8(5):525–538, 1998.

27. N. Paragios and V. Ramesh. A MRF-based approach for real-time subway monitoring. In *IEEE Conference on Computer Vision and Pattern Recognition, CVPR 2001*, pages 1034–1040, 2001.

28. M. Piccardi. Background subtraction techniques: A review. In *IEEE International Conference on Systems, Man and Cybernetics, SMC 2004*, pages 3099–3104, 2004.

29. Y. Raja, S. McKenna, and S. Gong. Segmentation and tracking using color mixture models. In Roland T. Chin and Ting-Chuen Pong, editors, *Third Asian Conference on Computer Vision (ACCV)*, volume 1351 of *Lecture Notes in Computer Science*, pages 607–614, 1998.

30. C. Robert and G. Casella. *Monte Carlo Statistical Methods.* Springer-Verlag, 1999.

31. J. Sethuraman. A constructive definition of Dirichlet priors. *Statistica Sinica*, 4:639–650, 1994.

32. C. Stauffer and W. Grimson. Adaptive background mixture models for real-time tracking. In *IEEE Conference on Computer Vision and Pattern Recognition, CVPR 1999*, pages 246–252, 1999.

33. B. Stenger, V. Ramesh, N. Paragios, F. Coetzee, and J. Buhmann. Topology free hidden markov models: Application to background modeling. In *IEEE International Conference on Computer Vision, ICCV 2001*, pages 294–301, 2001.

34. C. Stiller. Object-based estimation of dense motion fields. *IEEE Transactions on Image Processing*, 6(2):234 –250, 1997.

35. Y. Teh, M. Jordan, M. Beal, and D. Blei. Hierarchical Dirichlet processes. *Journal of the American Statistical Association*, 101:705–711, 2004.

36. K. Toyama, J. Krumm, B. Brumitt, and B. Meyers. Wallflower: Principles and practice of background maintenance. In *IEEE International Conference on Computer Vision, ICCV 1999*, pages 255–261, 1999.

37. J. Wang and E. Adelson. Representing moving images with layers. *IEEE Transactions on Image Processing*, 3(5):625–638, 1994.

38. Y. Weiss and E. Adelson. A unified mixture framework for motion segmentation: incorporating spatial coherence and estimating the number of models. In *IEEE Computer Society Conference on Computer Vision and Pattern Recognition, CVPR 1996*, pages 321–326, 1996.

39. Z. Zivkovic and F. van der Heijden. Recursive unsupervised learning of finite mixture models. *IEEE Transactions on Pattern Analysis and Machine Intelligence*, 26:651–656, 2004.

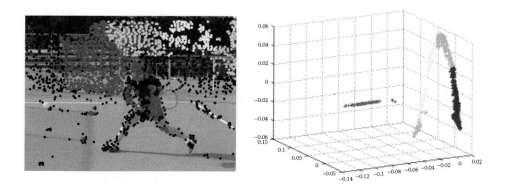

FIGURE 4.5 Example trajectories in the image and their corresponding embedding coordinates, from Elqursh and Elgammal [9]. Each cluster is marked with a different color. Black dots are short trajectories not yet added to the embedding. The foreground cluster is marked with two red concentric circles around its trajectories.

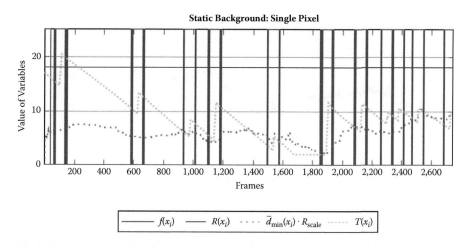

FIGURE 6.7 Inspection of a pixel ($x = 280$, $y = 120$) in the video *sofa* of the Change Detection competition.

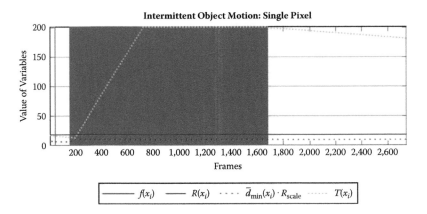

FIGURE 6.8 Inspection of a pixel ($x = 60$, $y = 200$) in the video *sofa* of the Change Detection competition.

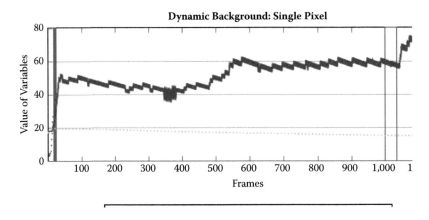

FIGURE 6.9 Inspection of a pixel ($x = 40$, $y = 200$) in the video *canoe* of the Change Detection competition.

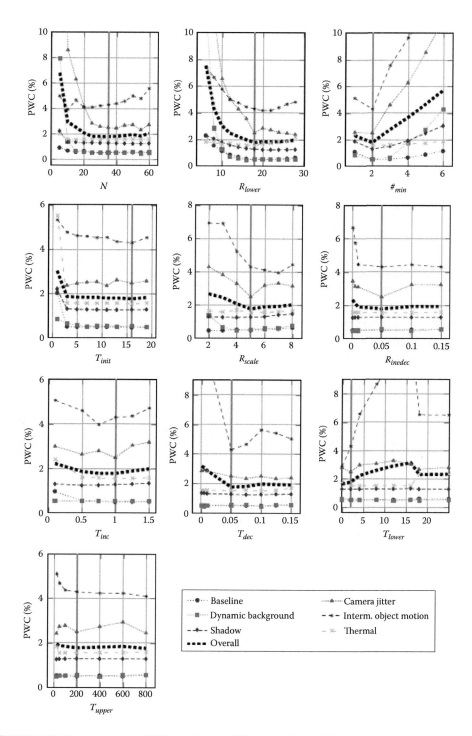

FIGURE 6.10 Impact on the *PWC* metric for different values of the parameters of the PBAS.

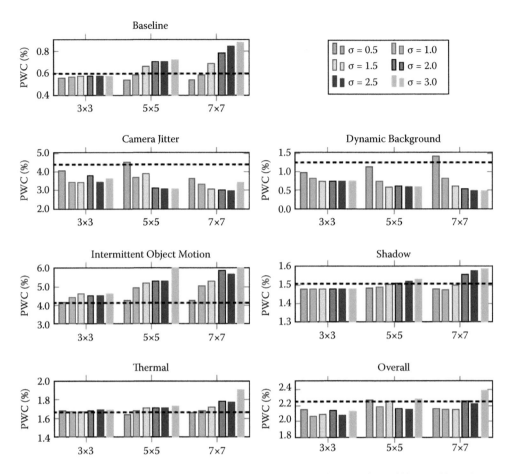

FIGURE 6.12 An overview of the impact on the *PWC* metric by using different filter sizes and standard deviations σ in a Gaussian pre-filter step.

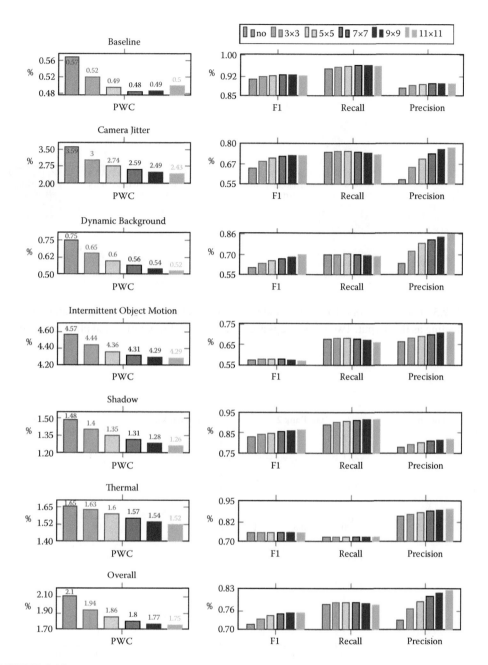

FIGURE 6.13 Results of each category for *PWC*, Recall, Precision and *F*1 measure with different sizes of median filter for post-processing.

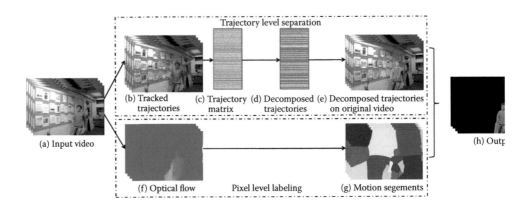

FIGURE 9.5 The framework. LRGS method takes a raw video sequence as input, and produces a binary labeling as the output. Two major steps are trajectory level separation and pixel level labeling.

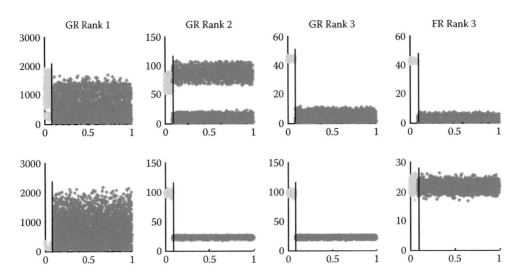

FIGURE 9.8 $\|F_i\|_2$ distribution of the GR and the FR model. Green means foreground and blue means background. Separation means good result. Top row is from moving camera, and bottom row is from stationary camera. The four columns are GR–1, GR–2, GR–3 and FR.

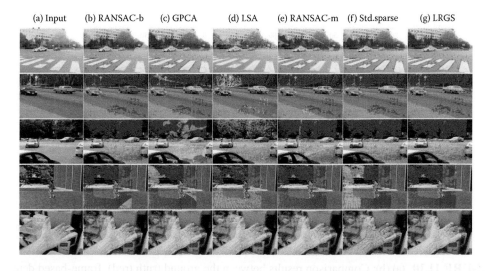

(a) Input   (b) RANSAC-b   (c) GPCA   (d) LSA   (e) RANSAC-m   (f) Std.sparse   (g) LRGS

FIGURE 9.9 Results at trajectory level labeling. Green: foreground; purple: background. From top to bottom, five videos are *"VCars"*, *"cars2"*, *"cars5"*, *"people1"* and *"VHand"*.

FIGURE 11.9 Results using the proposed method (a) and frame-based (b). Final tracking results (c).

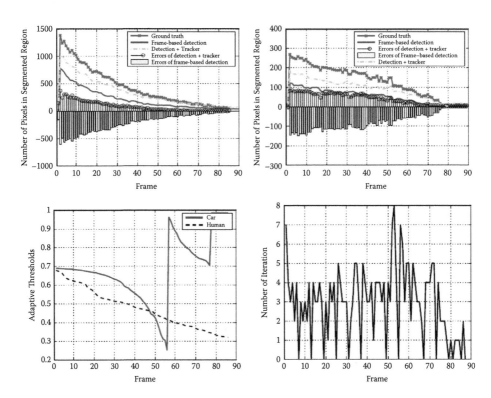

FIGURE 11.10 (a),(b): Comparison results between the ground truth (red), frame-based detection (blue) and the proposed approach (green). (c): Adaptive threshold. (d): Average number of iterations (© (2008) Springer).

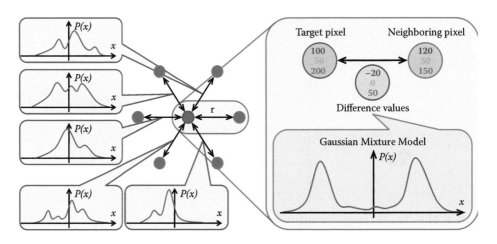

FIGURE 13.3 Background model based on *Statistical Local Difference Pattern*: Local Difference (LD) is a local feature, and is defined by the difference between a target pixel and a neighboring pixel. LD is modeled using a GMM to represent its distribution, making it a statistical local feature called the Statistical Local Difference (SLD). Then, Statistical Local Difference Pattern (SLDP) is defined using several SLDs for the background model (this figure shows an example with six SLDs).

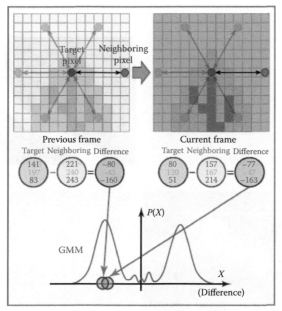

Adaptivity to illumination changes

(a)

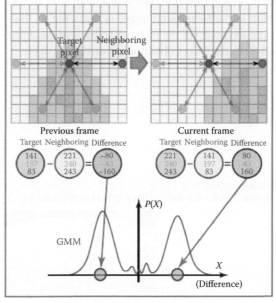

Adaptivity to dynamic changes

(b)

FIGURE 13.4 Adaptivities of our method to background fluctuation : (a) shows the case of illumi-nation changing suddenly. SLDP can adapt to illumination changes since LD has the ability to toler-ate the effects of illumination changes which affect the target pixel value in proportion with others. (b) shows the case where a background texture changes repeatedly. SLDP can adapt to this kind of dynamic background changes, since GMMs can learn the variety of background hypotheses.

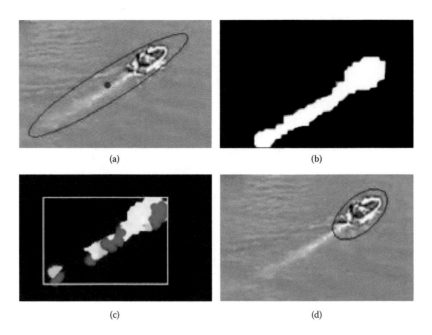

(a)   (b)

(c)   (d)

FIGURE 14.8 Example of optical flow (OF) use to deal with boat wakes. a) False positives caused by boat wakes. b) The blob in the foreground image is much bigger than the real boat. c) Optical flow mask. d) By clustering the points of the blob on the basis of the direction of the OF points, the boat can be distinguished from the wakes obtaining a correct detection.

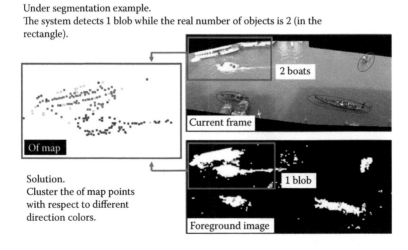

FIGURE 14.9 An example of ghost observation solved by clustering the optical flow points.

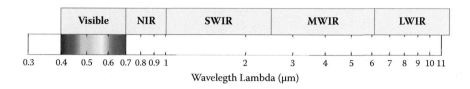

FIGURE 17.2 Visible and infra red spectrum bands.

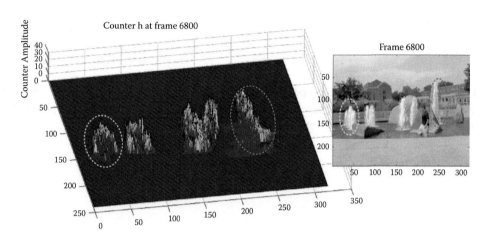

FIGURE 22.4 Original frame and the plot of the counter values $h(i, j)$ for different pixel locations (i, j). Higher values correspond to outer boundaries of multiple fountains, indicating regions with low reliability and non-salient motion (©(2010) Elsevier).

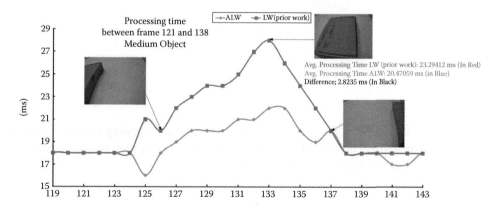

FIGURE 22.19 Processing time (ms) versus the frame number for two different versions of the algorithm when there is a foreground object in the scene (©(2010) Elsevier).

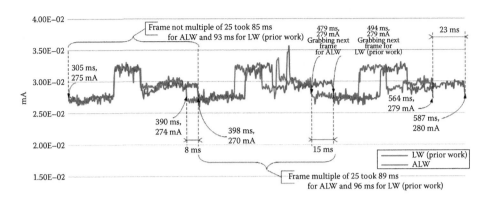

FIGURE 22.20 Variations in the operating current during the processing of three consecutive frames containing a foreground object. The method presented in this chapter (blue plot) is faster than the prior work [4] (red) (©(2010) Elsevier).

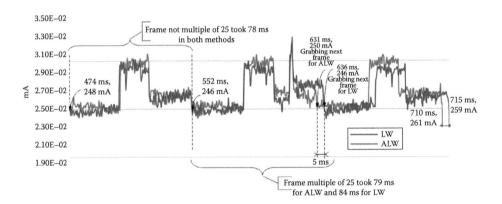

FIGURE 22.21 Variations in the operating current during the processing of three consecutive frames of an empty scene. The method presented in this chapter (blue plot) provides speed gain at frame numbers that are multiple of 25 (©(2010) Elsevier).

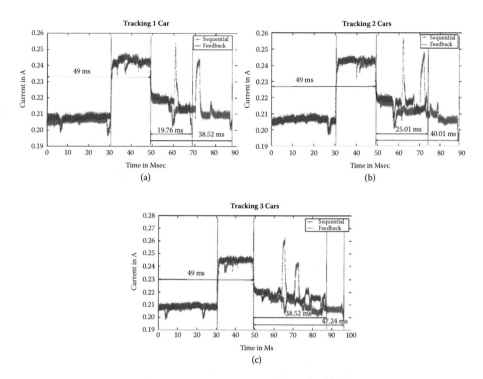

FIGURE 22.25 Operating current of the camera board with the feedback and sequential methods when tracking (a) one (b) two and (b) three remote-controlled cars (©[2011] IEEE).

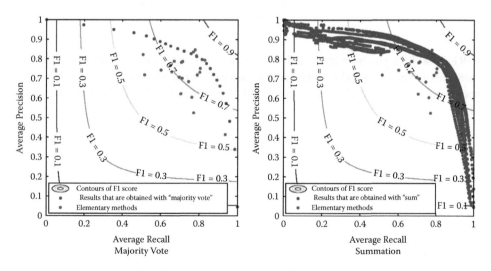

FIGURE 24.3 Results obtained from the combination of all (22) background subtraction methods with 2 combination rules. For the purpose of comparison, the precision and the recall of the 22 individual methods are displayed in red. The blue dots correspond to different decision thresholds (*Th*) as well as different estimations of the priors (**Y**).

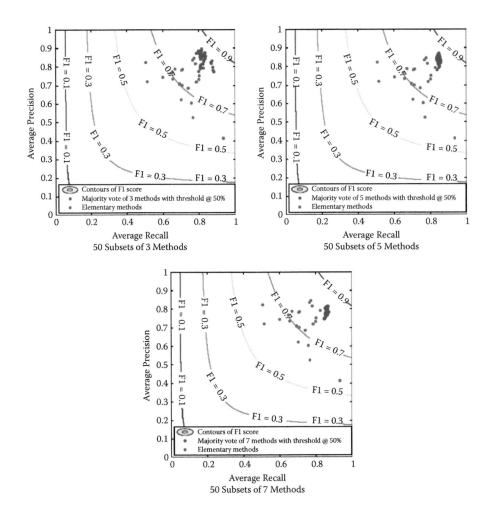

FIGURE 24.4 Real precision and recall of the majority vote combination rule (at the neutral decision threshold). The predicted performance is shown, in blue, for 50 combinations of 3, 5, and 7 methods, selected theoretically. The precision and the recall of the 22 individual methods are shown in red.

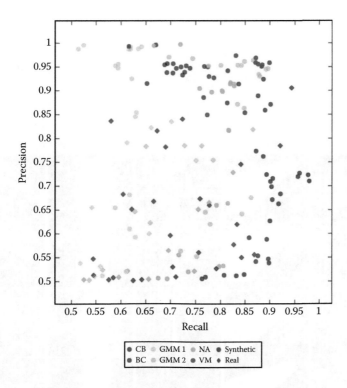

FIGURE 25.5 Precision/recall values for all the videos of the benchmark.

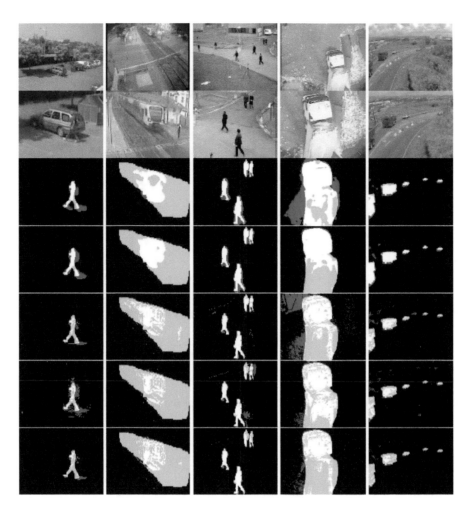

FIGURE 25.6 Foreground masks obtained from the top five BS algorithms on sequences 001, 006, 003, 002, 008 of the real videos. Input frames are depicted on the first row, followed by the region of interest chosen, and the five BS outputs, according to the order of Table 25.4. In these images, the TP pixels are in white, TN pixels in black, FP pixels in red and FN pixels in green.

13

Spatio-temporal Background Models for Object Detection

Satoshi Yoshinaga
LIMU, Kyushu University, Japan

Yosuke Nonaka
LIMU, Kyushu University, Japan

Atsushi Shimada
LIMU, Kyushu University, Japan

Hajime Nagahara
LIMU, Kyushu University, Japan

Rin-ichiro Taniguchi
LIMU, Kyushu University, Japan

13.1   Introduction

One of the fundamental problems in computer vision is detecting regions or objects of interest from an image sequence. Background subtraction, which removes a background image from the input image, is widely used for detecting foreground objects in practical applications, since it enables us to detect foreground objects without any previous knowledge of them. However, simple background subtraction often detects not only foreground objects but also a lot of noise regions, because it is quite sensitive to background changes. In general, background changes which occur in outdoor scenes can be mainly classified into two types:

- **Illumination changes** – changes caused by lighting conditions such as the sun rising, setting, or being blocked by clouds,

- **Dynamic changes** – changes caused by the swaying motion of tree branches, leaves and grass, fleeting cloud, waves on water and so on.

To handle these background changes, many researchers have proposed background modeling approaches [1–4, 6–13, 15–17]. They are classified into three categories: statistical, local feature-based and extrapolation-based approaches.

Statistical background modeling: In general, the occurrence probability of pixel value of the foreground objects is smaller than that of background, suppose that foreground objects do not remain in the same place. This above property holds even when we consider the effect of the dynamic changes of the background, because the dynamic changes usually occur repeatedly. Therefore, statistical approaches, which model this probabilistic characteristic to adapt to dynamic changes, have been proposed [1,8,10–12]. Most of these approaches are intended for on-line surveillance, and they model the background using a probability distribution of the previously observed intensity values of each pixel. By using a multi-modal distribution, statistical models can learn the multiple different appearances of the background which are caused by dynamic changes. A Gaussian mixture model has been used for representing the multiple distributions in the literature [10,11]. Non-parametric statistical methods [1,8,12] which use kernel density estimation have also been proposed. While these statistical approaches can model the effects of dynamic changes, they have difficulty correctly adapting to sudden illumination changes, which are not observed in the previous frames.

Local feature-based background modeling: Since illumination changes affect not only a target pixel but also its neighboring pixels, local feature-based approaches which use this spatial characteristic have been proposed [3,4,6,7,9,15,17]. In early research, the similarities of a edge magnitude and a block histogram between a current and a background images are used for object detection [4,7]. Local Binary Pattern (LBP) and Radial Reach Correlation (RRC) use magnitude relations between a target pixel and its neighboring pixels, and detect foreground objects based on whether the current feature is similar to the feature of background [3,6,9]. Some researches extend RRC by adaptively determining the neighboring pixels based on off-line analysis [15,17]. These approaches assume that local features are not affected by the illumination changes. However, surveillance scenes also often include dynamic changes, and then the features of not only foreground regions but also of background regions vary significantly. Therefore, it is difficult for local feature-based background models to handle dynamic changes in the background.

Extrapolation-based background modeling: To handle illumination changes, an extrapolation-based approach which extrapolates the current background image from original background image has also been proposed [2]. In the literature [2], background candidate regions are extracted from current image, and then a background generation function is estimated based on the correspondence between the brightness of a current region and that of its corresponding original background images. However, this method assumes that the illumination changes uniformly in the whole image, and therefore non-uniform illumination changes cannot be handled.

As mentioned above, each background modeling approach has merits and demerits. Statistical approaches can adapt to dynamic changes, but they cannot handle illumination changes. On the other hand, local feature-based and extrapolation-based approaches can adapt to illumination changes, but they cannot handle dynamic changes.

To adapt to various background changes such as illumination changes, dynamic changes, etc., we should use a spatio-temporal information by combining or integrating background models which have different characteristics. In this chapter, we present three types of spatio-temporal background models, which have been addressed in our laboratory, to detect foreground objects robustly against various background changes. The first model, which is described in Section 13.2, is a combinational background model in which the detection results from multiple different background models are combined adaptively [13]. On the contrary, in Section 13.3 and 13.4, we do not combine the results from multiple different background models, but we integrate their methodologies into a single framework. The second model uses a spatio-temporal feature which is established by integrating a statistical framework into an illumination-invariant local feature defined based on a spatiality. While the second model uses a spatial information only for defining an illumination-invariant feature to adapt to illumination changes, the third model uses the spatiality to handle not only illumination but also dynamic changes. The third model defines a spatio-temporal feature by considering the similarity of intensity changes among pixels. By combining/integrating background models which have different characteristics, our spatio-temporal background models can handle various background changes more robustly compared to previous ones. To verify the effectiveness of our approaches, we report their evaluation results using the database of Background Models Challenge[31] (BMC). Furthermore, we also discuss the characteristics of each background model, which are different according to the combining/integrating methodologies.

13.2 Combinational Background Model Based on Spatio-temporal Feature

In this section, we present a combinational background model based on a spatio-temporal feature. By combining a spatial and a local feature-based models, each of which can handle illumination and dynamic changes respectively, we can handle both changes. However, it is difficult for a local feature-based model to handle non-uniform illumination changes, which influence pixel values depending on the positions, since local features of background regions are also influenced by them. Meanwhile, these kinds of illumination changes can be extrapolated by using an extrapolation based on non-linear prediction. Therefore, in addition to a statistical and a local feature-based models, we combine an extrapolation-based model with them adaptively so that the combinational background model can adapt to various background changes [13]. First, from Section 13.2.1 to Section 13.2.3, we present each component of our combinational background model. Then, in Section 13.2.4, we explain how to combine three different background models.

13.2.1 Fast Non-parametric Background Model

It is quite important for statistical background modeling to rapidly and accurately estimate the probability density functions (PDFs) of each pixel values. Kernel density estimation (KDE) is quite effective for the PDF estimation, since it can provide accurate PDF estimation by using a sufficient number of samples. However, a typical method [1] requires a lot of computation, which is proportional to the number of samples. Therefore, it can be hardly applied to real-time processing. To solve this problem, we have designed a fast algorithm of PDF estimation [12]. In our algorithm, we have used a rectangular function as the kernel

---

[31] 1st ACCV Workshop on Background Models Challenge: http://bmc.univ-bpclermont.fr/

function W, instead of Gaussian function, which is often used in KDE.

$$W(u) = \begin{cases} \frac{1}{h^d} & \text{if } |u| \leq \frac{h}{2}, \\ 0 & \text{otherwise,} \end{cases} \tag{13.1}$$

where h is a parameter representing the width of the kernel, i.e., some smoothing parameter and d is the dimension of color space. Using this kernel, the PDF at time t is represented as follows:

$$P^t(\boldsymbol{X}) = \frac{1}{S}\sum_{i=1}^{S} W\left(|\boldsymbol{X} - \boldsymbol{X}^{t-i}|\right), \tag{13.2}$$

where $|\boldsymbol{X} - \boldsymbol{X}^{t-i}|$ means the chess-board distance of pixel values in d-dimensional space, and S is the number of samples.

Thus, we estimate the PDF based on Eq.(13.2), and $P^t(\boldsymbol{X})$ is calculated by enumerating pixels whose values are inside of the kernel located at $\boldsymbol{X}$. However, if we calculate the PDF in a naive way (i.e., by enumerating the particular pixels), the computational time is still proportional to the number of samples S. Therefore, we have designed a fast algorithm of PDF estimation, whose computation cost does not depend on S.

In background modeling, we estimate the PDF at time t (i.e., $P^t(\boldsymbol{X})$) by referring to the pixel values $\{\boldsymbol{X}^{t-1}, \dots, \boldsymbol{X}^{t-i}, \dots, \boldsymbol{X}^{t-S}\}$ observed in the latest S frames. Then, at time $t + 1$, we can estimate an updated PDF $P^{t+1}(\boldsymbol{X})$ referring to a new sample $\boldsymbol{X}^t$. Basically, the essence of PDF estimation is accumulation of the kernel estimator, and when the new value $\boldsymbol{X}^t$ is acquired, the kernel estimator corresponding to $\boldsymbol{X}^t$ should be accumulated. At the same time, the oldest kernel estimator corresponding to $\boldsymbol{X}^{t-S}$ at frame $t - S$ should be discarded, since the length of the pixel process is constant, S. This idea leads to reduction of the PDF computation into the following incremental computation:

$$P^{t+1}(\boldsymbol{X}) = P^t(\boldsymbol{X}) + \frac{1}{S}W\left(|\boldsymbol{X} - \boldsymbol{X}^t|\right) - \frac{1}{S}W\left(|\boldsymbol{X} - \boldsymbol{X}^{t-S}|\right). \tag{13.3}$$

This equation means that when a new pixel value is observed, the PDF is updated by:

- increasing the probabilities of pixel values which are inside of the kernel located at the new pixel value $\boldsymbol{X}^t$ by $\frac{1}{h^d}$ (see Figure 13.1 red parts),

- decreasing those which are inside of the kernel located at the oldest pixel value $\boldsymbol{X}^{t-S}$ by $\frac{1}{h^d}$ (see Figure 13.1 blue parts).

In other words, the new PDF is acquired by local operations of the previous PDF as shown in Figure 13.1. Please refer to the paper [12] for experimental results of the fast algorithm.

13.2.2   Local Feature-based Background Model

To realize robust local feature-based background model, we have improved Radial Reach Correlation (RRC) [9] so that the background model is updated according to the background changes of the input image frames, and we call it Adaprive RRC.

RRC is defined to evaluate local texture similarity without the influence of illumination changes, and calculated at each pixel (x, y). At first, pixels whose intensity differences to the intensity of the center pixel (x, y) exceed a threshold T_P are searched for in every radial reach extension in 8 directions around the pixel (x, y). The searched 8 pixels are called as peripheral pixels hereafter. Then, the signs of intensity differences (positive difference or negative difference) of the 8 pairs, each of which is a pair of one of eight peripheral pixels and the center pixel (x, y), are represented in a binary code. The basic idea is that the

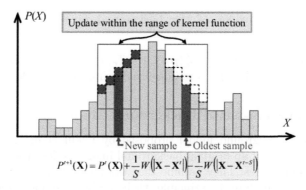

FIGURE 13.1 Update mechanism of our statistical background model based on kernel density estimation.

binary code represents intrinsic information about local texture around the pixel, and that it does not change under illumination changes. To make this idea concrete, the correlation value of the binary codes extracted from the observed image and the reference background image is calculated to evaluate their similarity. For further details, we refer the readers to the paper [9].

Using RRC, the similarity of the incremental codes between a background image and the observed image is calculated. Then, the pixels whose codes are "not similar" to their corresponding pixels in the background image are detected as foreground pixels. In principle, if the background does not change, we can prepare adequate codes of the background image in advance. However, usually, due to the illumination changes and various noises, it is almost impossible to prepare them. Even if we manage to prepare such fixed background codes, accurate results cannot be acquired, and therefore we should update the background codes adaptively. Here, we have developed a mechanism to update the background codes according to the following formula:

$$b_k^{t+1}(\boldsymbol{p}) = (1 - \alpha) \cdot b_k^t(\boldsymbol{p}) + \alpha \cdot b_k'^t(\boldsymbol{p}) \tag{13.4}$$

where $b_k^t(\boldsymbol{p})(k = 0, 1, ..., 7)$ represents the incremental code of a pixel $\boldsymbol{p}$ in the background image at time t, and $b_k'^t(\boldsymbol{p})$ is the code calculated from the input image in the same manner. α is a learning rate, and when it is large enough, the above code can be quickly adapted to the current input image, i.e., adapted to sudden background changes. The range of $b_k^t(\boldsymbol{p})$ is $[0, 1]$, and when $b_k^t(\boldsymbol{p})$ is close to either 0 or 1, it means that the magnitude relation of intensity between the center pixel and its peripheral pixel does not change. Otherwise, i.e., if $b_k^t(\boldsymbol{p})$ is close to 0.5, the magnitude relation is not stable. According to this consideration, a peripheral pixel is sought again when $T_r \le b_k^t(\boldsymbol{p}) \le 1 - T_r$ holds. T_r is a threshold value to invoke re-searching of peripheral pixels. Then, we define the similarity of the incremental codes as follows:

$$B^t(\boldsymbol{p}) = \sum_{k=0}^{7} |b_k^t(\boldsymbol{p}) - b_k'^t(\boldsymbol{p})| \tag{13.5}$$

Whether the pixel $\boldsymbol{p}$ belongs to background or foreground is judged by comparing the similarity $B^t(\boldsymbol{p})$ value with a threshold T_B. When $B^t(\boldsymbol{p})$ is smaller than T_B, we regard the pixel $\boldsymbol{p}$ as background.

Detailed procedure of background modeling is summarized as follows:

Step1. The incremental codes of the current frame t are calculated, and foreground pixels are discriminated from background pixels according to the similarity (defined by Eq.(13.5)) of the codes of the input image pixels and those of the background pixels.

Step2. The incremental codes of the background pixels are updated according to Eq.(13.4).

Step3. When $b_k^t(\boldsymbol{p})$ goes into $T_r \leq b_k^t(\boldsymbol{p}) \leq 1 - T_r$, its peripheral pixel is sought again in the current frame, and $b_k^t(\boldsymbol{p})$ is re-initialized using the newly found peripheral pixel.

13.2.3   Extrapolation-based Background Model

Extrapolation-based background model proposed by Fukui *et al.* [2], which is based on brightness normalization of an original background image, is designed to be robust against sudden illumination changes. They assume that the illumination changes occur uniformly in the entire image. Therefore, in principle, their method cannot detect objects robustly under non-uniform illumination changes. In addition, since the original background image must be prepared in advance, it cannot handle unexpected background changes as well.

In our method, on the contrary, the background image is adaptively generated referring to recent pixel values, which can deal with background changes. Then, the brightness of the generated background image is normalized based on on-line training, and the influence of non-uniform illumination changes is reduced. Finally, objects can be simply detected by subtracting the normalized background image from an observed image. To realize the robust brightness normalization, we have designed a multi-layered perceptron, by which the brightness mapping between the background image and an observed image is established. To reflect the locality of the brightness distribution, the input vector of the perceptron consists of the coordinates (x, y) and the pixel value (R, G, B) of a pixel, and this combination can handle the non-uniform illumination changes. Detailed procedure of the extrapolation-based background modeling is as follows:

Step 1 (Learning): the mapping between an input vector (x, y, R, G, B) of a pixel in the background image and an output vector, (R', G', B') of the pixel at the same position in the observed image is learned. To achieve on-line training, we also have to acquire training data on-line, which is achieved in the integration process of the three background modelings. Details will be presented in the next section.

Step 2 (Normalization): the brightness of the background image is normalized using the perceptron learned in Step1, which means that the background image corresponding to the observed image is estimated.

Step 3 (Object detection): subtraction of the normalized background image, which is estimated in Step2, from the observed image gives us the object detection result. That is, pixels whose pixel values differ more than a given threshold T_{det} from the normalized pixel values are detected as foreground.

13.2.4   Combinational Background Model

Each background model presented above is effective for dynamic, illumination and non-uniform changes, respectively. Therefore, by combining them adaptively, our combinational background model can detect foreground objects robustly against various background changes. Here, we present our major contribution, i.e., combinational background model based on a spatio-temporal feature. The flow of the combination process is shown in Figure 13.2 and the details are as follows.

Step1. Foreground objects are detected based on the **statistical** background model.

Step2. Foreground objects are detected based on the **local feature-based** background model.

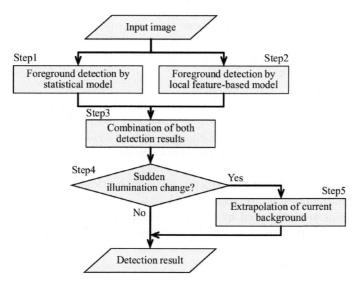

FIGURE 13.2 Processing flow of combinational background modeling.

Step3. Detection results of Step1 and Step2 are combined. That is, pixels which are judged as foregrounds by both of the above models are judged as foreground objects, and other pixels are judged as the background. Then, the parameters of the background models are updated. First, the PDF of the pixel value of the input images, which is maintained in the statistical background model, is updated. In addition, for the pixels judged as the background, the parameters of local feature-based background model are updated.

Step4. In the current frame, more than certain number of pixels whose pixel values differ considerably from the previous frame, then we establish TTL (Time To Live) to the extrapolation-based background model. Here, TTL represents the duration where the extrapolation-based model is activated. By using TTL, we activate the extrapolation-based model only when the illumination condition suddenly changes.

Step5. If TTL > 0, foreground objects are finally detected based on the **extrapolation-based** background model and TTL is decreased. Otherwise, object detection result acquired in Step3 is regarded as the final result. The extrapolation-based background model is achieved as follows:

Substep5-1. A background image is generated so that each pixel has the most frequent pixel value in its PDF, which is maintained in the statistical background model.

Substep5-2. Training samples for brightness mapping are selected from pixels judged as background in Step3. This is because, we have found experimentally that the object detection result of Step3 has little false negatives, and pixels judged as background (called as background candidate pixels (BCPs)) have little misidentification. Therefore, at BCPs, the correspondences of pixel values of the generated background image and those of the observed image can become adequate training samples for background normalization. In practice, at each frame, a few percent of BCPs are randomly sampled and used as training samples.

Substep5-3. The background image is normalized by using the multi-layered perceptron trained in Step5-2 at each frame. Finally, the subtraction of the normalized background image from the observed image becomes the final object detection result.

Note that the extrapolation-based background model is used only when the illumination condition suddenly changes. This is because statistical and local feature-based background models can adapt to other background changes such as dynamic and gradual illumination changes. Therefore, we selectively use the extrapolation-based background model from the viewpoint of computational cost.

By combining three different components, our combinational background model can benefit from a spatio-temporal information and adapt to various background changes. However, when some regions are influenced by dynamic and illumination changes at the same time, these regions cause a problem as we will discuss in Section 13.5.2. That is, the combinational model also has a difficulty in handling these regions, since each of their components falsely detects these regions as foreground.

13.3   Background Model Based on Statistical Local Feature

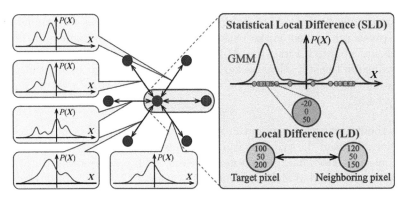

FIGURE 13.3   (See color insert.) Background model based on Statistical Local Difference Pattern: Local Difference (LD) is a local feature, and is defined by the difference between a target pixel and a neighboring pixel. LD is modeled using a GMM to represent its distribution, making it a statistical local feature called the Statistical Local Difference (SLD). Then, Statistical Local Difference Pattern (SLDP) is defined using several SLDs for the background model (this figure shows an example with six SLDs).

In Section 13.2, we combined the detection results of the multiple different background models in order to solve the following problems: statistical approaches cannot handle illumination changes, local feature-based approaches cannot deal with multi-modal backgrounds caused by dynamic changes. However, the combinational model cannot handle the regions which are influenced by dynamic and illumination changes at the same time.

To address this problem, instead of combining the results of multiple different background models, we have integrated their methodologies into a single framework so that a single background model can handle various background changes [16]. Hence, in this section, we present an integrated background model based on a spatio-temporal feature by applying a statistical framework to a local feature-based approach. In practice, as shown in Figure 13.3, we apply a Gaussian mixture model (GMM) to a local feature called the *Local Difference* (LD) to get a statistical local feature called the *Statistical Local Difference* (SLD). Finally, we define *Statistical Local Difference Pattern* (SLDP) for the background

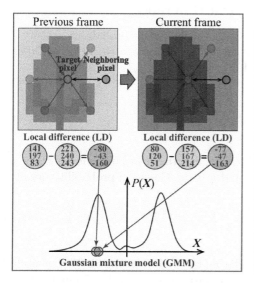

(a) Adaptivity to illumination changes

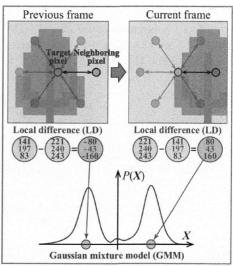

(b) Adaptivity to dynamic changes

FIGURE 13.4   (See color insert.) Adaptivities of our method to background fluctuation : (a) shows the case of illumination changing suddenly. SLDP can adapt to illumination changes since LD has the ability to tolerate the effects of illumination changes which affect the target pixel value in proportion with others. (b) shows the case where a background texture changes repeatedly. SLDP can adapt to this kind of dynamic background changes, since GMMs can learn the variety of background hypotheses.

model by using several SLDs. Figure 13.4 shows the advantages of using SLDP. In most cases where the illumination condition changes, the values of pixels in a localized region increase or decrease proportionally, and therefore there are small changes in the difference between a target pixel and its neighboring pixel. Due to the invariance of the difference value with respect to illumination changes, SLDP has the ability to tolerate the changes as shown in Figure 13.4(a), since it uses the difference value (LD) as a local feature. Furthermore, SLDP can also cope with dynamic changes, since it can learn the multiple different appearances of the background caused by dynamic changes as shown in Figure 13.4(b). This is because a GMM, which can handle a multi-modal distribution, is applied to LD which is an important component of SLDP. Thus, SLDP can integrate the concepts of both statistical and local feature-based approaches into a single framework.

13.3.1   Construction of Local Difference

A target pixel and its neighboring pixel in an observed image are described by the vectors $\boldsymbol{p}_c = (x_c, y_c)^T$ and $\boldsymbol{p}_j = (x_j, y_j)^T$ respectively. We can then represent a local feature $\boldsymbol{X}_j$, called the *Local Difference* (LD), by using the difference between the target and its neighboring pixel as $\boldsymbol{X}_j = f(\boldsymbol{p}_c) - f(\boldsymbol{p}_j)$, where $f(\boldsymbol{p})$ is the pixel value at pixel $\boldsymbol{p}$.

In cases where illumination changes occur, the changes in the LD are smaller than the pixel values, since the pixels in the localized region show a similar change. Therefore, the value of LD is stable under the illumination changes as shown in Figure 13.4(a).

13.3.2   Construction of Statistical Local Difference

We apply a Gaussian mixture model (GMM) to LD to represent probability density functions (PDF) for LD. This gives a statistical local feature called *Statistical Local Difference* (SLD). We define the SLD $P(\boldsymbol{X}_j^t)$ (PDF for LD) at time t by:

$$P(\boldsymbol{X}_j^t) = \sum_{m=1}^{M} w_{j,m}^t \eta(\boldsymbol{X}_j^t | \boldsymbol{\mu}_{j,m}^t, \boldsymbol{\Sigma}_{j,m}^t), \tag{13.6}$$

where $w_{j,m}^t$, $\boldsymbol{\mu}_{j,m}^t$ and $\boldsymbol{\Sigma}_{j,m}^t$ are the weight, the mean and the covariance matrix of the m-th Gaussian in the GMM at time t respectively, and η is the Gaussian probability density:

$$\eta(\boldsymbol{X}_j^t | \boldsymbol{\mu}_j^t, \boldsymbol{\Sigma}_j^t) = \frac{1}{(2\pi)^{\frac{d}{2}} |\boldsymbol{\Sigma}|^{\frac{1}{2}}} \exp\left(-\frac{1}{2}(\boldsymbol{X}_j^t - \boldsymbol{\mu}_j^t)^T \boldsymbol{\Sigma}^{-1}(\boldsymbol{X}_j^t - \boldsymbol{\mu}_j^t)\right). \tag{13.7}$$

We construct the background model by updating the GMM (in the SLD). The updating method for the GMM is based on the statistical approach proposed by Shimada *et al* [10]. This method allows automatic changes of M (the number of Gaussian distributions) in response to background changes. That is, M increases when the background has many hypotheses because of motion changes, for example. On the other hand, when pixel values are constant for a while, some Gaussian distributions are eliminated or integrated, and M consequently decreases.

13.3.3   Object Detection Using Statistical Local Difference Pattern

In our method, each pixel has a pattern of SLD in the background model, and we call it *Statistical Local Difference Pattern* (SLDP) [16]. The SLDP at time t is defined as $\boldsymbol{S}^t = \{P(\boldsymbol{X}_1^t), \dots, P(\boldsymbol{X}_j^t), \dots, P(\boldsymbol{X}_N^t)\}$ by using a target pixel $\boldsymbol{p}_c$ and N neighboring pixels $\boldsymbol{p}_j$ which radiate out from $\boldsymbol{p}_c$. Here, N represents the number of neighboring pixels (Figure 13.3 and 13.4 show an example for $N = 6$), and each neighboring pixel is defined as follows:

$$\boldsymbol{p}_j = \boldsymbol{p}_c + r\boldsymbol{a}_j, \tag{13.8}$$

$$\boldsymbol{a}_j = \left(\cos\frac{j-1}{N}2\pi, \sin\frac{j-1}{N}2\pi\right)^T, \tag{13.9}$$

where r is a radial distance from the target pixel $\boldsymbol{p}_c$, and all of the neighboring pixels lie on a circle of radius r centered at a target pixel $\boldsymbol{p}_c$.

Foreground detection using SLDP uses a voting method to judge whether a target pixel $\boldsymbol{p}_c$ belongs to the background or the foreground. When the pattern of N LDs is given as $\boldsymbol{D}^t = \{\boldsymbol{X}_1^t, \dots, \boldsymbol{X}_j^t, \dots, \boldsymbol{X}_N^t\}$, foreground detection based on SLDP is decided according to:

$$\Phi(\boldsymbol{p}_c) = \begin{cases} \text{background} & \text{if} \quad \sum_j \phi(\boldsymbol{D}_j^t, \boldsymbol{S}_j^t) \geq T_D, \\ \text{foreground} & \text{otherwise}, \end{cases} \tag{13.10}$$

where T_D is a threshold for determining whether a target pixel $\boldsymbol{p}_c$ belongs to the background or the foreground. In Eq.(13.10), $\phi(\boldsymbol{D}_j^t, \boldsymbol{S}_j^t)$ is a function which returns 0 or 1, depending on whether or not the LD $\boldsymbol{X}_j^t$ matches the SLD $P(\boldsymbol{X}_j^t)$ at time t. For further details, we refer the readers to the literature [10].

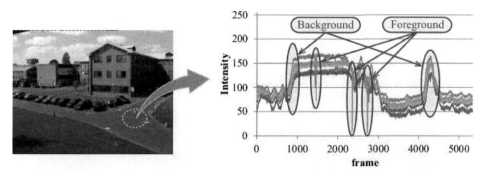

FIGURE 13.5 Similarity in the intensity changes among pixels: The pixels which are geographically and chromatically close to each other show similar intensity changes against illumination changes.

13.4   Background Model Considering Intensity Change Similarly among Pixels

As presented in Section 13.2 and 13.3, the combinational and the integrated (SLDP) background models employ a statistical framework to adapt to dynamic changes in the background. However, a statistical framework is not enough to handle heavy dynamic changes, because the heavy dynamic changes often give us a complicated distribution of pixel values, which is far from typical multi-modality. Even in such cases a target pixel is likely to observe similar changes occurring around it, and therefore, spatial information can be exploited to handle those dynamic changes. While the combinational and SLDP models use the spatial information only to define their local features to adapt to illumination changes, here, we use the spatial characteristic to handle dynamic changes more robustly in addition to illumination changes. In this section, we propose another integrated background model based on a spatio-temporal information, where the spatiality is introduced into a statistical approach. In other words, region-level statistical information is introduced, instead of pixel-level statistical information, for the integrated background model.

Previous statistical approaches [1, 8, 10–12] model the background using a probability distribution of the previously observed pixel values of each pixel (i.e. at the pixel-level). For this reason, their pixel-level statistical background models cannot adapt to illumination changes which are not observed in the previous frames, and they have difficulty handling heavy dynamic changes which give complicated distributions of pixel values far from typical multi-modality. To solve these problems, we use the spatial information, and in particular, by considering pixel clusters where all the pixels are close to one another and are similar in intensity change as shown in Figure 13.5. Here, background changes, i.e., dynamic changes and illumination changes, are handled as follows:

Dynamic changes: Each of such clusters usually consist of pixels which have a high probability to observing the same characteristic of a heavy dynamic change, since the characteristic of each dynamic change is associated with the place (e.g., in cases where trees shiver in the wind, pixels around the tree are likely to observe tree leaves). Then, by modeling the observation probability of features of the pixel, such as pixel value, within each cluster, each pixel can estimate the effects of heavy dynamic changes. Therefore, each pixel can adapt to heavy dynamic changes more robustly. This is why spatial information is effective for adapting to heavy dynamic changes robustly.

Illumination changes: By referring to the pixel value distributions of such clusters, we can easily identify whether each pixel of the same cluster belongs to foreground objects

(a) Background image   (b) Visualization of pixel clustering

FIGURE 13.6 Example of a result of the pixel clustering using K-means algorithm ($K = 100$): The clustering result at least shows that each cluster contains the pixels which belong to the same part of the scene context (e.g., grass, road, sky, walls of the buildings and so on).

or background (illumination changes) as shown in Figure 13.5. Since illumination changes affect not only the target pixel but also its neighboring pixels, all the pixels belonging to the same cluster observe similar intensity changes. Therefore, by defining an illumination-invariant feature (e.g., difference among the pixels), the feature is not influenced by illumination changes similarity, even if illumination changes vary the pixel values of each pixel significantly. This is why spatial information is effective for adapting to illumination changes robustly.

In Section 13.4.1, to use these spatial characteristics described above, we classify the pixels into pixel clusters where all the pixels are close to one another and are similar in intensity change. Then, we explain how to introduce the spatial characteristics into a statistical background model in Section 13.4.2.

13.4.1   Pixel Clustering Based on Intensity Change Similarity

By classifying the pixels into several clusters based on the similarity in intensity change and coordinate as shown in Figure 13.5, we can easily identify whether each pixel belongs to background or foreground. Furthermore, by referring the changes observed at pixels belonging to the same cluster, we can detect foreground objects more robustly. In terms of the similarity in the intensity changes, it is known that scene points having similar surface normals show similar intensity changes against a smoothly moving distant light source. Koppal *et al.* divide a complex scene into geometrically consistent clusters (scene points that have the same or very similar surface normals) irrespective of their material properties and lighting [5]. The results of the literature [5] at least show that the pixels which are positionally and chromatically close to each other belong to the same cluster. Therefore, we assume that the pixels which are geographically and chromatically close to one another show similar changes against illumination changes. Under this assumption, we classify pixels into several clusters, in each of which all the pixels have similar coordinates and show similar changes against illumination changes. We employ K-means clustering to acquire the clusters.

For K-means algorithm, we use a chromatic and positional feature. We assume that an initial background image without any foreground objects is given. The chromatic and positional feature is defined as the following quintet $\boldsymbol{F} = (Nx, Ny, NY, NU, NV)$, where Nx and Ny are the image coordinate (x, y) normalized by the image size, and NY, NU and NV are the normalized pixel values in YUV color space. Based on this feature, we classify the pixels into K clusters $\boldsymbol{C} = \{\boldsymbol{c}_1, \ldots, \boldsymbol{c}_k, \ldots, \boldsymbol{c}_K\}$ as shown in Figure 13.6. Figure 13.6 shows that each cluster contains the pixels which belong to the same part of the scene context (e.g., grass, road, sky, walls of the buildings and so on). Then, according to the literature [5], we regard the pixels belonging to the same cluster show similar changes against illumination changes.

13.4.2   Spatio-temporal Similarly of Intensity Change

We consider a target pixel at (x, y) belonging to a certain cluster $\boldsymbol{c}_k$, and define an illumination-invariant feature $\boldsymbol{X}^t = (D^t, U^t, V^t)$ in YUV color space. Here, D^t is the difference defined as $D^t = Y^t - Y_r^t$, where Y^t and Y_r^t are the target pixel intensity and the representative intensity of its cluster at time t respectively. We use a median intensity of the cluster as its representative intensity Y_r^t. Then, we can estimate the probability density function (PDF) $P(\boldsymbol{X})$ at time t by kernel density estimation (KDE) using past S samples. To rapidly and accurately estimate the PDF, we have used a rectangular function for the kernel function W as is the case in Section 13.2.

$$P^t(\boldsymbol{X}) = \frac{1}{S} \sum_{i=1}^{S} W\left(|\boldsymbol{X} - \boldsymbol{X}^{t-i}|\right). \tag{13.11}$$

By accumulating the PDFs of the pixels belonging to the cluster $\boldsymbol{c}_k$, we also estimate the PDF of the cluster as follows:

$$P_{\boldsymbol{c}_k}^t(\boldsymbol{X}) = \frac{1}{S|\boldsymbol{c}_k|} \sum_{(x,y) \in \boldsymbol{c}_k} \sum_{i=1}^{S} W\left(|\boldsymbol{X} - \boldsymbol{X}_{(x,y)}^{t-i}|\right), \tag{13.12}$$

where $|\boldsymbol{c}_k|$ is the number of pixels belonging to the cluster $\boldsymbol{c}_k$. By using these two PDFs (i.e., $P^t(\boldsymbol{X})$ and $P_{\boldsymbol{c}_k}^t(\boldsymbol{X})$), we can construct the background model based on "*Spatio-temporal Similarly of Intensity Change* (StSIC)." The updating method for the PDFs is based on a fast computation method described in Section 13.2.1. Foreground detection using StSIC uses a thresholding method as follows:

$$\Phi(\boldsymbol{X}^t) = \begin{cases} \text{background} & \text{if } P^t(\boldsymbol{X}^t) \geq T_S \text{ or } P_{\boldsymbol{c}_k}^t(\boldsymbol{X}^t) \geq T_S, \\ \text{foreground} & \text{otherwise}, \end{cases} \tag{13.13}$$

where T_S is a threshold for determining whether the target pixel (x, y) belongs to the background or the foreground.

By modeling the PDF of each cluster, in addition to modeling the PDF of each pixel, StSIC can adapt to heavy dynamic changes more robustly compared to previous statistical approaches which only use the PDFs of each pixel. This is because, by using the PDFs of the clusters, each pixel can refer to the features observed at the pixels having the same characteristics of dynamic changes, and then each pixel can be classified either background or foreground correctly even in heavy dynamic changes. This is the reason why StSIC is more robust against heavy dynamic changes compared to previous statistical approaches. Furthermore, StSIC can tolerate the effects of illumination changes, since it uses an illumination-invariant feature defined by difference between the target and its cluster representative. Each cluster is designed to contain the pixels which show similar changes against illumination changes, then the difference features are not influenced by illumination changes significantly. Therefore, StSIC is also robust against illumination changes.

13.5   Evaluation

To adapt to various background changes such as illumination changes, dynamic changes, etc., we have presented three different spatio-temporal background models by combining or integrating the concepts of background models which have different characteristics. Each of our spatio-temporal background models has different characteristic depending on the combining/integrating methodology. Hence, in this section, we evaluate the effectiveness our spatio-temporal background models and discuss each of their characteristics. We have used a database provided for the Background Models Challenge (BMC) for performance evaluation. BMC database provides human annotated ground truth for all videos, and therefore it is suitable for exhaustive competitive comparison of background models.

13.5.1   Experimental Condition

BMC database consists of 10 synthetic videos for "learning" and 10 synthetic and 9 real videos for "evaluation." When comparing the performance of background models, their parameters are tuned based on 10 synthetic videos for learning. Then, for 10 synthetic and 9 real videos for evaluation, a unique set of parameters is used for object detection. Since each of our spatio-temporal background models has several parameters, here lists their parameter settings as follows.

Combinational background model (CMB):

  For statistical model based on KDE, the width of the rectangular kernel is $h = 9$ and the number of samples is $S = 250$ for the details.

  For local feature-based model based on Adaptive RRC, the thresholds for selecting peripheral pixels is $T_P = 10$, object detection threshold is $T_B = 1.5$, threshold for re-searching peripheral pixels is $T_r = 0.1$, and learning rate of incremental codes is $\alpha = 0.05$.

  Extrapolation-based model is used for a certain period ($TTL = 20$) when the number of pixels whose pixel values differ more than 10 from the previous frame exceeds the half number of total pixels. The detection threshold is $T_{det} = 30$.

Integrated background based on SLDP:

  The radial distance is $r = 20$, the number of neighboring pixels is $N = 6$ and the detection threshold for SLDP is $T_D = 5$. Although the details of GMM are not explained in Section 13.3, we also indicate the parameter settings in GMM for reproducibility: the learning rate is $\alpha = 0.01$, the initial weight is $W = 0.05$ and the threshold of choosing the background model $T = 0.7$. For the details of GMM, we refer the readers to the paper [10].

Integrated background based on StSIC:

  the number of pixels clusters is $K = \dfrac{\text{image size}}{35 \times 35}$ (e.g., if image size $= 320 \times 240$, then $K = 62$), the number of past samples is $S = 250$ and the width of rectangular kernel is $h = 9$.

13.5.2   Performance Evaluation Using BMC Database

To evaluate the effectiveness of spatio-temporal background modeling approaches, we have compared the accuracy of foreground detection of ours with conventional approaches: statistical and local feature-based approaches. For conventional approaches, we have employed a

TABLE 13.1   Evaluation results using 5 synthetic videos of "Street" for evaluation

Method	Measure	112(no noise)	212(with noise)	312(sunny)	412(foggy)	512(windy)
GMM [10]	Recall	0.927	0.927	0.897	0.861	0.921
	Precision	0.866	0.868	0.580	0.526	0.619
	F-measure	**0.896**	0.896	0.705	0.653	0.740
Adaptive RRC	Recall	0.843	0.889	0.866	0.848	0.878
	Precision	0.840	0.760	0.745	0.726	0.560
	F-measure	0.841	0.820	0.801	0.782	0.684
Ours (CMB)	Recall	0.928	0.927	0.905	0.832	0.920
	Precision	0.857	0.878	0.875	0.855	0.660
	F-measure	0.891	**0.902**	**0.890**	0.844	0.769
Ours (SLDP)	Recall	0.857	0.857	0.827	0.822	0.852
	Precision	0.883	0.894	0.876	0.773	0.643
	F-measure	0.870	0.875	0.851	0.797	0.733
Ours (StSIC)	Recall	0.922	0.892	0.863	0.832	0.886
	Precision	0.868	0.907	0.891	0.921	0.805
	F-measure	0.894	0.899	0.877	**0.874**	**0.843**

statistical background model using Gaussian Mixture Model (GMM [10]) and a local feature-based background model using Adaptive Radial Reach Correlation (Adaptive RRC), which is a revised version of RRC [9] as described in Section 13.2.2.

For performance evaluation, we have used three static quality metrics [14]: Recall, Precision and F-measure. For a given frame i, the numbers of true and false positive pixels are denoted by TP_i and FP_i, and the true and false negative ones are denoted by TN_i and FN_i. Then, the metrics are calculated as follows:

$$\text{Recall} = \frac{1}{n}\sum_{i=1}^{n} \text{Rec}_i = \frac{1}{n}\sum_{i=1}^{n} \frac{\text{Rec}_i(P) + \text{Rec}_i(N)}{2}, \qquad (13.14)$$

$$\text{Precision} = \frac{1}{n}\sum_{i=1}^{n} \text{Prec}_i = \frac{1}{n}\sum_{i=1}^{n} \frac{\text{Prec}_i(P) + \text{Prec}_i(N)}{2}, \qquad (13.15)$$

$$\text{F-measure} = \frac{1}{n}\sum_{i=1}^{n} \frac{2 \times \text{Rec}_i \times \text{Prec}_i}{\text{Rec}_i + \text{Prec}_i}, \qquad (13.16)$$

where $\text{Rec}_i(P)$, $\text{Rec}_i(N)$, $\text{Prec}_i(P)$ and $\text{Prec}_i(N)$ are $\text{Rec}_i(P) = TP_i/(TP_i + FN_i)$, $\text{Rec}_i(N) = TN_i/(TN_i + FP_i)$, $\text{Prec}_i(P) = TP_i/(TP_i + FP_i)$ and $\text{Prec}_i(N) = TN_i/(TN_i + FN_i)$, respectively.

Performance Evaluation Using Synthetic Videos of BMC database

We discuss the characteristics of each background modeling approach based on evaluation results using 10 synthetic videos, each scene of which corresponds to one of following five event types:

1. cloudy scene without acquisition noise,

2. cloudy scene with noise,

3. sunny scene with noise, where illumination condition changes,

4. foggy scene with noise, where the fog is coming and after a while it is cleared,

5. windy scene with noise, where dynamic changes such as swaying motion of tree branches are observed.

TABLE 13.2   Evaluation results using 5 synthetic videos of "Rotary" for evaluation

Method	Measure	122(no noise)	222(with noise)	322(sunny)	422(foggy)	522(windy)
GMM [10]	Recall	0.923	0.931	0.897	0.843	0.934
	Precision	0.886	0.890	0.626	0.535	0.840
	F-measure	0.904	0.910	0.738	0.655	0.884
Adaptive RRC	Recall	0.856	0.897	0.853	0.836	0.894
	Precision	0.867	0.836	0.830	0.756	0.726
	F-measure	0.861	0.865	0.841	0.794	0.801
Ours (CMB)	Recall	0.921	0.921	0.892	0.845	0.923
	Precision	0.878	0.902	0.896	0.840	0.874
	F-measure	**0.907**	0.912	**0.894**	0.842	0.898
Ours (SLDP)	Recall	0.915	0.920	0.885	0.854	0.924
	Precision	0.894	0.906	0.888	0.794	0.870
	F-measure	0.904	**0.913**	0.886	0.823	0.896
Ours (StSIC)	Recall	0.905	0.888	0.855	0.818	0.899
	Precision	0.907	0.932	0.922	0.946	0.921
	F-measure	0.903	0.910	0.887	**0.877**	**0.910**

Each video is numbered according to presented event type (from 1 to 5 described above), the scene number (1:Street or 2:Rotary) and the use-case (1: for "learning" or 2: for "evaluation"). For example, the video 312 corresponds to a sunny street for evaluation phase. Table 13.1 and Table 13.2 show evaluation results using 5 synthetic videos of "Street" and "Rotary" respectively. As we can see from Table 13.1 and Table 13.2, in cases of event type $= 1, 2$, every approach can achieve high accuracy since the scenes include no significant background changes. On the other hand, in cases where the background changes significantly (event type $= 3, 4, 5$), our spatio-temporal background models can adapt to the changes robustly, but either of the conventional models cannot. To demonstrate the effectiveness of our approaches, we show some examples of foreground detection results using the videos whose event type is from 3 to 5, in Figure 13.7:

- The first and the second rows correspond to a frame of video 312 and 322 respectively, where illumination condition is changing rapidly.

- The third and the fourth rows correspond to a frame of video 412 and 422 respectively, where the fog is collecting rapidly.

- The last two rows correspond to a frame of video 512 and 522 respectively, tree branches are swaying in the wind.

Figure 13.7 (from the first to the fourth rows) shows that GMM method [10] detects a lot of false positive pixels influenced by illumination changes and the fog, which cause changes of pixel values which are not observed in the previous frames. The reason why GMM method cannot adapt to illumination changes and the fog is because its background model refers only to the previous information. Meanwhile, by using the spatiality or extrapolating the current background, the other background models can adapt to illumination changes and the effect of the fog more robustly compared to GMM. This is also shown in Table 13.1 and Table 13.2 (videos whose event type is 3 and 4), where the precision values of Adaptive RRC, Combinational (CMB), SLDP and StSIC models are much higher than that of GMM. These are typical evidence of the effectiveness of local feature-based and extrapolation-based approaches to adapt to illumination changes and fogs. Figure 13.7, Table 13.1 and Table 13.2 also show, in the videos 412 and 422, StSIC model can adapt to the fog more accurately compared to Adaptive RRC, CMB and SLDP models. The fogs also cause sudden intensity changes as well as illumination changes, but they influence the scenes more non-uniformity compared to illumination changes, since the densities of the fogs depend on the place in the background (e.g., road, sky, walls of the buildings, etc.). This is the reason why Adaptive RRC, CMB and SLDP models, whose local features are independent of the scene context of

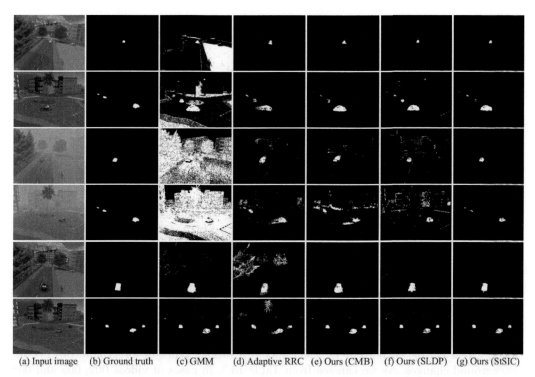

(a) Input image　(b) Ground truth　(c) GMM　(d) Adaptive RRC　(e) Ours (CMB)　(f) Ours (SLDP)　(g) Ours (StSIC)

FIGURE 13.7 Examples of foreground detection results using synthetic videos: The first and second rows are sunny scenes, the third and fourth rows are foggy scenes, and the last two rows are windy scenes.

the background, become less effective for the fogs. Even in cases of an extrapolation-based approach used in CMB model, it is difficult to extrapolate the current background when the changes have a high degree of non-uniformity. In contrast, by using pixel clustering which enables to classify the pixels based on the similarity in intensity change and pixel coordinate, StSIC model can consider the scene context of the background and then StSIC can reduce the effects of non-uniform background changes.

Figure 13.7 (the last two rows) also shows that Adaptive RRC falsely detects the movement of tree leaves, and this is also shown in Table 13.1 (video 512) and Table 13.2 (video 522), where the precision values of Adaptive RRC is much lower than that of the others. This is because Adaptive RRC is easily disturbed by texture changes of the background caused by dynamic changes. On the other hand, since dynamic changes are repeatedly observed, the other method can adapt to dynamic changes by using a statistical framework, which enables to refer the previous information. This proves the effectiveness of the statistical approaches against dynamic changes in the background.

As discussed above, our spatio-temporal background models can adapt to various background changes more robustly compared to conventional ones. These results are typical evidence of the effectiveness of our spatio-temporal background modeling approaches against various background changes. The reason why our background models can detect foreground objects robustly against various background changes is because they can use the multiple different characteristics of multiple different background modeling approaches.

Performance Evaluation Using Real Videos of BMC database

As discussed above, by using synthetic videos, we have confirmed our background models can adapt to various background changes if the scene includes only one type of background

TABLE 13.3   Evaluation results using 9 real videos for evaluation

Method	Measure	Real Applications								
		001	002	003	004	005	006	007	008	009
GMM [10]	Recall	0.949	0.680	0.959	0.929	0.854	0.880	0.791	0.823	0.928
	Precision	0.782	0.646	0.880	0.680	0.535	0.736	0.703	0.595	0.890
	F-measure	0.857	0.662	0.918	0.785	0.658	0.802	0.744	0.691	0.909
Adaptive RRC	Recall	0.849	0.819	0.870	0.894	0.835	0.832	0.722	0.764	0.756
	Precision	0.824	0.889	0.820	0.812	0.657	0.794	0.823	0.609	0.914
	F-measure	0.837	**0.853**	0.844	0.851	0.735	0.813	0.769	0.678	0.828
Ours (CMB)	Recall	0.933	0.799	0.941	0.944	0.871	0.902	0.757	0.827	0.903
	Precision	0.828	0.857	0.867	0.914	0.663	0.832	0.845	0.711	0.921
	F-measure	0.878	0.827	0.903	**0.929**	**0.753**	**0.866**	**0.799**	0.765	0.912
Ours (SLDP)	Recall	0.926	0.671	0.954	0.916	0.823	0.856	0.790	0.824	0.909
	Precision	0.818	0.862	0.913	0.891	0.597	0.825	0.780	0.829	0.920
	F-measure	0.869	0.754	**0.933**	0.904	0.692	0.841	0.785	0.827	0.914
Ours (StSIC)	Recall	0.927	0.689	0.915	0.911	0.824	0.814	0.688	0.875	0.933
	Precision	0.847	0.879	0.905	0.893	0.601	0.871	0.802	0.801	0.901
	F-measure	**0.885**	0.773	0.910	0.902	0.695	0.842	0.741	**0.836**	**0.917**

changes under ideal circumstances. In the real scenes, sometimes more than two types of background changes are observed at the same time, e.g., illumination changes and dynamic changes. Moreover, in the real scenes, various types of foreground objects (e.g., big objects, objects whose color is similar to the background and so on) are observed, and some of them are hard to detect. Here, as shown in Table 13.3, we evaluate each background model using 9 real videos, each scene of which includes various types of foreground objects and background changes. As we can see from Table 13.3, in cases of videos 002, 005 and 008, there are differences in the accuracy among the combinational (CMB) and the integrated (SLDP and StSIC) models. For each of these videos, we show the examples of success and failure in foreground detection in Figure 13.8.

In case of video 002, in the uniform background with no texture, two types of large objects are observed: the one with non-uniform color and the other with uniform color. Our background models can detect the first one robustly against background changes as shown in the first row of Figure 13.8. However, it is difficult for SLDP and StSIC models to detect the second one correctly (see the second row of Figure 13.8), and Table 13.3 shows that the recall values of them are lower than that of CMB model. The is because they detect foreground objects based on a relationship between a target and its neighboring pixels, and then the relationship dose not vary before and after the second type of object appears. Therefore, SLDP and StSIC models confuse the second type of objects with background changes sometimes.

The video 005 is a scene where the snow is falling in the night and sometimes a light is projected. In usual case, each background model can detect foreground objects correctly as in the third row of Figure 13.8. On the other hand, when the projection of the light is observed, SLDP and StSIC models detect more false positive pixels (i.e., snow shines by reflecting the light) compared to Adaptive RRC and CMB model. As a result, the precision values of SLDP and StSIC models are lower than those of Adaptive RRC and CMB models in Table 13.3. The reason why Adaptive RRC and CMB models are more robust is because they use a binary codes which are not influenced significantly in this kind of low contrast video. These kind of low contrast videos are an exception, where the binary codes work better than SLDP and StSIC.

In case of video 008, both illumination and dynamic changes are observed, and sometimes some regions are influenced by both changes. When illumination and dynamic changes affect different regions, then our spatio-temporal background models can detect foreground object robustly against background changes as shown in the fifth row of Figure 13.8. However, when illumination and dynamic changes affect the same region at the same time, then CMB model

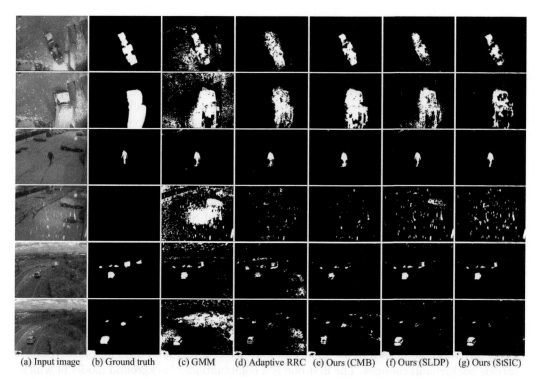

(a) Input image   (b) Ground truth   (c) GMM   (d) Adaptive RRC   (e) Ours (CMB)   (f) Ours (SLDP)   (g) Ours (StSIC)

FIGURE 13.8 Examples of foreground detection results using real videos: The first and second rows correspond to video 002, the third and fourth rows correspond to video 005, and the last two rows correspond to video 008.

falsely detect such regions as foreground, and its precision value is lower than those of SLDP and StSIC model in Table 13.3. By combining the results of multiple different background models, CMB enables to handle background changes. Then, as a result, CMB falsely detects regions which are judged as foreground by both of statistical and local feature-based models under the influence of background changes. On the other hand, SLDP and StSIC integrate the methodologies of multiple different background models into a single framework so that a single background model can handle various background changes. This is why if some regions are influenced by dynamic and illumination changes at the same time, SLDP and StSIC can judge these regions as background.

13.5.3 Discussion of the Adaptivity to Background Changes

Each of our spatio-temporal background models has different characteristics depending on its combining/integrating methodology. Therefore, they have advantages and disadvantages as discussed above, and the most appropriate background model is different depending on the scenes. Here, according the videos whose scene context is different, we present how best to choose a background model.

CMB can provide more correct detection results compared to SLDP and StSIC in the following two cases. The first case is "close-up" scenes where big foreground objects are observed, and then both the background and the foreground regions often have uniform colors. Then, in such close-up scenes, SLDP and StSIC sometimes confuse foreground objects with background changes, since they use the relations between a target and its neighboring pixels. The other case is "low contrast" scenes, where SLDP and StSIC which use the magnitude of the difference among pixel values for object detection are more sensitive to background changes compared to CMB which use the magnitude relations among pixel values.

On the other hand, in cases where illumination and dynamic changes affect the same region at the same time, SLDP and StSIC can provide more correct detection results compared to CMB. This is because CMB falsely detects regions which are judged as foreground by both of statistical and local feature-based models under the influence of background changes. Furthermore, we can also choose either SLDP or StSIC according to whether we can have adequate background images without foreground object for pixel clustering, which images are, in other words, initial training data for background model generation. StSIC can consider the scene context of the background by classifying the training image into pixel clusters, and then StSIC can adapt to various background changes more robustly than SLDP. However, an unsuitable training image including foreground objects provides unsuitable pixel clusters, which reduce the accuracy of StSIC. Therefore, in cases where the background image including no foreground objects cannot be available, SLDP can provide more correct detection results compared to StSIC. This is because SLDP uses a fixed localized region controlled by radial distance r instead of considering the scene context.

13.6 Conclusion

In this chapter, we have presented three spatio-temporal background modeling approaches, which have been addressed in our laboratory, to detect foreground objects robustly against various background changes. The first one is a combinational background model in which a statistical, a local feature-based and an extrapolation-based background model are combined adaptively. The second one is an integrated background model based on SLDP where a statistical framework is applied to an illumination-invariant feature. The third one is an integrated background model StSIC realized by considering a similarity of intensity changes among pixels. Each of our background model can use a spatio-temporal feature, they can adapt to various background changes robustly. The combinational model is suitable for "close-up" and "low contrast" scenes, and SLDP and StSIC are suitable for the scenes where illumination and dynamic changes affect the same region at the same time. In terms of the adaptivity to the background changes, StSIC is the most robust because it can use the spatial information not only for illumination changes but also for dynamic changes in the background. As discussed in Section 13.5.2, in a particular scene where each of the background and large foreground objects is uniform color, SLDP and StSIC sometimes cannot detect the insides of the objects. However, these mis-detections can be reduced by post processing such as Graph-cuts.

References

1. A. Elgammal, R. Duraiswami, D. Harwood, and L. Davis. Background and Foreground Modeling using Non-parametric Kernel Density Estimation for Visual Surveillance. *Proceedings of the IEEE*, 90:1151–1163, 2002.
2. S. Fukui, T. Ishikawa, Y. Iwahori, and H. Itoh. Extraction of Moving Objects by Estimating Background Brightness. *Journal of the Institute of Image Electronics Engineers of Japan*, 33:350–357, 2004.
3. M. Heikkila, M. Pietikainen, and J. Heikkila. A texture-based method for detecting moving objects. *British Machine Vision Conference*, 2004.
4. S. Jabri, Z. Duric, and H. Wechsler. Detection and location of people in video images using adaptive fusion of color and edge information. *15th International Conference on Pattern Recognition*, 4:627– 630, 2000.

5. S. Koppal and S. Narasimhan. Appearance Derivatives for Isonormal Clustering of Scenes. *IEEE Transactions on Pattern Analysis and Machine Intelligence*, 31:1375–1385, 2009.

6. H. Marko and P. Matti. A Texture-Based Method for Modeling the Background and Detecting Moving Objects. *IEEE Transactions on Pattern Analysis and Machine Intelligence*, 28:657–662, 2006.

7. M. Mason and Z. Duric. Using histograms to detect and track objects in color video. *In 30th Applied Imagery Pattern Recognition Workshop*, pages 154 – 159, 2001.

8. E. Monari and C. Pasqual. Fusion of Background Estimation Approaches for Motion Detection in Non-static Backgrounds. *International Conference on Advanced Video and Signal Based Surveillance*, 2007.

9. Y. Satoh, S. Kaneko, Y. Niwa, and K. Yamamoto. Robust object detection using a Radial Reach Filter (RRF). *Systems and Computers in Japan*, 35:63–73, 2004.

10. A. Shimada, D. Arita, and R. Taniguchi. Dynamic Control of Adaptive Mixture-of-Gaussians Background Model. *IEEE International Conference on Advanced Video and Signal Based Surveillance*, 2006.

11. C. Stauffer and W. Grimson. Adaptive background mixture models for real-time tracking. *IEEE International Conference on Computer Vision and Pattern Recognition (CVPR)*, 2:246–252, 1999.

12. T. Tanaka, A. Shimada, D. Arita, and R. Taniguchi. A Fast Algorithm for Adaptive Background Model Construction Using Parzen Density Estimation. *IEEE International Conference on Advanced Video and Signal Based Surveillance*, 2007.

13. T. Tanaka, S. Yoshinaga, A. Shimada, R. Taniguchi, T. Yamashita, and D. Arita. Object Detection Based on Combining Multiple Background Modelings. *IPSJ Transactions on Computer Vision and Applications*, 2:156–168, 2010.

14. A. Vacavant, T. Chateau, A. Wilhelm, and L. Lequievrevre. A Benchmark Dataset for Outdoor Foreground/Background Extraction. *1st ACCV Workshop on Background Models Challenge (BMC)*, 2012.

15. K. Yokoi. Probabilistic BPRRC: Robust Change Detection against. Illumination Changes and Background Movements. *IAPR Conference on Machine Vision Applications*, 148-151, 2009.

16. S. Yoshinaga, A. Shimada, H. Nagahara, and R. Taniguchi. Statistical Local Difference Pattern for Background Modeling. *IPSJ Transactions on Computer Vision and Applications*, 3:198–210, 2011.

17. X. Zhao, Y. Satoh, H. Takauji, S. Kaneko, K. Iwata, and R. Ozaki. Object detection based on a robust and accurate statistical multi-point-pair model. *Pattern Recognition*, 44:1296–1311, 2011.

14

Background Modeling and Foreground Detection for Maritime Video Surveillance

Domenico Bloisi
Sapienza University of Rome, Italy

14.1 Introduction

Surveillance systems for the maritime domain are becoming more and more important. The increasing threats coming from illegal smuggling, immigration, illegal fishing, oil spills, and, in some part of the world, piracy make the protection of coastal areas a necessary requirement. Moreover, the control of vessel traffic is often correlated to environment protection issues, since vessels carrying dangerous goods (e.g., oil-tankers) can cause huge environmental disasters.

There exist various surveillance systems for the maritime domain, including Vessel Monitoring Systems (VMS) [35], Automatic Identification System (AIS) [25], ship- and land-based radars [2], air- and space-born Synthetic-Aperture Radar (SAR) systems [47], harbor-based visual surveillance [41], and Vessel Traffic Services (VTS) systems [9]. Data from several information sources generated by multiple heterogeneous sensors are fused in order to provide vessel traffic monitoring. Examples are systems combining AIS data with SAR-imagery [43], buoy-mounted sensors with land radars [21,31], visual- with radar-based surveillance [42], and multiple ship-based sensors [60].

A widely used solution for maritime surveillance consists in combining radar and AIS information [10]. However, radar sensors and AIS data are not sufficient to ensure a complete solution for the vessel traffic monitoring problem, due to the following limitations.

- The AIS signal may not be available for cooperative targets (e.g., AIS device not activated or malfunctioning).

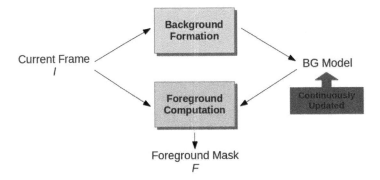

FIGURE 14.1 Background subtraction. The frames in input are used to create a model of the observed scene (BG model) and to detect the foreground pixels (FG mask).

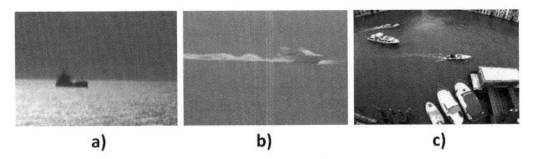

FIGURE 14.2 Challenges in the maritime scenario. a) Sun reflections. b) Boat wakes. c) Waves on the water surface. The images in this figure are taken from the MAR data set [32].

- The recognition task for non-cooperative (non-AIS) targets can be addressed only by analyzing visual features (e.g., color, shape, plates).

- Radar-based systems are not suitable for vessel traffic monitoring in populated areas, due to high electro-magnetic radiation emissions.

Replacing radars with cameras can be a feasible and low-cost solution for addressing the problem of maritime surveillance, without the need of placing radar antennas in populated areas [4].

In this chapter, the use of background subtraction (BS) for detecting moving objects in the maritime domain is discussed. BS is a popular method that aims at identifying the moving regions (the foreground, FG) by comparing the current frame with a model of the scene background (BG) as shown in Fig. 14.1. The possibility of achieving real-time performance made BS appealing for being the initial step for a number of higher-level tasks, such as tracking, object recognition, and detection of abnormal behaviors [14].

Maritime domain is one of the most challenging scenarios for automatic video surveillance due to the complexity of the scene to be observed. Indeed, sun reflections (Fig. 14.2a), boat wakes (Fig. 14.2b), and waves on the water surface (Fig. 14.2c) contribute to generate a highly dynamic background. In addition, weather issues (such as heavy rain or fog), gradual and sudden illumination variations (e.g., clouds), motion changes (e.g., camera jitter due to winds), and modifications of the background geometry (e.g., parked boats) can provoke false detections.

An appropriate BG model has to deal with all the above mentioned issues. In particular, the model has to provide an approximation for a multi-modal probability distribution, that can address the problem of modeling an inherently dynamic and fast changing background. Solutions based on a predefined distribution (e.g., Gaussian) for creating the BG model can prove ineffective, due to the need of modeling non-regular patterns [52].

A method for producing a "discretization" of an unknown distribution that can model the highly dynamic background that is typical of the maritime domain is provided in the following. Furthermore, a publicly available data set of videos and images called MAR - Maritime Activity Recognition data set [32], that contains data coming from real systems and from multiple maritime scenarios is presented. MAR includes video sequences with varying light and weather conditions, together with ground truth information.

The remainder of the chapter is organized as follows. The state-of-the-art is analyzed in the next Section 14.2, while a multi-modal approach for creating a robust BG model, including a noise removal module, designed to filter out false positives generated by shadows, reflections, and boat wakes, is presented in Section 14.3. A quantitative evaluation on publicly available data is reported in Section 14.4. Section 14.5 provides the conclusions.

14.2   State of the Art

BS is generally composed by two stages

1. The background initialization, where the first model of the background is generated.

2. The model updating, where the information in the BG model is updated to take into account the changes in the observed scene.

Since BS is highly dependent on the creation of the BG model to be effective, how to obtain an accurate model is the key issue. A series of variable aspects, depending also on the kind of environment considered, such as illumination changes (gradual and sudden) [36], shadows [15], camera jitter (due to winds) [19], movement of background elements (e.g., trees swaying in the breeze) [57], and changes in the background geometry (e.g., parked cars, moved furniture) [53] make the BG model generation a hard problem.

14.2.1   Classification of BG Modeling Methods

Different classifications of BS methods have been proposed in literature. Cristani *et al.* in [14] organize BS algorithms in four classes

(i) *Per-pixel.* The class of per-pixel approaches (e.g [15, 49]) consider each pixel signal as an independent process. This class of approaches requires the minimum computational effort and can achieve real-time performance.

(ii) *Per-region.* Region-based algorithms (e.g., [27, 33]) usually divide the frames into blocks and calculate block-specific features in order to obtain the foreground. This provides a more robust description of the visual appearance of the observed scene. Indeed, information coming from a set of pixels can be used to model parts of the background scene which are locally oscillating or moving slightly, like leaves or flags. Moreover, considering a region of the image instead of a single pixel allows to compute histograms and to extract edges that can be useful for filtering out false positives.

(iii) *Per-frame.* Frame-level methods look for global changes in the scene (e.g., [37, 51]). Per-frame approaches can be used to deal with sudden illumination changes involving the entire frame, such as switching the lights on and off in an indoor scene or cloud passing by the sun in an outdoor one.

TABLE 14.1 Three possible classifications of background subtraction methods

Cristani et al. [14]	Cheung and Kamath [13]	Mittal and Paragios [33]
Per-pixel		
Per-region	Recursive	Predictive
Per-frame		
Multi-stage	Non-Recursive	Non-Predictive

(iv) *Multi-stage.* Multi-stage methods (e.g., [54,58]) combine the previous approaches in a serial process. Multiple steps are performed at different levels, in order to refine the final result.

Cheung and Kamath in [13] identify two classes of BS methods, namely *recursive* and *non-recursive*.

(i) Recursive algorithms (e.g., [49]) recursively update a single background model based on each new input frame. As a result, input frames from distant past could have an effect on the current background model [19].

(ii) Non-recursive approaches (e.g., [15,37]) maintain a buffer of previous video frames (created using a sliding-window approach) and estimate the background model based on a statistical analysis of these frames.

The simplest, but rather effective, recursive approaches are Adjacent Frame Difference (FD) and Mean-Filter (MF) [54]. As its name suggests, FD operates by subtracting a current pixel value from its previous value, marking it as foreground if absolute difference is greater than a threshold. MF computes an on-line estimate of a Gaussian distribution from a buffer of recent pixel values either in the form of single Gaussians (SGM) [28,61] or mixture of Gaussians (MGM) [18,22,50] or by using different approaches (e.g., median [15] or minimum-maximum values [24]).

A third classification [33] divides existing BS methods in *predictive* and *non-predictive*.

(i) Predictive algorithms (e.g., [17]) model the scene as a time series and develop a dynamical model to recover the current input based on past observations.

(ii) Non-predictive techniques (e.g., [18,49]) neglect the order of the input observations and build a probabilistic representation of the observations at a particular pixel.

14.2.2 Background Modeling on Water Background

While there are many general approaches for background modeling in dynamic environments, only a few of them have been tested on water scenarios.

As suggested by Ablavsky in [1], for dealing with water background it is fundamental to integrate a pixel-wise statistical model with a global model of the movement of the scene (for example, by using optical flow). Indeed, the BG model has to be sensitive enough to detect the moving objects, adapting to long-term lighting and structural changes (e.g., objects entering the scene and becoming stationary). It needs also to rapidly adjust to sudden changes. By combining the statistical model with the optical flow computation it is possible to satisfy simultaneously both the sensitivity to foreground motion and the ability to model sudden background changes (e.g., due to waves and boat wakes).

Monnet *et al.* in [34] present an on-line auto-regressive model to capture and predict the behavior of dynamic scenes. A prediction mechanism is used to determine the frame to be observed by using the k latest observed images. Ocean waves are considered as an example of a scene to be described as dynamic and with non-stationary properties in time.

Spencer and Shah in [48] propose an approach to determine real world scale as well as other factors, including wave height, sea state, and wind speed, from uncalibrated water frames. Fourier transforms of individual frames are used to find the energy at various spatial frequencies, while Principal Component Analysis (PCA) of the whole video sequence followed by another Fourier transformation are used to find the energy at various temporal frequencies. The approach works only for water waves in the open ocean due to some assumptions on wavelengths that can be not valid in a different scenario (e.g., a water channel).

Zhong and Sclaroff [65] describe an algorithm that explicitly models the dynamic, textured background via an Auto-Regressive Moving Average (ARMA) model. A Kalman filter algorithm is used for estimating the intrinsic appearance of the dynamic texture. Unfortunately, this approach is not usable in real time application, since the authors claim a computational speed of 8 seconds per frame.

Related to the detection of reflective surfaces, two approaches have been proposed by Rankin *et al.* in [40] and by He and Hu in [26]. Both solutions are based on the analysis of the HSV color space of the images. Indeed, the single components H (hue), S (saturation), and V (value or brightness) have specific behaviors that allow to detect in which part of the image the reflective surface is.

14.2.3   Open Source BS Algorithms

The possibility of having the source code of the BS methods proposed in literature represents a key point towards the generation of more and more accurate foreground masks and towards a wider application of this technology. Usually, the approach by Stauffer and Grimson [49] is used as a gold-standard, given that implementations of more recent algorithms are not always available [56]. In the following, two libraries containing open source BS methods are presented.

OpenCV
OpenCV library [38] provides the source code for a set of well-known BS methods. In particular, OpenCV library version 2.4.4 includes

  (i) MOG [30]. A Gaussian mixture-based BS algorithm that provides a solution for dealing with some limitations of the original approach by Stauffer and Grimson [49] related to the slow learning rate at the beginning, especially in busy environments.

 (ii) MOG 2 [66]. An improved adaptive Gaussian mixture model similar to the standard Stauffer and Grimson one [49] with additional selection of the number of the Gaussian components. The code is very fast and performs also shadow detection.

(iii) GMG [23]. A BS algorithm for dealing with variable-lighting conditions. A probabilistic segmentation algorithm identifies possible foreground regions by using Bayesian inference with an estimated time-varying background model and an inferred foreground model. The estimates are adaptive to accommodate variable illumination.

BGSLibrary
BGSLibrary [46] is an OpenCV based C++ BS library containing the source code for both native methods from OpenCV and several approaches proposed in literature. The library offers more than 30 BS algorithms and a JAVA graphical user interface (GUI), that can be used for comparing the different methods, is also provided.

14.2.4   Data Sets for BS Benchmarking

A complete and updated list of BS methods and publicly available data sets can be found in the "Background Subtraction - Site web" [8]. A section of such website is dedicated to the available implementations of both traditional (e.g., statistical methods [7]) and recent emerging approaches (e.g., fuzzy background modeling [7]). Another section contains links and references for available data sets.

Image Sequences

Some image sequences with water background have been used in literature for evaluating the performance of BS methods.

- Jug [29]: A foreground jug floats through the background rippling water.

- Water Surface [59]: A person walks in front of a water surface with moving waves.

- UCSD Background Subtraction Data set [55]. The data set consists of a total of 18 video sequences, some of them containing water background. In particular "Birds", "Boats", "Flock", "Ocean", "Surf", "Surfers", and "Zodiac" sequences are related to maritime scenarios.

Changedetection.net Video Data Set

Changedetection.net [11] is a benchmark data set containing several video sequences that can be used to quantitatively evaluate BS methods. Five sequences containing water background are included in the category "Dynamic Background".

MAR - Maritime Activity Recognition Data Set

MAR [32] is a collection of maritime video streams and images, with ground truth data, that can be used for evaluating the performance of automatic video surveillance systems and of computer vision techniques such as tracking and video stabilization. In particular, for each video the following details are provided:

- Sensor type (electro-optical EO or infrared IR);

- Camera type (static or moving);

- Location;

- Light conditions;

- Foreground masks to evaluate foreground segmentation;

- Annotations with the bounding box vertexes to evaluate detection;

- Identification numbers to evaluate data association and tracking over time.

In addition, the videos have been recorded with varying observing angles and weather conditions. At this moment the data set contains

(i) A selection of images containing boats from the publicly available VOC data base [20].

(ii) EO and IR videos recorded in a VTS center in Italy.

(iii) EO and IR videos recorded in a VTS center in the North of Europe.

(iv) EO videos from the ARGOS system [4] monitoring the Grand Canal in Venice.

All the above video sequences are available for downloading from the MAR - Maritime Activity Recognition data set homepage [32].

14.2.5   Discussion

Although state-of-the-art approaches can deal with dynamic background, a real-time, complete, an effective solution for maritime scenes does not yet exist. In particular, water background is more difficult than other kinds of dynamic background since waves on the water surface do not belong to the foreground even though they involve motion. Moreover, sun reflections on the water surface do not have the same behavior of a reflective surface.

As stated by Cheung *et al.* in [12], FD is the worst approach for computing dynamic background, since the key point for solving the problem of modeling a water background is to take into account the time-series nature of the problem.

Per-pixel approaches (e.g., [49]) usually fail because the rich and dynamic textures typical of the water background cause large changes at an individual pixel level [16]. A non-parametric approach (e.g., [18]) cannot learn all the changes, since on the water surface the changes do not present any regular patterns [52]. More complex approaches (e.g., [44,64,65]), can obtain better results at the cost of increasing the computational load of the process.

In the next section, a multi-stage (per-pixel, per-region), non-recursive, non-predictive, and real-time BS approach is described. The method has been specifically designed for dealing with water background, but can be successfully applied to every scenario [5]. The algorithm is currently in use within a real 24/7 video surveillance system[32] for the control of naval traffic in Venice [4].

The key aspects of the method are

1. An on-line clustering algorithm to capture the multi-modal nature of the background without maintaining a buffer with the previous frames.

2. A model update mechanism that can detect changes in the background geometry.

3. A noise removal module to help in filtering out false positives due to reflections and boat wakes.

Quantitative experiments show the advantages of the developed method over several state-of-the-art algorithms implemented in the well-known OpenCV [38] library and its real-time performance.

The approach has been implemented in C++ by using OpenCV functions and it is released as an open source code[33]. All the data used for the evaluation phase can be downloaded from the publicly available MAR - Maritime Activity Recognition data set [32].

14.3   Multi-modal Background Modeling

The analysis of water background highlights the presence of non-regular patterns. In Fig. 14.3, RGB and HSV color histograms of a pixel from the *Jug* sequence [29] over 50 frames are reported. RGB values (shown in the three histograms on the top of Fig. 14.3) are distributed over a large part of the spectrum $[0, 255]$. In particular, red values range from 105 to 155. HSV values (shown in the three histograms on the bottom of Fig. 14.3) also cover a large part of the spectrum and are characterized by a strong division in several bins. A more marked distribution over the whole spectrum can be observed analyzing a pixel location affected by reflections (see Fig. 14.4). In this second example, the red values range from 85 to 245.

---

[32]http://www.argos.venezia.it
[33]http://www.dis.uniroma1.it/~bloisi/software/imbs.html

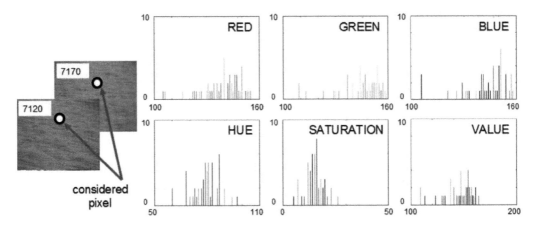

FIGURE 14.3 RGB and HSV values of a pixel (white dot) from frame 7120 to frame 7170 of *Jug* sequence [29]. The x-axis represents the color component values, while on the y-axis the number of occurrences are reported. HSV values are scaled to fit to $[0, 255]$.

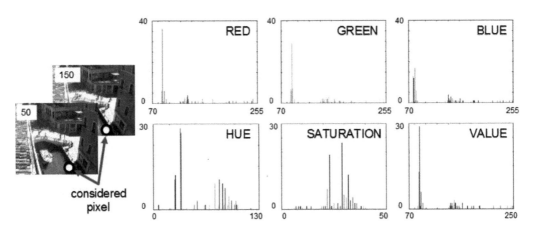

FIGURE 14.4 RGB and HSV values of a pixel (white dot) from frame 50 to frame 150 of sequence 1 from the MAR dataset [32]. The x-axis represents the color component values, while on the y-axis the number of occurrences are reported. HSV values are scaled to fit to $[0, 255]$.

In order to cope with such a strongly varying behavior, a possible solution is to create a discretization of an unknown distribution by using an on-line clustering mechanism. Each pixel of the background is modeled by a set of bins of variable size, without any *a priori* assumption about the observed distribution. To further validate the hypothesis that the color value distributions from the two examples reported in Fig. 14.3 and Fig. 14.4 are not Gaussian, the Anderson-Darling test [62] has been applied to those color values and the hypothesis of normality has been rejected for all the color components.

The proposed method is called Independent Multi-modal Background Subtraction (IMBS) and is described in Algorithm 14.1. The main idea behind IMBS is to assign to each pixel p a set of values representing the discretization of the background color distribution in p over a period of time R. The BG model $\mathfrak{B}$ is computed through a per-pixel, on-line statistical analysis of N frames from the video sequence in input (called *scene samples* S_k with $1 \leq k \leq N$) that are selected on the basis of a sampling period P.

Each element $\mathfrak{B}(i, j)$ of the BG model is a set of pairs $\{c, f(c)\}$, where c is the discrete

Algorithm 14.1 Independent Multi-modal Background Subtraction (IMBS).

Input: $I, P, N, D, A, \alpha, \beta, \tau_S, \tau_H, \omega, \theta$
Output: F
Data Structure: $\mathfrak{B}$
Initialize: $k = 1$, $t_s = 0$, $\forall\, i, j$ $\mathfrak{B}_{(i,j)}(t_s) = \oslash$
for each $I(t)$ **do**
    if $t - t_s > P$ **then**
        $S_k = I(t)$
        $BGFormation(S_k, D, A, k)$
        $t_s = t$
        $k = (k + 1)$
    end if
    $F_{\text{mask}}(t) \leftarrow FGComputation(I(t), A)$
    $F_{\text{filtered}}(t) \leftarrow NoiseSuppression(F_{\text{mask}}(t), \alpha, \beta, \tau_S, \tau_H, \omega, \theta)$
    $F(t) \leftarrow ModelUpdate(F_{\text{filtered}}(t), S_k);$
    if $k = N$ **then**
        $k = 1$
    end if
end for

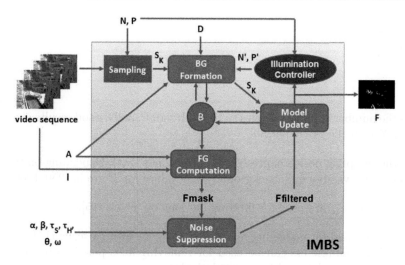

FIGURE 14.5   The modular architecture of IMBS approach.

representation of a set of adjacent colors in a given color space (e.g., a range in RGB or HSV space) and $f(c)$ is the number of occurrences of c (i.e., of colors within the range represented by c) in the sample set. After processing all the sample set, only those colors that have enough occurrences are maintained in the background model. This per-pixel BG model generation, that allows to achieve a high computational speed, is followed by a per-region analysis aiming at removing false positives due to possible compression artifacts, shadows, reflections, and boat wakes.

A diagram illustrating the IMBS algorithm is reported in Fig. 14.5. $I(t)$ is the current image at time t, $\mathfrak{B}$ is the BG model, S_k is the last processed scene sample and t_s its time stamp. The following are the input parameters.

- A is the association threshold. A pixel p in position (i, j) is considered a foreground point if the condition $dist(c, \mathfrak{B}(i, j)) \geq A$ holds. Typical values for A are in the interval $[5, 20]$.

- D is the minimal number of occurrences to consider a color value c to be a significant background value. Typical values for D are in $[2, 4]$.

Algorithm 14.2 Background Formation.

Input: S_k, D, A, k, N, P
Data Structure: $\mathfrak{B}$
for each *pixel* $p \in S_k$ *having color value* v **do**
  if $k = 1$ **then**
    add couple $\{v, 1\}$ *to* $\mathfrak{B}(i, j)$
  else
    for each *couple* $T := \{c, f(c)\} \in \mathfrak{B}(i, j)$ **do**
      if $|v - c| \leq A$ **then**
        $c' \leftarrow \left\lfloor \frac{c \cdot f(c) + v}{f(c+1)} \right\rfloor$
        $T \leftarrow \{c', f(c+1)\}$
        break;
      else
        add couple $\{v, 1\}$ *to* $\mathfrak{B}(i, j)$
      end if
    end for
    if $k = N$ **then**
      for each *couple* $T := \{c, f(c)\} \in \mathfrak{B}(i, j)$ **do**
        if $f(c) < D$ **then**
          delete T;
        end if
      end for
    end if
  end if
end for

- N is the number of scene samples to be analyzed. Usually, good results can be achieved when N is in the range $[20, 30]$.

- P is the sampling period expressed in milliseconds (ms). Depending on the monitored scene (static vs. dynamic background), the values for P are in $[250, 1000]$.

- α, β, τ_S, and τ_H are used for filtering out shadow pixels [15].

- θ and ω are used for managing reflections. If the monitored scene is affected by reflections, these parameters can be used for removing false positives on the basis of the brightness of the detected foreground pixels.

14.3.1   Background Formation

The procedure *BG-Formation* (see Algorithm 14.2) creates the background model, while the routine *FG-Computation* (see Algorithm 14.3) generates the binary FG image F_{mask}.

Let $I(t)$ be the $W \times H$ input frame at time t, and $F_{\text{mask}}(t)$ the corresponding foreground mask. The BG model $\mathfrak{B}$ is a matrix of H rows and W columns. Each element $\mathfrak{B}(i, j)$ of the matrix is a set of couples $\{c, f(c)\}$, where c is a value in a given color space (e.g., RGB or HSV) and $f(c) \rightarrow [1, N]$ is a function returning the number of pixels in the scene sample $S_k(i, j)$ associated with the color component values denoted by c. Modeling each pixel by binding all the color channels in a single element has the advantage of capturing the statistical dependencies between the color channels, instead of considering each channel independently.

Each pixel in a scene sample S_k is associated with an element of $\mathfrak{B}$ according to A (see Algorithm 14.2). Once the last sample S_N has been processed, if a couple T has a number $f(c)$ of associated samples greater or equal to D, namely $T := \{c, f(c) \geq D\}$, then its color value c becomes a significant background value. Up to $\lfloor N/D \rfloor$ couples for each element of $\mathfrak{B}$ are considered at the same time, thus approximating a multi-modal probability distribution.

Algorithm 14.3 Foreground Computation.

Input: I, A
Output: F_{mask}
Data Structure: $\mathfrak{B}$
Initialize: $\forall\ i, j\ F_{mask}(i, j) = 1$
for each *color value* v *of pixel* $p \in I$ **do**
   if $|\mathfrak{B}(i, j)| \neq \oslash$ **then**
      for each *couple* $T := \{c, f(c)\} \in \mathfrak{B}(i, j)$ **do**
         if $|v - c| < A$ **then**
            $F_{mask}(i, j) \leftarrow 0$
            break;
         end if
      end for
   end if
end for

In this way, the problem of modeling waves, gradual illumination changes, noise in sensor data, and movement of small background elements can be addressed [5]. Indeed, the adaptive number of couples for each pixel can model non-regular patterns that cannot be associated with any predefined distribution (e.g., Gaussian). This is a different approximation with respect to the popular Mixture of Gaussians approach [30,49,66], since the "discretization" of the unknown background distribution is obtained considering each BG point as composed by a varying number of couples $\{c, f(c)\}$ without attempting to force the found values to fit in a predefined distribution.

F_{mask} is computed according to the thresholding mechanism shown in Algorithm 14.3, where $|\mathfrak{B}(i, j)|$ denotes the number of couples in $\mathfrak{B}(i, j)$, with $|\mathfrak{B}(i, j) = \oslash| = 0$. The use of a set of couples instead of a single BG value makes IMBS robust with respect to the choice of the parameter A, since a pixel that presents a variation in the color values larger than A will be modeled by a set of contiguous couples.

The first BG model is built after $R = NP$ *ms*, then new BG models, independent from the previous ones, are built continuously every R *ms*. The independence of each BG model is a key aspect of the algorithm, since it permits to adapt to fast changing environments avoiding error propagation and it does not affect the accuracy for slow changing ones. Moreover, the on-line model creation mechanism allows for avoiding to store the N scene samples, that is the main drawback of the non-recursive BS techniques [39].

An *Illumination Controller* module manages sudden illumination changes. N and P values are reduced ($N' = N/2$ and $P' = P/3$) if the percentage of foreground pixels in F_{mask} is above a certain threshold (e.g., 50 %), in order to speed up the creation of a new BG model. To further increase the speed of the algorithm, the foreground computation can be paralleled, dividing the FG mask in horizontal slices.

14.3.2   Noise Removal

In the maritime domain, when BS is adopted to compute the foreground mask, sun reflections on the water surface and shadows generated by boats (and sometimes buildings, as in Fig. 14.7) can affect the foreground image producing false positives. In order to deal with the erroneously classified foreground pixels, that can deform the shape of the detected objects, a noise suppression module is required. In the following, two possible solutions for dealing with reflections and shadows are described.

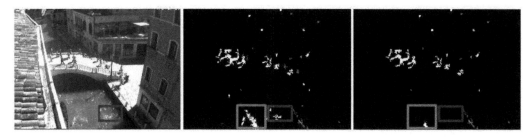

FIGURE 14.6   Reflection suppression example. Original frame (left), foreground extraction without reflection removal (center), and with reflection suppression (right).

Reflections

In order to characterize reflections (see Fig. 14.6), a region level analysis based on the HSV color space can be carried out. As reported in [26] and [40], the H, S, and V components in a reflective surface assume specific values through which it is possible to model, detect, and describe these particular surfaces.

Hue specifies the base color, while the other two values of saturation and brightness (or value) specify intensity and lightness of the color. For a reflective surface the H component assumes similar values of subtractive colors (e.g., cyan, yellow, and magenta), that are more likely to reflect sun lights than dark primary colors (e.g., red, green and blue). Instead, saturation and brightness respectively assume low and high values in the range [0..1].

However, through these information it is impossible to characterize the behavior of a single reflection. Indeed, the saturation does not assume low values, but it has the same behavior of the brightness. Therefore, the analysis can be limited to the brightness component. Starting from the RGB values of the video sequence in input, an HSV conversion is carried out to overcome the information loss due to the image compression. The conversion is computed according to [23]:

$$
H = \begin{cases}
60 \times \left(\dfrac{g-b}{i_{\max} - i_{\min}} \right) \ mod\,360 & if \ \ i_{\max} = r \\[2mm]
60 \times \left(2 + \dfrac{b-r}{i_{\max} - i_{\min}} \right) \ mod\,360 \ if \ \ i_{\max} = g \\[2mm]
60 \times \left(4 + \dfrac{r-b}{i_{\max} - i_{\min}} \right) \ mod\,360 \ if \ \ i_{\max} = b
\end{cases}
\tag{14.1}
$$

$$
S = \begin{cases}
\dfrac{i_{\max} - i_{\min}}{i_{\max} + i_{\min}} & l < 0.5 \\[3mm]
\dfrac{i_{\max} - i_{\min}}{2 - i_{\max} - i_{\min}} & l \geq 0.5
\end{cases}
\tag{14.2}
$$

$$
V = \frac{i_{\max} - i_{\min}}{2}
\tag{14.3}
$$

where:

$$
i_{\max} = max(r,g,b), \ \ i_{\min} = min(r,g,b) \ \ and \ \ 0 \leq r,g,b \leq 1
$$

IMBS makes a region analysis of the image parts detected as foreground (blobs). For each region (blob) a percentage based on the evaluation of the brightness of each point is calculated, in order to establish if the blob has been generated by reflections. Given a blob, all the pixel p for which the following condition hold

FIGURE 14.7 Shadow suppression example. Original frame (left), foreground extraction without shadow removal (center), and with shadow suppression (right). The shadows in the red box are correctly removed.

$$V > \theta \tag{14.4}$$

are considered affected by reflections. If the percentage of such points with respect to the total points of the blob under analysis is greater than a predefined threshold ω, than the region (i.e., the blob) is classified as a reflection area and filtered out. The parameters θ and ω are defined by the user and in our experiments we set $\theta = 0.7$ and $\omega = 60\%$.

Shadows

To detect and suppress shadows, IMBS adopts a pixel-level strategy that is a slight modification of the HSV based method proposed by Cucchiara *et al.* in [15].

Let $I^c(i,j)$, $c = \{H, S, V\}$ be the HSV color values for the pixel (i,j) of the input frame and $B_T^c(i,j)$ the HSV values for the couple $T \in \mathfrak{B}(i,j)$. The shadow mask M value for each foreground point is:

$$M(i,j) = \begin{cases} 1 \ if \ \exists \ T : \alpha \le \frac{I^V(i,j)}{B_T^V(i,j)} \le \beta \ \wedge \\ \quad |I^S(i,j) - B_T^S(i,j)| \le \tau_S \ \wedge \\ \quad |I^H(i,j) - B_T^H(i,j)| \le \tau_H \\ 0 \ otherwise \end{cases} \tag{14.5}$$

The parameters $\alpha, \beta, \tau_S, \tau_H$ are user defined and can be found experimentally. In all the experiments show in this paper $\alpha = 0.75$, $\beta = 1.15$, $\tau_S = 30$, and $\tau_H = 40$. Setting β to a value slightly greater than 1 permits to filter out light reflections.

It is worth to note that shadow suppression is essential for increasing the accuracy of the algorithm. The HSV analysis can effectively remove the errors introduced by a dynamic background, since it is a more stable color space with respect to RGB [63].

After the removal of the shadow pixels, a F_{filtered} binary image is obtained (see Fig. 14.7). It can be further refined by exploiting the opening and closing morphological operators. The former is particularly useful for filtering out the noise left by the shadow suppression process, the latter is used to fill internal holes and small gaps.

Wakes

Optical Flow (OF) can be used to filter out false positives due to wakes. OF correlates two consecutive frames, providing for every feature which is present in both the frames a motion vector (including direction, versus, and value) that is not null if the position of the feature in the two frames is different.

We used the OpenCV implementation of the pyramidal Lucas-Kanade OF algorithm [6, 45], that generates a sparse map (see the left side of Figure 14.9), to obtain the OF

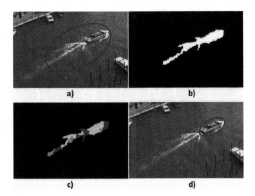

FIGURE 14.8 (See color insert.) Example of optical flow (OF) use to deal with boat wakes. a) False positives caused by boat wakes. b) The blob in the foreground image is much bigger than the real boat. c) Optical flow mask. d) By clustering the points of the blob on the basis of the direction of the OF points, the boat can be distinguished from the wakes obtaining a correct detection.

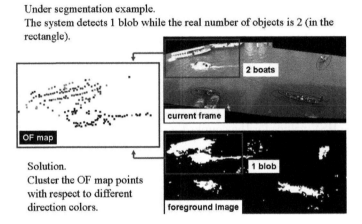

FIGURE 14.9 (See color insert.) An example of ghost observation solved by clustering the optical flow points.

points. The points in the map are associated with one of four predefined directions and are characterized by a color. Thus, the OF map consists of a set of colored points in which different colors indicate different moving directions.

Calculating the OF can be useful when a boat has a long wake on the back. An example of a false detection error is reported in Fig. 14.8a. The initial number of FG points (Fig. 14.8b) is over estimated with respect to the real one. Since the direction for each OF point is known, it is possible to cluster (details can be found in [45]) the OF points (Fig. 14.8c) in order to detect the biggest cluster that corresponds to the boat and to discard the outliers corresponding to the water waves (Fig. 14.8d).

OF analysis also helps in detecting false positives due to changes in the background geometry. Not moving (e.g., parked boats) or slow objects (e.g., gondolas) can be included incorrectly in the background, producing the so called ghost observations (i.e., false positives are generated when the static object that has been absorbed in the BG model starts to move). Although the background model can be corrupted by those elements, it is possible to filter out them by analyzing the OF points. Indeed, since ghosts do not produce motion, their OF is null and can be filtered out (see the example in Fig. 14.9).

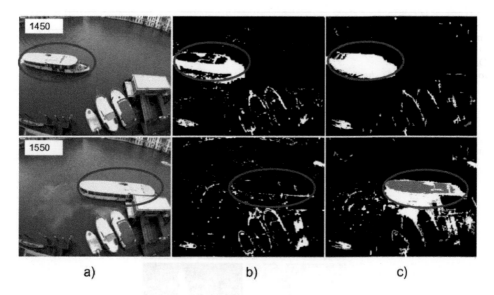

FIGURE 14.10   Model update. a) A vaporetto enters the monitored scene and remains in the same position over several frames. b) A blind update obtained by using the algorithm from [23] implemented in OpenCV. The model includes incorrectly the vaporetto as part of the scene background. c) IMBS model update. The vaporetto is identified as a potential foreground region (grey pixels).

However, this post-processing analysis based on OF has also some limitations. It fails when a boat is involved in particular maneuvers. As an example, when a boat turns around itself, the optical flow may detect different directions for the different parts of the boat (e.g., one for the prow and another for the stern), leading to a wrong suppression for that observation. Nevertheless, from an analysis of the performance of the BS process on many live video streams from a real application scenario [4], it is possible to state that situations where optical flow worsen the performance of the segmentation process are very limited, with respect to the advantages it introduces.

14.3.3   Model Update

Elgammal *et al.* in [18] proposed two alternative strategies to update the background

1. *Selective update.* Only pixels classified as belonging to the background are updated.

2. *Blind update.* Every pixel in the background model is updated.

The selective (or *conditional*) update improves the detection of the targets since foreground information are not added to the BG model, thus solving the problem of ghost observations [15]. However, when using selective updating any incorrect pixel classification produces a persistent error since the BG model will never adapt to it.

Blind update does not suffer from this problem since no update decisions are taken, but it has the disadvantage that the values not belonging to the background are added to the model.

A different solution, aiming at solving the problems of both selective and blind update, can be adopted. As shown in Algorithm 14.4, given a scene sample S_k and the current foreground mask after noise removal F_{filtered}, if $F_{\text{filtered}}(i, j) = 1$ and $S_k(i, j)$ is associated to a couple T in the BG model under development (namely, $\exists \ c \in \mathfrak{B}(i, j) : \{|v - c| \leq A\}$, where v is the color value of $S_k(i, j)$), then T is labeled as a "foreground couple". When computing

---

Algorithm 14.4 Model Update.

Input: S_k, F_{filtered}
Output: F
Data Structure: $\mathfrak{B}$
for each *pixel* $p \in S_k$ *having color value* v **do**
   for each *couple* $T := \{c, f(c)\} \in \mathfrak{B}(i, j)$ **do**
     if $F_{\text{filtered}}(i, j) = 1 \wedge |v - c| \leq A$ **then**
       T *is a foreground couple*
     end if
   end for
end for

FIGURE 14.11   IMBS output for frames 1516 and 1545 of the *water surface* sequence [59].

the foreground, if $I(i, j)$ is associated with a foreground couple, then it is classified as a potential foreground point. Such a solution allows for identifying regions of the scene that represent not moving foreground objects, as shown in Fig. 14.10, where the pixel belonging to a boat that stops for several frames are detected as potential foreground points.

The decision about including or not the potential foreground points as part of the background is taken on the basis of a *persistence map*. If a pixel is classified as potential foreground consecutively for a period of time longer than a predefined value (e.g., $R/2$ or 10 seconds), then it becomes part of the BG model. Furthermore, the labeling process provides additional information to higher level modules (e.g., a visual tracking module) helping in reducing ghost observations.

14.4   Results

Two well-known publicly available sequences involving water background, namely *water surface* [59] and *jug* [29], have been used to qualitatively test the results obtained by IMBS. The two examples, that are shown in Fig. 14.11 and Fig. 14.12, demonstrate the capacity of IMBS to correctly model the background in both the situations, extracting the foreground masks with great accuracy.

To quantitatively test IMBS on water background, some of the video sequences available in the MAR data set containing reflections have been used. The sequences come from a real video surveillance system for the control of naval traffic in Venice [4].

IMBS has been compared with three BS algorithms implemented in the OpenCV Library [38]: MOG [30], MOG2 [66], GMG [23]. The scenes used for the comparison present varying light conditions and different camera heights and positions. All the ten ground truth frames, together with the results of the compared methods, are reported in Fig. 14.13.

Table 14.2 reports the results in terms of false negatives (the number of foreground

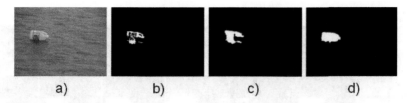

a)   b)   c)   d)

FIGURE 14.12 IMBS results on the *jug* sequence [29]. a) Frame number 36. b) Foreground mask obtained by using the approach presented in [65]. c) Results obtained in [16]. d) IMBS results.

TABLE 14.2 A comparison of IBMS with OpenCV methods on the MAR data set.

Seq.	Error	MOG	MOG2	GMG	IMBS	IMBS*
	f. neg.	14418	6106	**1152**	3625	5061
1	f.pos.	**15595**	514082	108899	46272	30058
	tot. e.	**30013**	520188	110051	49897	35119
	f. neg.	80959	11910	22252	**9077**	9604
2	f.pos.	**7432**	141628	43820	25259	23306
	tot. e.	88391	153538	66072	34336	**32910**
	f. neg.	29659	4761	13416	**8526**	11161
3	f.pos.	**5153**	207511	35204	34219	44587
	tot. e.	**34812**	212272	48620	42745	55748
	f. neg.	45634	**9759**	17148	9848	10153
4	f.pos.	**9339**	151342	40220	34301	29516
	tot. e.	54973	161101	57368	44149	**39669**
	f. neg.	72158	**9946**	45873	25162	21567
5	f.pos.	**5730**	151103	43661	44640	21131
	tot. e.	77888	161049	89534	69802	**42698**
	f. neg.	46741	**7197**	26856	7265	10311
6	f.pos.	**1825**	153352	13102	13005	10020
	tot. e.	48566	160549	39958	20270	**20331**
	f. neg.	35130	9225	10574	7682	**7279**
7	f.pos.	25394	361048	76184	53041	**24729**
	tot. e.	60524	370273	86758	60723	**32367**
	f. neg.	80959	11910	22252	9077	**7638**
8	f.pos.	**7432**	141628	43820	25259	24729
	tot. e.	88391	153538	66072	34336	**32367**
	f. neg.	16384	**1011**	3568	1135	1060
9	f.pos.	**5677**	182025	37908	24989	20676
	tot. e.	22061	183036	41476	26124	**21736**
	f. neg.	16707	10337	15499	1807	**1737**
10	f.pos.	666	97782	9298	11831	**11549**
	tot. e.	17373	108119	24797	13638	**13286**
	Tot. Err.	522.992	2.183.663	630.706	396.020	**348.985**

points detected as background, FN) and false positives (the number of background points detected as foreground, FP). IMBS has been analyzed both with (column IMBS*) and without (column IMBS) the activation of the reflection suppression mechanism. The results show the effectiveness of the IMBS approach, that performs better than the other considered methods in terms of total error.

To further validate the IMBS approach, three additional metrics have been considered

- *F-measure*

- *Detection Rate*

- *False Alarm Rate*

F-measure is computed as shown in [56]:

$$F\text{-}measure = \frac{1}{n} \sum_{i=1}^{n} 2 \frac{Prec_i \times Rec_i}{Prec_i + Rec_i} \tag{14.6}$$

where i represents the current frame, TP the true positives observations, TN the true negatives ones, and

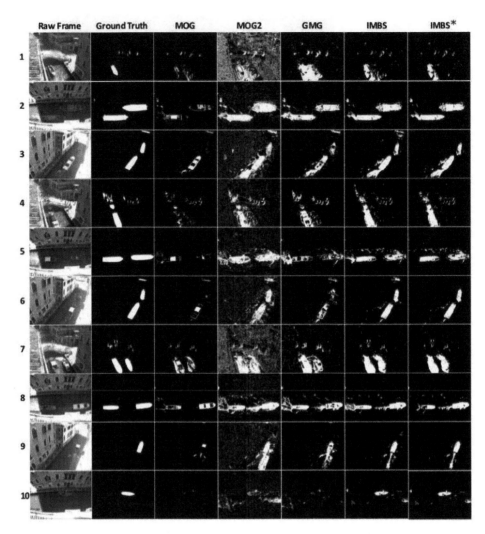

FIGURE 14.13 IMBS qualitative comparison with OpenCV BS algorithms. The first two columns illustrate the analyzed frame and the ground truth, respectively. The remaining columns show the foreground obtaining by using OpenCV algorithms and IMBS method (last two columns).

$$Rec_i\left(P\right) = TP_i/\left(TP_i + FN_i\right); \qquad Prec_i\left(P\right) = TP_i/\left(TP_i + FP_i\right)$$
$$Rec_i\left(N\right) = TN_i/\left(TN_i + FP_i\right); \qquad Prec_i\left(N\right) = TN_i/\left(TN_i + FN_i\right)$$
$$Rec_i = \left(1/2\right)\left(Rec_i\left(P\right) + Rec_i\left(N\right)\right); \; Prec_i = \left(1/2\right)\left(Prec_i\left(P\right) + Prec_i\left(N\right)\right)$$

Detection Rate (DR) and *False Alarm Rate* (FAR) are computed as follows.

$$DR = \frac{TP}{TP + FN} \tag{14.7}$$

$$FAR = \frac{FP}{TP + FP} \tag{14.8}$$

Table 14.3 shows the results in terms of *F-measure*, DR and FAR obtained by using the ten ground truth in Fig. 14.13.

The computational speed of the algorithm has been tested with different sequences coming from publicly available data sets [3,32,57] and with multiple frame dimensions. The

TABLE 14.3 Computation of F-measure, Detection Rate and False Alarm Rate.

Measure	MOG	MOG2	GMG	IMBS	IMBS*
F-measure	0.72	0.70	0.77	0.85	**0.86**
DR	0.26	**0.84**	0.67	0.83	**0.84**
FAR	**0.34**	0.78	0.55	0.41	0.37

results obtained by using a single-thread implementation are reported in Table 14.4 and show the real-time performance of the proposed approach. It is worth noting that, for all the test sequences, IMBS parameters have been set as: $A = 12$, $N = 30$, $D = 2$, $P = 500$ *ms*, and the OF map has not been activated.

TABLE 14.4 IMBS computational speed in terms of frame per seconds.

Data	Frame Dim.	FPS
Wallflower [57]	160×120	115
ATON [3]	320×240	70
MAR [32]	320×240	38.6
MAR	640×480	22.4
MAR	1024×768	12
MAR	1280×960	9.4

14.5   Conclusion

In this chapter, the use of background subtraction (BS) for detecting moving objects in the maritime domain has been discussed.

Maritime domain is a challenging scenario for automatic video surveillance, due to its inherently dynamic background. Indeed, waves on the water surface, boat wakes, weather issues (such as heavy rain), gradual and sudden illumination variations (e.g., clouds), motion changes (e.g., camera jitter), modification of the background geometry (e.g., parked boats), and reflections can provoke false positives.

A fast and robust background subtraction algorithm, specifically conceived for being effective in maritime scenarios, has been proposed to deal with the above issues. It has been quantitatively compared by using data coming from a real system [4]. The data are publicly available from the MAR - Maritime Activity Recognition data set [32].

Thanks to an on-line clustering algorithm to create the model and a conditional update mechanism, the method can achieve good accuracy while maintaining real-time performance.

The accuracy of the foreground detection is increased by a noise suppression module, that can deal with reflections on the water surface, shadows, and boat wakes.

The results of the comparison with several other state-of-the-art methods implemented in the well-known OpenCV library demonstrate the effectiveness of the proposed approach.

References

1. V. Ablavsky. Background models for tracking objects in water. In *ICIP 2003*, pages 125–128, 2003.
2. F. Amato, M. Fiorini, S. Gallone, and G. Golino. Fully solid state radar for vessel traffic services. In *IRS*, pages 1–5, 2010.
3. ATON. Autonomous Agents for On-Scene Networked Incident Management. `http://cvrr.ucsd.edu/aton/testbed`.

4. D. Bloisi and L. Iocchi. ARGOS - a video surveillance system for boat traffic monitoring in Venice. *International Journal of Pattern Recognition and Artificial Intelligence*, 23(7):1477–1502, 2009.

5. D. Bloisi and L. Iocchi. Independent multimodal background subtraction. In *CompIMAGE*, pages 39–44, 2012.

6. J. Bouguet. Pyramidal implementation of the Lucas Kanade feature tracker. *Intel Corporation, Microprocessor Research Labs*, 2000.

7. T. Bouwmans. Recent advanced statistical background modeling for foreground detection: A systematic survey. *Recent Patents on Computer Science*, 4(3):147–176, 2011.

8. T. Bouwmans. Background Subtraction Website. `https://sites.google.com/site/backgroundsubtraction/`, 2013.

9. Canadian Coast Guard. Vessel traffic services (vts) update study. *OM: Canadian Coast Guard, Marine Navigation Services*, 159, 1991.

10. C. Carthel, S. Coraluppi, and P. Grignan. Multisensor tracking and fusion for maritime surveillance. In *FUSION*, pages 1–6, 2007.

11. ChangeDetection.net. Video Database. `http://www.changedetection.net/`, 2012.

12. L. Cheng, S. Wang, D. Schuurmans, T. Caelli, and S. V. N. Vishwanathan. An online discriminative approach to background subtraction. In *AVSS*, 2006.

13. S. Cheung and C. Kamath. Robust techniques for background subtraction in urban traffic video. In *Visual Comm. and Image Proc.*, volume 5308, pages 881–892, 2004.

14. M. Cristani, M. Farenzena, D. Bloisi, and V. Murino. Background subtraction for automated multisensor surveillance: A comprehensive review. *EURASIP J. Adv. Sig. Proc.*, 2010:1–24, 2010.

15. R. Cucchiara, C. Grana, M. Piccardi, and A. Prati. Detecting moving objects, ghosts, and shadows in video streams. *PAMI*, 25(10):1337–1342, 2003.

16. G. Dalley, J. Migdal, and W. Grimson. Background subtraction for temporally irregular dynamic textures. In *IEEE Workshop on Applications of Computer Vision*, pages 1–7, 2008.

17. G. Doretto, A. Chiuso, Y. N. Wu, and S. Soatto. Dynamic textures. *IJCV*, 51(2):91–109, 2003.

18. A. M. Elgammal, D. Harwood, and L. S. Davis. Non-parametric model for background subtraction. In *ECCV*, pages 751–767, 2000.

19. S. Elhabian, K. El-Sayed, and S. Ahmed. Moving object detection in spatial domain using background removal techniques - state-of-art. *Recent Patents on Computer Science*, 1:32–54, 2008.

20. M. Everingham, L. Van Gool, C. Williams, J. Winn, and A. Zisserman. The Pascal visual object classes (VOC) challenge. *IJCV*, 88(2):303–338, 2010.

21. S. Fefilatyev, D. Goldgof, M. Shreve, and C. Lembke. Detection and tracking of ships in open sea with rapidly moving buoy-mounted camera system. *Ocean Engineering*, 54(0):1–12, 2012.

22. N. Friedman and S. Russell. Image segmentation in video sequences: A probabilistic approach. In *Conf. on Uncertainty in Artificial Intelligence*, pages 175–181, 1997.

23. A. Godbehere, A. Matsukawa, and K. Goldberg. Visual tracking of human visitors under variable-lighting conditions for a responsive audio art installation. In *American Control Conference*, pages 4305–4312, 2012.

24. I. Haritaoglu, D. Harwood, and L. David. W4: Real-time surveillance of people and their activities. *PAMI*, 22(8):809–830, 2000.

25. I. Harre. AIS Adding New Quality to VTS Systems. *The Journal of Navigation*, 3(53):527–539, 2000.

26. Q. He and C. Hu. Detection of reflecting surfaces by a statistical model. In *SPIE*, volume 7251, 2009.

27. M. Heikkila and M. Pietikainen. A texture-based method for modeling the background and detecting moving objects. *PAMI*, 28(4):657–662, 2006.

28. S. Jabri, Z. Duric, H. Wechsler, and A. Rosenfeld. Detection and location of people in video images using adaptive fusion of color and edge information. *ICPR*, 4:627–630, 2000.

29. Jug. Video sequence. `http://www.cs.bu.edu/groups/ivc/data.php`.

30. P. Kaewtrakulpong and R. Bowden. An improved adaptive background mixture model for realtime tracking with shadow detection. In *European Workshop on Advanced Video Based Surveillance Systems*, pages 135–144, 2001.

31. W. Kruger and Z. Orlov. Robust layer-based boat detection and multi-target-tracking in maritime environments. In *WSS*, pages 1 –7, 2010.

32. MAR. Maritime Activity Recognition Data set. `http://labrococo.dis.uniroma1.it/MAR`.

33. A. Mittal and N. Paragios. Motion-based background subtraction using adaptive kernel density estimation. In *CVPR*, pages 302–309, 2004.

34. A. Monnet, A. Mittal, N. Paragios, and V. Ramesh. Background modeling and subtraction of dynamic scenes. In *ICCV*, pages 1305–1312, 2003.

35. C. Nolan. *International Conference on Integrated Fisheries Monitoring*. Food and Agriculture Organization of the United Nations (FAO), Sydney, Australia, 1999.

36. P. Noriega and O. Bernier. Real time illumination invariant background subtraction using local kernel histograms. In *BMVC*, pages 100.1–100.10, 2006.

37. N. M. Oliver, B. Rosario, and A. P. Pentland. A Bayesian computer vision system for modeling human interactions. *PAMI*, 22(8):831–843, 2000.

38. OpenCV. Open Source Computer Vision. `http://opencv.org`.

39. M. Piccardi. Background subtraction techniques: a review. In *IEEE Int. Conf. on Systems, Man and Cybernetics*, pages 3099–3104, 2004.

40. A. Rankin, L. Matthies, and A. Huertas. Daytime water detection by fusing multiple cues for autonomous off-road navigation. *24th Army Science Conference*, 1(9), 2004.

41. B. Rhodes, N. Bomberger, M. Seibert, and A. Waxman. Seecoast: Automated port scene understanding facilitated by normalcy learning. *MILCOM*, 0:1–7, 2006.

42. S. Rodriguez, D. Mikel, and M. Shah. Visual surveillance in maritime port facilities. *SPIE*, 6978:11–19, 2008.

43. G. Saur, S. Estable, K. Zielinski, S. Knabe, M. Teutsch, and M. Gabel. Detection and classification of man-made offshore objects in terrasar-x and rapideye imagery: Selected results of the demarine-deko project. In *Proceedings of IEEE Oceans*, Santander, June 2011.

44. Y. Sheikh and M. Shah. Bayesian object detection in dynamic scenes. In *CVPR*, pages 74–79, 2005.

45. J. Shi and C. Tomasi. Good features to track. In *CVPR*, pages 593–600, 1994.

46. A. Sobral. BGSLibrary: A OpenCV C++ Background Subtraction Library. `http://code.google.com/p/bgslibrary/`, 2013.

47. M. Soumekh. *Synthetic aperture radar signal processing*. Wiley New York, 1999.

48. L. Spencer and M. Shah. Water video analysis. In *ICIP*, pages 2705–2708, 2004.

49. C. Stauffer and W. Grimson. Adaptive background mixture models for real-time tracking. *CVPR*, 2:246–252, 1999.

50. C. Stauffer and W. Grimson. Learning patterns of activity using real-time tracking. *PAMI*, 22(8):747–757, 2000.

51. B. Stenger, V. Ramesh, N. Paragios, F. Coetzee, and J. Buhmann. Topology free hidden markov models: application to background modeling. In *ICCV*, volume 1, pages 294–301, 2001.

52. A. Tavakkoli, M. Nicolescu, and G. Bebis. Robust recursive learning for foreground region detection in videos with quasi-stationary backgrounds. In *ICPR*, pages 315–318, 2006.

53. Y. Tian, R. Feris, H. Liu, A. Hampapur, and M. Sun. Robust detection of abandoned and removed objects in complex surveillance videos. *IEEE Transactions on Systems, Man, and Cybernetics, Part C*, 41(5):565–576, 2011.

54. K. Toyama, J. Krumm, B. Brumitt, and B. Meyers. Wallflower: principles and practice of background maintenance. In *ICCV*, volume 1, pages 255–261, 1999.

55. UCSD. Background Subtraction Data set. `http://www.svcl.ucsd.edu/projects/background_subtraction/ucsdbgsub_dataset.htm`.

56. A. Vacavant, T. Chateau, A. Wilhelm, and L. Lequievre. A benchmark dataset for outdoor foreground/background extraction. *In ACCV 2012, Workshop: Background Models Challenge*, 2012.

57. Wallflower Sequence. Test Images for Wallflower Paper. `http://research.microsoft.com/en-us/um/people/jckrumm/wallflower/testimages.htm`.

58. H. Wang and D. Suter. Background subtraction based on a robust consensus method. In *ICPR*, pages 223–226, 2006.

59. Water surface. Video sequence. `http://perception.i2r.a-star.edu.sg/bk\textunderscoremodel/bk\textunderscoreindex.html`.

60. H. Wei, H. Nguyen, P. Ramu, C. Raju, X. Liu, and J. Yadegar. Automated intelligent video surveillance system for ships. *SPIE*, 73061:1–12, 2009.

61. C. Wren, A. Azarbayejani, T. Darrell, and A. Pentland. Pfinder: Real-time tracking of the human body. *PAMI*, 19(7):780–785, 1997.

62. J. Zhao, X. Xu, and X. Ding. New goodness of fit tests based on stochastic EDF. *Commun. Stat., Theory Methods*, 39(6):1075–1094, 2010.

63. M. Zhao, J. Bu, and C. Chen. Robust background subtraction in HSV color space. In *SPIE: Multimedia Systems and Applications*, pages 325–332, 2002.

64. B. Zhong, H. Yao, S. Shan, X. Chen, and W. Gao. Hierarchical background subtraction using local pixel clustering. In *ICPR*, pages 1–4, 2008.

65. J. Zhong and S. Sclaroff. Segmenting foreground objects from a dynamic textured background via a robust kalman filter. In *ICCV*, pages 44–50, 2003.

66. Z. Zivkovic. Improved adaptive Gaussian mixture model for background subtraction. In *ICPR*, volume 2, pages 28–31, 2004.

15

Hierarchical Scene Model for Spatial-color Mixture of Gaussians

Christophe Gabard
CEA, LIST, France

Catherine Achard
Univ. Pierre et Marie Curie, Paris 06, France

Laurent Lucat
CEA, LIST, France

15.1 Introduction

Classical motion-detection methods, such as those presented by Stauffer and Grimson [7] or Elgammal *et al.* [3], use statistical temporal modeling for each background pixel. Afterwards, optional post-processing steps are applied with the aim to introduce spatial consideration such as object compactness. Moreover, most approaches from the literature only use a background model. In such approaches, motion detection is performed independently for each pixel, by evaluating whether each pixel is sufficiently similar to the background model. Furthermore, when the background and the target (foreground) appearance are close (similar colors, for instance), it becomes difficult to discriminate them without *a priori* knowledge. If we want an efficient and generic method, this information can then only be obtained with an updatable target model in addition to the background one. By looking for the closest model for each new incoming data, the detection process is therefore much more accurate.

A solution consists to model the scene globally, by gathering similar neighbor pixels [1, 2, 5, 8–11]. The whole image is thus represented by a mixture of Gaussian modes. This model is commonly referred to *SMOG* in the literature, meaning for *Spatial-color Mixture Of Gaussians* or *SGMM* for *Spatial Gaussian Mixture Models*. The method described in this chapter extends this approach and is based on a hierarchical scene model. It combines a global spatial modeling with a temporal pixel modeling of the scene which takes into account the spatial consistency between pixels. Detection decision is no longer local but performed on consistent pixel sets and is thus more robust. Global models allow describing the whole scene, both background and foreground areas. The target model is dynamically created during the detection process, allowing for a better Background versus Foreground separation and provides data information of a higher level than pixel. This can help a tracking step by introducing a natural combination between the detection and the tracking module. Our proposed method is described following the subsequent presentation: in the next section, the related works are presented. Section 15.3 will describe the proposed approach and particularly the methodology used to monitor the modeling quality and the automatic component labeling. The detection results are then presented in Section 15.4 where the proposed *SMOG* approach outperforms other tested methods especially under difficult conditions. Finally, we will provide conclusions on the proposed method and describe some interesting potential perspectives this kind of method allows at a higher level like tracking.

15.2 SMOG Overview

15.2.1 Basic Principle

The proposed approach uses the principle of $SGMM^{34}$, also known as $SMOG^{35}$, proposed, among other, in [1,2,5,8–11]. In this work, a mixture of Gaussian distributions is used, but unlike the well-known Stauffer and Grimson method [7], each Gaussian Mode represents a group of pixels instead of only a single pixel. Then a single list of modes is created and updated for each frame to represent the whole scene.

To introduce spatial and color consistency, each pixel observation is described by both spatial and color coordinates: $\mathbf{x}_t = [x, y, R, G, B]$ where (x, y) are the spatial coordinates of the pixel and (R, G, B) its color components.

From these observations, the scene is characterized by a set of Gaussian distributions in a five-dimensional space. A simplified representation of the so-obtained model is shown in Figure 15.1.

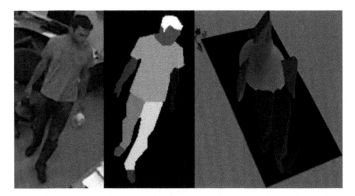

FIGURE 15.1 Simplified model representation: Left: the source image. Center: the perfect segmentation. Right: a schematic representation of the model. For representation purposes, only the average color of each mode is shown.

At each time t, all Gaussian distributions are characterized by the parameters $\theta_{(j,t)} = \{\omega_{(j,t)}, \mu_{(j,t)}, \Sigma_{(j,t)}\}$ where $\omega_{(j,t)}$ is the weight of the Gaussian mode j at time t while $\mu_{(j,t)}$ and $\Sigma_{(j,t)}$ are, respectively, the mean and the covariance matrix of the data distribution.

---

[34] Spatial Gaussian Mixture Models
[35] Spatial-Color Mixture Of Gaussian

The probability density function of each component is given by:

$$p(\mathbf{x}_t|\theta_{(j,t)}) = \frac{\omega_{(j,t)}}{\sqrt{(2\pi)^d.|\Sigma_{(j,t)}|}} e^{-\frac{1}{2}(\mathbf{x}_t-\mu_{(j,t)})^T(\Sigma_{(j,t)})^{-1}(\mathbf{x}_t-\mu_{(j,t)})} \tag{15.1}$$

where d is the problem dimension (5 here).

This model is clearly composed of two parts:

- The spatial components that are used to characterize the position, size and orientation of the pixel set.

- The colorimetric components that are used to characterize the color of the set.

In addition, an interpretation can be associated with the Gaussian components. Taking the example of the spatial components:

- The mean vector of the pixel set provides centroid position information.

- The eigenvectors of the covariance matrix define the main axes of the distribution and determine the mode orientation.

- The eigenvalues of the covariance matrix provide information on the extent along the principal axes.

A schematic diagram is given in Figure 15.2 where the Gaussian distribution is represented by an ellipse[36].

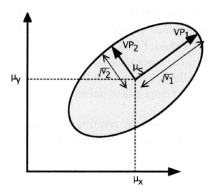

FIGURE 15.2 **Geometric interpretation of the Gaussian characteristics :** The ellipse represents the Gaussian distribution where $\mu_S = (\mu_x, \mu_y)$ is the mean vector, VP_1 and VP_2 are both eigenvectors and (v_1, v_2) the associated eigenvalues.

The image modeling principle is extended to video sequences by updating the model temporally: with a model parameter set $\theta_{(j,t-1)}$, each observed pixel value of a new image is classified by assigning it to the model component providing the maximum *a posteriori* probability C_{map}. For efficiency purposes, a log-likelihood formulation is generally used:

---

[36]The ellipse is the shape obtained by thresholding the probability density, introduced in equation 15.1, or the iso-value curve.

$$C_{map} = argmax_j\{log(p(\mathbf{x}_t|\theta_{(j,t-1)}))\} \tag{15.2}$$

The model components $\theta_{(j,t)}$ are then updated before processing the next image. In practice, the model is usually simplified by assuming that color and spatial components are independent and then uncorrelated. Then the probability density function (equation 15.2) can be re-expressed as the product of a spatial two-dimensional Gaussian distribution and a color three-dimensional Gaussian mode, with their respective parameters set $\theta^S_{(j,t)}$ and $\theta^C_{(j,t)}$. Each pixel is represented by a space vector $\mathbf{x}^S_t = [x, y]$ and a color vector $\mathbf{x}^C_t = [R, G, B]$. Thus equation (15.2) can be re-expressed:

$$C_{map} = argmax_j\{log(p(\mathbf{x}^S_t|\theta^S_{(j,t)})) + log(p(\mathbf{x}^C_t|\theta^C_{(j,t)}))\} \tag{15.3}$$

Then, the model defined on the entire image is composed of a set of Gaussian modes in a joint spatial-colorimetric space. It has been used in the literature for various purposes as shown in the next section.

15.2.2   Various Existing Work

The *SMOG* principle is mainly used for two applications in the literature:

Tracking: The *SGMM* model is used to characterize only one or several moving targets. Then, a set of modes represents each targeted moving object and is temporally tracked [5, 9, 10].

Background subtraction: The *SMOG* can also be used for background subtraction topics, modelling the whole scene (both background and foreground elements). The advantage of this work is twofold. First, this model allows a better discrimination between foreground and background since each of them is modelled. Secondly, it leads to higher level information that allows to combine both segmentation and target tracking since moving target modes are temporally maintained [1, 2, 11].

SMOG initialization, which does not require a learning step, is performed on the first sequence frame. Model initialization is often performed using EM[37] algorithm, such as in [9–11]. As such algorithms may slowly converge, the initial model can also be evaluated by analyzing the statistical characteristics of the spatial-colorimetric distribution. The modes are then successively cut and merged like in [1, 2, 8].

When initialization is completed, the modes are temporally updated in order to represent the different scene parts: for each incoming image, each pixel is classified to the most probable mode (if the distance is not too large). Target modes are thus intrinsically tracked during the segmentation process. To improve monitoring, some authors such as [11] pre-estimate the target's mode position, thanks to a simplified EM algorithm.

Finally, while an initial handmade segmentation of object is often provided [5, 9–11], some works such as [1] allow the dynamic creation of a target model during the process. This creation is performed when no available mode does properly characterize a region of space. From these data a new Gaussian mode is created and is labelled as an object because it was not previously observed.

---

[37] Exception maximization

15.2.3 Previous Work Limitations

So, the objective is to develop a generic and low constraint motion detection algorithm which can adapt to any kind of complex sequences without manual intervention during initialization. In this context, the work of [1] is the most interesting because it allows background and target patterns updating and also, the dynamic creation of new modes, unlike methods such as [5, 9–11] that need a provided *a priori* model. However, various problems detailed below appeared during the experiments of this approach, leading to significant evolutions of the original algorithm.

Initialization Problems

The initialization proposed in [1], although faster, is mainly based on a set of split and merge heuristics in the eigen-space. This model creation leads to several problems:

Initialization and update inconsistency: The models first created during the initialization are largely modified during the updating step. In fact, the model provided by initialization constraints is relatively inconsistent compared to the final model reached during the updating process. Thus, several images are required to stabilize the initial model.

Arbitrary and, sometimes, inconsistent split of modes the proposed system switches between spatial and color cutting stages. Although the process is relatively efficient, no warranty is provided on the quality of the final model. Especially, highly inconsistent modes appear, such as scattered and sparse clusters of pixels.

Initialization is not the sole cause of this mismatch between model and data. As once initialized the system is totally unconstrained, the updating process generates some temporal drifts and the model moves away from the optimal solution. Then distorted modes appear and lead to a results fall down. These drift problems appear both on the color and spatial spaces.

Temporal Color Drifts Problems

A totally unconstrained system generates a mode shifting whose covariance matrices become widely disparate. Indeed, some modes are then represented by an accurate covariance matrix, allowing only little color deviations from the mean value, while other modes, in contrast, are very diffuse. A too large covariance matrix produces a mode that can represent all observations, even too far ones. This kind of mode can aggregate all pixels, even the new object entering the scene. This defect is particularly embarrassing because the model becomes too uninformative to allow proper segmentation and therefore accurate detection. Instead, too focused and specialized modes are not robust and do not manage the slightest noise. Again, the quality of detection is affected. Finally, the presence of these two types of modes in a single mode pile significantly degrades performances. Wide modes have a trend to become larger and larger while specific modes will be more and more specific. The drift phenomenon is self-sustaining. Therefore, it is necessary to control the color covariance matrix to keep it within a given template and ensure the consistency between modes.

Temporal Spatial Drifts Problems

The spatial domain is also affected by the drift phenomenon during the unconstrained updating step. However, the size is not an important factor in this space. Indeed, it is necessary to allow all modes sizes to correctly and efficiently model all scene elements. Small size modes are required to represent details while major modes represent large and homogeneous surfaces such as floors or walls. However, the patterns need to be consistent and particularly they must be compact. Indeed, each mode is supposed to represent a colorimetrically consistent physical object-part. However, in the commonly used methods, a drift appears with time and some modes become highly sparse. For example, a mode could represent several spatially separated objects with close colors. Another example is the modeling of a single homogeneous area by the superposition of two spatially scattered modes. These modes are spatially equivalent, but each of them is specialized on a specific color. So, in addition to the control of the covariance matrix in the color space, the spatial density must be managed to keep homogeneous and compact modes.

Background/Foreground labeling

Finally, the labeling system proposed by [1] has an important inherent flaw and becomes completely unusable after a long time. Indeed, when a mode is created after the initialization period, it is automatically labelled as belonging to a target. The authors allow a mode to become "background" only if it does not move during a given period after its creation. However, in many cases, mislabelled modes can appear and then remain mislabelled. In particular, defects drift presented above lead to new modes creation that are labelled "object" if they move during the convergence time ensuring its stability. The opposite effect can also appear if an object with a color similar to the background enters into the scene. Then, a mode previously labelled as "background" will move with the object but will remain labelled as "background". This case also appears when a fixed object starts to move (a car parked in a parking lot for example).

All these problems make the final segmentation results not satisfactory. Then, after being identified, we tried to solve them and have developed a new approach presented in the next section.

15.3   Proposed Approach

The proposed system aims to reach a moving object detection with a pixel-size precision, and robust to realistic difficult environment conditions like acquisition noise or camera vibration. It is based on the work presented in [1] with several major extensions. First, the scene is modeled in a hierarchical way, with a pixel-wise temporal modeling (as Stauffer and Grimson [7] for example) and a global representation encoding the spatial consistency between pixels. This allows one to overcome some problems encountered with the *SMOG* approach and secondly, to make detection decisions on consistent sets of pixels, leading to robustness. Another contribution is the control of mode creation and evolution. It avoids color dispersion of modes enforces spatial density and so eliminates important drift with time. Finally, as the scene is modeled as an entity, the Gaussian mixture represents both background and foreground areas and a decision step is implemented and combined with a local shadow detection process to label each mode as background or object.

15.3.1   General Method Overview

As indicated in Section 15.2.1, the model used to represent the scene to be analyzed consists of a list of five-dimensional modes. This list is created on the first image and then maintained from frame to frame. A general view of the algorithm is presented in Figure 15.3, it is then detailed in the following sections.

For each new image, pixels are initially independently analyzed and assigned to the most likely modes (Section 15.3.2). By using a pseudo default mode (an uniform probability distribution), we allow pixels to be assigned to any modes. These pixels, not accurately

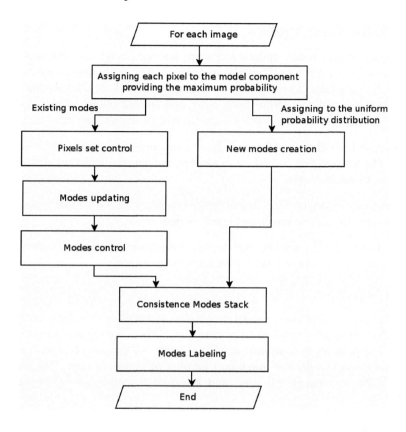

FIGURE 15.3   General method overview.

modeled by existing modes, and therefore probably not previously observed, are used to create new modes (Section 15.3.5). Contrary to the work proposed in [1], all modes (background and foreground) are considered in the same way. This allows avoiding discontinuities in a components model. A mode may represent a background area or a real moving object that can be tracked. Thus the algorithm manages in the same way new objects and moving objects present at initialization. Mode updating is performed using its associated pixels (see Section 15.3.3). To maintain an accurate modelization of the scene, the consistency of the mode pile is checked every image (Section 15.3.4). So, despite mode evolution, spatial and color consistencies are maintained. Finally it is necessary to label each mode as background or foreground in order to generate the segmentation result (Section 15.3.7).

15.3.2   Pixel to Mode Assignment

A five-dimensional mode list is built from the first image to model the scene. The creation of this initial model is detailed in Section 15.3.6. As already mentioned in Equation 15.3, each pixel of the current image is described with a feature vector $\mathbf{x}_t = [x, y, R, G, B]$ and is simply assigned to the Gaussian mode providing the highest likelihood.

When a new object appears in the scene, it is necessary to adapt the Gaussian mixture and create new modes for these new data. Thus an uniform distribution is added to the mode stack. Then, all pixels away from existing Gaussian modes will be assigned to this uniform distribution and are specifically addressed in a future step for new modes creation.

15.3.3   Modes Stack Update

If pixel assignment and mode update steps are unconstrained, models may derive as explained before (Section 15.2.3). Thus the update can gradually distort modes that deviate from their initial position. Then, before updating the model, it is important to check that each mode remain consistent, from both color and spatial points of view as presented in Sections 15.2.3 and 15.2.3. This control is performed at several levels:

On the assigned pixel set: All pixels have been independently tested in the previous stage. This step is not robust as mode can include outliers that disturb the system and lead to mode derive.

On the color mode part: This step keeps the modes within a template and avoids a drift towards too specific or too broad color models.

On the mode stack: During the update, an additional check is performed to ensure the consistency of the whole model with homogeneous modes.

Pixels Set Control

In order to have an accurate and consistent pixellic description, and since modes are parts of physical objects, they are supposed to be compact. This property is ensured by introducing a connected components decomposition of the pixels set assigned to each mode. Only the main component part (with the maximum number of pixels) is kept. The other pixels are reassigned to the uniform distribution and are used to create new Gaussian modes (see Section 15.3.5).

Modes Updating

Let us consider the updating of the jth mode. The first step is to find all pixels x that have been assigned to this mode L_j. The new mode components estimated from current data are evaluated by the following equations:

$$\omega_{(j,sm)} = \frac{n_j}{N} \qquad \mu_{(j,sm)} = \frac{1}{n_j} \sum_{x \in L_j} x \qquad \Sigma_{(j,sm)} = \frac{\sum_{x \in L_j} x^T x}{n_j} - \mu_j^T \mu_j \quad (15.4)$$

where n_j is the number of pixels in L_j and N is the total number of pixels in the image.

These data constitute the parameter vector $\theta_{(j,sm)} = \{\omega_{(j,sm)}, \mu_{(j,sm)}, \Sigma_{(j,sm)}\}$ which is used to update the parameter vector of the jth mode. From the parameters of the previous model $\theta_{(j,t-1)}$, the new values $\theta_{(j,t)}$ are calculated using an adaptive learning (α_j is the learning rate) :

$$\theta_{(j,t)} = \alpha_j.\theta_{(j,sm)} + (1 - \alpha_j)\theta_{(j,t-1)} \tag{15.5}$$

Color Modes Control

Control of the mode color size must be performed to keep all modes in an homogeneous template and avoid extreme drifts. As explained in Section 15.2.1, the color covariance matrix Σ_j^C represents the shape and mode size in the color space. So, its components must be checked to prevent excess. The control is not done by changing the pixels assignment but by "thresholding" the color size of the covariance matrix to avoid reaching extreme values. To this end, let us define the size of a covariance matrix.

Covariance is useful but its components are not easily interpretable. By definition, it is a symmetric diagonalizable positive definite matrix, with an endomorphism entirely determined by its eigenvectors and eigenvalues:

The eigenvectors allow determining the orientation of the data.

The eigenvalues allow determining the mode extent in each direction defined by the eigenvectors.

The mode size, that is independent of data orientation, is defined by the product of the square roots of the eigenvalues (to remain consistent according to dimension):

$$S\left(\Sigma_j^C\right) \propto \prod_{k=1}^{K} \sqrt{\mathrm{vp}_j^C\left[k\right]} \tag{15.6}$$

with $\mathrm{vp}_j^C\left[k\right]$ the kth eigenvalue of the color covariance matrix Σ_j^C of dimension $K * K$ ($K = 3$ in our case).

As already mentioned at the beginning of this section, it is necessary to maintain a consistent color size to avoid, for example, a mode specialization on a particular color. It is thus necessary to limit the size but also to control the proportion of the model components. Indeed, if a component is disproportionate, it is specialized in a color range and causes the same bad effects as a too specialized mode. So, consecutive eigenvalues (assuming that they are classified in a descending order) should have the same order of magnitude. Thus, their ratio should remain below a threshold representing the maximum allowed deviation. If this is not the case, the eigenvalues of the matrix are modified to satisfy this condition and ensure that the matrix remains inside a fixed template. The principle of this constraint is given in the following pseudo-code algorithm:

```
INPUT  : all k eigenvalues -> vp[]
INPUT  : the maximum ratio allowable  between two consecutive
         eigenvalues -> rlim
OUTPUT : The k new eigenvalues -> vp[]

FOR i = 1 TO k-1 DO :
    ract = vp[i] / vp[i+1]
    IF ract > rlim THEN :
        vp[i+1] = vp[i] / rlim
    END IF
END FOR
```

In a second step, the size of the matrix is controlled to avoid too specialized or too wide modes. Thus, if Th_S^C is the ideal size of the matrix, we impose to all the color covariance matrix sizes to remain inside the template $\left[(1 - p_S^C) * Th_S^C; (1 + p_S^C) * Th_S^C\right]$ where p_S^C ($0 < p_S^C < 1$) is a parameter set by the user to provide some flexibility.

The matrix control is quite simple: if the size is out of the template, all the eigenvalues are reduced (or increased) by the same ratio to bring the equation 15.6 to the perfect size Th_S^C. The advantage of processing the eigenvalues is obvious: all distributions will have similar sizes while maintaining their own orientation. The principle is illustrated with the following algorithm:

```
INPUT  : all k eigenvalues -> vp[]
INPUT  : Threshold th_sc
INPUT  : The acceptance rate p_sc
OUTPUT : The k new eigenvalues -> vp[]

// Extremes threshold rate
```

```
th_min = (1 - p_sc) * th_sc
th_max = (1 + p_sc) * th_sc

// Mode size evaluation
mode_size = 1
FOR i = 1 TO k DO :
    mode_size = mode_size * vp[i]
END FOR

// Eigenvalues updating
// Reduction rate
IF mode_size < th_min THEN :
    f_red = (mode_size / th_min)^(2/k)
ELSE IF mode_size > th_max THEN :
    f_red = (mode_size / th_max)^(2/k)
ELSE :
    f_red = 1
END IF
// Rate apply
FOR i = 1 TO k DO :
    vp[i] = vp[i] / f_red
END FOR
```

Once the new eigenvalues have been determined, the covariance matrices are reconstructed. If $M^C_{\{j,VP\}}$ is the matrix containing the eigenvectors of Σ^C_j and $N^C_{\{j,VP\}}$ the diagonal matrix containing the associated eigenvalues, the new covariance matrix Σ^C_j is reconstructed by:

$$\Sigma^C_j = M^C_{\{j,VP\}} N^C_{\{j,VP\}} \left(M^C_{\{j,VP\}} \right)^{-1} \tag{15.7}$$

Actually, the control is performed twice:

- the first time on the covariance matrix of the pixels set so that outliers do not disturb the model

- a second time after the updating step to ensure a consistent modes stack.

Interest of This Control

The proposed system allows having consistent modes stack, improving results and avoiding the drifts presented at the beginning of this section. It provides also a simple way to control the complexity of the modeling thanks to the threshold Th^C_S combined with the updating and mode creation processes (Section 15.3.5). A low value of this threshold results in small components, relatively specialized on their color. Conversely a high value allows more flexibility with some outliers and results in larger modes. This characteristic is particularly interesting because it can be fixed regardless of the scene. For a given threshold value Th^C_S, a scene with large homogeneous areas will automatically have less modes than a highly textured scene. The number of modes fits to obtain the same description level.

15.3.4 Consistence of Mode Stack

Finally, in order to maintain a consistent mode stack, several operations are performed after the updating step:

- Deleting modes having too weak weight ω_j (~ 0), or those that are no longer present in the scene;

- Deleting modes with a low density (the ratio between the weight ω_j and the spatial size obtained identically to the equation 15.6);

- Normalising weight to satisfy their sum equals to one $\left(\sum_{j=1}^{N} \omega_j = 1 \right)$.

These operations lead to a clean stack mode, consistent and well balanced.

15.3.5 New Mode Creation

As previously mentioned, some pixels are assigned to the uniform distribution added to the modes stack. Three cases may be distinguished:

Local Singularity: This case occurs occasionally and locally. Some pixels do not match to the model due to noise acquisition, color artifacts, compression default,... These data should be ignored.

New object appearance: When an new object enters the scene, the corresponding pixels do not match to an existing model. It is necessary to adapt the Gaussian mixture and create new modes for these new pixels.

Pixel group splitting: If a single mode is associated to two physical objects that are being distant, the process will separate this mode in two parts and the one with the smallest number of pixels will be assigned to the uniform distribution. As in the previous case, these pixels represent an important group of connected pixels that will be used to create new modes.

As it is important to create modes consistent with the constraints imposed by the update process, several steps are executed.

Cutting into Connected Components

The set of pixels assigned to the uniform distribution is decomposed into connected components and only the components with a sufficient number of pixels are stored. Groups with too few pixels are considered as *local singularities* and remain assigned to the uniform distribution. Remaining pixel groups are used to initialize new modes using equations 15.4.

At this stage, all modes satisfy the compacity constraint, but are not necessarily in the color template. In this case, they should be cut into smaller modes respecting the template.

Color-Space Constraints

The size of each color covariance matrix is analyzed:

- If it fits in the color template, the mode is added to the final stack mode model.

- If the size is too small, the color constraints are applied to artificially increase the mode size and prevent its specialization on a particular color. Indeed, by construction we know that this mode contains enough pixels. It is then also added to the mixture of Gaussian modes.

- Finally, if the size is too large, the mode is split in two parts in the direction associated with the largest eigenvalue. Then as the two newly formed groups may be not consistent, they are studied again by the previous steps.

These operations are iterated until reaching the stabilization of the mode stack. In the general case, except for large scene changes, only a few iterations are needed.

15.3.6   Initial Model Construction

With the presented system, no special action is required to create the initial model. In the first image, all pixels are attributed to the uniform distribution (the only distribution in the initial mode stack). The modes stack is created thanks to the step of modes creation.

15.3.7   Modes labeling

The above steps maintain a consistent statistical model of the scene over time. Each mode may represent a background area or be attached to a real moving object part. It is then necessary for the detection step to label each mode as *background* or *foreground*. This decision is difficult to take at the mode level, for example by analyzing temporal evolution of μ_j^S. Indeed background elements modes are likely to move. This is for example the case on large homogeneous surfaces where several modes exist and move on this surface to suit the changing environment. Moreover, background modes are distorted during a partial occlusion due to the passage of an object.

Background/Foreground Segmentation

The decision, which is taken for each mode, involves pixellic information coming from a local temporal modelling obtained with the classical Stauffer and Grimson algorithm [7] whose limitations are the generation of over-detections due to acquisition noise or brightness changes and absence of detection when objects look like background part. The decision offered by the proposed approach, is performed globally, on a pixels set. It is thus more robust and partially avoids local errors (pixellic). A simple threshold on the average of the probability map generated with the Stauffer and Grimson method on all pixels inside the *SMOG* mode allows deciding whether globally the pixels belong to the background mode or not. Note that the use of all local probabilities is more robust than a simple majority vote around each pixel, and that this approach provides more robust decisions than a simple low-pass filtering of pixellic decision as shown in the results. Finally, for the isolated pixels that have not been assigned to any mode, the final decision is the one provided by the Stauffer and Grimson method, locally.

Shadow Removal

To further improve the results, an additional algorithm has been implemented. Based on the work of [6], it allows to obtain a new score characterizing the probability that each pixel is inside a shadow area. Similarly as above, this score is averaged over all the pixels of each mode in order to make a global decision.

15.3.8   Frame-to-Frame Object Tracking

Many systems implement a frame to frame object tracking after the detection algorithm. In the *SMOG* approach, the tracking is integrated within the model updating process. Assuming that the movement of objects from one image to the next one is relatively small, the modes corresponding to the moving objects will evolve over time to track them. If objects have larger displacements, the corresponding "target" component disappears and a new component is recreated automatically. The segmentation and labeling procedures remain effective but the temporal correspondence is lost in this case.

15.4 Validation and Results

To evaluate the results of the presented method, many tests were performed. At first, some results showing the internal structure of the *SMOG* system are presented to highlight the contribution of the *SMOG* approach compared to the pixel-size approach of Stauffer and Grimson [7], both in terms of detection results and shadows removing [6] point of views. Then, quantitative results are presented to measure performance on real sequences, compared to other methods from the literature. Finally, in order to emphasize the robustness to difficult conditions, while keeping pixel-wise accuracy, results in more severe sequences (camera shake, color noise) are presented.

15.4.1 Test Sequences

Tests on several sequences were performed and evaluated, in Section 15.4.3, thanks to the associated handmade ground-truths. Three sequences are presented here (Figure 15.4):

"PETS2006" [4]: This is a sequence captured within a railway station. The environment and lighting conditions introduce many reflections and shadows and some targets are difficult to extract because of a textured background.

"Blue Room": This is an indoor sequence where the environment and lighting conditions are controlled. The main sequence difficulty is some clothes colors close to the ground and walls ones.

"Yellow Room" : This is an indoor sequence with direct outdoor lighting changes and a light flicker.

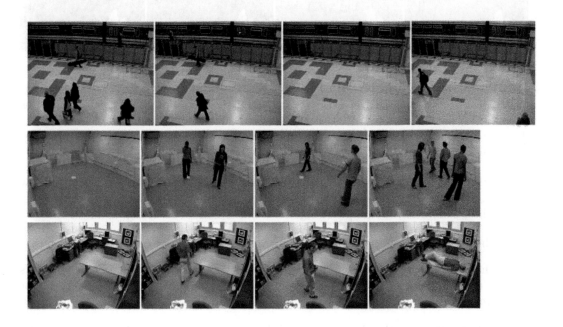

FIGURE 15.4   First row: PETS2006 sequence [4]; Second row: "Blue Room" sequence; Third row: "Yellow Room" sequence.

15.4.2   Qualitative Evaluation

To illustrate the main steps of the algorithm, key-frames are shown from the processing of an image coming from the PETS2006 dataset [4] in Figure 15.5 and from the "Blue Room" sequence in Figure 15.6. A component modeling is shown: each mode is associated with a single random color and all pixels assigned to this mode are displayed in this color. First, despite the important number of shadows and reflections in this sequence, we observe that the targets are properly extracted with both methods. On the SMOG representation, we observe that targets and background do not use the same modes (clean and separated regions). This feature is important to achieve good segmentations results.

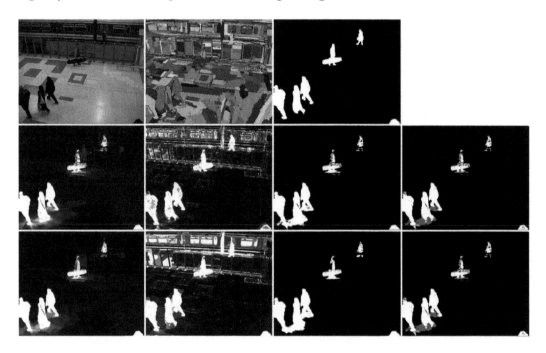

FIGURE 15.5   Presentation of a PETS2006 frame processing First line: source image, *SMOG* modeling and associated ground truth. 2nd line: step of [7] algorithm processing. 3rd line: the processing steps of the proposed *SMOG* implementation. 1st column: probability map of belonging to the foreground; the second column: distance map of the shadows removing algorithm; 3rd column: raw segmentation results; 4th column: shadow removing application.

Another interest of the global model is shown in Figure 15.7. In this sequence ("Yellow Room"), a person is wearing clothes whose color is similar to a background-object one. Under these conditions, the proposed algorithm has two main advantages:

- The overall decision, performed globally on a Gaussian mode, is much more robust.

- The coupled background and objects modelization allows an optimal detection. This is equivalent to an automatically adjusted detection threshold based on the color proximity (or not) between background and foreground.

Finally, to illustrate the shortcomings of the literature *SMOG* method [1], two results are presented in Figure 15.8 and 15.9. The first result (Figure 15.8) shows the labeling at the end of the sequence. While the second SMOG method considers rightly, like Stauffer & Grimson one, that no objects are present, the method presented in [1] has created several

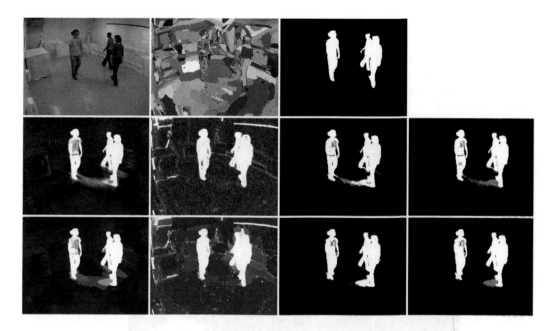

FIGURE 15.6 **Presentation of a "Blue Room" frame processing** (see Figure 15.5 for frame description).

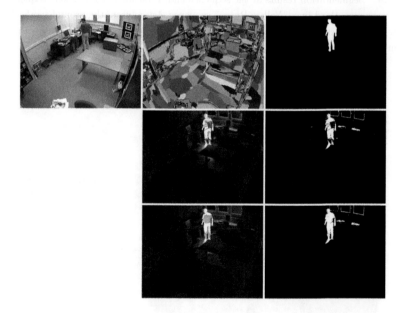

FIGURE 15.7 **Presentation of a "Yellow Room" frame processing** First line: source image, $SMOG$ modeling and associated ground truth. 2nd line: step of [7] algorithm processing. 3rd line: the processing steps of the proposed $SMOG$ implementation. Second column: probability map of belonging to the foreground; 3rd column: raw segmentation results.

modes, labelled as objects, that are mislabelled until the end of the sequence. The second result (Figure 15.9) shows some evolution drift over time of the [1] SMOG model method unlike the presented one. In [1], foreground elements, that are updated quickly, have a major drawback: the modes components are not informative enough. This is the case when a mode is assigned to two spatially disconnected people, when the same mode is used to represent a head with both dark hair and fair face or when a green shirt and human arms are associated. Under these conditions, the drifting effect is obvious: this kind of mode can equally match with all the scene objects, both foreground or background ones, resulting in a poor modeling.

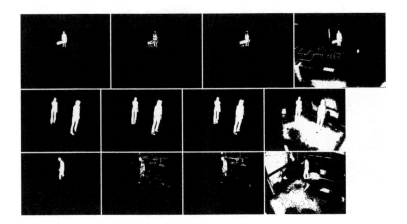

FIGURE 15.8   Segmentation results at the sequence end. First line: "PETS2006" sequence, second line: "Blue Room" sequence, third line: "Yellow Room" sequence. 1st column: the ground truth; Second column: the Gaussian mixture method [7]; 3rd column: the previously described *SMOG* algorithm; 4th column: *SMOG* from the literature [1].

FIGURE 15.9   Illustrations of the SMOG models. First row: *SMOG* from the literature [1], second row: the previously described *SMOG* algorithm.

15.4.3   Quantitative Results on Real Sequences

Metrics

All the results of this chapter (Figure. 15.10 to 15.15) are illustrated with two curves: the ROC[38] curve on the left side of the figures and the precision/recall curves on the right. The illustrated concepts by these two kinds of representation are defined below.

ROC curves : ROC curve represents the true positive rate versus the false positive rate. Then, the ideal point is $(0,1)$. This kind of curve is used to compare algorithms performance independently of the selected detection threshold.

Precision/recall curves: Although ROC curves are frequently used to compare methods, they have the drawback of not taking into account the proportion of positive and negative examples. So, if positive data are more numerous than negative ones (as it is the case in this application[39]), ROC curves do not represent the real quality of results. Indeed, in extreme cases, 1% of negative examples wrongly considered as positive (so false positive) may actually represent 10-20 or 30% of the actual target size.

To take into account this disparity, precision-recall curves were added[40,41]. In this case, the ideal point is (1.1). These curves are much more specific to the tested sequence but allow a better evaluation of perceived performance.

Using these metrics, three methods of the literature are compared:

Pixel-wise modeling by a mixture of Gaussian: this is the classic algorithm as proposed by Stauffer and Grimson [7].

SMOG modeling reference: this is the approach proposed in [1], for a *SMOG* reference.

The SMOG approach: this is the approach described in this paper.

The parameters of all these methods have been adjusted to reach the best compromise on all processed sequences. Furthermore, in a first step, the shadow removing algorithm has not been activated in order to compare raw results of the different methods.

Raw Detections Results

Results on the PETS2006 dataset [4]

On this sequence, the best results are obtained with the modified SMOG algorithm (Figure 15.10). The main difficulties are introduced by shadows and highlights. Thanks to the global decision at the mode level, the described approach compensates some local defects generated by the approach of [7] and due to noise. These strong reflections degrade the first *SMOG* approach [1], by creating many aberrant modes labelled as objects.

Results on the "Blue Room" sequence

Results have relatively few differences. However, a person whose clothing color is close to the background one is present in the scene. The modeling of both foreground and background

---

[38] Receiver Operating Characteristic

[39] Targets are generally smaller than background and sometimes any target is present in the scene

[40] Accuracy: ratio between the number of correctly classified positive examples(true positives) and all classified positive data (true + false positives)

[41] Recall : ratio between the number of correctly classified positive examples (true positives) and all real positive data (true positives + false negatives)

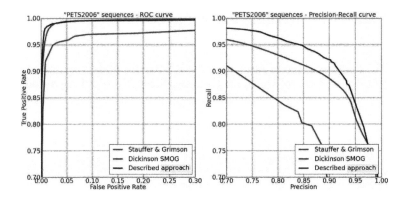

FIGURE 15.10   PETS2006 results. Left: ROC curve; Right : Precision-Recall curve. Red: Gaussian mixture method [7]; Green: $SMOG$ from the literature [1]; Blue the previously described $SMOG$ algorithm.

areas allows a better extraction of the target than with the classical Stauffer and Grimson [7] approach. With a well controlled environment, the SMOG method from the literature [1] is quite stable than in the previous test.

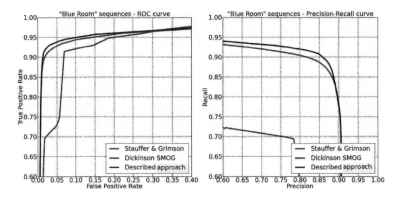

FIGURE 15.11   "Blue Room" results. Left: ROC curve; Right: Precision-Recall curve. Red: Gaussian mixture method [7]; Green: $SMOG$ from the literature [1]; Blue: the previously described $SMOG$ algorithm.

Results on the "Yellow Room" sequence

The difficult lighting conditions lead to similar results between the local Gaussian Mixture model [7] and the proposed SMOG approaches. In particular, the modified SMOG method is penalized by its grouping into modes. When the proportion of pixels with high distances to the local Gaussian mixture model becomes too large, the entire Mode is mislabeled, which penalizes the results. This method is then sensitive to excessive error rates provided by the local algorithm [7]. Nevertheless, it provides better results over a wide threshold range. The $SMOG$ method [1] is highly affected by these difficult conditions. New mislabeled modes appear and remain throughout the sequence.

The presented results show the relevance and efficiency of the described method on three real sequences having different characteristics. The gains are particularly significant on accuracy, and confirm the improvements due to a pixel-set analysis for decision. The poor results achieved by [1] show the limitations already presented in Section 15.2.3 and in

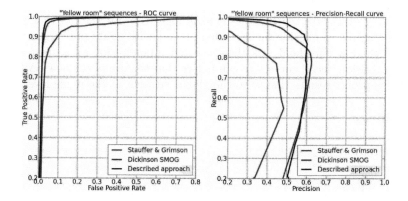

FIGURE 15.12   "Yellow Room" results. Left: ROC curve; Right: Precision-Recall curve. Red: Gaussian mixture method [7]; Green: $SMOG$ from the literature [1]; Blue: the previously described $SMOG$ algorithm.

particular model and label drift over long sequences.

Shadow Removal

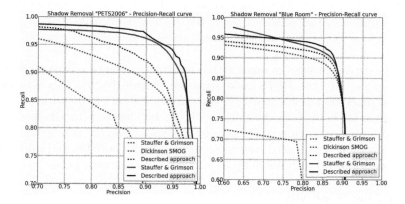

FIGURE 15.13   Precision-Recall curve for shadow removal algorithm on PETS2006 sequence (left) and "Blue Room" sequence (right). Red: the local Gaussian mixture method [7]; Green: $SMOG$ from the literature [1]; Blue: the new $SMOG$ algorithm. Solid lines: with shadows removing. Dashed curves: without shadows removing.

To illustrate another advantage of the mode-level approach, we have implemented the algorithm [6] for shadow removing as presented in Section 15.3.7. It provides for each pixel a distance between the color of the pixel and the background model taking into account the shadows effect. The PETS2006 and "Blue Room" sequences have many shadows and reflections and were used for these tests whose results are observable on Figure 15.13. The sequence "Yellow Room" is not considered as it does not have much shadow. Moreover, the $SMOG$ method proposed in [1] is based on a different labeling that does not easily allow the integration of the shadow removal algorithm. Thus, it is not tested here. The algorithm for shadows removal [6] is applied only on the pixels detected as belonging to the target in order to possibly classify them as shadow. Therefore, it can only reduce data classified as positive (a real shadow removal decreases the false positive rate while an error will decrease the true positive rate). To compare the shadow removal algorithm for the different methods, they have been configured to provide the same high true positive rate (all targets are detected). The removal shadow threshold was then selected uniquely for all sequences in order to draw a new precision/recall curve. We can notice that the use of the new SMOG algorithm significantly improves results thanks to the global decision at the mode level.

Noise Sensitivity

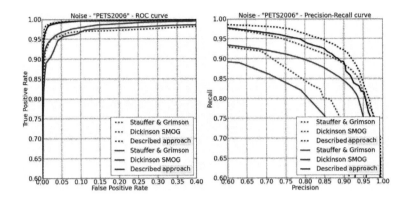

FIGURE 15.14   Results on PETS2006 sequence degraded by a Gaussian noise with standard deviation of 10. Red: local Gaussian mixture method [7]; Green: *SMOG* from the literature [1]; Blue: the new *SMOG* algorithm. Solid lines: with shadows removing. Dashed curves: without shadows removing.

The next experiment aims to quantify the sensitivity of the methods according to noise by adding a noise into the *PETS 2006* sequence (Figure 15.14). This Gaussian noise is spatially and timely independent with a standard deviation of 10. Without noise, results are quite similar for both methods. However, it can be observed that the proposed SMOG algorithm performance remains stable as noise increases while the Stauffer & Grimson algorithm loses precision. The spatial clustering allows a better stability and robustness thanks to the joint decision and distance analysis over a set of consistent pixels.

Camera Shaking Sensitivity

Finally, with a global spatial and color modeling, the described method is less sensitive to camera-shake. Then, the *PETS 2006* frames have been distorted with a sinusoidal move around the true position with a Gaussian random on period and amplitude (Figure 15.15). It can be noticed that, like on noise test, our method remains stable while the Stauffer & Grimson algorithm loses precision under camera vibration.

15.5   Conclusion

Local modeling performed independently on each pixel is typically proposed by conventional approaches of motion detection. Using a descriptor defined on blocks of pixels of pre-determined shape and size leads to more robustness, but penalizes the detection precision. To overcome these limitations, the described approach combines the accuracy of the pixel information with the robustness of a consistent pixels set decision. Particularly, the consistent model allows improving both detection and shadow removal methods thanks to the spatial consistency and pixel group decision with respect to both the Stauffer and Grimson method and the first *SMOG* principle.

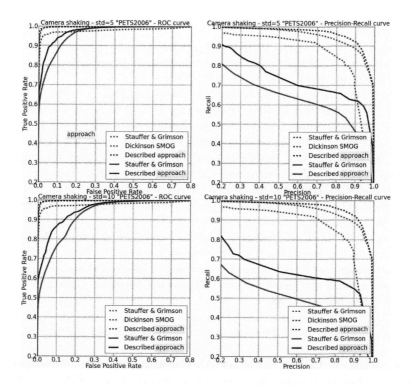

FIGURE 15.15 Results on PETS2006 sequence with vibration. First row: standard deviation of the amplitude on the Gaussian random of 5, second row: standard deviation of the amplitude on the Gaussian random of 10. Red: local Gaussian mixture method [7]; Green: $SMOG$ from the literature [1]; Blue: the new $SMOG$ algorithm. Solid lines: with shadows removing. Dashed curves: without shadows removing.

The complete scene modeling (background + foreground) also allows to manage some ambiguities when a moving object has an appearance similar to the one of the background. Where conventional algorithms have only a background model, the proposed method introduces two models (background and object). This allows to transform the detection task into a classification problem with two classes.

The experimental results show that the described approach improves both methods from which it derives thanks to the rigorous control of mode consistency, avoiding model drifts.

In addition to the improvement of the detection, the method provides a spatial and colorimetric modeling of modes that is continuously updated. As a result, each mode tracks a part of the object (or background) they characterize. We then obtain high-level information that can be directly integrated to perform a complete target tracking. Using this information allows interesting tracking perspectives. Indeed this approach would combine at the same time complete scene modeling, segmentation task, informative object descriptors and local tracking.

References

1. P. Dickinson, A. Hunter, and K. Appiah. A spatially distributed model for foreground segmentation. *Image and Vision Computing*, 27(9):1326–1335, 2009.
2. P. Dickinson, A. Hunter, and K. Appiah. Segmenting video foreground using a multi-class MRF. In *International Conference on Pattern Recognition, ICPR 2010*, pages 1848–

1851. IEEE, 2010.

3. A. Elgammal, R. Duraiswami, D. Harwood, and L. S. Davis. Background and foreground modeling using nonparametric kernel density estimation for visual surveillance. *Proceedings of the IEEE*, 90(7):1151–1163, 2002.

4. J. Ferryman. http://www.cvg.rdg.ac.uk/PETS2006/data.html.

5. J. Gallego, M. Pardas, and G. Haro. Bayesian foreground segmentation and tracking using pixel-wise background model and region based foreground model. In *IEEE International Conference on Image Processing, ICIP 2009*, pages 3205–3208, 2009.

6. K. Lo, M. Yang, and R. Lin. Shadow removal for foreground segmentation. In *Advances in Image and Video Technology*, pages 342–352. Springer, 2006.

7. C. Stauffer and WEL Grimson. Adaptive background mixture models for real-time tracking. In *IEEE International Conference on Computer Vision and Pattern Recognition, CVPR 1999*, 1999.

8. C. Trunong, N. Dung, L. Khoudour, C. Achard, and A. Flancquart. Adaptive model for object detection in noisy and fast-varying environment. In *ISAP*, 2011.

9. H. Wang, D. Suter, and K. Schindler. Effective appearance model and similarity measure for particle filtering and visual tracking. *European Conference Computer Vision, ECCV 2006*, pages 606–618, 2006.

10. H. Wang, D. Suter, K. Schindler, and C. Shen. Adaptive object tracking based on an effective appearance filter. *IEEE Transactions on Pattern Analysis and Machine Intelligence*, pages 1661–1667, 2007.

11. T. Yu, C. Zhang, M. Cohen, Y. Rui, and Y. Wu. Monocular video foreground/background segmentation by tracking spatial-color Gaussian mixture models. In *IEEE Workshop on Motion and Video Computing, WMVC 2007*, page 5, 2007.

Online Robust Background Modeling via Alternating Grassmannian Optimization

Jun He
Nanjing University of Information Science and Technology, China

Laura Balzano
University of Michigan, Ann Arbor, USA

Arthur Szlam
City University of New York, USA

16.1 Introduction

Low-rank subspaces have long been a powerful tool in data modeling and analysis. In particular, they have proven very useful in computer vision: Subspace models are of great interest in computer vision for background subtraction [52], object tracking [5, 19], and to represent a single scene under varying illuminations [9, 20]. Other applications in communications [40], source localization and target tracking in radar and sonar [33], and medical imaging [3] all leverage subspace models in order to recover the signal of interest and reject noise. In these classical signal processing problems, a handful of high-quality sensors are co-located such that data can be reliably collected.

The challenges of modern data analysis and large-scale computer vision breach this standard setup. A first difference, one that cannot be overstated, is that data are being collected everywhere, on a more massive scale than ever before, by cameras, sensors, and people. Consider just a few examples: There are an estimated minimum 10,000 surveillance cameras in the city of Chicago and an estimated 500,000 in London [4, 11]. Netflix collects ratings from 25 million users on tens of thousands of movies [16]. On its peak day of the holiday season in 2008, Amazon.com collected data on 72 items purchased every second [51]. The Large Synoptic Survey Telescope, which will be deployed in Chile and will photograph the whole sky visible to it every three nights, will produce 20 terabytes of data every night [49].

A second and equally important difference is that, in all these examples mentioned, the data collected may be unreliable or an indirect indicator of what one really wants to know. The data are collected from many possibly distributed sensors or even from people whose responses may be inconsistent, and the data may be missing or corrupted.

In order to address both these issues, algorithms for data analysis must be computationally fast as well as robust to corruption, unknown data transformations, and missing data. This fact is the motivation for the work we describe in this chapter. Recent developments at the interface of optimization and statistical signal processing have resulted in the theory of matrix completion [15], which shows that a very large yet low-rank matrix can be reconstructed from a very small number of its entries. For example, a million-by-million rank-50 matrix could be reconstructed from a uniform random sample of only 3% of its entries with probability at least .96[42]. The theory of Robust PCA [1, 14, 17, 53, 54] takes this further to say that a low-rank matrix can be reconstructed even when the matrix has corrupted entries. We may also turn this around: since the subspace can be identified from incomplete vectors with corruptions, we can subsample in order to improve on computational efficiency, and we still retain subspace estimation accuracy in the face of corruptions.

In this chapter we present the Grassmannian Robust Adaptive Subspace Tracking Algorithm , or GRASTA [24], an online algorithm for robust subspace tracking that is inspired by these results in optimization, and its variant t-GRASTA [25, 26] that incorporates geometric transforms on the data. For GRASTA, we seek a low-rank model for data that may be corrupted by outliers and have missing data values; for t-GRASTA, we seek a low-rank model for misaligned images that may be corrupted by outliers. Both GRASTA and t-GRASTA use the natural ℓ^1-norm cost function for data corrupted by sparse outliers, and both perform incremental gradient descent on the Grassmannian, the manifold of all d-dimensional subspaces for fixed d. For each subspace update, we use the gradient of the augmented Lagrangian function associated to this cost. Our GRASTA-type algorithms operate only one data vector at a time, making them faster than other state-of-the-art algorithms and amenable to streaming and real-time applications.

We show the application of two versions of the GRASTA algorithm to background subtraction in particular. Background subtraction is a useful technique for preprocessing image and video data, especially when there are active, moving objects of interest and a relatively more static background. It is has been used as a crucial component in human activity recognition and the analysis of video from surveillance cameras. Computational cost and real-time processing are of utmost importance here: for example, the video datasets mentioned above are almost entirely used post-hoc for analysis after a disruptive or criminal event. The goal of preprocessing is to enable technologies that can analyze video data in real-time.

---

[42]See [48] and ignore the constant; we see empirically that this is a good rule of thumb.

16.2 Related Work

16.2.1 Robust PCA

If we wish to learn the background subspace dynamically from the video to be analyzed, classical least squares low-rank approximation is insufficient, precisely because we expect the "foreground" objects to not follow the low-rank model; the pixels corresponding to these objects will not be well approximated by the model. It is well known that even small numbers of gross coordinate corruptions lead to large changes in a least squares low-rank model [29].

Due to the vast applicability of principle components analysis, robust versions have been studied for some time. Precipitated by a new understanding of convex surrogates for cost functions minimizing rank and sparsity of a matrix, a great deal of recent work has focused on theoretical results showing that convex programming may exactly solve the Robust PCA (RPCA) problem under a variety of assumptions and problem formulations. Much of the work was concurrent or nearly so; we do our best to describe the results here.

The work of [17] provided breakthrough theory for decomposing a matrix into a sum of a low-rank matrix and a sparse matrix; it defined a notion of rank-sparsity incoherence under which the problem is well-posed. What's more, using this as an assumption on one's data allows one to use convex programming in order to solve for the true decomposition. The requirements on the rank-sparsity incoherence are deterministic, but the authors also show how randomly drawn low-rank and sparse matrices satisfy the assumptions of their theory. Similarly, [14] gives guarantees for the same convex program to recover both the sparse and the low-rank matrix; these guarantees are with high probability with respect to a randomly drawn support for the sparse matrix. The authors in [53] again provide very similar guarantees, with high probability for particular random models: the random orthogonal model for the low-rank matrix and a Bernoulli model for the sparse matrix. The work of [1] is slightly more general in the sense that it proves results about matrix decompositions that are the sum of a low-rank matrix and a matrix with complementary structure, of which sparsity is only one example. This paper studies a family of M-estimators which are the solution to regularized least-squares convex programs. They also show that these estimators are minimax optimal for the Frobenius norm error. In [27], the authors again follow a similar story as [17], providing guarantees for the low rank + sparse model for deterministic sparsity patterns. Since the sparsity pattern is allowed to be deterministic, they do pay a small price in terms of the size of the support of the sparse matrix relative to [14], however, their guarantees allow the support to be larger than in [17].

Another formulation of the Robust PCA problem considers that *entire columns* (or rows) of the low-rank matrix may be corrupted. That is, as opposed to [14,17,53], the corruptions are column-wise or row-wise sparse as opposed to entry-wise sparse. The paper [55] presents an outlier pursuit algorithm, and their supporting theory states that as long as the fraction of corrupted points is small enough, their algorithm will recover the low-rank subspace as well as identify which columns are outliers. Their theoretical results hold in both the noise-free and the noisy measurement case. Like [55], the work of [54] supposes that a constant fraction of the observations (rows or columns) are outliers. In [54] the authors provide an algorithm for Robust PCA and give very strong guarantees in terms of the *breakdown point* of the estimator [29]. Two algorithms are proposed in [39] for Robust PCA. The first is an instance of what is known as projection pursuit [34]. This algorithm takes the view of PCA as the directions of maximum variance in the data. Their theory here is based on a formulation of RPCA which replaces maximization of the empirical standard deviation of PCA, which is not robust, with maximization of a robust scale measuring the variance of the observations. The second algorithm takes the view of PCA as the optimal low-rank model for the data. Their guarantees for this second algorithm come in the form of bounds on the leverage of the points in the low-rank estimate. This paper especially has a nice related work section discussing the history of their particular formulation of RPCA as well as several algorithms that have been proposed.

Several of the papers above proposed the use of semi-definite program (SDP) generalized solvers, but SDP solvers are known to be slow in practice though they run in polynomial time. Some of the already mentioned papers proposed their own algorithms, and many more algorithms have also been proposed, which we attempt to outline now.

The work in [37] provides Augmented Lagrange Multiplier algorithms for recovery of the low-rank and sparse matrices under the same model as [53]. They provide convergence theory of their algorithms: they can show that their exact method converges to the global optimal point with a rate given, and the theory for their inexact method converges in the limit without a rate given. The paper derives stopping conditions for their algorithms and gives applications of their algorithms to the matrix completion problem as well. A non-convex reformulation is proposed in [50] which also uses the Augmented Lagrangian formulation and alternating estimation of the factors of the low-rank matrix. They show that any limit point of a sequence of iterations of their algorithm satisfies the KKT conditions.

Many previous Robust PCA algorithms [14] operate in batch[43]. Those batch algorithms which are based on convex optimization can estimate the low-dimensional subspace with subsampled data, taking a cue from low-rank matrix completion theory. The majority of algorithms use SVD (singular value decomposition) computations to perform Robust PCA. The SVD is too slow for many real-time applications, and consequently many online SVD and subspace identification algorithms have been developed.

Online algorithms for Robust PCA were presented in [23,30,38], but these do not support estimation from subsampled data. The work of [47] applies principle components pursuit to video processing; the authors wish to exploit correlation between the sparse part in two successive video frames. Their algorithm is online, though they require an accurate initial estimate of the principle components.

16.2.2   Transformed Robust PCA

In many applications, the low-dimensional subspace assumption which is central to Robust PCA may only hold after an unknown transformation is applied to the data. In computer vision in particular, images lie collectively in a low-dimensional subspace only when the images are exactly aligned. The algorithm "Robust Alignment by Sparse and Low-rank decomposition," or RASL [28,42], poses the robust image alignment problem as a transformed version of Robust PCA. The transformed batch of images can be decomposed as the sum of a low-rank matrix of recovered aligned images and a sparse matrix of errors. RASL seeks the optimal domain transformations while trying to minimize the rank of the matrix of the vectorized and stacked aligned images and while keeping the gross errors sparse. While the rank minimization and sparsity regularization can be relaxed to their convex surrogates as in Robust PCA, the relaxation proposed in RASL is still highly non-linear due to the domain transformation. However, the authors apply others' convergence results to argue that RASL converges to a local optimum as long as the transformations satisfy a continuity assumption. Additionally, they have demonstrated that RASL works well for batch aligning the linearly correlated images despite large illumination variations and occlusions.

Though the convex programming methods used in [43] are polynomial in the size of the

---

[43]For example, [14] used batch size 200 for the "Airport hall" video, see Section 16.5.1.

problem, that complexity can still be too demanding for very large databases of images. In order to improve the scalability of robust image alignment for massive image datasets, especially for streaming video data, we propose Transformed GRASTA, or t-GRASTA for short, which is an extension of GRASTA and RASL: We extend GRASTA to transformations, and extend RASL to the incremental gradient optimization framework.

16.2.3   Foreground and Background Separation

We investigate the performance of our algorithms through the application of foreground and background separation in video. A common assumption for background subtraction is that certain attributes of the objects of interest change more rapidly than the background scene. It is very natural to think that objects with attributes that change rapidly should be analyzed in a different manner from objects changing slowly. However, deciding which attributes are the correct ones and how to measure change can be subtle. For example, one might wish to ignore rapid changes to objects from the sun going behind the clouds, but notice slow changes like an increasing puddle on the kitchen floor from a leaky faucet.

The simplest methods of background subtraction keep a running mean or median of previous frames as a model for the background. While this is efficient and simple to implement, and can be effective in sufficiently regular environments, changes in lighting or slow moving objects can cause trouble. Using the median (or mean) gives a 0-dimensional model for the background: we measure the residual of a given frame as a distance to a point, the median. We can make this model more powerful by keeping a d dimensional subspace or affine set with $d > 0$ [41,45,46]. Suppose we have T frames and $V = [v_1 \ldots v_T]$, where each length-mp vector v_i corresponds to the $m \times p$ pixels of the ith image in the sequence. A d-dimensional subspace model is then a factorization $V \approx LR^T$, where L is an $mp \times d$ basis matrix, and R is a $T \times d$ coefficient matrix. In the following we define $n := mp$ for the vectorized image dimension. It has been noticed experimentally [22] that the vectorized images corresponding to a given object under different lighting conditions approximately lie on a low dimensional subspace. A theoretical justification for this is described in [9].

16.2.4   The Contributions of GRASTA

The algorithms presented in this chapter, GRASTA (Grassmannian Robust Adaptive Subspace Tracking Algorithm) [24] and t-GRASTA [25], are *online* algorithms which are therefore amenable to real-time processing such as in video processing applications. They both perform iterative gradient descent on the Grassmannian, the manifold of all fixed-dimensional subspaces, and that leads to simple subspace update equations that are fast and easy to implement. GRASTA performs Robust PCA when matrix elements are missing, and t-GRASTA performs Robust PCA when an unknown transformation has been applied to the data. Our algorithms do not have theoretical convergence guarantees, though combining the convergence studies in [8,31] may allow such a theory to be developed.

16.3   Realtime Robust Background Modeling from a Stream of Highly Subsampled Video Frames

In this section, we derive the algorithm GRASTA, or Grassmannian Robust Adaptive Subspace Tracking Algorithm, which performs Robust PCA in an incremental fashion. We start by discussing the model and cost function, we derive the augmented Lagrangian formulation, and derive alternating minimization steps for this particular choice of cost function.

16.3.1   Model

At each time step t, we assume that v_t is generated by the following model:

$$v_t = U_t w_t + s_t + \zeta_t \tag{16.1}$$

where w_t is the $d \times 1$ weight vector, s_t is the $n \times 1$ sparse outlier vector whose nonzero entries may be arbitrarily large, and ζ_t is the $n \times 1$ zero-mean Gaussian white noise vector with small variance. We observe only a small subset of entries of v_t, denoted by $\Omega_t \subset \{1, \ldots, n\}$.

Conforming to the notation of GROUSE [6], we let U_{Ω_t} denote the submatrix of U_t consisting of the rows indexed by Ω_t; also for a vector $v_t \in \mathbb{R}^n$, let v_{Ω_t} denote a vector in $\mathbb{R}^{|\Omega_t|}$ whose entries are those of v_t indexed by Ω_t. A critical problem raised when we only partially observe v_t is how to quantify the subspace error only from the incomplete and corrupted data. GROUSE [6] uses the natural Euclidean distance, the ℓ^2-norm, to measure the subspace error from the subspace spanned by the columns of U_t to the observed vector v_{Ω_t}:

$$F_{grouse}(\mathbb{S}; t) = \min_w \|U_{\Omega_t} w - v_{\Omega_t}\|_2^2 . \tag{16.2}$$

It was shown in [7] that this cost function gives an accurate estimate of the same cost function with full data ($\Omega = \{1, \ldots, n\}$), as long as $|\Omega_t|$ is large enough. In [7] the authors show that $|\Omega_t|$ must be larger than $\mu(\mathbb{S}) d \log(2d/\delta)$, where $\mu(\mathbb{S})$ is a measure of incoherence on the subspace and δ controls the probability of the result. See the paper for details. However, if the observed data vector is corrupted by outliers as in Equation (16.1), an ℓ^2-based best-fit to the subspace can be influenced arbitrarily with just one large outlier; this in turn will lead to an incorrect subspace update in the GROUSE algorithm.

16.3.2   Subspace Error Quantification by ℓ^1-Norm

In order to quantify the subspace error robustly, we use the ℓ^1-norm as follows:

$$F_{grasta}(\mathbb{S}; t) = \min_w \|U_{\Omega_t} w - v_{\Omega_t}\|_1 . \tag{16.3}$$

With U_{Ω_t} known (or estimated, but fixed), this ℓ^1 minimization problem is the classic least absolute deviations problem; Boyd [12] has a nice survey of algorithms to solve this problem and describes in detail a fast solver based on the technique of ADMM (Alternating Direction Method of Multipliers)[44]. More references can be found therein.

According to [12], we can rewrite the right hand of Equation (16.3) as the equivalent constrained problem by introducing a sparse outlier vector s:

$$\min \|s\|_1 \tag{16.4}$$
$$s.t. \ U_{\Omega_t} w + s - v_{\Omega_t} = 0$$

The augmented Lagrangian of this constrained minimization problem is then

$$\mathcal{L}(s, w, y) = \|s\|_1 + y^T (U_{\Omega_t} w + s - v_{\Omega_t}) + \frac{\rho}{2} \|U_{\Omega_t} w + s - v_{\Omega_t}\|_2^2 \tag{16.5}$$

where y is the dual vector. Our unknowns are s, y, U, and w. Note that since U is constrained to a non-convex manifold ($U^T U = I$), this function is not convex (neither is Equation (16.2)). However, note that if U were estimated, we could solve for the triple (s, w, y)

---

[44]http://www.stanford.edu/~boyd/papers/admm/

using ADMM; also if (s, w, y) were estimated, we could refine our estimate of U. This is the alternating approach we take with GRASTA. We describe the two parts in detail in Sections 16.3.3 and 16.3.4.

As we have said, GRASTA alternates between estimating the triple (s, w, y) and the subspace U. Here we discuss those two pieces of our algorithm. Section 16.3.3 describes the update of (s, w, y) based on an estimate $\widehat{U}_t$ for the subspace variable. Section 16.3.4 describes the update of our subspace variable to $\widehat{U}_{t+1}$ based on the estimate of (s^*, w^*, y^*) resulting from the first step.

16.3.3 ℓ^1 Regression from Subsampled Video Frame

Given the current estimated subspace $\widehat{U}_t$, the partial observation v_{Ω_t}, and the observed entries' indices Ω_t, the optimal (s^*, w^*, y^*) of Equation (16.4) can be found with the following minimization of the augmented Lagrangian.

$$(s^*, w^*, y^*) = \arg \min_{s,w,y} \mathcal{L}(\widehat{U}_{\Omega_t}, s, w, y) \tag{16.6}$$

Equation (16.6) can be efficiently solved by ADMM [12]. That is, s, w, and the dual vector y are updated in an alternating fashion:

$$\begin{cases} w^{k+1} = \arg \min_w \mathcal{L}(\widehat{U}_{\Omega_t}, s^k, w, y^k) \\ s^{k+1} = \arg \min_s \mathcal{L}(\widehat{U}_{\Omega_t}, s, w^{k+1}, y^k) \\ y^{k+1} = y^k + \rho(\widehat{U}_{\Omega_t} w^{k+1} + s^{k+1} - v_{\Omega_t}) \end{cases} \tag{16.7}$$

Specifically, these quantities are computed as follows. In this paper we always assume that $U_{\Omega_t}^T U_{\Omega_t}$ is invertible, which is guaranteed if $|\Omega_t|$ is large enough [6]. We have:

$$w^{k+1} = \frac{1}{\rho}(\widehat{U}_{\Omega_t}^T \widehat{U}_{\Omega_t})^{-1} \widehat{U}_{\Omega_t}^T (\rho(v_{\Omega_t} - s^k) - y^k) \tag{16.8}$$

$$s^{k+1} = \mathsf{S}_{\frac{1}{1+\rho}} (v_{\Omega_t} - \widehat{U}_{\Omega_t} w^{k+1} - y^k) \tag{16.9}$$

$$y^{k+1} = y^k + \rho(\widehat{U}_{\Omega_t} w^{k+1} + s^{k+1} - v_{\Omega_t}) \tag{16.10}$$

where $\mathsf{S}_{\frac{1}{1+\rho}}$ is the elementwise soft thresholding operator [13].

16.3.4   Subspace Update via Incremental Gradient Descent over Grassmannian Manifold

The set of all subspaces of $\mathbb{R}^n$ of fixed dimension d is called *Grassmannian*, which is a compact Riemannian manifold, and is denoted by $\mathcal{G}(d, n)$. Edelman, Arias and Smith (1998) have a comprehensive survey [21] that covers how both the Grassmannian geodesics and the gradient of a function defined on the Grassmannian manifold can be explicitly computed.

GRASTA achieves online robust subspace tracking by performing incremental gradient descent on the Grassmannian step by step. That is, we first compute a gradient of the loss function, and then follow this gradient along a short geodesic curve on the Grassmannian. Figure 16.1 illustrates the basic idea of gradient descent along a geodesic.

Augmented Lagrangian as the Loss Function

It seems that it would be natural to use Equation (16.3) as the robust loss function. However, there is a critical limitation of this approach: when regarding U as the variable, this loss function is not differentiable everywhere.

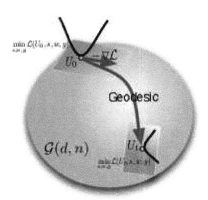

FIGURE 16.1   Illustration of the gradient descent along geodesic on the Grassmannian manifold.

Here we propose to use the augmented Lagrangian as the subspace loss function once we have estimated (s^*, w^*, y^*) from the previous $\widehat{U}_{\Omega_t}$ and v_{Ω_t} by Equation (16.7). The new loss function is stated as Equation (16.11):

$$\mathcal{L}(U) = \|s^*\|_1 + y^{*T}(U_{\Omega_t}w^* + s^* - v_{\Omega_t}) + \frac{\rho}{2}\|U_{\Omega_t}w^* + s^* - v_{\Omega_t}\|_2^2 \qquad (16.11)$$

This new subspace loss function is differentiable.

Grassmannian Geodesic Gradient Step

In order to take a gradient step along the geodesic of the Grassmannian, according to [21], we first need to derive the gradient formula of the real-valued loss function Equation (16.11) $\mathcal{L} : \mathcal{G}(d, n) \to \mathbb{R}$.

From Equation (2.70) in [21], the gradient $\nabla\mathcal{L}$ can be determined from the derivative of $\mathcal{L}$ with respect to the components of U. Let χ_{Ω_t} is defined to be the $|\Omega_t|$ columns of an $n \times n$ identity matrix corresponding to those indices in Ω_t; that is, this matrix zero-pads a vector in $\mathbb{R}^{|\Omega_t|}$ to be length n with zeros on the complement of Ω_t. The derivative of the augmented Lagrangian loss function $\mathcal{L}$ with respect to the components of U is as follows:

$$\frac{d\mathcal{L}}{dU} = [\chi_{\Omega_t}(y^* + \rho(U_{\Omega_t}w^* + s^* - v_{\Omega_t}))]w^{*T} \qquad (16.12)$$

Then the gradient $\nabla\mathcal{L}$ is $\nabla\mathcal{L} = (I - UU^T)\frac{d\mathcal{L}}{dU}$ [21]. Here we introduce three variables Γ, Γ_1, and Γ_2 to simplify the gradient expression:

$$\Gamma_1 = y^* + \rho(U_{\Omega_t}w^* + s^* - v_{\Omega_t}) \qquad (16.13)$$
$$\Gamma_2 = U_{\Omega_t}^T\Gamma_1 \qquad (16.14)$$
$$\Gamma = \chi_{\Omega_t}\Gamma_1 - U\Gamma_2 \qquad (16.15)$$

Thus the gradient $\nabla\mathcal{L}$ can be further simplified to:

$$\nabla\mathcal{L} = \Gamma w^{*T} \qquad (16.16)$$

From Equation (16.16), it is easy to verify that $\nabla\mathcal{L}$ is rank one since Γ is a $n \times 1$ vector and w^* is the optimal $d \times 1$ weight vector. Then it is trivial to compute the singular value decomposition of $\nabla\mathcal{L}$, which will be used for the following gradient descent step along the geodesic according to Equation (2.65) in [21]. The sole non-zero singular value is $\sigma = \|\Gamma\|\|w^*\|$, and the corresponding left and right singular vectors are $\frac{\Gamma}{\|\Gamma\|}$ and $\frac{w^*}{\|w^*\|}$ respectively. Then we can write the SVD of the gradient explicitly by adding the orthonormal set $x_2, \ldots, x_d$ orthogonal to Γ as left singular vectors and the orthonormal set $y_2, \ldots, y_d$ orthogonal to w^* as right singular vectors as follows:

$$\nabla\mathcal{L} = \left[\frac{\Gamma}{\|\Gamma\|} \quad x_2 \quad \cdots \quad x_d \right] \times diag(\sigma, 0, \ldots, 0) \times \left[\frac{w^*}{\|w^*\|} \quad y_2 \quad \cdots \quad y_d \right]^T \tag{16.17}$$

Finally, following Equation (2.65) in [21], a gradient step of length η in the direction $-\nabla\mathcal{L}$ is given by

$$U(\eta) = U + \left((cos(\eta\sigma) - 1)\frac{Uw_t^*}{\|w_t^*\|} - sin(\eta\sigma)\frac{\Gamma}{\|\Gamma\|} \right) \frac{w_t^{*T}}{\|w_t^*\|} . \tag{16.18}$$

16.3.5   The Algorithm - GRASTA

The discussion of previous sections can be summarized into our algorithm as follows. For each time step t, when we observe an incomplete and corrupted data vector v_{Ω_t}, our algorithm will first estimate the optimal value (s^*, w^*, y^*) from our current estimated subspace U_t via the ℓ^1 minimization ADMM solver (16.7); then compute the gradient of the augmented Lagrangian loss function $\mathcal{L}$ by Equation (16.16); then select a proper step-size η_t, for example constant step-size or diminishing step-size; and finally do the rank one subspace update via Equation (16.18). We state our main algorithm GRASTA (Grassmannian Robust Adaptive Subspace Tracking Algorithm) in Algorithm 16.1 and validate its performance in Section 16.5.1.

Algorithm 16.1 - Grassmannian Robust Adaptive Subspace Tracking

---

Require: An $n \times d$ orthogonal matrix U_0. A sequence of corrupted vectors v_t, each vector observed in entries $\Omega_t \subset \{1, \ldots, n\}$. A structure OPTS that holds parameters for ADMM.
Return: The estimated subspace U_t at time t.

1: **for** $t = 0, \ldots, T$ **do**
2:    Extract U_{Ω_t} from U_t: $U_{\Omega_t} = \chi_{\Omega_t}^T U_t$
3:    Estimate the sparse residual s_t^*, weight vector w_t^*, and dual vector y_t^* from the observed entries Ω_t via ℓ^1 regression ADMM algorithm using OPTS:
       $(s_t^*, w_t^*, y_t^*) = \arg\min_{w,s,y} \mathcal{L}(U_{\Omega_t}, w, s, y)$
4:    Compute the gradient of the augmented Lagrangian $\mathcal{L}$, $\nabla\mathcal{L}$ as follows:
       $\Gamma_1 = y_t^* + \rho(U_{\Omega_t}w_t^* + s_t^* - v_{\Omega_t}), \quad \Gamma_2 = U_{\Omega_t}^T\Gamma_1, \quad \Gamma = \chi_{\Omega_t}\Gamma_1 - U\Gamma_2$
       $\nabla\mathcal{L} = \Gamma w_t^{*T}$
5:    Select a proper step-size η_t.
6:    Update subspace: $U_{t+1} = U_t + ((cos(\eta_t\sigma) - 1)U_t\frac{w_t^*}{\|w_t^*\|} - sin(\eta_t\sigma)\frac{\Gamma}{\|\Gamma\|})\frac{w_t^{*T}}{\|w_t^*\|}$
       where $\sigma = \|\Gamma\|\|w_t^*\|$
7: **end for**

16.4 Robust Background Modeling from Jittered Camera Surveillance Images

16.4.1 Motivations

When applying Robust PCA to video streams for foreground and background separation, the central assumption is that the background of several frames together lie in a low-dimensional subspace. Unfortunately in the very common case of camera jitter, the background is no longer low-rank; while the aforementioned algorithms work very well for a stationary camera, they fail quite dramatically with camera jitter [2, 18, 32]. Robustly and efficiently detecting moving objects from an unstable camera is a challenging problem, since we need to accurately estimate both the background and the transformation of each frame. In this section we present a t-GRASTA, an extension of GRASTA to include geometric transforms of the data such as translations and rotations caused by camera jitter. Fig. 16.2 shows that for a video sequence generated by a simulated unstable camera, GRASTA fails to do the separation, but t-GRASTA can successfully separate the background and moving objects despite camera jitter.

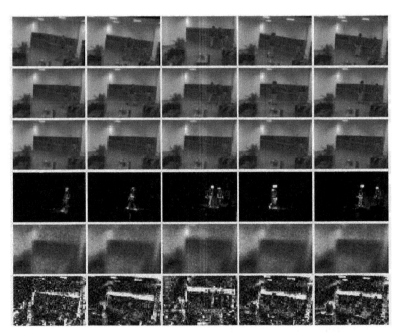

FIGURE 16.2 Video background and foreground separation by t-GRASTA despite camera jitter. 1st row: misaligned video frames by simulating camera jitters; 2nd row: images aligned by t-GRASTA; 3rd row: background recovered by t-GRASTA; 4th row: foreground separated by t-GRASTA; 5th row: background recovered by GRASTA; 6th row: foreground separated by GRASTA (©(2014) Elsevier).

16.4.2 Iterative Subspace Learning to Approximate the Nonlinear Transform

Inspired by the recent proposed transformed version of Robust PCA, named RASL [43], we have merged the merits of both RASL and GRASTA to design an efficient approach that can handle image misalignment frame by frame. For each video frame I, we may model the

ℓ^1 minimization problem as follows:

$$\min_{U,w,e,\tau} \|e\|_1 \tag{16.19}$$
$$s.t.\ I \circ \tau = Uw + e$$
$$U \in \mathcal{G}(d,n)$$

Note that with the constraint $I \circ \tau = Uw + e$ in the above minimization problem, we suppose for each frame the transformed image is well aligned to the low-rank subspace U. However, due to the nonlinear geometric transform $I \circ \tau$, directly exploiting online subspace learning techniques [6, 44] is not possible.

Here we approach this as a manifold learning problem, supposing that the low-dimensional image subspace under nonlinear transformations forms a nonlinear manifold. We propose to learn the manifold approximately using a union of subspaces model U^ℓ, $\ell = 1, \ldots, L$. The basic idea is illustrated in Fig. 16.3, and the locally linearized model for the nonlinear problem (16.19) is as follows:

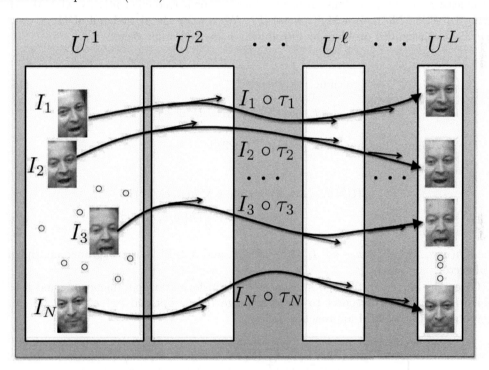

FIGURE 16.3   The illustration of iteratively approximating the nonlinear image manifold using a union of subspaces (©(2014) Elsevier).

$$\min_{w,e,\triangle\tau} \|e\|_1 \tag{16.20}$$
$$s.t.\ I \circ \tau^\ell + J^\ell \triangle \tau = U^\ell w + e .$$
$$U^\ell \in \mathcal{G}(d^\ell, n)$$

Intuitively, from Fig. 16.3, it is reasonable to think that the initial misaligned image sequence should be high rank; then after iteratively approximating the nonlinear transform with a locally linear approximation, the rank of the new subspaces U^ℓ, $\ell = 1, \ldots, L$, should be decreasing as the images become more and more aligned. Then for each misaligned image I and the unknown transformation τ, we iteratively update the union of subspaces U^ℓ, $\ell = 1, \ldots, L$, and estimate the transformation τ.

16.4.3   ADMM Solver for the Locally Linearized Problem

Since we iteratively approximate the nonlinear transform with a locally linear approximation, the key problem is how to quantify the subspace error robustly for the locally linearized problem. At iteration k, given the i-th image I_i, its estimate of transformation τ_i^k, the Jacobian J_i^k, and the current estimate of U_t^k, we use the ℓ^1 norm as follows:

$$F(S; t, k) = \min_{w, \triangle \tau} \| U_t^k w - (I_i \circ \tau_i^k + J_i^k \triangle \tau) \|_1 \qquad (16.21)$$

With U_t^k known (or estimated, but fixed), this ℓ^1 minimization problem is a variation of the least absolute deviations problem, which can be solved efficiently by ADMM (Alternating Direction Method of Multipliers) [12]. We rewrite the right hand of (16.21) as the equivalent constrained problem by introducing a sparse outlier vector e:

$$\min_{w, e, \triangle \tau} \| e \|_1 \qquad (16.22)$$

$$s.t. \; I_i \circ \tau_i^k + J_i^k \triangle \tau = U_t^k w + e \;.$$

The augmented Lagrangian of problem (16.22) is

$$\mathcal{L}(U_t^k, w, e, \triangle \tau, \lambda) = \| e \|_1 + \lambda^T h(w, e, \triangle \tau)$$
$$+ \frac{\mu}{2} \| h(w, e, \triangle \tau) \|_2^2 \qquad (16.23)$$

where $h(w, e, \triangle \tau) = U_t^k w + e - I_i \circ \tau_i^k - J_i^k \triangle \tau$, and $\lambda \in \mathbb{R}^n$ is the Lagrange multiplier or dual vector.

Given the current estimated subspace U_t^k, transformation parameter τ_i^k, and the Jacobian matrix J_i^k with respect to the i-th image I_i, the optimal $(w^*, e^*, \triangle \tau^*, \lambda^*)$ can be computed by the ADMM approach as follows:

$$\begin{cases} \triangle \tau^{p+1} = (J_i^k J_i^{k^T})^{-1} J_i^{k^T} (U_t^k w^p + e^p - I_i \circ \tau_i^k + \frac{1}{\mu} \lambda^p) \\ w^{p+1} = (U_t^k U_t^{k^T})^{-1} U_t^{k^T} (I_i \circ \tau_i^k + J_i^k \triangle \tau^{p+1} - e^p - \frac{1}{\mu^p} \lambda^p) \\ e^{p+1} = \mathsf{S}_{\frac{1}{\mu}} (I_i \circ \tau_i^k + J_i^k \triangle \tau^{p+1} - U_t^k w^{p+1} - \frac{1}{\mu^p} \lambda^p) \\ \lambda^{p+1} = \lambda^p + \mu^p h(w^{p+1}, e^{p+1}, \triangle \tau^{p+1}) \\ \mu^{p+1} = \rho \mu^p \end{cases} \qquad (16.24)$$

where $\mathsf{S}_{\frac{1}{\mu}}$ is the elementwise soft thresholding operator [13], and $\rho > 1$ is the ADMM penalty constant enforcing $\{\mu^p\}$ to be a monotonically increasing positive sequence. The iteration (16.24) indeed converges to the optimal solution of the problem (16.22) [10]. We summarize this ADMM solver as Algorithm 16.2 in Section 16.3.5.

Algorithm 16.2 is the ADMM solver for the locally linearized problem (16.22). From our extensive experiments, if we set the ADMM penalty parameter $\rho = 2$ and the tolerance $\epsilon^{tol} = 10^{-7}$, Algorithm 16.2 has always converged in fewer than 20 iterations.

Algorithm 16.2 - ADMM Solver for the Locally Linearized Problem (16.22)

Require: An $n \times d$ orthogonal matrix U, a wrapped and normalized image $I \circ \tau \in \mathbb{R}^n$, the corresponding Jacobian matrix J, and a structure OPTS which holds four parameters for ADMM: ADMM penalty constant ρ, the tolerance ϵ^{tol}, and ADMM maximum iteration K.
Return: weight vector $w^* \in \mathbb{R}^d$; sparse outliers $e^* \in \mathbb{R}^n$; locally linearized transformation parameters $\triangle\tau^*$; and dual vector $\lambda^* \in \mathbb{R}^n$.

1: Initialize $w, e, \triangle\tau, \lambda, and \ \mu$: $e^1 = 0, w^1 = 0, \triangle\tau^1 = 0,\ \lambda^1 = 0,\ \mu = 1$
2: Cache $P = (U^T U)^{-1} U^T$ and $F = (J^T J)^{-1} J^T$
3: **for** $k = 1 \to K$ **do**
4:      Update $\triangle\tau$: $\triangle\tau^{k+1} = F(Uw^k + e^k - I \circ \tau + \frac{1}{\mu}\lambda^k)$
5:      Update weights: $w^{k+1} = P(I \circ \tau + J\triangle\tau^{k+1} - e^k - \frac{1}{\mu}\lambda^k)$
6:      Update sparse outliers:
     $e^{k+1} = \mathsf{S}_{\frac{1}{\mu}}(I \circ \tau + J\triangle\tau^{k+1} - Uw^{k+1} - \frac{1}{\mu}\lambda^k)$
7:      Update dual: $\lambda^{k+1} = \lambda^k + \mu h(w^{k+1}, e^{k+1}, \triangle\tau^{k+1})$
8:      Update μ: $\mu = \rho\mu$
9:      **if** $\|h(w^{k+1}, e^{k+1}, \triangle\tau^{k+1})\|_2 \leq \epsilon^{tol}$ **then**
10:          Converge and break the loop.
11:      **end if**
12: **end for**
13: $w^* = w^{k+1}$, $e^* = e^{k+1}$, $\triangle\tau^* = \triangle\tau^{k+1}$, $\lambda^* = y^{k+1}$

16.4.4   The Algorithm - t-GRASTA

In Section 16.4.2, we propose to tackle the difficult nonlinear online subspace learning problem by iteratively learning online a union of subspaces U^ℓ, $\ell = 1, \ldots, L$. For a sequence of video frames $I_i, i = 1, \ldots, N$, the union of subspaces U^ℓ are updated iteratively as illustrated in Fig. 16.4.

Specifically, at i-th frame I_i, for the locally approximated subspace U_i^1 at the first iteration, given the initial roughly estimated transformation τ_i^0, the ADMM solver Algorithm 16.2 gives us the locally estimated $\triangle\tau_i^1$, and the updated subspace U_{i+1}^1 is obtained by taking a gradient step along the geodesic of the Grassmannian $\mathcal{G}(d^1, n)$ which is similar to GRASTA as discussed in Section 16.3.4. The transformation τ_i^1 of the next iteration is updated by $\tau_i^1 = \tau_i^0 + \triangle\tau_i^1$. Then for the next locally approximated subspace U_i^2, we also estimate $\triangle\tau_i^2$ and update the subspace along the geodesic of the Grassmannian $\mathcal{G}(d^2, n)$ to U_{i+1}^2. Repeatedly, we will update U_i^ℓ in the same way to get U_{i+1}^ℓ and the new transformation $\tau_i^\ell = \tau_i^{\ell-1} + \triangle\tau_i^\ell$. After completing the update for all L subspaces, the union of subspaces $U_{i+1}^\ell(\ell = 1, \ldots, L)$ will be used for approximating the nonlinear transform of the next video frame I_{i+1}.

We summarize the above statements as Algorithm 16.3, and we call this approach the *fully online mode* of t-GRASTA[45]. The performance of t-GRASTA is validated in Section 16.5.2.

---

[45]The interested reader may also find algorithm descriptions for a *batch mode* of t-GRASTA and a *trained online mode* of t-GRASTA in the paper [25].

Algorithm 16.3 - Transformed GRASTA - *Fully Online Mode*

Require: The initial L $n \times d^\ell$ orthonormal matrices U^ℓ spanning the corresponding subspace $\mathbb{S}^\ell$, $\ell = 1, ..., L$. A sequence of unaligned images I_i and the corresponding initial transformation parameters τ_i^0, $i = 1, \ldots, N$.

Return: The estimated iteratively approximated subspaces U_i^ℓ, $\ell = 1, \ldots, L$, after processing image I_i. The transformation parameters τ_i^L for each well-aligned image.

1: **for** unaligned image $I_i, i = 1, \ldots, N$ **do**

2:      **for** the iterative approximated subspace $U^\ell, \ell = 1, \ldots, L$ **do**

3:         Update the Jacobian matrix of image I_i:

$$J_i^\ell = \frac{\partial(I_i \circ \zeta)}{\partial \zeta}\Big|_{\zeta = \tau_i^\ell}$$

4:         Update the wrapped and normalized images:

$$I_i \circ \tau_i^\ell = \frac{vec(I_i \circ \tau_i^\ell)}{\|vec(I_i \circ \tau_i^\ell)\|_2}$$

5:         Estimate the weight vector w_i^ℓ, the sparse outliers e_i^ℓ, the locally linearized transformation parameters $\triangle\tau_i^\ell$, and the dual vector λ_i^ℓ via the ADMM algorithm 16.2 from $I_i \circ \tau_i^\ell$, J_i^ℓ, and the current estimated subspace U_i^ℓ

$$(w_i^\ell, e_i^\ell, \triangle\tau_i^\ell, \lambda_i^\ell) = \arg\min_{w,e,\triangle\tau,\lambda} \mathcal{L}(U_i^\ell, w, e, \lambda)$$

6:         Compute the gradient $\nabla\mathcal{L}$ as follows:
        $\Gamma_1 = \lambda_i^\ell + \mu h(w_i^\ell, e_i^\ell, \triangle\tau_i^\ell),$
        $\Gamma = (I - U_t^\ell U_t^{\ell T})\Gamma_1, \qquad \nabla\mathcal{L} = \Gamma w_i^{\ell T}$

7:         Compute step-size η_i^ℓ.

8:         Update subspace:
        $U_{i+1}^\ell = U_i^\ell + \left((\cos(\eta_i^\ell \sigma) - 1) U_t \frac{w_i^\ell}{\|w_i^\ell\|} - \sin(\eta_i^\ell \sigma) \frac{\Gamma}{\|\Gamma\|} \right) \frac{w_i^{\ell T}}{\|w_i^\ell\|},$
        where $\sigma = \|\Gamma\| \|w_i^\ell\|$.

9:         Update the transformation parameters:

$$\tau_i^{\ell+1} = \tau_i^\ell + \triangle\tau_i^\ell$$

10:      **end for**

11: **end for**

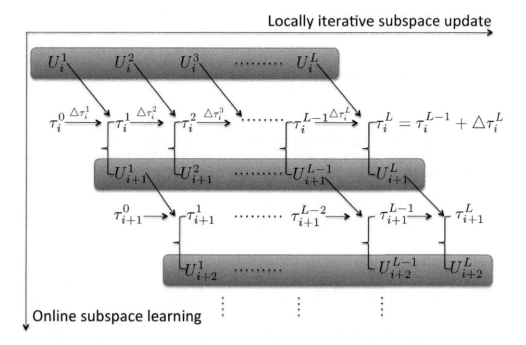

FIGURE 16.4   The diagram of the *fully online mode* of t-GRASTA (ⓒ(2014) Elsevier).

16.5   Experimental Validation

16.5.1   Realtime Separation - GRASTA

In this subsection we discuss the application of GRASTA to the prominent problem of real-time separation of foreground objects from the background in video surveillance. Imagine we had a video with only the background: When the columns of a single frame of this video are stacked into a single column, several frames together will lie in a low-dimensional subspace. In fact if the background is completely static, the subspace would be one-dimensional. That subspace can be estimated in order to identify and separate the foreground objects; if the background is dynamic, subspace tracking is necessary. GRASTA is uniquely suited for this burgeoning application.

Here we consider three scenarios in the video tasks, with a spectrum of challenges for subspace tracking. In the first we have a video with a static background and objects moving in the foreground. In the second, we have a video with a still background but with changing lighting. In the third, we simulate a panning camera to examine GRASTA's performance with a dynamic background. The results are summarized in Table 16.1.

Static Background

If the video background is known to be static or near static, we can use GRASTA to track the background and separate the moving foreground objects in real-time. Since the background is static, we use GRASTA first to identify the background, and then we use only the ℓ^1 regression ADMM algorithm to separate the foreground from the background. More precisely we do the following:

1. Randomly select a few frames of the video to train the static low-rank subspace U. In

our experiments, we select frames randomly from the entire video; however for real-time processing these frames may be chosen from initial piece of the video, as long as we can be confident that every pixel of the background is visible in one of the selected frames. The low-rank subspace U is then identified from these frames using partial information. We use 30% of the pixels, select 50 frames for training, and set RANK = 5 in all the following experiments.

2. Once the video background BG has been identified as a subspace U, separating the foreground objects FG from each frame can be simply done using Equation (16.25), where the weight vector w_t can be solved via the ADMM Algorithm, again from a small subsample of each frame's pixels.

$$\begin{cases} BG = Uw_t \\ FG = video(t) - BG \end{cases} \qquad (16.25)$$

Table 16.1 shows the real-time[46] video separation results. From the first experiment, we use the "Hall" dataset from [35][47], which consists of 3584 frames each with resolution 144×176. We let GRASTA cycle 10 times over the 50 training frames just from 30% random entries of each frame to get the stationary subspace U. Training the subspace costs 11.3 seconds. Then we perform background and foreground separation for all frames in a streaming fashion, and when dealing with each frame we only randomly observe 1% entries. The separation task is performed by Equation (16.25), and the separating time is 20.9 seconds, which means we achieve 171.5 FPS (frames per second) real-time performance. Figure 16.5 shows the separation quality at $t = 1, 230, 1400$. In order to show GRASTA can handle higher resolution video effectively, we use the "Shopping Mall" [35] video with resolution 320×256 as the second experiment. We also do the subspace training stage with the same parameter settings as "Hall". We do the background and foreground separation only from 1% entries of each frame. For "Shopping Mall" the separating time is 27.5 seconds for a total of 1286 frames. Thus we achieve 46.8 FPS real-time performance. Figure 16.6 shows the separation quality at $t = 1, 600, 1200$. The details of each tracking set-up are described in Table 16.2.

Dataset	Resolution	Total Frames	Training Time	Tracking and Separating Time	FPS
Airport Hall	144×176	3584	11.3 sec	20.9 sec	171.5
Bootstrap	120×160	3055	13.8 sec	15.5 sec	197.1
Shopping Mall	320×256	1286	33.9 sec	27.5 sec	46.8
Lobby	144×176	1546	3.9 sec	71.3 sec	21.7
Hall with Virtual Pan (1)	144×88	3584	3.8 sec	191.3 sec	18.7
Hall with Virtual Pan (2)	144×88	3584	3.7 sec	144.8 sec	24.8

TABLE 16.1 Real-time video background and foreground separation by GRASTA. Here we use three different resolution video datasets, the first two with static background and the last three with dynamic background. We train from 50 frames; in the first two experiments they are chosen randomly, and in the last three they are the first 50 frames. In all experiments, the subspace dimension is 5.

---

[46]We comment here that to call something "real-time" processing of course will depend on one's application requirements and hardware (camera frame capture rate, in the example of video processing). For example, standard 35mm film video uses 24 unique frames per second. The maximum frame rate for most CCTVs is 30 frames per second.

[47]Find these along with the videos at http://perception.i2r.a-star.edu.sg/bk_model/bk_index.html.

Dataset	Training Sub-Sampling	Tracking Sub-Sampling	Separation Sub-Sampling	Tracking/Separation Alg.
Hall	30%	-	1%	ℓ^1 ADMM + Eqn 16.25
Bootstrap	30%	-	1%	ℓ^1 ADMM + Eqn 16.25
Shopping Mall	30%	-	1%	ℓ^1 ADMM + Eqn 16.25
Lobby	30%	30%	100%	Full GRASTA Alg 16.1
Hall with Virtual Pan (1)	100%	100%	100%	Full GRASTA Alg 16.1
Hall with Virtual Pan (2)	50%	50%	100%	Full GRASTA Alg 16.1

TABLE 16.2 Here we summarize the approach for the various video experiments. For the initial training stage, we use the full GRASTA Algorithm 16.1. When the background is dynamic, we use the full GRASTA for tracking. We used K=10 iterations of the ADMM algorithm for static video experiments and K=20 for dynamic video.

$$t = 1 \qquad\qquad t = 230 \qquad\qquad t = 1400$$

FIGURE 16.5 Real-time video background and foreground separation from partial information. We show the separation quality at t=1, 230, 1400. The resolution of the video is 144 × 176. The first row is the original video frame at each time; the middle row is the recovered background at each time only from 5% information; and bottom row is the foreground calculated by Equation (16.25).

Dynamic Background: Changing Lighting

Here we want to consider a problem where the lighting in the video is changing throughout. We use the "Lobby" dataset from [35], which has 1546 frames, each 144 × 176 pixels. In order to adjust to the lighting changes, GRASTA tracks the subspace throughout the video; that is, unlike the last two experiments, we run the full GRASTA Algorithm 16.1 for every frame. We use 30% of the pixels of every frame to do this update and 100% of the pixels to do the separation. Again, see the numerical results in Table 16.1. The results are illustrated in Figure 16.7.

Dynamic Background: Virtual Pan

In the last experiment, we demonstrate that GRASTA can effectively track the right subspace in video with a dynamic background. We consider panning a "virtual camera" from left to right and right to left through the video to simulate a dynamic background. Periodically, the virtual camera pans 20 pixels. The idea of the virtual camera is illustrated cleanly with Figure 16.8.

We choose "Hall" as the original dataset. The original resolution is 144 × 176, and we

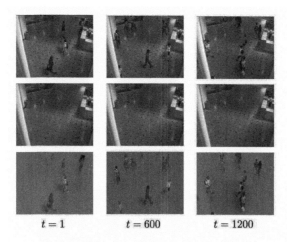

$$t = 1 \qquad t = 600 \qquad t = 1200$$

FIGURE 16.6   Real-time video background and foreground separation from partial information. We show the separation quality at t=1, 600, 1200. The resolution of the video is 320 × 256. The first row is the original video frame at each time; the middle row is the recovered background at each time only from 1% information; and bottom row is the foreground calculated by Equation (16.25).

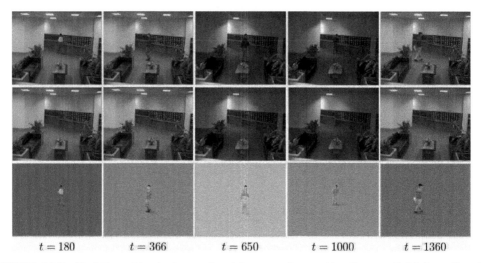

$$t = 180 \qquad t = 366 \qquad t = 650 \qquad t = 1000 \qquad t = 1360$$

FIGURE 16.7   Real-time video background and foreground separation from partial information. The first row is the original video frame at each time; the middle row is the recovered background at each time only from 30% information; and bottom row is the foreground estimated using all pixels. The differing background colors of the bottom row is simply an artifact of colormap in Matlab.

set the scope of the virtual camera to have the same height but half the width, so the resolution of the virtual camera is 144 × 88. We set the subspace $RANK = 5$. Figure 16.9 shows how GRASTA can quickly adapt to the changed background in just 25 frames when the virtual camera pans 20 pixels to the right at $t = 101$. We also let GRASTA track all the 3584 frames and do the separation task for all frames. When we use 100% of the pixels for the tracking and separation, the total computation time is 191.3 seconds, or 18.7 FPS, and adjusting to a new camera position after the camera pans takes 25 frames as can be

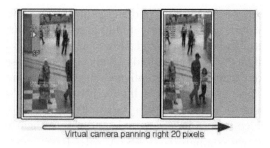

Virtual camera panning right 20 pixels

FIGURE 16.8 Demonstration of panning the "virtual camera" right 20 pixels.

seen in Figure 16.9. When we use 50% of the pixels for tracking and 100% of the pixels for separation, the total computation time is 144.8 seconds or 24.8 FPS, and the adjustment to the new camera position takes around 50 frames.

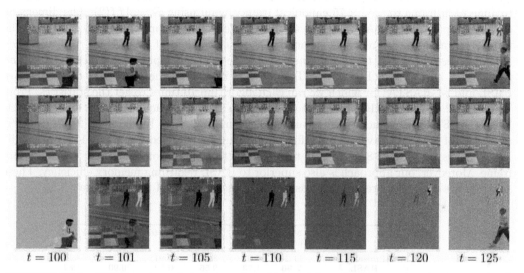

$t = 100$ $t = 101$ $t = 105$ $t = 110$ $t = 115$ $t = 120$ $t = 125$

FIGURE 16.9 Real-time dynamic background tracking and foreground separation. At time t=101, the virtual camera slightly pans to right 20 pixels. We show how GRASTA quickly adapts to the new subspace by t=125. The first row is the original video frame at each time; the middle row is the tracked background at each time; the bottom row is the separated foreground at each time.

16.5.2   Robust Separation despite Camera Jitter - t-GRASTA

In this section, we apply t-GRASTA to separation problems made difficult by video jitter. Here we apply both the *fully online mode* Algorithm 16.3 and the *trained online mode* t-GRASTA [25] to different datasets. We show the superiority of t-GRASTA regarding both the speed and memory requirement of the algorithms.

Hall

Firstly, we apply t-GRASTA to the task of separating moving objects from static background in the video footage recorded by an unstable camera. In the previous section, we simulate a virtual panning camera to show that GRASTA can quickly track sudden changes in the background subspace caused by a moving camera. The low-rank subspace tracking model is well-defined, as the camera after panning is still stationary, and thus the recorded video frames are accurately pixelwise aligned. However, for an unstable camera, the recorded frames are no longer aligned; the background cannot be well represented by a low-rank subspace unless the jittered frames are first aligned. In order to show that t-GRASTA can tackle this separation task, we consider a highly jittered video sequence generated by a simulated unstable camera. To simulate the unstable camera, we randomly translate the original well-aligned video frames in x- / y- axis and rotate them in the plane.

In this experiment, we compare t-GRASTA with RASL and GRASTA. We use the first 200 frames of the "Hall" dataset, each 144×176 pixels. We first perturb each frame artificially to simulate camera jitter. The rotation of each frame is random, uniformly distributed within the range of $[-\theta_0/2, \theta_0/2]$, and the ranges of x- and y-translations are limited to $[-x_0/2, x_0/2]$ and $[-y_0/2, y_0/2]$. In this example, we set the perturbation size parameters $[x_0, y_0, \theta_0]$ with the values of [20,20,10°].

For comparing with RASL, we just let RASL run its original batch model without forcing it into an online algorithm framework. The task we give to RASL and t-GRASTA is to align each frame to a 62×75 canonical frame, again using $\mathbb{G} = Aff(2)$. The dimension of the subspace in t-GRASTA is set to be 10. We first randomly select 30 frames of the total 200 frames to train the subspace by the batch mode of t-GRASTA [25] and then align the rest using the *trained online mode*. The visual comparison between RASL and t-GRASTA is shown in Fig. 16.10. Table 16.5.2 illustrates the numerical comparison of RASL and t-GRASTA, for which we ran each algorithm 10 times to get the statistics. From Table 16.5.2 and Fig. 16.10 we can see that the two algorithms achieve a very similiar effect, but t-GRASTA runs much faster than RASL: On a PC with Intel P9300 2.27GHz CPU and 2 GB of RAM, the average time for aligning a newly arrived frame is 1.1 second, while RASL needs more than 800 seconds to align the total batch of images, or 4 seconds per frame. Moreover, our approach is also superior to RASL regarding memory efficiency. These superiorities become more dramatic as one increases the size of the image database.

	Max error	Mean error	X1 std	Y1 std	X2 std	Y2 std
Initial misalignment	11.24	5.07	3.35	3.01	3.34	4.17
RASL	**2.96**	1.73	0.56	**0.71**	0.90	1.54
t-GRASTA	6.62	**0.84**	**0.48**	1.11	**0.57**	**0.74**

TABLE 16.3 Statistics of errors in two pixels P_1 and P_2, selected from the original video frames and traced through the jitter simulation process to the RASL and t-GRASTA output frames. Max error and mean error are calculated as the distances from the estimated P_1 and P_2 to their statistical center $E(P_1)$ and $E(P_2)$. Std are calculated as the standard deviation of four coordinate value (X_1, Y_1) for P_1 and (X_2, Y_2) for P_2 across all frames.

(a)   (b)   (c)   (d)   (e)

FIGURE 16.10 Comparison between t-GRASTA and RASL. (a) Average of initial misaligned images; (b) average of images aligned by t-GRASTA; (c)average of background recovered by t-GRASTA; (d) average of images aligned by RASL; (e) average of background recovered by RASL.

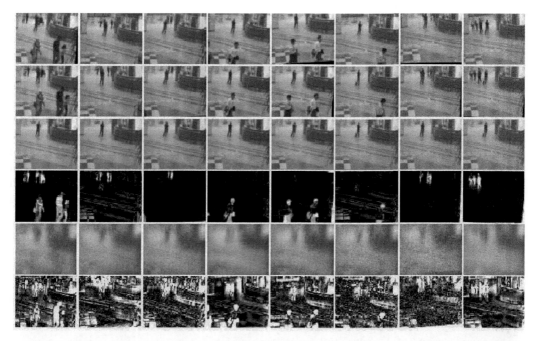

FIGURE 16.11   Video background and foreground separation with jittered video. 1st row: 8 misaligned video frames randomly selected from artificially perturbed images; 2nd row: images aligned by t-GRASTA; 3rd row: background recovered by t-GRASTA; 4th row: foreground separated by t-GRASTA; 5th row: background recovered by GRASTA; 6th row: foreground separated by GRASTA (ⓒ(2014) Elsevier).

In order to compare with GRASTA, we use 200 perturbed images to recover the background and separate the moving objects for both algorithms; Fig. 16.11 illustrates the comparison. For both GRASTA and t-GRASTA, we set the subspace rank = 10 and randomly selected 30 images to train the subspace first. For t-GRASTA, we use the affine transformation $\mathbb{G} = Aff(2)$. From Fig. 16.11, we can see that our approach successfully separates the foreground and the background and simultaneously aligns the perturbed images. But GRASTA fails to learn a proper subspace, thus, the separation of background and foreground is poor. Although GRASTA has been demonstrated to successfully track a dynamic subspace, e.g. the panning camera, the dynamics of an unstable camera are too fast and unpredictable for the GRASTA subspace tracking model to succeed in this context without pre-alignment of the video frames.

Sidewalk

In the last experiment, we use misaligned frames caused by real camera jitter to test t-GRASTA. Here we align all 1200 frames of "Sidewalk" dataset[48] to 50×78 canonical frames, again using $\mathbb{G} = Aff(2)$ and subspace dimension 5. We also use the first 20 frames to train the initial subspace using the batch mode, and then use the *fully online mode* to align the rest of the frames. Here we can see that aligning the total 1200 frames is a heavy task for RASL – for our PC with Intel P9300 2.27GHz CPU and 2 GB of RAM, it was

---

[48]Find it along with other datasets containing misaligned frames caused by real video jitters at `http://wordpress-jodoin.dmi.usherb.ca/dataset`.

necessary to divide the dataset into four parts each containing 300 frames. We then let RASL separately run on each sub-dataset. The total time needed by RASL was around 1000 seconds for 1.2 frames per second, while t-GRASTA achieved more than 4 frames per second without partitioning the data.

Compared to the *trained online mode*, the *fully online mode* can track changes of the subspace over time. This is an important asset of the *fully online mode*, especially when it comes to large streaming datasets containing considerable variations. We see that we usually need no more than 20 frames for *fully online mode* to adapt to the changes of the subspace, such as illumination changes or dynamic background caused by the motion of the subspace. Moreover, if the changes are slow, i.e the natural illumination changes from daylight or the camera moving slowly, then t-GRASTA needs no extra frames to track such changes; it incorporates such information with each iteration during the slowly changing process.

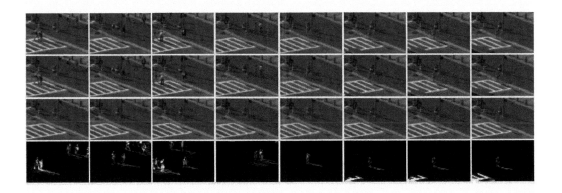

FIGURE 16.12　Video background and foreground separation with jittered video. 1st row: 8 original misaligned video frames caused by video jitter; 2nd row: images aligned by t-GRASTA; 3rd row: background recovered by t-GRASTA; 4th row: foreground separated by t-GRASTA.

16.6　Conclusion

In this chapter we have presented two variants of a robust online subspace tracking algorithm, GRASTA and t-GRASTA. The algorithms estimate a low-rank model from noisy, corrupted, and either transformed or incomplete data, even when the best low-rank model may be time-varying. Though this work presents successful algorithms, many questions remain.

The cost function for GRASTA in Equation (16.3) has the subspace variable U, which is constrained to a non-convex manifold; therefore the resulting optimization is non-convex. A proof of convergence to the global minimum of this algorithm is of great interest. It may be possible to extend the local convergence analysis of [8], for ℓ_2 subspace estimation, to an analysis of the alternating directions method herein for minimizing the ℓ_1 cost function.

GRASTA uses alternating minimization, alternating first to estimate (s, w, y) and then fixing this triple of variables to estimate U. Observe that if (s, w, y) are correct estimates, we could then estimate U *without* the robust cost function. This would be quite useful in situations when speed is of utmost importance, as the GROUSE subspace update is faster than the GRASTA subspace update. Of course, knowing when (s, w, y) are accurate is a very tricky business. Exploring this tradeoff is part of our future work.

We have shown that one of the very promising applications of GRASTA is that of separating background and foreground in video surveillance. We are very interested to apply GRASTA to more videos with dynamic backgrounds: for example, natural background scenery which may blow in the wind. In doing this we will study the resulting trade-off between the kinds of movement that would be captured as part of the background and the movement that would be identified as foreground.

A foremost remaining problem is how to scale the proposed t-GRASTA approach to a very large streaming dataset such as is typical in real-time video processing. The fully online t-GRASTA algorithm presented here is a first step towards a truly large-scale real-time algorithm, but several practical implementation questions remain, including online parameter selection and error performance cross-validation. Another question of interest is regarding the estimation of d^k for the subspace update. Though we fix the rank d in this paper, estimating d^k and switching between Grassmannians is a very interesting future direction.

While preparing the conference version of this work [26], we noticed an interesting alignment approach proposed in [36]. Though the two approaches of ours and [36] are both obtained via optimization over a manifold, they perform alignment for very different scenarios. For example, the approach in [36] focuses on semantically meaningful videos or signals, and then it can successfully align the videos of the same object from different views; t-GRASTA manipulates the set of misaligned images or the video of unstable camera to robustly identify the low-rank subspace, and then it can align these images according to the subspace. An intriguing future direction would be to merge these two approaches.

A final direction of future work is toward applications which require more aggressive background tracking than is possible by a GRASTA-type algorithm. For example, if a camera is following an object around different parts of a single scene, even though the background may be quickly varying from frame to frame, the camera will get multiple shots of different pieces of the background. Therefore, it may be possible to still build a model for the entire background scene using low-dimensional modeling. Incorporating camera movement parameters and a dynamical model into GRASTA would be a natural way to solve this problem, merging classical adaptive filtering algorithms with modern manifold optimization.

16.7 Acknowledgements

This work of Jun He is supported by NSFC (61203273). Laura Balzano would like to acknowledge 3M for funding her Ph.D. work.

References

1. A. Agarwal, S. Negahban, and M. Wainwright. Noisy matrix decomposition via convex relaxation: Optimal rates in high dimensions. *The Annals of Statistics*, 40(2):1171–1197, 2012.

2. S. Apewokin, B. Valentine, L. Wills, S. Wills, and A. Gentile. Multimodal mean adaptive backgrounding for embedded real-time video surveillance. *IEmbedded Computer Vision Workshop, ECVW 2007*, June 2007.

3. M. Audette, F. Ferrie, and T. Peters. An algorithmic overview of surface registration techniques for medical imaging. *Medical Image Analysis*, 4(3):201 – 217, 2000.

4. D. Babwin. Cameras make Chicago most closely watched U.S. city, April 6 2010.

5. F. El Baf and T. Bouwmans. Comparison of background subtraction methods for a multimedia learning space. *International Conference on Signal Processing and Multimedia, SIGMAP*, July 2007.

6. L. Balzano, R. Nowak, and B. Recht. Online identification and tracking of subspaces from highly incomplete information. In *Communication, Control, and Computing (Allerton), 2010 48th Annual Allerton Conference on*, pages 704–711, 2010.

7. L. Balzano, B. Recht, and B. Nowak. High-dimensional matched subspace detection when data are missing. In *Information Theory Proceedings (ISIT), 2010 IEEE International Symposium on*, pages 1638–1642. IEEE, 2010.

8. L. Balzano and S. Wright. Local convergence of an algorithm for subspace identification from partial data. Available at `http://arxiv.org/abs/1306.3391`.

9. R. Basri and D. Jacobs. Lambertian reflectance and linear subspaces. *Pattern Analysis and Machine Intelligence, IEEE Transactions on*, 25(2):218–233, 2003.

10. D. Bertsekas. *Nonlinear Programming*. Athena Science, 2004.

11. T. Bouwmans. Background subtraction for visual surveillance: A fuzzy approach. *Handbook on Soft Computing for Video Surveillance, Taylor and Francis Group*, 5, March 2012.

12. S. Boyd, N. Parikh, E. Chu, B. Peleato, and J. Eckstein. Distributed optimization and statistical learning via the alternating direction method of multipliers. *Found. Trends Mach. Learn.*, 3(1):1–122, January 2011.

13. S. Boyd and L. Vandenberghe. *Convex optimization*. Cambridge university press, 2004.

14. E. Candès, X. Li, Y. Ma, and J. Wright. Robust principal component analysis? *J. ACM*, 58(3):11:1–11:37, June 2011.

15. E. Candès and B. Recht. Exact matrix completion via convex optimization. *Foundations of Computational Mathematics*, 9(6):717–772, 2009.

16. Netflix Media Center. Accessed August 2011 at `http://www.netflix.com/MediaCenter`.

17. V. Chandrasekaran, S. Sanghavi, P. Parrilo, and A. Willsky. Rank-sparsity incoherence for matrix decomposition. *SIAM Journal on Optimization*, 21(2):572–596, 2011.

18. L. Cheng, M. Gong, D. Schuurmans, and T. Caelli. Real-time discriminative background subtraction. *IEEE Transaction on Image Processing*, 20(5):1401–1414, December 2011.

19. J. Costeira and T. Kanade. A multibody factorization method for independently moving objects. *International Journal of Computer Vision*, 29, 1998.

20. M. Cristani, M. Farenzena, D. Bloisi, and V. Murino. Background subtraction for automated multisensor surveillance: A comprehensive review. *EURASIP Journal on Advances in Signal Processing*, 2010:24, 2010.

21. A. Edelman, T. Arias, and S. Smith. The geometry of algorithms with orthogonality constraints. *SIAM Journal on Matrix Analysis and Applications*, 20(2):303–353, 1998.

22. R. Epstein, P. Hallinan, and A. Yuille. 5+/-2 eigenimages suffice: An empirical investigation of low-dimensional lighting models. In *PBMCV95*, page SESSION 4, 1995.

23. B. Han and L. Davis. Density-based multi-feature background subtraction with support vector machine. *IEEE Transactions on Pattern Analysis and Machine Intelligence, PAMI 2012*, 34(5):1017–1023, May 2012.

24. J. He, L. Balzano, and A. Szlam. Incremental gradient on the Grassmannian for online foreground and background separation in subsampled video. In *IEEE Conference on Computer Vision and Pattern Recognition, CVPR 2012*, pages 1568–1575, 2012.

25. J. He, D. Zhang, L. Balzano, and T. Tao. Iterative Grassmannian optimization for robust image alignment, 2013. arXiv preprint arXiv:1306.0404.

26. J. He, D. Zhang, L. Balzano, and T. Tao. Iterative online subspace learning for robust image alignment. In *Conference on Face and Gesture Recognition, FG 2013*, 2013.

27. D. Hsu, S. Kakade, and T. Zhang. Robust matrix decomposition with sparse corruptions. *Information Theory, IEEE Transactions on*, 57(11):7221–7234, 2011.

28. J. Hu and T. Su. Robust background subtraction with shadow and highlight removal for indoor surveillance. *International Journal on Advanced Signal Processing*, pages 1–14, 2007.

29. P. Huber and E. Ronchetti. *Robust Statistics*. Wiley, 2009.

30. O. Javed, K. Shafique, and M. Shah. A hierarchical approach to robust background subtraction using color and gradient information. *IEEE Workshop on Motion and Video Computing, WMVC 2002*, December 2002.

31. P. Jimenez, S. Bascon, R. Pita, and H. Moreno. Background pixel classification for motion detection in video image sequences. *International Work Conference on Artificial and Natural Neural Network, IWANN 2003*, pages 718–725, 2003.

32. P. Jodoin, J. Konrad, V. Saligrama, and V. Veilleux-Gaboury. Motion detection with an unstable camera. In *International Conference on Image Processing, ICIP 2008*, pages 229–232, 2008.

33. H. Krim and M. Viberg. Two decades of array signal processing research: the parametric approach. *Signal Processing Magazine, IEEE*, 13(4):67–94, July 1996.

34. G. Li and Z. Chen. Projection-pursuit approach to robust dispersion matrices and principal components: primary theory and monte carlo. *Journal of the American Statistical Association*, 80(391):759–766, 1985.

35. L. Li, W. Huang, I. Gu, and Q. Tian. Statistical modeling of complex background for foreground object detection. *IEEE Transactions on Image Processing*, 13(11):1459–1472, November 2004.

36. R. Li and R. Chellappa. Spatiotemporal alignment of visual signals on a special manifold. *IEEE Transactions on Pattern Analysis and Machine Intelligence*, 35(3):697–715, 2013.

37. Z. Lin, M. Chen, L. Wu, and Y. Ma. The augmented lagrange multiplier method for exact recovery of corrupted low-rank matrices. Technical Report Technical Report UILU-ENG-09-2215, University of Illinois, Urbana-Champaign, October 2009. Available at http://arxiv.org/abs/1009.5055.

38. G. Mateos and G. Giannakis. Sparsity control for robust principal component analysis. In *Signals, Systems and Computers (ASILOMAR), 2010 Conference Record of the Forty Fourth Asilomar Conference on*, pages 1925–1929, 2010.

39. M. McCoy and J. Tropp. Two proposals for robust PCA using semidefinite programming. *Electronic Journal of Statistics*, 5(0):1123–1160, June 2011.

40. E. Moulines, P. Duhamel, J. Cardoso, and S. Mayrargue. Subspace methods for the blind identification of multichannel FIR filters. *Signal Processing, IEEE Transactions on*, 43(2):516–525, February 1995.

41. N. Oliver, B. Rosario, and A. Pentland. A bayesian computer vision system for modeling human interactions. *IEEE Transactions on Pattern Analysis and Machine Intelligence*, 22(8):831–843, 2000.

42. Y. Peng, A. Ganesh, J. Wright, W. Xu, and Y. Ma. Rasl: Robust alignment by sparse and low-rank decomposition for linearly correlated images. In *Computer Vision and Pattern Recognition (CVPR), 2010 IEEE Conference on*, pages 763–770, 2010.

43. Y. Peng, A. Ganesh, J. Wright, W. Xu, and Y. Ma. Rasl: Robust alignment by sparse and low-rank decomposition for linearly correlated images. *Pattern Analysis and Machine Intelligence, IEEE Transactions on*, 34(11):2233–2246, 2012.

44. M. Piccardi. Background subtraction techniques: a review. *IEEE International Conference on Systems, Man and Cybernetics*, October 2004.

45. D. Pokrajac and L. Latecki. Spatiotemporal blocks-based moving objects identification and tracking. *IEEE Visual Surveillance and Performance Evaluation of Tracking and*

      Surveillance (VS-PETS 2003), pages 70–77, October 2003.

46. F. Porikli. Detection of temporarily static regions by processing video at different frame rates. *IEEE International Conference on Advanced Video and Signal based Surveillance, AVSS 2007*, 2007.

47. C. Qiu and N. Vaswani. Real-time robust principal components' pursuit. In *Communication, Control, and Computing (Allerton), 2010 48th Annual Allerton Conference on*, pages 591–598. IEEE, 2010.

48. B. Recht. A simpler approach to matrix completion. *Journal of Machine Learning Research*, 2009. Preprint available at `http://arxiv.org/abs/0910.0651`.

49. A. Schechtman-Rook. Personal Communication, 2011. Details at `http://www.lsst.org/lsst/science/concept_data`.

50. Y. Shen, Z. Wen, and Y. Zhang. Augmented lagrangian alternating direction method for matrix separation based on low-rank factorization. *Optimization Methods and Software*, pages 1–25, 2012.

51. T. Stynes. Amazon lauds its holiday sales. *Wall Street Journal*, January 2009. Available at `http://online.wsj.com/article/SB123029910235635355.html`.

52. L. Wang, L. Wang, M. Wen, Q. Zhuo, and W. Wang. Background subtraction using incremental subspace learning. In *ICIP 2007*, 2007.

53. J. Wright, A. Ganesh, S. Rao, Y. Peng, and Y. Ma. Robust principal component analysis: Exact recovery of corrupted low-rank matrices via convex optimization. *Advances in neural information processing systems*, 22:2080–2088, 2009.

54. H. Xu, C. Caramanis, and S. Mannor. Outlier-robust PCA: The high dimensional case. *IEEE Transactions on Information Theory*, 59(1):546–572, 2013.

55. H. Xu, C. Caramanis, and S. Sanghavi. Robust PCA via outlier pursuit. *IEEE Transactions on Information Theory*, 58(5):3047–3064, 2012.

IV

Sensors, Hardware and Implementations

17

Ubiquitous Imaging (Light, Thermal, Range, Radar) Sensors for People Detection: An Overview

Zoran Zivkovic
NXP Semiconductors

17.1   Introduction

Smart environments that respond to humans are expected to become a part of our everyday life [1,32,44]. Low cost and low power sensors that enable detecting humans and their movements are seen as the major trend in the technology development that is needed to enable the future human-responsive environments. This chapter presents an overview and analysis of the "imaging" sensors that can be used for sensing movement and presence of people. Main sensing principles are described for the major available technologies. The "video" data obtained with each sensor will be very different. The main problems and the typical algorithms for detecting movement and people are presented for each sensor. Finally, the costs, the development trends and other practical aspects are highlighted and the sensors are compared.

Among the "imaging" sensors, the most widely used one is the regular camera that senses the reflected visible light. In this chapter we will consider the "imaging" sensors in a broader sense, where a sensor is an imaging sensor if it produces at least a two dimensional representation of the space. Such sensors typically provide enough resolution to monitor a certain area, e.g. a room, and detect, count and localize the people in the area. Another approach for detecting movement and people is to have many small distributed sensors, but this goes beyond the scope of this chapter. A nice overview of other approaches and sensors can be found in [40].

There are many sensors that fall under the category of imaging sensors as defined here, but we will focus here on the main sensor technologies currently in use. The current imaging sensors operate in different parts of the electromagnetic spectrum and the three main categories are:

- **Reflected (visible, near infra red) light cameras** - these are inexpensive regular cameras measuring the reflected light in the visible spectrum. It also turns out that the common imaging sensors are also sensitive to the near infra red light which is invisible to the human eye. This is often used, e.g. in many surveillance applications, to provide invisible near infra red illumination of the scene.

- **Thermal cameras (long wavelength infra red)** - cameras measuring the emitted thermal radiation of the objects. Such cameras operate in a different part of the light spectrum and do not need any additional light source. This part of the spectrum is detected using different principles than the regular cameras, e.g. bolometers.

- **Phased array radar (millimeter wavelengths)** - radar operates a very different part of the electromagnetic spectrum. It is typically an active technique where the electromagnetic waves are emitted and the time of flight of the reflection is measured. With the new developments and operation at higher frequencies, reasonably sized phased array antennas can be made and the radar can become an interesting imaging solution that provides very different types of data than other more common imagers.

Besides the three main sensors operating in different bands of the electromagnetic radiation described above, another important category that is widely used are the "range cameras" which are the cameras generating depth map of the scene in various ways. There are three most widely used methods:

- **Stereo (multifocal):** Systems consisting usually of multiple lenses and visible light cameras.

- **Structured light:** Systems using structured light (usually near infra red light).

- **Time of flight:** Systems actively emitting modulated light (usually near infra red light) and measuring time of flight of the emitted light to determine the depth.

Phased array radar is also measuring depth similar to the "time of flight" cameras but since it is a very specific sensor it is considered separately. The radar is also not commonly labeled as "range camera". The laser range scanners that mechanically scan the environment are mentioned under the "time of flight" range imaging category.

The chapter starts with an overview of the complex human sensing field landscape mostly following the lines presented in [40]. The role of imaging sensors in the immense field of human sensing is highlighted. Each of the defined imaging sensor categories is analyzed in detail next. In the analysis we explain first the basic operating principles and achievable resolution. Having in mind possible ubiquitous use, each sensor is also evaluated in terms of power consumption, its form factor and current cost and trends. The chapter ends with a discussion comparing the technologies and conclusions.

17.2   Overview of Human Sensing

Figure 17.1 presents an overview of the human sensing landscape, inspired by the recent report [40]. The taxonomy of the human sensing properties is presented in the upper part of the figure. The human properties that are typically of interest are divided into four groups: physiological (e.g. heart rate), spatio-temporal, behavioral (e.g. gesture [5]), and psychological (e.g. emotions [33]).

The video surveillance and home-building automation applications focus mainly on the spatio-temporal properties, namely:

- **Presence:-** "Is there a person present?"

- **Count:-** "How many persons are there?"

- **Location:-** "Where is each person?"

- **Track:-** "Where was this person before?"

- **Identity:-** "Who is the person?"

Besides the spatio-temporal properties, there is an increasing trend toward extracting also the higher level behavioral concepts such as the human activities, gestures, behaviors [5, 51]. On top of that, knowing the psychological state, i.e. how the person feels at the moment, and then reacting properly is the ultimate goal of designing the future human friendly smart environments [33]. Measuring the physiological properties (such as heart rate, breathing rate, etc...) is often required for medical purposes. Besides, extracting the emotions can also benefit from knowing the physiological properties. In some cases the physiological properties can be used for extracting the spatio-temporal properties, for example remote detection of the heart and breathing rate can be used to detect and localize people.

The smart systems extract the human properties based on various sensor reading. The sensors are the interface between the physical world, and the digital world where the sensor readings are processes and information about humans and their movement extracted. This is illustrated in Figure 17.1. The lower part of the image presents a taxonomy of the physical cues that can be used for detecting people. There are various types of sensors for various type of physical phenomena. As a result there is an enormous number of possible ways to detect presence and movement of people. One type of classification is according to the amount of the data produced by the sensor:

- **Binary sensors:** - these are typically lowest cost sensors which typically respond to the presence or movement of people in close proximity to the sensor or in a certain area. Examples are: contact sensors, break-beam sensors, pyroelectric movement sensors. Combining many of such responses can provide more information about the people [3, 11].

- **"Medium data size" sensors:** - there is a huge set of sensors that provide more information than a binary sensors but still less than what could be called an imaging sensor. Examples are: vibration sensors, electric/magnetic field sensors, laser scanners, ultrasound, radio (tomographic) sensors.

- **Imaging sensors:** - we will classify a sensor as an imaging sensor if it produces at least a two dimensional representation of the space. This means that counting and localizing people in the certain area of the sensor could be possible, provided the sensor has enough resolution. The imaging sensors can vary considerably in the number of pixels provided but also in the type of information contained in each pixel. This will be discussed in more detail in the next section.

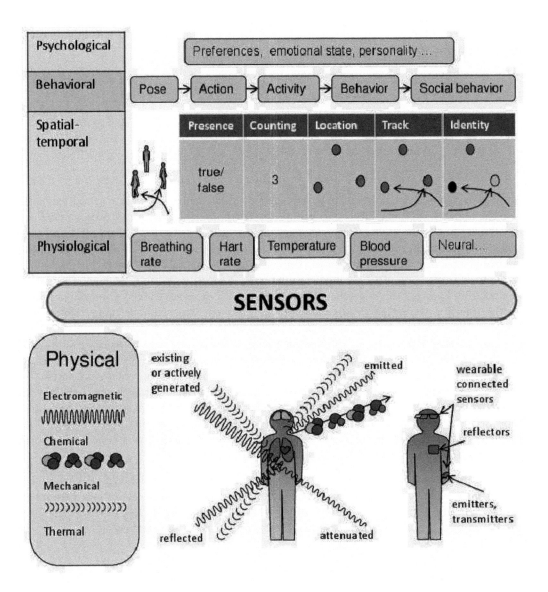

FIGURE 17.1   Human sensing landscape. Taxonomy of properties and the sensing cues. Inspired by the recent report [40].

Before moving to the next section, for completeness we should also mention another aspect of human sensing. We are living in a connected world with more people continuously wearing various devices that contain sensors. From this point of view the sensors can be divided into:

- **External sensors (non-instrumented scenarios)** - where the sensors are part of the environment and there are no restrictions put on people wearing some special devices.

- **Wearable sensors (instrumented scenarios)** - where a person wears special tags or devices that could complement the environment infrastructure. The wearable devices can, in certain scenarios, co-operatively share their identity and possibly also their local sensor information [42].

The future human sensing systems will probably be some combination of different sensors of different complexity, where some sensors are part of the environment and some are part of the devices that people are wearing [40]. Imaging sensors are expected to play an important role.

17.3   Imaging Sensors Algorithms and Architecture

The most widely used "imager" is the regular camera that senses the reflected visible light. The data from all "imagers" considered here will share many similar properties. Two important properties commonly linked to the imaging sensors that we will mention here are: the large amount of data and the complex algorithms.

17.3.1   Complex Algorithms

If there is enough resolution, people and other objects can be recognized based on their shape. While this task is often easy for our brain and eyes when looking at the visible light images, complexity of the human shape and possible complex environments make it very hard to design an algorithm for robust people detection from various viewing angles. Furthermore, in many surveillance situations, people are obscured or just partially visible in a complex scene that it is hard even for the human eye to spot them. Therefore, in most practical systems, motion is the basic cue for detecting the presence of people [12]. Just detecting the movement by e.g. detecting large changes between two consecutive images, is usually of limited use. Common approach is the background modeling, where a statistical model of the scene is constructed and continuously updated. The foreground objects are detected as parts of the image that do not fit the static scene model. Eventually, after detecting the foreground objects, some shape cues are used to classify the detected objects and remove some wrong detections. For each of the selected imaging sensors we explain also the basic motion detection and background modeling principles. See Figure 17.8 for some examples of basic background modelling and detection processing steps for different types of imaging signals. Other possible cues for people presence detection are also mentioned. However, a detailed analysis of the huge field of methods for people detection in images is beyond the scope of this chapter.

17.3.2   Smart Imagers

Large arrays of sensors provide large amounts of data. Widely available and cheap high definition visible light cameras generate gigabytes of data per second. Phased array radar typically has much less antennas for the angular resolution than a regular camera pixels.

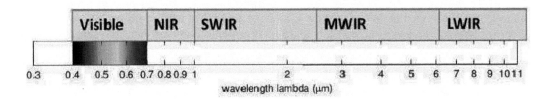

FIGURE 17.2   (See color insert.) Visible and infra red spectrum bands.

However, radar signals are sensitive to motion and depth. As a result radar is providing actually a four dimensional images of the space [41]. The large amount of data produced by most imagers becomes problematic in practice both for storage and transmission. Therefore, the concept of "smart imagers" is often considered. The idea is to process the data in the sensor and then transmit or store only the relevant information. In scenarios where there are many imagers it will also become important to offload the central processor in such a way. Often just video compression is applied locally. However, there is increasingly more interest for implementing also the complex processing and detection algorithms locally embedded in the camera module. An overview of smart cameras for wireless networks can be found in [24, 50].

17.4   Reflected (Visible, Near Infrared) Light Cameras

Operating principles: Most current cameras are using complementary metal-oxide-semiconductor (CMOS) active-pixel sensors. Older and still used technology are digital charge-coupled devices (CCD). In both cases, an array of photo sensors is placed behind an optical system (i.e. lens). The incident light is transformed to an electrical signal by the photo sensors. The signal depends on the light intensity, the sensor area and the sensor sensitivity [14]. In surveillance applications the captured images are often watched by human operators. As a result these sensors are often tuned to mimic the response of the human eye in the "visible part of the spectrum". The photo sensors are usually sensitive to a wider part of the spectrum than the visible [8]. In Figure 17.3 a typical response of a silicon based photo sensor is shown. The required sensitivity that is close to the human eye is typically achieved by using some light filtering materials, but the same sensor could be used for detecting other parts of the spectrum, i.e. near infra red.

The common definition of the relevant spectral bands is depicted in Figure 17.2. The "near infra red" (NIR) and "short wavelength infra red" (SWIR) are near the visible bands and their behavior is quite similar to the more commonly used visible light. Energy in these bands must be reflected from the scene objects in order to produce an image. This means that there must be some external illumination source. The Visible and NIR systems can take advantage of sunlight and other light sources. Typically those systems require some type of artificial illumination during night. Arrays of infrared Light Emitting Diodes (LEDs) often provide a very cost effective solution for illumination and are often used for surveillance purposes. Such illumination is usually good enough only for short ranges.

Resolution: Standard image sensors typically have a huge amount of sensors (pixels) and combined with a good quality lens, large angular resolutions can be achieved. The images do not provide depth information, but some limited information about the distance of the detected objects can be obtained using some geometric constraints, e.g. known size of the object or knowing that the object is moving over a known surface which is often the case. The other limitation of the visible, NIR and SWIR systems is that they rely on

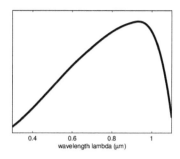

0.4 0.6 0.8   1
wavelength lambda (μm)

FIGURE 17.3 Typical response of a silicon based light diode.

existing light sources. In many surveillance applications, the cameras are equipped with a light source. Typical light sources give limited depth range of 10's of meters.

Detection Cues and principles: People detection is one of the most challenging problems in computer vision [19] due to large variations caused by articulation, viewpoint and appearance. Visual people detection has been researched intensively for a number of years [12]. In surveillance applications motion is one of the main cues used for detecting humans and background modeling is a widely studied subject [7,22,35]. The images depend on the external illumination that is usually not controlled by the detection system, e.g. sun. The changes of the illumination are one of the major challenges of background modeling based on the visual and near infra red images. Other common problems are the shadows casted by the foreground objects [34,37].

Power, cost and size: Being a mainstream widely used technology, the standard imaging sensors are available at a very low cost. The power consumption is low, however battery powered operation is still difficult for a very long time. The size of the sensor array is very small and typically the main limiting factor is the optical system. There is always inherent trade off between the size and the performance where the smaller sensors usually have signal to noise ratio. Small but good quality optical systems are also a challenge. However, using imaging sensors in the mobile and portable devices leads to a huge demand for further size reduction. As a result, low power, small size and low cost imaging sensors are widely available today and they are the dominating imaging technology.

17.5 Thermal Images

Operating principles: The word "infrared" usually refers to a broad part of the electromagnetic spectrum. Everything between visible light and microwaves might be denoted as "infrared". Much of the infrared range is blocked by the atmosphere and the portions of spectrum that are not blocked are often called "atmospheric transmission windows". The following infrared bands are defined: Near Infrared (NIR), Short-Wave Infrared (SWIR), Medium-Wave Infrared (MWIR), and Long-Wave Infrared (LWIR).

The MWIR and LWIR bands are often called "thermal" bands because warm objects in a typical scene emit radiation in these ranges. An imaging system that operates in these ranges can be completely passive, requiring no external illumination because sensing the energy that is radiated directly from objects also during the night. Two major factors that determine how bright an object appears in a thermal image are: the object's temperature and its emissivity. As an object gets hotter, it radiates more energy and appears brighter to a thermal imaging system. Emissivity is a physical property of materials that describes how efficiently it radiates. A black body is a theoretical object which will radiate infrared

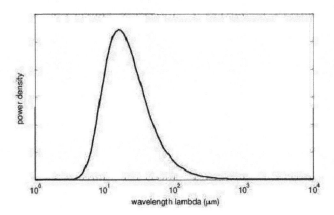

FIGURE 17.4   Spectral radiation of a human body using the black body model and the Planck's law.

radiation at its contact temperature with an emissivity of 1. The fraction of radiation compared to the ideal black body is called emissivity. For example, if human skin and cloth the person is wearing are at exactly the same temperature, the cloth will appear darker since they have typically a lower emissivity than skin.

The Planck's law describes the spectrum of blackbody radiation, which depends on the object's temperature:

$$P = \frac{2hf^3}{c^2} \frac{1}{e^{\frac{hf}{k_B T}} - 1} \tag{17.1}$$

where P denotes the object spectral radiance, f is the spectral frequency, T its absolute temperature, k_B the Boltzmann constant, h the Planck constant, and c the speed of light. Human skin temperature is about $33^\circ C$ and the spectrum of the radiation is shown in Figure 17.4. The peak of the radiation is in the LWIR part of the spectrum around wavelength of $10\mu m$. Simple silicon sensors can not be used for these wavelengths, see the typical photodiode response in Figure 17.3 and usually more exotic materials are used. The most widely used detection principle is "bolometry" [36]. A bolometer consists of an absorptive material such that radiation impinging on the absorptive element raises its temperature. The more absorbed power, the higher the temperature will get. The very small temperature change is measured directly with an attached thermometer device, or the resistance change of the absorptive element itself can be used as a measure of temperature.

At the long wavelengths, infrared radiation behaves differently from visible light. For example, glass is not transparent for the LWIR band, and blocks most energy in the MWIR band radiation. Consequently, LWIR and MWIR systems cannot use inexpensive glass lenses, but are forced to use other materials like silicon or germanium.

Images in general exhibit different behavior than visible light images most people are familiar with. An illustration is given in Figure 17.5. Glass windows are not transparent, so they appear brighter or darker according to their temperature. Many materials become reflective, e.g. metal objects. Objects can get warmed up by other warm objects or various heat sources and start appearing brighter. For example, a person sitting on a chair will make the chair warm, or a laptop waking up from its sleep mode will produce heat and warm up.

For outdoor recording, the LWIR is influenced by rain since radiation in the LWIR band is not transmitted by water. As a result the thermal contrast in the thermal images is reduced if it is raining. On the other hand, longer wavelengths are reasonably tolerant to smoke, dust and fog since the light gets less scattered in this part of the spectrum.

Humans
as warm
objects
appear
bright

Other warm
objects are
also bright,
here the chair
where the
person was
sitting

Reflection
on metal
objects e.g.
white
board in
this image

FIGURE 17.5 Example image in the LWIR band to illustrate some typical behavior that is different than the visual light.

Resolution: Similar to the standard image sensors large angular resolutions are possible. However, the thermal sensing elements are typically larger than the imaging pixels. Furthermore, it is not possible to use standard glass lenses but more exotic materials. As a result the thermal imaging sensors are usually lower in spatial resolution. While in the standard visible spectrum imaging world VGA resolution of 640×480 pixels is considered low resolution, it is typically considered to be a high resolution for the thermal imaging.

Detection Cues and Principles: Thermal infrared video cameras detect relative differences in the amount of thermal energy emitted and reflected from objects in the scene. As long as the thermal properties of a foreground object are slightly different (higher or lower) from the background radiation, the corresponding region in a thermal image appears at a contrast from the environment.

The use of thermal imagery alleviates several classic computer vision problems such as the presence of shadows and sudden illumination changes. The shadows can appear in the thermal domain but only when an object is stationary long enough for the shadow to cool down the background. Furthermore, the thermal imaging works also during night without any light sources present. The thermal imagery has its own unique challenges. Other objects that warm up, might appear in the image, e.g. computer screen. The heat is transferred between the objects, e.g. the sun will warm up the scene, a person will warm up the chair and other objects it is touching. Finally many materials will reflect the LWIR radiation, e.g. metal walls in an office building.

Many methods for foreground detection take also into account the belief that the objects of interest are warmer than their surroundings [16,20,23,49]. Even though this is common in cooler nighttime environments, during winter and in air-conditioned indoor environments, it is not true in general. Other methods try to solve a more general case assuming that surroundings can also be warmer, as for example in outdoor summer scenes [9].

Power, cost and size: The thermal imaging sensors are mostly used for military applications. The operating principles and calibration are quite a bit more complex than for the regular imagers. One of the reasons is that the sensing is based on very small temperature changes and any other unwanted heating, e.g. by processor, needs to be compensated for. As a result the thermal imagers are typically much more expensive than the regular visible light cameras. However, there are more low-cost thermal imagers appearing recently. The low cost sensors are typically also low resolutions starting from 4x4 pixels for the Omron D4T sensor, 8x8 for Panasonic Grid Eye, going up to around 100x100 pixels. It is expected that new low cost low resolution thermal sensors will continue to emerge.

17.6   Range Images

Range imaging is the name for a collection of techniques and devices which are used to produce a 2D image showing the distance to points in a scene from a specific viewing point. The resulting image, the range image, has pixel values which correspond to the distance, e.g., brighter values mean shorter distance, or vice versa. If the sensor which is used to produce the range image is properly calibrated, the pixel values can be translated into physical units such as meters. The range cameras are known under many names: ranging camera, lidar camera, time-of-flight (ToF) camera, and RGB-D camera. The underlying sensing mechanisms can be very different. There are three most widely used methods that are discussed separately.

17.6.1   Detection Cues and Principles

Similar to the thermal imaging, depth map image removes several classic computer vision problems such as the presence of shadows and sudden illumination changes. In most applications the goal is to detect the movement of the physical objects and therefore the depth images are a good choice to detect movement of the physical objects. Besides that the spatial information provided by the depth maps is very useful for extracting location and other spatial information about the detected objects, e.g human pose for detected people. Many robust systems have been developed based on range imaging for detecting humans, their pose, gestures, etc. In most cases the range cameras are made using same or similar photo detectors as the regular cameras. As consequence it is often possible to provide the traditional images, either in visible or near infra red spectrum, in combination with the depth information. Combining the two sources of information is common to increase the robustness of the detection.

17.6.2   Stereo (multifocal) Cameras

Operating principles: A multifocal camera is a type of camera with a special optical element, e.g. two lenses, that are used to create projection of the world from different viewing angles. Stereo cameras usually have 2 regular lenses. Either a separate image sensor is used for each lens or special optical systems that provide multiple views with a single sensor. Imaging sensors that include optical solutions to provide many viewing angles are also available usually under name of "light filed photography" or "plenoptic cameras" [2,29].

Viewing the scene from different angles enables a stereo (mutifocal) camera system to determine the depth to points in the scene. In order to solve the depth measurement problem using a stereo camera system, it is necessary to first find corresponding points in the different images. Solving the correspondence problem is one of the main problems when using this type of technique. For instance, it is difficult to solve the correspondence problem for image points which lie inside regions of homogeneous intensity or color. As a consequence, range imaging based on the stereo triangulation can usually produce reliable depth estimates only for a subset of all points visible in the multiple cameras [38]. Furthermore, as with all optical methods, reflective or transparent surfaces raise difficulties.

In the basic case of two cameras that are at a baseline distance of d_B. If cameras are looking forward, a world 3D point at distance r from the camera will be detected in the two images at different horizontal positions. The difference in the positions is denoted as disparity d. The disparity d is related to the depth r by the following equation:

$$d = d_B F \frac{1}{r} \tag{17.2}$$

where F is the focal length. This equation indicates that for the same depth the disparity is proportional to the baseline, or that the baseline length d_B acts as a magnification factor in measuring d in order to obtain depth r. That further means that the estimated depth is more precise if the two cameras are farther apart from each other - cameras with a longer baseline d_B. A longer baseline implies larger devices and also other problems. For example, because a longer disparity range must be searched, matching is more difficult. Therefore there is a greater possibility of false matches and typically there is a trade-off between precision and accuracy (correctness) in matching between different views [31].

The advantage of this technique is that the measurement is passive; it does not require special conditions in terms of scene illumination. However, it assumes enough texture details in the scene.

Resolution: Since stereo systems are based on regular image sensors large angular resolutions are possible. However, the depth map is computed by matching the points from different views and typically a texture pattern from a number of points is needed for reliable matching. As a result, the depth image effective resolution is lower and often not available for the image points where there is not enough details. The depth accuracy is typically a few centimeters. The range typically goes up to tens of meters. The accuracy and maximum range depend on the distance between the cameras/lenses, but also on the scene, as mentioned above.

Detection cues and principles: Similar detection cues apply as for other depth cameras. A difference is that the stereo cameras obtain at the same time the regular visible light camera images. The information from the regular images and the estimated depth can be combined to achieve better performance [15, 21, 27].

Power, cost and size: Stereo captured images can be used to simulate human binocular vision. With the new generation of 3D displays there is more demand for low cost stereo cameras. Even some mobile phone devices include stereo cameras and 3D diplays today. These devices are usually not meant for extracting depth and the depth extraction would require also careful calibration and synchronization of the cameras. Power consumption for recording data is small. However, complex processing is needed to extract the depth map from two images. The size of the devices is determined mainly by the needed baseline between the cameras/lenses. Most devices are 5-10cm long, except the mobile phone devices but the small stereo cameras in the mobile phones would have poor depth resolution.

17.6.3 Structured Light Cameras

Operating principles: By illuminating the scene with a specially designed light pattern, depth can be determined using only a single image of the reflected light. The structured light can be in many forms, horizontal and vertical lines, points, checker board patterns [18]. In practical applications usually, invisible NIR light is used. The Kinect-Camera from Microsoft is the first consumer-grade example. It uses a complex dotted pattern of near infrared light to generate a dense 3D-Image [48].

As with all optical methods, reflective or transparent surfaces are difficult to handle. Double reflections and inter-reflections can cause the stripe pattern to be overlaid with unwanted light. Reflective cavities and concave objects are therefore difficult to handle. Similar to the stereo imaging, usually there is a need to have a larger baseline between the illuminator and the camera. Larger baseline leads to better accuracy but also larger devices. On the positive side the structured light principle requires just a single camera with a regular lens and the illumination can be realized by arrays of infrared LEDs in a cost effective way. As a result the total system costs are usually smaller than for the stereo/multifocal systems.

The depth map quality is usually better than the stereo cameras since the patterns are provided by the active illumination while the stereo matching depends on the existence of the patterns in the scene. As with all active methods, there is a possibility of interference. This means that two cameras cannot observe the same scene unless special care is taken to remove the interference, e.g. time multiplexing or using a different wavelength for the light source and filters.

Resolution: Similar to the stereo systems the structured light systems are based on regular image sensors and large angular resolutions are possible. The depth map is computed by matching the points from different views and typically a texture pattern from a number of points is needed for reliable matching. As a result the depth map resolution is usually not very high. The Kinect-Camera from Microsoft gives a 320×240 depth map. The depth accuracy is typically a few centimeters and range goes up to 10s of meters. The Kinect-Camera from Microsoft works for distances up to 4m. The accuracy and maximum range depend on the distance between the cameras and illuminator as mentioned above. The depth range is additionally limited by the strength of the structured light source.

Detection cues and principles: Similar detection cues apply as for other depth cameras. Since the projected light patterns are usually in the near infra red band, a filter is used to remove the visible light before reaching the imaging sensor. As a result regular visible light images of the scene are not directly available. The Kinect from Microsoft includes an additional visible light color camera.

Power, cost and size: The Kinect-Camera from Microsoft is the first camera in the $100 price range. Compared to stereo cameras, the devices are usually cheaper and depth map more accurate. The principle cannot work at very low power since still considerable power is needed to project the structured pattern on the scene. In addition, complex processing is needed to extract the depth map from the image with structured light patterns. The Kinect device contains a dedicated application specific IC for this purpose to be able to achieve the high frame rate. As for the stereo cameras, the size of the devices is determined mainly by the needed baseline between the camera and the illuminator. Most devices are 5-10cm long.

17.6.4 Time of Flight Cameras

Operating principles: A time-of-flight camera (ToF camera) is a range imaging camera system that resolves distance by measuring the time-of-flight of a light signal between the camera and the subject for each point of the image. The principles are similar to the "Lidar" which is a remote sensing technology that measures distance by illuminating a target with a laser, or radar where mm-waves are used. Illumination is provided by LEDs, usually in the invisible near infra red spectrum, rather than a laser or mm-Waves.

There is number of different principles that can be used e.g. range-gated ToF or modulated ToF [25]. The range gated imagers have a built-in electronic or mechanical shutter in front of the image sensor that opens and closes in some relation to the rate the light pulses are sent out. The base principle is that part of every returning pulse is blocked by the shutter according to its time of arrival, the amount of light received relates to the distance the pulse has traveled. The ZCam by 3DV Systems [46] is a range-gated system (purchased in 2009 by Microsoft). Range gated imagers can also be used in 2D imaging to suppress reflections outside a specified distance range. A pulsed light source provides illumination, and an optical gate allows light to reach the imager only during the desired time period.

Modulated light sources with phase detectors work by modulating the outgoing light beam. Measuring the phase shift of the modulation can be used to determine time of flight and in this way distance. This approach has a modular error challenge; ranges are modulo half the RF carrier wavelength. The Swiss Ranger [30] is a compact, short-range device, with ranges of 5 or 10 meters. With phase unwrapping algorithms, the maximum range can be increased. Initial device had resolution of with 176×144 range pixels. Photonic Mixer Devices (PMD) [45] camera is another similar device.

For the time of flight principle cameras both the illumination unit and the image sensor timing needs to be very accurate to obtain a high resolution [39]. As the structured light approach, the time of flight cameras suffer from the interference problems and special measures need to be taken to have more than one device observing the same scene.

Resolution: The available systems cover ranges of a few meters up to about 60 m. The distance resolution can be reasonably high often about 1 cm. The lateral resolution of time-of-flight cameras is generally low compared to standard 2D video cameras. Most commercially available devices provide at 320×240 pixels or less resolution.

Detection cues and principles: Similar detection cues apply as for other depth cameras. The depth estimates are obtained directly or after some relatively simple per pixel processing. The depth estimates are usually more robust than other range cameras.

Power, cost and size: Existing commercial products, such as the Swiss Ranger and PMD Tech products, are still very expensive. In contrast to stereo vision or structured light, the whole system is very compact: the illumination is placed just next to the lens, whereas the other systems need a certain minimum base line. In contrast to laser scanning systems, no mechanical moving parts are needed. It is very easy to extract the distance information out of the output signals of the TOF sensor, therefore this task uses only a small amount of processing power, again in contrast to stereo vision or structured light. The principle can work at reasonably low power since emitting light pulses requires typically less power than structured patterns on the scene, and there is no need for complex processing to extract depth map.

17.7   Phased Array Radar

Radar is in principle similar to the time-of-flight camera, but where, instead of light, radio waves are used. The radio waves are a type of electromagnetic radiation with wavelengths much longer than the infrared light. Millimeter range wavelengths are the most commonly used for remote sensing for various reasons described later. The sensor is an antenna that can be often used for both emitting and detecting the radio waves. Instead of using a lens for getting the information about the direction from which the signal is coming, phase differences between the signals measured at multiple antennas provide this information. The multiple antennas are also called "phased array" antennas [41, 43].

17.7.1   Passive Radar - Radiometry

Passive radar or "radiometer" is a device measuring the radiant flux (power) of existing electromagnetic radiation [10]. It was shown many times that the existing communication radio waves can be used for detecting movement [47] and localization of devices [4]. Even the thermal radiation of the human body can be measured in the radio spectrum. The energy distribution shown Figure 17.4. indicates that there will be still some small part of energy in the radio frequencies. Various systems were made to show that detection of the thermal radiation is possible in the mm wavelengths band [28]. Since these topics are still in research phase and often do not fall into category of real "imaging" sensor, we will focus in this chapter only on the more common active radar described next.

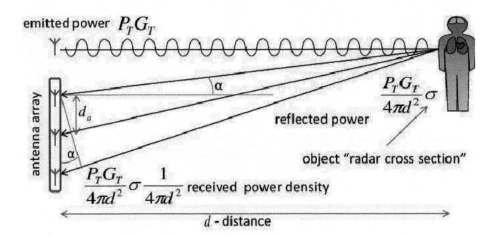

FIGURE 17.6   Basic principles of phased array radar.

17.7.2   Active Radar

Operating principles: The basic operating principle is illustrated in Figure 17.6. The radio wave with power P_T is emitted by an antenna with gain G_T. The wave is reflected from an object at distance d. The reflection will depend on the object and the factor describing the reflected power is denoted as "radar cross section" σ. The power density that is received back can be calculated as:

$$\frac{P_T G_T \sigma}{(4\pi)^2 d^4} \tag{17.3}$$

An effective area of a receiver antenna is calculated as:

$$A_{eff} = \frac{G_R \lambda^2}{4\pi} \tag{17.4}$$

where G_R is the antenna gain and λ is the radiation wavelength. For making effective antennas physical antenna size is determined mainly by the wavelength of the radiation. This is one of the main reasons why mm-wavelengths are especially interesting, as they allow small antenna sizes.

Combining the previous two equations gives the so called "radar equation" for the received power:

$$\frac{P_T G_T G_R \lambda^2 \sigma}{(4\pi)^3 d^4} \tag{17.5}$$

The equation combined with the noise characteristics of the radar receiver system can be used to calculate the transmitted power needed for reliable detection. Large amount of power is needed for the typical radar airborne applications. However, for a shorter distance typical for surveillance applications the amount of power is reasonably small and in the mW range.

Let $s(t)$ denote emmitted radar signal where t is time. The general form of the reflected signal can be written as:

$$s(t - t_d) e^{j2\pi f_D t} e^{-j\phi} e^{-j\psi} \tag{17.6}$$

where t_d is the time of flight delay of the emitted signal $s(t)$, f_D is the Doppler frequency and the factor $e^{j2\pi f_D t}$ is the resulting frequency shift described in the complex domain. There are two typical additional delays: ϕ is the delay introduced by the physically different positon of multiple radar antennas and ψ is phase shift introduced by the object.

The time of flight delay t_d is typically estimated from the refelected signal and used to calculate the distance to the object d. If c is the speed of light we have:

$$t_d = \frac{2d}{c} \qquad (17.7)$$

For moving objects, the reflected signal will also have the second component due to the Doppler's shift. If the object is moving with speed of v and the radial part of the speed is v_r, the Doppler's frequency shift measured in the radar signal will have the following frequency:

$$f_D = \frac{2f_0}{c} v_r \qquad (17.8)$$

where c is speed of light and f_0 is the radar operating frequency. For higher operating frequencies, higher Doppler frequency is observed. For example if $f_0 = 60GHz$ we have $f_D = 400 * v_r$. As a result it becomes easier and faster to detect small movements.

In case of multiple antennas an additional delay ϕ will be introduced between antennas due to their different positions in space. If the object that reflected the signal is far away and the distance between two antennas is d_a, as in Figure 17.6, the additional delay at the second antenna becomes:

$$\phi = \frac{d_a sin(\alpha)}{c} \qquad (17.9)$$

where α is the angle corresponding to the position of the object with respect to the antennas. It is important to notice also that this angular position related phase delay can be introduced by multiple physical posions of either transmit or receive antennas or combinations of the two. Many practical systems therefore include combinations of multiple transmit or receive antennas in specially chosen spatial configurations usually known as "Multiple Input Multiple Output" (MIMO) radar systems [17]. In a MIMO system an angular resolution of $N \times N$ can be realized using only N transmit and N receive antennas.

Various types of radar signals $s(t)$ are used by various radar system. Examples are the pulse radar where the signal is a short pulse, or continuous wave radars where much longer signals are used. Typically many signals are transmitted and their reflections are measured. Complex processing is then applied to extract the depth, speed and angle from the signals that have the general form presented above.

Resolution: The ranges that can be covered depend on the emitted power and also reflectance properties of the objects that need to be detected. With limited transmitted power of few mW distances of up to about 50 m could be covered for detecting people. The distance resolution can be very high and often about 1 cm. The resolution depends on the signal used by the radar system. Signals that occupy large bandwidth of the spectrum lead to high range resolution. The lateral resolution is generally low and it is determined by the number of transmit and receive antennas. For example, current systems used in automotive applications go up to just around 10 antennas. The low lateral resolution is often compensated by the very high depth resolution and also by the sensitivity to movement, i.e. Doppler shift. Possibility to measure the radial movement directly is unique for radar when compared to other imaging sensors. The resolution in radial speed measurements using the Doppler effect can be very high. An example are the radar systems used for remote vital signs monitoring that are able to detect even the very small movements of human breathing and the heart beat [13, 26].

Detection cues and principles: As mentioned above, radar signals typically require complex processing steps to extract information about the radial distance, the radial speed, and the angle corresponding to the position of the objects. Since the movement is directly detected by the Doppler's frequency shifts, many radar based surveillance systems use only the Doppler effect. Distance and angle can give additional information about the object.

As other active systems, the radar systems can interfere with each other. Different measures typical for radio communications systems can be used to reduce the chance for the interference or avoid it completely. Since the signals are similar to the typical signals in radio communication and acoustics, many techniques are borrowed from those areas. For example the common task of background modeling and subtraction is for radar signals more similar to the so called "spectral subtraction" used to remove background noise in acoustic systems [6]. See Figure 17.8 for an example of background spectral subtraction for radar signals. The power of the reflected signal corresponding to different distances is shown. The system had four different antennas. The background can be modelled in this domain and then used to extract the parts of the signal corresponding to the foreground objects.

Power, cost and size: Existing commercial products are still very expensive. However, further integration of the radar components becomes possible with new technologies and cost can be heavily reduced. Wider application would also drive down the costs. Recent research effort on miniaturized integrated radar has large potential. Advanced technology enables low cost devices, e.g in standard CMOS, operating at high frequencies where the wavelengths approach millimeters. Since the needed antenna size depends on the wavelength of the radiation it becomes possible to arrange many antennas on a small space. The antenna arrays are typically flat and can be mounted under non-transparent materials, e.g. plastic, such that the device is invisible.

17.8 Conclusion

An overview of "imaging" sensors is presented here. Regular cameras, thermal cameras, range cameras and phased array radar are selected as the most promising imaging sensors to be widely used for people and movement detection. These imaging technologies are operating in different parts of the spectrum, as illustrated in Figure 17.7, and have different operating principles and characteristcs. Each type of sensor is described and discussed. Strong and weak points were highlighted. Practical aspects were included such as power consumption and costs. Table in Figure 17.9 presents a short summary of the main properties for each technology.

Because of their very low cost the regular visible and near infra red light imagers are currently dominating the market. Recent introduction of the consumer grade range camera by Microsoft opened up the opportunities for the range cameras. The range cameras typically give much more reliable detection results and are much more suitable than the regular cameras for extracting additional information about the humans, such as pose, gestures, etc. The thermal imaging and phased array radar can be seen as new emerging technologies where it is very difficult to predict if they will be widely used. In cases when there is a difference in temperature between the detected objects and the environment the thermal imaging becomes a good option. An especiall interesting property of the thermal imaging is that it is a passive technique measuring just the emitted thermal radiation. The radar is quite different than the other technologies, especially since it is very good in detecting movement and it can be hidden under visually non-transparent cover. Different part of the spectrum and principle of detection make it also interesting to combine with some other imager. Low cost radar solutions could make radar a very interesting sensor in the future.

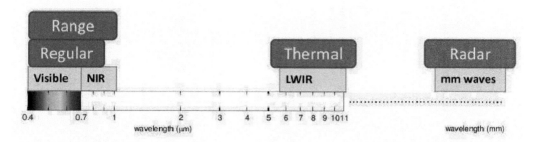

FIGURE 17.7 Different imaging sensors and their spectrum bands.

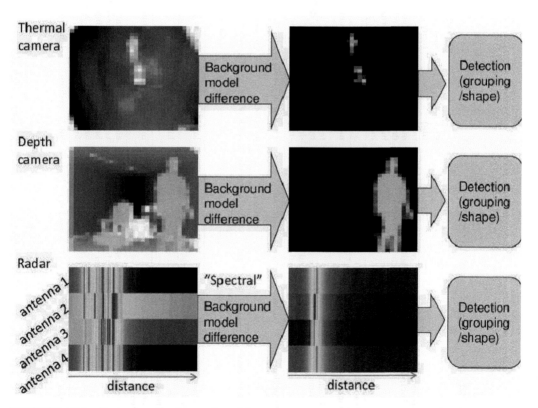

FIGURE 17.8 Examples of background model segmentation typical processing steps. Similar to regular images, thermal images contain observed thermal radiation values and depth images contain depth values per pixel. For the radar case the power of the reflected signal for different distances and different antennas is shown.

TABLE 17.1 Imaging Sensors: An Overview

	Visible	Thermal	Range cameras			Phased Array Radar
			Stereo (Visible)	Structured light (NIR)	Time of Flight (NIR)	
Operating Principle	detecting reflected light from existing light sources	detecting thermal radiation of the objects	range from matching reflected patterns in two camera images	range from emitted invisible NIR light patterns	range from time of flight of emitted invisible NIR light	antennas emitting radio-waves and detecting reflections
Resolution						
-angle (pixels)	high	medium / low, e.g. 640x480	medium	medium	medium	low, e.g. 4x4 (4 + 4 antennas in MIMO randar)
-distance	~	~	few cm	~cm	~cm	~cm
-motion	by difference - medium lateral - very low radial	by difference - low lateral -very low radial	by difference -low lateral -medium radial	by difference -low lateral -medium radial	by difference -low lateral -medium radial	by difference -very low lateral -high radial (Doppler)
Detection Cues	-reflectance contrast -motion	-thermal contrast -motion	-depth contrast -motion	-depth contrast -motion	-depth contrast -motion	-depth contrast -motion
Limitations	-sudden light changes -shadows	-other warm objects	-lack of texture	-interference -reflections	-interference -reflections	-interference -reflections -angular resolution
Power (Sensing)	low	low/medium	low	medium	medium	medium
Base Local Processing	low	low/medium	high	high	medium	high
Detection Processing	complex	medium	medium	medium	medium	medium
Form factor	small (determined by the lens, e.g. 5mm x 5mm x 5mm)	small (determined by the lens, e.g. 10mm x 10mm x 10mm)	elongated (determined by the baseline, e.g. 5mm x 5mm x 5mm)	elongated (determined by the baseline, e.g. 5mm x 5mm x 5mm)	small/medium (needs light source, e.g. 10mm x 10mm x 10mm)	medium and flat (determined by the wavelength, e.g. 3cm x 3cm x 3mm)
Costs	very low	medium / high	low / medium	low / medium	medium / high	high

References

1. E. Aarts. *Ambient Intelligence*. AH, 2004.
2. E. Adelson and J. Wang. Single lens stereo with a plenoptic camera. *IEEE Transactions on Pattern Analysis and Machine Intelligence*, 14(2):99–106, 1992.
3. J. Aslam, Z. Butler, F. Constantin, V. Crespi, G. Cybenko, and D. Rus. Tracking a moving object with a binary sensor network. In *SenSys*, pages 150–161, 2003.
4. P. Bahl and V. Padmanabhan. RADAR: An in-building RF-based user location and tracking system. In *INFOCOM 2000. Nineteenth Annual Joint Conference of the IEEE Computer and Communications Societies*, volume 2, pages 775–784, 2000.
5. A. Bobick. Movement, activity, and action: The role of knowledge in the perception of motion. *Royal Society Workshop on Knowledge-based Vision in Man and Machine*, 352:1257–1265, 1997.
6. S. Boll. Suppression of acoustic noise in speech using spectral subtraction. *Acoustics, Speech and Signal Processing, IEEE Transactions on*, 27(2):113–120, 1979.
7. T. Bouwmans, F. El Baf, and B. Vachon. Background modeling using mixture of Gaussians for foreground detection-a survey. *Recent Patents on Computer Science*, 1(3):219–237, 2008.
8. G. Brooker. *Introduction to Sensors for Ranging and Imaging*. ScitTech Publishing, 2009.
9. J. Davis and V. Sharma. Background-subtraction in thermal imagery using contour saliency. *International Journal of Computer Vision*, 71(2):161–181, 2007.
10. R. Dicke. The measurement of thermal radiation at microwave frequencies. *Review of Scientific Instruments (AIP)*, 17(7):268275, 1946.
11. P. Djuric, M. Vemula, and M. Bugallo. Target tracking by particle filtering in binary sensor networks. *IEEE Transactions on Signal Processing*, 56(6):2229–2238, 2008.
12. P. Dollár, C. Wojek, B. Schiele, and P. Perona. Pedestrian detection: An evaluation of the state of the art. *IEEE Trans. Pattern Anal. Mach. Intell.*, 34(4):743–761, 2012.
13. A. Droitcour, V. Lubecke, J. Lin, and O. Boric-Lubecke. A microwave radio for doppler radar sensing of vital signs. In *Microwave Symposium Digest, 2001 IEEE MTT-S International*, volume 1, pages 175–178. IEEE, 2001.
14. R. Dyck and G. Weckler. Integrated arrays of silicon photodetectors for image sensing. *IEEE Trans. Electron Devices*, 15(4):196201, 1968.
15. C. Eveland, K. Konolige, and R.C. Bolles. Background modeling for segmentation of video-rate stereo sequences. In *IEEE Conference on Computer Vision and Pattern Recognition*, pages 266–271. IEEE, 1998.
16. A. Fernández-Caballero, J.C. Castillo, J. Serrano-Cuerda, and S. Maldonado-Bascón. Real-time human segmentation in infrared videos. *Expert Systems with Applications*, 38(3):2577–2584, 2011.
17. E. Fishler, A. Haimovich, R.Blum, D. Chizhik, L. Cimini, and R. Valenzuela. MIMO radar: an idea whose time has come. In *Radar Conference, 2004. Proceedings of the IEEE*, pages 71–78. IEEE, 2004.
18. D. Fofi, T. Sliwa, and Y. Voisin. A comparative survey on invisible structured light. In *SPIE Electronic Imaging Machine Vision Applications in Industrial Inspection XII*, page 9097, 2004.
19. D. Forsyth and J. Ponce. *Computer Vision, A Modern Approach*. Prentice Hall, 2003.
20. T. Gandhi and M. Trivedi. Pedestrian protection systems: Issues, survey, and challenges. *IEEE Transactions on Intelligent Transportation Systems*, 8(3):413–430, 2007.
21. M. Harville, G. Gordon, and J. Woodfill. Foreground segmentation using adaptive mixture models in color and depth. In *IEEE Workshop on Detection and Recognition of Events in Video*, pages 3–11. IEEE, 2001.
22. W. Hu, T. Tan, L. Wang, and S. Maybank. A survey on visual surveillance of object motion

and behaviors. *IEEE Transactions on Systems, Man, and Cybernetics, Part C: Applications and Reviews*, 34(3):334–352, 2004.

23. S. Iwasawa, K. Ebihara, J. Ohya, and S. Morishima. Real-time estimation of human body posture from monocular thermal images. In *IEEE Conference on Computer Vision and Pattern Recognition*, pages 15–20. IEEE, 1997.

24. B. Kisacanin. Examples of low-level computer vision on media processors. In *IEEE CVPR 2005 Workshop on Embedded Computer Vision*, 2005.

25. A. Kolb, E. Barth, R. Koch, and R. Larsen. Time-of-flight sensors in computer graphics. In *Proc. Eurographics (State-of-the-Art Report)*, pages 119–134, 2009.

26. J. Lin. Non-invasive microwave measurement of respiration. In *IEEE*, volume 63, page 1530, 1975.

27. R. Muñoz-Salinas, E. Aguirre, and M. García-Silvente. People detection and tracking using stereo vision and color. *Image and Vision Computing*, 25(6):995–1007, 2007.

28. J. Nanzer and R. Rogers. Human presence detection using millimeter-wave radiometry. *IEEE Transactions on Microwave Theory and Techniques*, 55(12):2727–2733, 2007.

29. R. Ng, M. Levoy, M. Brédif, G. Duval, M. Horowitz, and P. Hanrahan. Light field photography with a hand-held plenoptic camera. *Computer Science Technical Report CSTR*, 2(11), 2005.

30. T. Oggier, B. Buttgen, F. Lustenberger, G. Becker, B. Ruegg, and A. Hodac. Swissranger SR3000 and first experiences based on miniaturized 3DToF cameras. In *First Range Imaging Research Day at ETH Zurich*, 2005.

31. M. Okutomi and T. Kanade. A multiple-baseline stereo. *IEEE Transactions on Pattern Analysis and Machine Intelligence*, 15(4):353–363, 1993.

32. M. Pantic, A. Pentland, A. Nijholt, and T. Huang. Human computing and machine understanding of human behavior: A survey. *Artificial Intelligence for Human Computing, T.S. Huang, A. Nijholt, M. Pantic, A. Pentland, Eds., Lecture Notes in A.I.*, 4451:47–71, 2007.

33. R. Picard. *Affective computing*. MIT Press, Cambridge, MA, USA, 1997.

34. A. Prati, I. Mikic, M. Trivedi, and R. Cucchiara. Detecting moving shadows: algorithms and evaluation. *IEEE Transactions on Pattern Analysis and Machine Intelligence*, 25(7):918–923, 2003.

35. R. Radke, S. Andra, O. Al-Kofahi, and B. Roysam. Image change detection algorithms: a systematic survey. *IEEE Transactions on Image Processing*, 14(3):294–307, 2005.

36. P. Richards. Bolometers for infrared and millimeter waves. *Journal of Applied Physics*, 76:1–36, 1994.

37. A. Sanin, C. Sanderson, and B.C. Lovell. Shadow detection: A survey and comparative evaluation of recent methods. *Pattern recognition*, 45(4):1684–1695, 2012.

38. D. Scharstein and R. Szeliski. High-accuracy stereo depth maps using structured light. In *IEEE Conference on Computer Vision and Pattern Recognition*, volume 1, pages I–195. IEEE, 2003.

39. S. Schuon, C. Theobalt, J. Davis, and S. Thrun. High-quality scanning using time-of-flight depth superresolution. In *IEEE Conference on Computer Vision and Pattern Recognition Workshops*, pages 1–7. IEEE, 2008.

40. T. Teixeira, G. Dublon, and A. Savvides. A survey of human-sensing: Methods for detecting presence, count, location, track, and identity. Technical report, ENALAB, 2010.

41. H. Van Trees. *Detection, estimation and modulation theory. 3, Radar-sonar signal processing and Gaussian signals in noise*. John Wiley, 2001.

42. R. Want, V. Falcao, and J. Gibbons. The active badge location system. *ACM Transactions on Information Systems*, 10:91–102, 1992.

43. D. Wehner. High resolution radar. *Norwood, MA, Artech House, Inc., 1987*, 1:484, 1987.

44. M. Weiser. The computer for the twenty-first century. *Scientific American*, 265(3):94–104,

1991.

45. Z. Xu, R. Schwarte, H. Heinol, B. Buxbaum, and T. Ringbeck. Smart pixel photonic mixer device (pmd). In *International Conference on Mechatronics & Machine Vision*, page 259264, 1998.

46. G. Yahav, G.J. Iddan, and D. Mandelbaum. 3D imaging camera for gaming application. In *Digest of Technical Papers of Int. Conf. on Consumer Electronics*, 2007.

47. M. Youssef, M. Mah, and A. Agrawala. Challenges: device-free passive localization for wireless environments. In *Annual ACM International Conference on Mobile Computing and Networking*, pages 222–229. ACM, 2007.

48. Z. Zalevsky, A. Shpunt, A. Maizels, and J. Garcia. Method and system for object reconstruction. In *Patent WO2007043036*, 2007.

49. L. Zhang, B. Wu, and R. Nevatia. Pedestrian detection in infrared images based on local shape features. In *IEEE Conference on Computer Vision and Pattern Recognition*, pages 1–8. IEEE, 2007.

50. Z. Zivkovic and R. Kleihorst. Smart cameras for wireless camera networks: Architecture overview. *In A. Cavallaro, H. Aghajan, editors, Multi-Camera Networks: Concepts and Applications, Elsevier*, pages 497–511, 2009.

51. Z. Zivkovic, N. Sebe, B. Kisacanin, and H. Aghajan. Human computer interaction: Real-time vision aspects of natural user interfaces, (special issue introduction). *International Journal on Computer Vision*, 101:401–402, 2013.

18

RGB-D Cameras for Background-Foreground Segmentation

Massimo Camplani
Grupo de Tratamiento de Imágenes - ETSIT, Universidad Politécnica de Madrid, Spain

Luis Salgado
Grupo de Tratamiento de Imágenes - ETSIT, Universidad Politécnica de Madrid, Spain.
Video Processing and Understanding Lab, Universidad Autónoma de Madrid, Spain

18.1 RGB-D Cameras: Applications, Characteristics and Challenges

A new generation of RGB-D cameras have strongly influenced the research activities in the computer vision field. Due to their affordable price and their good performance in terms of frame rate and image resolution, cameras such as Microsoft's Kinect or the Asus's Xtion Pro have gained popularity and have been employed in many different applications. Originally proposed for controller-free game platforms, these devices have been applied for more complex human body motion analysis, as presented in the recent review [8]: in general, the depth data is used to identify and segment human users to track their body parts. RGB-D cameras have been also successfully applied for object segmentation purposes [36], in indoor video surveillance systems as recently proposed in [9], home care activity monitoring [30] or in robot-based application as presented by [12]. A detailed review of RGB-D cameras applications can be found in [16].

Microsoft Kinect is the most popular RGB-D camera. Depth data from the scene is obtained by means of a structured light sensor. In fact, the device is equipped, apart from a color sensor, with an infrared projector and an infrared camera to implement active triangulation [32] to estimate depth data. In particular, the projector emits a known dot pattern to the surrounding environment and the reflected infrared light is registered by the infrared camera; depth measurements can be computed by estimating the shift between the received pattern and the projected one [16]. It is worth noting that locally the dot pattern is unique, thus it is possible to apply local matching strategies between the two patterns. A scheme of the Kinect depth measurements acquisition system is presented in Figure 18.1.

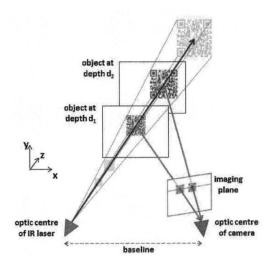

FIGURE 18.1   Kinect depth measurements system, image taken for the survey on Kinect-based applications [16] (©(2013) IEEE).

An example of depth and color data provided by the Kinect is shown in Figure 18.2. As it can be noticed, there are large areas for which the depth sensor is not able to provide depth information (marked in black in the image). These *non-measured depth* pixels (*nmd* pixels) are mainly due to occlusions (typically around object boundaries) that obviously affect the triangulation system previously mentioned. Moreover, depth data cannot be recovered in case of particular reflective surfaces (such as the monitors in the lab in Figure 18.2), or particular scattering surfaces (such as human hairs). Furthermore, *nmd* pixels may appear when either natural or artificial illumination negatively affect the received infrared pattern [32].

Regarding the computed depth measurements, the noise in the pattern projection and acquisition process, and the interference between illumination and the patterns affect the depth data stability over time and space. In particular, measurements show temporal random fluctuations, so different depth values are obtained for spatially neighboring pixels that correspond to points situated at the same distance from the camera. This noise varies with the distance; in fact, as shown in [24], the theoretical dispersion of depth measurements varies with the object-camera distance following a quadratic law. This noise-measurement relationship has been confirmed experimentally in several studies [6, 19, 31]; the results of our tests are shown in Figure 18.3, where the standard deviation (dgray line), σ_{noise}, of the measured values is reported as a function of the object-camera distance (depth). As expected, σ_{noise} increases with the distance following a quadratic function (black line). More information about Kinect depth measurements can be found in [24, 31, 32].

In addition, the computed depth measurements are also corrupted by a high level of noise at object boundaries: strong discontinuities in depth produce misleading reflection patterns that result in rough and inaccurate depth measurements, misaligned with respect to the actual object boundaries (see Figure 18.2 and other examples in [6, 32]).

FIGURE 18.2   Kinect color data (a) and the corresponding depth map (b) where pixel depth values are grey level coded.

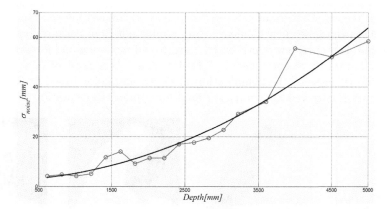

FIGURE 18.3   Depth measurements dispersion as function of the object-camera distance.

18.2   Background-Foreground Segmentation with Depth Data

Many of the abovementioned applications, such as human computer interaction systems or video surveillance systems, aim at detecting moving objects and separating them from the static environment to be further processed and analyzed. Their performance relies on the efficiency of the underlying background-foreground segmentation algorithms, which using the available depth information, in general, overperform traditional color-based approaches. The test proposed in the recent review [16] on a long run sequence is an example of the very high performance of depth-based foreground detectors.

Depth data are particularly attractive and suitable for these applications, since they are not affected by several problems typical of color-based imagery. Let us consider the detection example proposed in Figure 18.4, where the Mixture of Gaussians (MoG) [38] algorithm has been used to obtain the foreground mask by using as feature either color or depth. Color-based segmentations are affected by several problems: color camouflage leads to a fragmented foreground masks, due to similar color of foreground and background regions. Shadows cast by the foreground objects generate several false positive foreground pixel detections. Other problems that affect color-based background-foreground segmentation algorithms are illumination changes causing modification of the background regions that can be erroneously detected as foreground ones and the problem of moved background objects, that occurs when a background object is moved and the new *empty* space is wrongly detected as foreground. A more detailed description of the background-foreground segmentation problems can be found in [11].

The impact of these open problems can be dramatically reduced by considering depth data. In fact, the foreground detection obtained with the MoG based on the depth feature (Figure 18.4 (d)) is not influenced by the presence of shadows or the color camouflage; and it is clear that depth data is not affected by illumination changes.

However, depth data suffer from other types of problems, such as depth-camouflage (foreground objects get too close to the modeled background) or depth sensor noisy measurements, which bound the efficiency of depth-only based background-foreground detection approaches. Distance dependent noise, noisy measurements at object boundaries and the presence of *nmd* pixels (see their impact on the foreground mask in Figure 18.4 (d)) have to be considered in the design of the foreground detection strategy in order to improve its performance. Additionally, the depth sensor short range and the interferences with sunlight make it only applicable indoor environments and for objects close enough to the camera.

The complementary nature of color and depth synchronized information acquired with these RGB-D sensors poses new challenges and opportunities of background-foreground algorithms design. However, few depth-based background modeling and foreground detection works are referenced in recent reviews [4, 11, 16]. There is a need of new strategies that explore the effectiveness of the combination of depth and color based features, and their joint incorporation into well-known background-foreground segmentation frameworks.

FIGURE 18.4 Kinect color data (a) and the corresponding depth map (b), MoG detection based on color feature (c), MoG detection based on color feature (d).

The strategy proposed in [14] adapts the MoG algorithm to process jointly depth (obtained with a stereo device) and color data. In particular, a per-pixel background model is built using a four dimensional mixture of Gaussians distribution, where color and depth data are considered independent. Mixture components' parameters are updated by following the same strategy of the original MoG algorithm.

A similar approach has been proposed in [40], where depth and infrared data are combined for moving objects detection. Two independent background models are estimated, and the corresponding foreground regions are identified when the classification based on these two models agrees. Performance of this approach is limited since a failure of one of the models affects the final pixel classification.

A multi-camera system composed by color and ToF cameras is used for video segmentation in [26]. The Vibe algorithm [3] is applied independently to the color and the depth data such that the final foreground detection is obtained by combining with logical and morphological operations the independent foreground masks.

Recently, in the Microsoft Kinect based surveillance system proposed in [9], a per pixel background subtraction technique is presented. A four dimensional Gaussian distribution (using color and depth) is used to model the background. The main drawback of this approach is that it cannot manage multimodal backgrounds, and cannot address the depth-data noise issues associated to the Kinect.

In the gesture recognition system proposed [29], the moving objects are extracted by applying to depth map the Otsu segmentation algorithm [33]. This approach is efficient in very controlled environments characterized by a constant background, and with the restriction that there can be only a single user well separated from the background.

Other examples based on new RGB-D devices, such as [39] and [30], rely only on the depth data without considering a possible color and depth data integration.

In [7], a per-pixel background modeling approach is presented and fused different statistical classifiers based on depth and color data by means of a weighted average combiner that takes into account the characteristics of depth and color data. A mixture of Gaussians distribution is used to model the background pixels, and a uniform distribution is used for the modeling of the foreground.

18.3   Color and Depth Data Fusion

In this chapter, we propose an efficient combination of depth and color data provided by RGB-D cameras for background-foreground segmentation in indoor environments. The proposed approach fuses several independent classifiers in a mixture-of-experts fashion to improve the final foreground detection performance. For each pixel, each independent classifier estimates its probability to belong to the scene foreground and background. Classification is based on computed background models of the scene. In particular, four independent background models are considered which provide a scene description at pixel and region level based on color and depth features. The pixel-level background models are obtained considering the temporal evolution of depth and color data, using the MoG parametric model [38]. Background regions' depth information is obtained through the Mean-Shift algorithm [10]; MoG is again used to model these regions' depth characteristics. The background regions description considering color is obtained through Local Binary Patterns (LBP) [17]; in particular the LBP histogram based approach proposed in [18] has been employed. The combination of these classifiers is based on a weighted average that allows to adaptively modify the *importance* of each classifier in the ensemble. In this way, it is possible to reduce false detections due to critical issues that cannot be tackled by the individual classifiers.

The block diagram of the proposed background-foreground segmentation approach is presented in Figure 18.5. It is composed by three phases: features extraction, classification and classifiers combination. In the first phase, pixel-based and region-based features are extracted to be processed by the independent classifiers; mean-shift (*MShift* block) is applied to the depth data to obtain the corresponding segmented maps $MS - D$, the *LBP* block extracts the local binary pattern histograms that gather color data texture information at region level. In the second phase, the four independent classifiers (included in *BgMOD* in Figure 18.5) process the extracted features and classify the image pixels. Finally, in the last phase, the classifiers' output are fused together in a mixture of experts' fashion by the *Mixture of Experts* block. In particular, a weighted average scheme that considers depth discontinuities, previous foreground detections and the analysis of regions characterized by non-measured depth (nmd) pixels distribution is employed

The problem of background and foreground segmentation can be considered as a two-class problem [1] in which for each pixel the label of the foreground class (ω_{fg}) or of the background one (ω_{bg}) has to be assigned. As described above, we have introduced four

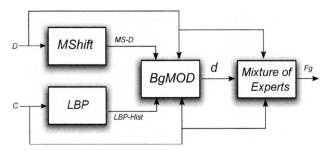

FIGURE 18.5   Scheme of the proposed approach. D and C are respectively the depth and the color data. Depth region map and data obtained with mean-shift is indicated with $MS - D$. Color region information based on LBP-histogram is indicated with $LBP - Hist$. BgMOD includes the different classifiers, their outputs are indicated with d. The Mixture of Experts block combines classifiers' outputs and characteristics of depth and color data to identify the foreground regions (indicated with Fg).

different classifiers based on the analysis of different features (color and depth data at pixel and region level). They are, individually, well suited for this task, each showing different - but to some extent complementary- advantages and drawbacks. In our approach we propose an efficient solution to combine these features based on a mixture of experts model.

The combination of classifiers, also referenced as mixture of experts, is a popular and efficient strategy employed to solve pattern recognition tasks. In particular, it has been demonstrated that complex problems can be efficiently tackled by combining more simple classifiers that work on a *local* level instead of training and building a unique sophisticated classifier that processes all the features. In our case the mixture of experts is useful to efficiently solve a data fusion problem, as underlined by [34], where the data provided by different sensors has to be combined to obtain the final classification. In these cases, poor performance is generally obtained if a single classifier is used to learn the information contained in all the data. This research area has been extensively treated in the literature, for a complete review see [13, 25, 34].

Let us consider the pixel s at position (x, y), and the corresponding measured data $\mathbf{x}_s = [D, C, MS - D, LBP - Hist]$, Ω the set of c classes $\{\omega_1, \ldots, \omega_c\}$ and CL the set of l classifiers $\{CL_1, \ldots, CL_l\}$. Each classifier CL_i gives as output a support d_{ij} for all the c classes. The support d_{ij} can be considered as an estimation (obtained with the i_{th} classifier) of the posterior probability that $\mathbf{x}_s$ belongs to the j_{th} class. It has to be noticed that we are considering classifiers that produce a continuous output (see [13, 25] for more details). The decision profile DP [25] for $\mathbf{x}_s$ is:

$$DP(\mathbf{x}_s) = \begin{bmatrix} d_{1,1} & \cdots & d_{1,j} & \cdots & d_{1,c} \\ \vdots & \cdots & \vdots & \cdots & \vdots \\ d_{i,1} & \cdots & d_{i,j} & \cdots & d_{i,c} \\ \vdots & \cdots & \vdots & \cdots & \vdots \\ d_{l,1} & \cdots & d_{l,j} & \cdots & d_{l,c} \end{bmatrix} \tag{18.1}$$

where the i_{th} row represents the output of the classifier CL_i, and the j_{th} column represents the overall support $M_j(\mathbf{x}_s)$ of the classifiers to the class ω_j. The decision profile $DP(\mathbf{x}_s)$ can be used to find the overall support $\mathbf{M}(\mathbf{x}_s)$ for all the classes; the label is assigned to the class with the greatest $M_j(\mathbf{x}_s)$.

Different combination techniques of the data contained in $DP(\mathbf{x}_s)$ can be considered: *class conscious* approaches, in which each column is processed independently without combining the $M_j(\mathbf{x}_s)$ of different classes together; on the contrary, in *class indifferent* approaches the obtained $M_j(\mathbf{x}_s)$ are used to generate new features processed by another classifier that completes the final classification task. *Class conscious* combiners are non trainable combiners since the overall supports $M_j(\mathbf{x}_s)$ are obtained with arithmetic operations; hence, the computational and memory cost of the classifiers' combination is kept low. On the contrary, *class indifferent* combiners are more complex since they add new parameters in the classification system that have to be tuned and initialized after an additional training phase. For these reasons, in our approach we select a *class conscious* combiner to reduce its complexity and computational requirements. Typical choices of the combination are average, median, maximum, etc. In our case, we use a weighted average in order to extract different information from the different feature sets and to adapt efficiently the *contribution* of each classifier to the final classification. $M_j(\mathbf{x}_s)$ is estimated as:

$$M_j(\mathbf{x}_s) = \sum_{i=1}^{l} W_i(\mathbf{x}_s) d_{i,j} \tag{18.2}$$

where the weight W_i is chosen as a function of the input $\mathbf{x}_s$ in order to increase the support of the most reliable classifier. In the following sections a description of the classifiers employed in the proposed system and the weights selection strategy will be presented.

18.3.1 Depth-based Classifiers

Depth-based classifiers included in the *BgMOD* block are both based on the MoG background modeling algorithm on raw depth data D and on the region information $MS - D$ obtained with the mean-shift algorithm. From now on, we will refer to the classifiers built on these models as MoG_D and MoG_{MS-D}. In this section, we first briefly introduce the MoG algorithm adapted to the depth data and we finish highlighting the differences between the pixel-based and the region-based one.

The MoG algorithm is a popular per-pixel background modeling technique (see the review presented in [5] for more details), which is able to model multi-modal backgrounds, and to adapt the model to gradual changes. The pixels are modeled as a mixture of Gaussians distribution and their parameters are iteratively updated.

In the MoG model, the probability to find a pixel at position s at time t of value $\mathbf{x}$ is defined as a mixture of Gaussians:

$$P\left(\mathbf{x}_{s,t}|\omega_{bg}, \omega_{fg}\right) = \sum_{i=1}^{K} v_{i,t} \cdot \mathcal{N}\left(\mathbf{x}_{s,t}, \mu_{i,t}, \mathbf{\Sigma}_{i,t}\right) \tag{18.3}$$

where K is the number of Gaussians used in the model, $v_{i,t}$ is the weight associated to the i_{th} Gaussian $\mathcal{N}$ at the time t with mean $\mu_{i,t}$ and covariance matrix $\mathbf{\Sigma}_{i,t}$. It is worth noting that the statistics of both models, background and foreground, are described with the same mixture, which is indicated in eq. 18.3 by inserting the two classes symbols, ω_{bg} (for the background) and ω_{fg} (for the foreground), in the left term of the equation. Depth-based distributions have a single dimension.

The background model is formed by those distributions of the mixture that are characterized by a high ratio (r) between the weight and the variance. In the case of a univariate distribution r, the ratio for the i_{th} Gaussian is estimated as $r_i = v_i/\sigma_i$. A high value for r_i means that the corresponding distribution is very stable. The distributions are ordered by considering the factor r and the background is composed by the first B distributions that exceed the threshold T:

$$B = \arg\min_b \left(\sum_{i=1}^{b} v_{i,t} \geq T \right) \tag{18.4}$$

T indicates the minimum portion of the data that should be accounted for by the background.

The Mahalanobis distance is used to check if the pixel belongs to one of the distributions:

$$\sqrt{(\mathbf{x}_{s,t+1} - \mu_{i,t})^T \Sigma_{i,t}^{-1} (\mathbf{x}_{s,t+1} - \mu_{i,t})} < \lambda, \tag{18.5}$$

where λ is usually set to 2.5 (see for example [5]). If eq. 18.5 is satisfied for one of the background distributions, the pixel will be classified as a background pixel, otherwise it is classified as a foreground pixel.

As mentioned in the introduction, the depth measurement variation follows a quadratic relationship with the depth [24]. Therefore, depth measurements corresponding to regions located far from the camera are characterized by a higher value of σ than those regions situated closer. This fact could introduce a bias in the estimation of the ranking parameter r. For this reason, we normalize its value with σ_{noise}, as presented in [6], that is selected according to the noise-measurements relationship.

The parameters of the background distributions are updated as proposed in [38]. In particular, in the case of a single dimension feature space (i.e. depth):

$$v_{i,t+1} = v_{i,t}(1 - \alpha) + \alpha * \Gamma \tag{18.6}$$

$$\rho = \alpha \cdot \mathcal{N}(x_{s,t}, \mu_{i,t}, \sigma_{i,t}) \tag{18.7}$$

$$\mu_{i,t+1} = \mu_{i,t}(1 - \rho) + \rho x_{s,t+1} \tag{18.8}$$

$$\sigma_{i,t+1}^2 = \sigma_{i,t}^2(1 - \rho) + \rho(x_{s,t+1} - \mu_{i,t+1})^2 \tag{18.9}$$

where α is the *learning rate*, which influences the speed of background modifications due to new objects in the scene or gradual changes. For the unmatched Gaussians, all the parameters remain unchanged, except their weight that is updated with $\Gamma = 0$ in equation 18.6.

In the case that eq. 18.5 has not been satisfied by all the distributions, the one with the lowest ratio r is substituted by a new one with low weight, high variance, and a mean equal to the current value. It is worth noting that in each iteration the weights are normalized such that their sum is equal to one.

Instead of using a fixed learning rate value to update the distribution parameters, we employ a variable learning rate, as proposed by [23]. Its value is decreased at each iteration until a minimum fixed value of α is reached, thus limiting the absorption of moving objects to the background model at the beginning of the sequence.

The MoG_{MS-D} uses the same scheme of MoG_D presented above, but the model is built by considering the region characteristics of the background. In [22], it has been demonstrated that the distribution of neighbor pixel values (spatial information) can be successfully used to build a robust background model of the analyzed scene, which has similar characteristics to the one obtained by considering the per-pixel temporal evolution. MoG_{MS-D} has been designed by following these results; however, we modify the neighborhood definition with respect to [22]. In fact, instead of considering a fixed region area, we select the different regions as the ones that are identified by the mean-shift algorithm proposed by [10]. The mean-shift algorithm is a non parametric approach that is well suited to identify the principal modes in a complex multi modal feature space. Thanks to the mean-shift, we are able to identify in the data local regions that are characterized by similar properties in the depth or in the color feature space. In this way we are able to describe each pixel not only by considering its temporal evolution (pixel level model), but considering also the temporal evolution of its neighborhood. In particular, for each pixel we built MoG-like classifiers

(MoG_{MS-D}) by using the region principal mode (identified by the mean-shift) in the calculation of the Gaussian distribution mean value and by estimating the mode variance among all the pixels that belong to the region corresponding to the pixel. The model parameters are updated as presented above, and similarly to the pixel-level models, the ranking parameters normalization is applied.

18.3.2 Color-based Classifiers

The color-based classifier that works at pixel level is based on the MoG algorithm, we will refer to this classifier as MoG_C. The same strategy used for the depth-based classifiers is used. Moreover, it is necessary to reduce the effect of sudden changes of illumination that can lead to wrong pixel classification. In this case, we propose to use the frame-level control strategy proposed by [41], where the fraction of the pixels detected as foreground is computed and compared with a predefined threshold. If the computed fraction exceeds this threshold, the MoG_C Gaussian parameters and the corresponding learning rate are re-initialized. It is worth noting that low cost RGB-D cameras do not allow controlling acquisition parameters such as aperture time, shutter time, white balance, etc. Due to the automatic modification of these parameters, the color data is affected by sudden changes of the illumination condition (also in very controlled environments) that affect large portions of the image.

For the color-based classifier that works at region level, we use the background-foreground classifier presented in [18] (we will refer to this technique with the acronym CL_{LBP}). The algorithm proposed in [18] use histograms of LBP to extract for each pixel the characteristics of the local texture. LBP [17] is a gray-scale invariant texture primitive statistic. The operator labels the pixels of an image region by thresholding the neighborhood of each pixel with the center value and considering the result as a binary number (binary pattern). LBP is a powerful feature for background subtraction since it is invariant to monotonic gray-level variation in the images. Moreover, it is a nonparametric method and its computation is not expensive. Its main drawback is that it is not very robust on uniform and flat image regions. In [18] for each pixel, a LBP histogram is computed over a circular region and used as a feature vector. The corresponding background model is composed by K adaptive LBP histograms. Each histogram of the model has associated a weight. The comparison between the LBP histogram of the current pixel and the background model ones is performed by using the histogram intersection:

$$\cap \left(\overline{a}, \overline{b} \right) = \sum_{n=0}^{N-1} min \left(a_n, b_n \right) \qquad (18.10)$$

where $\overline{a}$ and $\overline{b}$ are two histograms with N bins. This measure calculates the common part of two histograms. Its main advantages are that it explicitly excludes features which only occur in one of the histograms, and its very low complexity. In our case, we normalize the obtained value of the intersection (with respect to the number of pixels in the region) in order to obtain an estimation of the background-foreground probabilities that are used by the mixture of experts. In [18], the background histograms and the corresponding weights are updated following a similar strategy to that of the MoG algorithm. They are sorted by considering only the weight, and the less probable one is substituted by the new ones.

18.3.3 Mixture of Experts Weights Selection

The block diagram of the weights selection strategy is presented in Figure 18.6. It is worth noting that in this section we refer to W_D as the weights associated to the depth-based classifiers, and to W_C as the ones associated with the color-based classifiers. First of all, the presence or not of nmd values is checked and used to update the temporal consistency score t_c; this factor is used to reduce the values of W_D (see the last part of this section). For all the nmd pixels, the weights of the classifiers are set as $W_D(\mathbf{x}_s) = 0$ and $W_C(\mathbf{x}_s) = 1$. It is clear that in the case that either the depth measurement or the depth-based background model is not available, only color-based classifiers are considered for the final pixel classification.

For all those pixels that do not belong to the nmd set, we assign the weights as proposed in [7], by considering the edge regions in the depth domain, to reduce the noisy depth values at object boundaries, and the previous detections to reduce the effect of depth camouflage.

Depth data typically guarantees compact detections of moving object regions but with errors in accuracy and definition close to object boundaries. Therefore, we propose selecting $W_i(\mathbf{x}_s)$ as a function of the pixel edge-closeness probability in the depth and color images. The idea is to give a higher weight to the color based classifiers for those pixels that simultaneously belong to edge regions in the depth map and are located close to edges in the color image. We estimate a global edge-closeness probability P_G^e for each pixel, as the product between depth edge-closeness probability (P_D^e) and the color edge-closeness probability (P_C^e). The values of P_D^e and P_C^e are computed as a function of the distance between the pixel and the closest edge weighted with a Gaussian function. The value of P_G^e is high for those regions for which the color edges correspond to depth edges; in these regions it is necessary to assign a higher value to W_C. It should be noticed that the product between the two edge-closeness probability functions limits the impact of MoG_C where there is not an edge in the color domain. The weights are assigned as $W_C(\mathbf{x}_s) = P_G^e(x_s)$ and $W_D(\mathbf{x}_s) = 1 - W_C(\mathbf{x}_s)$. The weights values are bounded to a minimum and a maximum value (W_{min} and W_{max}) to ensure the contribution of both classifiers in the final classification.

For those pixels that have a too low value of edge-closeness probability $(P_g^e(x_s) < W_{low})$, weights are assigned to reduce the effect of depth camouflage. In particular, if previously the pixel has been classified as foreground, new W_D and W_C are computed. The normalized depth distance $\delta(\mathbf{x}_s, t)$ between the foreground pixel and the corresponding pixel in the background model is computed and used to set the weights for the pixel classification for the following frame $(\mathbf{x}_s, t + 1)$:

$$W_D(\mathbf{x}_s, t+1) = W_{min} + \frac{W_{max} - W_{min}}{\left(1 + Qe^{-B(\delta(\mathbf{x}_s,t)-M)}\right)^{1/\nu}} \tag{18.11}$$

The weight of CL_C is calculated as $W_C(\mathbf{x}_s, t+1) = 1 - W_D(\mathbf{x}_s, t+1)$. The function in Equation 18.11 is the generalized logistic function [35]. Regarding the normalized distance δ, it is defined as:

$$\delta(\mathbf{x}_s, t) = |(D(\mathbf{x}_s, t) - \mu_s(t)| / \sigma_s(t) \tag{18.12}$$

where $D(\mathbf{x}_s, t)$ is the depth value of the pixel, $\mu_s(t)$ is the depth value of the most representative sample of the corresponding background model, and $\sigma_s(t)$ its interval of confidence. It is worth noting that this definition of δ allows to adapt the weights in Equation 18.11 to the time varying characteristics of the background distributions. The logistic curve has been selected because it allows to smoothly select classifiers' weights and bind their values to the minimum and maximum value W_{min} and W_{max}: in this way, it is guaranteed that all the classifiers' support are included in the final classification stage. A low (high) weight is assigned to the depth-based (color-based) classifiers (color-based) for those foreground pixels that are very close in depth to that of the background model and vice versa, thus limiting the effect of depth camouflage.

The estimated values of W_D have to be finally scaled by the temporal consistency score t_c, a parameter that measures the reliability of the depth model, and helps reducing the

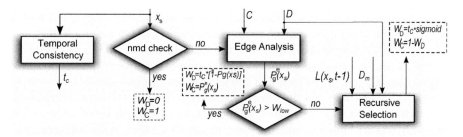

FIGURE 18.6 Weights selection strategy.

depth-based background model regions affected by the presence of *nmd* pixels. Let us consider a region that contains several *nmd* pixels for which the depth based model is not available. If a moving object passes throughout this region, the depth-based model is initialized with the object depth values that do not correspond to the real background one, thus leading to errors in further pixels classification. In order to reduce the impact of this situation on the final classification, we multiply for each pixel the element W_D by the factor tc calculated as:

$$tc = \frac{1}{1 + e^{\beta \# Hit}}, \tag{18.13}$$

where $\# Hit$ represents a counter that is incremented when a valid depth measurement is obtained, and is decremented when a *nmd* pixel is registered. Hit_{max} is the value for the counter $\# Hit$ for which the tc value is equal to one, in this case, the depth based background model is considered to be reliable. The value of the parameter β is selected according to the chosen value of $\# Hit$. Thanks to the temporal consistency score, it is possible to give a different weight to the background model pixels in regions that are strongly affected by the presence of *nmd* pixels.

18.4   Results

In this section, we present the results obtained with the proposed strategy and other state-of-the-art background modeling techniques based on depth and color features. We use indoor sequences from three different datasets: the *GenSeq* sequence, from the dataset [7], that shows an indoor environment containing moving objects, cast shadows and illumination changes; one sequence (called *seq0*) from the dataset [37], which records a crowded university building hall where strong illumination changes and objects located out of the measurement range heavily affect color and depth data; one last sequence (called *cam1*), from the dataset presented in [2], where the camera is placed on the ceiling looking forward with a significant pitch angle. The camera position in the last sequence differentiates it from the others where the cameras are positioned parallel to the floor at a height between 1 and 1.5 meters.

To analyze and evaluate the algorithm performance, we use several measurements proposed in the literature: False Positive Ratio (FPR), the fraction of the *Bg* pixels that are marked as *Fg*; False Negative Ratio (FNR), the fraction of *Fg* pixels that are marked as *Bg*; Total Error (TE), the total number of misclassified pixels, normalized with respect to the image size. Moreover, we consider also the similarity measure S defined in [21], which was proposed to evaluate the accuracy of background-foreground segmentation algorithms in [27]. This metric fuses together the concepts of FPR and FNR, and its value is close to 1 when the detected foreground regions are similar to the ground truth, and close to 0 if they significantly differ. Finally, we use also the overall metric proposed in [15] to rank the accuracy of the analyzed methods. We combine the performance across different metrics

(RM) and sequences into a single rank that is indicative of how well a method performs with respect to the others, computing an average ranking RC across all sequences.

The proposed strategy is compared with the following approaches in the state of the art: $Vibe_{Bin}$ [26], which is based on a binary combination of foreground masks obtained by the ViBe algorithm; $PBAS$ [20], which first models the background using the recent history of pixel values, and then computes the foreground using a decision threshold calculated dynamically for each pixel; $SOBS$ [28], which adopts a neural-network based approach to detect foreground objects without making any assumption about the pixel distribution; CL_W [7], that proposes a probabilistic classifier to fuse a set of foreground masks, which are computed by a Mixture of Gaussian approach using color and depth information; and the Mixture of Gaussian algorithm adapted to the RGB-D data presented by [14] (MOG_{RGB-D}). Regarding the $PBAS$ and $SOBS$ algorithms, they have been extended to use RGB-D imagery, since originally they were designed to be used only with color. It is worth noting that we will refer to the proposed approach as CL_{Reg}.

In the first part of this section we use the sequence *genSeq* to demonstrate how the proposed mixture of experts system guarantees an improvement of the detection accuracy with respect to the individual classifiers. In the second part of this section we compare the proposed algorithm considering only the state of the art approaches mentioned above.

In Table 18.1 are reported the results obtained using the individual classifiers and the proposed mixture of experts approach. The proposed approach CL_{Reg} guarantees the lowest value for TE and the higher value for S. The classifiers MoG_D and MoG_C lead individually to inaccurate results: the color-based classifier is dramatically affected by color camouflage problems that lead to a high value of FNR; on the contrary, the depth-based one leads to a higher value of FPR, that is due mainly to the noisy detections at objects boundaries. A similar behavior is observed for the classifiers that operate at region level. As it can be noticed, the fusion of these classifiers balance the ratio of FPR and FNR, thus reducing the TE value and obtaining a very high value for S. The lowest value for RM is also obtained with the proposed mixture of experts system.

TABLE 18.1   Detection accuracy obtained by analyzing the *genSeq* sequence of the dataset [7], individual classifiers and mixture of experts.

	TE	FNR	FPR	S	RM
CL_{Reg}	0.92	1.88	0.80	0.87	1.25
MoG_C	2.13	8.41	1.35	0.74	3.50
MoG_D	1.61	3.70	1.36	0.81	2.50
CL_{LBP}	2.50	5.65	0.65	0.69	4.00
MoG_{MS-D}	2.49	3.86	2.32	0.75	3.75

In Figure 18.7, an example of the foreground masks obtained with the individual classifiers and the proposed CL_{Reg} approach is presented. The detection masks provided by the independent classifiers are affected by different types of errors. Depth based classifiers (see Figure 18.7 (c) and (e)) show masks with irregular and inaccurate object boundaries, lacking compactness due, mainly, to the areas that contain *nmd* pixels. On the contrary, the foreground masks provided by the color-based classifiers (see Figure 18.7 (d) and (f)) are affected by the color camouflage problem. These errors are dramatically reduced with the proposed approach CL_{Reg}, where the adaptive balance of the different classifiers contribution result in more accurate object silhouettes and compact foreground regions. It is worth noting that also the false detections in the background regions that appear with the individual classifiers have been eliminated.

Table 18.2 reports the results obtained using the *genSeq* sequence with the proposed approach and the other algorithms from the state of the art. As it can be noticed, the proposed approach guarantees the higher value for S and the lowest value for TE, thus

FIGURE 18.7 Foreground masks for the *genSeq* sequence of the dataset [7]: ground truth mask (a), CL_{Reg} (b), MoG_D (c), MoG_C (d), MoG_{MS-D} (e), CL_{LBP} (f).

obtaining the best value for RM. Similar results are obtained with CL_W, especially if the value of S is considered, however the proposed approach leads a lower value of FNR and FPR. Other approaches such as MoG_{RGB-D} and $ViBe_{bin}$ lead to lower values for FNR, although they show a high value for FPR; the proposed approach guarantees a good tradeoff between false positive and false negative detections, thus leading to a lower value for total number of errors, TE. The performance of SOB and $PBAS$ algorithms are fair, but they cannot reach the same accuracy than the proposed method, especially if we consider the values of TE and S.

TABLE 18.2 Detection accuracy obtained by analyzing the *genSeq* sequence of the dataset [7].

	TE	FNR	FPR	S	RM
CL_{Reg}	0.92	1.88	0.80	0.87	1.75
CL_W	1.13	2.26	0.99	0.85	3.00
MoG_{RGB-D}	1.93	0.63	2.09	0.79	4.00
$ViBe_{bin}$	12.39	0.65	13.85	0.44	5.00
SOB	1.91	1.34	1.98	0.80	3.25
$PBAS$	1.93	2.10	1.91	0.80	4.00

The results obtained by the proposed approach and the state of the art algorithms by processing the *seq0* sequence of the dataset [37] are reported in Table 18.3. The lowest value of RM is obtained also in this case with the proposed approach CL_{Reg}, moreover this method guarantees the highest value for the similarity measure S. A similar value of S is obtained with CL_W, although this method leads to a higher value for TE and FPR. As far as the other approaches are concerned, they are heavily affected by the strong illumination changes in the scene and the noisy depth data (especially the presence of large regions containing *nmd* pixels) of this sequence. In fact, their performance is quite poor, and characterized by high values for TE and a low values for S.

An example of the foreground masks obtained with these algorithms is reported in Figure 18.8. The most accurate silhouette is obtained by the proposed approach (see Figure 18.8 (d)). As expected, also the foreground region obtained with CL_W (Figure 18.8 (e)) is quite accurate, but is negatively affected by the presence of a large region with *nmd* pixels. In this case, it is clear that the use of region information and the temporal consistency score results in a more accurate detection. The foreground region masks obtained with the other approaches are affected by different types of errors due to the strong illumination changes and depth data noise.

Table 18.4 reports the results obtained by analyzing the *cam1* sequence of the [2] dataset. For this sequence, the proposed approach leads to very similar results to the CL_W method. In particular, both methods obtain the same value of RM, the proposed approach with a higher value for S but with also a higher value of TE. The other algorithms show, in general, worst performance with very high values of TE and low values of S.

Figure 18.9 reports the foreground masks obtained with the different algorithms. The

TABLE 18.3 Detection accuracy obtained by
analyzing the *seq0* sequence of the [37] dataset.

	TE	FNR	FPR	S	RM
CL_{Reg}	4.82	24.68	3.02	0.36	1.50
CL_W	7.06	18.06	6.06	0.36	2.25
MoG_{RGB-D}	48.04	9.21	51.56	0.13	4.50
$ViBe_{bin}$	25.24	41.31	23.79	0.12	5.00
SOB	13.42	46.10	10.46	0.19	4.00
$PBAS$	8.44	66.46	3.18	0.17	3.75

FIGURE 18.8 RGB-D data and foreground masks for the *seq0* sequence of the dataset [37]: color data
(a), depth data (b), ground truth mask (c), CL_{Reg} (d), CL_W (e), MoG_{RGB_D} (f), $ViBe_{Bin}$ (g), SOM
(h), $PBAS$ (i).

foreground masks obtained with CL_{Reg} and CL_W are quite similar, showing good accuracy:
the foreground objects are well detected and the moving object boundaries are well refined.
In this example also the foreground mask produced by MoG_{RGBD} is characterized by a
good accuracy level. On the contrary, the other algorithms provide fragmented foreground
masks.

TABLE 18.4 Detection accuracy obtained by
analyzing the *cam1* sequence of the [2] dataset.

	TE	FNR	FPR	S	RM
CL_{Reg}	6.79	7.15	6.80	0.61	1.75
CL_W	6.50	9.50	5.98	0.60	1.75
MoG_{RGB-D}	32.91	0.89	37.36	0.33	4.50
$ViBe_{bin}$	13.76	48.59	8.97	0.30	5.00
SOB	10.49	25.59	8.42	0.47	3.25
$PBAS$	14.74	43.53	10.78	0.34	4.75

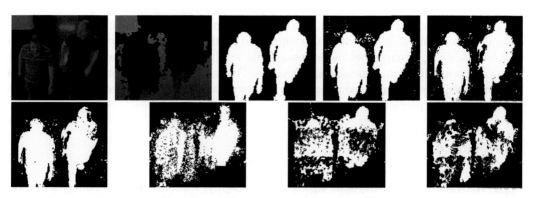

FIGURE 18.9 RGB-D data and foreground masks for the *cam1* sequence of the [2] dataset: color data (a), depth data (b), ground truth mask (c), CL_{Reg} (d), CL_W (e), MoG_{RGB_D} (f), $ViBe_{Bin}$ (g), SOM (h), $PBAS$ (i).

TABLE 18.5 RC values for the analyzed algorithms.

	CL_{Reg}	CL_W	MoG_{RGB-D}	$ViBe_{bin}$	SOB	$PBAS$
RC	1.67	2.33	4.33	5.00	3.50	4.17

18.5   Conclusion

A new generation of RGB-D cameras have strongly influenced the research activities and applications in the computer vision field. In particular, these devices are very promising to improve the performance of background-foreground segmentation algorithms in indoor environments. However, the combined use of depth and color data poses new challenges in the design of background-foreground segmentation algorithms. In particular, efficient strategies to combine the complementary color and depth features are required to obtain more accurate and reliable background-foreground segmentations, thus reducing the impact of the errors that affect both sensors.

In this chapter, we have presented a background-foreground segmentation algorithm based on mixture of expert scheme that improves the background-foreground segmentation accuracy through a weighted combination of independent classifiers that use features based on color and depth data. Moreover, we propose combining pixel-based and region based-background modeling, thus extracting also characteristics of the pixels' neighbor regions. Four different classifiers are combined through a weighted average that adapts, for each pixel, the support of each classifier to the final classification by considering depth and color images edge information and the previous foreground detections. The main idea is to give a higher influence to the color-based classifiers near object borders, thus reducing the impact in the detection of noisy depth measurements at object boundaries. On the contrary, the depth-based classifiers contribution become dominant within the mixture of experts for pixels located in low gradient areas of the depth map, thus guaranteeing compact detected foreground regions. Moreover, an iterative estimation of depth-based classifier weights is proposed: the reliability of the depth-based model is checked to modify the influence of the color classifiers to mitigate the depth camouflage errors. A temporal consistency score guarantees a low impact on the background model of large regions lacking depth information. It has to be remarked that the proposed mixture of experts scheme can be applied to a heterogeneous set of classifiers, MoG and the LBP-based approach proposed in [18], and by using different features such as depth, segmented depth regions centroids and LBP.

The results section demonstrates that the proposed classifier CL_{Reg} allows to obtain more reliable and accurate detections than other state of the art algorithms. In fact, the proposed approach efficiently combines the independent statistical classifiers improving the overall performance. This result is also supported by the performance score RC reported in Table 18.5: CL_{Reg} guarantees the lowest RC value, thus highlighting the robustness of the approach with different benchmark sequences. Furthermore, the high standard results of the proposed classifier are also apparent considering a qualitative analysis of the examples presented in the previous sections.

References

1. T. Aach and A. Kaup. Bayesian algorithms for adaptive change detection in image sequences using markov random fields. *Signal Processing: Image Communication*, 7(2):147–160, 1995.
2. A. Albiol, J. Mossi, and J. Oliver. Who is who at different cameras: people re-identification using depth cameras. *IET Computer Vision*, 6(5):378–387, September 2012.
3. O. Barnich and M. Van Droogenbroeck. ViBe: a universal background subtraction algorithm for video sequences. *IEEE Transactions on Image Processing*, 20(6):1709–24, 2011.
4. T. Bouwmans. Recent Advanced Statistical Background Modeling for Foreground Detection - A Systematic Survey. *Recent Patents on Computer Science*, 4(3):147–176, 2011.
5. T. Bouwmans and F. El Baf. Background modeling using mixture of Gaussians for foreground detection-a survey. *Recent Patents on Computer Science*, 3:219–237, 2008.
6. M. Camplani, T. Mantecon, and L. Salgado. Depth-Color Fusion Strategy for 3D scene modeling with Kinect. *IEEE Transactions on Cybernetics*, 2013.
7. M. Camplani and L. Salgado. Background Foreground segmentation with RGB-D Kinect data: an efficient combination of classifiers. *Journal of Visual Communication and Image Representation*, 2013.
8. L. Chen, H. Wei, and J. Ferryman. A survey of human motion analysis using depth imagery. *Pattern Recognition Letters*, 34(15):1995 – 2006, 2013.
9. A. Clapés, M. Reyes, and S. Escalera. Multi-modal user identification and object recognition surveillance system. *Pattern Recognition Letters*, 34(7):799–808, December 2013.
10. D. Comaniciu and P. Meer. Mean shift: a robust approach toward feature space analysis. *IEEE Transactions on Pattern Analysis and Machine Intelligence*, 24(5):603–619, May 2002.
11. M. Cristani, M. Farenzena, D. Bloisi, and V. Murino. Background Subtraction for Automated Multisensor Surveillance: A Comprehensive Review. *EURASIP Journal on Advances in Signal Processing*, 2010:1–24, 2010.
12. G. Doisy, A. Jevtic, E. Lucet, and Y. Edan. Adaptive person-following algorithm based on depth images and mapping. In *IEEE IROS Workshop on Robot Motion Planning*, 2012.
13. R. Duda, P. Hart, and D. Stork. *Pattern Classification*. Wiley-Interscience, 2001.
14. G. Gordon, T. Darrell, M. Harville, and J. Woodfill. Background estimation and removal based on range and color. In *IEEE Conference on Computer Vision and Pattern Recognition*, volume 2, page 464, 1999.
15. N. Goyette, P.-M. Jodoin, F. Porikli, J. Konrad, and P. Ishwar. changedetection.net: A new change detection benchmark dataset. In *IEEE Computer Society Conference on Computer Vision and Pattern Recognition Workshops (CVPRW)*, pages 1–8. IEEE, 2012.
16. J. Han, L. Shao, D. Xu, and J. Shotton. Enhanced computer vision with Microsoft Kinect sensor: A review. *Cybernetics, IEEE Transactions on*, 43(5):1318–1334, 2013.
17. D. He and L. Wang. Texture unit, texture spectrum, and texture analysis. *Geoscience and Remote Sensing, IEEE Transactions on*, 28(4):509–512, 1990.

18. M. Heikkila and M. Pietikainen. A texture-based method for modeling the background and detecting moving objects. *IEEE Transactions on Pattern Analysis and Machine Intelligence*, 28(4):657–662, 2006.

19. D. Herrera, J. Kannala, and J. Heikkila. Joint Depth and Color Camera Calibration with Distortion Correction. *IEEE transactions on pattern analysis and machine intelligence*, pages 1–8, May 2012.

20. M. Hofmann, P. Tiefenbacher, and G. Rigoll. Background segmentation with feedback: The pixel-based adaptive segmenter. In *Computer Vision and Pattern Recognition Workshops (CVPRW), 2012 IEEE Computer Society Conference on*, pages 38–43, 2012.

21. P. Jacard. Distribution de la flore alpine dans le bassin des dranses et dans quelques regions voisines. *Bulletin de la Societe Vaudoise des Sciences Naturelles*, 37:241–272, 1901.

22. P. Jodoin, M. Mignotte, and J. Konrad. Statistical Background Subtraction Using Spatial Cues. *IEEE Transactions on Circuits and Systems for Video Technology*, 17(12):1758–1763, December 2007.

23. P. KaewTraKulPong and R. Bowden. An Improved Adaptive Background Mixture Model for Real-time Tracking with Shadow Detection. *European Workshop on Advanced Video Based Surveillance Systems*, pages 149–158, September 2001.

24. K. Khoshelham and S. Elberink. Accuracy and Resolution of Kinect Depth Data for Indoor Mapping Applications. *Sensors*, 12(2):1437–1454, 2012.

25. L. Kuncheva. *Combining Pattern Classifiers: Methods and Algorithms*. Wiley-Interscience, 2004.

26. J. Leens, O. Barnich, S. Piérard, M. Van Droogenbroeck, and J. Wagner. Combining color, depth, and motion for video segmentation. *Computer Vision Systems*, 5815:104–113, 2009.

27. L. Li, W. Huang, I. Gu, and Q. Tian. Statistical modeling of complex backgrounds for foreground object detection. *IEEE Transactions on Image Processing*, 13(11):1459–1472, 2004.

28. L. Maddalena and A. Petrosino. A self-organizing approach to background subtraction for visual surveillance applications. *IEEE transactions on Image Processing*, 17(7):1168–77, July 2008.

29. U. Mahbub, H. Imtiaz, T. Roy, S. Rahman, and A. Ahad. A template matching approach of one-shot-learning gesture recognition. *Pattern Recognition Letters*, 34(15):1780 – 1788, 2013.

30. G. Mastorakis and D. Makris. Fall detection system using Kinect's infrared sensor. *Journal of Real-Time Image Processing*, pages 1–12, 2012.

31. F. Menna, F. Remondino, R. Battisti, and E. Nocerino. Geometric investigation of a gaming active device. In *Proceedings of SPIE*, volume 8085, page 80850G, 2011.

32. C. Mutto, P. Zanuttigh, and G. Cortelazzo. *Time-of-Flight Cameras and Microsoft Kinect(TM)*. Springer Publishing Company, Incorporated, 2012.

33. N. Otsu. A threshold selection method from gray-level histograms. *Systems, Man and Cybernetics, IEEE Transactions on*, 9(1):62–66, 1979.

34. R. Polikar. Ensemble based systems in decision making. *IEEE Circuits and Systems Magazine*, 6(3):21–45, 2006.

35. F. Richards. A Flexible Growth Function for Empirical Use. *J. Exp. Bot.*, 10(2):290–301, June 1959.

36. A. Richtsfeld, T. Morwald, J. Prankl, M. Zillich, and M. Vincze. Learning of perceptual grouping for object segmentation on RGB-D data. *Journal of Visual Communication and Image Representation*, 2013.

37. L. Spinello. People detection in RGB-D data. *IEEE Robots and Systems*, 2011.

38. C. Stauffer and W. Grimson. Adaptive background mixture models for real-time tracking. In *Conference on Computer Vision and Pattern Recognition*, pages 246–252. IEEE,

1999.

39. E. Stone and M. Skubic. Evaluation of an inexpensive depth camera for in-home gait assessment. *Journal of Ambient Intelligence and Smart Environments*, 3(4):349–361, 2011.

40. A. Stormer, M. Hofmann, and G. Rigoll. Depth gradient based segmentation of overlapping foreground objects in range images. In *IEEE Conference on Information Fusion*, pages 1–4, 2010.

41. K. Toyama, J. Krumm, B. Brumitt, and B. Meyers. Wallflower: principles and practice of background maintenance. *IEEE International Conference on Computer Vision, ICCV 1999*, pages 255–261, 1999.

19

Non-Parametric GPU Accelerated Background Modeling of Complex Scenes

Ashutosh Morde
intuVision Inc., USA

Sadiye Guler
intuVision Inc., USA

19.1 Introduction

Modeling a scene background and finding the discrepancies between each new frame and the model to determine the foreground objects, i.e. background subtraction, is a common approach to detecting moving objects in a video scene and it is a critical component of many video object tracking systems. Segmentation of the foreground objects from the scene background is a challenging task, especially with the dynamically changing nature of most real-world scene backgrounds due to variations in illumination, dynamic scene elements such as wind blown trees, water ripples and reflections, etc. Many methods have been proposed in literature to model a video scene background, from modeling the image pixels as a Mixture of Gaussians (MOG) [36], to the use of non-parametric kernel density estimation or artificial neural network based approaches for background subtraction. In [36], where the MOG approach was proposed, the number of mixture components was kept fixed to an experimentally determined value. Various updates and refinements have been proposed for this method. In [20] the MOG model is improved by using a faster initialization and updating process and introducing a chromatic color space based shadow removal step. The number of mixture components, up to a maximum value of 4, is estimated using Dirichlet priors in [43]. Alternatively a non-parametric kernel density estimation (KDE) approach for background subtraction, which adapts both short-term and long-term models to handle quick and slow changes in background was proposed in [8]. The local texture around a pixel is taken into account in [16] where the pixels are modeled as adaptive local binary pattern histograms calculated over a circular area around the pixels. A recent approach that takes the neighborhood pixel values into account to build a background model was proposed in [1]. In [1] the foreground pixel determination is treated as a classification process and the background pixels are not modeled with a probability density function. A random sampling strategy is used to update the exemplars for each pixel. This method is not sensitive to small camera displacements, noise or ghost objects. The algorithm presented in [25] takes a self organizing approach for background subtraction based on artificial neural networks. Each pixel is modeled with a neuronal map of weight vectors and a weight vector closest to an incoming pixel, along with its neighborhood, is updated over time.

Our method attempts to provide a simple yet robust and flexible approach, fit for real-world video tracking applications, for foreground background segmentation using a Chebyshev probability inequality based background model without any assumptions of normality on the scene background pixels. The foreground object detection relies on system components such as peripheral and stationary object detectors to account for the slow and fast changing elements of real-world scenes as well as the intermittent objects that may become part of a scene background for periods of time [15]. Additionaly recurrent motion and shadow filters and relevance feedback from object tracking provide means for eliminating false foreground detections. The proposed background model is used as part of our GPU accelerated, end-to-end video tracking system benchmarked using the ChangeDetection2012 dataset [27].

The remainder of the chapter is organized as follows. Section 19.2 provides a general overview of our proposed method for foreground background segmentation. In Section 19.3 we provide the details of our Chebyshev probability inequality background model followed by details of the use of relevance feedback in determining the foreground objects in Section 19.4. In Section 19.5 we discuss static and abandoned object detection. We describe the GPU acceleration implementation of proposed algorithms in Section 19.6. The accuracy performance results of the system on the ChangeDetection 2012 dataset and computational performance of GPU acceleration is presented in Section 19.7. We present a summary of our approach in Section 19.8.

19.2   Algorithm Overview

In this section we provide a general overview of the end-to-end object tracking system with the Chebyshev probability inequality background model. Details of the individual components can be found in later sections. Figure 19.1 shows the various components, their interactions and the process flow involved in generating the background model and determining the foreground objects. The thumbnail images aligned with the functional blocks in the diagram denote the resultant intermediate images at each step of the foreground background segmentation process. As illustrated in the figure each new input image frame is used to update a Chebyshev probability model for the background as well as to update the motion regions based on the frame differences calculated over a few consecutive frames. We call this preliminary moving object region detector the "peripheral motion detector" due to its functional resemblance to biological peripheral motion detection. The foreground pixel candidate probability map based on the background model is thresholded and then validated based on the peripheral detector output. This ensures that any pixels with very little motion in the recent past do not get tagged as candidate foreground pixels consistently. The peripheral motion validated foreground pixel candidates pass through a recurrent motion filter to remove the pixels that alternate as foreground and background due to light effects such as reflections. This stage is followed by the removal of shadow pixels from the candidate foreground pixels using pixel intensity change ratios in each color channel. The surviving foreground pixel candidates pass through the connected components stage to be combined into labeled foreground object region candidates. In addition to the moving object detector, a stationary object detector keeps track of moving objects that have come to a stop on their own or left behind. The final foreground object pixels thus consist of both moving and static objects. These foreground object candidates are tracked and further filtered using relevance feedback based on the object track history and the object classifiers in use. The foreground regions that qualify as foreground objects are then used to inhibit the background model update at the corresponding pixel locations.

19.3   Scene Background Model

In building our scene background we avoid making assumptions of normality of pixel distributions and use a probability inequality to model the background. As mentioned earlier, this model is augmented by the peripheral detector to enhance the sensitivity to moving objects for foreground detection, and a recurrent motion detector to handle dynamic scenes. In the following subsections we provide details of these components and discuss how our background models handle illumination issues, dynamic backgrounds and unstable cameras.

19.3.1   Chebyshev Inequality Probability Model

In our model, the intensity of every pixel in a new frame is assigned a probability of not belonging to the background model based on the Chebyshev inequality. Let X be a random variable corresponding to the pixel intensity, $E[X]$ be the mean and σ_x the standard deviation of the distribution then the Chebyshev's inequality is given by equation (19.1)

$$P(|X - E[X]| > \lambda\sigma_x) \leq 1/\lambda^2 \qquad (19.1)$$

Thus the Chebyshev's inequality provides upper and lower tail bounds on X with the tails decaying as $1/\lambda^2$. Besides requiring the mean and the standard deviation, this inequality does not make any assumptions for the underlying distribution (e.g. Mixture of Gaussians) of the pixel intensities. The foreground probability is inversely proportional to the standard deviation normalized distance of the observed pixel from the mean value.

19.3.2   Color Spaces for Background Modeling

The Chebyshev inequality provides a metric to differentiate a foreground object from the background in the chosen feature space. Common feature spaces include the various color spaces [36] [9] [4] and texture measures [24] [31] [41] [16] [39]. Our approach allows the use of multiple color spaces to handle illumination changes. The use of color spaces as features has the additional advantage of being fast to process due to the limited transformation of the frame data as compared to other features like texture. The background model can be trained in either grayscale, RGB color space or the CIELa*b* color space. The RGB channels are correlated and are equally affected by illumination changes. As the scene gets brighter the RGB values increase in unison and vice versa. On the other hand the CIELa*b* color space decorrelates the color and illumination information. It approximates the human vision system with the a* and b* channels capturing color information based on the opponent color axis while L captures a lightness value that is closer to human perception. This allows

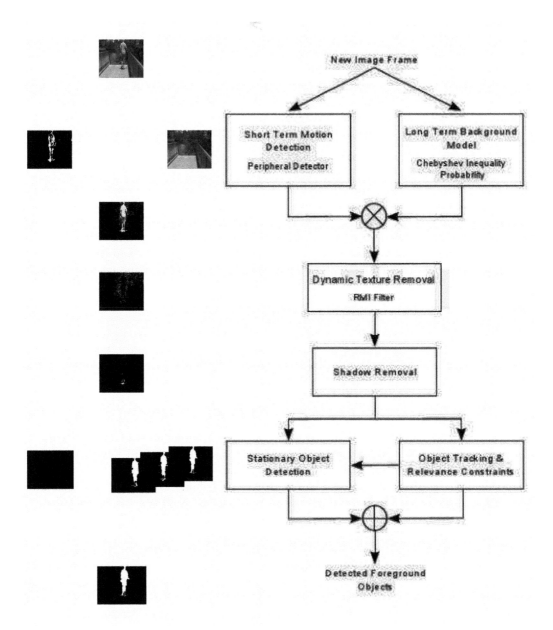

FIGURE 19.1   Non-parametric scene background modeling algorithm overview. Left) Intermediate results at various stages Right) Interaction between various scene background modeling components.

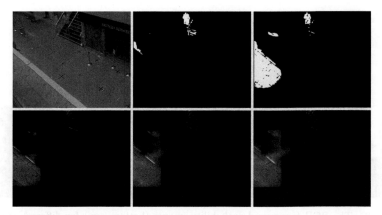

FIGURE 19.2 Comparison of CIELa*b* and RGB background models to illumination change. The top row displays, from left to right, the input image from a PETS2007 video, the foreground detected based on a CIELa*b* background model and one based on RGB background model. Bottom row displays, from left to right, the absolute difference between the Red, Green and Blue channels for the two approaches. For comparison, the CIELa*b* background image was converted to RGB

the detection of any changes in illumination and the ability to control the response of the background model when such a situation occurs. Figure 19.2 shows the background subtraction results for a frame from a PETS 2007 [10] sequence. In this video a passing cloud cover causes a drop in the ambient illumination of the scene. The RGB background model picks up the illumination change as a foreground region (top row right image) while the CIELa*b* based background model correctly handles the change and does not result in any false positive foreground pixel detections (top row middle image). The second row of Figure 19.2 shows the difference between RGB and CIELa*b* background images, for comparison the CIELa*b* images were converted to RGB. The difference between the two background models get reflected in the detected foreground regions. Figure 19.3 shows the foreground probability images based on the RGB color space background model for frames from different videos in the ChangeDetection dataset [12]. The top row of the figure shows the original input image while the bottom row shows the corresponding foreground probability image. In the first column, we see some shadow pixels are determined to be part of the foreground with low probability while in the second column parts of the water fountain jets come up in the probability image. In the thermal video in column 3 the foreground probability image does not detect most of the reflections of the person on the floor and the side walls while in another indoor video, in column 4, the person in the subway station gets assigned high probability to be part of the foreground. Overall, the foreground probability images are clean and segment the moving object from the background.

19.3.3   Peripheral Motion Detection

In a typical scene the background subtraction is required to detect both very slow and fast moving objects. The conflicting background model update rate requirements for these conditions cannot be handled by a single model. To provide inherent stability to the background model being generated, the background probability needs to be updated rather slowly but a very slow background update caused problems in detecting slow moving objects. We incorporate a peripheral detector that performs a coarse but very efficient moving region detection to satisfy the fast model update requirements. By augmenting the background probability model with the peripheral detector, the motion regions are emphasized for fore-

FIGURE 19.3 The RGB foreground probability images (bottom row) for different video frames (top row). In the probability images, brighter is the pixel greater is the probability that it is part of the foreground.

ground pixel detection. The peripheral detector compares an incoming frame image I_t with a few past frame images as indicated in equation 19.2 to generate the Motion History Image (MHI). The motion history image has a memory of N frames over which to accumulate the current difference image defined by k. Typically both these variables are set to 3-5. The constant c is inversely proportional to k and controls the contribution of the motion pixels to the MHI. To minimize small flicker noise influencing the difference image, only pixels with absolute difference above a threshold m are used to determine the motion regions. During the update, for display purposes, the MHI pixels are mapped to [0, 255] with 0 indicating a background pixel and 255 indicating a motion region pixel. The MHI image is thresholded based on desired motion sensitivity and filtered with morphological dilate and erode operations to provide the peripheral detector output.

$$
\begin{aligned}
&\text{Initialize } MHI_{t=0:k}(x,y) \text{ to zeros} \\
&D_t(x,y) = c \text{ if } |I(x,y,t) - I(x,y,t-k)| > m \\
&\qquad\qquad = 0 \text{ otherwise} \\
&MHI_t(x,y) = MHI_{t-1}(x,y) - D_{t-N}(x,y) \\
&\qquad\qquad \text{if } MHI_{t-1}(x,y) > 0, t >= k + N \\
&\text{followed by} \\
&MHI_t(x,y) = MHI_t(x,y) - D_t(x,y) + D_t(x,y) \\
&\qquad\qquad \text{if } MHI_t(x,y) < 255, t > k
\end{aligned}
\tag{19.2}
$$

Figure 19.4 shows the peripheral motion history images for the videos from the ChangeDetection dataset. As the peripheral detector is based on accumulation of frame differences, in column 1 we see significant peripheral motion as the person is close to the camera and is walking towards the photocopying machine. In the *subway* video, in column 4, there is very limited peripheral motion detected as the person did not move around a lot. As the peripheral detector output is used to validate moving object detections, the person gets detected as a foreground by the stationary object detection component once he comes to a halt.

19.3.4   Recurrent Motion Filter

Most background models have problems with dynamic background elements that create frequent motion in parts of a scene such as swaying trees and ripples on water. To circumvent

FIGURE 19.4   The peripheral motion history image (bottom row) for different video frames (top row).

this problem we use a Recurrent Motion Image (RMI) to filter out false positive foreground pixels due to dynamic scene elements. RMI was introduced in [18] as a measure of periodic object motion for object classification; we use RMI not in object regions as in [18] but over the entire frame to detect the flickering of pixels due to dynamic backgrounds.

Let $F(x, y, t)$ be the binary foreground image pixel at position (x, y) obtained at time t based on the background probability model and the peripheral detector. The RMI is then defined by equation 19.3 in which $\oplus$ is an XOR operation and the time duration t is long enough to cover the temporal extent of object motion.

$$RMI(x, y, t) = \sum_{k=1}^{\tau} F(x, y, t - k - 1) \oplus F(x, y, t - k) \qquad (19.3)$$

The RMI will have low values for pixels that get consistently triggered as a part of foreground object and a high value for pixels that get intermittently tagged as foreground due to flickering effects of specular reflections on water or leaves moving in the wind. Additionally, the average strength of the recurrent motion image is used to determine the presence of camera jitter. The peripheral detector will result in false motion regions due to jittering of an unstable camera, as it would be difficult to distinguish between true motion of the moving objects in the scene and the induced motion caused by the camera jitter. These false motion regions would then result in false positive foreground pixel detections. The back and forth jittering of the camera causes some pixels to be detected as foreground for a few frames and then going back to being part of the background. In our system when the camera jitter is detected based on the RMI strength, then the peripheral detector memory is reduced accordingly.

Figure 19.5 shows the results for recurrent motion detection. In the *fountain* video the water stream gets detected as recurrent motion and thus gets removed from the final foreground image. The *badminton* video has significant camera jitter, this results in the court markings occasionally coming up in the foreground probability image and getting correctly detected as recurrent motion.

FIGURE 19.5 Recurrent motion images (bottom) highlighting the dynamic texture regions for various scenes (top).

19.3.5 Shadow Detection

The shadow is considered as a semi transparent region in which the image retains underlying appearance properties [17] and the shadows cast by moving objects as they move in the scene are often detected as part of the foreground. We assign a probability to each pixel for being a shadow pixel based on the RGB channel ratios, ψ_R, ψ_G, ψ_B [26] as given by equation 19.4. To identify and remove darker cast shadows, the current pixel RGB values are compared with the corresponding means, μ_R, μ_G, μ_B, from the background model and a pixel is tagged as a shadow pixel when the current RGB pixel values are less than the corresponding background model means. We only focus on the case where there is consistent darkening across the 3 color channels. In calculating the tolerance, ξ, we use an empirically determined scale factor of 0.1 for the average channel ratio for the maximum deviation between two channels ratios. The probability that a pixel belongs to a shadow region is determined by comparing the absolute R-G and G-B channel ratio differences with this tolerance factor. If neither of these ratios are less than 1 then the probability that the pixel is a shadow is set to 0. Thus the absolute channel ratio difference is allowed to be up to 10 percent of the average channel ratio for a shadow pixel. The identified shadow pixels are then marked as part of the background but are not included in the background update as shadows are considered a temporary effect.

$$\Psi_R = \frac{R_{current}}{\mu_R}; \Psi_G = \frac{G_{current}}{\mu_G}; \Psi_B = \frac{B_{current}}{\mu_B}$$

$$\xi = 0.1\frac{(\Psi_R + \Psi_G + \Psi_B)}{3}$$

$$\upsilon_{RG} = |\Psi_R - \Psi_G|; \upsilon_{BG} = |\Psi_B - \Psi_G| \qquad (19.4)$$

$$P(shadow) = max\left(\frac{\upsilon_{RG}}{\xi}, \frac{\upsilon_{BG}}{\xi}\right) \text{ if } \upsilon_{RG} < \xi, \upsilon_{BG} < \xi$$

$$= 0 \text{ otherwise}$$

19.4 Relevance Feedback from Tracking and Classification

The use of relevance feedback in the form of spatiotemporal information of foreground objects, obtained by higher level analysis such as object tracking and object classification, has been found effective in updating the background model [22] [37]. We use such information to eliminate the foreground blobs that are due to reflections, 'ghost' objects created by the motion of objects in the background, and illumination changes or dynamic scene elements that did not get removed by previously described components of our algorithm. Transient foreground object candidates due to these effects usually have noisy short tracks. Moreover, we generally expect certain classes of objects, like people, vehicle, etc., to be present in a given scene. Using object classification feedback is especially useful in the removal of false foreground candidates due to ghost objects. Once the false positive foreground detections are filtered based on these constraints, the background model gets updated in those regions thereby improving its quality.

19.4.1   Object Tracking

Our object tracking matches up foreground objects obtained in each frame after background subtraction and labeling, and establishes a spatial and temporal motion history for each object. This information is fed back into the background model updating, providing:

1. Removal of false objects that appear only in one or two frames, usually caused by noise, or insufficient background model update rate. At the pixel or blob level, a very fast moving foreground object might cause a large area of pixels to change. If the background update rate is not fast enough, this creates a ghost object in the background image. With tracking information available, the speed of foreground objects can be estimated to adjust the background update rate, and these ghost objects can easily be removed by feature matching (color, edge, etc.).

2. Maintaining moving objects that come to a stop and remain motionless. Without feedback from tracking, the background model would eventually absorb the stopped objects as background, due to no change detection at the pixel level. With tracking feedback, the proper object metadata can be maintained and object region can be classified as foreground.

3. Resurrecting objects that are occluded by other moving objects or scene elements. This helps maintain an object's motion history even after the object has been occluded for a number of frames. Without the high-level tracking feedback, a foreground object after being occluded might be labeled as a new object.

19.4.2   Object Classification

In addition to object tracking, use of object classification can help validate the foreground object detections. There is an infinitely large set of moving objects in the real world, however, only a small number of objects are of interest to automated video surveillance, and statistically they appear more often in videos, such as person, group of people, vehicle, animal, etc. We use Foreground candidates that survive dynamic texture and shadow removal get evaluated by the SVM classifiers. The confidence for an object class is updated over its track based on the per frame classification scores. The objects that do not get classified into one of the prespecified categories can be tagged for removal. However, candidate objects that have poor track properties may still be considered as valid objects if they are confidently classified by object classification. Figure 19.6 shows the example usage of relevance feedback. In the depicted frames some parts of the background get detected as foreground but the use of relevance feedback from tracking and classification correctly removes the false detections.

Support Vector Machine (SVM) [6] based object classification to evaluate foreground object candidates. SVM is a supervised learning method that simultaneously minimizes the

FIGURE 19.6   Examples of false positive regions selected for removal, marked with a circle, by object tracking and classification relevance feedback.

classification error while maximizing the geometric margin. Hence SVMs perform better than many other classifiers and have been used in a wide range of computer vision applications for object classification, object recognition [42] [38] and action recognition [15] [19] [34]. We use oriented edges [2], aspect ratios and relative object sizes for person and vehicle classification. Oriented edges have shown to be robust to changes in illumination, scale and view angle [7]. In most surveillance applications, robustly classifying objects without having a large set of sample data to train the classifier is of particular interest. SVM classifiers are well suited for such applications as they can be trained with few samples and generalize well to novel scenes.

19.5   Stationary Object Detection

While moving objects are detected and tracked by the object tracker, a stationary object detector runs in parallel using the same background model as the moving object tracker to determine the regions of stationary objects. Our stationary object detection algorithm introduced in [14] looks for foreground pixels that deviate from the background for a long duration of time. The algorithm produces an intermediate image called a stationary object confidence image, S, in which each pixel value represents the confidence in that pixel belonging to a stationary object. These confidence values are mapped to [0, 255], 0 indicating a background pixel and 255 indicating a stationary object pixel. At the start of processing all pixels, $S(x, y)$, of the stationary object confidence image are initialized to 0. Equation 19.5 gives the details for generating the stationary object confidence image. I denotes the original frame image and B denotes the background model. For each new frame, every pixel $S(x, y)$ is updated based on a comparison of the corresponding channel of the original frame image pixel $I(x, y)$ with the corresponding background model $B_{x,y}(m, s)$ at pixel (x, y) and a duration based update function. An increment counter $C(x, y)$ at each pixel keeps track of number of consecutive $I(x, y)$ observations that do not fit the background model at that pixel. This counter is reset to zero when the observed image pixel is deemed to be part of the background model. Similarly a decrement counter $D(x, y)$ is started each time $C(x, y)$ is reset to keep track of number of consecutive image pixel observations that fit the background model at (x, y). The response time for an object pixel to get incorporated into the stationary object confidence map is controlled by t_s while the parameter r controls how quickly a stationary pixel gets removed from the stationary object confidence map. Both the update and decay of the confidence are inversely proportional to the video frame rate, FPS.

Initialize $S_{t-1}(x, y)$ to zero

$$S_t(x, y) = S_{t-1}(x, y) + C(x, y)\frac{255}{t_s * FPS}$$

$$\text{if } I_t(x, y) \notin B_{x,y}(\mu, \sigma) \tag{19.5}$$

$$S_t(x, y) = S_{t-1}(x, y) + rD(x, y)\frac{255}{t_s * FPS}$$

$$\text{if } I_t(x, y) \in B_{x,y}(\mu, \sigma)$$

Pixels that get tagged as belonging to a stationary object will not fit the background model and remain different from the background for as long as that object stays in place, causing the corresponding $S(x, y)$ pixels to be incremented at every frame and to reach the maximum confidence value (255) by the time the object has been stationary for t_s seconds. Figure 19.7 shows the stationary object detection component in action. The black car comes to a stop on the left side of the road. The sequence of images on the top row shows the stationary object confidence map that getting built over time. Early on the pixels on the black car have a low confidence for belonging to a stationary object and as time progresses their confidence increases. The momentary slowing down of the cars behind the black car causes more pixels getting accumulated into the stationary object image as seen in the 2nd and 3rd frames in the top row of the figure, but when they continue to move those pixels get erased from the stationary object image. In the final image we see the black car has been correctly detected as having come to a stop.

19.5.1 Left Object Detection

The stationary object detector quickly identifies potential stationary foreground objects while the object tracker keeps track of all object-split events, such as someone putting down a carried object which may lead to an "abandoned object". The detected stationary objects are filtered by correlating with abandoned objects and also by using the distance of the owner from the object to confirm an abandoned object. This dual detection mechanism helps reduce the effect of occlusions by strengthening the abandoned object hypothesis using cues from both moving and stationary object analysis.

19.6 GPU Performance Enhancement

Today's commodity graphics hardware contains powerful programmable coprocessors i.e., General Purpose Graphic Processing Units (GPUs) designed to independently process pixels in parallel providing great potential for acceleration of computer vision and image processing algorithms [3]. Over the last decade GPUs have evolved faster than CPU's and they continue to do so with NVIDIA's newer Fermi and Kepler architecture [29]. GPUs provide a more attractive and affordable alternative to dedicated special purpose hardware used for speeding up computationally intense algorithms in many applications. In better part of the last decade GPUs have attracted a lot of attention from the computer vision community for speeding up parallel operations in video analysis algorithms [11]. Pixel based parallel operations of object detection and frame pre-processing stages of video analysis tasks are well suited to GPU platform and several image and video processing algorithms have been implemented on GPU platforms [13] [21] [28] [35] [40]. In recent years there has been an increased interest in proposing background subtraction algorithms on the GPU [5] [32] [33] and speeding up of existing GMM variants [23] [30].

The most computationally intense operations of our video object detection and tracking

FIGURE 19.7 As the car comes to stop the system starts increasing its confidence in the stationary object map and finally detected as a stationary object, as indicated by the dark bounding box around the car.

system described here are implemented on the GPUs resulting in significant speed-ups. Parallel operations such as background model generation, background model update and new frame to background comparisons, as well as image filtering operations are all accelerated on GPUs resulting in large savings in processing time. By performing only the inherently non-parallel operations on the CPU and utilizing multi-threaded processing within the CPU, the overall computational performance of our video tracking system is significantly boosted.

The conventional CPU architecture uses few general purpose processors capable of multithreading that are optimized for instruction level parallelism while the GPU architecture makes use of many small specialized processors, executing a single instruction on multiple data (SIMD), optimized for data level parallelism. In the GPU architecture the data parallel portions of an application are executed on the device as kernels with multiple kernels executed in parallel. The CUDA processing model uses a grid of 1D, 2D or 3D blocks with each block running multiple threads. In this model it is important to minimize the control flow divergence as each thread executes the same kernel and any divergence results in the slowing down of all the threads. A group of 32 threads is defined as a warp in the CUDA model and it corresponds to the smallest executable unit of parallelism on the GPU device.

The CUDA memory model contains multiple memory spaces, viz. thread local, block shared, global, constant and texture memory spaces. The global, constant and texture memory spaces are persistent across kernel launches by the same application. Of these memory spaces global, local and texture memory have the greatest access latency and these are followed by constant and shared memory. Considering the access latency, especially for global memory, it is necessary to ensure memory coalescing. Memory coalescing enforces consecutive threads in the same warp to concurrently request consecutive logical addresses from global memory thereby allowing to minimize the number of global memory segments requested. We improved the memory access pattern of the background subtraction algorithm by using a structure of arrays for the background model instead of treating the models of each pixel as an array of structures. The number of memory accesses were also reduced by combining multiple 1 byte pixel values into a single 4 byte multi-pixel data. We also take advantage of CUDA textures which, on devices of compute capability of 1.x, are cached

FIGURE 19.8 Output at various stages of the pipeline. top row) input image, background model, foreground probability image, peripheral motion image. bottom row) recurrent motion image, shadow image, final foreground image.

unlike the global memory.

19.7 Experimental Results

We benchmarked the real time video tracking system employing the foreground/background segmentation algorithm described here on the ChangeDetection dataset. The dataset consists of 6 video categories, viz Baseline, Camera Jitter, Dynamic Background, Intermittent Object Motion, Shadow and Thermal, with 31 videos totaling over 80,000 frames.

Figure 19.8 depicts the results from the various components of the system on the *Overpass* video from the Dynamic Background category of the ChangeDetection 2012 dataset. The second column in the top row shows a generated background model, it is blurred in the regions with water and tree movement. This is followed by the foreground probability image which contains non zero probability values for the trees swaying around the person and finally the peripheral image. The recurrent motion image, in the bottom row, captures the swaying tree motion while the shadow detection identifies the shadow region at the foot of the person. The final foreground image is much cleaner and is devoid of most of the dynamic background artifacts.

Figure 19.9 shows few examples of foreground object detection results for various video clips from the ChangeDetection dataset [12]. All the categories were processed with a single setting file. Table 19.1 summarizes the average scores for the video clips across the 6 categories. Our system did exceptionally well with dynamic scenes, with the best result for *fountain02* video at f-measure of 0.91. The system also handled the strong shadows and the transitions between shadow and light well. The best performer was the backdoor video with f-measure of 0.88. Among the 5 videos in the Thermal category, *library* video achieved the best performance with f-measure of 0.95. Figure 19.10 depicts the challenging reflections on the floor and the side walls in the *corridor* video that were nicely eliminated with our system. The worst performer was the *lakeside* video due to the people emerging from the cold lake water. For the Baseline category, the PETS2006 subway video had the best f-measure of 0.93 while for Camera Jitter category it was the *badminton* video with an f-measure of 0.90. The Intermittent Object Motion category was our worst performer due to the competing system component requirements; specifically this category needs the

FIGURE 19.9 Sample foreground detections on the ChangeDetection 2012 dataset. Top Row) Input frame, Middle Row) Ground truth with foreground pixels having a value 255, Bottom Row) Detected foreground pixels.

TABLE 19.1 Performance metrics for the ChangeDetection 2012 dataset

Category	Recall	Sp	FPR	FNR	PBC	FMeasure	Precision
Baseline	0.8266	0.9970	0.0030	0.0058	0.8304	0.8646	0.9143
Camera Jitter	0.7223	0.9725	0.0275	0.0106	3.6203	0.6416	0.5960
Dynamic Background	0.8182	0.9976	0.0024	0.0017	0.4086	0.7520	0.7339
Intermittent Object Motion	0.3570	0.9807	0.0193	0.0515	6.4700	0.3863	0.7688
Shadow	0.8670	0.9887	0.0113	0.0051	1.5561	0.8333	0.8103
Thermal	0.6887	0.9963	0.0037	0.0121	1.4283	0.7230	0.8906

Stationary Object Detector which inherently conflicts with the RMI filter. This constraint results in poor settings for the intermittent objects which would not be the case in a real deployment scenario.

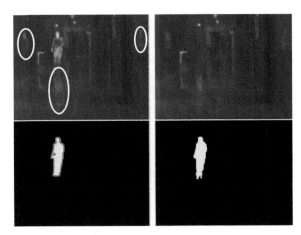

FIGURE 19.10 Top Left: A frame from the corridor video with marked reflections, Bottom Left: Ground truth foreground, Top Right: proposed background model, Bottom Right: foreground result.

TABLE 19.2   Total frame processing time, in milliseconds, for Gray, RGB and CIELa*b* background models on the CPU and GPU

Video Resolution	CPU			GPU		
	Gray	RGB	CIELa*b*	Gray	RGB	CIELa*b*
320 x 240	6.86	8.32	17.73	4.86	5.16	5.72
640 x 480	16.50	21.05	58.05	10.66	10.21	12.82
1280 x 720	39.06	56.80	184.23	16.10	16.07	23.63

TABLE 19.3   Average peripheral detection processing time, in milliseconds, on the CPU and GPU

Video Resolution	CPU Grayscale	GPU Grayscale
320 x 240	0.83	0.72
640 x 480	1.72	0.88
1280 x 720	5.53	1.00

19.7.1   GPU Performance Results

The per frame processing time of the system was tested on a Core i7 CPU with 4 cores and a NVIDIA GTX690 video card with 3092 cores. Table 19.2 shows the total processing time, per frame, in milliseconds for different video resolutions and for different background models on the CPU and the GPU. For all resolutions the GPU implementation is faster with the speed improvements becoming more pronounced with increasing video resolution. For a 720p video with the CIELa*b* color background model, the GPU version is about 7.8 times faster than the CPU version. As expected, the gray scale background model is the fastest at all resolutions. Up to 640×480 resolution the GPU implementation of the 3 background models have comparable total frame processing time, only at higher resolution does the CIELa*b* model starts becoming more expensive on the GPU.

Tables 19.3 and 19.4 show the processing times for some of the components of the system on the CPU and the GPU. As can be seen in 19.3 peripheral detection is very fast on both the CPU and the GPU. It benefits from the use of GPU only at video resolution of 720p. Background compare and recurrent motion image processing times are presented in 19.4, the GPU benefits becoming apparent at video resolutions of 640 x 480 and higher.

The accelerated processing with the GPU improves real time video tracking. Figure 19.11 depicts snapshots from the processing of a far field HD video scene with very small tracking targets. The top frame shows a snapshot from the processing of this video without GPU acceleration while the bottom frame shows a snapshot from the processing with the GPU on the same hardware hence the same CPU. Without the GPU acceleration the generated background model is not optimal due to a lower processing frame rate resulting in missed detections and poor tracking of small targets with low contrast. With GPU enabled processing, several tasks are off-loaded to GPU in addition to the CPU processing. When

TABLE 19.4   Background compare and Recurrent Motion Image processing time, in milliseconds, for Gray, RGB and CIELa*b* background models on the CPU and GPU

Video Resolution	CPU			GPU		
	Gray	RGB	CIELa*b*	Gray	RGB	CIELa*b*
320 x 240	4.28	6.06	4.39	1.85	2.15	2.59
640 x 480	11.06	15.22	16.09	2.12	1.91	2.48
1280 x 720	25.15	43.42	50.17	2.17	2.40	3.04

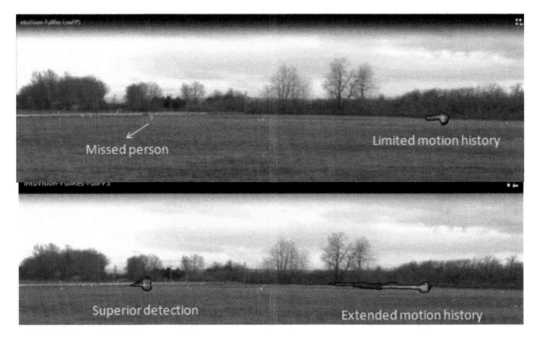

FIGURE 19.11   Processing of an HD video on a low cost hardware platform misses low contrast small targets (top); GPU accelerated processing on the otherwise same hardware detects and tracks all targets reliably (bottom).

processed at full frame rate an optimal background model is maintained, and even the very small and low contrast targets are detected and tracked very reliably.

GPU enabled analytics processing makes it possible to perform high quality real-time tracking, on high resolution video such as this example, or multiple video streams on low cost computers without the need for special-purpose hardware solutions.

19.8   Conclusion

We presented a non-parametric background model as part of a real time video object detection and tracking system. In our approach the long and short term background model components are aided by recurrent motion filter, shadow removal, stationary object detection and relevance feedback from tracking and classification. This allows the detection and tracking of moving objects in various operational deployment scenarios from shoreline tracking of vessels in rough water, to tree branches blown in high winds, with cast shadows from moving objects and other scene elements. Most of the components make use of GPU acceleration to get better than real time processing speed and support for handling multiple camera streams on a single machine. The improved processing speed of GPU accelerated components results in a higher frame rate background update. This allows for better segmentation of the moving objects from the background.

Our design philosophy in building this system has been to include a range of algorithms with parameters that allow easy configuration and customization of the system components to a wide range of intended use case scenarios. Customizing the component configuration and flow the system can be easily tuned to the specific viewing conditions and a range of real-world applications.

References

1. O. Barnich and M. Van Droogenbroeck. Vibe: A universal background subtraction algorithm for video sequences. *IEEE Transactions on Image Processing*, 20(6):1709–1724, 2011.
2. S. Bileschi and L. Wolf. Image representations beyond histograms of gradients: The role of gestalt descriptors. In *Computer Vision and Pattern Recognition, CVPR 2007*, pages 1–8, 2007.
3. K. Bjorke. NVIDIA corporation, image processing using parallel GPU units. In *Proceedings of SPIE*, volume 6065, 2006.
4. A. Chan, V. Mahadevan, and N. Vasconcelos. Generalized stauffer-grimson background subtraction for dynamic scenes. *Machine Vision and Applications*, 22(5):751–766, 2011.
5. L. Cheng, M. Gong, D. Schuurmans, and T. Caelli. Real-time discriminative background subtraction. *IEEE Transactions on Image Processing*, 20(5):1401–1414, 2011.
6. C. Cortes and V. Vapnik. Support-vector networks. *Machine learning*, 20(3):273–297, 1995.
7. N. Dalal and B. Triggs. Histograms of oriented gradients for human detection. In *Computer Vision and Pattern Recognition, CVPR 2005*, volume 1, pages 886–893, 2005.
8. A. Elgammal, D. Harwood, and L. Davis. Non-parametric model for background subtraction. *European Conference on Computer Vision, ECCV 2000*, pages 751–767, 2000.
9. R. Evangelio, M. Patzold, and T. Sikora. Splitting Gaussians in mixture models. In *International Conference on Advanced Video and Signal-Based Surveillance (AVSS 2012)*, pages 300–305, 2012.
10. J. Ferryman and D. Tweed. An overview of the PETS 2007 dataset. In *IEEE International Workshop on Performance Evaluation of Tracking and Surveillance, PETS*, 2007.
11. M. Gong, A. Langille, and M. Gong. Real-time image processing using graphics hardware: A performance study. *Image Analysis and Recognition*, pages 1217–1225, 2005.
12. N. Goyette, P. Jodoin, F. Porikli, J. Konrad, and P. Ishwar. Changedetection.net: A new change detection benchmark dataset. In *IEEE Computer Society Conference on Computer Vision and Pattern Recognition Workshops, CVPRW 2012*, pages 1–8, 2012.
13. A. Griesser, S. De Roeck, A. Neubeck, and L. Van Gool. GPU-Based Foreground-Background Segmentation using an Extended Colinearity Criterion. In G. Greiner, J. Hornegger, H. Niemann, and M. Stamminger, editors, *Vision, Modeling, and Visualization (VMV 2005)*, pages 319–326. IOS Press, November 2005.
14. S. Guler and M. Farrow. Abandoned object detection in crowded places. In *PETS 2006*, pages 18–23, 2006.
15. S. Guler, K. Garg, and A. Silverstein. Video scene assessment with unattended sensors. In *Proceedings of SPIE Vol*, volume 6741, pages 674108–1, 2007.
16. M. Heikkila and M. Pietikainen. A texture-based method for modeling the background and detecting moving objects. *IEEE Transactions on Pattern Analysis and Machine Intelligence*, 28(4):657–662, 2006.
17. T. Horprasert, D. Harwood, and L. Davis. A statistical approach for real-time robust background subtraction and shadow detection. In *IEEE ICCV*, volume 99, pages 256–261, 1999.
18. O. Javed and M. Shah. Tracking and object classification for automated surveillance. *European Conference on Computer Vision, ECCV 2002*, pages 439–443, 2002.
19. I. Junejo, E. Dexter, I. Laptev, and P. Pérez. View-independent action recognition from temporal self-similarities. *IEEE Transactions on Pattern Analysis and Machine Intelligence*, 33(1):172–185, 2011.
20. P. KaewTraKulPong and R. Bowden. An improved adaptive background mixture model for real-time tracking with shadow detection. In *European Workshop on Advanced Video Based Surveillance Systems*, volume 25, pages 1–5, 2001.
21. P. Labatut, R. Keriven, and J. Pons. A GPU implementation of level set multiview stereo.

Computational Science-ICCS 2006, pages 212–219, 2006.

22. L. Li, I. Gu, M. Leung, and Q. Tian. Adaptive background subtraction based on feedback from fuzzy classification. *Optical Engineering*, 43(10):2381–2394, 2004.

23. Y. Li, G. Wang, and X. Lin. Three-level GPU accelerated Gaussian mixture model for background subtraction. In *IS&T/SPIE Electronic Imaging*, pages 829514–829514, 2012.

24. S. Liao, G. Zhao, V. Kellokumpu, M. Pietikainen, and S. Li. Modeling pixel process with scale invariant local patterns for background subtraction in complex scenes. In *Computer Vision and Pattern Recognition, CVPR 2010*, pages 1301–1306, 2010.

25. L. Maddalena and A. Petrosino. A self-organizing approach to background subtraction for visual surveillance applications. *IEEE Transactions on Image Processing*, 17(7):1168–1177, 2008.

26. B. Mitra, P. Birch, I. Kypraios, R. Young, and C. Chatwin. On a method to eliminate moving shadows from video sequences. *SPIE Photonics Europe Optical and Digital Image Processing, Strasbourg, France*, 7000:700012–1, 2008.

27. A. Morde, X. Ma, and S. Guler. Learning a background model for change detection. In *Computer Vision and Pattern Recognition Workshops (CVPRW 2012)*, pages 15–20, 2012.

28. L. Mussi, S. Cagnoni, and F. Daolio. GPU-based road sign detection using particle swarm optimization. In *Intelligent Systems Design and Applications, ISDA 2009*, pages 152–157, 2009.

29. NVIDIAs next generation CUDA compute architecture : Kepler GK 110. White Paper, 2012.

30. V. Pham, P. Vo, and V. Hung et al. GPU implementation of extended Gaussian mixture model for background subtraction. In *International Conference Computing and Communication Technologies, Research, Innovation, and Vision for the Future, RIVF 2010*, pages 1–4, 2010.

31. J. Pilet, C. Strecha, and P. Fua. Making background subtraction robust to sudden illumination changes. *European Conference on Computer Vision, ECCV 2008*, pages 567–580, 2008.

32. M. Poremba, Y. Xie, and M. Wolf. Accelerating adaptive background subtraction with GPU and CBEA architecture. In *IEEE Workshop on Signal Processing Systems, SIPS 2010*, pages 305–310, 2010.

33. D. Schreiber and M. Rauter. GPU-based non-parametric background subtraction for a practical surveillance system. In *International Conference on Computer Vision Workshops, ICCV Workshops 2009*, pages 870–877, 2009.

34. C. Schuldt, I. Laptev, and B. Caputo. Recognizing human actions: a local SVM approach. In *International Conference on Pattern Recognition, ICPR 2004*, volume 3, pages 32–36, 2004.

35. S. Sinha, J. Frahm, M. Pollefeys, and Y. Genc. GPU-based video feature tracking and matching. In *EDGE, Workshop on Edge Computing Using New Commodity Architectures*, volume 278, page 4321, 2006.

36. C. Stauffer and W. Grimson. Adaptive background mixture models for real-time tracking. In *Computer Vision and Pattern Recognition, CVPR 1999*, volume 2, 1999.

37. L. Taycher, J. Fisher, and T. Darrell. Incorporating object tracking feedback into background maintenance framework. In *Application of Computer Vision, WACV/MOTIONS 2005*, volume 2, pages 120–125, 2005.

38. K. van de Sande, T. Gevers, and C. Snoek. Evaluating color descriptors for object and scene recognition. *IEEE Transactions on Pattern Analysis and Machine Intelligence*, 32(9):1582–1596, 2010.

39. L. Wang, H. Wu, and C. Pan. Adaptive ε lbp for background subtraction. *ACCV 2010*, pages 560–571, 2011.

40. R. Yang and G. Welch. Fast image segmentation and smoothing using commodity graphics hardware. *Journal of graphics tools*, 7(4):91–100, 2002.

41. J. Yao and J. Odobez. Multi-layer background subtraction based on color and texture. In *Computer Vision and Pattern Recognition, CVPR 2007*, pages 1–8, 2007.

42. J. Zhang, M. Marszałek, S. Lazebnik, and C. Schmid. Local features and kernels for classification of texture and object categories: A comprehensive study. *International journal of computer vision*, 73(2):213–238, 2007.

43. Z. Zivkovic. Improved adaptive Gaussian mixture model for background subtraction. In *International Conference on Pattern Recognition, ICPR 2004*, volume 2, pages 28–31, 2004.

20

GPU Implementation for Background-Foreground-Separation via Robust PCA and Robust Subspace Tracking

Clemens Hage
Technische Universitat Munchen, Germany

Florian Seidel
Technische Universitat Munchen, Germany

Martin Kleinsteuber
Technische Universitat Munchen, Germany

20.1 Introduction

Any method for automatic background-foreground-separation requires a robust model of the video background. Subspace estimation methods, such as algorithms derived from Principal Component Analysis (PCA) can be applied to a wide range of data [6] and have been reported to be especially successful at modeling lighting conditions and repetitive motion in the background [13]. On the downside they are reported to be very costly compared to competing methods. Many use cases for background subtraction methods, such as surveillance applications require algorithms that are realtime capable and can be implemented on concentrated hardware such as an off-the-shelf graphics processing unit (GPU) in a desktop PC. Thus, a way needs to be found to implement such methods in a time and memory efficient way.

Robust PCA (RPCA) methods rely on the assumption that a video sequence can be separated into a low-dimensional background and foreground objects that are sparse in space and time, i.e. they occupy a small part of the scene and are not persistent over more than a certain number of frames. From an algorithmic point of view, a video sequence is observed as a set of m-dimensional feature vectors which are constructed by vectorizing each frame with each pixel serving as an independent feature. If the background of a video sequence is static or exhibits repetitive motion it is possible to find a low-rank model for these features. This is done by fitting a low-dimensional subspace to the data. For each pixel in a frame the residual error of this approximation is computed and if it exceeds a certain threshold it is labeled as a foreground object.

A possible drawback of common RPCA methods is the fact that a whole video sequence needs to be accumulated and is then batch-processed to determine a low-rank approximation of the background in the whole sequence. Depending on the application, the resulting processing delay and increased memory requirements may introduce some difficulties. If rather long sequences are considered the background model may become quite complex, as even static parts of the scene (e.g. shadows) might change gradually over time and have to be represented in the low-rank approximation. It may therefore be desirable to find a way of estimating and maintaining a low-rank background model in an online setting, i.e. in a frame-per-frame manner. While the batch-processing case leads to a subspace estimation or regression problem, the online scenario can be addressed by so-called subspace tracking methods. As the name indicates, subspace tracking methods are able to detect and to deal with temporal changes in the data by adapting a low-dimensional subspace over the time of observation, which allows for a more flexible use in unknown conditions. As a desirable side effect, sample-wise processing can also reduce the computational complexity significantly.

In this chapter we will analyze how the tasks of subspace estimation and subspace tracking can be solved with optimization methods on manifolds and how such methods can efficiently be implemented on a GPU. Starting with a brief introduction to some basic concepts of manifold optimization we will state the Robust PCA problem in the manifold context, discuss which cost functions are suitable and how their optimization can practically be performed in batch and online mode. In order to keep the discussion close to real-world applications we discuss the recently proposed methods *GRASTA* [15] and *pROST* [24], which both perform online background subtraction on video sequences. We introduce the reader to common tools and illustrate some steps of realtime video processing on the GPU at the example of a *pROST* implementation.

20.2   Background Modeling with Subspaces

Any video sequence can be written as a matrix, where the columns represent the video frames over time. If successive frames exhibit certain similarities, which the human eye intuitively interprets as a video background, this matrix becomes low-rank. The key idea of subspace approaches for background modeling is now to search for the low-dimensional subspace which contains the background of a given video sequence. Once this low-dimensional subspace has been determined, a foreground-background segmentation can be performed by comparing each video frame to the background model. Areas in a video frame that cannot be explained by the background model are labeled as foreground in this process, while the remaining pixels are classified as background.

We first outline the batch case, in which a video sequence is processed at once. The online setting will be discussed in Section 20.4. Consider a video sequence of n frames, each consisting of m pixels. If the video frames are vectorized and concatenated, a video data matrix $X \in \mathbb{R}^{m \times n}$ is obtained. Even for low-resolution images m will take on large values, since each pixel is observed as an independent feature. However, as is often the case in high-dimensional feature spaces, the actual dimension of the data is much lower and it is possible to approximate the data with a low-dimensional model. The image sequence can well be approximated by a low-rank matrix $L \in \mathbb{R}^{m \times n}$ with $\mathrm{rk}(L) = k \ll m, n$. L can be composed as the product of a matrix $U \in \mathbb{R}^{m \times k}$ with $U^\top U = I_k$ and a matrix $Y \in \mathbb{R}^{k \times n}$, i.e. $X \approx L = UY$. The columns of the matrix U span a low-dimensional subspace and Y contains the subspace coordinates which are in some sense closest to the original video frames. For appropriately chosen U and Y and given the sparsity assumptions made on the video foreground, the matrix $S = X - UY$ will therefore be sparse and the non-zero entries identify foreground pixels in the video frames. In the following sections we will address the question of how U and Y can efficiently be found via optimization on matrix manifolds.

20.3 The Manifold Setting

The m-dimensional Euclidean space is spanned by the columns of any matrix from the space of m-dimensional orthogonal matrices $O(m)$:

$$Q \in O(m) := \{Q \in \mathbb{R}^{m \times m} \mid Q^\top Q = I_m\}. \tag{20.1}$$

Consequently, a k-dimensional subspace within this surrounding space is spanned by the k orthogonal columns of a matrix $U^{m \times k}$ from the so-called *Stiefel manifold*:

$$U \in \mathrm{St}_{k,m} := \{U \in \mathbb{R}^{m \times k} \mid U^\top U = I_k\}. \tag{20.2}$$

Once the coordinate frame of such a subspace is defined, all that remains is to find the particular representation $Y \in \mathbb{R}^{k \times n}$, which leads to the low-rank approximation

$$L = UY \quad \text{with } U \in \mathrm{St}_{k,m}, Y \in \mathbb{R}^{k \times n}. \tag{20.3}$$

The elements of the Stiefel manifold are not unique representations of a low-dimensional subspace. Namely, whenever (U, Y) defines a particular background solution then so does $(UQ, Q^\top Y)$ for any orthogonal matrix $Q \in O(k)$. In other words, one can find many pairs (U, Y) resulting in the same L, and a particular low-rank model is represented by a set of Us in a many-to-one mapping. One way to resolve this ambiguity is to consider the so-called *Grassmannian* $\mathrm{Gr}_{k,m}$, the set of all k-dimensional subspaces in a surrounding space of dimension m. As pointed out in [9], however, there exists no explicit matrix representation.

There are two possible ways of identifying the elements of $\mathrm{Gr}_{m,k}$ with matrices. One is via symmetric rank-k-projectors and one uses equivalence classes. In the first definition a subspace is uniquely identified by an orthogonal projector that maps any element in the surrounding space to the closest point on this k-dimensional subspace, and the Grassmannian is defined as

$$\mathrm{Gr}_{k,m} := \{P \in \mathbb{R}^{m \times m} \mid P^2 = P, P^\top = P, \operatorname{tr} P = k\}. \tag{20.4}$$

Exploiting that

$$UU^\top = P \in \mathrm{Gr}_{k,m} \quad \text{and} \quad UQ(UQ)^\top = UQQ^\top U^\top = UU^\top \tag{20.5}$$

one can establish a one-to-one mapping between an element of the Grassmannian and the set of all $U \in \mathrm{St}_{k,m}$ which create the same projector $P = UU^\top$.

Another way of resolving the ambiguity is to define the Grassmannian as the quotient manifold

$$\mathrm{Gr}_{k,m} = \mathrm{St}_{k,m} / O(k). \tag{20.6}$$

In this concept, all Stiefel elements $\tilde{U}$ that correspond to an element of $\mathrm{Gr}_{k,m}$ are represented by one representative $U \in \mathrm{St}_{k,m}$. Formally, an equivalence relation between all members of a so-called equivalence class

$$[U] = \{\tilde{U} \in \mathrm{St}_{k,m} \mid \tilde{U} \sim U\} \in \mathrm{Gr}_{k,m} \tag{20.7}$$

is established, where any member $\tilde{U}$ in the class is equivalent to its class representative $[U]$ if and only if there exists a $Q \in O(k)$ such that $\tilde{U} = UQ$.

A great advantage of defining elements of the Grassmannian as projectors is their unique matrix representation, which allows an intuitive way of optimizing on the manifold. However, the size of a projector is much larger than its Stiefel component ($m \times m$ compared to $m \times k$), which results in significant overhead concerning computation and storage. A way to reduce the computational complexity and the storage requirements is choosing a suitable retraction, cf. [14].

The concept of equivalence classes does not lead directly to numerical algorithms. There exists no unique matrix representation of an equivalence class, but storing one of its members is much cheaper than storing a full projector. Establishing and understanding an optimization method that alternates between learning the subspace and the coordinates using the equivalence class view of the Grassmannian needs more care due to the aforementioned ambiguities. Since the focus of this work is on discussing efficient numerical schemes for subspace tracking, we will use the equivalence class representation in the following and recall the required concepts.

20.3.1   Geometric Optimization

Geometric Optimization is an elegant and effective framework for solving optimization problems. The core of the approach is to define the geometry of the set of all possible solutions as a so-called (sub-)manifold and to apply optimization techniques on this space. Starting at some element and moving along the shortest paths on this manifold in reasonable steps, one finally reaches an element which leads a locally or globally optimal solution to the problem. The difficulties are that many concepts of the common Euclidean space such as directions and operations between the elements cannot be used in the same manner on a manifold, but need to be explicitly defined, if possible at all. A variety of literature has therefore been dedicated to the field of Geometric Optimization. To keep things as simple as possible we will restrict ourselves to the most relevant concepts and refer the interested reader to [9] and [1]. Some more insight on Grassmannian optimization and a thorough discussion of conjugate gradient methods on the Grassmannian are provided in [19].

Tangent space and projections

The tangent space at a certain point on the manifold is the Euclidean approximation of the local manifold surface, in a similar manner as e.g. a linear approximation of a nonlinear function. For the quotient manifold definition, the tangent space of the Grassmannian $\mathrm{Gr}_{k,m}$ at a point $[U]$ can be defined by

$$T_{[U]} \mathrm{Gr}_{k,m} := \{B \in \mathbb{R}^{m \times k} \mid U^\top B = 0\}, \tag{20.8}$$

where B represents any possible tangent direction. While certain measures, such as the inner product of two elements might not be defined on the manifold, the tangent space now allows to use the inner product of the surrounding space, i.e.

$$\langle Z_1, Z_2 \rangle = \mathrm{tr}(Z_2^\top Z_1). \tag{20.9}$$

Any element from the surrounding space $H \in \mathbb{R}^{m \times n}$ can be projected onto the tangent space $T_{[U]} \mathrm{Gr}_{k,m}$ via

$$\pi(H) = (I - UU^\top)H. \tag{20.10}$$

Geodesics

In order to move along the manifold one approach to define the shortest paths from one point on the manifold to another point, such as great circles on a ball. When the topology of the manifold becomes more difficult than that of a ball, these so-called geodesics may become costly to compute. In practical applications one seeks to reduce the computational effort, which can be done by introducing *retractions* based e.g. on the QR decomposition. According to [9], the exponential map and thus the geodesics for a point $[U]$ on the Grassmannian can also be described as

$$U(t) = (UV\cos(\Sigma t) + \Theta\sin(\Sigma t))V^\top, \qquad (20.11)$$

with the compact Singular Value Decomposition of a tangent direction $B = \Theta\Sigma V^\top$.

Gradient descent with line search

If $G \in \mathbb{R}^{m\times k}$ is the gradient of a cost function (see Section 20.4), its projection $\Gamma := \pi(G)$ onto the tangent space of a submanifold is denoted as the *Riemannian gradient*. Notice that we require the Riemannian metric defined by the scalar product in the tangent space to be compliant with the scalar product of the surrounding space $\mathbb{R}^{m\times k}$. The projection is assumed to be orthogonal with respect to this product.

Now a simple gradient descent method on the Grassmannian can be established, which is illustrated in Figure 20.1. Starting at an initial point, one can walk along the geodesics of the manifold in the direction of the Riemannian gradient in order to reduce the value of the cost function at the next iteration. The length t of this step, which is commonly referred to as the *step size*, needs to be determined such that sufficient progress is made. Finding the optimum step size in an optimization method is a problem on its own and countless sophisticated methods have been proposed to solve it, cf. [22]. In our application, however, it has been observed that a simple condition such as the *Armijo rule* or even a fixed step size lead to acceptable results if some additional safeguards such as a minimum and maximum value for t are imposed.

In the following we will discuss potential cost functions and how the subproblem of optimizing on the Grassmannian takes part in the overall subspace tracking problem.

20.4   The Optimization Problem

Since Robust PCA has been defined as the decomposition of a matrix $X \in \mathbb{R}^{m\times n}$ into a low-rank and a sparse matrix [8], many methods have been developed to solve this task. Formally, the decomposition can be written as

$$X = L + S \quad \mathrm{rk}(L) \le k, \|S\|_0 \le s, \qquad (20.12)$$

where $\mathrm{rk}(\cdot)$ denotes the rank operator and the so-called zero norm $\|\cdot\|_0$ counts the number of non-zero entries in a matrix. The intuitive formulation as an optimization problem would be

$$\hat{L} = \arg\min_{\mathrm{rk}(L)\le k} \|X - L\|_0 \qquad (20.13)$$

which, however, is impractical for several reasons. Besides the discrepancy between the data model and statistical properties of noisy real-world data the problem itself is NP-hard, i.e. no efficient solver is at hand. A practical way to approach this problem is to solve its convex relaxation, where the rank constraint is approached by a minimization of the nuclear norm and the zero norm is approximated with the Manhattan distance, resulting in the problem

$$(\hat{L}, \hat{S}) = \arg\min_{L,S} \|L\|_* + \lambda\|S\|_1 \quad \text{s.t. } X = L + S \qquad (20.14)$$

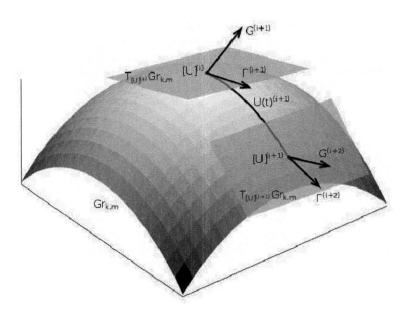

FIGURE 20.1　Illustration of a Gradient Descent method on the Grassmannian.

Many efficient solvers have been proposed for this problem [7] [28] [21] [8] [25], and they differ in the way the data is modelled, how the sparsity is enforced and how different optimization terms are combined and solved. For example, in some cases an additional Gaussian noise term N is considered ($X = L + S + N$, cf. [29]) or data is observed only partially ($\min \|X_\Omega - L_\Omega\|$, where Ω denotes the index set of the respective observations, cf. [27]). While most methods evaluate convex loss functions such as the ℓ_1-norm, it has also been shown how non-convex sparsifying functions can be used to penalize the residual $S = X - L$, cf. [14]. One possibility is to consider a smoothed version of the ℓ_p-quasi norm,

$$h_\mu : \mathbb{R}^{m \times n} \to \mathbb{R}^+, \quad S \mapsto \sum_{j=1}^{n} \sum_{i=1}^{m} \left(s_{ij}^2 + \mu \right)^{\frac{p}{2}}, \quad 0 < p < 1. \tag{20.15}$$

In the limit $p \to 0$ and $\mu \to 0$ this loss function resembles the zero-norm, which penalizes all non-zero elements regardless of their magnitude, and the introduced smoothing parameter $\mu > 0$ makes the function differentiable and, despite the zero-norm behavior, more tolerant against very small residuals such as measurement noise.

Convex rank minimization can be performed with Singular Value Thresholding techniques [7], and Augmented Lagrangian Multipliers [21] couple the rank minimization problem with the sparsity condition. Another way is to model the low-rank component as a product of two matrices, as proposed e.g. in [25], and to minimize a loss function of the residual in an alternating fashion. Convexity brings many advantages, most of all considering global convergence guarantees. However, non-convex methods might be preferable from a numerical point of view or because of their efficiency. Manifold optimization exhibits reliable performance despite its non-convex nature and, as outlined in Section 20.3, it is an elegant way of accounting for the rank constraint in equation (20.13). If a cost function is defined and optimized over the Grassmannian this guarantees an inherent upper bound on the rank, as the search space is explicitly constructed such as to generate low-rank matrices only. In [5] it has furthermore been shown how manifold optimization allows for tracking a subspace that changes slowly over time. For the application of background modeling this means that a subspace approximation is initially learned and represented by an element of the Grassmannian, and if the background should change over time a reasonable algorithm can follow these changes by moving slowly over the manifold in order to constantly adjust the low-rank representation to the particular conditions. Thus, in the following we will make a distinction between estimating a subspace for a whole data set $X^{m \times n}$ at once (batch mode) and analyzing a data set sample per sample (online subspace tracking). The latter is the special case of $n = 1$ for the sparsifying function defined in equation (20.15) and the subspace model of the past observation is updated using the current data sample $x \in \mathbb{R}^m$. Comparing the low-rank approximation of a single video frame

$$l = Uy \quad \text{with } U \in \text{St}_{k,m}, y \in \mathbb{R}^k \tag{20.16}$$

with the batch approximation in equation (20.3) one can see that in either case a single Stiefel element U represents the coordinate frame for the background model, while the length of the observed sequence corresponds to the number n of coordinate vectors y. The low-rank background results from the product of U and y (or respectively Y), which can be computed in an alternating manner. Considering a loss function l, e.g. h_μ from equation (20.15), the required cost functions and their gradients are

$$f_1 \colon \text{Gr}_{k,m} \to \mathbb{R}, \quad U \mapsto l(X - UY); \quad \nabla f_1 = -\nabla l(X - UY) \, Y^\top \tag{20.17}$$

$$f_2 \colon \mathbb{R}^{k \times n} \to \mathbb{R}, \quad Y \mapsto l(X - UY); \quad \nabla f_2 = -U^\top \nabla l(X - UY) \tag{20.18}$$

and their (common) gradients are denoted by ∇f_1 and ∇f_2, respectively. The minimization scheme for the batch mode is straightforward, as the following steps are repeated until convergence:

- Compute the gradient with respect to Y

- Minimize the loss function over Y to obtain $Y^{(i+1)}$

- Compute the gradient of the loss function with respect to U

- Project the gradient onto the tangent space to obtain the Riemannian gradient Γ

- Compute the geodesics $U(t)$ from the SVD of Γ

- Determine a suitable step size and obtain the next iterate $U^{(i+1)}$

The formal procedure is listed in Algorithm 20.1.

Algorithm 20.1 - Robust subspace estimation (batch mode)

Require: Video sequence $X^{m \times n}$
Ensure: Low-rank background sequence $L^{m \times n}$
  1: Initialize $U^{(0)}$, $Y^{(0)}$
  2: **repeat**
  3:    $\nabla f_2^{(i+1)} = \nabla f_2(U^{(i)}, Y^{(i)})$

  4:    $Y^{(i+1)} = \arg\min_Y f_2$, using $\nabla f_2^{(i+1)}$

  5:    $\nabla f_1^{(i+1)} = \nabla f_1(U^{(i)}, Y^{(i+1)})$

  6:    $\Gamma^{(i+1)} = \pi(\nabla f_1^{(i+1)}) = (I - U^{(i)}U^{(i)\top})\nabla f_1^{(i+1)}$

  7:    $U(t)^{(i+1)} = (U^{(i)}V\cos(\Sigma t) + \Theta\sin(\Sigma t))V^\top, \quad \Theta\Sigma V^\top = \mathrm{svd}(\Gamma^{(i+1)})$

  8:    $t^{(i+1)} = \arg\min_t f_1(U(t)^{(i+1)}, Y^{(i+1)}) \quad$ s.t. $t_{min} < t < t_{max}$

  9:    $U^{(i+1)} = U(t^{(i+1)})^{(i)}$
 10: **until** converged
 11: $L = U^{(i+1)}Y^{(i+1)}$

Subspace tracking algorithms require an initial subspace estimate $U^{(0)}$. The approach taken in GRASTA is to have a batch initialization phase, in which frames are sampled uniformly from a short video sequence over several rounds. By sampling uniformly, foreground objects are rarely ever at the same location in consecutive frames and will thus not be incorporated into the background model. The pROST algorithm on the other hand uses additional weighting factors $w_{ij} \in \mathbb{R}^+$ for each pixel in the loss function (20.15). Their purpose is to reduce the loss incurred by likely foreground pixels, which allows the method to avoid a batch initialization phase. The weighting for a particular pixel is set to a small positive value if it was labeled a foreground pixel in the last frame, whereas the weight is set to one if the pixel was in the background before. Thus the integration of foreground objects into the background model can be avoided and it becomes possible to run the algorithm entirely online.

The optimization over Y is an ordinary minimization problem of a smooth cost function in Euclidean space. Several methods for solving this problem are conceivable. In [15] an ADMM scheme is proposed, while Conjugate Gradient (CG) methods [17] are used in [14] and for pROST [24]. Advantages of CG methods are their simple and efficient implementation and their applicability to non-convex optimization problems. With a line-search method and appropriate conditions for determining a reasonable step size (e.g. the *Armijo rule*) fast and reliable convergence behavior is observed in practice. In each iteration of Algorithm 20.1 the search direction is updated using the updated gradient of the cost function and an appropriate update rule for the search direction (e.g. the *Hestenes-Stiefel* rule). Then an appropriate step size is determined and one optimization step is performed before the algorithm moves on to the optimization of U.

In an online setting the crucial question is how changes in the background model can be tracked and how subsequent estimates $U^{(i)}$ over time are related to the corresponding input frames $x^{(i)}$. Clearly, minimizing the loss function $l(x^{(i+1)} - Uy^{(i+1)})$ of a current frame $x^{(i+1)}$ over the whole Grassmannian would not be sensible, as this would lead to overfitting the background model to the content of that particular frame. This could destroy the temporal consistency between successive background estimates $l^{(i)}$ and the rank of a concatenated background sequence L would be undefined. One possible way to avoid this is to compute the (Riemannian) gradient over a certain window of frames and to update the gradient with each new incoming frame while controlling the adaptation rate with a forgetting factor [14]. In [5] it has been shown that the gradient can actually be computed from the current frame only and continuity in the subspace tracking process is achieved through taking only a small step on the Grassmannian.

We will now show how the second approach can reduce the computational complexity. The cost functions in the online settings and the respective gradients are

$$g_1 \colon \operatorname{Gr}_{k,m} \to \mathbb{R}, \quad U \mapsto l(x - Uy); \qquad \nabla g_1 = -\nabla l(x - Uy)\, y^\top \tag{20.19}$$

$$g_2 \colon \mathbb{R}^k \to \mathbb{R}, \quad y \mapsto l(x - Uy); \qquad \nabla g_2 = -U^\top \nabla l(x - Uy). \tag{20.20}$$

Here the loss function of the residual $l(x - Uy)$ is a vector-valued function, and also ∇g_2, the gradient with respect to y is a vector. The derivative ∇g_1 with respect to U is still a matrix of dimension $m \times k$ in the online setting, but it has a rank of one. This property holds for any differentiable loss function $l(x - Uy)$ and reduces the cost for computing the geodesics of the Grassmannian (cf. equation (20.11)) tremendously. As outlined in [5], the singular value decomposition of the Riemannian gradient Γ can easily be computed from the product of the gradient ∇g_1 projected onto the tangent space:

$$\Gamma = \pi(\nabla g_1) = \pi\left(-\nabla l(x - Uy)\, y^\top\right) =: \theta \sigma_1 v^\top \tag{20.21}$$

with the one non-zero singular value

$$\sigma_1 = \|\pi\left(\nabla l(x - Uy)\right)\|_2 \|y\|_2 \tag{20.22}$$

and the left and right singular vectors

$$\theta = \frac{-\pi\left(\nabla l(x - Uy)\right)}{\|\pi\left(\nabla l(x - Uy)\right)\|_2} \quad \text{and} \quad v = \frac{y}{\|y\|_2}. \tag{20.23}$$

Thus, the cost of computing the SVD of a full m-by-k matrix is avoided and the subspace update becomes feasible for realtime processing. The online subspace tracking scheme is listed in Algorithm 20.2.

Algorithm 20.2 - Robust subspace tracking (online mode)

1: **while** video data available **do**
2:      Vectorized video frame $x^{(i+1)} \in \mathbb{R}^m$, previous subspace estimate $U^{(i)} \in \operatorname{St}_{k,m}$
3:      Low-rank background frame $l^{(i+1)}$

4:      $y^{(i+1)} = \arg\min_y g_2$

5:      $\nabla g_1^{(i+1)} = \nabla g_1(U^{(i)}, y^{(i+1)})$

6:      $\sigma^{(i+1)} = \|(I - U^{(i)} U^{(i)\,\top}) \nabla l(x - U^{(i)} y^{(i+1)})\|_2 \|y^{(i+1)}\|_2$

7:      $\theta^{(i+1)} = \dfrac{(U^{(i)} U^{(i)\,\top} - I) \nabla l(x - U^{(i)} y^{(i+1)})}{\|(I - U^{(i)} U^{(i)\,\top}) \nabla l(x - U^{(i)} y^{(i+1)})\|_2}$

8:      $v^{(i+1)} = \dfrac{y^{(i+1)}}{\|y^{(i+1)}\|_2}$

9:      $U(t)^{(i+1)} = (U^{(i)} \cos(\sigma^{(i+1)}t) + \theta^{(i+1)} \sin(\sigma^{(i+1)}t)) v^{(i+1)\,\top}$

10:     $t^{(i+1)} = \arg\min_t f_1(U(t)^{(i+1)}, Y^{(i+1)}) \quad$ s.t. $t_{min} < t < t_{max}$

11:     $U^{(i+1)} = U(t^{(i+1)})^{(i)}$
12:     $l^{(i+1)} = U^{(i+1)} y^{(i+1)}$

13: **end while**

The minimization over y in Algorithm 20.2 is a much less complex optimization problem than the respective minimization over Y in the batch scenario (Algorithm 20.1). A few iterations are usually sufficient for the CG algorithm to converge, and instead of alternating between y and U all iteration steps of the CG minimization are performed before the gradient descent step on the Grassmannian is undertaken.

20.5   Implementation

A mathematical and algorithmic description of a computer vision algorithm is seldomly detailed enough to deduce a feasible implementation. Besides developing an algorithm a major part of engineering lies in practical issues like pre- and post-processing and in tweaking the computational efficiency. After having outlined a possible algorithmic framework we now want to discuss these topics in the context of background modeling and foreground detection. We will do this at the example of the pROST algorithm [24], which is a realtime robust online subspace tracking algorithm implemented for *NVIDIA* GPUs. Figure 20.2 depicts its processing pipeline with the order of processing steps and their distribution between CPU and GPU. A certain number of preprocessing steps is performed on each incoming video frame. Then the data is transferred to the GPU, which is used to solve the respective optimization problems and to perform a thresholding operation that outputs a binary segmentation mask. After post-processing this segmentation mask it is used to update the foreground weights of the cost function and it is transferred to the host as the result of the segmentation process. Although the detailed discussion of the building blocks in this section is specific for the particular implementation of pROST, many of the aspects generalize to other methods for realtime foreground-background-segmentation.

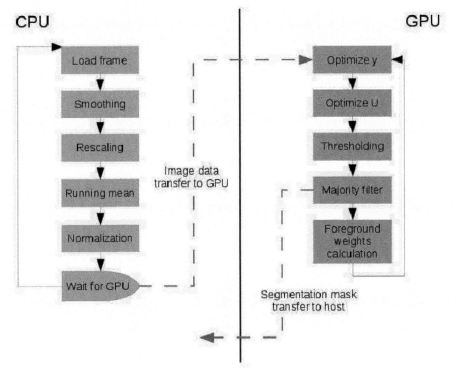

FIGURE 20.2   Processing flow for pROST. Preprocessing for frame $(i + 1)$ is done on the CPU during the optimization and post-processing steps for iteration i on the GPU.

20.5.1 GPU Programming for Foreground Detection and Background Modeling

The recent advancements of graphic processing units (GPUs) have made high-performance computing available to the masses, and GPUs have become the computational device of choice for many computer vision applications. The main reason is that computer vision algorithms on the one hand are often computationally demanding, but can be parallelized on the other hand. Moreover, GPUs can handle images efficiently and provide functionality that can be used to greatly improve performance and to accelerate program development. Apart from the large number of processing units, the most important component of a GPU for image processing applications is the *texture memory*.

Texture memory is a means of storage that can be accessed very efficiently by CUDA kernels and it provides functionality like interpolation and clamping (i.e. boundary management) in hardware. Moreover, its cache is optimized for two-dimensional spatial locality. Usually caches are optimized for a linear memory access pattern, i.e. it is assumed that programs access memory locations in a consecutive order and an access to the main memory causes a whole 1D *cache line* to be loaded to the cache. In contrast to this, image processing and computer vision applications usually need to access memory in a 2D spatial pattern. Filtering operations for example compute a weighted sum of the intensities of neighboring pixels around a target pixel in an image. The texture memory cache is organized in such a way that a pixel in the image is accessed by loading its 2D spatial neighborhood into the cache. Consecutive access to this neighborhood can then be serviced from the faster cache. This makes texture memory uniquely suited for computer vision and image processing. In Section 20.5.3 we show how to access texture memory for implementing a weighted majority filter.

Due to the specialized hardware the implementation of algorithms on the GPU using CUDA is not trivial and the CUDA programming model is somewhat unintuitive to many programmers. However, there are a number of generic building blocks for efficiently implementing computer vision algorithms on a GPU in CUDA, and quite reasonable performance can be achieved by using high-level libraries that shield the programmer from the complexities of GPU programming.

Robust PCA-based image processing algorithms use a number of linear algebra operations like matrix-matrix multiplication and matrix-vector multiplication, and some algorithms require matrix factorizations like the SVD or the QR decomposition. As they are mostly generic and might be part of many different algorithms, it is reasonable to implement them in dedicated libraries and make them available for reuse. This has been recognized a long time ago and has lead to the definition of a generic linear algebra interface called *BLAS* (Basic Linear Algebra Subprograms) [20] and implementations for the CPU [11]. LAPACK [3] is a software package containing highly optimized solvers for linear systems of equations, but also decompositions like SVD or QR. Virtually all linear algebra operations defined in BLAS are amenable to parallelization. Consequently, *NVIDIA* has published a freely available BLAS implementation for its GPUs, which is called *CUBLAS* [23]. Given an appropriate problem scale, CUBLAS can lead to a considerable speed-up compared to its CPU counterparts. The matrices encountered in PCA based background modeling are usually tall and skinny matrices, i.e. they have many more rows than columns. For these types of matrices an efficient scheme for computing the SVD of a matrix G is to compute the reduced QR decomposition $G = QR$ in a first step, then the SVD $R = USV$, and finally the SVD of G is obtained via $G = QR = \tilde{U}SV$ with $\tilde{U} = QU$. [4] have presented a communication avoiding QR decomposition scheme for computing the QR decomposition of tall matrices on the GPU in a highly-efficient way. Commercial libraries like *CULA* [18] and open-source libraries like *MAGMA* [26] provide functions for computing the SVD and

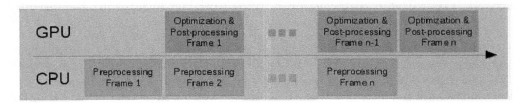

FIGURE 20.3   Interleaving of the processing steps on CPU and GPU for consecutive frames.

QR decomposition on the GPU. The CULA library takes approximately 42 ms for computing the SVD of a 57600×15 matrix (i.e. a 15-dimensional subspace background model at a resolution of 160×120) on a GTX-660 GPU. For a 5-dimensional subspace it takes approximately 30 ms.

20.5.2   Processing Steps

The pROST pipeline is designed for heterogeneous computing environments with a fast CPU and an auxiliary GPU. Wherever possible, the computational workload is distributed between CPU and GPU in order to achieve maximum performance. Although it would be possible in general to execute preprocessing steps like image acquisition, scaling and noise removal efficiently on the GPU, all these tasks are delegated to the CPU instead. The reason for this is that operations of this kind are relatively cheap compared to the more expensive subspace tracking (Optimization of U) and model fitting (Optimization of y) tasks of the pROST pipeline, which are computed on the GPU. Since the optimization steps of the pROST pipeline are dependent on the preprocessing steps, high-level parallelism can only be achieved by executing the preprocessing steps of Frame $(i + 1)$ in parallel to the optimization steps of Frame i. This is illustrated in Figure 20.3.

Scaling and Smoothing

Robust PCA-based methods have difficulties with backgrounds like waving trees or water fountains that exhibit chaotic and unpredictable dynamics, because these backgrounds do not lie on a low-dimensional linear subspace. Also, the higher the subspace dimension and the video resolution the more computationally demanding the algorithms become. Even with optimized code it is currently not possible to process videos in VGA resolution at real-time frame rates on off-the-shelf hardware. Both problems can be circumvented by preprocessing the video frames appropriately. The first step in the pROST pipeline performs Gaussian smoothing to reduce noise and to prevent artifacts before the image is downsampled to a lower resolution. This decreases the time required for subsequent steps and also mitigates problems arising from chaotic dynamics in the background. Of course, reducing the resolution of an image also makes small structures vanish and edges of foreground objects are less precise on an upsampled segmentation mask. As the resources on a GPU are limited the tradeoff between spatial and spectral resolution is obvious, and the permissible amount of granularity in the segmentation result depends on the application. One way to alleviate the artifacts arising from scaling and smoothing is to employ an edge aware smoothing filter on the segmentation mask. Possible choices are *Guided Image Filtering* [16], which can be computed efficiently on a CPU, or *Domain Transform Image Filtering* [10], which allows fast GPU implementations. In order to show how a GPU can be used for edge aware filtering, we use a simple weighted majority filter, which allows a concise implementation. Its purpose and its implementation are discussed in Section 20.5.3.

Mean Subtraction

A preprocessing step which we found extremely beneficial is to estimate the running mean (i.e. the average over a window of frames) and to subtract it from each incoming video frame. It has been observed that this eases the task of modeling backgrounds with repetitive dynamics.

RGB Channels

Many computer vision algorithms only work with gray scale images in order to reduce computational complexity. As this discards information it can only be beneficial to process color images. Still, the question is how to use the color information practically. The approach taken in pROST is to learn one background model for each of the channels and to search at each pixel for the maximum residual among the three color channels during the thresholding operation. In Section 20.5.3 we outline how to implement this step on a GPU via atomic intrinsics.

20.5.3 Implementation Spotlights

In this section we want to discuss particular steps in the pROST pipeline and list a selection of CUDA kernels, which have been selected to explain the most important programming techniques by example. The core elements of the pipeline are of course the model fitting step (*Optimize y*) and the subspace tracking step (*Optimize U*). For the model fitting step a nonlinear Conjugate Gradient method is used. The most expensive aspect of this step is to compute the gradient and to evaluate the loss function in a line search method. While this can be done on the GPU, the computation of conjugate search directions is best done on the CPU. The tracking step can be implemented very efficiently on a GPU. We will discuss and compare two possible implementations, one using only CUBLAS functionality and one using a CUDA kernel for the retraction in Section 20.5.3.

Line Search along Geodesics on the *Grassmannian*

Projection onto the tangent space
    As explained in Section 20.3 doing a line search on the *Grassmannian* requires a projection of the ambient space gradient onto the tangent space at the current subspace estimate. Using BLAS, this can be achieved with only two calls to the $GEMM$ (General Matrix Multiply) routine. The $GEMM(transposeA, transposeB, \alpha, A, B, \beta, C)$ function performs the operation $C \leftarrow \alpha * A * B + \beta * C$ and the parameters *transposeA* and *transposeB* indicate if the matrices A and B are to be transposed. The actual signature of the CUBLAS routine contains arguments for the matrix sizes and memory pitches which we omit here for clarity. Algorithm 20.5.3 shows how GEMM operations can be used to implement the projection of the ambient space gradient G (equation (20.17) or equation (20.19)) onto the tangent space at the current subspace estimate U.

Algorithm 20.3 - Projection of the ambient space gradient G onto the tangent space at U

1: allocate buffer B
2: GEMM('t', 'n', 1.0, U, G, 0.0, B)
3: GEMM('n', 'n', -1.0, U, B, 1.0, G)

Line 2 of Algorithm 20.5.3 computes $B \leftarrow U^T G$ and in line 3 $G \leftarrow G - UB$ is calculated, which completes the projection. In the online case the rank-one gradient can be computed as $G = gy^T$ with $g \in \mathbb{R}^{n \times 1}$ and $y \in \mathbb{R}^{1 \times k}$, and due to associativity of matrix multiplication, only g has to be projected onto the tangent space.

Efficient implementation of the exponential map retraction on the *Grassmannian*

In batch mode the exponential map retraction onto the *Grassmannian* requires the calculation of the singular value decomposition of the search direction. This can be done efficiently using one of the SVD implementations of the libraries mentioned in Section 20.5.1. After the SVD of the descent direction has been obtained, the next step is to do a line search along the geodesics of the *Grassmannian*. The form of the retraction implemented here is the form given in [15],

$$U(t) = U + ((cos(t\sigma) - 1)Uv - sin(t\sigma)\theta)v^T. \tag{20.24}$$

It is possible to calculate the retraction using the functionality provided in CUBLAS and this will already provide a speed-up over a CPU implementation. Algorithm 20.4 shows the respective pseudo code. The SCAL(α, A) (SCALing) operation computes $A \leftarrow \alpha A$ and the $AXPY(\alpha, X, Y)$ (Alpha X Plus Y) operation computes $Y \leftarrow \alpha * X + Y$. Notice that three calls to CUBLAS functions are required and that a buffer for an intermediate result has to be allocated. This leads to additional reads and writes from and to global memory. The product $B = Uv$, which is independent of the step size, can be computed before the step size is determined in a line-search loop.

Algorithm 20.4 - Calculation of online retraction using GEMM operations.

1: $U' \leftarrow$ U;
2: $\theta' \leftarrow \theta$;
3: SCAL(-sin($t\sigma$), θ');
4: AXPY(cos($t\sigma$) - 1, B, θ');
5: GEMM('n', 't', 1.0, θ', v, 1.0, U');

In the first and second line of Algorithm 20.4 a copy of the current subspace estimate and the left singular vector is made. In line 3 the operation $\theta' \leftarrow -sin(t\sigma)\theta'$ is computed, in line 4 the operation $\theta' \leftarrow \theta' + (cos(t\sigma) - 1)B$ and in line 5 the operation $U' \leftarrow \theta'v + U'$. A specialized CUDA implementation can be even more efficient, especially in the case of the rank-one retraction. The reason is that we avoid the use of an intermediary buffer and multiple function calls which each read and write intermediary results from and to global device memory. In Listing 20.1 we outline the body of a CUDA kernel for the online retraction.

Listing 20.1   Body of online retraction

```
1   //Calculate  global  row  index
2   int  row = blockIdx.x*blockDim.x + threadIdx.x;
3   if(row>=rows)
4           return;
5   //compute  Uprime+=(B*(cos(step)-1) - U*sin(step))*v^T
6   const  float  g=(B[row]*(cos(sigma*t)-1)
7           - U[row]*sin(sigma*t));
8   for(int  c=0;c<cols;c++)
9   {
10          Uprime[c*pitchOut + row] +=g*v[c];
11  }
```

CUDA kernels are executed on a multi-dimensional grid. The grid is partitioned in blocks of threads which are running concurrently. When designing a CUDA kernel a crucial question is how to structure this grid. In the case of the retraction, our strategy for implementing the kernel is to allocate a grid of $n \times 1$ threads, i.e. one thread per row of $B = Uv$ and to do the multiplication with $(cos(t\sigma) - 1)B - sin(t\sigma)U)v^T$ in a loop in the kernel. An alternative and maybe more intuitive approach would be to omit the loop in the kernel and to allocate one thread per element of U'. However, the grid size has to be a multiple of the warp size, which is currently 32, and the number of columns of U is rarely ever a multiple of 32. Thus, many of the threads in such a grid would be idle. Allocating a one dimensional grid is therefore more efficient in practice, especially because the number of rows of B is usually large enough to utilize all of the GPU's resources. In line 2 of Listing 20.1 the row of B is calculated to which the current thread is assigned. Since the number of threads allocated in total can be greater than the number of actual rows of B, a safeguard is imposed in line 3, which returns if the the thread is allocated to a row index greater or equal to the number of rows in B. Lines 6 to 10 contain the loop in which $(cos(t\sigma) - 1)B - sin(t\sigma)U)v^T$ is calculated for a single row of the matrix U'. This approach is approximately 1.6 times faster than a CUBLAS implementation for a subspace dimension of 15 and a frame size of 160×120. Even greater speed-ups over the CUBLAS implementation (≈ 1.9) can be attained in the case where the gradient has a rank larger than one.

Postprocessing

Thresholding

For three-channel images a pixel i is classified as foreground if the difference between the reconstructed background of either of the channels is large enough, i.e. if

$$\max\{|x_{R,i} - b_{R,i}|, |x_{G,i} - b_{G,i}|, |x_{B,i} - b_{B,i}|\} \geq \delta. \tag{20.25}$$

Here $x_{.,i}$ and $b_{.,i}$ denote the pixel values of the channels of foreground and background and δ is a threshold. Of course this can be implemented on the GPU in many different ways and most of them lead to similar performance. We will focus on the implementation for RGB images with interleaved packing order, as this is the format commonly used in many image processing libraries like OpenCV. Interleaved packing order means, that the RGB values for a pixel are stored after another in memory, i.e. [RGB RGB RGB], instead of storing the channels separately, as is the default in MATLAB. The thresholding operation is a type of reduction in which three consecutive elements of an array have to be reduced to one value in another array. Our implementation strategy is to spawn one thread per element in the image array and to use atomic intrinsics to synchronize the reduction process. The body of the kernel is listed in Listing 20.2.

Listing 20.2 Body of thresholding operation

```
1  //Check if index is withing array bounds
2  if (x<size)
3  {
4  //calculate linear index into segmentation
5  //mask
6          const int maskIdx =
7          ((x/3)/maskCols)*maskPitch + (x/3)%maskCols;
8          //thresholding check
9          const float d =
10         (fabsf((img[x]-background[x]))>threshold);
11         //atomic reduction
12         atomicCAS(mask+maskIdx,0,d);
13 }
```

(a) input data and segmentation ground truth

(b) 3x3 majority filter

(c) 31x31 majority filter

(d)   31x31 weighted majority filter, $\sigma = 0.1$

FIGURE 20.4   Comparison of post-processing steps at the example of frame # 1069 of *traffic* sequence from *changedetection.net* [12]

The grid allocated for this purpose is of size $n \times 1$. The segmentation mask pointed to by `mask` in Listing 20.2 is a pitched 2D float array. The memory pitch is stored in `maskPitch` and the number of columns in `maskCols`. The array is pitched, i.e. every `maskCols*sizeof(float)` bytes are aligned with a certain memory pitch. Therefore, in lines 8 and 9 the linear index into the `mask` array is computed from the linear index into the image array to which the thread is allocated to. In lines 11 and 12 the thresholding operation is computed. If the absolute difference between image and estimated background is greater than some threshold, then the variable `d` is set to 1.0. Notice that three threads have to modify the same memory location, based on a conditional expression depending on the content of the memory location. If not handled properly this would cause a race condition. CUDA provides so-called *atomic intrinsics*, which can be used to set values of shared or global memory without interference from other threads, thus preventing race conditions. The *atomicCAS* operation used here checks if the element at memory location `mask+maskIdx` is set to 0 and if so, sets it to `d`.

Weighted majority filter

In order to remove small holes and spurious pixels a median filter is commonly applied to the segmentation mask, which needs to sort at least half of the elements in a window. For binary segmentation masks, however, the filter only needs to count the number of set pixels and set the center pixel if the majority of the pixels in the window are set. A straight-forward extension of this majority filter is to count every pixel in the window with a different weight. This weight can depend for example on the color or intensity similarity and on the distance to the center pixel. If such a weighting is used the filter will tend to respect object boundaries and over-smoothing can be avoided as seen in Figure 20.4. This can also overcome camouflaging effects in some cases, but may also lead to the removal of small foreground structures.

We define the weighted majority filter as:

$$S(x,y) \leftarrow \sum_{-r \le k \le r} \sum_{-r \le l \le r} w(x,k,y,l)S(x+k,y+k) \tag{20.26}$$

$$\ge \sum_{-r \le k \le r} \sum_{-r \le l \le r} w(x,k,y,l)(1 - S(x+k,y+k))$$

where $S \in \{0,1\}^{m \times n}$ is the segmentation mask and $w(x,k,y,l) = \exp\left(-\frac{\|I(x,y)-I(x+k,y+l)\|_2^2}{2\sigma^2}\right)$ is a Gaussian weight function, which measures the intensity or color difference between the pixels.

An efficient implementation would make use of an efficient data structure for high-dimensional filtering like the permutohedral lattice [2]. Here we will restrict the discussion to a simpler implementation, which is sufficient to illustrate several important programming techniques. The implementation strategy is to store the image as well as the segmentation mask in texture memory. A two dimensional grid of thread blocks is allocated and each thread processes one element of the mask. To efficiently handle three-channel float textures with CUDA texture memory, the image has to be copied to a 4 channel representation first.

The complete kernel is listed in listing 20.3. In line 6 and 7, the 2D index into the image and mask are calculated from the thread and block indices. Since more threads may be allocated than elements are contained in the image in line 8 the indices are checked against the image extents. In line 12 the RGB values of the pixel that the thread is allocated to are read from the texture memory. Notice that texture memory requires center pixel addressing, i.e. pixel (x, y) is accessed with $(x + 0.5, y + 0.5)$. Lines 14 to 28 contain the filter loop. In line 19 the RGB value of pixel $(x + j, x + i)$ is loaded and in lines 20 to 24 its weight is calculated. In line 25 the weighted sum of foreground pixels in the window is updated and in the following line the total sum of all weights in the window. Finally, in line 30 the current pixel in the output mask is set to 1 if the weighted foreground sum `fgWeight` is larger than half of the sum of all weights. On a GTX 660 GPU a 31x31 weighted majority filter implemented as in Listing 20.3 takes about 9 ms to execute for a 160×120 image.

Listing 20.3 Body of weighted majority filter

```
1   //scale of weighting function
2   #define sigma 0.1
3   //radius of filter
4   #define r 7
5   //global memory coordinates
6   const int x = blockIdx.x * blockDim.x + threadIdx.x;
7   const int y = blockIdx.y * blockDim.y + threadIdx.y;
8   if(x<cols && y<rows)
9   {
10    float fgWeight=0;
11    float wsum=0;
12    float4 centerPixelC = tex2D(imageTex, x + 0.5, y + 0.5);
13    for(int i=-r;i<=r;i++)
14    {
15      for(int j=-r;j<=r;j++)
16      {
17      //skip if outside of mask
18      if(x+j>=cols || x+j <0 || y+i>=rows || y+i<0) continue;
19      float4 otherPixelC = tex2D(imageTex, x+j+0.5, y+i+0.5);
20      const float rd=(centerPixelC.x-otherPixelC.x);//Difference in R channel
21      const float gd=(centerPixelC.y-otherPixelC.y);//Difference in G channel
22      const float bd=(centerPixelC.z-otherPixelC.z);//Difference in B channel
23      const float w= __expf(-0.5*(rd*rd+gd*gd+bd*bd)/(sigma*sigma));
24      //look up mask element in texture memory and compute weighted sums
25      fgWeight+=w*tex2D(maskTex,x+j,y+i);
26      wsum+=w; //sum over all weights
27      }
28    }
29    //set mask element
30    maskOut[y*pitch + x]=fgWeight>wsum/2.0?1.0:0.0;
31  }
```

20.6   Conclusion

In this chapter the required concepts for Robust PCA and Robust Subspace Tracking on manifolds have been recalled and it has been shown how to implement such an algorithm on a GPU. We have outlined how an efficient interplay between CPU and GPU can be achieved and several spotlights of the implementation have shed some light on the use of the hardware and some popular software libraries. Apart from the specialities of manifold optimization many discussed building blocks are also valid for any other algorithm for realtime background-foreground-separation.

References

1. P. Absil, R. Mahony, and R. Sepulchre. *Optimization Algorithms on Matrix Manifolds.* Princeton University Press, Princeton, NJ, 2008.
2. A. Adams, J. Baek, and M. Davis. Fast high-dimensional filtering using the permutohedral lattice. *Computer Graphics Forum*, 29(2):753–762, 2010.
3. E. Anderson, Z. Bai, C. Bischof, S. Blackford, J. Demmel, J. Dongarra, J. Du Croz, A. Greenbaum, S. Hammarling, A. McKenney, and D. Sorensen. *LAPACK Users' Guide.* Society for Industrial and Applied Mathematics, Philadelphia, PA, third edition, 1999.
4. M. Anderson, G. Ballard, J. Demmel, and K. Keutzer. Communication-Avoiding QR Decomposition for GPUs. In *Proceedings of the 2011 IEEE International Parallel & Distributed Processing Symposium*, IPDPS '11, pages 48–58, Washington, DC, USA, 2011. IEEE Computer Society.
5. L. Balzano, R. Nowak, and B. Recht. Online identification and tracking of subspaces from highly incomplete information. In *Allerton Conference on Communication, Control, and Computing*, pages 704–711, 2010.
6. T. Bouwmans. Subspace learning for background modeling: A survey. *Recent Patents on Computer Science*, 2(3):223–234, 2009.
7. J. Cai, E. J. Candes, and Z. Shen. A singular value thresholding algorithm for matrix completion. *SIAM J. on Optimization*, 20:1956–1982, 2010.
8. E. Candes, X. Li, Y. Ma, and J. Wright. Robust principal component analysis? *Journal of ACM*, 58(3):1–37, 2011.
9. A. Edelman, T. Arias, and S. Smith. The geometry of algorithms with orthogonality constraints. *SIAM Journal on Matrix Analysis and Applications*, 20(2):303–353, 1998.
10. E. Gastal and M. Oliveira. Domain transform for edge-aware image and video processing. *ACM Transactions on Graphics (TOG)*, 30(4):69, 2011.
11. K. Goto and R. Van De Geijn. High-performance implementation of the level-3 BLAS. *ACM Transactions on Mathematical Software (TOMS)*, 35(1):4, 2008.
12. N. Goyette, P. Jodoin, F. Porikli, J. Konrad, and P. Ishwar. Changedetection.net: A new change detection benchmark dataset. In *Computer Vision and Pattern Recognition Workshops*, pages 1–8, June 2012.
13. C. Guyon, T. Bouwmans, and E. Zahzah. Robust Principal Component Analysis for Background Subtraction: Systematic Evaluation and Comparative Analysis. In *Principal Component Analysis*, chapter 12, pages 223–238. INTECH, 2012.
14. C. Hage and M. Kleinsteuber. Robust PCA and subspace tracking from incomplete observations using ℓ_0-surrogates. *Computational Statistics (accepted for publication)*, 2013.
15. J. He, L. Balzano, and A. Szlam. Incremental gradient on the Grassmannian for online foreground and background separation in subsampled video. In *Computer Vision and Pattern Recognition*, pages 1568–1575, 2012.
16. K. He, J. Sun, and X. Tang. Guided image filtering. In *International Conference on*

Computer Vision, ECCV 2010, pages 1–14. Springer, 2010.

17. R. Hestenes and E. Stiefel. Methods of Conjugate Gradients for Solving Linear Systems. *Journal of Research of the National Bureau of Standards*, 49:409–436, December 1952.

18. J. Humphrey, D. Price, K. Spagnoli, L. Paolini, and E. Kelmelis. CULA: hybrid GPU accelerated linear algebra routines. In *SPIE Defense and Security Symposium (DSS)*, 2010.

19. K. Hüper, M. Kleinsteuber, and H. Shen. Averaging complex subspaces via a Karcher mean approach. *Signal Process.*, 93(2):459–467, February 2013.

20. L. Lawson, R. Hanson, D. Kincaid, and T. Krogh. Basic linear algebra subprograms for Fortran usage. *ACM Transactions on Mathematical Software (TOMS)*, 5(3):308–323, 1979.

21. Z. Lin, M. Chen, L. Wu, and Y. Ma. The augmented lagrange multiplier method for exact recovery of corrupted low-rank matrices. *Arxiv preprint arXiv:1009.5055*, 2010.

22. J. Nocedal and S. Wright. Numerical optimization, series in operations research and financial engineering. *Springer, New York*, 2006.

23. NVIDIA. CUBLAS library. *NVIDIA Corporation, Santa Clara, California*, 15, 2008.

24. F. Seidel, C. Hage, and M. Kleinsteuber. pROST : A Smoothed ℓ_p-norm Robust Online Subspace Tracking Method for Realtime Background Subtraction in Video. *Technical Report, Technische Universität München*, 2013.

25. Y. Shen, Z. Wen, and Y. Zhang. Augmented Lagrangian alternating direction method for matrix separation based on low-rank factorization. *Optimization Methods and Software*, 2011.

26. P. Tomov, S. Du, and R. Nath. MAGMA: Matrix Algebra on GPU and Multicore Architectures, 2011.

27. A. Waters, A. Sankaranarayanan, and R. Baraniuk. SpaRCS: Recovering Low-Rank and Sparse Matrices from Compressive Measurements. In *Advances in Neural Information Processing Systems*, 2011.

28. J. Wright, A. Ganesh, S. Rao, Y. Peng, and Y. Ma. Robust Principal Component Analysis: Exact Recovery of Corrupted Low-Rank Matrices via Convex Optimization. *Advances in Neural Information Processing Systems*, pages 2080–2088, 2009.

29. T. Zhou and D. Tao. GoDec: Randomized low-rank and sparse matrix decomposition in noisy case. In *International Conference on Machine Learning*, pages 33–40, 2011.

21

Background Subtraction on Embedded Hardware

Enrique J. Fernandez-Sanchez
CITIC, Univ. of Granada, Spain

Rafael Rodriguez-Gomez
CITIC, Univ. of Granada, Spain

Javier Diaz
CITIC, Univ. of Granada, Spain

Eduardo Ros
CITIC, Univ. of Granada, Spain

21.1 Introduction

Extracting background from a video sequence is the very first stage in many applications related to video surveillance: vehicle traffic control, intruders' detection, suspicious objects, etc. The most usual approach to segment moving objects is known as background subtraction. This technique consists of building a reference model which represents the static background of the scene during a certain period of time. Multiple factors and events may affect the scene, making this first background subtraction a non-trivial task: sudden and gradual illumination changes, presence of shadows, or background repetitive movements (such as waving trees), among many others.

Although there are many different algorithms to perform background subtraction, in general it is a computationally expensive task. Whilst this could be solved by using powerful processing units in a centralized network, in multi-camera video analytics systems that would not be a scalable solution. When the number of cameras increases, having a centralized surveillance network leads to issues related to the network topology, due to high bandwidth requirements. For that reason, our interest lies in the deployment of background subtraction techniques on embedded hardware, which allows for distributed and decentralized camera networks. Thus, the use of FPGAs [1,3,11,17] and DSPs [8,18] is justified by requirements of scalability, size and low power consumption which are key features that other technologies are not able to achieve. There are also other real-time approaches using GPUs [4,19], but these ones are not suitable yet for embedded devices. In the case of embedded systems, commodity processor implementations are not usually utilized although latest devices, such as Intel Atom, could soon address this market.

Oliveira et al. [17] presented an FPGA implementation for the Horprasert algorithm, although the throughput reached by this approach was fairly low. Jiang et al. [11] presented a compression scheme for Mixture of Gaussians model [27] which allows reaching a high frame rate. However, this approach is not explained in detail, and no results are shown about accuracy or power consumption. Appiah et al. [1] proposed an implementation based on a simplified MOG, which offers fairly good throughput and acceptable accuracy. Bravo et al. [3] recently proposed an FPGA implementation based on Principal Component Analysis (PCA), getting a good throughput and specifying the resource consumption. Nevertheless, there is no data about accuracy with a standardized dataset. Carr [4] and Pham et al. [19] presented GPU implementations based on MOG, with very different results in performance. Despite the approach described in [19] it has a higher frame rate than any of the other mentioned hardware implementations, being a GPU implementation is an impediment for embedded systems and low power constraints. Strictly speaking, there are new GPU families oriented to embedded devices that could solve that problem. However, the performance of these families is considerably lower than the ones of GPUs used on standard PCs or laptops.

We have proposed two FPGA architectures based on the method described by Horprasert [7, 12], with the extension that allows for shadow detection [20], and also the Codebook method described by Kim et al. [13].

The first method [7] has been selected since it requires less memory to store the model while keeping fairly good accuracy, hence being more suitable for implementation in low cost FPGAs [17]. This algorithm builds a static background model, which means that the model is obtained at an initial training phase. There are other methods which build dynamic background models [13,27], which can adapt themselves to changes in the scene. The main difference between these models, as far as required hardware resources are concerned, is that the latter have much higher memory consumption requiring external memory with an important bandwidth. Furthermore, the shadow detection capabilities increase the accuracy of the object shape detection, which helps to achieve a better object classification and reduces the errors due to shadows artifacts.

The second implementation [22] is based on the algorithm proposed by Kim et al. [13]. This algorithm has been classified by several authors [9, 29] as a good trade-off between accuracy and efficiency. This has motivated the choice of this model to be implemented even though it is a much more complex model than the previous hardware implementations available in the literature. This model requires much higher memory than the one proposed by Horprasert et al. [7], requiring thus both optimization and simplifications as well as external memory with an important bandwidth.

The rest of this chapter is organized as follows. In Sections 21.2 and 21.3 we explain the two background algorithms studied and the architectures developed on reconfigurable hardware. Section 21.4 shows the experimental results obtained by the proposed architectures. Finally, we summarize in Section 21.5 the conclusions of this work.

21.2 Horprasert's Algorithm

As previously mentioned, our first implementation [23] is based on the algorithm proposed by Horprasert et al. [7]. This algorithm basically obtains a reference image to model the background of the scene so that it can perform automatic threshold selection, subtraction operation and, finally, pixel-wise classification.

Horprasert et al. [7] proposed a color model in the three-dimensional RGB color space. The main idea is that chromaticity and brightness are perceived separately, since humans tend to assign a constant color to an object despite illumination changes.

In this section we describe the original model, explaining the background modeling and the subtraction operation and classification steps. We also give an overview of the architecture we have developed on FPGA and the modifications made to the model towards a hardware-friendly implementation.

21.2.1 Background Model

In order to build a reference image which represents the background, a number N of images will be used, whose color space is given in RGB. Each pixel i from the image is modeled by a 4-tuple $< E_i, S_i, a_i, b_i >$, where each element is defined as follows:

- E_i the expected color value, defined as $E_i = [\mu_R(i), \mu_G(i), \mu_B(i)]$, being $\mu_R(i), \mu_G(i), \mu_B(i)$ the arithmetic means of each color channel for pixel i.

- S_i the value of the color standard deviation for each channel, defined as $S_i = [\sigma_R(i), \sigma_G(i), \sigma_B(i)]$.

- a_i the variation of the brightness distortion, computed as the root mean square (RMS) of the brightness distortion α_i, given by equation (21.1).

- b_i the variation of chromaticity distortion, the RMS of the chromaticity distortion CD_i, which is described in equation (21.2).

$$\alpha_i = \frac{\frac{I_R(i)\mu_R(i)}{\sigma_R^2(i)} + \frac{I_G(i)\mu_G(i)}{\sigma_G^2(i)} + \frac{I_B(i)\mu_B(i)}{\sigma_B^2(i)}}{\left(\frac{\mu_R(i)}{\sigma_R(i)}\right)^2 + \left(\frac{\mu_G(i)}{\sigma_G(i)}\right)^2 + \left(\frac{\mu_B(i)}{\sigma_B(i)}\right)^2} \tag{21.1}$$

$$CD_i = \sqrt{\left(\frac{I_R(i) - \alpha_i\mu_R(i)}{\sigma_R(i)}\right)^2 + \left(\frac{I_G(i) - \alpha_i\mu_G(i)}{\sigma_G(i)}\right)^2 + \left(\frac{I_B(i) - \alpha_i\mu_B(i)}{\sigma_B(i)}\right)^2} \tag{21.2}$$

More detailed information about how the background model is built can be found in [7].

21.2.2 Subtraction Operation and Classification

In this stage, the difference between the background model and the current image is evaluated. This difference consists of two components: brightness distortion α_i and chromaticity distortion CD_i. In order to use a single threshold for all pixels, it is necessary to normalize α_i and CD_i as indicated in expressions (21.3) and (21.4):

$$\hat{\alpha}_i = \frac{\alpha_i - 1}{a_i} \tag{21.3}$$

$$\hat{CD}_i = \frac{CD_i}{b_i} \tag{21.4}$$

After the normalization of brightness and chromaticity distortions, the given pixel can be classified into one of four categories, according to Equation (21.5), that is the analytical representation derived from the model presented in Figure 21.1.

$$C(i) = \begin{cases} \text{Foreground:} & \hat{CD}_i > \tau_{CD}, \text{ or } \hat{\alpha}_i < \tau_{\alpha lo} \text{ else} \\ \text{Background:} & \hat{\alpha}_i > \tau_{\alpha_1}, \text{ and } \hat{\alpha}_i \geq \tau_{\alpha2} \text{ else} \\ \text{Shadowed background:} & \hat{\alpha}_i < 0, \text{ else} \\ \text{Highlighted background:} & \text{otherwise} \end{cases} \tag{21.5}$$

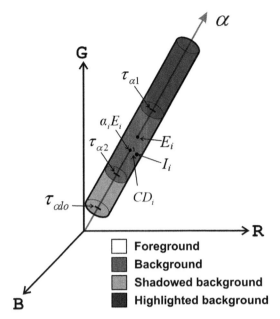

FIGURE 21.1 Graphic representation of the model used to classify the pixels in the categories. This model is oriented to shadow and highlights detection, taking into account chromaticity lines as well as brightness changes.

The thresholds τ_{CD}, $\tau_{\alpha1}$, $\tau_{\alpha2}$ are automatically selected from the information obtained during the training stage, as explained in [7]. $\tau_{\alpha lo}$ is a lower bound used to avoid misclassification of dark pixels.

This approach to shadow detection is considered a Statistical non-parametric (SNP) method [20], which means that the approach uses probabilistic functions to describe the class membership, and it is non-parametric since the thresholds are automatically determined by means of a statistical learning procedure.

This model has been selected for its implementation on reconfigurable hardware. For that reason, several modifications have been made in order to reduce the hardware complexity of the architecture. These simplifications towards a hardware friendly model generate some degradation on the original model's quality that will be evaluated in subsequent sections. These modifications are described in Section 21.2.3.

21.2.3   Model Simplifications

The foreground/background segmentation is executed by a hardware module with an independent access to the memory, where the current image and the background model are stored. Considerable reduction of the hardware complexity of the architecture is achieved through precalculating and storing several constants during the training stage and avoiding division operations by substituting them for multiplications, which require less hardware resources. In the case of brightness distortion α_i, these constants are computed according to Equation (21.6):

$$A_i = \left(\frac{\mu_R(i)}{\sigma_R(i)}\right)^2 + \left(\frac{\mu_G(i)}{\sigma_G(i)}\right)^2 + \left(\frac{\mu_B(i)}{\sigma_B(i)}\right)^2$$

$$B_i = \left(\frac{\mu_R(i)}{A_i\sigma_R^2(i)}\right)$$

$$C_i = \left(\frac{\mu_G(i)}{A_i\sigma_G^2(i)}\right) \tag{21.6}$$

$$D_i = \left(\frac{\mu_B(i)}{A_i\sigma_B^2(i)}\right)$$

The brightness distortion α_i will remain as in Equation (21.7), making use of the constants B_i, C_i, D_i.

$$\alpha_i = B_iI_R(i) + C_iI_G(i) + D_iI_B(i) \tag{21.7}$$

In order to remove the divisions in the computation of the chromaticity distortion CD_i, we store $(S_i)^{-1}$, $(a_i)^{-1}$ and $(b_i)^{-1}$ instead of S_i, a_i and b_i. Besides, the training stage is done with $N = 128$ images to facilitate the computation of the mean, standard deviation and root mean square, avoiding divisions. Previously, the model had a 4-tuple for each pixel, composed by $< E_i, S_i, a_i, b_i >$, whereas now a 7-tuple will have to be stored$< E_i, B_i, C_i, D_i, (S_i)^{-1}, (a_i)^{-1}, (b_i)^{-1} >$. The hardware complexity has been reduced considerably, but at the cost of increasing memory consumption, since now we also have to store the constants B_i, C_i and D_i.

TABLE 21.1 Bit-width of each variable taking part in the calculation of brightness and chromaticity distortions.

Variable	Bits
$< B_i, C_i, D_i >$	[18 8]
$< E_i, (S_i)^{-1} >$	[8 8]
$< (a_i)^{-1}, (b_i)^{-1} >$	[8 10]
α_i	[28 8]
CD_i	[18 8]
$\acute{\alpha}_i$	[36 10]
$\acute{C}D_i$	[26 10]

The software implementation has been developed using double floating-point representation. This allows reaching a higher degree of accuracy at the expense of a worse performance on embedded devices. In order to develop a hardware implementation on FPGA with constrained resources, a fixed-point representation is usually adopted since it adjusts itself better to the type of available resources, although a detailed study is required in order to optimize the trade-off between accuracy and hardware consumption. It is important to take into account that an insufficient number of bits may lead to inaccurate results with high quantification noise. On the contrary, the use of too many bits can increase the hardware resources consumption, making the system implementation on a moderate cost FPGA unfeasible. In order to determine the appropriate number of bits for the fractional part of the variables, we have measured the error between the results obtained with different bit-width configurations and the floating-point representation. This comparison has been performed by using the Wallflower dataset [28]. The selected bit-width configuration is shown in Table 21.1, where the first value represents the integer part and the second value represents the fractional part. The bit-width values have been determined as the minimum values of the fractional part for which the quantization error is approximately stable. More details about this study can be found in [23].

21.2.4 Hardware Architecture

An optimized hardware architecture has been developed using novel ideas that allow for a high degree of algorithm tuning for optimized digital hardware implementation. They can be summarized as follows:

1. Hardware/software co-design. The use of a mixed hardware/software architecture allows us to share its resources to solve many algorithm stages as the ones related with communication, system initialization, basic control, system debugging, *etc.* It is not necessary to develop custom datapaths for these stages because no critical real-time restrictions are imposed to them. This permits to reduce hardware resources, to extend the system flexibility and to significantly reduce development time.

2. Superscalar and pipelined architecture. Multiple functional units run in parallel to take full advantage of the intrinsic algorithm parallelism. The whole implementation has been carefully pipelined in order to increase the throughput. These strategies allow us to keep the pixel-rate very high and to achieve a significant performance.

3. Adaptable fixed-point arithmetics. The bit-width of the different processing stages has been tuned according to the accuracy requirements of each processing element. This approach is very different from the one used on many DSPs or digital hardware implementations that has a basic bit-width for all the processing stages. Our approach allows us to keep resources always tuned to the required accuracy at the cost of increasing the complexity of the system design. Hopefully, the use of high-level description languages helps to reduce the development time and make this option feasible with acceptable design time.

4. Proper utilization of the right level of abstraction for description of the different algorithm modules. The processing stages mainly require a DSP-based design flow which is well described using high-level description languages (as provided by ImpulseC [10]) whilst basic controllers such as the ones required for memory interfaces or low level communications are better described in RTL (for instance using VHDL or Verilog). In addition, sequential operations such as the ones required to build communication packages are well described by software code. Our implementation uses different descriptions based on the previous considerations. This enables to get the maximum output out of each description level in terms of performance or development time.

As it could be understood from the previous items, the advantage of our implementation relies on the combination of the latest design methodologies, seldom addressed together in the same design. The drawback of this novel approach is that it requires a high degree of competences at many different design levels, languages and tools. Nevertheless, the advantage is that it allows highly optimized designs that completely fit the target application.

In order to address this implementation, we have used EDK (Embedded Developer's Kit) of Xilinx Inc. [30]. The EDK environment facilitates the design of complex and completely modular SoC architectures able to support embedded microprocessors (MicroBlaze, PowerPC,...), peripheral and memory controllers (Ethernet, DDR2, ZBT,...), and interconnecting buses (PLB, NPI, MCH, FSL,...), whilst IP cores for specific processing can be designed using HDL languages through the ISE tool. As a prototyping board we use the ViSmart4 video processing board from Seven Solutions [24].

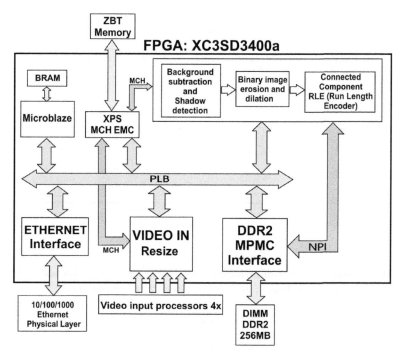

FIGURE 21.2 Scheme of the complete architecture and connections between modules, peripherals, memory and processor (©(2012) Springer).

This architecture consists of several modules and interconnect buses, as shown in Figure 21.2. Processing modules, peripherals and a Microblaze processor are connected to a PLB bus. The VIDEOIN module captures images from four independent analog inputs and stores them in a ZBT SSRAM external memory, through the MCH port of the XPS MCH EMC memory interface module. Through the PLB bus, Microblaze has access to: memory regions (ZBT or DDR2), configuration registers of the peripherals and the ethernet interface for data and image sending/receiving. The Background subtraction, shadow detection and blob detection (erosion, dilation and RLE) module performs an intensive processing on the pixels of each image in order to separate foreground and background, and then proceeds to the blob extraction of the different objects. This module uses ImpulseC [10] in order to develop a DSP-based design flow system. The Microblaze processor is programmed in C/C++ for initialization and communications tasks, and the rest of peripherals are described in VHDL. The MPMC module (DDR2 memory controller) offers an easy access to the external DDR2 memory, which stores the background model. This memory offers efficient high bandwidth access, thus providing a feasible use for applications requiring real-time processing.

Figure 21.3 shows a basic scheme of the proposed architecture for the IP core that performs the background subtraction, pixel classification and blob detection processing stages [2, 6]. This architecture consists of a pipelined structure divided into several basic stages which work in parallel. In addition, it is controlled by means of the embedded Microblaze processor. It is important to note that memory has a key role in the system performance and requires an efficient memory accessing scheme. This has motivated the use of high-performance multiport memory controllers (Xilinx MPMC for DDR2) as well as very specific and optimized memory ports (NPI).

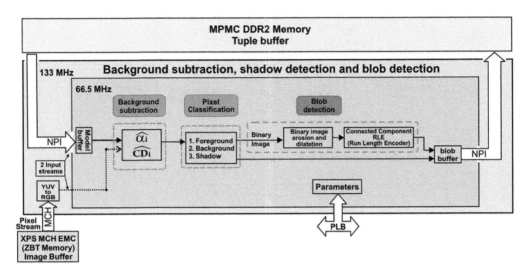

FIGURE 21.3 Simplified datapath architecture for background subtraction and blob detection core. The IP core can process streams from up to four cameras (©(2012) Springer).

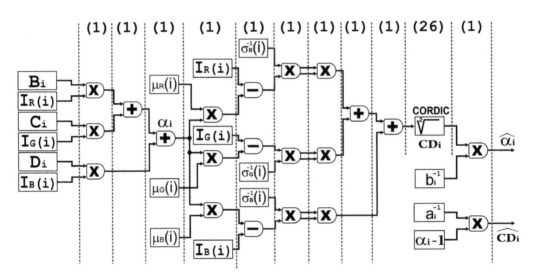

FIGURE 21.4 Fine grain pipelined datapath for background subtraction stage. The number of clock cycles is indicated on top. The different operations are indicated by X sign (multiplication) and $+$ sign (addition), registers by rectangles while routing paths are indicated by arrows (©(2012) Springer).

Since the background model is built only in the beginning, the construction is made by a software application running in the Microblaze processor. Data of the tuple $< E_i, B_i, C_i, D_i, (S_i)^{-1}, (a_i)^{-1}, (b_i)^{-1} >$ are computed in floating point notation by Microblaze, although they are stored in memory in fixed point to be used by the hardware module. The size of a fixed point tuple will determine the required memory space for the background model, in our case $< E_i$ (48 bits), B_i (26 bits), C_i (26 bits), D_i (26 bits), $(S_i)^{-1}$ (48 bits), $(a_i)^{-1}$ (18 bits), $(b_i)^{-1}$ (18bits)$>$ in total 210 bits. In order to maintain a background model for a 1024x1024 image, we will need 26.25 MB, which is affordable for current FPGA platforms. After the background model has been built, the background subtraction stage is executed by the IP core, designed with ImpulseC [10]. Figure 21.4 shows the fine-grain pipelined datapath for this hardware module.

After conducting background subtraction, the system generates a binary mask image in which 0 and 1 represent background and foreground respectively. Subsequently, morphological filtering and connected component stages are executed, making use of the architectures described by Herdberg et al. [6] and Appiah et al. [2].

21.3 Codebook Algorithm

The Codebook algorithm, as proposed by Kim et al. [13], is based on the construction of a background model adopting a quantization/clustering technique described by Kohonen [14] and Ripley [21]. The above-mentioned work shows that the background model for each pixel is given by a codebook consisting of one or more codewords. Codewords are data structures that contain information about pixel colors, color variances and information about how frequently each codeword is updated or accessed.

The different stages of the Codebook algorithm are described below:

21.3.1 Background Model

Given a set of N time steps (frames), a training sequence S is used for each pixel consisting of N RGB vectors. Each pixel has a different codebook, represented as $C = \{c_1, c_2, c_3, ...c_L\}$, consisting of L codewords, where L can be different for each pixel. Each codeword $c_i, i = 1...L$; consists of a RGB vector $v_i = (\bar{R}_i, \bar{G}_i, \bar{B}_i)$ and a 6-tuple $aux_i = \left\langle \check{I}_i, \hat{I}_i, f_i, \lambda_i, p_i, q_i \right\rangle$, described as follows:

- $v_i = (\bar{R}_i, \bar{G}_i, \bar{B}_i)$, average value of each color component.

- $\check{I}_i, \hat{I}_i$, min and max brightness, respectively, of all pixels assigned to codeword c_i.

- f_i, the frequency (number of frames) with which codeword c_i has been updated.

- λ_i, the *maximum negative run-length* (MNRL), defined as the longest interval of time during which codeword c_i has not been updated.

- p, q, the first and last updating access times of codeword c_i.

The detailed algorithm for codebook construction is given in Fig. 21.1.

Conditions (a) and (b) in step 2 (ii) must be evaluated in order to determine if a pixel $x_t = (R, G, B)$ matches the codeword c_m. These two conditions and λ parameter are explained in detail in [13]. Summarizing, we can say that the evaluation of color distortion basically consists of determining the distance between the color of an input pixel $x_t = (R, G, B)$ and $v_i = (\bar{R}_i, \bar{G}_i, \bar{B}_i)$ of codeword c_i, as indicated in (21.8).

$$
\begin{aligned}
\|x_t\|^2 &= R^2 + G^2 + B^2 \\
\|v_i\|^2 &= \bar{R}_i^{\,2} + \bar{G}_i^{\,2} + \bar{B}_i^{\,2} \\
\langle x_t, v_i \rangle^2 &= \left(\bar{R}_i R + \bar{G}_i G + \bar{B}_i B \right)^2
\end{aligned}
\tag{21.8}
$$

Color distortion δ is calculated as indicated in (21.9).

$$
p^2 = \|x_t\|^2 cos^2\theta = \frac{\langle x_t, v_i \rangle^2}{\|v_i\|^2}
\tag{21.9}
$$

$$
colordist\,(x_t, v_i) = \delta = \sqrt{\|x_t\|^2 - p^2}
$$

Algorithm 21.1 - Algorithm for Codebook Construction

$C \leftarrow \emptyset$

for $t = 1 \rightarrow N$ **do**

$\quad x_t = (R, G, B), I \leftarrow \sqrt{R^2 + G^2 + B^2}$

$\quad$ Find the codeword c_m in C matching to x_t based on two conditions:

$\quad$ (a) $colordist(x_t, v_m) \leq \epsilon_1$

$\quad$ (b) $brightness\left(I, \left\langle \check{I}_m, \hat{I}_m \right\rangle\right) = $ **true**

$\quad$ **if** $C = \emptyset$ **or** there is no match **then**

$\qquad$ {Create new codeword and add it to C}

$\qquad v_i = (R, G, B)$

$\qquad aux_i = \langle I, I, 1, t - 1, t, t \rangle$

$\quad$ **else**

$\qquad$ {Update matched codeword c_m consisting on $v_m = \left\langle \bar{R}_m, \bar{G}_m, \bar{B}_m \right\rangle$ and $aux_m = \left\langle \check{I}_i, \hat{I}_i, f_m, \lambda_m, p_m, q_m \right\rangle$ by setting}

$\qquad v_m = \left(\frac{f_m \bar{R}_m + R}{f_m + 1}, \frac{f_m \bar{G}_m + G}{f_m + 1}, \frac{f_m \bar{B}_m + B}{f_m + 1} \right)$

$\qquad aux_m = \left\langle min(I, \check{I}_m), max(I, \hat{I}_m), f_m + 1, max(\lambda_m, t + q_m), p_m, t \right\rangle$

$\quad$ **end if**

end for

Now, condition (b) evaluates how brightness change of $x_t = (R, G, B)$ lies within $[I_{low}, I_{hi}]$ range for codeword c_i. In this way, we allow the brightness change to vary in a certain range, as defined in (21.10).

$$I_{low} = \alpha \hat{I} \qquad (21.10)$$

$$I_{hi} = min\left\{ \beta \hat{I}, \frac{\check{I}}{\alpha} \right\}$$

Typically, α is in the interval $[0.4, 0.8]$, and β is in the interval $[1.1, 1.5]$. The brightness function is defined in (21.11).

$$brightness\left(I, \left\langle \check{I}, \hat{I} \right\rangle\right) = \begin{cases} true & if\, I_{low} \leq \|x_t\| \leq I_{hi} \\ false & otherwise \end{cases} \qquad (21.11)$$

The set of codebooks obtained from the previous step may include some moving foreground objects as well as noise. In order to obtain the true background model, it is necessary to separate the codewords containing foreground objects from the true background codewords. This true background includes both static pixels and background pixels with quasi-periodic movements (for instance waving trees in outdoor scenarios). The background model M obtained from the initial set of codebooks C is given in (21.12).

$$M = \{ c_m | c_m \in C \wedge \lambda_m \leq T_M \} \qquad (21.12)$$

λ_M (MNRL) is defined as the maximum interval of time that the codeword has not been updated during the training period. T_M is the time threshold set equal to half the number of training frames, $N/2$. Thus, codewords having a large λ_m (larger than T_M) will be eliminated from the corresponding codebook.

21.3.2 Foreground Detection

Extracting the foreground from the current image is straightforward once we have obtained the background model M. The algorithm performing this task is detailed in Fig. 21.2.

Algorithm 21.2 - Algorithm for Background Subtraction

$x_t = (R, G, B), I \leftarrow \sqrt{R^2 + G^2 + B^2}$

For all codewords in M, find the codeword c_m matching to x_t based on two conditions:

(a) $colordist(x_t, v_m) \leq \epsilon_2$

(b) $brightness\left(I, \left\langle \check{I}_m, \hat{I}_m \right\rangle\right) = \textbf{true}$

Update matched codeword as in algorithm 1

$BGS(x) = \begin{cases} \text{foreground} & \text{if there is no match} \\ \text{background} & \text{otherwise} \end{cases}$

21.3.3 Background Model Update

Kim et al. [13] propose a layered modeling and detection scheme which updates the background model M. This can be done by defining an additional model H, called *cache* [13,25], where the new codewords are stored. Three new parameters $(T_H, T_{add}, T_{delete})$ are also defined. The periodicity of a codeword h_i stored in cache H is filtered by T_H, as we did previously with T_M in the background model M in eq. (21.12). The codewords h_i remaining in cache H for a time interval larger than T_{add} are added to the background model M. Codewords of M not accessed for a period of time (T_{delete}) will be deleted from the background model. The detailed procedure is given below:

Algorithm 21.3 - Algorithm for Background Model Update

After training, the background model M is obtained. Create a new model H as cache.

For an incoming pixel x, find a matching codeword in M. If found, update the codeword. Otherwise, find a matching codeword in H and update it. If not found, create a new codeword h and add it to H.

Filter out the cache codewords based on T_H:

$H \leftarrow H - \{h_i | h_i \in H \wedge \lambda_i > T_H\}$

Move the cache codewords staying for enough time to M:

$M \leftarrow M \cup \{h_i | h_i \in H \wedge h_i \text{ stays longer than } T_{add}\}$

Delete the codewords not accessed for a long time from M:

$M \leftarrow M - \{c_i | c_i \in M \wedge \lambda_i > T_{delete}\}$

Repeat process from Step 2.

21.3.4 Model Simplifications

The original model by Kim et al. [13] needs to be simplified in order to arrive at an affordable high-performance hardware system. The main stages suitable for simplification issues can be summarized as follows:

- Storage and memory management. The amount of memory required to store the codebooks may change because the total number of codewords may increase or decrease dynamically. Our system uses a DDR2 memory, which provides high bandwidth, but it needs regular access to reach the maximum performance, which complicates the dynamic management of the set of codebooks.

- Color distortion and brightness distortion computation. Equation (21.9) requires a square root and division operations, which are expensive on resources-constrained hardware devices.

- Model accuracy degradation due to fixed point arithmetic. Customized hardware systems normally use fixed-point data representation to reduce hardware resources utilization. However, this strategy requires careful analysis to avoid any degradation in accuracy.

In order to adapt the algorithm to the hardware implementation, allowing an easier use of the external memory containing the codewords, we have limited the number of codewords in each pixel. If we limit the number of codewords to 2 or 3 in sequences with non-static backgrounds (i.e. waving trees), our architecture performs similarly to unimodal models. After an in-depth analysis of accuracy and resources consumption, our choice of optimal number of codewords for a wide range of possible scenes is set to 5. With this simplification, it may happen that the memory space of certain codebooks is already full (maximum limit reached) when new codewords are created. Then, it is necessary to replace an existing codeword. If this is the case, the replaced codeword will be the one not having been accessed for the longest time period. This condition will be added in Alg. 1 (Fig. 21.1), when new codewords are created. A proper study of the optimal number of codewords and the effect over accuracy is shown in [22].

Regarding the computation of color and brigthness distortions, all the codewords c_m are compared with the incoming pixel $x_t = (R, G, B)$ in parallel, using a *colordist*(x_t, v_i) and *brightness* $\left(I, \langle \check{I}, hatI \rangle\right)$ block for each codeword c_m. Both division and square root operations implemented in these blocks represent a considerable consumption of hardware resources. Therefore, it is desirable to avoid these calculations in the FPGA implementation without significantly affecting the accuracy of the algorithm, using in some cases multipliers which are optimized with the embedded resources of the FPGA DSP48 for a Xilinx Spartan-3A DSP [30].

With respect to color distortion δ, we have implemented the modifications indicated in eq. (21.13).

$$\|v_i\|^2 p^2 = \langle x_t, v_i \rangle^2 \tag{21.13}$$
$$\|v_i\|^2 \delta^2 = \|v_i\|^2 \|x_t\|^2 - \|v_i\|^2 p^2 = \|v_i\|^2 \|x_t\|^2 - \langle x_t, v_i \rangle^2$$

Condition (a) in Alg. 2 (Fig. 21.2) will remain as eq. (21.14).

$$\|v_i\|^2 \|x_t\|^2 - \langle x_t, v_i \rangle^2 \leq \epsilon^2 \|v_i\|^2 \tag{21.14}$$

With respect to brightness, we have established values $\alpha = 0.5$ and $\beta = 1.25$, so that calculations in eq. (21.10) can be easily computed by means of bit shifts, as follows in eq. (21.15).

$$I_{low} = \alpha \hat{I} = \hat{I} >> 1 \tag{21.15}$$
$$I_{hi} = min\left\{\beta \hat{I}, \frac{\check{I}}{\alpha}\right\} = min\left\{\hat{I} + (\hat{I} >> 2), \check{I} << 1\right\}$$

In this way, we have reduced the consumption of hardware resources by avoiding the use of two multipliers and one division.

TABLE 21.2   Bit-width of each variable taking part in the calculation of colordist and brightness.

	Variable	Bits
	R_i, G_i, B_i	[8 0]
	$\bar{R}_i, \bar{G}_i, \bar{B}_i$	[8 5]
$\breve{I}, \hat{I}$	$\|x_t\|$	[9 3]
$\langle x_t, v_i \rangle^2$ $\|v_i\|^2 \|x_t\|^2 - \langle x_t, v_i \rangle^2$ $\|v_i\|^2 \|x_t\|^2$		[36 5]
$\|v_i\|^2$ $\dfrac{\|x_t\|^2}{\|x_t\|^2}$		[18 5]
	$\epsilon^2 \|v_i\|^2$	[26 5]

During the updating process of v_m, there is a division for each color component. In order to reduce the consumption of resources in the FPGA, we have approximated f to its nearest power of 2, so that shift operations can be used instead of divisions. The implementation details of this modification can be seen in Fig. 21.4.

Algorithm 21.4 - Simplified Algorithm for Updating v_m

---

  if $1 \leq f \leq 3$ **then**
    $sf = 2$
  else if $4 \leq f \leq 6$ **then**
    $sf = 4$
  else if $7 \leq f \leq 12$ **then**
    $sf = 8$
  else if $13 \leq f \leq 24$ **then**
    $sf = 16$
  else
    $sf = 32$
  end if
  $v_m = \left(\dfrac{sf \cdot \bar{R}_m - \bar{R}_m}{sf} + \dfrac{R}{sf}, \dfrac{sf \cdot \bar{G}_m - \bar{G}_m}{sf} + \dfrac{G}{sf}, \dfrac{sf \cdot \bar{B}_m - \bar{B}_m}{sf} + \dfrac{B}{sf} \right)$

The software implementation has been developed using double floating-point representation, which enables a high accuracy (at the expense of using high cost computing units with high resources consumption). For FPGA hardware implementation with constrained resources, a fixed-point data representation is usually adopted, as it is more suitable for the type of resources in FPGA devices. In addition, specific purpose architectures cannot afford floating point arithmetic when implementing long-datapath pipelined computing architectures that may have a large number of processing elements. In this case, a study has been performed using an analog method to the one used with Horprasert algorithm. We have measured the error between the results obtained with different bit-width configurations and the original floating-point approach. In order to evaluate the accuracy of each configuration, the Wallflower dataset [28] has been used.

Table 21.2 shows the main algorithm data structures (system registers) and the associated bit-width choices, fixed after the study performed in [22]. Along the pipeline datapath, each data structure has a different bit-width which is optimized to the type of performed operations (multiplications, additions, subtractions, etc...). This approach allows researchers to tune resources and accuracy of the system in a much finer way.

Once we have established the bit-width of the fractional part of $v_i = \left(\bar{R}_i, \bar{G}_i, \bar{B}_i \right)$ and $\left(\breve{I}, \hat{I} \right)$, we have also evaluated the degradation of our design combining all the modifications required to obtain a hardware-friendly model. These results are shown in Section 21.4.

21.3.5   Hardware Architecture

An optimized hardware architecture for the Codebook algorithm is described in this section, following the same ideas that motivated the development of the Horprasert architecture, summarized in Section 21.2.4.

The proposed architecture has been developed using EDK (Embedded Developer's Kit) and ISE Foundation of Xilinx Inc [30]. The EDK environment facilitates the design of complex and completely modular SoC architectures able to support embedded microprocessors (MicroBlaze, PowerPC...), peripheral and memory controllers (Ethernet, DDR2, ZBT...), interconnecting buses (PLB, NPI, MCH, FSL...) whilst IP cores for specific processing can be designed using HDL languages in the ISE tool. For a better understanding of the designed architecture using EDK, it is important to highlight the use of a ViSmart video processing board from Seven Solutions [24], including: two Xilinx XC3SD3400aFG676 FPGAs, two 256MB DDR2 DIMM memory modules, four independent analog video inputs, two gigabit ethernet connections, 485 connection, a 3G connection module, a 64 MB Flash memory, and two 1MBx36 bits ZBT memories. In our case, we have only used one of the FPGAs included in the ViSmart board. This architecture has already been described in Figure 21.2.

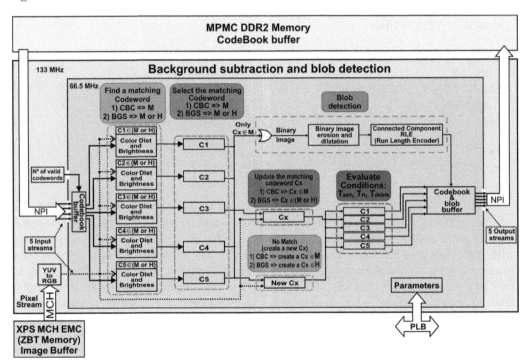

FIGURE 21.5   Simplified datapath architecture for background subtraction and blob detection core with 5 scalar units (C1, C2, C3, C4, C5).

The hardware description language that we have used to implement this IP core is ImpulseC [10], which allows us to work at a high level of abstraction, enabling the construction of a multi-stage pipelined architecture running in parallel. The parallel execution of these stages is the key point for the high performance obtained in our system. Figure 21.5 shows the proposed architecture for this IP core. Blob detection is performed in an analog way to the architecture designed for the Horprasert algorithm, which is explained in [2, 6]. It is important to remark that memory has a key role in the performance of the system and requires an efficient memory accessing scheme. The codebook model requires an intensive utilization of memory resources and poor system memory architectures drastically reduce the system performance. This has motivated the utilization of high performance multiport memory controllers (Xilinx MPMC for DDR2 and XPS_EMC_MCH for SSRAM) as well as very specific and optimized memory ports (NPI for DDR2 and MCH for SSRAM).

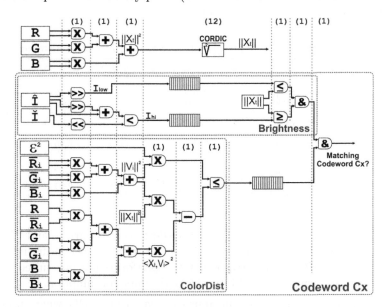

FIGURE 21.6 Fine-grain pipelined datapath for evaluating a codeword Cx. All these operations are computed using 18 pipelined stages. The number of clock cycles is indicated in brackets.

More details about this architecture are found in [22], including memory management and the memory map configuration for the storage of codebooks in order to enable an efficient use of memory.

21.4 Results

This section shows the results obtained by the proposed architectures. We have analyzed the performance of the algorithm in comparison with other hardware-oriented approaches, as well as an objective accuracy evaluation based on the Wallflower dataset [28].

21.4.1 Performance Comparison with Other Approaches

It is important to compare the current implementation with other approaches described in the literature (shown in Table 21.3). In order to evaluate the processing speed, we use the MegaPixels per Second measure (MPPS), multiplication of image size by frame rate. Table 21.3 summarizes the results obtained by previous works.

The algorithm by Horprasert [7] has been implemented by Oliveira et al. [17], reaching 30 fps with resolution 240×120, i.e. 0.824 MPPS. Our architecture for Horprasert presents a large improvement over this performance (32.8 fps, 1024x1024, i.e. 32.8 MPPS). The architecture implementing the Codebook approach [13] reaches 24 fps with resolution 1024x1024, obtaining much higher accuracy, as we show in Section 21.4.2.

Other authors have proposed different approaches, as in [11] and [1] based on MOG (Mixture of Gaussians). Jiang et al. [11] reach 38 fps with resolution 1024x1024 by applying a compression scheme, but with a considerable loss of accuracy. The system proposed by Appiah et al. [1] performs 145 fps for 768x576 frames, but obtaining worse results in terms of accuracy than our presented approach (Section 21.4.2). Bravo et al. [3] implement PCA algorithm on FPGA, which performs at maximum between 190 and 250 fps for 256x256 frames, i.e. between 11.875 and 15.625 MPPS. However, due to the lack of accuracy information, a deeper comparison is not possible.

TABLE 21.3   Comparison with other previous approaches described in the literature.

Approach	Method	Resolution	Frame Rate	MPPS	Processor Type
Oliveira et al. [17]	Horprasert	240x120	30	0.824	FPGA
Jiang et al. [11]	MOG	1024x1024	38	38	FPGA
Appiah et al. [1]	MOC	768x576	145	61.18	FPGA
Bravo et al. [3]	PCA	256x256	190-250	11.875-15.625	FPGA
Carr [4]	MOG	704x576	16.7	6.46	GPU
Pham et al. [19]	Zivkovic's Extended MOG	400x300	980	112.15	GPU
Ierodiaconou et al. [8]	MOG	352x288	21	2.03	DSP
Rodriguez-Gomez et al. [23]	Horprasert	1024x1024	32.8	32.8	FPGA
Rodriguez-Gomez et al. [22]	Codebook	1024x1024	24	24	FPGA

For standard GPU platforms, the approaches described in Carr [4] and Pham et al. [19] achieve high accuracy. Furthermore, Pham et al. [19] presents a high frame rate (980 fps, 400x300, i.e. 112.15 MPPS). Nevertheless, GPU platforms are not suited for embedded systems especially in terms of portability, size and power consumption.

Ierodiaconou et al. [8] proposes an approach based on MOG which uses TI DM642 DSP, reaching 2.03 MPPS with low power consumption (2.5 W, 0.8 MPPS per Watt). Our systems require 5.76 W for Horprasert and 5.13 W for Codebook, achieving 5.7 MPPS per Watt and 4.67 MPPS per Watt respectively, therefore our FPGA-based systems have better performance.

Finally, note that the processing performance is directly determined by the running clock frequency, and we are using low cost FPGAs. Therefore, migration to faster technologies as Virtex-7 FPGAs could directly represent an improvement of the system performance.

21.4.2   Evaluation of the Accuracy of the Background Model

Apart from the evaluation of system performance performed in the previous subsection, it is important to evaluate the quality of the segmentation obtained by the proposed architectures and to carry out a comparison with other background subtraction algorithms found in the literature. The algorithms which have been used for this comparison are MoG (Mixture of Gaussians) [27], a segmentation method based on Bayes decision rules [15], a simplification of MoG for FPGAs [1], and the original algorithms described in this chapter, that is, Horprasert [7] and Codebook [13]. These models have been selected since they represent different kinds of algorithms and they are among the most frequently used. The implementations of MoG and the Bayesian algorithm that have been used are versions from the OpenCV library, while the other approaches have been developed by ourselves from the information shown in their respective papers. In this Section, the methodology used to compare the different approaches is presented. We have evaluated the general performance as a background subtraction algorithm of the two proposed approaches and the behavior in the presence of shadows of the Horprasert implementation. The former has been performed by means of the dataset Wallflower [28], while the latter has been studied using the sequences presented in [20].

TABLE 21.4 Segmentation evaluation for dataset
Wallflower. The table shows F_1 results for six sequences.

Approach	B	C	FA	LS	TD	WT
Li et al. [15]	0,585	0,879	0,589	0,123	0,448	0,752
MOG [27]	0,401	0,447	0,387	0,506	0,793	0,859
Appiah et al. [1]	0,446	0,172	0,621	0,365	0,390	0,704
HOR Soft [7]	0,470	0,862	0,553	0,154	0,394	0,540
HOR Hard	0,462	0,858	0,540	0,146	0,345	0,533
CB Soft [13]	0,546	0,866	0,678	0,352	0,640	0,954
CB Hard	0,517	0,862	0,674	0,301	0,638	0,925

Background Subtraction Evaluation

The evaluation of the accuracy of the proposed architectures has been performed by using the dataset known as *Wallflower* [28]. The metrics used to evaluate background subtraction algorithms are based on True and False Positives and Negatives (TP,FP,TN,FN), and have been previously explained in this book. In this Section, we compare the accuracy of different approaches by means of F_1 measure.

However, the test *Moved Object* cannot be evaluated using these metrics, since the ground truth does not have any foreground pixel. As a result, there are not True Positives (TP) or False Negatives (FN), being impossible to compute *Precision* and thus F_1. For that reason, the performance in this test is only studied in a qualitative manner, by observing the resultant images (Figure 21.8).

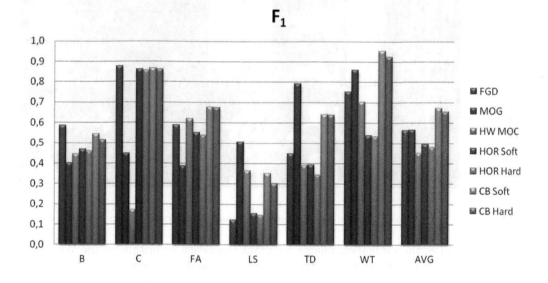

FIGURE 21.7 Overall performance evaluated using F_1. FGD is the Bayesian algorithm [15], MOG Mixture of Gaussians [27], HWMOC the FPGA implementation [1] for MOG, HOR Software the approach by Horprasert [7], CB Software [13] the original implementation of Codebook, and HOR Hardware and CB Hardware the proposed approaches.

Figure 21.7 shows the F_1 values obtained by each algorithm in each sequence, and the average accuracy in the entire dataset. From this figure, it can be seen that our proposed implementation for the algorithm by Horprasert [7] offers acceptable results, especially in comparison with the other hardware-oriented implementation [1]. Regarding the Codebook implementation, we can see that it gets very good results, being the best algorithm in two of the tests, and obtaining very high marks in the others. Concerning hardware approaches studied in this work, the accuracy decreases minimally so that they get slightly worse results than the original ones. Furthermore, the degradation caused by fixed-point limitations and additional modifications is fairly acceptable, obtaining by our architectures very good results, especially by the one based on the Codebook algorithm.

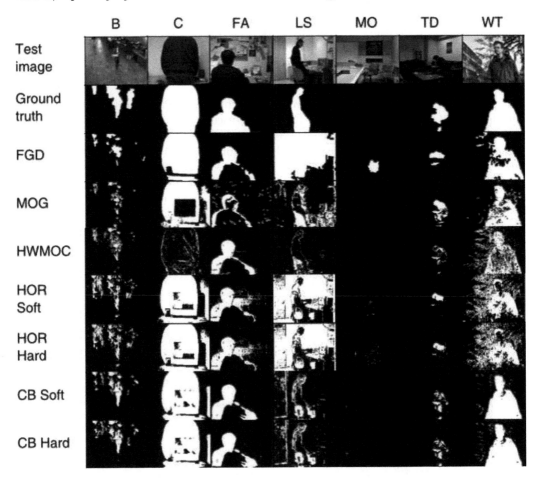

FIGURE 21.8   Wallflower evaluation frames, ground truth and resultant images from tested algorithms.

In Figure 21.8, the evaluation images resultant from each algorithm on the Wallflower dataset [28] are displayed, as well as the original frame and the ground truth to evaluate the quality of the segmentation. This comparison allows us to see in a qualitative manner the foreground/background segmentation quality of the different approaches. This subjective evaluation supports the conclusions of the quantitative analysis. Regarding the *Moved Object* test, the Horprasert algorithm, due to its static nature, produces more mistakes than MOG [27] or HW MOC [1]. In the Codebook implementation, only a small region of less than 10 pixels has been misclassified. In both cases, most of the misclassified pixels are sparse errors which could be removed by subsequent morphological filtering.

Shadow Detection Behavior

Despite the implementation proposed for the Codebook algorithm, it offers much higher accuracy than the Horprasert architecture, one of the benefits of the latter is that it is able to compute not only the background information of the scene but also information about the visible shadows. In order to evaluate these capabilities, we use the metrics proposed by Prati et al. [20], the *shadow detection accuracy* η and the *shadow discrimination accuracy* ξ.

TABLE 21.5   Shadow detection and discrimination accuracy.

Approach	η (%)	ξ (%)
SNP Soft	74.54	91.76
SNP Hard	71.14	88.13
SP	76.27	90.74
DNM1	78.61	90.29
DNM2	62.00	93.89

Table 21.5 shows the results obtained by the proposed architecture and the original software implementation in the *Intelligent Room* sequence as well as the results from other approaches found analyzed in [20]: SNP (Statistical Non Parametric, Horprasert, our approach), SP (Statistical Parametric) [16], DNM1 [5] and DNM2 [26] (Deterministic Non-Model-based approaches). Despite the degradation suffered by the hardware implementation (mainly due to the utilization of fixed-point arithmetics), it offers acceptable results, considering the greater complexity of the other approaches that makes them unsuitable for FPGAs with limited resources.

About the degradation between the software implementation and the proposed one, Figure 21.9 shows the results for the *Intelligent Room* sequence during a series of evaluation frames. In the worst of the scenarios, the loss of accuracy due to the restrictions of the hardware implementations is limited to 5%, offering fairly good results in both detection and discrimination metrics.

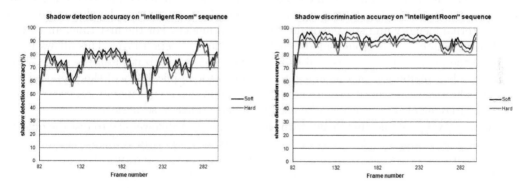

FIGURE 21.9   Shadow detection accuracy and discrimination accuracy of the original software model and the proposed approach, evaluated on Intelligent Room sequence.

This loss of accuracy can be easily seen in Figure 21.10. Images (b) and (c) show the segmentation obtained by the software and hardware implementation respectively. The degradation is noticeable in the higher dispersion of the shadow points in the hardware detection, whilst the shadow regions resultants from the software implementation are denser. However, the results are fairly accurate and the noise can be removed during the connected component stage, which was not included here in order to facilitate comparison with other approaches.

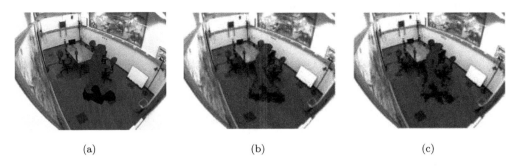

(a)   (b)   (c)

FIGURE 21.10   Frames 282 of the Intelligent Room sequence, (a) shows the ground truth for foreground and shadows, while (b) and (c) show the results of the software algorithm and hardware implementation respectively.

21.5   Conclusion

We have designed and analyzed two architectures to perform background subtraction in video sequences. Our first approach is based on the algorithm by Horprasert [7], a static model whose simplicity allows for a low cost FPGAs implementation and which has been extended to perform also shadow detection. The second one is based on the Codebook algorithm by Kim et al. [13], which allows us to model dynamic and multimodal backgrounds and is well known because of its robustness and good balance between accuracy and efficiency.

Two FPGA implementations of background subtraction algorithms have been developed which offer low degradation in comparison with the original ones. Since the hardware environment has limited resources, we have optimized memory access, low-level interfaces with external memory and storage of the background models. For the first time, an FPGA implementation of a background model includes shadow detection logic. This allows us to increase the model robustness as well as to improve object localization on the scene. This is a valuable contribution that significantly enhances the applicability of the proposed approach. The second approach, despite not having shadow detection capabilities, uses a much more accurate algorithm which can be used in a wider range of scenarios.

We have evaluated the approaches with the benchmark Wallflower [28] to test the quality of the segmentation. The first architecture offers good results (in terms of accuracy) in comparison with other hardware implementations found in the literature [1]. Furthermore, shadow detection performance has been analyzed by means of manually segmented video sequences [20]. The implementation is able to segment objects in complex sequences with resolution 1,024 × 1,024 at 32.8 fps (therefore 32.8 MPPS, Megapixels per Second) or from up to four cameras with less resolution. This represents a speed up over 35× with respect to the other approach [17] based on Horprasert. Concerning the cost of the system, the architecture has been designed for low cost FPGAs Spartan-3 by Xilinx, and it offers low power consumption (5.76 W). Therefore we achieve 5.7 MPPS per Watt. Our approach can be included in embedded systems, where parameters such as size and power are key elements that are not achievable by other approaches, such as commodity processors or GPU-based systems.

Regarding the second architecture, the results are excellent in comparison with other hardware-oriented approaches found in the literature, being similar to the ones offered by advanced software algorithms such as Bayesian and MoG. The implementation is able to segment objects in complex sequences with resolution 768x576 at 50 fps (therefore 21.1 MPPS) or at a higher speed with less resolution sources. Concerning the cost of the system, the architecture has been designed for low cost FPGAs Spartan-3 by Xilinx, with an estimated power consumption of 5.13 W (4.11 MPPS per Watt).

References

1. K. Appiah and A. Hunter. A single-chip FPGA implementation of real-time adaptive background model. In *IEEE International Conference on Field-Programmable Technology*, pages 95–102, December 2005.

2. K. Appiah, A. Hunter, P. Dickinson, and J. Owens. A run-length based connected component algorithm for FPGA implementation. In *International Conference on ICECE Technology*, pages 177–184, December 2008.

3. I. Bravo, M. Mazo, J. Lázaro, A. Gardel, P. Jiménez, and D. Pizarro. An intelligent architecture based on field programmable gate arrays designed to detect moving objects by using principal component analysis. *Sensors*, 10(10):9232–9251, 2010.

4. P. Carr. GPU accelerated multimodal background subtraction. In *Digital Image Computing: Techniques and Applications, DICTA 2008*, pages 279–286, December 2008.

5. R. Cucchiara, C. Grana, M. Piccardi, and A. Prati. Detecting objects, shadows and ghosts in video streams by exploiting color and motion information. In *International Conference on Image Analysis and Processing*, pages 360–365, September 2001.

6. H. Hedberg, F. Kristensen, P. Nilsson, and V. Owall. A low complexity architecture for binary image erosion and dilation using structuring element decomposition. In *IEEE International Symposium on Circuits and Systems, ISCAS 2005*, volume 4, pages 3431–3434, May 2005.

7. T. Horprasert, D. Harwood, and L. Davis. A statistical approach for real-time robust background subtraction and shadow detection. In *IEEE Frame-Rate Applications Workshop, Kerkyra, Greece*, 1999.

8. S. Ierodiaconou, N. Dahnoun, and L. Xu. Implementation and optimisation of a video object segmentation algorithm on an embedded DSP platform. In *The Institution of Engineering and Technology Conference on Crime and Security, 2006*, pages 432–437, June 2006.

9. A. Ilyas, M. Scuturici, and S. Miguet. Real time foreground-background segmentation using a modified codebook model. In *IEEE International Conference on Advanced Video and Signal Based Surveillance, 2009. AVSS 2009*, pages 454–459, September 2009.

10. Impulse accelerated technologies. `http://www.impulseaccelerated.com/`, 2011.

11. H. Jiang, H. Ardo, and V. Owall. Hardware accelerator design for video segmentation with multi-modal background modelling. In *International Symposium on Circuits and Systems, ISCAS 2005*, volume 2, pages 1142–1145, May 2005.

12. M. Karaman, L. Goldmann, D. Yu, and T. Sikora. Comparison of static background segmentation methods. In *SPIE 5960, 596069 (2005)*, volume 5960, 2005.

13. K. Kim, H. Thanarat, D. Harwood, and L. Davis. Real-time foreground-background segmentation using codebook model. *Special Issue on Video Object Processing, Real-Time Imaging*, 11(3):172–185, 2005.

14. T. Kohonen. Learning vector quantization. *Neural Networks*, 1:3 – 16, 1988.

15. L. Li, W. Huang, I. Gu, and Q. Tian. Foreground object detection from videos containing complex background. In *ACM International Conference on Multimedia, MULTIMEDIA 2003*, pages 2–10, New York, NY, USA, 2003.

16. I. Mikic, P. Cosman, G. Kogut, and M. Trivedi. Moving shadow and object detection in traffic scenes. *International Conference on Pattern Recognition*, 1:1321, 2000.

17. J. Oliveira, A. Printes, R. Freire, E. Melcher, and I. Silva. FPGA architecture for static background subtraction in real time. In *Annual Symposium on Integrated Circuits*

and Systems Design, SBCCI '06, pages 26–31, New York, NY, USA, 2006.

18. C. Peng. Video background/foreground detection implementation on TMS320C64/64x+DSP. Technical report, Texas Instruments, 2007.

19. V. Pham, P. Vo, V. Hung, and L. Bac. GPU implementation of extended Gaussian mixture model for background subtraction. In *International Conference on Computing and Communication Technologies, Research, Innovation, and Vision for the Future, RIVF 2010*, pages 1–4, November 2010.

20. A. Prati, I. Mikic, M. Trivedi, and R. Cucchiara. Detecting moving shadows: Algorithms and evaluation. *IEEE Transactions on Pattern Analysis and Machine Intelligence*, 25:918–923, 2003.

21. B. Ripley. *Pattern recognition and neural networks*. Cambridge: Cambridge University Press, 1996.

22. R. Rodriguez-Gomez, E. Fernandez-Sanchez, J. Diaz, and E. Ros. Codebook hardware implementation on FPGA for background subtraction. *Journal of Real-Time Image Processing*, pages 1–15, 2012.

23. R. Rodriguez-Gomez, E. Fernandez-Sanchez, J. Diaz, and E. Ros. FPGA implementation for real-time background subtraction based on Horprasert model. *Sensors*, 12(1):585–611, 2012.

24. Seven Solutions S.L. http://www.sevensols.com, 2013.

25. M. H. Sigari and M. Fathy. Real-time background modeling/subtraction using two-layer codebook model. In *Proceedings of the International MultiConference of Engineers and Computer Scientists*, volume 1, 2008.

26. J. Stander, R. Mech, and J. Ostermann. Detection of moving cast shadows for object segmentation. *IEEE Transactions on Multimedia*, 1(1):65–76, March 1999.

27. C. Stauffer and W. Grimson. Adaptive background mixture models for real-time tracking. In *IEEE International Conference on Computer Vision and Pattern Recognition, CVPR 1999*, volume 2, pages 637–663, 1999.

28. K. Toyama, J. Krumm, B. Brumitt, and B. Meyers. Wallflower: Principles and practice of background maintenance. In *IEEE International Conference on Computer Vision, ICCV 1999*, volume 1, page 255, Los Alamitos, CA, USA, 1999.

29. M. Wu and X. Peng. Spatio-temporal context for codebook-based dynamic background subtraction. *AEU - International Journal of Electronics and Communications*, 64(8):739–747, 2010.

30. Xilinx. http://www.xilinx.com, 2013.

22

Resource-efficient Salient Foreground Detection for Embedded Smart Cameras

Senem Velipasalar
Syracuse University, USA

Mauricio Casares
Syracuse University, USA

22.1 Introduction

The advances in VLSI technology and embedded computing have enabled the introduction of smart cameras, which are stand-alone units that combine sensing, processing and communication on a single embedded platform. With embedded smart cameras, it has now become viable to install many spatially-distributed cameras interconnected by wireless links. Yet, wireless and battery-powered, embedded smart camera networks introduce many additional challenges since they have very limited resources, such as energy, processing power, memory and bandwidth. Computer vision algorithms running on these camera boards should be lightweight and efficient. Considering the memory requirements of an algorithm and its portability to an embedded processor should be an integral part of the algorithm design in addition to the accuracy requirements.

Foreground detection is the first step in most of the object tracking applications. Existing methods for foreground detection can be broadly classified into two categories: temporal difference methods [1, 29], and background subtraction methods [12, 13, 15, 16, 19, 21, 23, 26, 32, 33, 35, 36]. Temporal difference methods subtract two consecutive frames and then apply a threshold to the output. These methods perform well when the background changes over time, however they cannot detect all the pixels of a moving object. Background subtraction methods build a model of the background and subtract this from the current image to detect objects in the scene. In order to adapt to changes in the environment, the background model is usually updated over time [12, 15, 16, 21, 26, 32, 33, 36]. The method described in Section 22.2[49] is a hybrid method, and it employs temporal difference to build the background model.

Horprasert et al. [17] obtain expected chromaticity by the arithmetic mean of the RGB values calculated over a number of background images. By using several thresholds, pixels are classified as foreground, background, shadow and highlighted background. Hidden Markov Models (HMMs) have been employed to represent the variations in the pixel intensity as discrete states [28] [34]. Nonparametric background models have been used in [11, 12, 31].

Oliver et al. [25] present an eigenbackground method, where images of a static background are collected, and PCA is employed to reduce the dimensionality of space.

Adaptive Mixture of Gaussians (MoG), introduced by Stauffer and Grimson [33], is one of the most commonly used background subtraction methods to model complex and non-static backgrounds. However, a few Gaussian distributions are usually not sufficient to accurately model backgrounds having fast variations. Methods have been introduced later that are based on Gaussian mixtures [2, 18, 22, 37]. Zivkovic [37] proposed an improved adaptive MoG model to constantly update the parameters of a Gaussian mixture and to simultaneously select the appropriate number of components for each pixel.

Kim et al. [20] proposed an algorithm for background modeling, where sample background values at each pixel are quantized into codebooks during training, which represent a compressed form of the background model. This algorithm performs well when background is non-static or there are lighting variations. However, its performance on different video sequences is dependent on the choice of multiple threshold values.

As mentioned above, many methods have been introduced for background subtraction and foreground object detection. However, most of these methods have been developed and tested on PCs instead of embedded smart cameras, and much less attention has been paid to the memory requirement and the portability of these algorithms to an embedded processor. Lighting variations and non-static backgrounds make the foreground detection problem even more challenging, since we are interested only in *salient* motion in tracking applications. We need to separate cases of uninteresting motion, such as swaying trees and water fountains, from the salient motion regions. The necessity of handling these challenging cases increases the algorithm complexity, and thus memory requirements. Most of the existing methods are too heavy-weight to implement on embedded platforms. Due to limited resources, most of the embedded smart camera systems [8, 27, 30] use relatively simple and sometimes less robust methods such as temporal difference and running average. In addition, traditional object tracking systems perform foreground object detection and object tracking at each frame independently and in a sequential manner.

In this chapter, we first present a resource-efficient background modeling and foreground detection algorithm* [6] that is highly robust against lighting variations and challenging non-static backgrounds including scenes with swaying trees, water fountains, strong wind and rain. Compared to many traditional methods, the memory requirement for the data saved for each pixel is very small in the proposed algorithm. For instance, Stauffer and Grimson [33] use multiple (three to five) Gaussian distributions per pixel to model non-static backgrounds. Kim et al. [20] form codewords for each pixel. Each codeword has nine entries, and on the average 6.5 codewords are needed for a pixel. The MoG method requires

---

[49]Reprinted from [6] ⓒ(2010) with permission from Elsevier.

23 to 32 bytes per pixel if three Gaussian distributions and one color channel are used. The codebook method requires 91 bytes on the average for one color channel. Whereas, in the proposed method, at most 6.25 bytes are needed per pixel. We provide a detailed comparison of the memory requirements in Section 22.2.6.

Moreover, the number of memory accesses and instructions are adaptive, and are decreased even more depending on the amount of activity in the scene and on a pixel's history. Each pixel is treated differently based on its history, and instead of requiring the same number of memory accesses and instructions for every pixel, we require less instructions for stable background pixels. The proposed method selectively updates the background model with an automatically adaptive rate, thus can adapt to rapid changes. Unlike many traditional methods treating each pixel individually, in the proposed method, information is obtained from neighboring pixels and incorporated into decision making, which increases accuracy and robustness. The results obtained with 10 challenging outdoor and indoor sequences are presented, and compared with the results of different state-of-the-art background subtraction methods. The Receiver Operation Characteristics (ROC) curves and memory comparison of different background subtraction methods are also provided. The experimental results demonstrate the success of the proposed lightweight salient foreground detection method in challenging situations such as scenes with water fountains, swaying trees, and strong wind and rain. Moreover, the proposed method has been implemented in its entirety on the microprocessor of an actual embedded smart camera, and the processing speed and the operating current of the camera board have been measured. To measure the current, we used a precise oscilloscope and a 1-ohm resistor configuration placed at the input of the supply source.

In this chapter, we also present a feedback method[50] [5] to increase the energy efficiency of the foreground object detection even further. Rather than following the traditional sequential approach in detection and tracking, and performing detection and tracking independently at each frame, the feedback method incorporates the information from the tracking stage into the foreground detection stage. This way, foreground detection is performed in smaller regions as opposed to the entire frame. The feedback method significantly reduces the processing time of a frame, and provides 48.7% decrease in the processing time when tracking one remote-controlled car. We take advantage of these savings by sending the microprocessor to idle state at the end of processing a frame without causing tracking failure. The feedback method also provides 10.44% savings in energy consumption compared to traditional sequential tracking when tracking one object. We present a detailed comparison of the feedback method and the sequential approach in terms of processing times and energy consumption.

22.2   Resource-efficient Background Modeling and Foreground Detection

The proposed algorithm employs a temporal difference method until a complete background model is built. It differentiates between salient and non-salient motion based on the history of a pixel's location, and by incorporating neighborhood information. At each frame, each pixel is classified either as a background or a foreground pixel, and its state is set to be 0 or 1, respectively. For a pixel at location (i, j), a counter $h(i, j)$ holds the number of changes in the state of this pixel during the last 100 frames, i.e. the counter $h(i, j)$ keeps

---

the number of times a pixel's state changes from 0 to 1 or vice versa. The stability of a pixel at location (i, j) is determined by this counter $h(i, j)$. The motivation is that the lower the value of $h(i, j)$, the more stable and reliable that location is, or vice versa. Until a complete background model is built, the state of a pixel is determined by using temporal difference.

The algorithm has an adaptive background model update rate. If a pixel location is determined to be consistently stable and very reliable, then the value of this pixel is incorporated to the background model with a higher weight. Instead of treating each pixel independently, information from neighboring pixels is used to differentiate between salient and non-salient motion, and in turn to classify a pixel as a foreground or background pixel. The details of the proposed algorithm will be explained by referring to the pseudo-code provided in Table 22.1.

22.2.1 Building the Background Model

A temporal difference-based method is used to build a complete background model, M. In order to detect slow motions or stopping objects, a weighted accumulation, I_t^{ac}, is used for temporal difference. At pixel location (i, j), I_t^{ac} is defined as:

$$I_t^{ac}(i, j) = (1 - w_{ac})I_{t-1}^{ac}(i, j) + w_{ac}|I_t(i, j) - I_{t-1}(i, j)| \tag{22.1}$$

where t is the current frame number, I_t is the current image frame, and w_{ac} is the weight. I_0^{ac} is set to be an empty image, and w_{ac} is set to be 0.5.

At the beginning, the background model is an empty array. In Table 22.1, M denotes the background model, and $s(i, j)$ denotes the state of a pixel at location (i, j), which is defined as:

$$s(i, j) = \begin{cases} 1 & I_{diff}(i, j) > \tau \\ 0 & Otherwise. \end{cases} \tag{22.2}$$

During the model building period, $I_{diff}(i, j)$ is set to be $I_t^{ac}(i, j)$, and τ is set to be $\tau_d = 15$. After the background model M is complete, τ is set to be $\tau_m = 25$, and $I_{diff}(i, j)$ is obtained by using the model M, as explained below. Since temporal difference is based on consecutive frames, and tends to give smaller differences, τ_d has a smaller value than τ_m.

When $s(i, j) = 1$, i.e. when the pixel is classified as foreground, this pixel location in the model $(M(i, j))$, is not updated/changed. On the other hand, if $s(i, j)$ is 0, the current value of $M(i, j)$ is checked. If $M(i, j)$ is not filled yet, $M(i, j)$ is set to be $I_t(i, j)$, which is the current pixel value. If $M(i, j)$ is already filled, its value is set to be $M(i, j) = 0.95M(i, j) + 0.05I_t(i, j)$. Thresholded temporal difference cannot detect all the pixels of a moving object as depicted in Fig. 22.1. Thus, existing model is given a 95% weight not to corrupt it by direct use of the values coming from the internal region of a moving object. As moving objects in the scene change their location, the M will gradually be filled as seen in Fig. 22.2. The process of building the background model ends when no empty location is left in M. When M is complete, temporal difference is not used anymore.

22.2.2 Updating the Counters

As stated previously, the stability of a pixel at location (i, j) is determined by a counter $h(i, j)$, which keeps the number of times a pixel's state changes from 0 to 1, or vice versa, in the last 100 frames. The motivation is that the lower the value of $h(i, j)$, the more stable and reliable that pixel location is.

Although it may look like an implementation detail, the computation of $h(i, j)$ for each pixel at each frame is worth emphasizing since we want fast processing, and we need to take the memory requirements into account. At any frame t, we want the number of changes in

TABLE 22.1   Salient foreground detection algorithm ($\copyright$(2010) Elsevier)

Set $M(i,j) = -1$ for all i,j;   Set $s(i,j) = 0$, $R(i,j) = 0$ for all i,j;
Set $I_1 = $ first frame; Set $model\_complete = false$;
for every frame $t > 1$
    Set $I_t = t^{th}$ frame, and set $I_{outp}(i,j) = 0$ for all i,j;
    if $\exists\ i,j$ for which $M(i,j) = -1$
      compute I_t^{ac}; set $I_{diff} = I_t^{ac}$; $\tau = \tau_d$;
    else
      set $model\_complete = true$;
      compute $I_t^{md} = |I_t - M|$; set $I_{diff} = I_t^{md}$; $\tau = \tau_m$;
    for all i,j
      if $I_{diff} > \tau$
        if $(s(i,j) == 0)$, set $s(i,j) = 1$; update CC_k, for $k \in \{1 \ldots 4\}$;
      else
        if $(s(i,j) == 1)$, set $s(i,j) = 0$; update CC_k, for $k \in \{1 \ldots 4\}$;
        if $model\_complete == false$
          if $M(i,j)$ is not equal to -1
            $M(i,j) = \alpha I_t(i,j) + (1-\alpha)M(i,j)$;
          else
            $M(i,j) = I_t(i,j)$;
      if $model\_complete == true$
        if $I_t^{md}(i,j) > \tau$
          if $R(i,j) == 0$
            Compute $h(i,j) = \sum_{i=1}^{4} CC_i$;
            if $h(i,j) < \tau_p$
              Set $I_{outp}(i,j) = 1$; Set $R(i,j) = 0$;
            else
              Set $neighb(i,j)$ to be 3×3 neighb. of $h(i,j)$
              if N $> 0.7(2w + 1)^2$
                $I_{outp}(i,j) = 1$; $R(i,j) = 0$;
              else
                $M(i,j) = \alpha I_t(i,j) + (1-\alpha)M(i,j)$;
          else
            $I_{outp}(i,j) = 1$; $R(i,j) = 0$;
        else
          Reset $FG\_duration(i,j) = 0$;
          if t is a multiple of 25
            if $R(i,j) == 0$
              Compute $h_{t-50}^{t}(i,j)$;
              if $(h_{t-50}^{t}(i,j) \leq 2)$, set $R(i,j) = 1$;
          if $R(i,j) == 1$
            $M(i,j) = 0.5I_t(i,j) + 0.5M(i,j)$;
          else $M(i,j) = \alpha I_t(i,j) + (1-\alpha)M(i,j)$;
        if $I_{outp}(i,j) == 1$ and t is a multiple of 100
          Create and/or increase $FG\_duration(i,j)$;
          if $100 \times FG\_duration(i,j) > T$
            $M(i,j) = 0.5 \times I_t(i,j) + 0.5 \times M(i,j)$;
    if $model\_complete == false$
      Set $I_{t-1} = I_t$;
return I_{outp}

---

a pixel's state between frames $t - 100$ and t. This requires saving the frame number each time a change occurs in a pixel's state. For locations with non-salient motion, this, in turn,

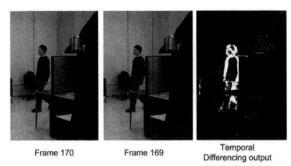

FIGURE 22.1   Output of the temporal difference after applying a threshold (©(2010) Elsevier).

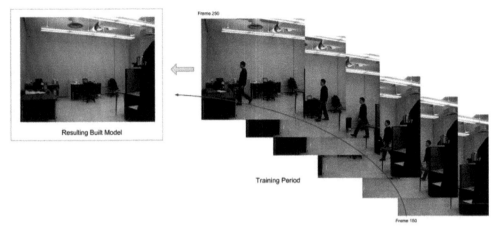

FIGURE 22.2   The background model is gradually built as moving objects change their location (©(2010) Elsevier).

requires an array with potentially high dimension for each pixel. Instead, we quantize the 100-frame window into 4 intervals, and keep a counter $CC_k(i,j), k \in \{1, \ldots, 4\}$, for each interval for pixel (i,j). The approach is illustrated in Fig. 22.3. Between frames 1 and 25, the counter CC_1 is increased each time the pixel's state changes, between frames 26 to 50 the counter CC_2 is increased, etc. At the end of the 100-frame period, the counter CC_1 is reset and its value is increased until frame 125 is reached, and the other counters are updated similarly. This avoids saving the frame instances of each change. Then,

$$h(i,j) = \sum_{k=1}^{4} CC_k(i,j).$$

Counters $h(i,j)$ are updated during the building of the model as well. Figure 22.4 shows a frame from a video containing a fountain, and a plot of the counter values $h(i,j)$ for different pixel locations (i,j). As can be seen, the counters are higher around the outer boundaries of the multiple fountains, where the water is constantly moving and splashing. The high counters indicate regions with low reliability and non-salient motion.

It should be noted that this approach provides only an approximation of the number of changes in a pixel's state without having to save the frame numbers of every state change. Other approaches can be used, and have been tried, that can give better approximations. However, they either require introducing additional variables, and/or additional instructions, and thus increase the memory requirement and decrease the algorithm speed. The presented approach is adapted for small memory requirement and better computational speed.

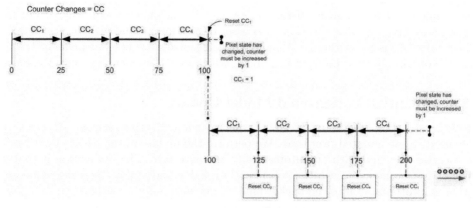

FIGURE 22.3 Illustration of how h(i,j) is computed (©(2010) Elsevier).

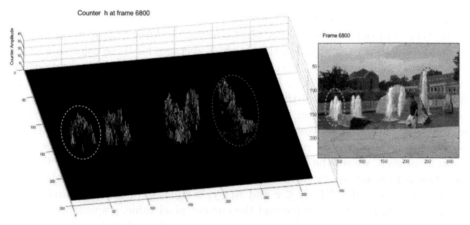

FIGURE 22.4 (See color insert.) Original frame and the plot of the counter values $h(i, j)$ for different pixel locations (i, j). Higher values correspond to outer boundaries of multiple fountains, indicating regions with low reliability and non-salient motion (©(2010) Elsevier).

22.2.3 Salient Foreground Detection

As can be seen in Table 22.1, after the background model is built, then the difference image is set to be $I_{diff} = I_t^{md} = |I_t - M|$.

If $I_t^{md}(i, j) \leq \tau$, then the pixel location (i, j) is classified as background. On the other hand, as opposed to many traditional model-based background subtraction approaches, in the proposed scheme, satisfying $I_t^{md}(i, j) > \tau$ is not enough for the pixel location (i, j) to be classified as foreground. Instead, reliability constraints are employed to differentiate between salient and non-salient motion. A pixel location satisfying $I_t^{md}(i, j) > \tau$ is classified as foreground only if its counter $h(i, j)$ satisfies $h(i, j) < \tau_p$, where $\tau_p = 15$ is the percentage threshold. The reasoning is that if $h(i, j) < 15$, then it means that the state of the pixel at this location changed less than 15% of the time during the last 100 frames making this location a reliable one. In other words, this location is not likely to be in a non-salient motion region. Thus, the intensity difference greater than τ is caused by a salient motion with high probability.

If $I_t^{md}(i, j) > \tau$ *and* $h(i, j) \geq \tau_p$, then we do not classify this location as background right away. We take a $(2w+1) \times (2w+1)$-window neighborhood, where $w = 1$, around location (i, j) and check the h counter for all the neighbors. In Table 22.1, N is the number of neighbors whose counter h is less than τ_p. If the majority of the neighbors (more than 70%) have a low counter, i.e. $h < \tau_p$, then location (i,j) is set to be a foreground pixel or vice versa. This way, we take into account the fact that neighboring pixels are not independent from each other. We obtain information from neighbors, which increases accuracy and robustness.

22.2.4 Adaptive Background Model Update

In order to adapt to changes in the environment, such as lighting changes, the background model needs to be updated over time. We perform the update of the background model M in a selective way, and with an automatically adaptive rate. The motivation is that when a pixel's location is deduced to be consistently reliable and stable, then the value at that location is incorporated into the background model with a higher weight.

If $I_t^{md}(i,j) \le \tau$, then we can conclude that, at this location, it is safe to update the background model. However, by looking at the *summary* of the recent past of a pixel, we can give a higher weight to the current pixel value, and better adapt to faster changes in the background. In other words, we have an automatically adaptive background update rate. The very compact *summary* of a pixel's history is formed as follows: Rather than saving many values for each pixel location, such as averages for three color values, multiple Gaussian distribution means and variances, multiple codewords with multiple entries, we use two of the four counters (CC_k, $k \in \{1, \dots, 4\}$) corresponding to the last 50 frames. Let $h_{t-50}^t(i,j)$ denote the sum of these two counters. Thus, $h_{t-50}^t(i,j)$ holds the number of state changes at pixel location (i,j) during the last 50 frames. If $h_{t-50}^t(i,j) \le 2$, it means that the state of this pixel has changed only two times or less during the last 50 frames, i.e. this location is very reliable. We perform this check every 25 frames, and if the condition is satisfied, we set the boolean variable $R(i,j)$, which is a reliability flag, to be 1. This location is then incorporated to the background model with a 50% weight.

On the other hand, if $I_t^{md}(i,j) \le \tau$ and $R(i,j)$ is equal to 0, then 95% and 5% weights are given to the existing model value and the current pixel value, respectively.

If a pixel at location (i,j) is classified as a foreground pixel, then $M(i,j)$ is not updated, which prevents corrupting the existing model. However, if a pixel location (i,j) is classified consecutively as foreground for a specified period of time (T) due to a static foreground object, then we start to push this location to the background by giving it 50% weight. T is set by the user, and determines how much time a stopped object should be static to be considered as part of the background.

22.2.5 Adaptive Number of Memory Accesses and Instructions

In Section 22.2.4, we described how we set the value of $R(i,j)$. If $I_t^{md} \le \tau$, and current $R(i,j)$ is 0, and the frame number is a multiple of 25, we compute the value of $h_{t-50}^t(i,j)$. A small $h_{t-50}^t(i,j)$ indicates that the pixel's state has not changed much in the last 50 frames, and thus $R(i,j)$ is set to be 1.

At the beginning, for each pixel, 1 byte is allocated for each CC_k, where $k \in \{1, \dots, 4\}$, 1 byte for the value saved in $M(i,j)$, 1 byte for the previous frame value, 1 bit for the state variable $s(i,j)$, and 1 bit for the reliability flag $R(i,j)$ making the total memory allocation 50 bits per pixel. After the background model is built, the pixel values of the previous frame are no longer needed. Instead, the memory allocated for the previous frame values is used for the *FG_duration* variable.

If the value of $R(i,j)$ is 1, this indicates that this pixel is a very reliable and stable background pixel. Compared to our prior work, the method presented here makes more and a better use of this information, which provides savings, in terms of number of memory accesses and number of instructions, in two different ways. With the presented method, first type of saving occurs when there is a foreground object in the scene covering reliable background pixels. In the presented method, when $I_t^{md}(i,j) > \tau$, $h(i,j)$ is not calculated for very reliable background pixels, i.e. pixels for which $R(i,j)$ is 1. The reasoning is the following: $h(i,j)$ is employed to determine the stability of a pixel by looking at its state changes in the last 100 frames. If $R(i,j)$ is 1, it is already known that this location is very reliable, thus we do not need to calculate and check the value of $h(i,j)$. In addition, we do not need to check the counters of the neighboring pixels either. This provides significant savings in terms of the number of memory accesses and instructions.

The second type of savings, over our prior work [4], occurs every 25 frames. If $R(i,j)$ is currently 1, then there is no need to compute $h_{t-50}^t(i,j)$, which provides additional savings. The detailed comparison of the method presented here and our prior work [4], in terms of the processing speed, will be presented in Section 22.2.6.

As described above, for very reliable and stable background pixels $R(i,j)$ is set to 1. Thus, the plot of the number of pixels, whose reliability flag $R(i,j)$ is 0, versus the frame number serves as a tool for activity summary. The changes and peaks in this plot will indicate the portions of the video with activity. Figures 22.5 and 22.6 show these plots obtained for different video sequences.

Figure 22.5 shows the number of pixels with $R(i,j) = 0$ for a video sequence of a fountain. Frames 1 through 100 correspond to the model building period, during which $R(i,j) = 0$ for all the pixels. After the model is built, and stable background pixels are determined, the number of pixels whose $R(i,j)$ is 0 drops significantly to about 2500 pixels per frame, and it remains around this value until some activity starts in the scene. For example, at frame 6800, there is a person walking in front of the camera. This creates a peak in the plot. A similar situation occurs at frame 7935, where the detected person is closer to the camera and thus its size is larger than the previous scenario. A bigger object covers more pixels, and causes them to be classified as foreground pixels. Thus, the $R(i,j)$ is set back to zero for these affected pixels. This is why the peak at frame 7935 is higher than the one at frame 6800.

The savings provided by the proposed method increases with increasing number of reliable background pixels, i.e. pixels whose $R(i,j)$ is 1. In Fig. 22.5, low values correspond to frames with a small number of unreliable pixels, and thus more number of reliable background pixels. Thus, in these portions of the video, the number of memory accesses, and the number of instructions will be less with the proposed method. More speed analysis will be provided in Section 22.2.6.

A more extreme example is presented in Fig. 22.6, in which sudden lightning causes a complete intensity change in the whole image at frame 369. As a result, a large peak is observed in the plot. After this, the total number of pixels with $R(i,j) = 0$ drops again.

22.2.6 Experimental Results

We compared the presented method with five other background subtraction methods, including our prior work, on 10 different video sequences with varying levels of difficulty. Henceforth, these algorithms will be referred to as follows; ALW: Adaptive LightWeight algorithm (the method presented in this chapter), LW: LightWeight algorithm [3], Org-MoG: Original MoG [33], Impr-MoG: Improved MoG [37], CB: Codebook [20], EB: Eigenbackground [25]. The method presented here modifies and optimizes our prior work [3] [4] in terms of the memory access, number of instructions, and thus, speed. The decision about whether a pixel is a foreground pixel is made differently and more efficiently. In addition, we provide a detailed comparison of the proposed (ALW) method with other state-of-the-art background subtraction algorithms in terms of their memory requirement, accuracy and processing time. We also present the Receiver Operation Characteristics (ROC) curves for

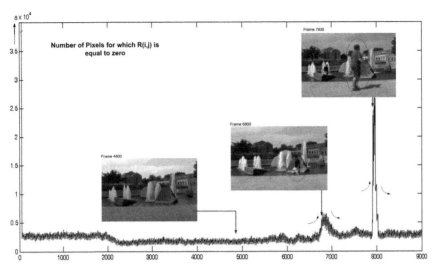

FIGURE 22.5   Video of a fountain: Number of pixels with $R(i, j) = 0$ vs. the frame number plot (©(2010) Elsevier).

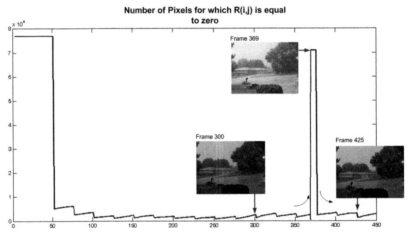

FIGURE 22.6   Rain Sequence: Number of pixels with $R(i, j) = 0$ vs. the frame number plot also serves as a tool for activity summary (©(2010) Elsevier).

different background subtraction algorithms.

Since embedded smart cameras have limited processing power and memory, it is very important to design lightweight algorithms that require less memory for storage. We first compared our algorithm with others in terms of the memory requirement for the data saved for each pixel. The proposed algorithm was run on the red channel, and its memory requirement is detailed in Section 22.2.5. For different background subtraction techniques, Table 22.2 lists the number of bytes necessary for the data saved for each pixel, for one color channel. The memory requirements for the other background subtraction methods are computed as follows. Let n denote the number of Gaussian distributions used in Org-MoG. Org-MoG requires two floating point numbers per Gaussian distribution, per color channel, per pixel (one for the mean and one for the variance of a Gaussian distribution). It also requires $n - 1$ many floating point numbers for the weights of distributions. Thus, if n is picked to be three, eight floating point numbers are needed per color channel. If 3 color channels are used, the memory required per pixel is 96 bytes. If one color channel is used, it is 32 bytes. Even if the mean for each distribution is rounded so that it can be represented by a byte, the memory requirement is still 23 bytes per color channel.

The codebook-based method (CB) uses three floating point numbers for the means of the RGB channels, 2 bytes for the minimum and maximum brightness values that the codeword accepted, one integer for the frequency of the codeword, one integer for the maximum negative run-length, and two integers for the first and last access times. Thus, the total memory needed is 22 bytes per codeword. If it is only one color channel, it is 14 bytes per codeword. In [20], it is stated an average of 6.5 codewords are needed per pixel codebook. Thus, the average memory requirement per pixel is 91 bytes.

For the EB method, the memory requirement per pixel is the number of the best eigen-backgrounds. During the training time the method requires allocation for all the training images. In general, 7 floating point numbers are required per pixel. Thus, the memory needed is 28 bytes.

The LW algorithm presented in [3] requires 7.25 bytes per pixel when the functionality of pushing the static foreground objects to the background is incorporated. For different methods, Fig. 22.7 shows a bar graph of the memory requirement (in bytes) per frame for a 240 × 320 frame.

TABLE 22.2   Memory requirement for the data saved for each pixel for different background subtraction methods (for one color channel)(©(2010) Elsevier)

	CB	Org-MoG	EB	LW [3]	ALW
Bytes per pixel	91	32	28	7.25	6.25

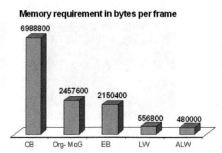

Memory requirement in bytes per frame

FIGURE 22.7   Per frame memory requirements of different background substraction methods when one color channel is used (©(2010) Elsevier).

We tested the proposed method on 10 challenging video sequences, and compared it with five other background subtraction methods including our previous work. It should be noted that all the displayed outputs below are the images obtained *without applying any* morphological or post-processing operations. All the results of our algorithm were obtained by using the same threshold values for all videos, specifically, $\tau_d = 15$, $\tau_m = 25$, $\tau_p = 15$, and $\alpha = 0.05$. Overall, the proposed method requires the least amount of memory per pixel while providing better or comparable outputs at the same time.

Figures 22.8 and 22.9 display the outputs obtained on videos of two different windy scenes. All the algorithms were run on one channel except the CB and Impr-MoG. As can be seen, the proposed method provides the least amount of noisy pixels, and good detection at the same time.

Figures 22.10 and 22.11 show the outputs for challenging videos of rainy scenes. Again, the proposed method provides comparable if not better outputs compared to the other algorithms, while requiring the least amount of memory at the same time.

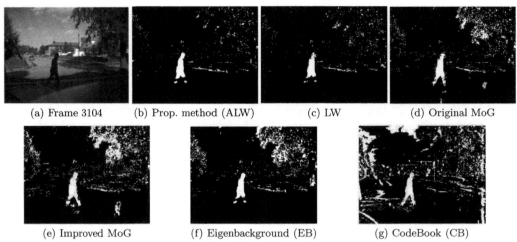

(a) Frame 3104   (b) Prop. method (ALW)   (c) LW   (d) Original MoG

(e) Improved MoG   (f) Eigenbackground (EB)   (g) CodeBook (CB)

FIGURE 22.8   Foreground detection results of different algorithms on a challenging video of a windy scene. Outputs are obtained without morphological operations (©(2010) Elsevier).

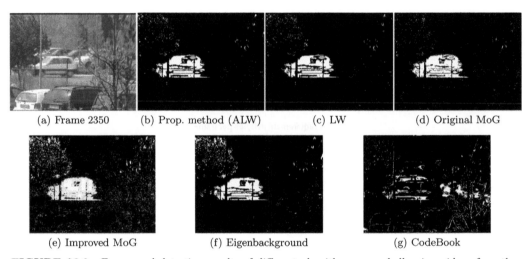

(a) Frame 2350   (b) Prop. method (ALW)   (c) LW   (d) Original MoG

(e) Improved MoG   (f) Eigenbackground   (g) CodeBook

FIGURE 22.9   Foreground detection results of different algorithms on a challenging video of another windy scene. Outputs are obtained without morphological operations (©(2010) Elsevier).

Figure 22.12 shows the outputs for another challenging video of a lake, where there are rippling water effects on the lake, and swaying trees in the background. Compared to Impr-MoG, EB and CB, the proposed method can differentiate the non-salient motion better. It gives the least amount of noisy pixels. The Org-MoG, on the other hand, has less noisy pixels than the proposed method. However, it misses the person and the dog, which should be detected as foreground objects. The outputs obtained on three other outdoor videos are shown in Figures 22.13, 22.14 and 22.15.

The results displayed in Fig. 22.16 were obtained from a video of a scene with a fountain, where the water level goes up and down. Moreover, during the video, lighting changes due to moving clouds, as seen in Figures 22.16(a) and 22.16(b). As the figure illustrates, since the eigenbackground method does not update the background model, it cannot handle the lighting change. The improved MoG method cannot detect most of the foreground pixels. The proposed method provides good detection, and can eliminate most of the non-salient motion caused by the fountains.

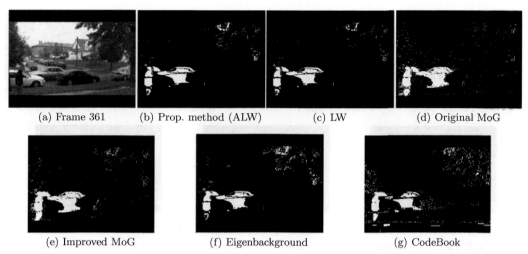

    (a) Frame 361   (b) Prop. method (ALW)   (c) LW   (d) Original MoG

    (e) Improved MoG   (f) Eigenbackground   (g) CodeBook

FIGURE 22.10   Foreground detection results of different algorithms on a video of a rainy scene. Outputs are obtained without morphological operations (ⓒ(2010) Elsevier).

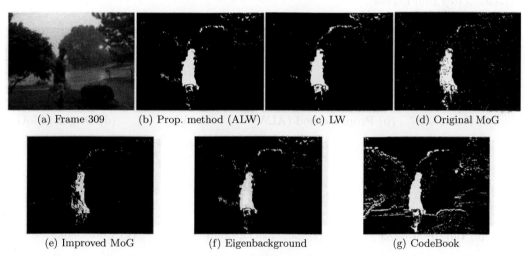

    (a) Frame 309   (b) Prop. method (ALW)   (c) LW   (d) Original MoG

    (e) Improved MoG   (f) Eigenbackground   (g) CodeBook

FIGURE 22.11   Foreground detection results of different algorithms on a video of another rainy scene. Outputs are obtained without morphological operations (ⓒ(2010) Elsevier).

Figure 22.17 displays the outputs obtained from an indoor sequence. Although the video was captured indoors, the flickering of the overhead lights affects the performance of the algorithms. The proposed method performs reliably, and like in the above examples, can differentiate the salient motion from a non-salient one.

We also compared the processing times of these algorithms on a PC. However, the codes for the ALW, EB, CB and Org-MoG are written in MATLAB, whereas the code for Impr-MoG is written in C. Also, these codes are not equally optimized. Hence, it is difficult to make a comparison of the processing times. We will list the frames/s rates to give the reader a general idea. The algorithms were run on a video with 240×320 frame size. ALW and EB run at 35 frames/s and 49.5 frames/s, respectively, in MATLAB. It should be noted that EB does not update the background model. The Org-MoG and the CB run at 0.2 frames/s and 0.24 frames/s, respectively, in MATLAB. The C++ version of the CB method runs at around 55 frames/s, and the Impr-MoG runs at 59 frames/s in C.

In addition, we performed a comparison of the different algorithms in terms of their

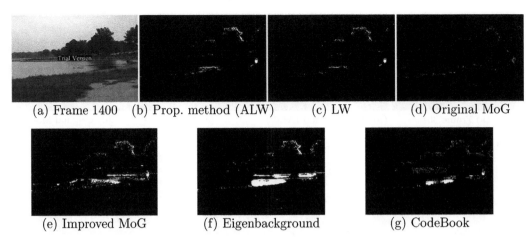

(a) Frame 1400   (b) Prop. method (ALW)   (c) LW   (d) Original MoG

(e) Improved MoG   (f) Eigenbackground   (g) CodeBook

FIGURE 22.12   Foreground detection results of different algorithms on a challenging video of a lake. Compared to (e), (f) and (g), the proposed method can eliminate the non-salient motion better. Although (d) has less noisy pixels, it misses the person and the dog. Outputs are obtained without morphological operations (ⓒ(2010) Elsevier).

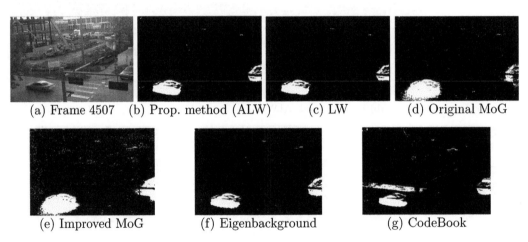

(a) Frame 4507   (b) Prop. method (ALW)   (c) LW   (d) Original MoG

(e) Improved MoG   (f) Eigenbackground   (g) CodeBook

FIGURE 22.13   Foreground detection results of different algorithms on a video of a street. Outputs are obtained without morphological operations (ⓒ(2010) Elsevier).

probability of detection (P_d) and probability of false alarm (P_{fa}) rates, and plotted their Receiver Operation Characteristics (ROC) curves [7, 10, 14, 24]. ROC curves are employed often when comparing background subtraction algorithms. Alongside the outputs obtained by different algorithms, ROC analysis provides a quantitative comparison. We obtained the ground truth for the foreground objects manually, and plotted the ROC curve for each algorithm. These curves are displayed in Fig. 22.18. As can be seen, for the same P_d rate, the proposed method has the least P_{fa}, and for the same P_{fa} rate it has the highest P_d.

As described above, compared to our prior work [4], the method presented here provides more savings, in terms of number of memory accesses and number of instructions, and thus speed and efficiency, in two different ways. To demonstrate these savings, we performed three different experiments.

First two experiments compare the processing speed of the presented (ALW) method and our prior work (LW) when a foreground object is in the scene. As discussed in detail above, with the method presented here, the first type of savings occurs when there is a foreground

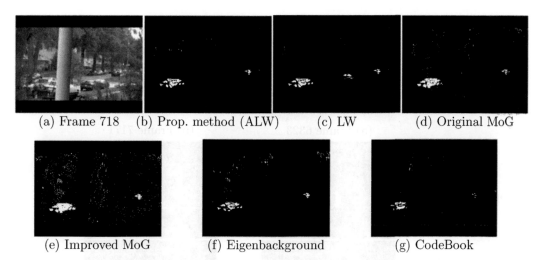

FIGURE 22.14   Foreground detection results of different algorithms on a video of a street. Outputs are obtained without morphological operations (ⓒ(2010) Elsevier).

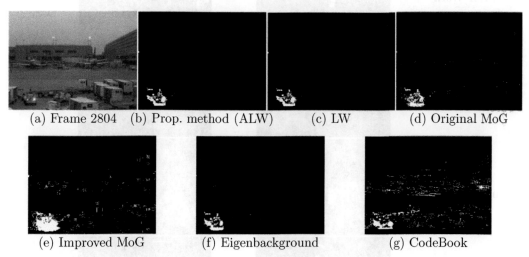

FIGURE 22.15   Foreground detection results of different algorithms on an airport sequence. Outputs are obtained without morphological operations.

object in the scene covering reliable background pixels. When $I_t^{md}(i,j) > \tau$ and $R(i,j) = 1$, $h(i,j)$ is not calculated for these reliable background pixels, i.e. pixels for which $R(i,j)$ is 1. In addition, we do not need to check the counters of the neighboring pixels either. This provides significant savings in terms of the number of memory accesses and instructions. For these experiments, we imported and implemented the ALW and LW algorithms on an embedded smart camera node. Figure 22.19 shows a plot of the processing time (in ms on the microprocessor of the camera) for the two algorithms during an interval when there is an object in the scene. The blue and red plots correspond to the methods presented in this chapter and in [4], respectively. As can be seen, on the average, the method presented here performs 2.82 ms faster per frame. Also, the speed gain provided by this method increases with increasing object size and also increasing the number of objects in the scene. When the foreground object is larger, the proposed method runs 4.5 ms faster per frame on the average. This gain is obtained in part by not accessing CC_k, $k \in \{1, ..., 4\}$, and not performing $\sum_{k=1}^{4} CC_k$ for reliable background pixels.

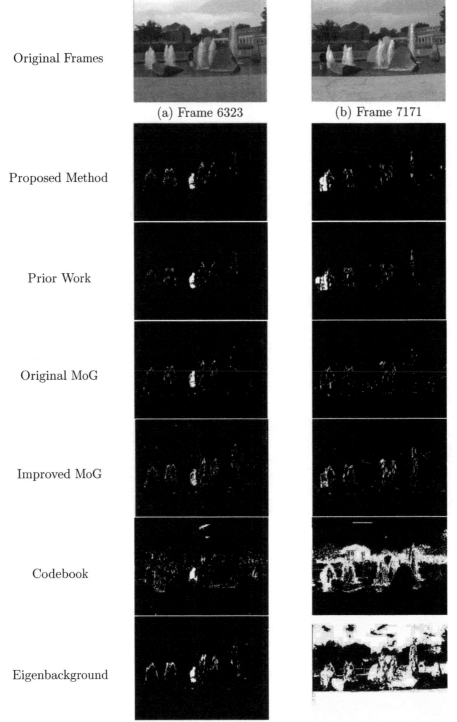

FIGURE 22.16 Comparison of foreground detection results of different algorithms on a video of a fountain. Two columns correspond to two different frames with significantly different lighting (©(2010) Elsevier).

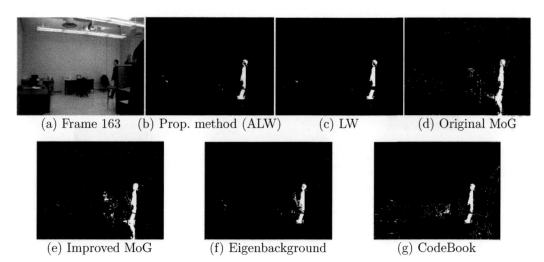

(a) Frame 163   (b) Prop. method (ALW)   (c) LW   (d) Original MoG

(e) Improved MoG   (f) Eigenbackground   (g) CodeBook

FIGURE 22.17 Foreground detection results of different algorithms on an indoor sequence, where there are flickering lights. Outputs are obtained without morphological operations (©(2010) Elsevier).

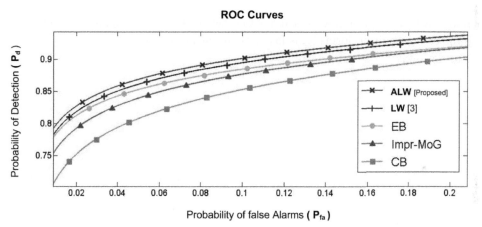

FIGURE 22.18 ROC curves of different background subtraction methods (©(2010) Elsevier).

In the second experiment, we ran the ALW and LW algorithms on the embedded smart camera platform, and measured the operating current of the board. The operating current increases or decreases based on the workload of the processor (number of instructions per task), the supply voltage source and the frequency at which the processor is working. To measure the current, we used a precise oscilloscope and a 1-ohm resistor configuration placed at the input of the supply source. Figure 22.20 shows the variations in the current during the processing of three consecutive frames containing a foreground object. As can be seen, the proposed method (blue plot) finishes processing the first frame 8 ms earlier than our prior work [4]. It also finishes processing the following two frames 7 and 8 ms faster.

The proposed method provides a second type of savings, over our prior work, every 25 frames. As described before, if $R(i,j)$ is currently 1, then there is no need to compute $h_{t-50}^{t}(i,j)$, which provides additional savings. In order to demonstrate these savings, we performed another experiment and measured the operating current of the camera board over time with an oscilloscope. To measure the gain obtained only from not calculating $h_{t-50}^{t}(i,j)$ at every 25 frames, we used an empty scene. Figure 22.21 shows the waveforms obtained. The blue and red plots correspond to the methods presented in this chapter and

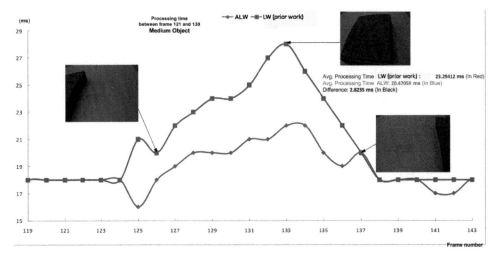

FIGURE 22.19   (See color insert.) Processing time (ms) versus the frame number for two different versions of the algorithm when there is a foreground object in the scene ($\copyright$(2010) Elsevier).

our prior work [4], respectively. As can be seen, when the frame number is a multiple of 25, the proposed method performs 5 ms faster than our prior work.

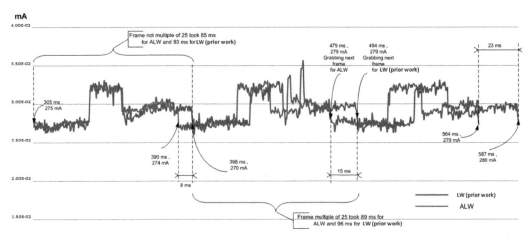

FIGURE 22.20   (See color insert.) Variations in the operating current during the processing of three consecutive frames containing a foreground object. The method presented in this chapter (blue plot) is faster than the prior work [4] (red) ($\copyright$(2010) Elsevier).

22.3   Feedback Method: Foreground Detection by Tracking Feedback

In the remainder of this chapter, two terms will be used to refer to different methods. In the *sequential method*, at every frame, we first run our foreground detection algorithm, described in Section 22.2, on the entire image to detect foreground pixels, group them together to form foreground blobs, and then match the foreground blobs to existing trackers. Most traditional tracking algorithms operate in this sequential manner. The method we present

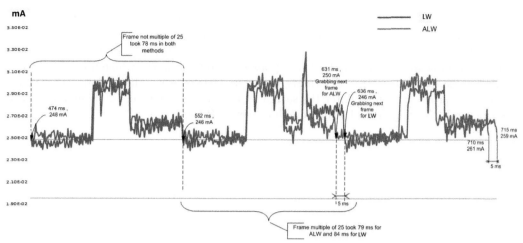

FIGURE 22.21 (See color insert.) Variations in the operating current during the processing of three consecutive frames of an empty scene. The method presented in this chapter (blue plot) provides speed gain at frame numbers that are multiple of 25 (©(2010) Elsevier).

in this section[51] will be referred to as the *feedback method*. Instead of performing foreground detection and tracking independently, the *feedback method* incorporates the information from the tracking stage into the foreground detection stage that employs the algorithm we presented in Section 22.2. This way, foreground detection is performed in smaller regions as opposed to the entire frame. As will be shown below, the feedback method significantly reduces the processing time of a frame. We take advantage of these savings by sending the microprocessor to idle state at the end of processing a frame without causing tracking failure.

When a foreground blob is detected in the scene, a bounding box is formed around it, and a new tracker is created. The intensity histogram of the foreground object is built and saved as the model histogram of the tracker (intensity histogram is used to keep the computational complexity low). The tracker also holds the coordinates of the bounding box of this object. Let $T = \{T^1(t-1), T^2(t-1)\dots T^n(t-1)\}$ denote the set of existing trackers at frame $t-1$. At frame t, a detected blob $B^i(t)$ will be matched to one of the trackers in the set T by using a matching criteria based on bounding box intersection and the Bhattacharyya coefficient [9]. The Bhattacharyya coefficient is derived from the sample data by using:

$$\hat{\rho}(y) \equiv \rho[\hat{\mathbf{p}}(\mathbf{y}), \hat{\mathbf{q}}] = \sum_{u=1}^{m} \sqrt{\hat{p}_u(\mathbf{y})\,\hat{q}_u} \tag{22.3}$$

where $\hat{\mathbf{q}} = \{\hat{q}_u\}_{u=1\dots m}$, and $\hat{\mathbf{p}}(\mathbf{y}) = \{\hat{p}_u(\mathbf{y})\}_{u=1\dots m}$ are the probabilities estimated from the m-bin histogram of the model in the tracker and the candidate blobs, respectively.

If the bounding box of a blob intersects with that of the tracker, the Bhattacharyya coefficient between the model histogram of the tracker and the histogram of the foreground blob is calculated by using (22.3). The tracker is assigned to the foreground blob which results in the highest Bhattacharyya coefficient. After blob $B^i(t)$ is matched to tracker $T^j(t-1)$ (which holds the bounding box location from frame $t-1$), the displacement of the centroid of the tracker's bounding box is calculated in x and y directions to obtain Δx and

---

[51]'SV'©[2011] IEEE. Reprinted, with permission, from [5].

Δy, respectively (Fig. 22.22). At frame $t + 1$, for each foreground object i, we determine a search region $R^i(t + 1)$ by using Δx, Δy, W and H, where W and H are the width and height of the bounding box of $B^i(t)$. Then, we perform the background subtraction and blob forming in the search regions $R^i(t+1)$ as opposed to doing it on the whole frame. As shown in Table 22.3, searching for and forming foreground blobs in smaller regions significantly reduces the processing time. After the search regions are determined, the bounding box of the tracker T^j is updated to be the bounding box of $B^i(t)$.

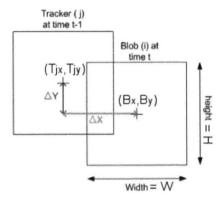

FIGURE 22.22   Displacement in the horizontal and vertical directions (©[2011] IEEE).

The center of the search region $R^i(t+1)$ is found by using (22.4), where $B_x^i(t)$ and $B_y^i(t)$ are the x and y coordinates of the center of the blob B^i at frame t. $\Delta x(t)$ and $\Delta y(t)$ are the displacements in the x and y directions calculated between frames $t - 1$ and t as shown in Fig. 22.22.

$$R_x^i(t + 1) = B_x^i(t) + \Delta x(t)$$
$$R_y^i(t + 1) = B_y^i(t) + \Delta y(t) \tag{22.4}$$

We determine the boundaries of the search region by using (22.5). Foreground detection at frame $t+1$ will be performed in the search regions formed around the estimated locations of objects that were detected at frame t.

$$R_{x\_min}^i(t + 1) = R_x^i(t + 1) - W$$
$$R_{x\_max}^i(t + 1) = R_x^i(t + 1) + W$$
$$R_{y\_min}^i(t + 1) = R_y^i(t + 1) - H$$
$$R_{y\_max}^i(t + 1) = R_y^i(t + 1) + H \tag{22.5}$$

The camera's capture rate is 15 fps. Since, the algorithm becomes localized around the regions R^i, we run the foreground detection on the whole frame every 500 ms to be able to detect new objects in the scene, and update the background model. Compared to the sequential method, this mechanism reduces the processing time significantly. We take advantage of these savings by sending the microprocessor to idle state at the end of processing a frame.

We also implemented an energy saving mechanism to be employed when the scene is empty. If no object is detected in the field of view of the camera, the microprocessor is sent to an idle state, and the camera captures only 2 fps. In Fig. 22.23, approximately the first seven seconds correspond to the operating current of the camera during this operation mode. During the idle state, the current drops from an average of 220mA to an average of 120mA, which is a 45% decrease in the current drawn. As soon as an object enters into the field of view of the camera, the system immediately resumes the full operation mode.

We also analyzed the cases of tracking targets that are close to or far from the camera. The bar graph in Fig. 22.24 shows the frame processing times when tracking an object in a

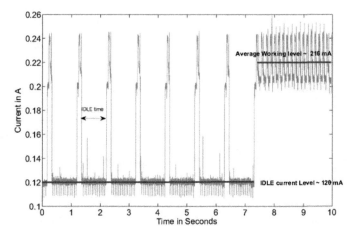

FIGURE 22.23 (See color insert.) The operating current of the camera board for a scenario where the scene is first empty, and then an object is detected in the scene, and the microprocessor resumes the full operation mode (©[2011] IEEE).

close, middle and far range from the camera together with the size of the object. The size of the bounding box of the object is displayed inside the bars. As expected, the processing time increases when the object is closer to the camera, since the object size, and thus, the area to be processed increases.

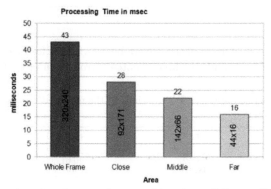

FIGURE 22.24 Processing time in ms. when an object is at different distances from the camera (©[2011] IEEE).

22.3.1   Comparison of the Processing Times of the Feedback Method and the Sequential Approach

We ran both the sequential method and the feedback method on embedded smart cameras to compare their processing times. We performed experiments where we tracked one, two and three remote-controlled cars by the two methods. The blue and red plots in Fig. 22.25(a) show the operating currents of the camera board when running the feedback method and the sequential method, respectively. The grabbing and buffering of a frame takes 49 ms. The feedback method and the sequential method finish the processing of the frame in 19.7 and 38.5 ms, respectively, and the feedback method provides 48.7% decrease in the processing time. Figures 22.25(b) and (c) show operating currents when tracking two and three cars, respectively. As expected, the gain in processing time decreases with an increasing number of tracked objects. But, the feedback method still outperforms the sequential method. The processing times and the results of the comparison are summarized in Table 22.3.

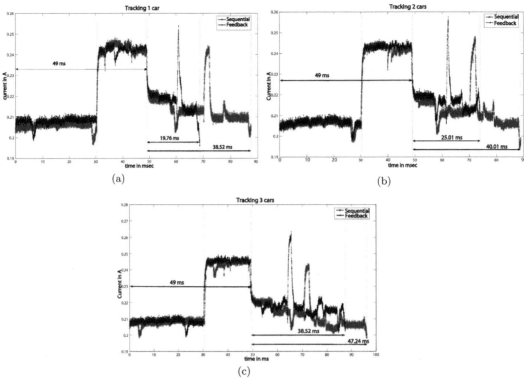

FIGURE 22.25   Operating current of the camera board with the feedback and sequential methods when tracking (a) one (b) two and (b) three remote-controlled cars (©[2011] IEEE).

TABLE 22.3   Comparison of the processing times of the proposed feedback method and the sequential approach (©[2011] IEEE).

Method	1 car (ms)	2 Cars (ms)	3 Cars (ms)
Feedback	19.76	25.01	38.52
Sequential	38.52	40.01	47.24
Savings	48.7%	37.49 %	18.45 %

22.3.2   Comparison of the Energy Consumptions of the Feedback Method and the Sequential Approach

We performed a set of experiments with three different tracking scenarios to measure the energy consumption of the camera when running the feedback method and the sequential method. In all three cases, remote-controlled cars are tracked for the same amount of time (five minutes) so that energy consumptions for different scenarios can be compared.

In the first scenario, a remote-controlled car is tracked continuously for five minutes. In other words, the car is always in the field of view, and the scene is never empty. When tracking one car, the feedback method finishes the processing of a frame, on the average, 18 ms earlier than the sequential method, and sends the microprocessor to idle state for 18 ms at the end of processing each frame. This way, the two methods process about the same number of frames during the 5-min period. Since the scene is never empty, the system never operates at 2 fps. It should also be noted that with the feedback method, the camera still processes the whole frame every 500 ms, to detect new objects and update the background model. Even in this case, using the feedback method provides 9.63% savings in energy consumption as seen in Table 22.4.

In the second scenario, the scene is empty for the first 100s. Then, a car enters the scene, and is tracked for 100s. Then, a second car enters the field of view of the camera, and two cars are tracked for another 100s. Table 22.5 shows the total energy consumptions while running each method during the 5-min experiment. The feedback method provides 17.34% savings in the energy consumption. Compared to the previous scenario, the savings in energy consumption increases, since the scene is empty for the first 100s.

TABLE 22.4   Energy consumptions for the feedback and the sequential methods when tracking one car continuously (©[2011] IEEE).

Method	Energy (J)
Feedback	304.25
Sequential	336.69
Savings	9.63%

TABLE 22.5   Energy consumptions for the feedback and the sequential methods when tracking one and then two cars (©[2011] IEEE).

Method	Energy (J)
Feedback	274.7057
Sequential	332.3419
Savings	17.34 %

The third scenario is the following: During the first 100s the scene is empty. Then, a remote-controlled car enters the scene, stays in the view of the camera for 50s, and leaves the field of view. After 100s, the car enters the view again, and stays there 50 more seconds. Table 22.6 shows the total energy consumptions while running each method during the 5-min experiment. The feedback method provides a 26.6% decrease in the energy consumption. Again, compared to the previous scenario, saving in the energy consumption is more, since the scene is empty for longer amount of time.

TABLE 22.6   Energy consumptions for the feedback & sequential methods when a car enters and leaves twice (©[2011] IEEE).

Method	Energy (J)
Feedback	242.6787
Sequential	330.8194
Savings	26.6%

22.4   Conclusion

A lightweight salient foreground detection algorithm, which is highly robust against challenging non-static backgrounds, has been presented in this chapter. Contrary to many traditional methods, the memory requirement for the data saved for each pixel is very small in the proposed algorithm, which is very important for portability to an embedded smart camera. Moreover, the number of memory accesses and instructions are adaptive, and are decreased even more depending on the amount of activity in the scene and on a pixel's history. Each pixel is treated differently based on its history, and instead of requiring the same number of memory accesses, and thus, instructions for every pixel, the algorithm requires less instructions for stable background pixels. This, in turn, increases the processing speed. This is achieved without sacrificing accuracy.

The proposed method selectively updates the background model with an automatically adaptive rate, thus can adapt to rapid changes. As opposed to traditional methods, pixels are not always treated individually, and information about neighbors is incorporated into decision making, which increases accuracy and robustness. The algorithm can use only intensity, or one color channel, and still provides very reliable results. The results obtained with 10 different challenging outdoor and indoor sequences have been presented, and compared with the results of different state-of-the-art background subtraction methods. All the results of our algorithm were obtained by using the same threshold values for all videos. The ROC curves of different background subtraction methods have also been provided. The memory requirements of the different algorithms have been compared as well, and it has been shown that the proposed method requires the least amount of memory per pixel. The experimental results demonstrate the success of the proposed lightweight method in challenging situations such as scenes with water fountains, swaying trees, and strong rain.

To increase the energy efficiency of the foreground object detection even further, a feedback method has also been presented in this chapter. Rather than following the traditional sequential approach, and performing detection and tracking independently at each frame, the feedback method incorporates the information from the tracking stage into the foreground detection. This way, foreground detection is performed in smaller regions as opposed to the entire image. The feedback method significantly reduces the processing time of a frame, and provides 48.7% decrease in the processing time when tracking one remote-controlled car. We take advantage of these savings by sending the microprocessor to idle state at the end of processing a frame without causing tracking failure. The feedback method also provides 10.44% savings in energy consumption compared to traditional sequential tracking when tracking one object.

22.5   Acknowledgment

This work has been funded in part by National Science Foundation (NSF) CAREER grant CNS-1206291 and NSF grant CNS-1302559.

References

1. C. Anderson, P. Burt, and G. van der Wal. Change detection and tracking using pyramid transform techniques. In *SPIE Intelligent Robots and Computer Vision*, pages 72–78, Cambridge, MA, 1985.

2. S. Bhandarkar and X. Luo. Fast and robust background updating for real-time traffic surveillance and monitoring. In *IEEE Workshop on Machine Vision for Intelligent Vehicles*, page 55, 2005.

3. M. Casares and S. Velipasalar. Light-weight salient foreground detection for embedded smart cameras. In *ACM/IEEE Int'l Conf. on Distributed Smart Cameras*, 2008.

4. M. Casares and S. Velipasalar. Light-weight salient foreground detection with adaptive memory requirement. In *IEEE International Conference on Acoustics, Speech, and Signal Processing, ICASSP 2009*, 2009.

5. M. Casares and S. Velipasalar. Adaptive methodologies for energy-efficient object detection and tracking with battery-powered embedded smart cameras. *IEEE Transactions on Circuits and Systems for Video Technology*, 21(10):1438–1452, October 2011.

6. M. Casares, S. Velipasalar, and A. Pinto. Light-weight salient foreground detection for embedded smart cameras. *Computer Vision and Image Understanding*, 114(11):1223–1237, 2010.

7. T. H. Chalidabhongse, K. Kim, D. Harwood, and L. Davis. A perturbation method for evaluating background subtraction algorithms. In *Joint IEEE International Workshop on Visual Surveillance and Performance Evaluation of Tracking and Surveillance*, 2003.

8. P. Chen and et al. Citric: A low-bandwidth wireless camera network platform. In *ACM/IEEE Int'l Conf. on Distributed Smart Cameras*, 2008.

9. D. Comaniciu, V. Ramesh, and P. Meer. Real-time tracking of non-rigid objects using mean shift. In *International Conference on Computer Vision and Pattern Recognition,* pages 142–149, 2000.

10. R. O. Duda and P. E. Hart. *Pattern Classification and Scene Analysis.* Wiley, New York, 1973.

11. A. Elgammal, R. Duraiswami, D. Harwood, and L. Davis. Background and foreground modeling using nonparametric kernel density estimation for visual surveillance. *Proc. of the IEEE,* 90(7):1151–1163, July 2002.

12. A. Elgammal, D. Harwood, and L. Davis. Non-parametric model for background subtraction. In *European Conference on Computer Vision,* pages 751–767, 2000.

13. N. Friedman and S. Russell. Image segmentation in video sequences: A probabilistic approach. In *Conference on Uncertainty in Artificial Intelligence,* 1997.

14. X. Gao, T. E. Boult, F. Coetzee, and V. Ramesh. Error analysis of background adaption. In *International Conference on Computer Vision and Pattern Recognition,* pages 503–510, 2000.

15. I. Haritaoglu, D. Harwood, and L. Davis. W^4: Real-time surveillance of people and their activities. *IEEE Transactions on Pattern Analysis and Machine Intelligence,* 22(8):809–830, August 2000.

16. M. Harville. A framework for high-level feedback to adaptive, per-pixel, Mixture-of-Gaussians background models. In *European Conference on Computer Vision, ECCV 2002,* pages 543–560, 2002.

17. T. Horprasert, D. Harwood, and L. Davis. A statistical approach for real-time robust background subtraction and shadow detection. In *IEEE ICCV Frame-Rate Workshop,* 1999.

18. P. Kaewtrakulpong and R. Bowden. An improved adaptive background mixture model for real-time tracking with shadow detection. In *Workshop on Advances in Vision-based Surveillance Systems,* 2001.

19. T. Kanade, R. T. Collins, A. J. Lipton, P. Burt, and L. Wixson. Advances in cooperative multi-sensor video surveillance. In *DARPA Image Understanding Workshop,* pages 3–24, 1998.

20. K. Kim, T. Chalidabhongse, D. Harwood, and L. Davis. Real-time foreground-background segmentation using codebook model. *Real-time Imaging,* 11(3):172–185, June 2005.

21. D. Lee. Effective Gaussian mixture learning for video background subtraction. *IEEE Transactions on Pattern Analysis and Machine Intelligence,* 27(5):827–832, 2005.

22. X. Luo and S. Bhandarkar. Real-time and robust background updating for video surveillance and monitoring. *Springer Lecture Notes in Computer Science,* 3656:1226–1233, 2005.

23. N. Nguyen, S. Venkatesh, G. West, and H. Bui. Multiple camera coordination in a surveillance system. *ACTA Automatica Sinica,* 29(3):408–422, 2003.

24. F. Oberti, A. Teschioni, and C. Regazzoni. Roc curves for performance evaluation of video sequences processing systems for surveillance applications. In *IEEE International Conference on Image Processing,* pages 949–953, 1999.

25. N. Oliver, B. Rosario, and A. Pentland. A bayesian computer vision system for modeling human interactions. *IEEE Transactions on Pattern Analysis and Machine Intelligence,* pages 831–834, 2000.

26. I. Pavlidis, V. Morellas, P. Tsiamyrtzis, and S. Harp. Urban surveillance systems: from the laboratory to the commercial world. *Proceedings of the IEEE,* 89(10):1478–1497, October 2001.

27. M. Rahimi, R. Baer, O. I. Iroezi, J. C. Garcia, J. Warrior, D. Estrin, and M. Srivastava. Cyclops: In situ image sensing and interpretation in wireless sensor networks. In *International Conference on Embedded Networked Sensor Systems,* pages 192–204, 2005.

28. J. Rittscher, J. Kato, S. Joga, and A. Blake. A probabilistic background model for tracking.

In *European Conference on Computer Vision*, volume 2, pages 336–350, 2000.

29. P. L. Rosin and T. Ellis. Image difference threshold strategies and shadow detection. In *British Machine Vision Conference*, pages 347–356, 1995.

30. S. S. Hengstler, D. Prashanth, S. Fong, and H. Aghajan. Mesheye: A hybrid-resolution smart camera mote for applications in distributed intelligent surveillance. In *International Symposium on Information Processing in Sensor Networks*, pages 360–369, 2007.

31. Y. Sheikh and M. Shah. Bayesian object detection in dynamic scenes. In *IEEE Conference on Computer Vision and Pattern Recognition*, pages 74–79, 2005.

32. A. Shimada, D. Arita, and R. Taniguchi. Dynamic control of adaptive Mixture-of-Gaussians background model. In *IEEE International Conference on Advanced Video and Signal Based Surveillance*, 2006.

33. C. Stauffer and W. E. L. Grimson. Learning patterns of activity using real-time tracking. *IEEE Transactions on Pattern Analysis and Machine Intelligence*, 22(8):747–757, August 2000.

34. B. Stenger, V. Ramesh, N. Paragios, F. Coetzee, and J. Bouhman. Topology free hidden markov models: Application to background modeling. In *International Conference on Computer Vision*, pages 294–301, 2001.

35. K. Toyama, J. Krumm, B. Brumitt, and B. Meyers. Wallflower: Principle and practice of background maintenance. In *International Conference on Computer Vision*, pages 255–261, 1999.

36. C. R. Wren, A. Azarbayejani, T. Darrell, and A. P. Pentland. Pfinder: Real-time tracking of the human body. *IEEE Transactions on Pattern Analysis and Machine Intelligence*, 19(7):780–785, July 1997.

37. Z. Zivkovic. Improved adaptive Gaussian mixture model for background subtraction. In *Proc. of the Int'l Conf. on Pattern Recognition*, pages 28–31, 2004.

V

Benchmarking and Evaluation

23

BGS Library: A Library Framework for Algorithms Evaluation in Foreground/Background Segmentation

Andrews Sobral
Federal University of Bahia, Brazil

Thierry Bouwmans
Lab. MIA, Univ. La Rochelle, France

23.1 Introduction

The background subtraction (BS) is an important task on computer vision applications such as moving object detection (e.g. to detect/segment vehicles [31] and peoples [27]), video surveillance [5] and multimedia applications [10]. Generally the BS consists of comparing an image with another image which represents an estimation of the background model. The foreground objects can be detected by the image regions that have a significant difference between the input image and the reference image (background model). Typically the BS process is defined by: a) background model initialization up to a certain threshold, b) background model maintenance (after the threshold) and c) foreground segmentation. The Figure 23.1 shows the block diagram of the background subtraction process described here.

Over the past years, several BS algorithms have been developed [4,5,7,8]. Some authors, such as Thierry Bouwmans [6], Donovan Parks [26], Zoran Zivkovic [37], Laurence Bender [1], Martin Hofmann [17], Jean Marc Odobez [24], Antoine Vacavant [32] and Domenico Bloisi [3], collaborate with the academic community providing free and open source implementation of some background subtraction methods. However several students, researchers and professionals find it difficult to use these implementations because each author codifies

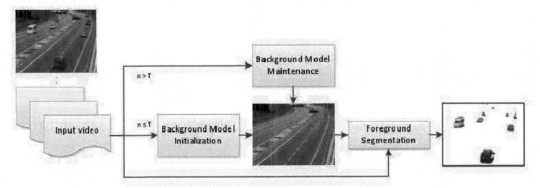

FIGURE 23.1 Block diagram of the background subtraction process.

their algorithms using some particularities such as programming language, operating system, additional libraries (sometimes proprietary), source code format and structure. Therefore an integrated and standardized library that is OS independent, free and open source is important. In this chapter, the authors propose the BGSLibrary [29] that was developed for this purpose.

The BGSLibrary has been designed to provide an easy-to-use C++ framework and tools to perform background subtraction. First released in March 2012, currently the library offers 32 background subtraction algorithms. The source code is available under GNU GPL v3 license and the library is free for non-commercial use, open source and platform independent. Note that the license of the algorithms included in BGSLibrary not necessarily have the same license of the library. Some authors do not allow that their algorithms will be used for a commercial purpose, first is needed to contact them to ask permission. However, by default, we decided to adopt the GPL-v3 license.

The BGSLibrary also provides one Java based GUI (Graphical User Interface) allowing the users to configure the input video-source, region of interest, and the parameters of each BS algorithm. A MFC-based GUI is also provided for a quick access in Windows computers. But, a QT-based GUI is coming for platform independent usage. To build/run the BGSLibrary, it is necessary to have the OpenCV[52] library installed previously. Everyone is invited to collaborate with the BGSLibrary. In this chapter some efforts have been made for how to make and add your contributions in the library. Feel free to contact us to suggest and share your ideas.

In the next sections, we present the main features and some remarks of the BGSLibrary such as available algorithms and tools. It is also explained how to integrate the BGSLibrary on an existing C++ project and how to contribute creating your own background subtraction algorithm. The BGSLibrary software architecture is also described in depth and how their respective components are integrated. Lastly, the conclusion of the present work and future developments are shown.

[52]http://opencv.org/

23.2   The BGSLibrary Algorithms

As initially presented, the BGSLibrary provides 32 background subtraction algorithms. The Table 1 shows the complete list of the algorithms with their respective authors. The algorithms were grouped by their similarity. The mathematical models and theory of each background subtraction method will not be described further in this chapter. For better information please see the references in the Author(s) column.

TABLE 23.1   Background subtraction algorithms available in BGSLibrary.

Method ID	Method Name	Author(s)
Basic methods, mean and variance over time:		
StaticFrameDifferenceBGS	Static Frame Difference	-
FrameDifferenceBGS	Frame Difference	-
WeightedMovingMeanBGS	Weighted Moving Mean	-
WeightedMovingVarianceBGS	Weighted Moving Variance	-
AdaptiveBackgroundLearning	Adaptive Background Learning	-
DPMeanBGS	Temporal Mean	-
DPAdaptiveMedianBGS	Adaptive Median	[23]
DPPratiMediodBGS	Temporal Median	[9]
Color and/or texture features:		
DPTextureBGS	Texture BGS	[16]
LbpMrf	Texture-Based Foreground Detection with MRF	[20]
Statistical methods with one gaussian:		
DPWrenGABGS	Gaussian Average	[33]
LBSimpleGaussian	Simple Gaussian	[2]
Statistical methods with multiple gaussians:		
DPGrimsonGMMBGS	Gaussian Mixture Model	[30]
MixtureOfGaussianV1BGS	Gaussian Mixture Model	[19]
MixtureOfGaussianV2BGS	Gaussian Mixture Model	[38]
DPZivkovicAGMMBGS	Gaussian Mixture Model	[38]
LBMixtureOfGaussians	Gaussian Mixture Model	[7]
Statistical methods with color and/or texture features:		
MultiLayerBGS	Multi-Layer BGS	[34]
Non-parametric methods:		
PixelBasedAdaptiveSegmenter	Pixel-Based Adaptive Segmenter	[18]
GMG	GMG	[14]
VuMeter	VuMeter	[15]
KDE	KDE	[13]
Eigenspace-based methods:		
DPEigenbackgroundBGS	Eigenbackground / SL-PCA	[25]
Fuzzy based methods:		
FuzzySugenoIntegral	Fuzzy Sugeno Integral	[35]
FuzzyChoquetIntegral	Fuzzy Choquet Integral	[11]
LBFuzzyGaussian	Fuzzy Gaussian	[28]
Type-2 Fuzzy based methods:		
T2FGMM_UM	Type-2 Fuzzy GMM-UM	[12]
T2FGMM_UV	Type-2 Fuzzy GMM-UV	[12]
T2FMRF_UM	Type-2 Fuzzy GMM-UM with MRF	[36]
T2FMRF_UV	Type-2 Fuzzy GMM-UV with MRF	[36]
Neural and neuro-fuzzy methods:		
LBAdaptiveSOM	Adaptive SOM	[21]
LBFuzzyAdaptiveSOM	Fuzzy Adaptive SOM	[22]

23.3   The BGSLibrary Tools

The BGSLibrary provides two simple tools for using the background subtraction methods available in Table 23.1. The instructions were divided into two subsections according to the user's operating system.

23.3.1   For Windows Users

For Windows users there are both 32 bits (**bgs_library_x86_vX.X.X_with_gui.7z**) and 64 bits (**bgs_library_x64_vX.X.X_with_gui.7z**) of the BGSLibrary, where vX.X.X is the current library version. Both versions include one Java GUI (Graphical User Interface) (see Figures 23.2 and 23.3) which enables the user to set and configure several parameters of the BGSLibrary. To run the Java GUI click on **run_gui.bat** file, or **run_gui_with_console.bat** for debug purposes. The Java GUI is optional, the user can run the BGSLibrary on windows console or command-line prompt by clicking on **bgslibrary.exe** file (see Section 23.3.2). Currently, the Java GUI runs only on Windows platform, this may seem strange at first because the Java language is multiplatform, but some functions will be adapted to make the Java GUI platform independent. Moreover, the source code of the Java GUI is currently closed, but in the next versions we will make it open source also. The BGSLibrary also provides a new, fast and easy-to-use MFC (Microsoft Foundation Class) GUI (see Figure 23.4). The MFC GUI is available currently only in 32 bits by **mfc_bgslibrary_x86_vX.X.X.7z**, were vX.X.X is the software version. The user can select the background subtraction method and set the input parameters. It is also possible to save the input image, the foreground mask and the background model. The MFC GUI is free and open source.

23.3.2   For Linux Users

For Linux users, the user needs to download the latest version of the BGSLibrary source code[53]. Firstly the user needs a Subversion client (SVN) and follow the instructions in the footnotes[54]. The BGSLibrary also provides a CMake and README.txt file with the instructions to build the library. After compilation, the binary files are available in the **build** folder. Currently there still isn't a GUI tool for linux users, but we plan to create it based on the Qt framework. Running the BGSLibrary on the command-line prompt, the following options are listed:

−**use_cam** Use the laptop's built-in webcam.

−**camera** Specify the camera index.

−**use_file** Use video file.

−**filename** Specify video file name.

−**stopAt** Stop at the specified frame number.

−**use_comp** Use the foreground mask comparator.

−**imgref** Specify the foreground mask reference.

---

[53]http://bgslibrary.googlecode.com/svn/trunk/
[54]http://code.google.com/p/bgslibrary/source/checkout

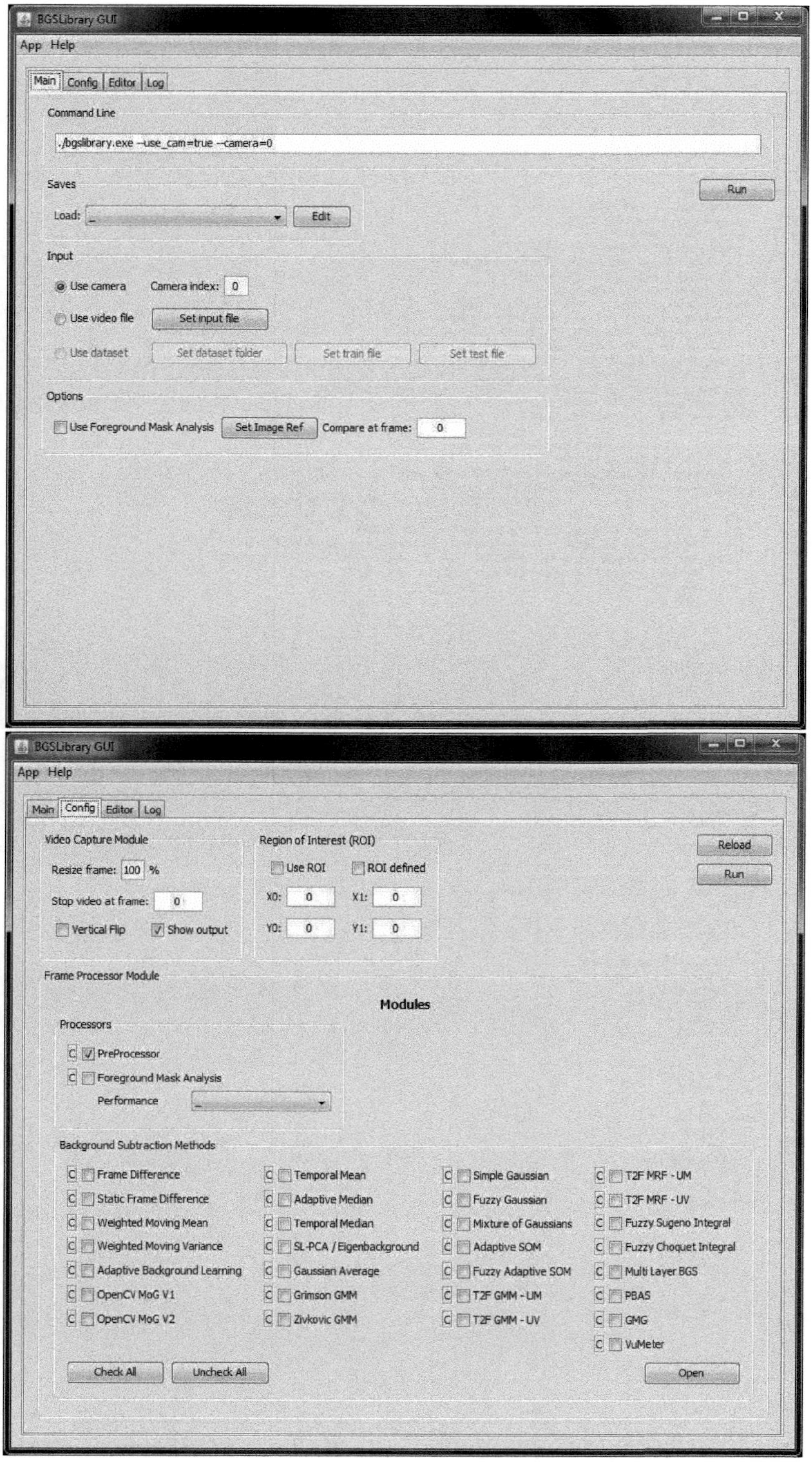

FIGURE 23.2 Java GUI for BGSLibrary (Part 1). The top image shows the main panel and the bottom image shows the configuration panel.

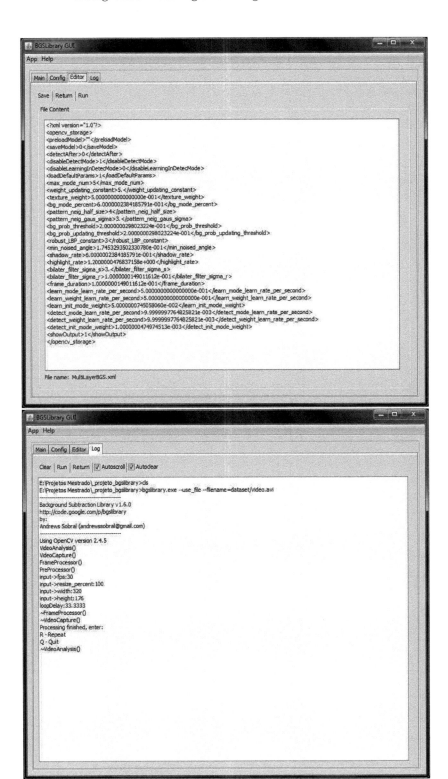

FIGURE 23.3 Java GUI for BGSLibrary (Part 2). The top image shows the editor panel and the bottom image shows the log/console panel.

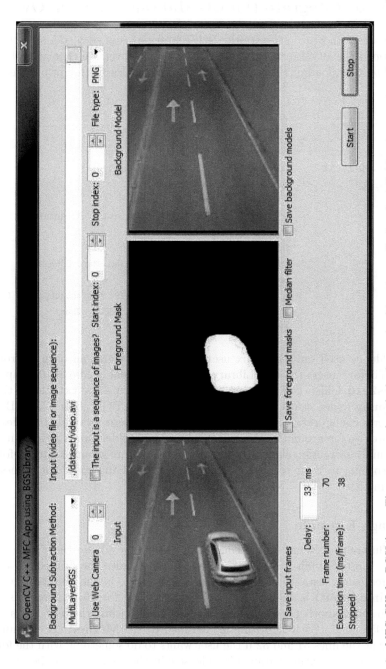

FIGURE 23.4 MFC GUI for BGSLibrary. The user can select the background subtraction method and set the input parameters. It is also possible to save the input image, the foreground mask and the background model.

Three batch files aere also provided. To run the webcam demo click on **run_camera.sh**, or **run_video.sh** and **run_demo.sh** for video file demo. Edit the **./config/FrameProcessor.xml** to enable/disable your chosen BS algorithm.

23.4   How to Integrate the BGSLibrary in Your Own Code

This section describes how to integrate the BGSLibrary in your own code. Firstly the user needs to download the BGSLibrary source code using a Subversion client (SVN) (see Section 23.3.2). After downloading the source code, the user will see the following directory tree:

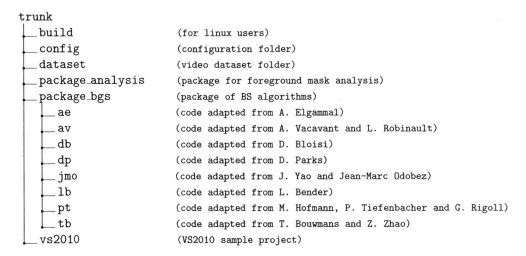

```
trunk
├── build              (for linux users)
├── config             (configuration folder)
├── dataset            (video dataset folder)
├── package_analysis   (package for foreground mask analysis)
├── package_bgs        (package of BS algorithms)
│   ├── ae             (code adapted from A. Elgammal)
│   ├── av             (code adapted from A. Vacavant and L. Robinault)
│   ├── db             (code adapted from D. Bloisi)
│   ├── dp             (code adapted from D. Parks)
│   ├── jmo            (code adapted from J. Yao and Jean-Marc Odobez)
│   ├── lb             (code adapted from L. Bender)
│   ├── pt             (code adapted from M. Hofmann, P. Tiefenbacher and G. Rigoll)
│   └── tb             (code adapted from T. Bouwmans and Z. Zhao)
└── vs2010             (VS2010 sample project)
```

The **build** folder is used only for linux users. This folder stores all outputs from cmake process. For Windows users the BGSLibrary provides a Visual Studio 2010 sample project available in **vs2010** folder (see also **bgslibrary_vs2010_opencv.txt** file in the **trunk** folder). The **config** folder stores all configuration files of the BGSLibrary. Each BS method also stores its own configuration file in this folder. The **package_analysis** provides a simple class to compare two foreground masks. The **package_bgs** folder stores all implementations of the background subtraction algorithms. Some algorithms are grouped into subfolders (ae, av, ..., tb) according to their respective authors (author of the code, not necessarily the author of the method).

23.4.1   Usage Example

Listing 23.1 illustrates a demo source code that shows how to use the Frame Difference method with a laptop's built-in webcam. The line 05 includes the desired BS algorithm. The line 17 creates an instance of the algorithm and lastly, in the line 29, the input image (variable **img_input**) is then processed. The resulting foreground mask is stored in **img_mask**. To select another background subtraction algorithm, the user needs to change only the lines 05 and 17. Each algorithm is identified by the **Method ID** column in Table 23.1. Uncomment the lines 31 and 32 if the user wants to do something with the foreground mask. To change the algorithm parameters, please edit the configuration file in the **config** folder (i.e. FrameDifferenceBGS.xml).

Listing 23.1   demo.cpp

```cpp
#include <iostream>
#include <cv.h>
#include <highgui.h>

#include "package_bgs/FrameDifferenceBGS.h"

int main(int argc, char **argv)
{
  CvCapture *capture = cvCaptureFromCAM(0);
  if(!capture)
  {
    std::cerr << "Cannot initialize camera!" << std::endl;
    return 1;
  }

  IBGS *bgs;
  bgs = new FrameDifferenceBGS;

  int key = 0;
  while(key != 'q')
  {
    IplImage *frame = cvQueryFrame(capture);
    if(!frame) break;

    cv::Mat img_input(frame);
    cv::imshow("input", img_input);

    cv::Mat img_mask;
    bgs->process(img_input, img_mask);

    //if(!img_mask.empty())
    //  do something

    key = cvWaitKey(1);
  }

  delete bgs;

  cvDestroyAllWindows();
  cvReleaseCapture(&capture);

  return 0;
}
```

23.5   BGSLibrary Architecture

In this section we show the BGSLibrary architecture. Firstly, the most important aspect is that all algorithms inherit a standard interface named IBGS. Listing 23.2 shows the IBGS interface code. In line 07, the process (...) function is defined. This function is responsible to process the input image (first parameter) and returns two parameters: **img_foreground**, the foreground mask, and **img_background** the background model. It is important to note that the **img_background** variable can be empty. This will depend on the chosen BS algorithm, some BS methods do not create a background model. Moreover, the **img_foreground** variable may be empty in the first video frames, because some of BS algorithms require some frames to create the background model.

The saveConfig() and loadConfig() functions are optional. These functions are responsible to store and load the BS methods parameters. The configuration file must be of the type XML (eXtensible Markup Language) and located at **config** folder.

Figure 23.5 shows the block diagram of the BGSLibrary architecture model. As can be seen in Figure 23.5, the main block (orange) is responsible for loading some configuration parameters and instantiate the Video Analysis (VA) component. The VA component, by its turn, process the user input parameters and instantiate two components, the Video Capture (VC) and Frame Processor (FP). The VC component is responsible to load the video stream (e.g. from video file or camera). The user can edit some configuration parameters (e.g. region of interest, input image/frame resize factor, etc...) of VC component editing the **Video-Capture.xml** file in **config** folder. To process each video frame / image from video stream, the VC component supports an IFrameProcessor interface. The FP component inherits this interface, and the VA component is responsible to instantiate the FP component and send it to VC component. The possibility to have many FP's with different purposes is the main advantage of this approach. In BGSLibrary, the FP component is responsible to instantiate the Pre Processor (PP) component and all user-selected background subtraction algorithms. Firstly, the PP component apply some preprocessing techniques in the input image/frame (e.g. histogram equalization, blur filters, conversion to gray scale, etc...). Next, the pre-processed image is sent to the all enabled BS algorithms. By default, all BS algorithms automatically shows the foreground mask image. The user can edit **FrameProcessor.xml** file in the **config** folder to enable/disable the BS algorithms.

Listing 23.2   IBGS.h

```
1   #pragma once
2   #include <cv.h>
3
4   class IBGS
5   {
6   public:
7     virtual void process(const cv::Mat &img_input, cv::Mat &img_foreground, cv
          ::Mat &img_background) = 0;
8     virtual ~IBGS() {}
9
10  private:
11    virtual void saveConfig() = 0;
12    virtual void loadConfig() = 0;
13  };
```

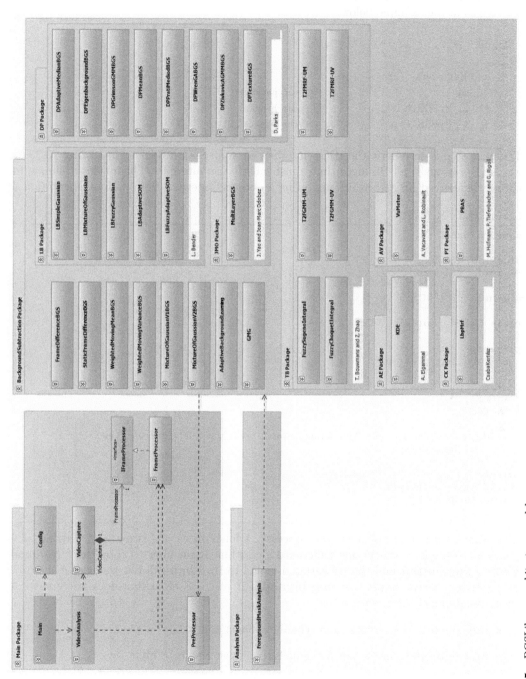

FIGURE 23.5 BGSLibrary architecture model.

23.6   How to Contribute

Every interested party is invited to cooperate by writing / sending your BS algorithm. In this section, some instructions and tips are presented. First you need to create your own package as in the example below:

```
trunk
├── ...
└── package_bgs
    └── my                    (your package)
        ├── MyBGS.h           (your header)
        └── MyBGS.cpp         (your source)
```

The next step is to inherit the IBGS interface in your class header as in Listing 23.3. As was said previously, all BS algorithms must inherit this interface. This makes it simple to choose the desired BS algorithm separating an objects interface from its implementation.

Listing 23.3   MyBGS.h

```cpp
1   #pragma once
2
3   #include <cv.h>
4   #include <highgui.h>
5
6   #include "../IBGS.h"
7
8   class MyBGS : public IBGS
9   {
10  private:
11      cv::Mat img_previous;
12
13  public:
14      MyBGS();
15      ~MyBGS();
16
17      void process(const cv::Mat &img_input, cv::Mat &img_output, cv::Mat &
            img_bgmodel);
18
19  private:
20      void saveConfig(){}
21      void loadConfig(){}
22  };
```

After that, you must implement the process(...) function as in Listing 23.4. Listing 23.4 shows an example of the Frame Difference algorithm. Remember to copy the foreground mask to **img_output** and the background model to **img_bgmodel** at the end of the process function. Leave empty the **img_bgmodel** if your BS algorithm does not creates a background model. Also, ensure that:

- **img_output** is an 8 bits single channel image; and

- **img_bgmodel** can be 1 or 3 channels, but must also be 8 bits.

The **img_input** is always an 8 bits 3 channels image.

Listing 23.4 MyBGS.cpp

```
1   #include "MyBGS.h"
2
3   MyBGS::MyBGS() {}
4   MyBGS::~MyBGS() {}
5
6   void MyBGS::process(const cv::Mat &img_input, cv::Mat &img_output, cv::Mat &
        img_bgmodel)
7   {
8     if(img_input.empty())
9       return;
10
11    if(img_previous.empty())
12      img_input.copyTo(img_previous);
13
14    cv::Mat img_foreground;
15    cv::absdiff(img_previous, img_input, img_foreground);
16
17    if(img_foreground.channels() == 3)
18      cv::cvtColor(img_foreground, img_foreground, CV_BGR2GRAY);
19
20    cv::threshold(img_foreground, img_foreground, 15, 255, cv::THRESH_BINARY);
21
22    img_foreground.copyTo(img_output);
23    img_previous.copyTo(img_bgmodel);
24
25    img_input.copyTo(img_previous);
26  }
```

Here are some basic tips and best practices you can use to help ensure the success of your code:

- Avoid using restricted or proprietary libraries. Give preference to free and open source libraries.

- Also, if possible, avoid using specific headers and functions of your OS. Write your BS algorithm with cross-platform compatibility.

- Please consider the use of name spaces carefully when you are writing your code. Some classes may have the same name in other package. This may cause problems when compiling the library.

- Check if your code has buffer overflows and memory leak occurrences. A memory leak is unnecessary memory consumption by a computer program. If a program with a memory leak runs long enough, it can eventually run out of usable memory. This issue can crash your BS algorithm for long stream videos. There are several memory leak detection tools available on the internet. The leaktracer[55] and deleaker[56] are simple and efficient memory-leak tracers for C++ programs.

---

[55] http://www.andreasen.org/LeakTracer/
[56] http://deleaker.com/

23.7   Conclusion

In this chapter, a background subtraction library is presented. This library was developed to serve as a framework for detection and segmentation of moving objects using background subtraction techniques. It is expected that the explanations presented here can collaborate with those who have an interest in this library. It also aims to evaluate all BS methods of the BGSLibrary with the most well-known public video data sets likes SABS[57], BMC[58] and ChangeDetection.net[59]. These evaluations can further help the user to choose the best background subtraction algorithm for his own situation.

As the majority of the researchers had more experience with scientific programming languages, we are joining efforts to create a compatible version of the BGSLibrary for MATLAB[60] users. The MATLAB software is widely used in academic and research institutions as well as industrial enterprises. Many researchers also write their BS algorithms in MATLAB which will make easier to add these algorithms in the BGSLibrary.

At the time of writing this chapter, the BGSLibrary's website has been viewed more than 30,000 times and the library files have been downloaded more than 10,000 times (only binaries, this does not includes the SVN source code checkouts). The top 10 countries that most used the library were: 1. USA ($>$3,226), 2. China ($>$3,195), 3. India ($>$2,427), 4. Brazil ($>$1,537), 5. Japan ($>$1,297), 6. Germany ($>$1,086), 7. France ($>$1,058), 8. South Korea ($>$987), 9. Russia ($>$963) and 10. Italy ($>$895), where ($>$X) is the number of accesses. This demonstrates how this library may be useful and famously known worldwide. The main goal of this work is to continuously improve the BGSLibrary, adding new features and new background subtraction methods. This library would not succeed without the collaboration of the academic community and professionals from around the world.

23.8   Acknowledgments

We would like to thank all those who have contributed in some way to the success of this library, especially the following people (in alphabetical order): Ahmed Elgammal (Rutgers University, USA), André Medeiros (Portugal), Antoine Vacavant (Université d'Auvergne, France) and Lionel Robinault (Foxstream, France), Csaba Kertész (Vincit Oy, Finland), Domenico Bloisi (Sapienza Università di Roma, Italy), Donovan Parks (Dalhousie University, Canada), Eduardo Barreto Alexandre (Lapix, Brasil), Iñigo Martínez, Jian Yao and Jean-Marc Odobez (Idiap Research Institute, Switzerland), Pierre-Marc Jodoin, Laurence Bender (Universidad Nacional de Tres de Febrero, Argentina), Luca Iocchi (Sapienza Università di Roma, Italy), Martin Hofmann, Philipp Tiefenbacher and Gerhard Rigoll (Technische Universitat München, Germany), Rim Trabelsi, Stefano Tommesani (Tommesani.com, Italy), Thierry Bouwmans (Université de La Rochelle, France), Vikas Reddy (University of Queensland, Australia), Yani Ioannou, Zhenjie Zhao (Nankai University, China), and Zoran Zivkovic (University of Amsterdam, Netherlands).

---

[57] http://www.vis.uni-stuttgart.de/index.php?id=sabs
[58] http://bmc.univ-bpclermont.fr/
[59] http://changedetection.net/
[60] www.mathworks.com/products/matlab/

References

1. L. Bender. Scene. `http://scene.sourceforge.net/`, 2013.
2. Y. Benezeth, P. Jodoin, B. Emile, H. Laurent, and C. Rosenberger. Review and evaluation of commonly-implemented background subtraction algorithms. *IEEE International Conference on Pattern Recognition, ICPR 2008*, pages 1–4, 2008.
3. D. Bloisi. IMBS. `http://www.dis.uniroma1.it/~bloisi/software/imbs.html`, 2013.
4. T. Bouwmans. Recent advanced statistical background modeling for foreground detection: A systematic survey. *Recent Patents on Computer Science*, 4(3):147–176, 2011.
5. T. Bouwmans. Background subtraction for visual surveillance: A fuzzy approach. In *Handbook on Soft Computing for Video Surveillance*. Taylor and Francis Group, 2012.
6. T. Bouwmans. Background Subtraction Website. `https://sites.google.com/site/thierrybouwmans/`, 2013.
7. T. Bouwmans, F. El Baf, and B. Vachon. Background modeling using mixture of Gaussians for foreground detection - a survey. *Recent Patents on Computer Science*, 1(3):219–237, 2008.
8. T. Bouwmans, F. El Baf, and B. Vachon. Statistical background modeling for foreground detection: A survey. In *Handbook of Pattern Recognition and Computer Vision*, volume 4, chapter 3, pages 181–199. World Scientific Publishing, 2010.
9. S. Calderara, R. Melli, A. Prati, and R. Cucchiara. Reliable background suppression for complex scenes. *ACM International Workshop on Video Surveillance and Sensor Networks*, 2006.
10. F. El Baf and T. Bouwmans. Comparison of background subtraction methods for a multimedia learning space. *International Conference on Signal Processing and Multimedia, SIGMAP 2007*, 2007.
11. F. El Baf, T. Bouwmans, and B. Vachon. Fuzzy integral for moving object detection. *IEEE International Conference on Fuzzy Systems, FUZZ-IEEE 2008*, pages 1729–1736, 2008.
12. F. El Baf, T. Bouwmans, and B. Vachon. Type-2 fuzzy mixture of Gaussians model: Application to background modeling. *International Symposium on Advances in Visual Computing (ISVC)*, 2008.
13. A. Elgammal, D. Harwood, and L. Davis. Non-parametric model for background subtraction. *European Conference on Computer Vision, ECCV 2000*, pages 751–767, 2000.
14. A. Godbehere, A. Matsukawa, and K. Goldberg. Visual tracking of human visitors under variable-lighting conditions for a responsive audio art installation. *American Control Conference, ACC 2012*, 2012.
15. Y. Goyat, T. Chateau, L. Malaterre, and L. Trassoudaine. Vehicle trajectories evaluation by static video sensors. *IEEE International Conference on Intelligent Transportation Systems*, 2006.
16. M. Heikkila and M. Pietikinen. A texture-based method for modeling the background and detecting moving objects. *IEEE Transactions on Pattern Analysis and Machine Intelligence*, 28, 2006.
17. M. Hofmann. PBAS Segmenter. `https://sites.google.com/site/pbassegmenter/home`, 2013.
18. M. Hofmann, P. Tiefenbacher, and G. Rigoll. Background segmentation with feedback: The pixel-based adaptive segmenter. *IEEE Computer Society Conference on Computer Vision and Pattern Recognition Workshops, CVPRW 2012*, pages 38–43, 2012.
19. P. Kaewtrakulpong and R. Bowden. An improved adaptive background mixture model for realtime tracking with shadow detection. *European Workshop on Advanced Video Based Surveillance Systems, AVSS 2001*, 2001.
20. C. Kertsz. Texture-based foreground detection. *International Journal of Signal Processing,*

Image Processing and Pattern Recognition, 4, 2011.

21. L. Maddalena and A. Petrosino. A self-organizing approach to background subtraction for visual surveillance applications. *IEEE Transactions on Image Processing*, 17(7):1168–1177, 2008.

22. L. Maddalena and A. Petrosino. A fuzzy spatial coherence-based approach to background/-foreground separation for moving object detection. *Neural Computing and Applications*, 19(2):179–186, 2010.

23. N. McFarlane and C. Schofield. Segmentation and tracking of piglets in images. *Machine Vision and Applications*, 8(3):187–193, 1995.

24. J. Odobez. `http://www.idiap.ch/~odobez/`, 2013.

25. N. Oliver, B. Rosario, and A. Pentland. A Bayesian computer vision system for modeling human interactions. *IEEE Transactions on Pattern Analysis and Machine Intelligence*, 22(8):831–843, 2000.

26. D. Parks. BGS Package. `http://dparks.wikidot.com/source-code/`, 2013.

27. J. Raheja, S. Kalita, P. Jyoti Dutta, and S. Lovendra. A robust real time people tracking and counting incorporating shadow detection and removal. *International Journal of Computer Applications*, 46(4):51–58, May 2012. Published by Foundation of Computer Science, New York, USA.

28. M. Sigari, N. Mozayani, and H. Pourreza. Fuzzy running average and fuzzy background subtraction: Concepts and application. *International Journal of Computer Science and Network Security*, 8(2):138–143, 2008.

29. A. Sobral. BGSLibrary: An opencv c++ background subtraction library. In *IX Workshop de Visão Computacional, WVC 2013*, Rio de Janeiro, Brazil, June 2013.

30. C. Stauffer and W. E. L. Grimson. Adaptative background mixture models for a real-time tracking. *IEEE International Conference on Computer Vision and Pattern Recognition, CVPR 1999*, 1999.

31. L. Unzueta, M. Nieto, A. Cortes, J. Barandiaran, O. Otaegui, and P. Sanchez. Adaptive multicue background subtraction for robust vehicle counting and classification. *IEEE Transactions on Intelligent Transportation Systems*, 13(2):527–540, 2012.

32. A. Vacavant. `http://isit.u-clermont1.fr/~anvacava/index.html`, 2013.

33. C. Wren, A. Azarbayejani, T. Darrell, and A. Pentland. Pfinder: Real-time tracking of the human body. *IEEE Transactions on Pattern Analysis and Machine Intelligence*, 19(7):780–785, 1997.

34. J. Yao and Jean marc Odobez. Multi-layer background subtraction based on color and texture. *IEEE Computer Vision and Pattern Recognition Conference (CVPR)*, 2007.

35. H. Zhang and D. Xu. Fusing color and texture features for background model. *International Conference on Fuzzy Systems and Knowledge Discovery*, 2006.

36. Z. Zhao, T. Bouwmans, X. Zhang, and Y. Fang. A fuzzy background modeling approach for motion detection in dynamic backgrounds. *International Conference Communications in Computer and Information Science*, 346:177–185, 2012.

37. Z. Zivkovic. `http://www.zoranz.net/`, 2013.

38. Z. Zivkovic and F. V. D. Heijden. Efficient adaptive density estimation per image pixel for the task of background subtraction. *Pattern Recognition Letters*, 27(7):773–780, 2006.

24

Overview and Benchmarking of Motion Detection Methods

Pierre-Marc Jodoin
Université de Sherbrooke, Canada

Sébastien Piérard
Université de Liège, Belgium

Yi Wang
Université de Sherbrooke, Canada

Marc Van Droogenbroeck
Université de Liège, Belgium

24.1 Introduction

Motion detection is closely coupled with higher level inference tasks such as detection, localization, tracking, and classification of moving objects, and is often considered to be a preprocessing step. Its importance can be gauged by the large number of algorithms that have been developed to-date and the even larger number of articles that have been published on this topic. A quick search for "motion detection" on IEEE Xplore© returns over 20,000 papers. This shows that motion detection is a fundamental topic for a wide range of video analytic applications. It also shows that the number of motion detection methods proposed so far is impressively large.

Among the many variants of motion detection algorithms, there seems to be no single algorithm that competently addresses all of the inherent real world challenges. Such challenges are sudden illumination variations, night scenes, background movements, illumination changes, low frame rate, shadows, camouflage effects (photometric similarity of object and background) and ghosting artifacts (delayed detection of a moving object after it has moved away) to name a few.

In this chapter, we provide an overview of the most highly cited motion detection methods. We identify the most commonly used background models together with their features, the kind of updating scheme they use, some spatial aggregation models as well as the most widely used post-processing operations. We also provide an overview of datasets used to validate motion detection methods. Please note that this literature review is by no means exhaustive and thus we provide a list of surveys that the reader can rely on for further details.

Together with this overview, we provide benchmarking results on different categories of videos, for different methods, different features and different post-processing methods. This should give the reader a perspective on some of the most effective methods available nowadays on different types of videos. We also report results on majority vote strategies that we use to combine the output of several methods. The goal being to identify complementary methods whose combination helps improving results. All benchmark results have been obtained on the changedetection.net dataset.

24.2   Motion Detection Methods

Motion detection is often achieved by building a representation of the scene, called background model, and then observing deviations from this model for each incoming frame. A sufficient change from the background model is assumed to indicate a moving object. In this document, we report on the most commonly-used models which we refer to as the *basic*, *parametric*, *non-parametric*, *data-driven* and *matrix decomposition* models. Other models for motion detection are also accounted for in this section such as the *prediction* model, the *motion segmentation* model, and the *machine learning* approaches. All these motion detection methods are summarized in Table 24.1.

Together with these 8 families of methods, we review commonly-used features, spatial aggregation techniques, updating scheme as well as post-processing methods.

24.2.1   Basic Models

The simplest strategy to detect motion is to subtract the pixel's color in the current frame from the corresponding pixel's color in the background model [7]. A temporal median filter can be used to estimate a color-based background model [55]. One can also generalize to other features such as color histograms [42, 98] and local self-similarity features [35]. In general, these temporal filtering methods are sensitive to compression artifacts, global illumination changes and incapable of detecting moving objects once they become stationary.

Frame differencing is another basic model. It aims to detect changes in the state of a pixel by subtracting the pixel's intensity (or color) in the current frame from its intensity (or color) in the previous frame. Although this method is computationally inexpensive, it cannot detect a moving object once it stops moving or when the object motion becomes small; instead it typically detects object boundaries, covered and exposed areas due to object motion.

Motion history images [8, 59] are also used as a basic model for motion detection. It is obtained by successive layering of frame differences. For each new frame, existing frame differences are scaled down in amplitude, subject to some threshold, and the new motion label field is overlaid using its full amplitude range. In consequence, image dynamics ranging from two consecutive frames to several dozen frames can be captured in a single image.

24.2.2   Parametric Models

In order to improve robustness to noise, parasite artifacts, and background motion, the use of a per-pixel Gaussian model has been proposed [91]. In a first step, the mean and standard deviation are computed for each pixel. Then, for each frame, the likelihood of each pixel color is determined and pixels whose probability is below a certain threshold are labeled as foreground pixels. Since pixels in noisy areas are given a larger standard deviation, a larger color variation is needed in those areas to detect motion. This is a fundamental difference with the basic models for which the tolerance is fixed for every pixel. As shown by Kim *et al.* [40], a generalized Gaussian model can also be used and Morde *et al.* [59] have shown that a Chebychev inequality can also improve results. With this model, the detection criteria depends on how many standard deviations a color is from the mean.

A single Gaussian, however, is not a good model for dynamic scenes [26] as multiple colors may be observed at a pixel due to repetitive object motion, shadows or reflectance changes. A substantial improvement is achieved by using multiple statistical models to describe background color. A Gaussian Mixture Model (GMM) [78] was proposed to represent each background pixel. GMM compares each pixel in the current frame with every model in the mixture until a matching Gaussian is found. If a match is found, the mean and variance of the matching Gaussian are updated, otherwise a new Gaussian with the mean equal to the current pixel color and some initial variance is introduced into the mixture. Instead of relying on only one pixel, GMM can be trained to incorporate extended spatial information [36].

Several papers [37] improved the GMM approach to add robustness when shadows are present and to make the background models more adaptive to parasitic background motion. A recursive method with an improved update of the Gaussian parameters and an automatic selection of the number of modes was presented in [101]. Haines *et al.* [29] also propose an automatic mode selection method, but with a Dirichlet process. A splitting GMM that relies on a new initialization procedure and a mode splitting rule was proposed in [23, 24] to avoid over-dominating modes and resolve problems due to newly static objects and moved away background objects while a multi-resolution block-based version was introduced in [73]. The GMM approach can also be expanded to include the generalized Gaussian model [2].

As an alternative to mixture models, Bayesian approaches have been proposed. In [69], each pixel is modeled as a combination of layered Gaussians. Recursive Bayesian update instead of the conventional expectation maximization fitting is performed to update the background parameters and better preserve the multi-modality of the background model. A similar Bayesian decision rule with various features and a learning method that adapt to both sudden and gradual illumination changes is used in [47].

Another alternative to GMM is background clustering. In this case, each background pixel is assigned a certain number of clusters depending on the color variation observed in the training video sequence. Then, each incoming pixel whose color is close to a background cluster is considered part of the background. The clustering can be done using K-means (or a variant of it) [13, 63] or codebook [41].

24.2.3 Non-Parametric Methods

In contrast to parametric models, non-parametric Kernel Density Estimation (KDE) fits a smooth probability density function to a time window with previously-observed pixel values at the same location [21]. During the change detection process, a new-frame pixel is tested against its own density function as well as those of pixels nearby. This increases the robustness against camera jitter or small movements in the background. Similar effects can be achieved by extending the support to larger blocks and using texture features that are less sensitive to inter-frame illumination variations. Mittal and Poggio [58] have shown that robustness to background motion can be increased by using variable-bandwidth kernels.

Although nonparametric models are robust against small changes, they are expensive both computationally and in terms of memory use. Moreover, extending the support causes small foreground objects to disappear. As a consequence, several authors worked to improve the KDE model. For instance, a multi-level method [61] makes KDE computationally independent of the number of samples. A trend feature can also be used to reliably differentiate periodic background motion from illumination changes [95].

24.2.4 Data-Driven Methods

Recently, pixel-based data-driven methods using random samples for background modeling have shown robustness to several types of error sources. For example, ViBe [3, 85] not only shows robustness to background motion and camera jitter but also to ghosting artifacts. Hofmann [32] improved the robustness of ViBe on a variety of difficult scenarios by auto-

Background model	References
Basic	Running average [7, 35, 42, 98] Temporal median [55] Motion history image [8, 59]
Parametric	Single Gaussian [91] Gaussian Mixture Model (GMM) [23, 24, 29, 36, 37, 73, 78, 101] Background clustering [13, 41, 63] Generalized Gaussian Model [2, 40] Bayesian [47, 69] Chebyshev inequality [59]
Non-Parametric	Kernel Density Estimation (KDE) [21, 58, 61, 95]
Data-driven	Cyclostationary [70] Stochastic K-nearest neighbors (KNN) [3, 32]. Deterministic KNN [101] Hidden Markov Model (HMM) [79]
Matrix Decomposition	Principal Component Analysis (PCA) [18, 48, 62, 75, 94] Sparsity and dictionary learning [68, 97]
Prediction Model	Kalman filter [39, 99] Wiener filter [84]
Motion Segmentation	Optical flow segmentation [51, 57, 90] GMM and Optical flow segmentation [63, 100]
Machine Learning	1-Class SVM [16] SVM [30, 31, 50] Neural networks [52, 53, 76]

TABLE 24.1   Overview of 8 families of motion detection methods.

matically tuning its decision threshold and learning rate based on previous decisions made by the system. In both [3, 32], a pixel is declared as foreground if it is not close to a sufficient number of background samples from the past. A deterministic K nearest neighbor approach has also been proposed by Zivkovic and van der Heijiden [101], and one for non-parametric methods by Manzanera [54].

A shortcoming of the above methods is that they do not account for any "temporal correlation" within video sequences, thus they are sensitive to periodic (or near-periodic) background motion. For example, alternating light signals at an intersection, a flashing advertisement board, the appearance of rotating objects, etc. A cyclostationary background generation method based on frequency decomposition that explicitly harnesses the scene dynamics is proposed in [70]. In order to capture the cyclostationary behavior at each pixel, spectral coefficients of temporal intensity profiles are computed in temporal windows and a background model that is composed of those coefficients is maintained and fused with distance maps to eliminate trail effects.

An alternative approach is to use a Hidden Markov Model (HMM) with discrete states to model the intensity variations of a pixel in an image sequence. State transitions can then be used to detect changes [79]. The advantage of using HMMs is that certain events, which may not be modeled correctly by unsupervised algorithms, can be learned using the provided training samples.

24.2.5   Matrix Decomposition

Instead of modeling the variation of individual pixels, the whole image can be vectorized and used in background modeling. In [62], a holistic approach using eigenspace decomposition is proposed. For a certain number of input frames, a background matrix (called *eigen background*) is formed by arranging the vectorized representations of images in a matrix where each vectorized image is a column. An eigenvalue decomposition via Principal Component Analysis (PCA) is performed on the covariance of this matrix. The background is then represented by the most descriptive eigenvectors that encompass all possible illuminations to decrease sensitivity to illumination. Several improvements of the PCA approach have been proposed. To name a few, Xu *et al.* [94] proposed a variation of the eigen background model which includes a recursive error compensation step for more accurate detection. Others [48, 75] proposed PCA methods with a computationally efficient background updating scheme, while Doug *et al.* [18] proposed an illumination invariant approach based on a multi-subspace PCA, each subspace representing different lighting conditions.

Instead of the conventional background and foreground definition, Porikli [68] decomposes a image into "intrinsic" background and foreground images. The multiplication of these images reconstructs the given image. Inspired by the sparseness of the intensity gradient, it applies spatial derivative filters in the log domain to a subset of the previous video frames to obtain intensity gradient. Since the foreground gradients of natural images are Laplacian distributed and independent, the Maximul Likelihood (ML) estimate of the background gradient can be obtained by a median operator and the corresponding foreground gradient is computed. These gradient results are used to reconstruct the background and foreground intensity images using a reconstruction filter and inverse log operator. This intrinsic decomposition is shown to be robust against sudden and severe illumination changes, but it is computationally expensive.

Another background subtraction approach based on the theory of sparse representation and dictionary learning is proposed in [97]. This method makes the following two important assumptions: (1) the background of a scene has a sparse linear representation over a learned dictionary; (2) the foreground is sparse in the sense that a majority of the pixels of the frame belong to the background. These two assumptions enable handling both sudden and gradual background changes.

24.2.6 Other Methods

Prediction Models

Early approaches use filters to predict background pixel intensities (or colors). For these models, each pixel whose observed color is far from its prediction is assumed to indicate motion. In [39] and [99], a Kalman filter is used to model background dynamics. Similarly, in [84] Wiener filtering is used to make a linear prediction at pixel level. The main advantage of these methods are their ability to cope with background changes (whether it is periodic or not) without having to assume any parametric distribution.

In [84] camouflage artifacts and small motionless regions (usually associated to homogeneous foreground regions) are filled in within a post-processing stage. In case most of the pixels in a frame exhibit sudden changes, the background models are assumed to be no longer valid at frame level. At this point, either a previously stored pixel-based background model is swapped in, or the model is reinitialized.

Motion Segmentation

Motion segmentation refers to the assignment of groups of pixels to various classes based on the speed and direction of their movements [51]. Most approaches to motion segmentation first seek to compute optical flow from an image sequence. Discontinuities in the optical flow can help in segmenting images into regions that correspond to different objects. In [90], temporal consistency of optical flow over a narrow time window is estimated; areas with temporally-consistent optical flow are deemed to represent moving objects and those exhibiting temporal randomness are assigned to the background.

Optical flow-based motion detection methods will be erroneous if brightness constancy or velocity smoothness assumptions are violated. In real imagery, such violations are quite common. Typically, optical flow methods fail in low-texture areas, around moving object boundaries, at depth discontinuities, etc. Due to the commonly imposed regularization term, most optical flow methods produce an over smooth optical flow near boundaries. Although solutions involving a discontinuity preserving optical flow function and object-based segmentation have been proposed [57], motion segmentation methods usually produce a halo artifact around moving objects. The resulting errors may propagate across the entire optical flow solution. As a solution, some authors [63, 100] use motion segmentation and optical flow in combination with a color-based GMM model.

Machine Learning

Motion detection methods in this category use machine learning discriminative tools such as SVM and neural networks to decide whether or not a pixel is in motion. The parameters of these functions are learned given a training video. Lin *et al.* [50] use a probabilistic SVM to initialize the background model. They use the magnitude of optical flow and inter-frame image difference as features for classification. Han and Davis [30] model the background with kernel density approximation with multiple features (RGB, gradient, and Haar) and use a Kernel-SVM as a discriminative function. A somewhat similar approach has also been proposed by Hao [31]. These approaches are typical machine learning methods that need positive and negative examples for training. This is a major limitation for any practical implementation since very few videos come with manually labeled data. As a solution, Chen *et al.* [16] proposed a GPU-based 1-class SVM method called SILK. This method does not need pre-labeled training data, but also allows for online updating of the SVM parameters. Maddalena and Petrosino [52, 53] model the background of a video with the weights of a neural network. A very similar approach but with a post-processing MRF stage has been proposed by Schick *et al.* [76]. Results reported in the paper show great compromise between processing speed and robustness to noise and background motion.

24.2.7   Features

Several features can be used to detect moving objects. The simplest one is certainly grayscale (or luminance) which is easy to interpret and has a well founded physical meaning [25]. Grayscale motion detection methods are normally used on mono-channel cameras like depth cameras, thermal cameras, or older grayscale surveillance cameras.

Nowadays, most motion detection methods rely on color. A color image consists of three channels per pixel (typically R, G, B) that can be processed separately or simultaneously. However, the physical meaning of these channels is less obvious than for mono-channel sensors. Ideally, color images are acquired using three spatially aligned sensors. But since this configuration increases the size and cost of the sensor and requires pixel registration, most color cameras use a single image sensor with a color filter array in front of it. The most widely implemented array is the Bayer color filter array [5]. Each location on the sensor measures one color and missing colors are interpolated from neighboring pixels. Suhr [80] proposes a GMM variant that conducts background modeling in a Bayer-pattern domain and foreground classification in an interpolated RGB domain. The authors argue that since performance is similar to that of the original GMM on RGB images, RGB video streams captured with one sensor are not 3 times more informative than their grayscale counterpart.

In practice though, most techniques exhibit a small performance increase for the classification task when using RGB instead of grayscale features [6]. Thus, from a classification perspective and despite that the computation time is more or less tripled, it is beneficial to use color images, even when colors have been interpolated in the image. In their survey paper, Benezeth *et al.* [6] compare six RGB color distance functions used for background subtraction, including the Euclidean distance, the L1 distance, and the Mahalanobis distance. They conclude that four of the six metrics had globally similar classification performances; only the simplest zero and first order distances were less precise.

Several motion detection techniques use other color spaces such as normalized color [58], cylindric color model [41], HSV [17], HSI [88], YCbCr [44], and normalized RGB [93]. From an application perspective, those colorspaces are believed to be more robust to shadows and illuminations changes than RGB or grayscale [17].

Other features, like edges [34], texture [49], optical flow [50, 58, 82], PCA-based features [62] are also used. Like the colorspace features, these features seem more robust to illumination changes and shadows than RGB features. Texture and optical flow features are also robust to noise and background motion. Since texture features integrate spatial information which often happens to be constant, a slight variation in the background does not lead to spurious false positives. For example, a bush with a uniform texture will be undetected when shaken by the wind. As for optical flow, since moving objects are assumed to have a smooth and coherent motion distribution [82], noise and random background motion can be easily decorrelated from actual moving objects.

In general, it seems like adding features improves performances. Parag *et al.* [64] even propose to select the best combination of features at each pixel. They argue that different parts of the image may have different statistics and thus require different features. But this comes at the price of both a complexity and a computation time increase.

24.2.8 Updating Strategies

In order to produce consistent results over time, background models need to be updated as the video streams in. From a model point of view, there are two major updating techniques [66] : the *recursive* and *non-recursive* techniques. The *recursive* techniques maintain a single background model that is updated with each new video frame. *Non-recursive* techniques maintain a buffer L of n previous video frames and estimate a background model based solely on the statistical properties of these frames. This includes median filtering and eigenbackgrounds [62]. The major limitation of this last approach is that computing the basis functions requires video clips void of foreground objects. As such, it is not clear how the basis functions can be updated over time if foreground objects are continuously present in the scene.

As mentioned by Elgammal *et al.* [22], other updating strategies use the output of the segmentation process. The *conditional* approach (also called *selective* or *conservative*) updates only background pixels in order to prevent the background model from being corrupted by foreground pixels. However, this approach is incapable of eliminating false positives as the background model will never adapt to it. Wang *et al.* [87] propose to operate at the blob level and define a mechanism to incorporate pixels in the background after a given period of time. As an alternative, the *unconditional* (or *blind*) approach updates every pixel whether it is identified as being active or not. This approach has the advantage of integrating new objects in the background and compensating for false detections caused, say, by global illumination changes or camera jitter. On the other hand, it can allow slowly moving objects to corrupt the background which leads to spurious false detections. Both conditional and unconditional techniques can be used, depending on the appropriateness to the model or on the requirements of the application.

Some authors introduce more nuances. For example, Porikli *et al.* [69] define a GMM method and a Bayesian updating mechanism, to achieve accurate adaptation of the models. A somewhat similar refinement method is proposed by Van Droogenbroeck *et al.* [85]. Both [69] and [85] distinguish between a *segmentation* mask, the binary output image which corresponds to the background/foreground classification result, and the *updating* mask. The updating mask corresponds to locations indicating which pixels have to be updated. The updating mask differs from the segmentation map in that it remains unknown to the user, and depends on updating strategies. For example, one can decide not to update the model inside of static blobs or, on the contrary, decide to erode foreground mask to progressively remove ghosts. Another recent updating strategy consists in spatial diffusion; it was introduced with ViBe [3]. Spatial diffusion is a mechanism wherein a background value is diffused in a neighboring model. This diffusion mechanism can be modulated to help remove ghosts or static objects.

24.2.9 Spatial Aggregation, Markovian Models and Post-Processing

Most motion detection techniques are local processes that focus on pixel-wise statistics and ignoring neighboring pixels (at least during the modeling phase). This is a well-founded approach from a statistical point of view since neighboring pixels might have very different underlying feature probability density functions. Nevertheless, there exist techniques that aggregate information from neighboring pixels into regular blocks or so-called superpixels. Block-based aggregation is a coherent approach for video encoder, as blocks and macroblocks are the fundamental spatial units in encoders.

Grouping pixels in blocks is motivated by several factors. First, statistics averaged over a rectangular region increases the robustness to non-stationary backgrounds, despite the fact that it blurs the object silhouette and that another method might be needed to refine edges as in [15]. Second, if sharp edges are not mandatory, processing blocks speeds up the motion detection process. Hierarchical methods, as proposed by Park et al. [65] or Chen et al. [14], are typical examples of methods that plays with different levels of pixel aggregation.

Pixels aggregation can also be achieved with the help of a Markovian model. Typical Markovian models are based on a maximum a posteriori formulation that is solved through an optimization algorithm such as iterative optimization scheme (ICM) or graphcut [1,56] which are typically slow. In [7] it was shown that simple Markovian methods (typically those using the Ising prior) produce similar results than simple post-processing filters.

Other Markovian methods have been proposed. In [33], Markov random fields are used to re-label pixels. First, a region-based motion segmentation algorithm is developed to obtain a set coherent regions. This serves to define the statistics of several Markovian random fields. The final labeling is obtained by maximizing the a posteriori energy of the Markov random fields, which can be seen as a post-processing step. The approach by Schick et al. [76] relies on similar ideas. A first segmentation is used to define a probabilistic superpixel representation. Then a post-processing is applied on the statistical framework to provide an enhanced segmentation map. It is interesting to note that Schick et al. [76] have successfully applied their post-processing technique to several motion detection techniques.

A more classical, simpler and faster way to re-label pixels is throughout a post-processing filter. For example, Parks and Fels [66] consider a number of post-processing techniques to improve the segmentation map. Their results indicate that the performance is improved by morphological filters (closings), noise removal filter (such as median filters), and area filters. Morphological filters are used to fill internal holes and small gaps, while area filters are used to remove small objects.

In Section 24.4.2, we present the results of some post-processing operations. It appears that simple post-processing operations, such as the median or close/open morphological operations always improve the segmentation map. It is thus recommended to include post-processing operations, even when comparing techniques. This was also the conclusion of Brutzer et al. [12] and Benezeth et al. [7]. Note that other filters can be used such as temporal filters, shadow filters [12], and complex spatio-temporal filtering techniques to re-label the classification results. However, it has been observed that not all post-processing filters do improve results [66].

24.3 Datasets and Survey Papers

Without aiming to be exhaustive, we list below 15 of the most widely used datasets for motion detection validation (see Table 24.2). Additional details regarding some of these

Dataset	Description	Ground truth
2012 CD.net[61]	31 videos in 6 categories : baseline, dynamic background, camera jitter, shadow, intermittent motion, and thermal.	Pixel-based labeling of 71,000 frames.
Wallflower [84]	7 short video clips, each representing a specific challenge such as illumination change, background motion, etc.	Pixel-based labeling of one frame per video.
PETS [96]	Many videos aimed at evaluating the performance of tracking algorithms	Bounding boxes.
CAVIAR[62]	80 staged indoor videos representing different human behaviors such as walking, browsing, shopping, fighting, etc.	Bounding boxes.
i-LIDS[63]	Very long videos meant for action recognition showing parked vehicle, abandoned object, people walking in a restricted area, and doorway	Not fully labeled.
ETISEO[64]	More than 80 videos meant to evaluate tracking and event detection methods.	High-level label such as bounding boxes, object class, event type, etc.
ViSOR 2009[65] [86]	Web archive with more than 500 short videos (usually less than 10 seconds)	Bounding boxes.
BEHAVE 2007[66]	7 videos shot by the same camera showing human interactions such as walking in group, meeting, splitting, etc.	Bounding boxes.
VSSN 2006[67]	9 semi-synthetic videos composed of a real background and artificially-moving objects. The videos contain animated background, illumination changes and shadows, however they do not contain any frames void of activity.	Pixel-based labeling of each frame.
IBM [68]	15 videos taken from PETS 2001 plus additional videos.	Bounding box around each moving object in 1 frame out of 30.
Karlsruhe[69]	4 grayscale videos showing traffic scenes under various weather conditions.	10 frames per video have pixel-based labeling.
Li *et al.* [47][70]	10 small videos (usually 160×120) containing illumination changes and dynamic backgrounds.	10 frames per video have pixel-based labeling.
Karaman *et al.* [38]	5 videos coming from different sources (the web, the "art live" project[71], etc.) with various illumination conditions and compression artifacts	Pixel-based labeling of every frame.
cVSG 2008 [83][72]	15 Semi-synthetic videos with various levels of textural complexity, background motion, moving object speed, size and interaction.	Pixel-based labeling obtained by filming moving objects (mostly humans) in front of a blue-screen and then pasted on top of background videos.
Brutzer *et al.* [12]	Computer-generated videos showing one 3D scene representing a street corner. The sequences include illumination changes, dynamic background, shadows, and noise.	Pixel-based labeling.

TABLE 24.2 Overview of 15 video datasets.

datasets can be found on a web page of the European CANTATA project[73].

Out of these 15 datasets, 7 were initially made to validate tracking and pattern recognition methods (namely PETS, CAVIAR, i-LIDS, ETISEO, ViSOR 2009, BEHAVE 2007, IBM). Although challenging for these applications, those datasets mostly contain day-time videos with fixed background, constant illumination, few shadows and no camera jitter. As a consequence, it is difficult to evaluate how robust motion detection methods are when looking at benchmarking results reported on these 7 datasets.

CD.net[74] [27] is arguably the most complete dataset devoted to motion detection benchmarking. It contains 31 videos in 6 categories with nearly 90,000 frames, most of it manually labeled. A unique aspect with this dataset is its web site which allows for performance evaluation and method ranking. As of today, 27 methods are reported on the website. CD.net replaced older datasets which had either few videos, partly labeled videos, very short video clips, or semi-synthetic (and yet not realistic) videos.

In parallel of these datasets, a number of survey papers have been written on the topic of motion detection. In this chapter, we list survey papers devoted to the comparison and ranking of motion detection algorithms. Note that some of these surveys use datasets mentioned previously while others use their own dataset. Details can be found in Table 24.3.

These survey papers often contain a good overview of state-of-the-art motion detection method. However, the reader shall keep in mind that statistics reported in some of these papers were not computed on a well-balanced dataset composed of real (camera-captured) videos. Typically, synthetic videos, real videos with synthetic moving objects pasted in, or real videos out of which only 1 frame was manually segmented for ground truth were used. Also, some survey papers report results from fairly simple and old motion detection methods.

24.4   Experimental Results

So far, we introduced eight families of motion detection methods, presented different features, several updating schemes and many spatial aggregation and post-processing methods. The goal of this section is to provide empirical results to validate which configuration performs best. Note that since the number of combinations of motion detection methods, features, updating schemes and post-processing methods is intractable, we provide benchmarks for each aspect independently.

The goal of this section is also to underline unsolved issues in motion detection and identify complementary methods whose combination can further improve results. Empirical results are obtained with the 2012 CD.net dataset. As mentioned previously, this dataset includes 31 videos divided in 6 categories namely dynamic background, camera jitter, shadow, intermittent motion, baseline, and thermal. With a manually labeled groundtruth for each video, to our knowledge this is the most complete dataset for motion detection validation.

But prior to present benchmarking results, we first describe and explain the evaluation metrics used in this section.

24.4.1   Metric Evaluation

As stated by Goyette *et al.* [28], it is not a trivial task to find the right metric to accurately measure the ability of a method to detect motion. If we consider background segmentation as a classification process, then we can recover the following 4 quantities for a processed video:

---

[73] www.hitech-projects.com/euprojects/cantata/datasets_cantata/
[74] www.changedetection.net

Survey	Description and Benchmark
Goyette *et al.*, 2012 [27]	Survey paper written in the wake of the CVPR 2012 Change Detection workshop. It surveys several methods and reports benchmark results obtained on the CD.net dataset.
Bouwmans *et al.*, 2011 [9]	Probably the most complete surveys to date with more than 350 references. The paper reviewed methods spanning 6 motion detection categories and the features used by each method. The survey also listed a number of typical challenges and gave insights into memory requirements and computational complexity. Benchmark on the Wallflower dataset.
Brutzer *et al.*, 2011 [12]	Report benchmarking results for 8 motion detection method on the computer-generated Brutzer dataset.
Benezeth *et al.*, 2010 [7]	Used a collection of 29 videos (15 camera-captured, 10 semi-synthetic, and 4 synthetic) taken from PETS 2001, the IBM dataset, and the VSSN 2006 dataset.
Bouwmans *et al.*, 2008 [10]	Survey of GMM methods. Benchmarking has been done on the Wallflower dataset.
Parks and Fels, 2008 [66]	Benchmark results for 7 motion detection methods and evaluation of the influence of post-processing on their performance. They used 7 outdoor and 6 indoor videos containing different challenges such as dynamic backgrounds, shadows and various lighting conditions.
Bashir and Porikli, 2006 [4]	Performance evaluation of tracking algorithms using the PETS 2001 dataset by comparing the detected bounding box locations with the ground-truth.
Nascimento and Marques, 2006 [60]	Report benchmarks obtained on a single PETS 2001 video with pixel-based labeling.
Radke *et al.* [72]	Extensive survey of several motion detection methods. Most of the discussion in the paper was related to background subtraction methods, pre- and post-processing, and methodologies to evaluate performances. Contains no quantitative evaluation.
Piccardi [67]	Reviewed 7 background subtraction methods and highlighted their strengths and weaknesses. Contains no quantitative evaluation.
Rosin and Ioannidis, 2003 [74]	Report results for 8 methods. Videos used for validation show two lab scenes with balls rolling on the floor.
Prati *et al.*, 2001 [71]	Used indoor sequences containing one moving person. 112 frames were labeled.

TABLE 24.3   15 motion detection surveys.

the number of true positives (TP) and false positives (FP), which accounts for the number of foreground pixels correctly and incorrectly classified, and the number of true negatives (TN) and false negatives (FN), which are similar measures but for background pixels. With these values, one can come out with the following seven metrics, often used to rank background subtraction methods. (1) The *True Positive Rate* (TPR), also named *sensitivity* and *recall*, is $Re = TPR = \frac{TP}{TP+FN}$ (2) The *False Negative Rate (FNR)* is $FNR = 1 - TPR$. (3) The *True Negative Rate* (TNR), also named *specificity*, is $TNR = \frac{TN}{TN+FP}$. (4) The *False Positive Rate* is $FPR = 1 - TNR$. (5) The *precision* is $Pr = \frac{TP}{TP+FP}$. (6) The *Probability of Wrong Classification* (also Error Rate) is $PWC = \frac{FP+FN}{TP+TN+FP+FN}$. (7) The *Accuracy* is $A = 1 - PWC$.

In the upcoming subsections, we will try to answer the question of which metric(s) should be used to rank methods.

The downside of TPR, TNR, Accuracy, and PWC.

For obvious reasons, TPR, TNR, FPR, and FNR cannot be used alone to rank methods. In fact, methods are typically adjusted to prevent FPR and FNR from being large simultaneously. Such trade-offs can be interpreted by showing a Receiver Operating Characteristic (ROC) graph. But ranking methods based on ROC curves is rather inconvenient due to the large number of results that need to be generated and which can be prohibitive in the context of large videos. Therefore, people often try to rank methods based on only one point in the ROC space. But, summarizing TPR and TNR into a single value remains difficult and this is highlighted by the following discussion.

Since most surveillance videos exhibit a low amount of activity (5 % on average for the CD.net video sequences), the TNR value will always dominate A (the *Accuracy*) and PWC. Actually, as one can see in Table 24.4, nearly all methods have a very low FPR (except for [35]) and a large FNR. As a consequence, when used alone, the accuracy A and the probability of wrong classification PWC will always favor methods with a low FPR and a large FNR. At the limit, a method that would detect no moving object at all would have a not-so-bad ranking score according to A and PWC alone. That is because only a small fraction of the pixels would be wrongly classified on average.

Another way of underlying the limitation of A (and PWC) is by rewriting the accuracy equation. If we denote the probabilities of observing a foreground pixel and a background pixel by p_{FG} and p_{BG} respectively, then one can show that the accuracy can be computed as follows $A = p_{FG} TPR + p_{BG} TNR$. Thus, with a low p_{FG} (as is the case with most surveillance videos), the TNR ends up having an overwhelming importance when computing A. As an alternative, one could consider the *Balanced Accuracy*, $BA = \frac{1}{2} TPR + \frac{1}{2} TNR$. However, since that metric is uncommon in the motion detection community, we decided not to use it.

So in conclusion, although the accuracy and the probability of wrong classification can be used to evaluate methods, they should not be used alone and one should keep in mind that they favor methods with a low FPR.

Metrics Derived from Pr and Re

Another trade-off for motion detection methods is to prevent Pr and Re from being small simultaneously, which is shown on a precision-recall curve. The classical difficulties encountered when interpreting such a curve are mainly related to an unachievable region in the precision-recall space [11]. There exists a lower bound on the precision which depends on the recall and p_{FG}. The size of this region grows with p_{FG}, and the precision varies significantly with p_{FG} [46]. Moreover, a purely random method is such that $Pr = p_{FG}$, and therefore

(a) ground truth	(b) method 1	(c) method 2	(d) method 3

FIGURE 24.1 Three methods with the same balanced accuracy (0.8) but with different F-Measures. For methods 1 and 2, $F_1 = 0.35$ while for method 3 $F_1 = 0.73$.

the range for valid Pr is not the same for all videos. Therefore, the metrics derived from Pr and Re should be interpreted with care.

But using precision-recall curves to rank methods is inconvenient for the same reasons ROC curves are. In practice, precision and recall must be combined into one single metric. The most frequent way of doing so is through the F-measure (F_1), which is the harmonic mean between Pr and Re: $F_1 = \left(\frac{1}{2}Pr^{-1} + \frac{1}{2}Re^{-1}\right)^{-1} = \frac{2TP}{2TP+FP+FN}$. When both Pr and Re are large, F_1 is approximately equal to the arithmetic mean of Pr and Re. Otherwise, it is approximately equal to $\min(Pr, Re)$.

The balanced accuracy and the F-Measure, although similar at the first glance, are not equivalent. Let's consider the case shown in Figure 24.1 for which it is difficult to identify the human silhouette based on the first two results. In that example, all three results have the same balanced accuracy but a much higher F-Measure for method 3. This is a strong indication that the F-measure is a better metric than the balanced accuracy in the context of motion detection and thus why we use it in our validation. Another reason for F_1 to be larger for method 3 is the fact that it does not take into account TN. As a consequence, F_1 is a metric that focuses more on the foreground than on the background.

Influence of Noise

The F-measure is not void of limitations. As will be shown in this section, it is sensitive to noise and thus should be used with care. In order to illustrate the impact of noise and the importance of post-filtering operations, let us add a "salt and pepper" noise to a segmentation map. Let α be the probability to switch the class of a pixel and TPR' and TNR' the estimates on the noisy segmentation maps. In that case, we have $TP' = TP(1-\alpha) + FN\alpha$, $FN' = FN(1-\alpha) + TP\alpha$, $TN' = TN(1-\alpha) + FP\alpha$, and $FP' = FP(1-\alpha) + TN\alpha$. Following some algebric manipulations, one can show that the relative ranking between 2 methods can change depending on the amount of noise in the data. This is illustrated in Figure 24.2 where F_1 for Spectral-360 goes below PBAS after noise has been added.

This sensitivity to noise leads us to conclude that it is preferable to filter noise with a post-processing filter before ranking background subtraction techniques according to F_1. This is what has been done for every method reported in Table 24.4.

Interestingly, the ranking provided by PWC, the accuracy and the balanced accuracy is not affected by noise as is the case for the F-Measure and here is why. After some algebraic manipulations, we find that $TPR' = \frac{TP'}{TP'+FN'} = (1-\alpha)TPR + \alpha(1-TPR)$, and $TNR' = \frac{TN'}{TN'+FP'} = (1-\alpha)TNR + \alpha(1-TNR)$. If $\alpha < \frac{1}{2}$, then the same amount of noise on the results of two segmentation algorithms does not change their respective ranking, if the ranking is determined by any linear combination Q of TPR and TNR. Indeed, $Q = \beta TPR + (1-\beta)TNR$ and $Q' = \beta TPR' + (1-\beta)TNR' = (1-\alpha)Q + \alpha(1-Q)$. Thus $Q_{algoritm\,1} \le Q_{algorithm\,2}$ if and only if $Q'_{algorithm\,1} \le Q'_{algorithm\,2}$. This implies that

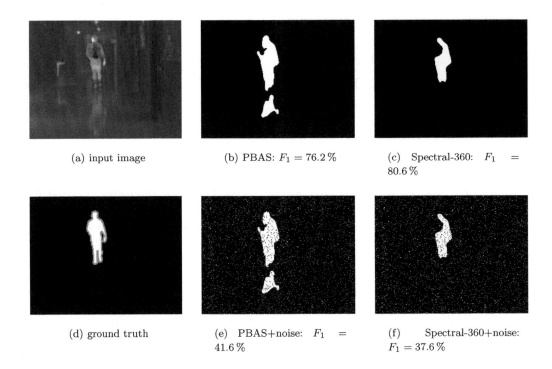

(a) input image   (b) PBAS: $F_1 = 76.2\%$ (c) Spectral-360: $F_1 = 80.6\%$

(d) ground truth   (e) PBAS+noise: $F_1 = 41.6\%$ (f) Spectral-360+noise: $F_1 = 37.6\%$

FIGURE 24.2 Ranking of the methods obtained according to F-measure is sensitive to noise. It is therefore important to filter out noise from the results before ranking methods with F_1.

the ranking according to the accuracy, the balanced accuracy, or the probability of wrong classification is stable as long as the same amount of noise is present on the output of the compared methods.

Evaluation and Ranking of Methods

The previous discussion made it clear that summarizing the performance of a background subtraction algorithm with a single metric is restrictive. Several metrics like FNR and FPR are complementary and cannot be used independently whereas others like PWC and A give an overwhelming importance to TNR. As for the F-measure, although widely used, it is sensitive to noise. This leads us to conclude that no metric is perfect and should thus be used with care.

The last question that we ought to answer before presenting benchmarking results, is how to compute evaluation metrics when considering more than one video sequence. Naively, one could add up the total number of TP, TN, FP and FN across all videos out of which metrics could be computed. But since videos have different sizes in space and time, large videos would end up having more influence on the ranking than smaller ones. As explained by Goyette *et al.* [28], a better solution is to compute the metrics for each video (that is $Re, FNR, FPR, Specificity, Pr, PWC, A,$ and F_1) and then average it across videos. CD.net also has a multi-criteria ranking which we do not retain in this chapter for the sake of simplicity.

In this chapter, we rank methods according to the average F_1 computed across all videos and categories of the CD.net dataset. Although sensitive to noise, the result of every method has been post-processed with a median filter to prevent the previously-mentioned ranking problems. Also, Goyette *et al.* [28] mentioned that the F_1 score is correlated with this

Method	Description	FPR	FNR	F-Measure
Spectral-360 [77]	Patent	0.008	0.22	0.77
DPGMM [29]	GMM + Dirichlet Process	0.014	0.17	0.77
SGMM-SOD [23]	Improved version of SGMM [24]	0.006	0.23	0.76
PBAS [32]	data-driven and stochastic method	0.010	0.21	0.75
PSP-MRF [76]	Probabilistic super-pixels with Neural Maps	0.017	0.19	0.73
SC-SOBS [53]	Improved version of SOBS [52]	0.016	0.19	0.72
SOBS [52]	Neural maps	0.018	0.21	0.71
SGMM [24]	GMM + new mode initialization, updating and splitting rule	0.009	0.29	0.70
Chebyshev Inequality [59]	Multistage method with Chebyshev inequality and object tracki	0.011	0.28	0.70
KNN [101]	Data-driven KNN	0.009	0.32	0.67
KDE Elgammal [21]	Original KDE	0.024	0.25	0.67
GMM Stauffer-Grimson [78]	Original GMM	0.014	0.28	0.66
GMM Zivkovic [101]	GMM with automatic mode selection	0.015	0.30	0.65
KDE Yoshinaga *et al.* [95]	Spatio-temporal KDE	0.009	0.34	0.64
KDE Nonaka *et al.* [61]	Multi-level KDE	0.006	0.34	0.64
Bayesian Multi layer [69]	Bayesian layers + EM	0.017	0.39	0.62
Mahalanobis distance [7]	Basic background subtraction	0.040	0.23	0.62
Euclidean distance [7]	Basic background subtraction	0.030	0.29	0.61
GMM KaewTraKulPong [37]	Self-adapting GMM	0.005	0.49	0.59
Histogram over time [98]	Basic method with color histograms	0.065	0.23	0.54
GMM RECTGAUSS-Tex [73]	Multiresolution GMM	0.013	0.48	0.52
Local-Self similarity [35]	Basic method with self- similarity measure	0.148	0.06	0.50

TABLE 24.4 Overall results for 22 methods. These results correspond to the average FPR, FNR and $F - Measure$ obtained on all 31 videos of the CD.net dataset.

multi-criteria ranking which is a good indication that F_1 is a well balanced metric.

Let us mentioned that the benchmarking results do not entirely capture the pros and cons of a method. Obviously, the complexity of an algorithm together with its processing speed and memory usage are to be considered for real-time applications.

24.4.2 Benchmarks
Motion Detection Methods

From the changedetection.net website, we retained results from 22 motion detection methods. Five methods are relatively simple as they rely on plain background subtraction, of which two use color features (Euclidean and Mahalanobis distance methods described in [7,9]), one uses RGB histograms over time [98], and one uses local self-similarity features [35].

We also report results for eight parametric methods, seven of which use a GMM model.

Category	1st	F-measure	2nd	F-measure	3rd	F-measure
Baseline	SC-SOBS	0.93	Spectral-360	0.93	PSP-MRF	0.92
Dynamic Back.	DPGMM	0.81	Spectral-360	0.79	Chebyshev Inequality	0.77
Shadows	Spectral-360	0.88	SGMM-SOD	0.86	PBAS	0.86
Camera Jitter	PSP-MRF	0.78	DPGMM	0.75	SGMM	0.75
Thermal	DPGMM	0.81	Spectral-360	0.78	PBAS	0.76
Interm. Motion	SGMM-SOD	0.72	SC-SOBS	0.59	PBAS	0.58

TABLE 24.5   Three highest ranked method for each category together with their F-measure.

This includes the well-known methods by Stauffer and Grimson [78], a self-adapting GMM by KaewTraKulPong [37], the improved GMM method by Zivkovic and Heijden [101], the multiresolution block-based GMM (RECTGAUSS-Tex) by Dora *et al.* [73], GMM method with a Dirichlet process (DPGMM) that automatically estimated the number of Gaussian modes [29] and the SGMM and SGMM-SOD methods by Evangelio *et al.* [23, 24] which rely on a new initialization procedure and novel mode splitting rule. We also included a recursive per-pixel Bayesian approach by Porikli and Tuzel [69] which shows good robustness to shadows according to [27].

We also report results on three KDE methods. The original method by Elgammal *et al.* [21], a multi-level KDE by Nonaka *et al.* [61], and a spatio-temporal KDE by Yoshinaga *et al.* [95]. Results for data-driven methods and machine learning methods are also reported. That is Hofmann's stochastic and self-adaptive method (PBAS) [32], a simple K-nearest neighbor method [101] and neural maps methods (SOBS and SC-SOBS) by Maddalena *et al.* [52, 53] and a neural network method with a region-based Markovian post-processing method (PSP-MRF) by Schick *et al* [76]. We also have results for two commercial products. One that does pixel-level detection using the Chebyshev inequality and peripheral and recurrent motion detectors by Morde *et. al.* [59] and one which has only been published in a pending patent so far and whose description is not available [77]. The false positives rate (FPR), false negative rate (FNR) and F-measure (F_1) for these 22 methods are reported in Table 24.4. Note that, as mentioned in [27], these are the *average* FPR, FNR and F_1 across all videos.

From these results, one can conclude that the top performing methods are mostly GMM methods (DPGMM, SGMM-SOD, SGMM), data-driven methods (KNN and PBAS) and machine learning methods (SOBS and PSP-MRF). As shown in Table 24.5, GMM methods (particularly DPGMM and SGM-SOD) seems robust to background motion, camera jitter and intermittent motion. This can be explained by the fact that these GMM methods come with mode initialization (and updating) procedures that react swiftly to changes in the background. Table 24.5 also shows that there is room for improvement on jittery sequences and intermittent motion which are the categories with the lowest F-measure. Another unsolved issue is robustness to shadows. Although the F-measure of the most effective methods is above 0.86, the FPR on shadows of the 3 best methods is above 58%. This means that even the most accurate methods wrongly classify hard shadows.

Features

Here, we report results for 8 of the most commonly-used features *i.e.*: grayscale, RGB, Normalized RGB, HSL, HSV, norm of the gradient, RGB+gradient and YCbCr. We tested these features with two different methods. The first one is a basic background subtraction method with a forgetting constant of 0.002 [7]. The second is a version of ViBe [3] (a stochastic data-driven method) that we adapted to the various color spaces and removed its postprocessing stage.

Results in Table 24.6 lead us to two main conclusions. First, using all three RGB color

Category	Gray	RGB	N-RGB	HSL	HSV	grad	RGB+grad	YCbCr
Basic Method	0.48	0.53	0.49	0.56	0.58	0.3	0.59	0.59
ViBe [3]	0.72	0.75	0.60	0.65	0.74	0.11	0.74	0.71

TABLE 24.6 F-Measure obtained for 8 different features and two motion detection methods.

Blind	Conservative	Soft-Conservative	Edge-Conservative
0.52	0.5	0.53	0.55

TABLE 24.7 F-measure for different background updating strategies.

channels when possible instead of grayscale only always improves results. Second, out of the "illumination-robust" features N-RGB, HSL, HSV and gradient (grad), only HSV seems to provide good results globally. That being said, combining gradient with RGB helps improving results, especially for the basic method. As mentioned by several authors, this suggests that for some methods, combining color and texture is a good way of improving results.

Updating Scheme

In this section, we tested different updating schemes on one method. We tested the blind, conservative, "soft" conservative and "edge" conservative updating schemes. Again, we implemented a simple background subtraction method with RGB color feature. The difference from one implementation to another is the forgetting constant α. For the blind scheme, $\alpha = 0.002$, for the conservative $\alpha = 0.002$ only for background pixels, the soft conservative $\alpha = 0.002$ for foreground pixels and $\alpha = 0.008$ for background pixels and edge conservative, $\alpha = 0.007$ for background pixel, $\alpha = 0.002$ for foreground edge pixels and $\alpha = 0$ for the other foreground pixels.

Results in Table 24.7 show that the edge-conservative strategy is the most effective one while the conservative strategy is the least effective, although by a small margin. The reason for this small difference between results comes from the fact that the CD.net videos are all relatively short (at most 6 minute-long videos) and thus do not exhibit major changes in the background as is the case when dealing with longer videos. Longer videos would certainly stretch the difference between each strategy.

Post-Processing

In this section, we compared different post-processing filters on the output of 3 methods. These methods are a basic background subtraction method with a forgetting constant of 0.002, ViBe [3] and ViBe+ [85]. Note that ViBe+ is a method which already has a post-processing stage. The post-processing methods are 3 median filters (3×3, 5×5, and 7×7), a morphological opening and closing operation, a closing operation followed by a region filling procedure (as suggested by Parks and Fels [66]) and a connected component analysis. The latter removes small isolated regions, whether they are active regions or not.

Results in Table 24.8 show that all post-processing filters improved the results of all 3 methods. Surprisingly, it also improved the results of ViBe+, a method which already had a post-processing stage! Of course, the improvement rate is more significant for a low ranked method than for a precise one. Given its positive impact on performance and noise removal, we recommend to use at least a 5x5 median, but also other filtering operations to fill gaps, smooth object shapes or remove small regions.

Method	No Post-processing	Med 3×3	Med 5×5	Med 7×7	Morph	Close +fill.	Connected component
Basic Method	0.53	0.56	0.63	0.60	0.55	0.54	0.58
ViBe [3]	0.67	0.68	0.68	0.69	0.70	0.70	0.68
ViBe+ [85]	0.71	0.72	0.73	0.73	0.74	0.74	0.72

TABLE 24.8 F-Measure obtained for 6 different postprocessing filters on the output of 3 motion detection methods.

24.4.3   Combining Methods

So far, we analyzed and compared the behavior of individual motion detection techniques. A further step consists of combining methods. From that point, at least two questions arise: how should methods be combined, and which methods, similar or dissimilar, should be combined?

There are two strategies to combine methods: (1) consider every available methods, regardless of their own performance, or (2) select a small subset of methods, based on their performance or on an optimization criterion. Here, we explore 3 different combination rules : two involving all $n = 22$ methods and one involving a subset of methods. Because it is difficult to model the correlation between individual methods and to take it into account, the combination rules considered here are based on the assumption that individual classifiers are independent of each other. An alternative would be to learn the combination rule [19], but this is out of the scope of this chapter.

The results obtained with the 3 different combination rules are shown on precision recall graphs with F_1 contour lines. It should be noted that the conclusions that can be drawn from the receiver operating characteristic space are different from those of the precision recall space. In this chapter, we only focus on the latter, and aim at maximizing the F_1 score. The following observations should therefore be interpreted with care.

Combination rule 1: majority vote among all methods.
We define a decision thresholding function $\mathcal{F}_{Th}$ as follows: $\mathcal{F}_{Th}(x) = 0$ if $x < Th$, and $\mathcal{F}_{Th}(x) = 1$ otherwise. Let us denote the output of the ith background subtraction method by $\hat{y}_i \in \{0,1\}$, the combined output by $\hat{y}_c \in \{0,1\}$, the ground truth by $y \in \{0,1\}$, and probabilities by $p(\cdot)$. The first combination rule considered in this chapter is

$$\hat{y}_c = \mathcal{F}_{Th}\left(\frac{1}{n}\sum_{i=1}^{n}\hat{y}_i\right), \qquad \text{with } Th \in [0,1]. \tag{24.1}$$

We refer to this technique as the "majority vote" rule, since it extends the classical unweighted majority vote (this one is obtained when n is odd, and the decision threshold is set to $Th = 0.5$). Note that Equation (24.1) defines $n + 1$ monotonic functions from $\{0,1\}^n$ into $\{0,1\}$. This combination rule supposes that the individual background subtraction algorithms are independent. Limits of what can be expected from such a combination are discussed in [45]. The results, obtained for every decision threshold, are shown in Figure 24.3(a).

Combination rule 2: summation.
Another combination rule which is often encountered in the literature is the summation rule [43], which is also known as the mean rule [81], or the averaged Bayes' classifier [92].

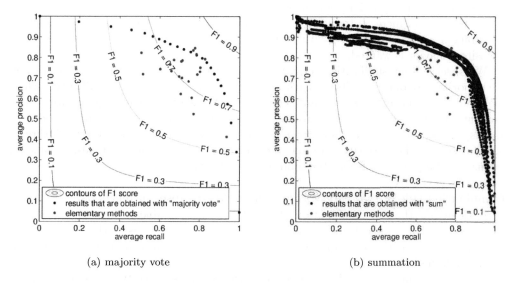

(a) majority vote　　　　　　　　　　　(b) summation

FIGURE 24.3 (See color insert.) Results obtained from the combination of all (22) background subtraction methods, with 2 combination rules. For the purpose of comparison, the precision and the recall of the 22 individual methods are displayed in red. The blue dots correspond to different decision thresholds (Th) as well as different estimations of the priors (Υ).

Adapted to our framework, it can be written as

$$\hat{y}_c = \mathcal{F}_{Th}\left(\frac{1}{n}\sum_{i=1}^{n} p\left(y = 1 | \hat{y}_i, \Upsilon\right)\right), \qquad \text{with } Th \in [0, 1], \tag{24.2}$$

where

$$p\left(y = 1 | \hat{y}_i = 0, \Upsilon\right) = \frac{FNR_i\, p\left(y = 1 | \Upsilon\right)}{TNR_i\, p\left(y = 0 | \Upsilon\right) + FNR_i\, p\left(y = 1 | \Upsilon\right)}, \tag{24.3}$$

$$p\left(y = 1 | \hat{y}_i = 1, \Upsilon\right) = \frac{TPR_i\, p\left(y = 1 | \Upsilon\right)}{FPR_i\, p\left(y = 0 | \Upsilon\right) + TPR_i\, p\left(y = 1 | \Upsilon\right)}. \tag{24.4}$$

Here Υ represents the knowledge about the context (Υ is sometimes named the *environment*, as in [92]). The context is, for example, an indoor video-surveillance application, a particular video stream, the other pixels in the same image, or some information related to the past. However, the choice should be carefully made, since it can have an important impact on the performance of the combination. The priors $p\left(y = 0 | \Upsilon\right)$ and $p\left(y = 1 | \Upsilon\right)$ are usually estimated on the basis of the decisions taken by the individual methods on the whole image, in order to adapt dynamically to the context. But, in some video-surveillance settings, some video regions are more likely to contain movement than others. In this case, it makes sense to estimate the priors on a neighborhood around the considered pixel, and also to take the history into account. This is somehow equivalent to the atlas used in [89], but in a dynamic setting.

The results for this combination rule are shown in Figure 24.3(b). We have considered the whole range of decision thresholds, and four ways of estimating the priors: (1) fixed priors $(p\left(y = 1\right) \in \{4\,\%,\, 8\,\%,\, 12\,\%,\, 16\,\%,\, 20\,\%\})$; (2) priors estimated on the whole image; (3) priors estimated on the whole image, with a temporal exponential smoothing applied on the estimated priors (with a smoothing factor $\alpha \in \{0.90, 0.37, 0.21, 0.14, 0.09, 0.05, 0.02\}$); (4) priors estimated per pixel, on a square neighborhood of size $s \in \{1, 7, 31, 127, 511\}$. Note that estimating the priors for a combination is an ill-posed problem since false positives (false negatives) tend to increase (reduce) the estimated prior of the foreground, and therefore to encourage a higher number of positives (negatives) in the combined output. Obviously, the opposite behavior is wanted.

We observe some similarities between the majority vote and the summation. However, the majority vote only permits to reach $n = 22$ points in the precision recall space, whereas the summation permits a fine tuning. The optimal threshold for the majority vote and the summation varies significantly from one video to another (this is not represented on the graphs). Thus, there is a trade-off when choosing the threshold. The best overall threshold is about 0.4 for the majority vote and the sum. We have obtained our best results when estimating the priors on a neighborhood of 31×31 pixels.

Combination rule 3: majority vote of a predefined subset of methods.

It turns out that no combination of the 22 methods is able to beat significantly the best individual methods. Carefully selecting a subset of methods is therefore necessary. Note that an alternative would be to assign a "good" weight to each individual background subtraction method.

Our third combination rule is the same as the previous majority vote one, except that it is applied on a subset of 3, 5 and 7 methods. Since computing the majority vote of every possible combination of methods is extremely time consuming, we first determined the 50 most promising subsets of methods. A prediction of the F_1 score has been obtained for every combination of 3, 5 and 7 methods, without the need to try them on the video sequences. This is possible under the assumption of independence, based on the values given in Table 24.4, and the knowledge of the proportion of foreground pixels in each video sequence[75].

The results obtained with the third combination rule are depicted in Figure 24.4. We used a decision threshold of 0.5. Whereas a blind combination of all methods together does not permit to beat significantly the best individual methods (see Figure 24.3), combining carefully selected subsets of methods leads to a higher performance than the methods independently (see Figure 24.4).

We have also observed how many times each method appears in the selected subsets of 3, 5, and 7 methods. We have noticed that, as expected, the methods which have the highest F_1 score are often taken into account, even if the ranking in Table 24.4 is not strongly correlated with the occurrences. Surprisingly, about one third of the methods are never selected. What is even more surprising is that the Local-Self similarity method [35], which has the worst ranking according to F_1 in Table 24.4, appears often in the selected combinations for 3 methods, and is systematically used in the top 50 subsets of 5 and 7 methods, with no exception. Note that it is not a side effect of the independence assumption, as taking this method into account does not harm to the performance when the errors are positively correlated, as the results shown in Figure 24.4 illustrate. What should be noted about the

---

[75] In a probabilistic framework, the classifier independence means that $p\left(\hat{y}_1, \ldots, \hat{y}_n | y\right) = \prod_{i=1}^{n} p\left(\hat{y}_i | y\right)$. The proportion of foreground pixels in a video sequence is $p\left(y = 1\right) = p_{FG}$. Table 24.4 gives $p\left(\hat{y}_i = 1 | y = 0\right) = FPR_i$ and $p\left(\hat{y}_i = 1 | y = 1\right) = 1 - FNR_i$. Let $\mathcal{C}$ denotes the combination function (the majority vote in this case) such that $\hat{y}_c = \mathcal{C}\left(\hat{\mathbf{y}}\right)$ with $\hat{\mathbf{y}} = \left(\hat{y}_1, \ldots, \hat{y}_n\right)$. Under the independence assumption, we expect $p\left(\hat{y}_c | y\right) = \sum_{\hat{\mathbf{y}} = \left(\hat{y}_1, \ldots, \hat{y}_n\right), \text{ s.t. } \mathcal{C}(\hat{\mathbf{y}}) = \hat{y}_c} \prod_{i=1}^{n} p\left(\hat{y}_i | y\right)$. The predicted precision and accuracy are derived from $p\left(\hat{y}_c = 1 | y = 1\right)$ and $p\left(\hat{y}_c = 0 | y = 0\right)$ thanks to the knowledge of $p\left(y = 1\right)$. They are averaged across all videos and categories, before deriving the predicted value of F_1. As we do not expect the independence assumption to hold in practice, the predicted values are only used for the selection of the most promising subsets of methods. The results presented in the figures are obtained experimentally.

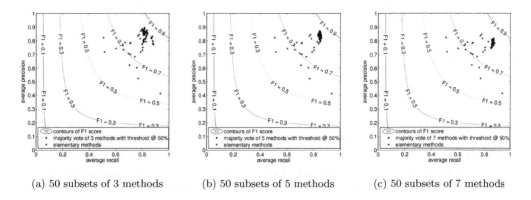

(a) 50 subsets of 3 methods   (b) 50 subsets of 5 methods   (c) 50 subsets of 7 methods

FIGURE 24.4   (See color insert.) Real precision and recall of the majority vote combination rule (at the neutral decision threshold). The predicted performance is shown, in blue, for 50 combinations of 3, 5, and 7 methods, selected theoretically. The precision and the recall of the 22 individual methods are shown in red.

Local-Self similarity method [35] is that it behaves differently from the other methods: it has the highest TPR, but also the highest FPR. Intuitively, a method that behaves differently may be useful in a combination, even if it has a bad ranking when evaluated alone, thanks to its complementarity with the other methods. This effect has already been observed by Duin *et al.* [20]. Therefore, if combining multiple background subtraction methods is possible, designing methods that are top-ranked, when they are evaluated alone, should not be the primary focus. Instead, designing complementary methods is preferable.

24.5   Conclusion

In this chapter, we presented a survey of 8 families of motion detection methods, presented different features, several updating schemes and many spatial aggregation and post-processing methods. We also provided several benchmarking results based on the CD.net dataset. These results lead us to the following conclusions :

1. **Methods** : As of today, GMM (DPGMM, SGMM-SOD, SGMM), data-driven methods (KNN and PBAS) and machine learning methods (SOBS and PSP-MRF) are among the most effective ones. That being said, none are performing best on every category.

2. **Remaining Challenges** : Intermittent motion, camera jitter and hard shadows are among the most glaring issues.

3. **Features** : HSV and RGB + gradient are the most effective features.

4. **Updating scheme** : The edge-conservative approach is the most effective scheme while the conservative approach is the least effective.

5. **Post-processing** : Every post-processing method that we have tested improved the results of our motion detection methods, especially for the simple low-ranked method. Post-processing should thus always be used.

6. **Combining methods** : One can beat the best performing methods by combining the output of several methods. The best results have been obtained with a majority vote of 3 and 5 methods and with a threshold of 50%. The best results are obtained

by not only combining top ranked methods, but by combining methods which are complementary by nature.

References

1. T. Aach and A. Kaup. Bayesian algorithms for adaptive change detection in image sequences using Markov random fields. *Signal Process., Image Commun.*, 7:147–160, 1995.
2. M.S. Allili, N. Bouguila, and D. Ziou. Finite general Gaussian mixture modelling and application to image and video foreground segmentation. *J. of Elec. Imaging*, 17:1–23, 2008.
3. O. Barnich and M. Van Droogenbroeck. ViBe: A universal background subtraction algorithm for video sequences. *IEEE Trans. Image Process.*, 20(6):1709–1724, June 2011.
4. F. Bashir and F. Porikli. Performance evaluation of object detection and tracking systems. In *Proc. IEEE Int. Workshop on Performance Evaluation of Tracking Systems*, 2006.
5. B. Bayer. Color imaging array, u.s. patent 3971065, 1976.
6. Y. Benezeth, P. Jodoin, B. Emile, H. Laurent, and C. Rosenberger. Review and evaluation of commonly-implemented background subtraction algorithms. In *Proc. Int. Conf. Pattern Recognition*, pages 1–4, December 2008.
7. Y. Benezeth, P-M. Jodoin, B. Emile, H. Laurent, and C. Rosenberger. Comparative study of background subtraction algorithms. *J. of Elec. Imaging*, 19(3):1–12, 2010.
8. A. F. Bobick and J. W. Davis. The recognition of human movement using temporal templates. *IEEE Trans. Pattern Anal. Machine Intell.*, 23(3):257–267, 2001.
9. T. Bouwmans. Recent advanced statistical background modeling for foreground detection: A systematic survey. *Recent Patents on Computer Science*, 4(3), 2011.
10. T. Bouwmans, F. El Baf, and B. Vachon. Background modeling using mixture of Gaussians for foreground detection: A survey. *Recent Patents on Computer Science*, 1(3):219–237, 2008.
11. K. Boyd, V. Costa, J. Davis, and C. Page. Unachievable region in precision-recall space and its effect on empirical evaluation. In *29th International Conference on Machine Learning (ICML)*, Edinburgh, UK, June-July 2012.
12. S. Brutzer, B. Höferlin, and G. Heidemann. Evaluation of background subtraction techniques for video surveillance. In *Proc. IEEE Conf. Computer Vision Pattern Recognition*, pages 1937–1944, June 2011.
13. D. Butler, S. Sridharan, and M. Bove. Real-time adaptive background segmentation. In *ICME*, pages 341–344, 2003.
14. S. Chen, J. Zhang, Y. Li, and J. Zhang. A hierarchical model incorporating segmented regions and pixel descriptors for video background subtraction. *IEEE Transactions on Industrial Informatics*, 8(1):118–127, February 2012.
15. Y. Chen, C. Chen, C. Huang, and Y. Hung. Efficient hierarchical method for background subtraction. *Pattern Recognition*, 40(10):2706–2715, 2007.
16. L. Cheng, M. Gong, D. Schuurmans, and T. Caelli. Real-time discriminative background subtraction. *IEEE Trans. Image Process.*, 20(5):1401–1414, 2011.
17. R. Cucchiara, C. Grana, M. Piccardi, and A. Prati. Detecting moving objects, ghosts, and shadows in video streams. *IEEE Trans. Pattern Anal. Machine Intell.*, 25(10):1337–1342, October 2003.
18. Y. Dong, T.X. Han, and G.N. Desouza. Illumination invariant foreground detection using multi-subspace learning. *Int. J. Know.-Based Intell. Eng. Syst.*, 14(1):31–41, 2010.
19. R. Duin. The combining classifier: to train or not to train? In *IEEE International Conference on Pattern Recognition (ICPR)*, volume 2, pages 765–770, Quebec City, Canada, August 2002.

20. R. Duin and D. Tax. Experiments with classifier combining rules. In *Multiple Classifier Systems*, volume 1857 of *Lecture Notes in Computer Science*, pages 16–29. Springer, 2000.

21. A. Elgammal, R. Duraiswami, D. Harwood, and L.S. Davis. Background and foreground modeling using nonparametric kernel density for visual surveillance. *Proc. IEEE*, 90:1151–1163, 2002.

22. A. Elgammal, D. Harwood, and L. Davis. Non-parametric model for background subtraction. *Proc. European Conf. Computer Vision*, pages 751–767, 2000.

23. R. Evangelio, M. Pätzold, and T. Sikora. Splitting Gaussians in mixture models. In *Proc. IEEE Int. Conf. on Advanced Video and Signal-based Surveillance*, 2012.

24. R. Evangelio and T. Sikora. Complementary background models for the detection of static and moving objects in crowded environments. In *Proc. IEEE Int. Conf. on Advanced Video and Signal-based Surveillance*, 2011.

25. N. Friedman and S. Russell. Image segmentation in video sequences: a probabilistic approach. In *Proc of Uncertainty in Artificial Intelligence*, pages 175–181, 1997.

26. X. Gao, T. Boult, F. Coetzee, and V. Ramesh. Error analysis of background adaption. *Proc. IEEE Conf. Computer Vision Pattern Recognition*, 2000.

27. N. Goyette, P.-M. Jodoin, F. Porikli, J. Konrad, and P. Ishwar. changedetection.net: A new change detection benchmark dataset. In *IEEE Workshop on Change Detection*, 2012.

28. N. Goyette, P.-M. Jodoin, F. Porikli, J. Konrad, and P. Ishwar. changedetection.net: A new change detection benchmark dataset. In *Change Detection Workshop (CDW)*, Providence, Rhode Island, June 2012.

29. T. Haines and T. Xiang. Background subtraction with dirichlet processes. In *Proc. European Conf. Computer Vision*, pages 97–111, 2012.

30. B. Han and L. Davis. Density-based multifeature background subtraction with support vector machine. *IEEE Trans. Pattern Anal. Machine Intell.*, 34(5), 2012.

31. Z. Hao, W. Wen, Z. Liu, and X. Yang. Real-time foreground-background segmentation using adaptive support vector machine algorithm. In *Proc of ICANN*, pages 603–610, 2007.

32. M. Hofmann. Background segmentation with feedback: The pixel-based adaptive segmenter. In *IEEE Workshop on Change Detection*, 2012.

33. S.-S. Huang, L.-C. Fu, and P.-Y. Hsiao. Region-level motion-based background modeling and subtraction using MRFs. *IEEE Transactions on Image Processing*, 16(5):1446–1456, May 2007.

34. O. Javed, K. Shafique, and M. Shah. A hierarchical approach to robust background subtraction using color and gradient information. In *Workshop on Motion and Video Computing*, pages 22–27, December 2002.

35. J-P. Jodoin, G. Bilodeau, and N. Saunier. Background subtraction based on local shape. Technical Report arXiv:1204.6326v1, Ecole Polytechnique de Montreal, 2012.

36. P-M. Jodoin, M. Mignotte, and J. Konrad. Statistical background subtraction methods using spatial cues. *IEEE Trans. Circuits Syst. Video Technol.*, 17(12):1758–1763, 2007.

37. P. KaewTraKulPong and R. Bowden. An improved adaptive background mixture model for realtime tracking with shadow detection. *European Workshop on Advanced Video Based Surveillance Systems*, 2001.

38. M. Karaman, L. Goldmann, D. Yu, and T. Sikora. Comparison of static background segmentation methods. In *Proc. SPIE Visual Communications and Image Process.*, 2005.

39. K. Karman and A. von Brandt. Moving object recognition using an adaptive background memory. In Capellini, editor, *Time-varying Image Processing and Moving Object Recognition*, volume II, pages 297–307, Amsterdam, The Netherlands, 1990. Elsevier.

40. H. Kim, R. Sakamoto, I. Kitahara, T. Toriyama, and K. Kogure. Robust foreground extraction technique using Gaussian family model and multiple thresholds. In *ACCV*, pages 758–

768, 2007.

41. K. Kim, T. Chalidabhongse, D. Harwood, and L. Davis. Real-time foreground-background segmentation using codebook model, real-time imaging. *Real-Time Imaging*, 11(3):172–185, 2005.

42. D. Kit, B. Sullivan, and D. Ballard. Novelty detection using growing neural gas for visuo-spatial memory. In *IEEE IROS*, 2011.

43. J. Kittler, M. Hatef, and R. Duin. Combining classifiers. In *IEEE International Conference on Pattern Recognition (ICPR)*, volume 2, pages 897–901, Vienna, Austria, August 1996.

44. F. Kristensen, P. Nilsson, and V. Öwall. Background segmentation beyond rgb. In *Proc of ACCV*, pages 602–612, 2006.

45. L. Kuncheva, C. Whitaker, C. Shipp, and R. Duin. Limits on the majority vote accuracy in classifier fusion. *Pattern Analysis & Applications*, 6(1):22–31, April 2003.

46. T. Landgrebe, P. Paclík, R. Duin, and A. Bradley. Precision-recall operating characteristic (P-ROC) curves in imprecise environments. In *IEEE International Conference on Pattern Recognition (ICPR)*, volume 4, pages 123–127, Hong Kong, August 2006.

47. L. Li, W. Huang, I. Y. Gu, and Q. Tian. Statistical modeling of complex backgrounds for foreground object detection. *IEEE Trans. Image Process.*, 13(11):1459–1472, 2004.

48. Y. Li. On incremental and robust subspace learning. *Pattern Recognit.*, 37:1509–1518, 2004.

49. C-C Lien, Y-M Jiang, and L-G Jang. Large area video surveillance system with handoff scheme among multiple cameras. In *MVA*, pages 463–466, 2009.

50. H.-H. Lin, T.-L. Liu, and J.-H. Chuang. A probabilistic SVM approach for background scene initialization. In *Proc. IEEE Int. Conf. Image Processing*, pages 893–896, 2002.

51. X. Lu and R. Manduchi. Fast image motion segmentation for surveillance applications. *Image Vis. Comput.*, 29(2-3):104–116, 2011.

52. L. Maddalena and A. Petrosino. A self-organizing approach to background subtraction for visual surveillance applications. *IEEE Transactions on Image Processing*, 17(7):1168–1177, 2008.

53. L. Maddalena and A. Petrosino. The SOBS algorithm: what are the limits? In *IEEE Workshop on Change Detection*, 2012.

54. A. Manzanera. Local jet feature space framework for image processing and representation. In *International Conference on Signal Image Technology & Internet-Based Systems*, pages 261–268, Dijon, France, 2011.

55. N. McFarlane and C. Schofield. Segmentation and tracking of piglets in images. *Mach. vis. and appl.*, 8:187–193, 1995.

56. J.M. McHugh, J. Konrad, V. Saligrama, and P.-M. Jodoin. Foreground-adaptive background subtraction. *IEEE Signal Process. Lett.*, 16(5):390–393, May 2009.

57. E. Memin and P. Perez. Dense estimation and object-based segmentation of the optical flow with robust techniques. *IEEE Trans. Image Process.*, 7(5):703–719, 1998.

58. A. Mittal and N. Paragios. Motion-based background subtraction using adaptive kernel density estimation. In *Proc. IEEE Conf. Computer Vision Pattern Recognition*, pages 302–309, 2004.

59. A. Morde, X. Ma, and S. Guler. Learning a background model for change detection. In *IEEE Workshop on Change Detection*, 2012.

60. J. Nascimento and J. Marques. Performance evaluation of object detection algorithms for video surveillance. *IEEE Trans. Multimedia*, 8(8):761–774, 2006.

61. Y. Nonaka, A. Shimada, H. Nagahara, and R. Taniguchi. Evaluation report of integrated background modeling based on spatio-temporal features. In *IEEE Workshop on Change Detection*, 2012.

62. N. M. Oliver, B. Rosario, and A. P. Pentland. A Bayesian computer vision system for modeling human interactions. *IEEE Trans. Pattern Anal. Machine Intell.*, 22(8):831–843, 2000.

63. S. K. Mitra P. Jaikumar, A. Singh. Background subtraction in videos using bayesian learning with motion information. In *British Mach. Vis. Conf*, pages 1–10, 2008.

64. T. Parag, A. Elgammal, and A. Mittal. A framework for feature selection for background subtraction. In *Proc. IEEE Conf. Computer Vision Pattern Recognition*, pages 1916–1923, New York, USA, June 2006.

65. J. Park, A. Tabb, and A. Kak. Hierarchical data structure for real-time background subtraction. In *IEEE International Conference on Image Processing (ICIP)*, pages 1849–1852, 2006.

66. D. Parks and S. Fels. Evaluation of background subtraction algorithms with post-processing. In *Proc. IEEE Int. Conf. on Advanced Video and Signal-based Surveillance*, pages 192–199, September 2008.

67. M. Piccardi. Background subtraction techniques: a review. In *Conf. IEEE Int. Conf. Sys., Man and Cyb.*, pages 3099–3104, 2004.

68. F. Porikli. Multiplicative background-foreground estimation under uncontrolled illumination using intrinsic images. In *Proc. of IEEE Motion Multi-Workshop*, Breckenridge, 2005.

69. F. Porikli and O. Tuzel. Bayesian background modeling for foreground detection. *Proc. of ACM Visual Surveillance and Sensor Network*, 2005.

70. F. Porikli and C. Wren. Change detection by frequency decomposition: wave-back. *Proc. of Workshop on Image Analysis for Multimedia Interactive Services*, Montreux, 2005.

71. A. Prati, R. Cucchiara, I. Mikic, and M. Trivedi. Analysis and detection of shadows in video streams: A comparative evaluation. In *Proc. IEEE Conf. Computer Vision Pattern Recognition*, pages 571–577, 2001.

72. R. Radke, S. Andra, O. Al-Kofahi, and B. Roysam. Image change detection algorithms: A systematic survey. *IEEE Trans. Image Process.*, 14:294–307, 2005.

73. D. Riahi, P. St-Onge, and G. Bilodeau. RECTGAUSS-tex: Block-based background subtraction. Technical Report EPM-RT-2012-03, Ecole Polytechnique de Montreal, 2012.

74. P. Rosin and E. Ioannidis. Evaluation of global image thresholding for change detection. *Pattern Recognit. Lett.*, 24:2345–2356, 2003.

75. J. Rymel, J. Renno, D. Greenhill, J. Orwell, and G.A. Jones. Adaptive eigen-backgrounds for object detection. In *Proc. IEEE Int. Conf. Image Processing*, pages 1847 – 1850, 2004.

76. A. Schick, M. Bäuml, and R. Stiefelhagen. Improving foreground segmentations with probabilistic superpixel markov random fields. In *IEEE Workshop on Change Detection*, 2012.

77. M. Sedky, M. Moniri, and C. Chibelushi. Object segmentation using full-spectrum matching of albedo derived from colour images, Patent application PCT GB2009/002829, 2009.

78. C. Stauffer and E. Grimson. Learning patterns of activity using real-time tracking. *IEEE Trans. Pattern Anal. Machine Intell.*, 22(8):747–757, 2000.

79. B. Stenger, V. Ramesh, N. Paragios, F. Coetzee, and J. Buhmann. Topology free Hidden Markov Models: application to background modeling. *Proc. IEEE Int. Conf. Computer Vision*, 2001.

80. J. Suhr, H. Jung, G. Li, and J. Kim. Mixture of Gaussians-based background subtraction for Bayer-pattern image sequences. *IEEE Trans. Circuits Syst. Video Technol.*, 21(3):365–370, March 2011.

81. D. Tax, M. van Breukelen, R. Duin, and J. Kittler. Combining multiple classifiers by averaging or by multiplying? *Pattern Recognition*, 33(9):1475–1485, 2000.

82. Y-L. Tian and A. Hampapur. Robust salient motion detection with complex background for real-time video surveillance. In *WACV/MOTION*, pages 30–35, 2005.

83. F. Tiburzi, M. Escudero, J. Bescos, and J. Martinez. A ground truth for motion-based video-object segmentation. In *Proc. IEEE Int. Conf. Image Processing*, pages 17–20, 2008.

84. K. Toyama, J. Krumm, B. Brumitt, and B. Meyers. Wallflower: Principles and practice of

background maintenance. In *Proc. IEEE Int. Conf. Computer Vision*, volume 1, pages 255–261, 1999.

85. M. Van Droogenbroeck and O. Paquot. Background subtraction: Experiments and improvements for ViBe. In *IEEE Workshop on Change Detection*, pages 32–37, Providence, Rhode Island, USA, June 2012.

86. R. Vezzani and R. Cucchiara. Video surveillance online repository (ViSOR): an integrated framework. *Multimedia Tools and Applications*, 50(2):359–380, 2010.

87. H. Wang and D. Suter. A consensus-based method for tracking: Modelling background scenario and foreground appearance. *Pattern Recognit.*, 40(3):1091–1105, March 2007.

88. W-H. Wang and R-C. Wu. Fusion of luma and chroma gmms for hmm-based object detection. In *Proc of. conference on Advances in Image and Video Technology*, pages 573–581, 2006.

89. S. Warfield, K. Zou, and W. Wells. Simultaneous truth and performance level estimation (STAPLE): An algorithm for the validation of image segmentation. *IEEE Transactions on Medical Imaging*, 23(7):903–921, July 2004.

90. L. Wixson. Detecting salient motion by accumulating directionally-consistent flow. *IEEE Trans. Pattern Anal. Machine Intell.*, 22, 2000.

91. C. R. Wren, A. Azarbayejani, T. Darrell, and A. P. Pentland. Pfinder: Real-time tracking of the human body. *IEEE Trans. Pattern Anal. Machine Intell.*, 19(7):780–785, 1997.

92. L. Xu, A. Krzyżak, and C. Suen. Methods of combining multiple classifiers and their applications to handwriting recognition. *IEEE Transactions on Systems, Man, and Cybernetics*, 22(3):418–435, May-June 1992.

93. M. Xu and T. Ellis. Illumination-invariant motion detection using colour mixture models. In *British Mach. Vis. Conf*, pages 163–172, 2001.

94. Z. Xu, P. Shi, and I. Yu-Hua Gu. An eigenbackground subtraction method using recursive error compensation. In *PCM*, pages 779–787, 2006.

95. S. Yoshinaga, A. Shimada, and Taniguchi R. Statistical background model considering relationships between pixels. In *IEEE Workshop on Change Detection*, 2012.

96. D. Young and J. Ferryman. PETS metrics: Online performance evaluation service. In *Proc. IEEE Int. Workshop on Performance Evaluation of Tracking Systems*, pages 317–324, 2005.

97. C. Zhao, X. Wang, and W-K. Cham. Background subtraction via robust dictionary learning. *EURASIP Journal on Image and Video Processing*, 2011.

98. J. Zheng, Y. Wang, N. Nihan, and E. Hallenbeck. Extracting roadway background image: A mode based approach. *J. of Transp. Research Report*, pages 82–88, 2006.

99. J. Zhong and S. Sclaroff. Segmenting foreground objects from a dynamic textured background via a robust kalman filter. In *Proc. IEEE Int. Conf. Computer Vision*, pages 44–50, 2003.

100. D. Zhou and H. Zhang. Modified GMM background modeling and optical flow for detection of moving objects. In *Conf. IEEE Int. Conf. Sys., Man and Cyb.*, pages 2224–2229, 2005.

101. Z. Zivkovic and F. Van Der Heijden. Efficient adaptive density estimation per image pixel for the task of background subtraction. *Pattern Recognit. Lett.*, 27:773–780, 2006.

25

Evaluation of Background Models with Synthetic and Real Data

Antoine Vacavant
ISIT, Université d'Auvergne, CNRS,
UMR6284, Clermont-Ferrand

Laure Tougne
LIRIS, Université Lyon 2, CNRS, UMR5205,
Lyon

Thierry Chateau
Pascal Institute, Blaise Pascal University,
CNRS, UMR6602, Clermont-Ferrand

Lionel Robinault
LIRIS/Foxstream, Université Lyon 2, CNRS,
UMR5205, Lyon

25.1 Introduction

As the other fields of computer vision [2, 19, 27], evaluating background subtraction (BS) algorithms is an important task to reveal the best approaches for a given application. Although the evaluation of BS techniques is an important issue, the impact of relevant papers that handle with both benchmarks and annotated dataset is limited [1, 6, 20]. Moreover, many authors that propose a novel approach compare their work with [25] or only a restricted part of the literature, but rarely with numerous recent related works.

Recently, this lack of durable reference has led to the emergence of several benchmarks, fully available on the Web, as ChangeDetection.net [9] or SABS (Stuttgart Artificial Background Subtraction Dataset) [3]. These datasets allow authors to download challenging videos, and to compare their work with both classical and recent contributions.

In this chapter, we first present in Section 25.2 the Background Models Challenge (BMC) [28], which is a benchmark proposing a set of both synthetic and real videos, together with several performance evaluation criteria. The first workshop BMC, organized within the ACCV (Asian Conference on Computer Vision) 2012 has been mainly dedicated to evaluate and rank background/foreground algorithms for intelligent video-surveillance applications. In this field, BS is of great importance in order to detect persons, vehicles, animals, *etc.*, within video streams. In Section 25.3, we present the most recent results obtained thanks to BMC, like the final ranking obtained for this very first challenge. We finally conclude this chapter by the Section 25.4, which draws possible evolutions for this benchmark, according to recent advances in BS and its potential applications.

25.2   The Background Models Challenge

25.2.1   Dataset

In the benchmark BMC (Background Models Challenge)[76], we have proposed a complete benchmark composed of both synthetic and real videos. They are divided into two distinct sets of sequences: learning and evaluation. The benchmark is first composed of 20 urban video sequences rendered with the SiVIC simulator [10]. With this tool, we are also able to render the associate ground truth, frame by frame, for each video (at 25 fps). We sum up in Table 25.1 the different situations we have proposed in BMC.

TABLE 25.1   Possible combinations of synthetic data generated for BMC.

Scenes	
1	Rotary
2	Street

Event types	
1	Cloudy, without acquisition noise, as normal mode
2	Cloudy, with salt and pepper noise during the whole sequence
3	Sunny, with noise, which generates moving cast shadows
4	Foggy, with noise, making both background and foreground hard to analyze
5	Wind, with noise, to produce a moving background

Use cases	
1	10 seconds without objects, then moving objects during 50 seconds
2	20 seconds without event, then event (*e.g.* sun uprising or fog) during 20 seconds, finally 20 seconds without event

The *learning* set is composed of the 10 synthetic videos representing the use case 1. Each video is numbered according to presented event type (from 1 to 5), the scene number (1 or 2), and the use case (1 or 2). For example, the video 311 of our benchmark describes a sunny street, under the use case 1, as illustrated in Figure 25.1. In the learning phase of the BMC contest, authors use these sequences in order to set the parameters of their BS method, thanks to the ground truth of each image that is available, and to a software of computation of quality criteria (see next section).

The *Evaluation* set first contains the 10 synthetic videos with use case 2. This set is also composed of real videos acquired from static cameras in video-surveillance contexts. This dataset has been built in order to test the algorithms reliability during time and in difficult situations such as outdoor scenes. Real long videos (about one hour and up to four hours) are available, and they may present long time change in luminosity with small density of objects in time compared to previous synthetic ones. This dataset allows to test the influence of some difficulties encountered during the object extraction phase. Those difficulties have been sorted according to:

1. the ground type (bitumen, ballast or ground);

2. the presence of vegetation (trees for instance);

---

[76]http://bmc.univ-bpclermont.fr

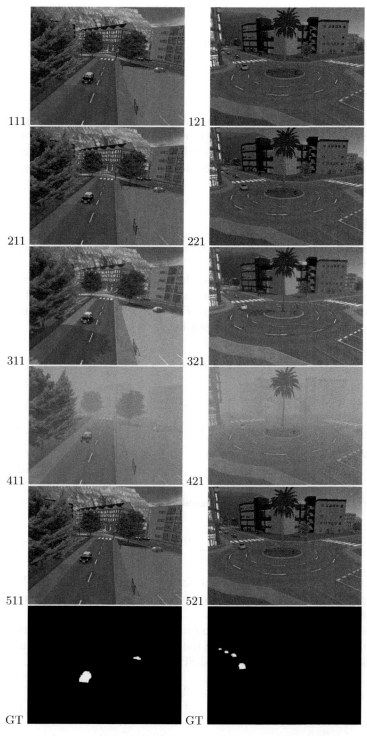

FIGURE 25.1 Images extracted from synthetic videos (learning mode) of BMC benchmark and their associated ground truth (GT).

3. cast shadows;

4. the presence of a continuous car flow near to the surveillance zone;

5. the general climatic conditions (sunny, rainy and snowy conditions);

6. fast light changes in the scene;

7. the presence of big objects.

Samples of images extracted from those real videos are presented in Figure 25.2. Each of them has been manually segmented some representative frames that can be used to evaluate a BS algorithm. In the evaluation phase of the BMC benchmark, no ground truth image is available, and authors should test their BS technique with the parameters they have set in the learning phase.

25.2.2 Metrics

In our benchmark, several criteria have been considered, and represents different kinds of quality of a BS method.

Let S be the set of n images computed thanks to a given BS algorithm A, and G be the ground truth video sequence. For a given frame i, we denote by TP_i and FP_i the true and false positive detections, and by TN_i and FN_i the true and false negative ones. We first propose to compute the F-measure, defined by:

$$F\text{-}measure(S, A, G) = \frac{1}{n}\sum_{i=1}^{n} 2\frac{Prec_i \times Rec_i}{Prec_i + Rec_i},\qquad(25.1)$$

with

$$Rec_i(P) = TP_i/(TP_i + FN_i) \; ; \qquad Prec_i(P) = TP_i/(TP_i + FP_i)\qquad(25.2)$$
$$Rec_i(N) = TN_i/(TN_i + FP_i) \; ; \qquad Prec_i(N) = TN_i/(TN_i + FN_i)\qquad(25.3)$$
$$Rec_i = (1/2)(Rec_i(P) + Rec_i(P)) \; ; \; Prec_i = (1/2)(Prec_i(P) + Prec_i(P)).\quad(25.4)$$

We also compute the PSNR (Peak Signal-Noise Ratio), defined by:

$$PSNR(S, A, G) = \frac{1}{n}\sum_{i=1}^{n} 10\log_{10}\frac{m}{\sum_{j=1}^{m}||S_i(j) - G_i(j)||^2}\qquad(25.5)$$

where $S_i(j)$ is the jth pixel of image i (of size m) in the sequence S (with length n). These two criteria should permit to compare the raw behavior of each algorithm for moving object segmentation.

We now consider the problem of background subtraction in a visual and perceptual way. To do so, we use the gray-scale images of the input and ground truth sequences to compute the perceptual measure SSIM (Structural SIMilarity), given by [29]:

$$SSIM(S, A, G) = \frac{1}{n}\sum_{i=1}^{n}\frac{(2\mu_{S_i}\mu_{G_i} + c_1)(2cov_{S_iG_i} + c_2)}{(\mu_{S_i}^2 + \mu_{G_i}^2 + c_1)(\sigma_{S_i}^2 + \sigma_{G_i}^2 + c_2)},\qquad(25.6)$$

where μ_{S_i}, μ_{G_i} are the means, $\sigma_{S_i}, \sigma_{G_i}$ the standard deviations, and $cov_{S_iG_i}$ the covariance of S_i and G_i. In our benchmark, we set $c_1 = (k_1 \times L)^2$ and $c_2 = (k_2 \times L)^2$, where L is the size of the dimension of the signal processed (that is, $L = 255$ for gray-scale images), $k_1 = 0.01$ and $k_2 = 0.03$ (which are the most used values in the literature).

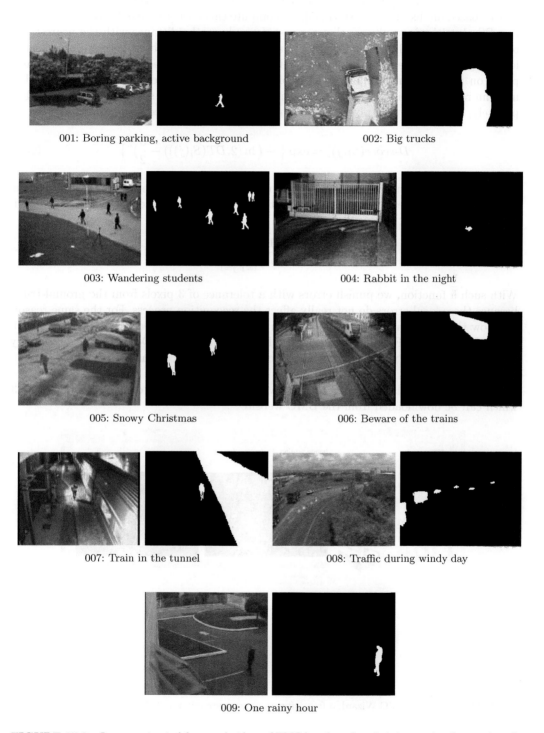

FIGURE 25.2 Images extracted from real videos of BMC benchmark and their associated ground truth.

We finally use the D-Score [16], which consists in considering localization of errors according to real object position. As Baddeleys distance, it is a similarity measure for binary images based on distance transform [5]. To compute this measure we only consider mistakes in a BS algorithm's results. Each error cost depends on the distance with the nearest corresponding pixel in the ground-truth. As a matter of fact, for object recognition, short or long range errors in segmentation step are less important than medium range error, because pixels on medium range impact greatly on object's shape. Hence, the penalty applied to medium range errors is heavier than the one applied to those in a short or large range. More precisely, the D-score of a pixel $S_i(j)$ in the sequence is computed as

$$D\text{-}score(S_i(j)) = \exp\left\{-\left(\ln\left(2.DT(S_i(j))\right) - \frac{5}{2}\right)^2\right\} \tag{25.7}$$

where $DT(x)$ is given by the minimal distance between x and the nearest reference point (obtained by distance transformation). We then compute the D-score measure by computing:

$$D\text{-}score(S, A, G) = \frac{1}{mn}\sum_{i=1}^{n}\sum_{j=1}^{m} D\text{-}score(S_i(j)) \tag{25.8}$$

With such a function, we punish errors with a tolerance of 3 pixels from the ground-truth, because these local errors do not really affect the recognition process. For the same reason, we allow the errors that occur at more than a 10 pixel distance. Details about such metric can be found in [16]. Few local/far errors will produce a near zero D-Score. On the contrary, medium range errors will produce high D-Score. A good D-Score has to tend to 0.

Criteria are calculated thanks to the BMC Wizard (BMCW), depicted in Figure 25.3, which can be downloaded from the BMC website[77].

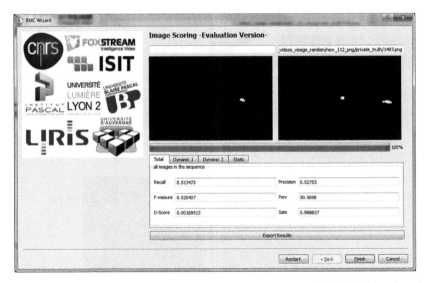

FIGURE 25.3   The BMC Wizard, a free software to compute criteria of the BMC benchmark.

---

[77]http://bmc.univ-bpclermont.fr/?q=node/7

25.3 Results, Rankings and Analysis

25.3.1 2012, First Edition

During the first wokrshop of BMC, we have evaluated 5 original algorithms [7,11,21,26,32], and we have compared them with 5 classic techniques from the literature [28].

In Table 25.2, we sum up the latter BS techniques, and their parameters, obtained thanks to a stochastic gradient descent to maximize the F-measure for the sequences of the learning phase. A first way to evaluate these algorithms is to study the distribution of each

TABLE 25.2 The methods extracted from the literature, with their associated references, tested with BMC.

Name	Description
NA	Naive approach, where pixels differing from the first image of the sequence (under a given threshold) are considered as foreground (*threshold* = 22).
GMM1	Gaussian mixture models from [12,25], improved by [14] for a faster learning phase.
GMM2	Gaussian mixture models improved with [33,34] to select the correct number of components of the GMM (*history size* = 355, *background ratio* = 16).
BC	Bayesian classification processed on feature statistics [17] ($L = 256$, $N_1 = 9$, $N_2 = 15$, $L^c = 128$, $N_1^c = 25$, $N_2^c = 25$, no holes, 1 morphing step, $\alpha_1 = 0.0422409$, $\alpha_2 = 0.0111677$, $\alpha_3 = 0.109716$, $\delta = 1.0068$, $T = 0.437219$, *min area* = 5.61266).
CB	Codewords and Codebooks framework [15].
VM	VuMeter [8], which uses histograms of occurences to model the background ($\alpha = 0.00795629$ and *threshold* = 0.027915).

score through the whole dataset, as illustrated in Figure 25.4. We can also visualize the precision/recall values like in Figure 25.5. Thanks to these plots, we are able to have an evaluation of the robustness of the tested BS algorithms. One of the best methods of this first test is BC, since its F-measure has the shortest range of values, with the highest values (from 0.65 to 0.93 approximately). The case of the VM method is interesting because its F-measure is focused around the interval [0.8; 0.85]. These observations can be confirmed by Figure 25.5, where BC and VM have the greatest numbers of points coming close to the (1, 1) point. GMM1 has also a similar behavior, around the 0.75 value, and a very good precision. GMM2 has a point of focus around the 0.9 value, but has also a wide interval of F-measures. The CB approach returns a very wide range of values, which could be induced by the high variability of the parameters of the method. Figure 25.5 informs us that the real videos of our benchmark are not correctly processed by CB, impacting a global bad result. This phenomenon can also be observed for the NA, in a more negative way. From a structural point of view, the values of SSIM and D-score lead to similar conclusions: CB and NA are not constant, and not efficient on the whole benchmark. It seems even better to choose NA (SSIM greater than 0.4) instead of CB (SSIM can be around 0.1 or 0.2). A more complete evaluation of these methods can be read in the supplementary material of [28].

We now consider all the BS algorithms of this first edition of BMC (from the literature and original submissions). Table 25.3 contains all the rankings for all the tested BS algorithms. Global rankings (right column) are obtained by summing ranks over all the evaluation sequences, for all the measures. We also show the ranks obtained only for synthetic or real data. We can notice that original BS algorithms submitted to this first edition are very competitive, since the ones proposed by [21,32] are the first of our global ranking. These techniques are very advanced methods, combining accurate image processing tools. Even if they are not at the top of this table, [26] have developed a promising method based on saliency. The interest of the BMC benchmark is also to help authors to improve their BS techniques, as [7,11], by comparing their work to the literature or to other original approaches. Moreover, Table 25.3 informs us that the rankings calculated only with real or synthetic data are quite different. We think that the combination of both of them is the best choice to evaluate the robustness of BS algorithms. We have observed that the ranking for

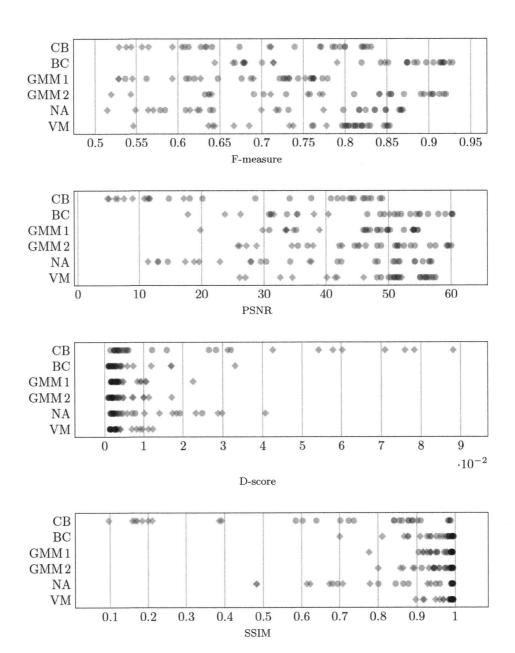

FIGURE 25.4   Distribution of each score, plotted for all the videos of the dataset.

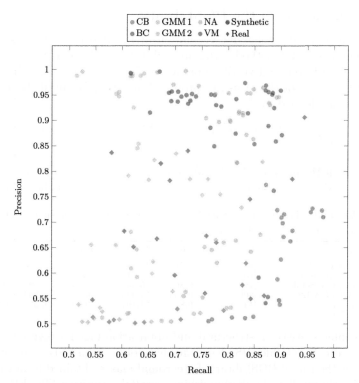

FIGURE 25.5 (See color insert.) Precision/recall values for all the videos of the benchmark.

TABLE 25.3 Final ranking of all evaluated methods for the first edition of BMC. Original submissions are highlighted in bold.

Name	Ref.	Rk, synth.	Rk, real	Rk, all
Statistical local difference pattern	[32]	**2nd**	**1st**	**1st**
GMM & SURF combination	[21]	**1st**	**2nd**	**2nd**
VM, or VuMeter	[8]	4th	3rd	3rd
BC, or Bayesian classification	[17]	3rd	6th	4th
Temporal saliency	[26]	**7th**	**4th**	**5th**
GMM2	[33]	5th	7th	6th
One-class classification	[7]	**6th**	**8th**	**7th**
GMM1	[14]	8th	5th	8th
Robust low rank matrix decomposition	[11]	**10th**	**9th**	**9th**
NA, or Naive approach	[28]	9th	10th	10th
CB, or Codebooks	[15]	11th	11th	11th

learning dataset (containing only synthetic videos) is very close to the one displayed within the middle row of this table.

25.3.2   Evaluation with the BGSLibrary

In [23], we have proposed a massive evaluation of BS algorithms thanks to the BGSLibrary [22], composed of 29 methods programmed in OpenCV. This work has permitted to extract the five most competitive BS approaches, listed in Table 25.4. To do so, we have calculated all the mean of the measures proposed in BMC, for all of them. We have then computed the score defined by:

$$FSD(A) = \frac{\overline{F - measure}(A) + \overline{SSIM}(A) + (1 - \overline{D\text{-}score}(A))}{3} \tag{25.9}$$

where A is a given BS method, $\overline{F\text{-}Measure}(A)$, $\overline{SSIM}(A)$ and $\overline{D\text{-}Score}(A)$ are the average F-measure, SSIM and D-score of the BS algorithm throughout the data set. This score

TABLE 25.4   The five first BS methods from BGSLibrary. Ranking is obtained with the FSD score calculated from Equation 25.9.

Name	Ref.	F-measure	PSNR	D-score	SSIM	FSD	Rk
Multi-layer BS	[31]	0.875	49.398	0.001	0.993	0.974	1st
Pixel-based segmenter	[13]	0.885	49.412	0.002	0.994	0.985	2nd
LB adaptive SOM	[18]	0.867	50.553	0.001	0.992	0.952	3rd
DP Wren GA-BS	[30]	0.853	51.394	0.001	0.993	0.922	4th
GMM V1	[14]	0.847	51.107	0.001	0.993	0.910	5th

permits to combine statistical, structural and application metrics together. We can notice that both very recent and older BS algorithms have obtained a FSD score greater than 0.9, ranking them at the top of BGSLibrary. Foreground masks obtained from these algorithms are depicted in Figure 25.6, by distinguishing true/false positive (TP/FP) and true/false negative (TN/FN) with different colors. Most algorithms, even the best techniques extracted from our study, critically fail for very complex situations, as illuminations due to sun rising (first column of Figure 25.6), casted shadows (fourth column of Figure 25.6), or very high lighting changes (second column of Figure 25.6). The learning of the background and foreground detection are very challenging in these scenes, and even the top 5 BS algorithms of our study are not completely able to handle them. New mathematical and algorithmic ideas have to be found to compute BS, and mostly for this kind of complicated configurations.

25.4   Conclusion

In this chapter, we have addressed the problem of evaluation for background modeling algorithms, by describing the BMC benchmark, composed of both synthetic and real videos. We have also shown the interest of such a dataset, by describing the first challenge organized within the ACCV 2012 conference, and by summing up the evaluation of many BS approaches thanks to the BGSLibrary. The visual results of the latter experience confirms that real data are still very challenging, because of many complex situations (moving trees, flashing lights, *etc.*). We think that the use of synthetic data cannot replace real ones, but we would like to keep on studying further the link between these two kinds of sequences, and mostly by using new convincing synthetic simulation tools, with high-quality 3D renderer. Up to now, we have focused our attention on the problems pointed out by intelligent video-surveillance. In the next steps of evolution of BMC, we would like to study the use of BS in other contexts, like animal behavior analysis or motion capture. Tracking fishes and other animals within underwater sequences propose very hard image conditions [24].

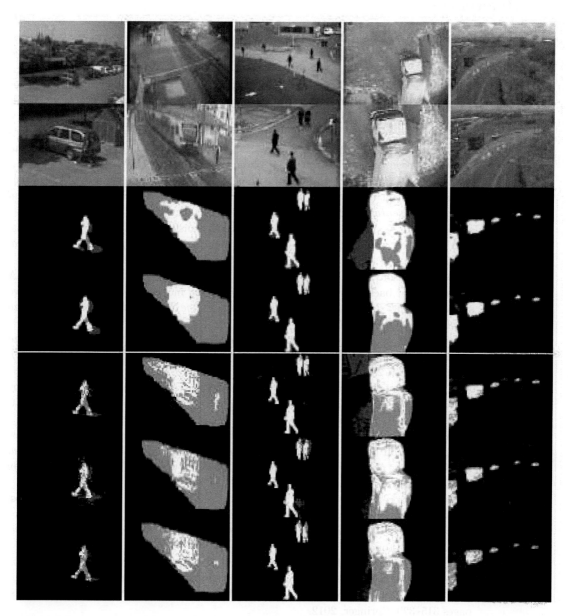

FIGURE 25.6 (See color insert.) Foreground masks obtained from the top five BS algorithms on sequences 001, 006, 003, 002, 008 of the real videos. Input frames are depicted on the first row, followed by the region of interest chosen, and the five BS outputs, according to the order of Table 25.4. In these images, the TP pixels are in white, TN pixels in black, FP pixels in red and FN pixels in green.

Motion capture using RGB-D data is still a challenging problem [4], since the BS of this kind of sequences has to be precise for further treatment, as motion analysis.

References

1. Y. Benezeth, P-M. Jodoin, B. Emile, H. Laurent, and C. Rosenberger. Review and evaluation of commonly-implemented background subtraction algorithms. In *IEEE International Conference on Pattern Recognition (ICPR 2008)*, Tampa, FL, USA, 2008.

2. K. W. Bowyer and P. J. Phillips. Overview of work in empirical evaluation of computer vision algorithms. In *Empirical Evaluation Techniques in Computer Vision*. IEEE Computer Society Press, 1998.

3. S. Brutzer, B. Höferlin, and G. Heidemann. Evaluation of background subtraction techniques for video surveillance. In *IEEE International Conference on Computer Vision and Pattern Recognition (CVPR 2011)*, Colorado Springs, CO, USA, 2011.

4. M. Camplani and L. Salgado. Background foreground segmentation with RGB-D Kinect data: An efficient combination of classifiers. *Journal of Visual Communication and Image Representation*, 2013. In press.

5. D. Coeurjolly and A. Vacavant. Separable distance transformation and its applications. In V. Brimkov and R. Barneva, editors, *Digital Geometry Algorithms. Theoretical Foundations and Applications to Computational Imaging*. Springer, 2012.

6. Y. Dhome, N. Tronson, A. Vacavant, T. Chateau, C. Gabard, Y. Goyat, and D. Gruyer. A benchmark for background subtraction algorithms in monocular vision: a comparative study. In *IEEE International Conference on Image Processing Theory, Tools and Applications (IPTA 2010)*, Paris, France, 2010.

7. A. Glazer, M. Lindenbaum, and S. Markovitch. One-class background model. In *Background Models Challenge (BMC) at Asian Conference on Computer Vision (ACCV 2012)*, volume LNCS 7728, pages 301–307. Springer, 2012.

8. Y. Goyat, T. Chateau, L. Malaterre, and L. Trassoudaine. Vehicle trajectories evaluation by static video sensors. In *IEEE Intelligent Transportation Systems Conference (ITSC 2006*, Toronto, ON, Canada, 2006.

9. N. Goyette, P.-M. Jodoin, F. Porikli, J. Konrad, and P. Ishwar. Changedetection.net: A new change detection benchmark dataset. In *Workshop on Change Detection (CDW) at the IEEE International Conference on Computer Vision and Pattern Recognition (CVPR 2012)*, Providence, RI, USA, 2012.

10. D. Gruyer, C. Royere, N. du Lac, G. Michel, and J.-M. Blosseville. SiVIC and RTMaps, interconnected platforms for the conception and the evaluation of driving assistance systems. In *World Congress and Exhibition on Intelligent Transportation Systems and Services*, London, UK, 2006.

11. C. Guyon, T. Bouwmans, and E.-H. Zahzah. Foreground detection via robust low rank matrix decomposition including spatio-temporal constraint. In *Background Models Challenge (BMC) at Asian Conference on Computer Vision (ACCV 2012)*, volume LNCS 7728, pages 315–320. Springer, 2012.

12. E. Hayman and J.-O. Eklundh. Statistical background subtraction for a mobile observer. In *International Conference on Computer Vision (ICCV 2003)*, Beijing, China, 2003.

13. M. Hofmann, P. Tiefenbacher, and G. Rigoll. Background segmentation with feedback: The pixel-based adaptive segmenter. *Change Detection Workshop at International Conference on Computer Vision and Pattern Recognition (CVPR 2012)*, 2012.

14. P. Kaewtrakulpong and R. Bowden. An improved adaptive background mixture model for realtime tracking with shadow detection. In *European Workshop on Advanced Video Based Surveillance Systems (AVBSS 2001)*, London, UK, 2001.

15. K. Kim, T. H. Chalidabhongse, D. Harwood, and L. Davis. Real-time foreground-background segmentation using codebook model. *Real-time Imaging*, 11(3):167–256, 2005.

16. C. Lallier, E. Renaud, L. Robinault, and L. Tougne. A testing framework for background subtraction algorithms comparison in intrusion detection context. In *IEEE International Conference on Advanced Video and Signal-based Surveillance (AVSS 2011)*, Klagenfurt, Austria, 2011.

17. L. Li, W. Huang, I. Y. H. Gu, and Q. Tian. Foreground object detection from videos containing complex background. In *ACM Multimedia*, Berkeley, CA, USA, 2003.

18. L. Maddalena and A. Petrosino. A self-organizing approach to background subtraction for visual surveillance applications. *IEEE Transactions on Image Processing*, 17(7):1168–1177, 2008.

19. R. J. Micheals and T. E. Boult. Efficient evaluation of classification and recognition systems. In *IEEE International Conference on Computer Vision and Pattern Recognition (CVPR 2001)*, Kauai, HI, USA, 2001.

20. A. Prati, I. Mikic, M. Trivedi, and R. Cucchiara. Detecting moving shadows: Algorithms and evaluation. *IEEE Transactions on Pattern Analysis and Machine Intelligence*, 25(7):918–923, 2003.

21. M. Shah, J. D. Deng, and B. J. Woodford. Illumination invariant background model using mixture of Gaussians and SURF features. In *Background Models Challenge (BMC) at Asian Conference on Computer Vision (ACCV 2012)*, volume LNCS 7728, pages 308–314. Springer, 2012.

22. A. Sobral. BGSLibrary: An opencv c++ background subtraction library. In *Workshop de Viso Computacional (WVC 2013)*, Rio de Janeiro, Brazil, 2013. Software available at `http://code.google.com/p/bgslibrary/`.

23. A. Sobral and A. Vacavant. A comprehensive review of background subtraction algorithms evaluated with synthetic and real videos. *Computer Vision and Image Understanding*, 2014.

24. C. Spampinato, Y.-H. Chen-Burger, G. Nadarajan, and R. B. Fisher. Detecting, tracking and counting fish in low quality unconstrained underwater videos. In *International Conference on Computer Vision Theory and Applications (VISAPP 2012)*, Rome, Italy, 2012.

25. C. Stauffer and W. E. L. Grimson. Adaptative background mixture models for a real-time tracking. In *IEEE International Conference on Computer Vision and Pattern Recognition*, Ft. Collins, CO, USA, 1999.

26. H. R. Tavakoli, E. Rahtu, and J. Heikkilä. Temporal saliency for fast motion detection. In *Background Models Challenge (BMC) at Asian Conference on Computer Vision (ACCV 2012)*, volume LNCS 7728, pages 321–326. Springer, 2012.

27. F. Tombari, L. Di Stefano, S. Mattoccia, and A. Galanti. Performance evaluation of robust matching measures. In *International Conference on Computer Vision Theory and Applications (VISAPP 2008)*, Funchal-Madeira, Portugal, 2008.

28. A. Vacavant, T. Chateau, A. Wilhelm, and L. Lequièvre. A benchmark dataset for foreground/background extraction. In *Background Models Challenge (BMC) at Asian Conference on Computer Vision (ACCV 2012)*, volume LNCS 7728, pages 291–300. Springer, 2012.

29. Z. Wang, A. C. Bovik, H. R. Sheikh, and E. P. Simoncelli. Image quality assessment: From error visibility to structural similarity. *IEEE Transactions on Image Processing*, 13(4):600–612, 2004.

30. C. Wren, A. Azarbayejani, T. Darrell, and A. Pentland. Pfinder: Real-time tracking of the human body. *IEEE Transactions on Pattern Analysis and Machine Intelligence*, 19(7):780–785, 1997.

31. J. Yao and Jean marc Odobez. Multi-layer background subtraction based on color and texture.

International Conference on Computer Vision and Pattern Recognition Conference (CVPR 2007), 2007.

32. S. Yoshinaga, A. Shimada, H. Nagahara, and R.-I. Taniguchi. Background model based on statistical local difference pattern. In *Background Models Challenge (BMC) at Asian Conference on Computer Vision (ACCV 2012)*, volume LNCS 7728, pages 327–332. Springer, 2012.

33. Z. Zivkovic. Improved adaptive Gaussian mixture model for background subtraction. In *IEEE International Conference on Pattern Recognition (ICPR 2004)*, Cambridge, UK, 2004.

34. Z. Zivkovic and F. V. D. Heijden. Efficient adaptive density estimation per image pixel for the task of background subtraction. *Pattern Recognition Letters*, 27(7):773–780, 2006.

Index